THE CITY READER
Third edition

The third edition of the highly successful *The City Reader* juxtaposes the very best of publications on the city. It has been extensively updated to reflect the latest thinking on globalization, information technology, and urban theory. Classic writings from such authors as Lewis Mumford, Jane Jacobs, and Le Corbusier meet the best contemporary writings of, among others, Peter Hall, Saskia Sassen, and Manuel Castells.

Fifty-four generous selections are included: a combination of thirty-nine readings from the second edition and fifteen entirely new selections. Structured to aid student understanding, the anthology features main and part introductions, as well as individual introductions to the selected articles. The latter contain a brief intellectual biography and a review of the author's writings and related literature. Furthermore, they provide an explanation of how the piece fits into the broader context of urban history and practice, competing ideological perspectives on the city and the major current debates concerning race and gender, global restructuring, sustainable urban development, and the impact of technology and postmodernism.

The City Reader provides *the* comprehensive mapping of the terrain of Urban Studies, old and new. It is illustrated with over forty photographs and is essential reading for anyone interested in the city.

Richard T. LeGates is Professor of Urban Studies at San Francisco State University. **Frederic Stout** is Lecturer in Urban Studies at Stanford University.

THE ROUTLEDGE URBAN READER SERIES

Series editors

Richard T. LeGates
Professor of Urban Studies, San Francisco State University

Frederic Stout
Lecturer in Urban Studies, Stanford University

The Routledge Urban Reader Series responds to the need for comprehensive coverage of the classic and essential texts that form the basis of intellectual work in the various academic disciplines and professional fields concerned with cities.

The readers focus on the key topics encountered by undergraduates, graduates and scholars in urban studies and allied fields. They discuss the contributions of major theoreticians and practitioners and other individuals, groups, and organizations that study the city or practice in a field that directly affects the city.

As well as drawing together the best of classic and contemporary writings on the city, each reader features extensive general, section and selection introductions prepared by the volume editors to place the selections in context, illustrate relations among topics, provide information on the author and point readers towards additional related bibliographic material.

Each reader will contain:

- ■ Approximately thirty-six *selections* divided into approximately six sections. Almost all of the selections will be previously published works that have appeared as journal articles or portions of books.
- ■ A *general introduction* describing the nature and purpose of the reader.
- ■ Two- to three-page *section introductions* for each section of the reader to place the readings in context.
- ■ A one-page *selection introduction* for each selection describing the author, the intellectual background of the selection, competing views of the subject matter of the selection and bibliographic references to other readings by the same author and other readings related to the topic.
- ■ A plate section with twelve to fifteen plates and illustrations at the beginning of each section.
- ■ An index.

The types of readers and forthcoming titles are as follows:

THE CITY READER

The City Reader: Third edition – an interdisciplinary urban reader aimed at urban studies, urban planning, urban geography and urban sociology courses – will be the *anchor urban reader*. Routledge published a first edition of *The City Reader* in 1996 and a second edition in 2000. *The City Reader* has become one of the most widely used anthologies in urban studies, urban geography, urban sociology and urban planning courses in the world.

URBAN DISCIPLINARY READERS

The series will contain *urban disciplinary readers* organized around social science disciplines. The urban disciplinary readers will include both classic writings and recent, cutting-edge contributions to the respective disciplines. They will be lively, high-quality, competitively priced readers which faculty can adopt as course texts and which will also appeal to a wider audience.

TOPICAL URBAN ANTHOLOGIES

The urban series will also include *topical urban readers* intended both as primary and supplemental course texts and for the trade and professional market.

INTERDISCIPLINARY ANCHOR TITLE

The City Reader: Third edition
Richard T. LeGates and Frederic Stout (eds)

URBAN DISCIPLINARY READERS

The Urban Geography Reader
Nick Fyfe and Judith Kenny (eds)

The Urban Sociology Reader
Jan Lin and Christopher Mele (eds)

The Urban Politics Reader
Elizabeth Strom and John Mollenkopf (eds)

The Urban and Regional Planning Reader
Eugenie Birch (ed.)

TOPICAL URBAN READERS

The City Cultures Reader, second edition
Malcolm Miles, Tim Hall with Iain Borden (eds)

The Cybercities Reader
Stephen Graham (ed.)

The Sustainable Urban Development Reader
Stephen M. Wheeler and Timothy Beatley (eds)

The Global Cities Reader
Neil Brenner and Roger Keil (eds)

For Further Information on The Routledge Urban Reader Series
please contact:

Andrew Mould
Routledge
11 New Fetter Lane
London EC4P 4EE
England
Andrew.mould@routledge.co.uk

Richard T. LeGates
Urban Studies Program
San Francisco State University
1600 Holloway Avenue
San Francisco, California 94132
(415) 338-2875
dlegates@sfsu.edu

Frederic Stout
Urban Studies Program
Stanford University
Stanford, California 94305-6050
(650) 725-6321
fstout@stanford.edu

The City Reader
Third edition

Edited by

Richard T. LeGates

and

Frederic Stout

Routledge
Taylor & Francis Group

LONDON AND NEW YORK

First published 1996
by Routledge
11 New Fetter Lane, London EC4P 4EE

Simultaneously published in the USA and Canada
by Routledge
29 West 35th Street, New York, NY 10001

Second edition 2000

Third edition 2003

Routledge is an imprint of the Taylor & Francis Group

Designed and typeset in Amasis MT Lt and Akzidenz Grotesk
by Keystroke, Jacaranda Lodge, Wolverhampton

Printed and bound in Spain by Graphos S.A. Arte Sobre Papel

British Library Cataloguing in Publication Data
A catalogue record for this book is available from the British Library

Library of Congress Cataloging in Publication Data
The city reader / [edited by] Richard T. LéGates & Frederic Stout. – 3rd ed.
    p. cm.— (Urban reader series)
Includes bibliographical references and index.
1. Urban policy. 2. Cities and towns. 3. City planning. I. LeGates, Richard T.
II. Stout, Frederic, 1943– III. Series: Routledge urban reader series.
HT151.C586 2003
307.76—dc21   2003002134

ISBN 0–415–27172–X (hbk)
ISBN 0–415–27173–8 (pbk)

To Courtney Elizabeth LeGates
and Amy Catherine Stout

They Are the Future

COMMENTS ON PREVIOUS EDITIONS

"Comprehensive and deep, this collection embodies the grand tradition, both classical and contemporary, of the urban field. It is a course itself; or a great lode for reference."

Robert J.S. Ross, Ph.D., *Professor of Sociology and Director of International Studies Stream, Clark University*

"This is the most useful reader on the market for students of cities. LeGates and Stout have refined the selections with each edition. My students tell me that the introductory notes and references make the readings more meaningful."

Ben Kohl, *Assistant Professor of Geography and Urban Studies, Temple University, Philadelphia*

"*The City Reader* brings together key works on the urban experience, problems, and policy alternatives in an engaging, accessibly structured and informative way. It draws together classic works and recent scholarship, capturing the dynamism of cities, urban processes and our interpretations of urban life. This is an impressive, comprehensive resource."

Dr Niall Majury, *School of Geography, Queen's University Belfast*

"*The City Reader* is a survival kit for practitioners who seek a better understanding of the towns and cities in which they work."

Steve Crocker, *Senior Skills Advisor, Neighbourhood Renewal Unit, Office of the Deputy Prime Minister, London*

"*The City Reader* weaves urban studies classics and modern writings in a masterful anthology. Editors' introductions to each section and piece make it an effective and accessible classroom tool."

Verrdie A. Craig, *Department of Geography, Rutgers University*

". . .remarkably well-chosen articles and excerpts . . . provides a solid introduction to the main currents of thought about urban form and processes in the twentieth century. . . ."

Environment and Planning

"An excellent, wide-ranging, stimulating reader; attractively presented and easy to read."

Brian Whalley, *Department of Built Environment, De Montfort University*

"A comprehensive mapping of the terrain of Urban Studies, old and new."

Jamie Peck, *Department of Geography, University of Manchester*

Contents

PART 8 THE FUTURE OF THE CITY

Plates

Acknowledgements

We received constant encouragement and many valuable suggestions, both for selections to include and approaches to critical commentary, from our colleagues. We wish particularly to thank Andrew Mould, our editor at Routledge, for his support, encouragement, and helpful suggestions. Ann Michael and Melanie Attridge at Routledge provided invaluable help in securing permissions. Nicola Cooper, our production editor, who sheparded the manuscript through production. Suzanne Fox, who meticulously copyedited the manuscript. Rufus Browning, Roger Crawford, Rich DeLeon, William Issel, Deborah LeVeen, Christopher McGee, Raquel Pinderhughes, Ayse Pamuk, Genie Stowers, and David Tabb at San Francisco State University; Paul Turner, Leonard Ortolano, and Gerry Gast at Stanford University; Cheyney Ryan at the University of Oregon, Dan Lewis of Northwestern University, Chester Hartman at the Poverty and Race Research and Action Council, and John Mollenkopf at the Graduate Center of the City University of New York gave us many valuable suggestions. Alexander Garvin of Yale University, Peter Calthorpe of Calthorpe Associates, and Chris McGee of San Francisco State University were generous in sharing their insights about what visual images to include and their own copies of images they had assembled over the years. Lisa Ryan contributed her artistic talents by creating visual images that appear at the beginning of the Prologue and other sections of the reader. Many others, too numerous to mention, made helpful suggestions. All errors and infelicities are, of course, ours.

A number of people helped us with technical support throughout the writing and editing process. Particularly helpful were Alex Keller, LaVonne Jacobsen, Thoreau Lovell, and Andrew Roderick at San Francisco State University and Lisa Marie Carlson, Milica Gurney, and Teresa Cameron of the Stanford University School of Education. Suzanne Radotic of Stanford was especially helpful, supportive, and resourceful in the provision of expert research assistance. Joanne Fraser LeGates and Lisa Ryan provided essential support and encouragement for our work.

INTRODUCTION

During the last thirty-five years our students in urban studies and city planning courses at San Francisco State University, Stanford University, and the University of California, Berkeley, have often asked us what is the best writing on a given topic or what single new writing will let them know what is happening in a given area right now. Since there was no one source to which we could refer them, each of us accumulated photocopies of what we consider to be *essential* writings and bibliographic references to many more. As time passed our colleagues began to come to us for suggested course readings, and we in turn added other selections they have found most useful to our list. We realized that a systematic organization of the best writings we use to meet both requests would make a good anthology to introduce students of urban studies and city planning to the field and to supplement course texts used in these and other courses concerned with cities. Accordingly we set to work on a first edition of *The City Reader* that was published by Routledge in 1996. The selections in the anthology represented both kinds of essential readings – enduring writings we use consistently and the most exciting new writings to which we point our students. The first edition was well received and we learned a great deal more about what readings students and faculty find most useful from using the first edition in our own courses and receiving feedback from faculty colleagues about what selections their students found most useful. In 2000 Routledge published an expanded and improved second edition of *The City Reader* that quickly established itself as required reading in courses in urban studies, city and regional planning, urban geography, and sociology worldwide.

The third edition of *The City Reader* continues and expands this tradition. Faculty and students familiar with the earlier editions will find all of the classic and contemporary selections which have proven most useful as well as exciting new material on green urbanism, Islamic cities, information technology, globalization, regionalism, city responses to terrorism, social capital, and much more. The third edition contains a series of first-person accounts of great cities in the past, a new plate section with images related to urban society, culture, and politics, and many new selections.

The City Reader: *Third edition*, has become the anchor text in a series of Routledge urban readers. Experts in the academic disciplines most related to the study of cities and in topical areas of greatest interest to urbanists will edit readers under our direction with a format similar to *The City Reader* to provide in-depth coverage of city-related material in their respective disciplines and topic areas. In addition to *The City Reader*: *Third edition*, Routledge will publish *The Cybercities Reader* edited by Stephen Graham and a second edition of *The City Cultures Reader* edited by Malcolm Miles, Tim Hall, and Iain Borden in 2003. Four more disciplinary and two more topical urban readers are scheduled to be published in 2004: *The Urban Geography Reader* edited by Nick Fyfe and Judith Kenny, *The Urban Sociology Reader* edited by Jan Lin and Christopher Mele, *The Urban Politics Reader* edited by Elizabeth Strom and John Mollenkopf, *The Urban and Regional Planning Reader* edited by Eugenie Birch, *The Sustainable Urban Development Reader* edited by Stephen M. Wheeler and Timothy Beatley, and *The Global Cities Reader* edited by Neil Brenner and Roger Keil.

Routledge plans to publish *The Urban History Reader*, *The Urban Design Reader*, *The Urban Community Services Reader*, and *The New Urbanism Reader* in 2005.

The City Reader and the other readings in the series focus on *essential* writings. We picked enduring issues in urban studies and planning across different cultures and times. In our courses, we have found that H.D.F. Kitto's "The Polis" raises fundamental questions about individuals' relations to their communities which are as relevant today as they were 2,400 years ago; that Louis Wirth's seventy-year-old essay on "Urbanism as a Way of Life" speaks to our students trying to understand contemporary urban violence, economic dislocation, homelessness, and anomie; and that our students are excited by William Julius Wilson's theories on the Black underclass and Manuel Castells's reflections on the rise of the network society. Most writings in this edition of *The City Reader* are from twentieth-century writers and more than half were written very recently.

This is an *international* anthology. In an increasingly global world students must learn from writers beyond the borders of their country of origin. In addition to writers from the USA, the third edition now contains writings by scholars from Austria, Australia, England, France, Germany, Italy, Scotland, Spain, and Morocco. Many of the writers included, from the great fourteenth-century Islamic wanderer Ibn Battuta to today's Manuel Castells, are truly world citizens. The third edition of *The City Reader* includes new material on cities in Asia and the Middle East, but space limitations prevented us from including as much international material as we would have liked. Many of the urban realities and urban processes described in the North American and European context are applicable everywhere precisely because they have become so internationalized. Other of the Routledge urban readers will include many more selections from everywhere in the world. Students studying globalization or comparative urbanism may particularly benefit from paired use of *The City Reader* and *The Global Cities Reader*.

The City Reader is an *interdisciplinary* anthology. The disciplines and professional fields represented in *The City Reader* include anthropology, architecture, archaeology, city planning, classics, culture studies, demography, economics, geography, history, landscape architecture, photography, political science, sociology, and urban design. Many of the writers blend insights from more than one discipline. Some of the best writing in *The City Reader*, such as Castells's writings about the informational city and Dolores Hayden's writings on non-sexist cities, don't fit in conventional disciplinary boxes at all.

We believe that cities are best studied from both interdisciplinary and disciplinary perspectives. The *disciplinary* Routledge urban readers will contain writings by scholars from the respective academic disciplines. Pairing *The City Reader* and one of the Routledge urban disciplinary readers will provide students in courses dealing with cities both the broad interdisciplinary perspective of *The City Reader* and the disciplinary perspective of the disciplinary reader. Thus, for example, using both *The City Reader* and *The Urban Geography Reader* as course texts will give students in urban geography courses both a broad interdisciplinary introduction to scholarship on cities *and* in-depth coverage of the best scholarship by geographers on topics of particular interest to urban geographers. Using both *The City Reader* and *The Urban Politics Reader* would give students in urban politics courses both an interdisciplinary understanding of cities and in-depth coverage of urban politics topics written by political scientists.

The City Reader emphasizes the connection between the built environment of cities and the natural environment. As the world's population soars and urbanization continues, the imperative to design sustainable cities becomes ever more important. Readings by Timothy Beatley, Stephen Wheeler, Peter Calthorpe and William Fulton, and Andres Duany and Elizabeth Plater-Zyberk introduce students to sustainable urban development, green urbanism, and the New Urbanism. Courses on sustainable urban development, green urbanism, ecological design, and related topics may benefit from pairing *The City Reader* with *The Sustainable Urban Development Reader*.

The writings in this anthology seek to combine *theory and practice*. "Urban Studies" is the term commonly used to refer to the academic study of cities. Knowledge about cities generated by social scientists and others is sometimes taught in a single program, sometimes dispersed among academic departments. The goal of these courses is primarily to teach students to *understand* cities, only secondarily to empower them to change cities. On the other hand, professional city planning, town planning, and regional planning courses explicitly train students to work as city planners. Often planning courses are taught as part of graduate or undergraduate professional degree programs; sometimes as part of geography, architecture, or departments in the social sciences. This anthology blends both the goal of *understanding* cities and the goal of *planning* and implementing policy for them. We feel planning and policy should be informed by understanding and that studying urban planning and policy can enhance understanding.

This anthology includes material on *race and gender* issues in cities, both in the selection of writings and introductions. Diversity characterizes cities throughout the world and students need to understand the situation, contributions, and perspective of women and people of color as an *integral* part of scholarship about cities. To produce balanced coverage of issues of race and gender, we have included essential writings by and about women and people of color throughout the anthology. We also include consideration of diverse groups in our introductions to the book, sections, and selections.

An anthology of essential writings on cities should have a *flexible organization*. There is no one best way to organize material on cities. Urban studies and city planning courses' content vary widely and courses are organized in as many different ways as there are courses. This dictates a flexible structure for the book. Readings are grouped into eight broad categories: The Evolution of Cities; Urban Culture and Society; Urban Space; Urban Politics, Governance, and Economics; Urban Planning History and Visions; Urban Planning Theory and Practice; Perspectives on Urban Design; and The Future of the City.

The section on the evolution of cities is chronological and works as a unit in the sequence in which the selections are presented. Some teachers may pick and choose selections from this section or use some selections as part of courses that do not have a chronological evolution section. The three sections on urban planning and design contain fifteen selections in all and could form the core readings for an entire course in urban planning. *The City Reader* and *The Urban and Regional Planning Reader* have been planned together to provide students beginning city and regional planning courses with the best possible theoretical and applied foundation for urban planning study. The other groupings work well if selections are read in the order in which they are presented, but different sequencing may work better in the context of a given course. Professors experiment; students enjoy!

One goal in picking the selections was to expose students to *great thinking and clear writing*. Almost everything written on the emergence of cities acknowledges a debt to the meticulous empirical research, creative theory building, and clear writing of Australian archaeologist V. Gordon Childe, on classical Greek cities to H.D.F. Kitto's delightfully written interpretation of the Greek polis, and on urban design to Kevin Lynch's brilliant and lucid writings on the image of the city. Students can learn a great deal from the way Childe, Kitto, Lynch, and other great urban scholars think and write beyond the substantive content of the selections.

The anthology begins with a section on the evolution of cities. We have found that even our brightest and most experienced students bring time- and place-specific cultural concepts to their study of cities. We warn them, and other readers of this anthology, against too quickly assuming what a "suburb," a "slum," an "ancient city," or any other urban settlement is. Some of our San Francisco Bay Area students who grew up in the affluent, predominantly white, residential suburb of Palo Alto think of Palo Alto when they think of a suburb. Material on the evolution of cities enriches their understanding that there were suburbs in medieval European cities composed of traders free from the medieval guilds, suburbs where Manchester's bourgeoisie fled the pollution and stench of industrial Manchester in the 1840s, and a proliferation of streetcar suburbs as nineteenth-century technology, entrepreneurship, and public tastes made them possible. These students may be surprised to realize there are a great range of suburbs today – including working-class suburbs, Black suburbs, and technoburbs. Studying the evolution of cities sharpens awareness of these differences. And understanding is the key to successful city planning. A new selection with writings by Marco Polo, Bernal Diaz, Ibn Battuta, and Albrecht Dürer provides students with first-hand accounts by contemporary observers of great cities in Europe, Asia, and the Middle East at different times in their past. *The City Reader* is appropriate to pair with *The Urban History Reader* for courses in urban history.

Part Two of *The City Reader* on Urban Culture and Society reflects our own interest in the relationship between urban studies and urban culture. It is deliberately placed near the beginning of the anthology because we feel materials on urban culture and society lay a good foundation for all that follows. If there was a single most important insight the great American intellectual Lewis Mumford emphasized in sixty years of tireless polemics against determinists of every stripe, it was that a city is an expression of the human spirit; not just a physical entity or a locus of economic activity. Accordingly, we begin Part Two with Lewis Mumford's essay on "What Is a City?" Other writings in this section review important contributions sociologists and anthropologists have made to our understanding of urban culture, explore the culture of poverty and underclass debates, and introduce students to material on social interaction. We close Part Two with Sharon Zukin's "Whose Culture? Whose City?" and an original essay and plate section by Frederic Stout on the city and the emergence of modern visual culture. We have added Robert Putnam's important essay "Bowling Alone" to this section. Students in culture studies and urban sociology courses may benefit from a pairing of *The City Reader* and *The City Cultures Reader* or the *Urban Sociology Reader*.

Part Three deals with Urban Space. Writings about urban space come from many disciplines and are essential to understanding cities. The physical form of cities is fundamental to urban planners, whose work involves the shaping of urban space. Understanding urban space is important for architects, landscape architects, urban designers, and anyone interested in cities. The great scholar of vernacular landscape, J.B. Jackson, spent a lifetime decoding the meaning of ordinary landscapes. Part Three contains Jackson's description of the logic underlying the physical organization of a hypothetical, ordinary, town in the Western United States both for its intrinsic interest and to illustrate the kind of intellectual inquiry Jackson pioneered. We have included in the section on Urban Space other writings by sociologists, architects, and planners, from Ernest Burgess's classic essay proposing a concentric zone model of the internal structure of the city to Mike Davis's rant against the oppressive design of "Fortress L.A." to Saskia Sassen's work on cities in a global economy. The third edition of *The City Reader* includes for the first time material on space and society in Islamic cities by Janet Abu-Lughod and on social exclusion in European cities by Ali Madanipour. Students in urban geography courses may particularly benefit from use of both *The City Reader* and *The Urban Geography Reader*.

Part Four of *The City Reader* is on Urban Politics, Governance, and Economics. Selections in this section explore the way in which political scientists, sociologists, and academics from other disciplines think about urban politics, governance, and economics. We particularly emphasize the debate between pluralists and structuralists, Marxist and non-Marxist theories, and liberal and conservative views on community development. Other selections in Part Four emphasize the impact of global restructuring on cities and interconnections between urban economics, politics, and society. We have added a new selection by Michael Sorkin dealing with the politics of rebuilding New York after the attack on the World Trade Center to this section. Students in urban politics courses may particularly benefit from use of both *The City Reader* and *The Urban Politics Reader*.

Part Five is on Urban Planning History and Visions. We tell our beginning students that the best city planning is slightly utopian and encourage them to envision urban futures that they would like to see. Part Five contains powerful visions of urban futures by Frederick Law Olmsted, Ebenezer Howard, Le Corbusier, and Frank Lloyd Wright that should challenge students to think through their own values about the importance of aesthetics, community, the relation of man to nature, social equity, efficiency, and other enduring issues. We hope our students won't lose the power to dream as they tackle research methodology, computers, finance, computer-assisted design, and other necessary professional coursework. Rather, we hope that they will see these technical skills as necessary knowledge to translate their own visions into reality. We have added a new selection by Peter Calthorpe and William Fulton to this section. A plate section containing photographs and line drawings illustrating a number of the classic planning visions by the authors follows Part Six. Students in courses in urban history and the history of urban planning may benefit from a pairing of *The City Reader* and *The Urban History Reader* or *The Urban and Regional Planning Reader*.

Part Six of *The City Reader* is on Urban Planning Theory and Practice. Peter Hall's masterful review, "The City of Theory," provides a fine introduction to urban planning theory. Edward J. Kaiser and David R. Godschalk's selection on twentieth-century land use planning provides an excellent introduction to the practice of city planning. Consistent with our emphasis on the importance of sustainable urban development, we close the section on urban planning theory and practice with a new selection by Timothy Beatley on green urbanism.

Part Seven covers Perspectives on Urban Design. It begins with a selection from Camillo Sitte's classic book on *The Art of Building Cities* and also contains seminal writings by Kevin Lynch, William Whyte, and Allan Jacobs and Donald Appleyard. Consistent with our emphasis on diversity we close this section with a selection by Dolores Hayden.

The final section of the anthology on The Future of the City picks up evolutionary themes from the first section and extends the visionary section of the sixth. We begin this section with Melvin Webber's prescient 1968 essay on "The Post-City Age." In this compact little piece, written when Bill Gates was in junior high school, Webber anticipated many of the debates now occurring about the future of cities in the information age. The section introduces students to Manuel Castells's theories about the network society and Stephen Wheeler's excellent summary of sustainable urban development literature. The selection closes with a new selection by William Mitchell on "The Teleserviced City" from *E-topia*.

We have said a good deal about the role of visions in urban studies and planning. We close with our own vision of how this anthology will be used. *The City Reader* is aimed primarily at students who will be encountering many of the writers and writings for the first time. We hope the writings touch responsive chords and will inspire all the students who use *The City Reader* to think more deeply and read more widely about cities. To that end, for each selection we point the way to other related writings by the same authors and other writers on the same subject matter. *The City Reader* will work well for students in general education courses who do not pursue urban studies further. For students who will take additional coursework in urban studies, city and regional planning, geography, sociology, political science, history, or other academic disciplines and professional fields we have designed *The City Reader* to provide an interdisciplinary overview of cities. In disciplinary courses *The City Reader* will work best if it is paired with another of the Routledge Urban Reader Series readers. *The City Reader* will provide a broad overview, an interdisciplinary perspective, and the best possible selections for each subject matter area that space permits. The disciplinary readers – edited by experts from the varying disciplines and containing many more writings by scholars from the respective discipline – will provide much more depth and focus in respective disciplinary areas.

We hope *The City Reader* will prove a book that professionally oriented students, professors, and practitioners will keep and periodically re-read. One test to which we put each of the essential writing included is that it should still be relevant to re-read and enjoy in twenty-five years.

Richard LeGates
Frederic Stout
San Francisco, February 2003

Prologue

"How to Study Cities"

adapted from Richard LeGates, "Urban Studies," *International Encyclopedia of Social and Behavioural Science*

Richard LeGates

Editors' Introduction

The City Reader contains just fifty-three selections written by writers, past and present, who have contributed most to our understanding of cities from literally millions of possible selections. The selections introduce readers to the best scholarship on urban history, geography, politics, economics, sociology, planning, and urban design. The following selection, written by one of the editors of *The City Reader*, steps back from the substantive material that follows to ask a single broad threshold question: "*how* do people study cities?"

Professor LeGates points out that colleges and universities today compartmentalize human knowledge, including knowledge about cities. Most colleges and universities rely on academic *departments* to teach material from the point of view of an academic *discipline*. Professors in a chemistry department who have Ph.D.s in chemistry, for example, teach chemistry using the accumulated knowledge from the discipline of chemistry while professors in math departments with Ph.D.s in mathematics teach math using the accumulated knowledge from their discipline. Professors are trained in and expected to know and to teach material in their respective academic *disciplines*.

The borderlines between the objects of study, methods, theory, and work product of different academic disciplines are fuzzy. Urban economists tend to study the economics of cities using quantitative methods and urban sociologists tend to study social aspects of the city using qualitative methods. But some urban economists use qualitative methods and their objects of interest extend beyond economics and some urban sociologists are very quantitative and interested in issues at or across the borderline with economics.

Some professors in academic disciplines value only theory and look down on applied research and writing intended to produce solutions to actual urban problems as derivative and inferior to theory building – not worthy of true scholars, and beneath their personal dignity. This is silly. Urban research lends itself well to applied research and good scholarship directed at problem solving can be just as theoretically subtle and methodologically sophisticated as pure theoretical research. Michael Porter's thoughtful prescriptions for urban poverty, James Q. Wilson and George L. Kelling's "broken window" theory and community policing ideas, and John Forester's insights about urban planning in the face of conflict are as intellectually rigorous as any of the less applied selections in *The City Reader*.

Most universities recognize that some topics are best understood by bringing the expertise of specialists from multiple disciplines together. They promote *interdisciplinary* programs where individual professors or groups of professors consciously combine material from more than one discipline.

Disciplines have the advantage that they are based on consistent methods for acquiring knowledge and a more or less agreed upon body of knowledge. All history professors, for example, in order to get their history Ph.D. degree must study methods of historical research that historians use. All history professors will have taken enough different history courses that they have a good overall knowledge of history in addition to their specialties in one or more specific time periods, issue areas, or methods of historical inquiry.

A disadvantage of disciplines is that they encourage rigid thinking within the four corners of the discipline itself. There is a danger that professors who are rigorously trained in economics, for example, will see only economic factors as important when they teach about urban issues. Because they have been trained in the importance of economics they may neglect political, social, and spatial aspects of the phenomena they are studying.

The strength of interdisciplinary approaches is that, done properly, they provide for a richer, more holistic, more varied understanding of multiple dimensions of the phenomena being studied. Urban issues do have multiple dimensions. There are social, spatial, and political aspects to urban poverty as well as economic ones.

The danger of interdisciplinary approaches is that they may become so loose and standardless that they lack intellectual rigor. Well-trained and specialized disciplinary scholars are often justifiably critical of colleagues who do wide but shallow teaching, research, and writing.

Many of the selections in *The City Reader* are written by recognized experts in the social science disciplines such as Kenneth Jackson (history), William Julius Wilson (sociology), and John Mollenkopf (political science). Other *City Reader* authors like Jane Jacobs, Lewis Mumford, Mike Davis, and Manuel Castells are true interdisciplinarians.

"How to Study Cities" provides some background on the evolution of scholarship about cities from the first half of the twentieth century, the nineteenth century, and before. This is helpful in understanding how we got to where we are now. In addition to placing many of *The City Reader* authors in context it mentions other scholars who have done and are now doing cutting edge research regarding the twelve most important issues areas for the study of cities.

Richard LeGates is a Professor of Urban Studies at San Francisco State University and the co-editor (with Frederic Stout) of *The City Reader*. He teaches the introductory Urban Studies course at San Francisco State and courses in research methods, land use law, and housing and community development policy.

Interdisciplinary academic urban programs are described in the Urban Affairs Association (UAA)'s *Directory of Programs in Urban Studies and Urban Affairs* (UAA, Cleveland, 1995).

Urban space is well described in Larry S. Bourne (ed.), *The Internal Structure of the City*, 2nd edn (New York: Oxford University Press, 1982), and Larry S. Bourne and J.W. Simmons (eds), *Systems of Cities* (New York: Oxford University Press, 1978).

Urban History is introduced in "New Perspectives on American History" in Raymond Mohl, *The Making of Urban America*, 2nd edn (New York: Scholarly Resources, Inc., 1997). Lewis Mumford's *The City in History* (New York: Harcourt Brace, 1961) and Sir Peter Hall's *Cities in Civilization* (New York: Pantheon, 1998) are excellent sources for sweeping historical accounts of the role of cities in civilization. John Reps's *The Making of Urban America* (Princeton, NJ: Princeton University Press, 1965) provides a clear history of the physical evolution of American cities. Spiro Kostof's *The City Shaped* (New York: Little Brown, 1991) is a visually exciting book that illustrates cities throughout the world at different periods in time.

John Mollenkopf's essay "How to Study Urban Political Power" from *A Phoenix in the Ashes: The Rise and Fall of the Koch Coalition in New York City Politics* (Princeton, NJ: Princeton University Press, 1992) which is included in *The City Reader* provides an excellent review of the urban politics and community power literature.

Jay Stein's anthology of *Classic Readings in Urban Planning* (New York: McGraw-Hill, 1995) collects in one volume classic scholarship on urban planning. A good way to get an overview of academic urban planning programs in North America is to look at the Association of Collegiate Schools of Planning (ACSP), *Guide to Graduate Programs in Urban and Regional Planning*, 10th edn (2000). Peter Katz, *The New Urbanism* (New York: McGraw-Hill, 1994) introduces important writings by people involved in this important contemporary urban planning movement.

Lewis Mumford's magisterial *The Culture of Cities* (New York: Harcourt Brace, 1938) was the first and remains the best book on the culture of cities. Richard Sennett's (ed.) *Classic Essays on the Culture of Cities* (New York: Appleton-Century-Crofts, 1969) contains more writings on the culture of cities.

Sakia Sassen's *Cities in a Global Society* (Thousand Oaks, CA: Pine Forge Press, 1994) is a readable summary of her own research and related research on globalization and cities.

Catharine Stimpson, Elsa Dixler, Martha Nelsen, and Kathryn Yatrakis (eds), *Women and the American City* (Chicago: University of Chicago Press) is an excellent introduction to feminist urban issues.

■ ■ ■ ■ ■ ■

Studying cities is a never-ending process – there is always more to learn. Fortunately many fine scholars, past and present, have focused their attention on cities and we now know a great deal about how cities evolved, their people and culture, how they are organized spatially, what economic functions they perform, how they are planned and governed, and their probable future. *The City Reader* contains the best classic and contemporary writings on all of these topics. But scholarship on cities has taught us something more. At the beginning of the twenty-first century we know a great deal more than ever in the past about *how to study cities*.

DISCIPLINARY AND INTERDISCIPLINARY TEACHING ABOUT CITIES

Almost all modern universities systematically organize research and teaching into academic units called *schools* or *colleges*. Schools or colleges contain academic *departments*. Professors trained in different academic *disciplines* are located within the departments: historians in the history department, economists in the economics department, and sociologists in the sociology department. Most universities also encourage research and teaching that cross disciplinary boundaries. Sometimes, for example, a history department may hire a faculty member whose Ph.D. is in ethnic studies or law, rather than history. A university may encourage a historian to teach a course that serves students in an urban studies department. Departments of urban studies and city planning are quite interdisciplinary. They may include faculty trained in city planning, architecture, geography, sociology, economics, political science, and other disciplines.

While professors from many different academic *disciplines* as well as *interdisciplinary* scholars study cities, most scholarship about cities – and most of the readings in *The City Reader* – come from *social scientists*: faculty who are trained to systematically study different aspects of human society. Most universities contain a school or college of social science. Schools of social science contain social science departments where

professors trained in the social science disciplines of geography, sociology, economics, political science, and anthropology teach. (History departments are sometimes located within schools of social science; sometimes within schools of humanities.) Within these social science departments professors interested in cities teach urban courses from the point of view of their disciplines: courses on *urban* geography, *urban* sociology, *urban* politics, and so on. They may include material from other academic disciplines in their courses. For example, a geography professor with an interdisciplinary orientation may be well read in economics and sociology and may use content and methods developed by economists and sociologists in her urban geography course.

Many universities have interdisciplinary urban studies programs specifically established to teach courses about cities that draw on multiple disciplines. Graduate and undergraduate urban planning departments are also interdisciplinary.

METHODS FOR STUDYING CITIES

Scholars who study cities use both *quantitative* and *qualitative* social science *research methods* to study cities. Both approaches can contribute to understanding cities and the best urban research often combines both quantitative and qualitative research. *Quantitative* methods involve analyzing data using statistical methods. Today most quantitative analysis is assisted by computers. A professor of urban politics doing statistical analysis of city voting data to see if recent immigrants feel differently about political priorities than longer-term residents would be doing quantitative urban research. Training in applied statistics is a regular part of most curricula that educate students to study cities. Students learn to use computerized statistical packages to do quantitative analysis. Because many urban phenomena have a spatial dimension, geographical information systems (GIS) software that permits users to map data is also very important in studying cities.

Qualitative research need not involve numbers or

statistical analysis. William Whyte's study of how people use urban parks and plazas is a notable example of *observation*. Urban sociologist William Julius Wilson conducted exhaustive *field research* in Black ghetto areas of Chicago. Urban planner John Forester *interviewed* dozens of urban planners to understand planning in the face of conflict. Urban designer Kevin Lynch and his students conducted *surveys* of residents of Boston and other cities to understand how they perceived the city environment.

An urban anthropologist interviewing Cambodian immigrants in New York City to see how they feel about local political issues would be doing qualitative analysis. If the political scientist who was doing statistical analysis of urban voting data and the anthropologist co-authored an academic paper reporting on what they found out using both quantitative and qualitative methods they would be doing interdisciplinary research using multiple methods to understand cities.

THEORY AND PRACTICE

Theory in the social sciences is intended to *explain* phenomena and provide a framework for understanding. Urban studies and urban planning are applied fields and much scholarship is directed towards the practical ends of solving problems. A professor like Saskia Sassen (p. 212) who is interested in the way in which information technology is affecting the size and relative importance of cities might use quantitative and qualitative methods to develop a body of *theory* explaining what is happening to the global system of cities. Policy makers and urban planners might draw on Sassen's theory to plan cities and implement policy for them. The relationship between theory and practice is reciprocal. Theory can inform practice and practice can inform theory.

ORGANIZATIONS AND JOURNALS DEVOTED TO THE STUDY OF CITIES

A number of academic associations organize conferences, set standards, publish academic journals, and work to advance scholarship related to understanding cities. In North America the academic association most directly concerned with cities is the Urban Affairs Association (UAA). The European counterpart organization is the European Urban Research Association (EURA). In North America, schools with urban planning programs are members of the Association of Collegiate Schools of Planning (ACSP); in Europe, of the Association of European Schools of Planning (AESOP), and in Asia, the Asian Planning Schools Association (APSA). Disciplinary academic organizations like the American Sociological Association, the American Political Science Association, and the American Geographical Association have specialized branches dealing with urban scholarship to help members of the disciplines with urban interests share their knowledge.

The main professional association of city planners in the United States is the American Planning Association (APA). In the UK an equivalent organization is the Royal Town Planning Institute. Practicing planners in these organizations meet to discuss their professional interests.

Journals like *The Journal of Urban Affairs*, *Urban Studies*, and *The International Journal of Urban and Regional Research* publish scholarly articles on cities. The *American Planning Association Journal* and *The Journal of Planning Education and Research* in North America and the *Town Planning Review* in the UK publish scholarly articles on urban planning.

THE EVOLUTION OF SCHOLARSHIP ABOUT CITIES

Cities have excited the interest of scholars since the earliest times. In *The Republic*, the Greek philosopher Plato envisioned the ideal polis in the fifth century BCE. The Roman architect Vitruvius formulated principles for planning cities. During the renaissance Florentine architect Leon Battista Alberti reminded his peers of how cities could contribute to the rebirth of humanism.

During the industrial revolution unplanned, congested, polluted slums jammed with a newly impoverished urban proletariat dominated nineteenth-century industrial cities. Friedrich Engels (p. 58) conducted empirical investigations of nineteenth-century Manchester – mainly by just walking through the city slums and observing the deplorable conditions. Ebenezer Howard (p. 309) envisioned humane, social, garden cities surrounded by greenbelts in place of cities of this time – and lived to see several garden cities built. Landscape architect Frederick Law Olmsted

(p. 302) led an urban parks movement. Austrian architect Camillo Sitte (p. 413) developed theories of plaza design to provide better urban civic space. Civil engineers like Baron Haussmann in Paris rebuilt the infrastructure and buildings of cities, and settlement house workers like Jane Adams in Chicago studied, theorized about, and agitated for the improvement of nineteenth-century industrial cities. But it was not until the twentieth century that academic courses and degree programs focusing on cities were created.

In the twentieth century, the University of Liverpool in England and the Massachusetts Institute of Technology in the United States began the first formal courses devoted to cities. Liverpool established a degree program in town and country planning in 1909 and MIT instituted a course in urban planning the same year.

Systematic study of cities took a giant leap forward a decade later when the University of Chicago created the world's leading center of urban sociology in the 1920s. The Chicago School of sociology attracted creative and talented scholars like Ernest W. Burgess (p. 156) and Louis Wirth (p. 97) who used neighborhoods surrounding the campus to study urban society. They developed the field of urban sociology and inspired other disciplines to study cities.

Coursework with cities as the unit of analysis slowly penetrated other social science disciplines. The number of urban planning programs grew during the 1930s, 1940s, and 1950s. Still, by the end of the 1950s there were only a handful of courses in urban politics, urban history, and urban anthropology, not one text in urban economics, and just a few dozen small design-oriented urban planning programs in the entire world.

Ghetto riots, and a massive increase in federal aid to cities in the United States in the 1960s greatly expanded academic study of cities. Jane Jacobs's *The Death and Life of Great American Cities* (p. 114) and Lewis Mumford's *The City in History* were both best sellers. Discipline-based and interdisciplinary urban courses proliferated during the 1960s and early 1970s and many of today's academic urban studies programs were created at that time.

CORE CONCERNS IN THE STUDY OF CITIES

Following is a list of twelve core concerns that have most intrigued scholars who study cities and an introduction to notable scholars and key scholarship in each area.

The evolution of cities

Historians have made the greatest contribution to understanding the evolution of cities. Archeologists, geographers, political scientists, economists, architects, and other scholars have also contributed to understanding of how cities evolved.

Archeologist V. Gordon Childe (p. 35) excavated the ancient Mesopotamian city of Ur in present-day Iraq. Mortimer Wheeler excavated Mohenjo Daro and Harrapa in the Indus River Valley, James Mellaart excavated Catal Hüyük in Turkey, and Dame Kathleen Kenyon excavated Jericho in Israel. These scholars developed *descriptions* of what the earliest cities were like and conflicting theories of *why* they arose.

Historians have approached the study of cities from many different perspectives. Raymond Mohl has proposed an eleven-category typology of American urban history that provides a useful introduction to all historical urban studies. Mohl's categories are: urban political history, suburbanization, city and region, sunbelt cities, technology and the city, planning and housing, the urban working class, immigration and ethnicity, African-American urban history, urban policy history, and urban culture.

Urban Planners such as John Reps, Eugenie Birch, Christine Boyer, and Mellior Scott in the United States and Sir Peter Hall and Steven Ward in England have developed the subfield of urban planning history. They have been joined by historians Carl Abbot, Mary Corbin, Christopher Silver, William H. Wilson, David Schuyler, and Stanley Buder, geographer Jay Vance, architectural historian Spiro Kostof, and many others who have written about urban planning history topics as varied as the building of Athens in the fifth century BCE, Michelangelo's work in renaissance Rome, Baron Haussmann's rebuild of Paris in the mid-nineteenth century, city beautiful planning for early twentieth-century Chicago, and the vision and reality of planned garden cities in early twentieth-century England.

Urban culture

Cities have always been centers of both high and low culture. Greek drama, poetry, art, and sculpture flourished in Athens in the fifth century BCE. Imperial Rome produced both high culture for elites and gladiatorial spectacles for the masses. Florence was an incubator of renaissance art, architecture, sculpture, poetry, and literature. Elizabethan London generated the theaters and dramatic troupes that made Shakespeare's plays possible.

Historian Arnold Toynbee, geographer/planner Peter Hall, sociologist Sharon Zukin (p. 136), polymath Lewis Mumford (p. 92), and many other scholars have studied urban culture.

Lewis Mumford pioneered the study of urban culture in the late 1930s in *The Culture of Cities*. Historians Warren I. Susman, Gunther Barth, Neil Harris, and others have since studied the history of urban culture. Richard Sennett has collected studies of urban culture.

The dominant strand in current culture studies of cities is postmodernism. Geographers David Harvey (p. 227) and Edward Soja (p. 189), sociologist Sharon Zukin (p. 136), social critic Mike Davis (p. 201) and many other scholars are fascinated and often horrified by the clash of cultures and values most manifest in contemporary cities.

Urban society

The Chicago school of sociology encouraged faculty and students to study Blacks in Bronzeville, prostitutes in the Gold Coast, hobos, Polish peasants, gangs, and many other Chicago groups. Urban ethnographies – detailed field research on urban subcultures – continue this tradition. Anthropologist Elliot Liebow's poignant descriptions of poor, Black streetcorner men and homeless women are exemplary of this tradition.

Another concern is with how rural migrants adapt to city settings. There are many studies of urban immigration, acculturation, and assimilation such as Oscar Handlin's studies of the Boston Irish and Ron Takaki's studies of Asian immigrants to the West Coast of the United States. Chicago school sociologists pioneered urban ecology to describe the structure and processes of urban ethnic neighborhood change, and the study of urban ecology continues.

Problems of poverty are central to urban studies.

Anthropologist Oscar Lewis's controversial theory that a distinct culture of poverty exists stimulated scholarly debate in the 1960s and 1970s. Harvard sociologist William Julius Wilson's studies of the urban underclass continue the debate (p. 126).

Yale professor Dolores Hayden (p. 448) and sociologists Lynn Lofland and Janet Abu-Lughod (p. 172) study the way in which humans use public space. Anthropologists and sociologists have studied the way in which human beings use the space around them. Edward T. Hall studied how people respond to intrusions into their space. Robert Sommers has observed behavioral aspects of how people use the space in such diverse settings as prisons, airports, and jury rooms. These behavioral and social science studies provide guidance to planners and designers.

Another enduring concern in urban studies is the nature of community. The pioneering nineteenth-century sociologists Emile Durkheim and Fernand Tönnies regretted a loss of community in large, impersonal nineteenth-century cities. Louis Wirth concluded that big city life created a distinct – and not very attractive – human personality type. Recently Robert Putnam's brilliant essay titled "Bowling Alone" (p. 105) has inspired a new generation of studies about the loss of community and declining social capital and a worldwide movement to engage young people in community service. Other social scientists have identified positive networks and supportive communities within cities. For example, sociologist Herbert Gans characterized Boston's West end as an urban village where migrants from southern Italy retained many of the relationships and folkways of their villages of origin. Michael Young and Peter Willmott's classic study of family and kinship in East London found positive networks in a London slum and far less community in new housing estates built for the relocated slum dwellers. The nature of community is one strand in the large and growing urban studies literature on suburbia by scholars such as historian Kenneth Jackson (p. 67), political scientist Robert Wood, and sociologist Mark Baldassare.

Urban politics and governance

The study of how cities are governed is a central concern for political scientists. A recent review of the literature on how to study urban politics by City University of New York political scientist John

Mollenkopf (p. 235) summarizes the evolution and current status of the subfield of urban politics.

Historians like Amy Bridges, Leo Hershkowitz, and Martin Schefter have provided historical studies of urban political bosses and political machines. Samuel Hayes and Roy Lubove have studied urban progressives and reformers. Economist James O'Connor, geographer David Gordon, social work professor Francis Fox Piven, and other urban political economists have brought Marxist and related critical theory to bear on understanding urban politics. Public choice theorists – mainly economists – have developed theory about how local governments do (and should) make the economic decisions that constitute the core of urban governance. Public administrators like David Osborne and Ted Gaebler study how cities are (and should be) managed. Journalist Robert Caro won a Pulitzer prize for his monumental study of Robert Moses and the politics of planning the twentieth-century New York region.

Sociologist Floyd Hunter and political scientist Robert Dahl, in classic studies of governance in Atlanta, Georgia, and New Haven, Connecticut, developed elitist and pluralist models of urban community power, respectively. Other pluralists like Edward Banfield, Herbert Kaufman, Martin Meyerson, Wallace Sayre, and Douglas Yates conflict with Peter Bachrach, Morton Baratz, John Manley, William G. Domhoff, and other theorists who continue to advance elitist explanations of community power.

Marxist political theorists like political scientists Ira Katznelsen, William Tabb, and David Gordon, economist James O'Connor, and French geographer Henri Lefebvre, posit structuralist theories of urban political power. Structuralist interpretations argue that underlying economic and class relationships predetermine most of what is important to decide at the city level. Structuralists see city governments as unable to control national and international capitalist forces. For them, urban politics is largely irrelevant.

Regime theory is an approach to studying community power, developed by political scientist Clarence Stone. Stone, Frank R. Baumgartner, Richard DeLeon, Steven Elkin, Brian D. Jones, Gerry Stoker, and other regime theorists conclude that complex, public–private regimes involve not only local elected officials, but also businessmen, church leaders, unions, the media, and other interest groups. Regime theorists do not deny the structural constraints local governments face, but they argue that complex, shifting public/private regimes are important to local political outcomes.

Urban economics, urban public finance, and regional science

The economics of cities is primarily the province of economists. Since Wilbur Thomson's *Preface to Urban Economics* (1985), urban economics has been recognized as a subfield of economics. Economic base theory, input–output analysis, location theory and other constructs, as well as numerous studies of specific city economies, enrich scholarly understanding of cities. Professors of business administration like Michael Porter (p. 227) contribute to this literature.

Economic geographers have studied the production and exchange of goods within and between cities. Historians interested in both economics and cities – like Belgian historian Henri Pierenne and French historian Fernand Braudel – have developed the subfield of urban economic history and have illuminated the relationship of capitalist institutions and trade to the rise of cities.

Two other interdisciplinary subfields closely related to economics help inform urban studies. Urban public finance is sometimes taught in economics departments; sometimes in business schools. It contributes to an understanding of how local governments raise revenue and the nature of their expenditures. The subfield of regional science includes economists, geographers, and others who study *regions* rather than cities. Regional scientists generally employ quantitative methodologies. They have their own professional associations and scholarly journal.

Urban and metropolitan space and city systems

Another core area of concern in studying cities is the *physical* organization of cities and regions and relationships among cities. This work is often classified into studies of the *internal structure of cities* and studies of *systems of cities*. Geographers are particularly interested in these issues, but sociologist Ernest W. Burgess (p. 156) and economist Homer Hoyt originally did pioneering work on the internal structure of cities. Writings about the internal structure of cities have been collected in an anthology edited by University of Toronto geographer Larry Bourne.

German geographer Walter Christaller's careful studies of the size and function of settlements in southern Germany during the 1930s caused him to formulate central place theory – a unified explanation of the way in which cities of different sizes are related to each other in a distinct hierarchy or system. Christaller's work led to many empirical studies of systems of cities. University of Toronto geographers Larry S. Bourne and J.W. Simmons have collected systems of cities studies (Bourne and Simmons, 1978).

Megacities and global city systems

The rise of enormous cities with populations of ten million or more people has captured the attention of urban studies scholars. So have the increasingly integrated world economic system and the close connections between cities and regions around the globe.

One strand of research and writing concerns the rise of *megacities* by Janet Abu-Lughod, Manuel Castells, Susan Fainstein, Peter Hall, Alan Gilbert, John Kasarda, Fu-chen Lo, Janice Perlman, Saskia Sassen, Yue-man Yeung, and authors of UN-commissioned city-specific megacity studies. The nature, causes, and appropriate policy response to megacities are particularly important in Asia and South America, where most of the megacities are emerging.

A related area concerns development of huge urbanized *regions* – what Scottish biologist Patrick Geddes in the early twentieth century termed conurbations and French geographer Jean Gottman in a study of the eastern seaboard of the United States named megalopolis.

A third area of inquiry involves the increasingly close relationship among cities caused by economic and political integration and by information technology. Spanish sociologist Manuel Castells (p. 475) uses the term informational city to describe advanced cities today and the term network society to describe the current global system in which information-rich multinational corporations are increasingly beyond the control of city government, citizens, and even nation-states. University of Chicago sociology professor Saskia Sassen (p. 212) is studying the current global system of cities.

Technology and cities

A number of urban studies scholars are interested in aspects of technology in cities. Transportation has received the most attention, but public infrastructure has also been of interest. The impact of the Internet and other information technologies on cities is attracting increasing attention and promises to be even more important in the twenty-first century.

There is a large literature on the impact of transportation technology on cities. Geographer Jay Vance has studied the relationship between transportation technologies and urban space. Historians like Sam Bass Warner, Charles W. Cheape, and Clifton Hood have documented the way in which the electric streetcar directly affected first the spatial structure of cities and metropolitan regions and, as a consequence, their social and economic structure. Kenneth Jackson (p. 67), Mark S. Foster, William Issel, Clay McShane, Mark H. Rose, and others have studied the impact of automobiles on the rise of suburbia. They are interested in public policy regarding freeways and automobiles and its impact on urban and regional physical form, social structure, and culture. Policy analysts and planners like Anthony Downs (p. 256) and Robert Cervero write about transportation policy and planning issues.

Scholars are interested in the way in which information technology is changing spatial and social relationships – particularly the issues of whether global access to information will lead to larger and more powerful global cities or disperse population and power and how information technology will impact social equity. Manuel Castells (p. 475), Saskia Sassen (p. 212), and William Mitchell (p. 497) write about these topics

Urban planning, urban design, landscape architecture, and architecture

Urban studies overlaps the professional fields of urban planning, urban design, architecture, and landscape architecture.

Professors of Urban Planning in departments that belong to the Association of Collegiate Schools of Planning (ACSP) in North America, the Association of European Schools of Planning (AESOP), and the Asian Planning Schools Association (APSA) study cities in order to help plan them. Twentieth-century land use planning books by Patrick Abercrombie,

Thomas Adams, Alan Altschuler, Edward Bassett, F. Stuart Chapin, Anthony Catanese, Barry Cullingworth, Alexander Garvin, Peter Hall, Jack Kent, John M. Levy, Edward J. Kaiser, David Gotschalk, and James C. Snyder combine material related to geography, economics, politics, sociology, and the design professions with applied material on plan preparation.

Ever since late nineteenth-century Austrian architect Camillo Sitte derived urban design principles from his careful observation of public spaces in European cities and how people use them, urban designers have been close observers of city space and contributors to urban studies. Kevin Lynch blended history, sociology, psychology, and humanistic material in *The Image of the City* (p. 424). Sociologist William Whyte's studies of urban park and plaza design (p. 429), planner Allan Jacobs's (p. 437) studies of streets and boulevards, Spiro Kostoff's studies of urban form and books by urban designers Edmund Bacon, Gary Hack, Clare Cooper Marcus, and other design scholar/practitioners have made major contributions to our understanding of cities.

Ever since the pioneering work of Frederick Law Olmsted in the nineteenth century (p. 302), landscape architects have contributed to understanding the relationship between the natural and built environments. Systematic study of how to plan human settlements in harmony with the natural environment were pioneered by Scottish biologist Patrick Geddes before World War I and revitalized by Scottish landscape architect Ian McHarg's approach to design with nature in the 1960s. This approach resonates with advocates of sustainable urban development such as Stephen Wheeler (p. 486), the New Urbanism such as Peter Calthorpe (p. 331), Andres Duany and Elizabeth Plater-Zyberk (p. 207), and green urbanists like Timothy Beatley (p. 399) today.

Architects like Frank Lloyd Wright (p. 325) in the United States, Raymond Unwin in England, and Le Corbusier (p. 317) in France moved far beyond design of individual structures to think and write about the design of entire communities and of society itself.

Race, ethnic, and gender relations in cities

W.E.B. Du Bois's 1899 study of *The Philadelphia Negro* (p. 119) began a tradition of studies by African-American scholars of urban Black society. Horace R. Cayton, Kenneth Clark, St Clair Drake, E. Franklin Frazier, and William Julius Wilson (p. 126) have continued that tradition.

Historian Oscar Handlin, political scientist Daniel Patrick Moynihan, and sociologist Nathan Glazer's early studies of the urban experience of Irish and Jewish immigrants to cities on the East Coast of America have been followed by many studies of Irish, Jewish, central European, Southern European, and other ethnic groups in cities of the East Coast and Great Lakes regions. Ron Takaki and other scholars are now building a similar body of scholarship about the experience of Asian and Hispanic migrants to cities in the American West and Southwest.

Sociologists, anthropologists, historians, planners, and women's studies scholars are generating an interdisciplinary literature about women in cities. Yale professor Dolores Hayden (p. 448) pioneered this subfield with studies on the relationship between gender, work roles, and the built environment highly critical of the male built environment. Eugenie Birch, Ann Forsyth, Jacqueline Leavitt, Lynn Lofland, Claire Cooper Markus, Ann Markusen, Risa Palm, Marsha Ritzdorf, Leonie Sandercock, Wendy Sarkissian, Catherine Simpson, Daphne Spain, and others have done important research on women in cities and helped inject feminist concerns into urban studies and planning.

Urban problems and policy

Urban studies is also concerned with urban problems and policy. It was poverty, overcrowding, congestion, disease, crime, vice, and other urban *problems* that inspired many nineteenth-century studies of cities. Urban pathology was a concern of the Chicago School of sociology. The focus on urban problems dominated urban studies after the ghetto riots of the 1960s. James Q. Wilson and George L. Kelling's classic "Broken Windows" article (p. 267) is in this tradition. The urban *problems* orientation is frequently balanced today with studies of *community assets and strengths*. Some studies like Michael Porter's work on the competitive advantage of the central city (p. 227) note the positive contributions neighborhoods can make and human capital and other assets they posses as well as negative aspects of these neighborhoods like poverty, crime, and low levels of education.

Most academic programs studying cities include applied coursework in policy areas such as land use, housing, community development, environment, transportation, and economic development. Some have coursework on crime, health, homelessness, education, welfare, and other policy areas. Scholars with applied urban interests include John DeGrove (land use), Anthony Downs (p. 256) (housing, transportation, race), Sim Van der Ryn (the environment), Robert Cervero (transportation), Michael Porter (p. 277) (economic development), James Q. Wilson (p. 267) (crime), James Coleman (education), and Francis Fox Piven (welfare).

Urban futurism

A final strand in the study of cities is reflection on probable, possible, and desirable urban futures. Virtually all demographers anticipate massive population growth and continuing urbanization – particularly in the third world. The "urbanization of the human population" that Kingsley Davis (p. 25) describes is likely to continue. There will likely be more congestion, sprawl, pollution, exhaustion of natural resources, extinction of species, and proliferation of megacities and vast urban conurbations unless governments intervene to plan and regulate city development far more than they have ever done in the past.

Studies of sprawl and ex-urban development raise the possibility that the rate of growth in cities may slow or even that counter-urbanization may occur in the future – at least in the developed world. Some writers believe information technologies will make cities unnecessary and a few even predict a withering away of cities, though empirical research by Saskia Sassen (p. 212) and other scholars has found no evidence that this is occurring.

In addition to *prediction*, urban futurists have formulated different *normative* visions for alternative urban futures. Ebenezer Howard (p. 309), Patrick Geddes, Lewis Mumford (p. 92), and Raymond Unwin opposed further growth of large cities and proposed regional systems of human-scale garden cities. Frank Lloyd Wright (p. 325) went further and advocated Broadacre City – a neo-Jeffersonian future in which each family would live on an acre. Soviet-era anti-urbanists took Marx's ideas of eliminating distinctions between the city and the countryside as a call for de-urbanization. Megastructuralists like Le Corbusier (p. 317) and Paolo Soleri took the opposite view and proposed building huge, self-contained structures.

CONCLUSION

Cities are civilization and the study of cities involves the study of humankind. No one can master all there is to know about cities and there will always be more to study and to understand as cities continue to evolve. But a great deal is known about cities and how to study them. Academic scholarship about cities is necessarily diverse, but it is increasingly well developed. *The City Reader* contains seminar writings on all of the main topics of greatest concern to students of the city. This selection provides a roadmap to the study of cities.

PART ONE

The evolution of cities

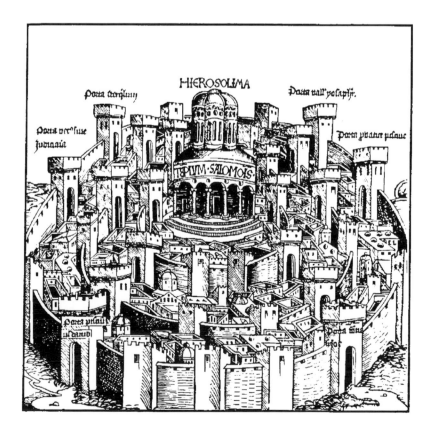

INTRODUCTION TO PART ONE

Cities are civilization. Humankind's rise to urban civilization took tens of thousands of years, but ever since the first true cities arose in Mesopotamia sometime between 4000 and 3000 BCE, the influence of city-based cultures and the steady spread and increase of urban populations around the world have been the central facts of human history.

As Kingsley Davis points out (p. 25) "urbanization" and "the growth of cities" are not the same thing. "Urbanization," as Davis defines it, is the *increase in the proportion* of a population that is urban as opposed to rural. That such an increase could take place without the growth of cities *per se* (for example, by the death of vast numbers of the rural population) or that city populations could grow without an increase in urbanization (as when the total population, urban and rural, increases at a similar rate) are important concepts that under-lie the history of urban life. This definition of urbanization helps to explain, for example, how immigration from the countryside to the city has repeatedly been the key factor in the history of urban development, as it continues to be today.

The first stage of urban history is what the Australian archaeologist V. Gordon Childe called "the Urban Revolution," that momentous shift from simple tribal communities and village-based agricultural production to the complex social, economic, and political systems that characterized the earliest cities of Mesopotamia, Egypt, and the Indus Valley. True, the earliest cities, in the ancient Near East and elsewhere, grew out of accumulated neolithic knowledge, and certain extensive neolithic communities (such as Catal Hüyük in Anatolia) pre-date the Mesopotamian cities by several millennia and may be regarded as at least proto-urban. For Childe, however, the development of writing was a crucial cultural element of true urbanism, and the emergence of the cities of the ancient Near East, where writing began, constituted the second of a series of massive transformations that gave shape to the whole of human evolutionary development. Although the successive stages overlapped, each of Childe's three "revolutions" (the agricultural, the urban, and the industrial) totally changed the world as it had been before.

In certain important respects, all the ancient cities are remarkably similar. Most are walled – except in Egypt, where the surrounding deserts may have been regarded as sufficient defenses – and all contain a distinct citadel precinct, separately walled, encompassing a temple, a palace, and the central granary. Most of the earliest cities also boasted some sort of pyramid or ziggurat. And, as Karl Wittfogel pointed out in *Oriental Despotism* (1957), almost all were located along major rivers and based their power (and that of their rulers) on the control of massive irrigation systems serving the surrounding countryside.

Thus, both the physical structure and socio-economic complexity of the earliest cities are unlike anything that had come before. Whereas the neolithic village had been ruled by a council of elders, the cities were mostly ruled by totalitarian god-kings and their attendant priests who formed a class totally apart from the rest of the citizenry. And whereas neolithic communities may have built earthen enclosures as ceremonial centers for ritual pageantry and hill forts for refuge and defense, the ancient cities – from Ur and Babylon on the Euphrates to Teotihuacan and Tenochtitlan in the Valley of Mexico – transformed these institutions into elaborate structures so massive that their remains are still visible today.

Many of the ancient cities elsewhere – in Shang and Chou China, sub-Saharan Africa, Southeast Asia, and Mesoamerica – arose quite independently of the cities of the ancient Near East. Still, what is remarkable is how similar ancient cities everywhere were in terms of social structure, economic function, political order, and architectural monumentality.

The cities of ancient Greece, on the other hand, developed on a very different model. Perhaps because they arose in narrow mountain valleys rather than on broad alluvial plains, the Greek cities that emerged around 1200 BCE and developed into an astonishing cultural efflorescence by 500 BCE were small (sometimes with a population of only a few thousand), economically self-contained, and almost village-like in their social and political institutions. It was the concept of urban citizenship and democratic self-government that was the distinctive contribution of the Greeks to the evolution of urban civilization. Greek democracy was by no means perfect and hardly inclusive. Women, slaves, and foreigners were all excluded from the rights and responsibilities of citizenship. But the cultural, artistic, and intellectual consequences of the democratic principle were extraordinary. "Within a couple of centuries," writes Lewis Mumford in *The City in History* (1961), "the Greeks discovered more about the nature and potentialities of man than the Egyptians or the Sumerians seem to have discovered in as many millennia."

If cities are civilization, they are also the cultural instrumentality by which humanity has attempted, since neolithic times, to achieve a higher, more inclusive concept of community. At the core of the Greek contribution to the history of urban civilization was the concept of the "polis." Sometimes translated as "city-state," at other times identified as the collective citizenry of a Greek city, the polis, as described so brilliantly by H.D.F. Kitto in *The Greeks* (p. 43), was both a community and a sense of community that helped to define the Greek citizen's relationship to his city and his fellow citizens, to the world at large, and to himself. In *The Politics*, Aristotle called man the "*zoon politikon*" (the "political animal" or, more properly, "the animal that belongs to a polis") and described the ideal city-state as one small enough so that a single citizen's voice could be heard by all his assembled fellow citizens. For the Greek citizen, social life was lived in the agora or public marketplace, and contact with rural nature was immediately at hand. In that sense, the polis was a reincarnation, in an urban context, of the face-to-face human relationships that characterized the pre-urban community of the neolithic village.

Marking another sharp break in the history of urban life, the city of Rome began as a cluster of villages along the Tiber in central Italy, emerged as a powerful republic similar to the earlier Greek cities, but then exploded into a giant metropolis and a city of world empire. Rome's contributions to civilization were considerable. Its roads, aqueducts, and sewers set new standards of engineering excellence. Its systems of military and colonial administration spread a common law, and established a common peace, throughout a large and populous area that extended from Persia to the borders of Scotland. Roman imperial expansion also spread Roman literature, philosophy, and art, establishing the basis for a widespread cultural hegemony. And Rome planted colonial towns wherever its legions marched, often leaving traces of an original *castrum* laid out along the cardinal points of the compass at the center of later medieval cities.

But if the administrative and infrastructural accomplishments of the Romans were impressive, their record in the field of social development is more problematic. In the place of the Greek conception of community and participation in the life of the polis, the Romans erected a citizenship of imperial privilege rooted in a rigid social hierarchy of patricians, clients, and plebeians. Beginning with Augustus, the Roman emperors proclaimed themselves gods, staged extravagant spectacles to awe the cowed populace, and ruled by "bread and circuses." Rome, at a population of one million, became a parasite on the entire Mediterranean world, and both city and empire eventually fell of their own weight.

For much of the medieval period that followed the fall of the Roman Empire in the West, Europe was a cultural backwater. In the early middle ages, self-contained monastery communities kept the larger world at bay, many provincial towns retreated inside the walls of the Roman amphitheaters, and the population of Rome itself dwindled to a few thousand. During this same period Hangzhou in China reached one million. Raided by Vikings from the North and invaded by North African Arabs on its southern flank, most of Europe reverted to rural conditions, and serfdom became widespread under a system of warlord feudalism. Meanwhile, the cities of Islam – Samara and Baghdad and Moorish Córdoba – were the real centers of power in what had been the Roman Empire. And other urban centers – the Khmer civilization at Angkor, Luoyang in Sung and Ch'in China, Great Zimbabwe in Africa – often surpassed Europe's cities in wealth and power.

After about the year 1000 CE, however, Europe began to revive, and the late medieval cities became true centers of commerce, culture, and community. As Henri Pirenne argued in *Medieval Cities* (1925), it was the economic function of the great trading towns that led inevitably to their growing power and political independence. Having used their wealth to win from the barons the right to self-government, the medieval towns became islands of freedom in a sea of feudal obligation.

The defensive walls of the medieval city provided a clear demarcation line between the urban and the rural, and the small size of most towns allowed for an easy reciprocity between urban industry and commerce, on the one hand, and agricultural pursuits on the other. Within the town walls, the guilds provided for the organization of economic and social life, while the church saw to the citizens' spiritual needs and established a framework for social ritual and communal unity. Cathedrals, guildhouses, charitable institutions, universities, and colorful marketplaces were all characteristic medieval institutions. Together, they established the perfect stage for what Lewis Mumford called "the urban drama," but as soon as "the unity of this social order was broken [with the advent of nation-statism and capitalist industrialization] everything about it was set in confusion . . . and the city became a battleground for conflicting cultures, dissonant ways of life."

As described in the first-person narratives of travelers like Marco Polo, Ibn Battuta, Bernal Diaz, and Albrecht Dürer, the world cities of the period between the ancient world and the modern world were extraordinarily diverse, and the urban civilizations of China and Mexico often rendered European visitors awestruck. In Europe itself, the slow decay of medieval urban unity began with the Renaissance and the rise of absolutist monarchies. The new national rulers built their royal palaces, such as Louis XIV's Versailles, outside of the traditional urban centers. Their interventions into the existing urban fabric included building broad boulevards and open squares fit for the display of baroque pomp and power. The Enlightenment and the Age of Revolutions brought down the divine right of kings and reestablished the political power of the urban commercial interests. And in the end, it was capitalism and the new industrial economic order that destroyed the last vestiges of the medieval city by separating the church from its social role and reducing the marketplace to its purely economic functions. The capitalist city, especially the city of the Industrial Revolution, created an entirely new urban paradigm and established the physical, social, economic, and political preconditions of all that was to follow. With the Industrial Revolution, we see the emergence of the modern city.

The political and economic consequences of the Renaissance had helped to spread European domination worldwide through extensive projects of exploration, discovery, and imperialist expansion. The forces of industrialization helped to complete that process of world domination by dividing the world between the advanced industrialized nations (originally Europe and North America) and the under-developed, non-industrialized nations. It also created a new social order based on property-owning capitalists and propertyless proletarians. And the cities, especially the new industrial centers, became dismal conurbations of factories and slums such as the world had never before seen.

One of the earliest and most acute observers of the new urban-industrial order was Friedrich Engels, himself the son of a major German industrialist. In *The Condition of the Working Class in England in 1844* (1845), Engels detailed the unrelenting squalor and misery that characterized the working-class districts of Manchester and the strategies employed by the capitalist bourgeoisie to protect themselves from the physical and social horror that was the source of their wealth. There were many responses to these horrifying conditions – the introduction of urban parks, systems of public hygiene, agitation for poor relief and model housing, even utopian visions of perfect societies – and all of these contributed to the development of modern urban planning, as described in Parts Five and Six of this volume. Yet another compelling strategy was middle-class flight to the suburbs. Suburbanization, with its consequent segregation by social class, became one of the continuing features of the modern city and one of the sources of its on-going social disharmony and class conflict.

Meanwhile, the model of middle-class suburbia has grown in size and influence to the point where it is no longer just an appendage to the central city. Instead, it now actually defines many cities, leaving the old inner cores to the poorest elements of the urban population and in need of massive efforts at renewal and redevelopment. Although the first modern suburbs were built along inter-urban railroad lines, the newer suburbs, especially those developments built after World War II, were automobile-based and created the "sprawl" that was first seen in North America and that characterizes more and more cities worldwide. The new tract-home developments have spawned a vast literature, much of it criticizing suburbia as a cultural wasteland and a segregated sanctuary of class privilege. In *The Levittowners*, Herbert Gans presented a sympathetic view of the community of tract-homes built by developer Arthur Levitt on Long Island, New York. He described a family-oriented community of skilled workers and mid-level managers – that is, a true *middle* class, not an upper-middle-class elite. But the more general view of automobile-dominated suburbia, a view that subjected sprawl to cultural as well as design criticism, is ably summarized in Kenneth T. Jackson's "The Drive-in Culture of Contemporary America" from *The Crabgrass Frontier*.

Although the Levittowns on the East Coast of the United States were clearly suburbs within the New York metropolitan region, a new city arose in California during the early decades of the twentieth century that signaled a new phase in the history of urbanism worldwide. Sometimes dismissed as a mere conurbation of suburbs in search of a city and derided as the ultimate in mindless, chaotic sprawl, Los Angeles did indeed break all the existing rules and natural boundaries of urban development but emerged finally as a new, radically decentralized urban paradigm: the contemporary multi-nucleated metropolis poised on the edge of post-modernity.

The essential characteristics of Los Angeles – a city that grew from less than 600,000 in 1920 to over ten million today – were present almost from the beginning, particularly its sense of "spatial freedom" and its preference for the middle-class single-family dwelling as an "expression of its design for living." For good or ill, these characteristics were further emphasized by a grid pattern of freeways and a reliance – many would say over-reliance – on the automobile that replaced a once-extensive network of streetcars and created a metropolis without a single downtown. Today, Los Angeles is a true world city, and its products – both industrial and cultural – are influential around the globe.

In the nineteenth century, middle-class suburbs developed outside major urban centers, spaced along commuter rail lines. In the twentieth century, the influence of the automobile turned once-attractive small-scale suburbs into an endless, congested sprawl. These first two stages in the development of suburbia depended on the existence of a vital central city, both as a center for production and employment and for cultural amenities. With the emergence of Los Angeles, however, that pattern began to change, and today the new "Edge City" suburban ring is clearly different from the earlier suburban developments in size, complexity, and even function. This is where most of the new houses, most of the new jobs, and even most of the new cultural centers are located. Increasingly, the major commute pattern is not from suburb to central city, but from suburb to suburb. Indeed, as Robert Fishman argues in *Bourgeois Utopias: The Rise and Fall of Suburbia* (1987), the new "Edge City" suburbs are not suburbs at all, but a fundamentally new kind of decentralized city that he calls "technoburbs."

What the future holds for urban civilization is infinitely debatable. Will central cities disappear? Will "Edge Cities" take over the primary urban functions? Will the urbanization process itself reverse direction (as Melvin Webber predicted in 1968 (p. 470) and lead to counter-urbanization and a general dispersal of the human population? Or will certain urban centers become worldwide command and control centers, internally characterized by the uneasy side-by-side coexistence of corporate power and service-sector marginality, as Saskia Sassen argues in "The Impact of the New Technologies and Globalization on Cities" (p. 212). No one knows for certain, but it increasingly appears that urban history is on the cusp of a major new transformation. Looming on the horizon is a new paradigm, a new discontinuity in urban history, tentatively described as post-modernism that will be characterized by telecommunications technologies, global systems of economic exchange, and new ecological constraints. Perhaps part-Los Angeles–part-technoburb, the new urban paradigm that will emerge promises to be a major new stage in the history of cities, in the history of civilization, and in the on-going evolution of the human community.

"The Urbanization of the Human Population"

Scientific American (1965)

Kingsley Davis

Editors' Introduction

The demographics of the urbanization process may be said to be the foundation of all urban history. Demography – from the Greek *demos*: "people" – is the study of human populations. Kingsley Davis (1908–1996) pioneered the study of historical urban demography and was particularly fascinated by the history of world urbanization: that is, the increase over time of the proportion of the total human population that is urban as opposed to rural.

The following selection synthesizes Davis's conclusions about how urbanization has occurred throughout the world during all of human history. He raises fundamental issues and lays out a clear framework for understanding population dynamics and urban growth. Davis's careful distinctions of possible sources of urbanization are fundamental. He concludes that urbanization is caused by rural–urban migration, not because of other possible factors such as differential birth and mortality rates.

Davis's extraordinary data on how tiny European urban settlements were after the fall of Rome, and how slowly they grew throughout the middle ages and early modern period, provides the demographic backdrop for the historic growth of European urbanization. During the long period of medieval urbanization the proportion of the population that was urban as opposed to rural changed very slowly. In sharp contrast, Davis concludes that as the Industrial Revolution occurred in England rapid population growth combined with rural–urban shifts changed both the proportion of the population living in cities and absolute city size very quickly. Friedrich Engels (p. 58) describes in horrifying detail what this revolution in urban demography meant to the impoverished urban proletariat of Manchester and other nineteenth-century industrial cities. His analysis is extremely relevant in assessing prospects for the twenty-first century as the advanced industrial societies and eventually the world reach what some environmental analysts regard as the full "carrying capacity" of the globe.

Davis argues that urbanization follows an attenuated S curve in which pre-industrial cities urbanize very slowly at the long bottom of the S, shoot up at the middle of the S as they industrialize, and then level off at the top of the S. He observes that advanced industrialized countries are now reaching the top part of an S curve, many rapidly urbanizing Third World countries are at the steep middle of the S, and other emerging countries are still moving along the long slowly rising bottom of the S.

The developing countries of Asia, South America, and Africa already have many huge and rapidly growing cities. As the twenty-first century progresses, it appears likely the human population will increasingly live in "megacities" of ten million inhabitants and more, often flowing together in vast urban conurbations.

Davis concludes that there will be an end to urbanization – but not necessarily to absolute population growth, the physical size of cities or the absolute number of people cities contain. He found that the rural population in third world countries today continues to grow as these countries urbanize, unlike European cities in the nineteenth century where industrialization led to depopulation of rural areas. His vision of third world societies unable to sustain their populations helps to explain Saskia Sassen's description of growing poverty and inequality worldwide and the growth of large, poorly-paid immigrant labor forces in the largest cities in the developed world (p. 212).

Research and scholarly debate continues on the nature and causes of world urbanization. Historians continue to shed light on the growth of cities, but because the records from which they work are often fragmentary and incomplete not everyone agrees with Davis or any other standard account. Debate continues on the relative importance of war, plague, medical advances, trade, technology, religion, and ideology on urban growth. And debate is even more intense in the normative area – about what, if anything, governments should do about population growth and urbanization.

Davis stressed the impact of overall population growth (which he saw as a real danger) on world urbanization and implies that family planning is essential if cities are to meet human needs. But many governments reject family planning on religious or policy grounds, and some European countries now face declining populations and are currently debating the desirability of enacting family-friendly policies to reward child-bearing.

Davis's other writings include many articles and studies on demographics and natural resources as well as two anthologies: *Cities: Their Origin, Growth and Human Impact* (San Francisco: W.H. Freeman, 1973) and, with Mikhail S. Bernstram, *Resources, Environment and Population: Present Knowledge, Future Options* (New York: Population Council, Oxford University Press, 1991).

Data on world urbanization are contained in Tertius Chandler and Gerald Fox, *3000 Years of Urban Growth* (New York: Academic Press, 1974). Further insight on demography and urbanization can be found in World Bank, *World Development Report: Infrastructure for Development* (Oxford: Oxford University Press, 1994), Ad van der Woude, Akira Hayami, and Jan de Vries (eds), *Urbanization in History: A Process of Dynamic Interaction* (Oxford: Oxford University Press, 1990), and the frequent revisions of *World Population Prospects* published by the Population Division of the United Nations.

For recent developments in Third World urbanization, see Alan Gilbert (ed.), *The Mega-City in Latin America* (New York: United Nations University Press, 1996), Carole Rakodi (ed.), *The Urban Challenge in Africa* (New York: United Nations University Press, 1997), and Fu-chen Lo and Yue-man Yeung (eds), *Emerging World Cities in Pacific Asia* (New York: United Nations University Press, 1996).

For an environmental view of world urbanization, consult Cedric Pugh (ed.), *Sustainability, the Environment, and Urbanization* (London: Earthscan, 1996). Lester R. Brown and Jodi L. Jacobson provide a somewhat alarming summary of recent world population studies and reflections on the future in *The Future of Urbanization: Facing the Ecological and Economic Constraints* (New York: Worldwatch Paper No. 77, 1987). The future of urbanization is of course an important issue for policy planners. For a fascinating review of population-related policy issues, including the possibility of a "world population implosion," see Nicholas Eberstadt, *Prosperous Paupers and Other Population Problems* (New Brunswick, NJ: Transaction, 2000).

▪ ▪ ▪ ▪ ▪ ▪

Urbanized societies, in which a majority of the people live crowded together in towns and cities, represent a new and fundamental step in man's social evolution. Although cities themselves first appeared some 5,500 years ago, they were small and surrounded by an overwhelming majority of rural people; moreover, they relapsed easily to village or small-town status. The urbanized societies of today, in contrast, not only have urban agglomerations of a size never before attained but also have a high proportion of their population concentrated in such agglomerations. In 1960, for example, nearly 52 million Americans lived in only 16 urbanized areas. Together these areas covered less land than one of the smaller counties (Cochise) of Arizona. According to one definition used by the U.S. Bureau of the Census, 96 million people – 53 percent of the nation's population – were concentrated in 213 urbanized areas that together occupied only 0.7 percent of the nation's land. Another definition used by the bureau puts the urban population at about 70 percent. The large and dense agglomerations comprising the urban population involve a degree of human contact and of social complexity never before known. They exceed in size the communities of any other large animal; they suggest the behavior of communal insects rather than of mammals.

Neither the recency nor the speed of this evolutionary development is widely appreciated. Before 1850 no society could be described as predominantly urbanized, and by 1900 only one – Great Britain – could be so regarded. Today, only 65 years later, all industrial nations are highly urbanized, and in the world as a whole the process of urbanization is accelerating rapidly.

Some years ago my associates and I at Columbia University undertook to document the progress of urbanization by compiling data on the world's cities and the proportion of human beings living in them; in recent years the work has been continued in our center – International Population and Urban Research – at the University of California at Berkeley. The data obtained in these investigations . . . show the historical trend in terms of one index of urbanization: the proportion of the population living in cities of 100,000 or larger. Statistics of this kind are only approximations of reality, but they are accurate enough to demonstrate how urbanization has accelerated. Between 1850 and 1950 the index changed at a much higher rate than from 1800 to 1850, but the rate of change from 1950 to 1960 was twice that of the preceding 50 years! If the pace of increase that obtained between 1950 and 1960 were to remain the same, by 1990 the fraction of the world's people living in cities of 100,000 or larger would be more than half. Using another index of urbanization – the proportion of the world's population living in urban places of all sizes – we found that by 1960 the figure had already reached 33 percent.

Clearly the world as a whole is not fully urbanized, but it soon will be. This change in human life is so recent that even the most urbanized countries still exhibit the rural origins of their institutions. Its full implications for man's organic and social evolution can only be surmised.

In discussing the trend – and its implications insofar as they can be perceived – I shall use the term "urbanization" in a particular way. It refers here to the proportion of the total population concentrated in urban settlements, or else to a rise in this proportion. A common mistake is to think of urbanization as simply the growth of cities. Since the total population

is composed of both the urban population and the rural, however, the "proportion urban" is a function of both of them. Accordingly, cities can grow without any urbanization, provided that the rural population grows at an equal or a greater rate.

Historically, urbanization and the growth of cities have occurred together, which accounts for the confusion. As the reader will soon see, it is necessary to distinguish the two trends. In the most advanced countries today, for example, urban populations are still growing, but their proportion of the total population is tending to remain stable or to diminish. In other words, the process of urbanization – the switch from a spread-out pattern of human settlement to one of concentration in urban centers – is a change that has a beginning and an end, but the growth of cities has no inherent limit. Such growth could continue even after everyone was living in cities, through sheer excess of births over deaths.

The difference between a rural village and an urban community is of course one of degree; a precise operational distinction is somewhat arbitrary, and it varies from one nation to another. Since data are available for communities of various sizes, a dividing line can be chosen at will. One convenient index of urbanization, for example, is the proportion of people living in places of 100,000 or more. In the following analysis I shall depend on two indexes: the one just mentioned and the proportion of population classed as "urban" in the official statistics of each country. In practice the two indexes are highly correlated; therefore either one can be used as an index of urbanization.

Actually the hardest problem is not that of determining the "floor" of the urban category but of ascertaining the boundary of places that are clearly urban by any definition. How far east is the boundary of Los Angeles? Where along the Hooghly River does Calcutta leave off and the countryside begin? In the past the population of cities and towns has usually been given as the number of people living within the political boundaries. Thus the population of New York is frequently given as around eight million, this being the population of the city proper. The error in such a figure was not large before World War I, but since then, particularly in the advanced countries, urban populations have been spilling over the narrow political boundaries at a tremendous rate. In 1960 the New York–Northeastern New Jersey urbanized area, as delineated by the Bureau of the Census, had more than 14 million people. That delineation showed it to be the largest city in the world and nearly twice as large as New York City proper.

As a result of the outward spread of urbanites, counts made on the basis of political boundaries alone underestimate the city populations and exaggerate the rural. For this reason our office delineated the metropolitan areas of as many countries as possible for dates around 1950. These areas included the central, or political, cities and the zones around them that are receiving the spillover.

This reassessment raised the estimated proportion of the world's population in cities of 100,000 or larger from 15.1 percent to 16.7 percent. As of 1960 we have used wherever possible the "urban agglomeration" data now furnished to the United Nations by many countries. The U.S., for example, provides data for "urbanized areas," meaning cities of 50,000 or larger and the built-up agglomerations around them.

. . . My concern is with the degree of urbanization in whole societies. It is curious that thousands of years elapsed between the first appearance of small cities and the emergence of urbanized societies in the nineteenth century. It is also curious that the region where urbanized societies arose – northwestern Europe – was not the one that had given rise to the major cities of the past; on the contrary, it was a region where urbanization had been at an extremely low ebb. Indeed, the societies of northwestern Europe in medieval times were so rural that it is hard for modern minds to comprehend them. Perhaps it was the non-urban character of these societies that erased the parasitic nature of towns and eventually provided a new basis for a revolutionary degree of urbanization.

At any rate, two seemingly adverse conditions may have presaged the age to come: one the low productivity of medieval agriculture in both per-acre and per-man terms, the other the feudal social system. The first meant that towns could not prosper on the basis of local agriculture alone but had to trade and to manufacture something to trade. The second meant that they could not gain political dominance over their hinterlands and thus become warring city-states. Hence they specialized in commerce and manufacture and evolved local institutions suited to this role. Craftsmen were housed in the towns, because there the merchants could regulate quality and cost. Competition among towns stimulated specialization and technological innovation. The need for literacy, accounting skills and geographical knowledge caused the towns to invest in secular education.

Although the medieval towns remained small and never embraced more than a minor fraction of each region's population, the close connection between industry and commerce that they fostered, together with their emphasis on technique, set the stage for the ultimate breakthrough in urbanization. This breakthrough came only with the enormous growth in productivity caused by the use of inanimate energy and machinery. How difficult it was to achieve the transition is agonizingly apparent from statistics showing that even with the conquest of the New World the growth of urbanization during three postmedieval centuries in Europe was barely perceptible. I have assembled population estimates at two or more dates for 33 towns and cities in the sixteenth century, 46 in the seventeenth and 61 in the eighteenth. The average rate of growth during the three centuries was less than 0.6 percent per year. Estimates of the growth of Europe's population as a whole between 1650 and 1800 work out to slightly more than 0.4 percent. The advantage of the towns was evidently very slight. Taking only the cities of 100,000 or more inhabitants, one finds that in 1600 their combined population was 1.6 percent of the estimated population of Europe; in 1700, 1.9 percent; and in 1800, 2.2 percent. On the eve of the industrial revolution Europe was still an overwhelmingly agrarian region.

With industrialization, however, the transformation was striking. By 1801 nearly a tenth of the people of England and Wales were living in cities of 100,000 or larger. This proportion doubled in 40 years and doubled again in another 60 years. By 1900 Britain was an urbanized society. In general, the later each country became industrialized, the faster was its urbanization. The change from a population with 10 percent of its members in cities of 100,000 or larger to one in which 30 percent lived in such cities took about 79 years in England and Wales, 66 in the U.S., 48 in Germany, 36 in Japan and 26 in Australia. The close association between economic development and urbanization has persisted: . . . in 199 countries around 1960 the proportion of the population living in cities varied sharply with per capita income.

Clearly, modern urbanization is best understood in terms of its connection with economic growth, and its implications are best perceived in its latest manifestations in advanced countries. What becomes apparent as one examines the trend in these countries is that urbanization is a finite process, a cycle through which nations go in their transition from agrarian to industrial society. The intensive urbanization of most of the advanced countries began within the past hundred years; in the underdeveloped countries it got under way more recently. In some of the advanced countries its end is now in sight. The fact that it will end, however, does not mean that either economic development or the growth of cities will necessarily end.

The typical cycle of urbanization can be represented by a curve in the shape of an attenuated S. Starting from the bottom of the S, the first bend tends to come early and to be followed by a long attenuation. In the United Kingdom, for instance, the swiftest rise in the proportion of people living in cities of 100,000 or larger occurred from 1811 to 1851. In the U.S. it occurred from 1820 to 1890, in Greece from 1879 to 1921. As the proportion climbs above 50 percent the curve begins to flatten out; it falters, or even declines, when the proportion urban has reached about 75 percent. In the United Kingdom, one of the world's most urban countries, the proportion was slightly higher in 1926 (78.7 percent) than in 1961 (78.3 percent).

At the end of the curve some ambiguity appears. As a society becomes advanced enough to be highly urbanized it can also afford considerable suburbanization and fringe development. In a sense the slowing down of urbanization is thus more apparent than real: an increasing proportion of urbanites simply live in the country and are classified as rural. Many countries now try to compensate for this ambiguity by enlarging the boundaries of urban places; they did so in numerous censuses taken around 1960. Whether in these cases the old classification of urban or the new one is erroneous depends on how one looks at it; at a very advanced stage the entire concept of urbanization becomes ambiguous.

The end of urbanization cannot be unraveled without going into the ways in which economic development governs urbanization. Here the first question is: where do the urbanites come from? The possible answers are few: the proportion of people in cities can rise because rural settlements grow larger and are reclassified as towns or cities; because the excess of births over deaths is greater in the city than in the country, or because people move from the country to the city.

The first factor has usually had only slight influence. The second has apparently never been the case. Indeed, a chief obstacle to the growth of cities in the past has been their excessive mortality. London's water in the middle of the nineteenth century came mainly from wells and rivers that drained cesspools, graveyards and tidal areas. The city was regularly ravaged by cholera. Tables for 1841 show an expectation of life of about 36 years for London and 26 for Liverpool and Manchester, as compared to 41 for England and Wales as a whole. After 1850, mainly as a result of sanitary measures and some improvement in nutrition and housing, city health improved, but as late as the period 1901–1910 the death rate of the urban counties in England and Wales, as modified to make the age structure comparable, was 33 percent higher than the death rate of the rural counties. As Bernard Benjamin, a chief statistician of the British General Register Office, has remarked: "Living in the town involved not only a higher risk of epidemic and crowd diseases . . . but also a higher risk of degenerative disease – the harder wear and tear of factory employment and urban discomfort." By 1950, however, virtually the entire differential had been wiped out.

As for birth rates, during rapid urbanization in the past they were notably lower in cities than in rural areas. In fact, the gap tended to widen somewhat as urbanization proceeded in the latter half of the nineteenth century and the first quarter of the twentieth. In 1800 urban women in the U.S. had 36 percent fewer children than rural women did; in 1840, 38 percent and in 1930, 41 percent. Thereafter the difference diminished.

With mortality in the cities higher and birth rates lower, and with reclassification a minor factor, the only real source for the growth in the proportion of people in urban areas during the industrial transition was rural–urban migration. This source had to be plentiful enough not only to overcome the substantial disadvantage of the cities in natural increase but also, above that, to furnish a big margin of growth in their populations. If, for example, the cities had a death rate a third higher and a birth rate a third lower than the rural rates (as was typical in the latter half of the nineteenth century), they would require each year perhaps 40 to 45 migrants from elsewhere per 1,000 of their population to maintain a growth rate of 3 percent per year. Such a rate of migration could easily be maintained as long as the rural portion of the population was large, but when this condition ceased to obtain, the maintenance of the same urban rate meant an increasing drain on the countryside.

Why did the rural–urban migration occur? The reason was that the rise in technological enhancement of human productivity, together with certain constant factors, rewarded urban concentration. One of the constant factors was that agriculture uses land as its prime instrument of production and hence spreads out people who are engaged in it, whereas manufacturing, commerce and services use land only as a site. Moreover, the demand for agricultural products is less elastic than the demand for services and manufactures. As productivity grows, services and manufactures can absorb more manpower by paying higher wages. Since nonagricultural activities can use land simply as a site, they can locate near one another (in towns and cities) and thus minimize the fraction of space inevitably involved in the division of labor. At the same time, as agricultural technology is improved, capital costs in farming rise and manpower becomes not only less needed but also economically more burdensome. A substantial portion of the agricultural population is therefore sufficiently disadvantaged, in relative terms, to be attracted by higher wages in other sectors.

In this light one sees why a large *flow* of people from farms to cities was generated in every country that passed through the industrial revolution. One also sees why, with an even higher proportion of people already in cities and with the inability of city people to replace themselves by reproduction, the drain eventually became so heavy that in many nations the rural population began to decline in absolute as well as relative terms. In Sweden it declined after 1920, in England and Wales after 1861, in Belgium after 1910.

Realizing that urbanization is transitional and finite, one comes on another fact – a fact that throws light on the circumstances in which urbanization comes to an end. A basic feature of the transition is the profound switch from agricultural to nonagricultural employment. This change is associated with urbanization but not identical with it. The difference emerges particularly in the later stages. Then the availability of automobiles, radios, motion pictures and electricity, as well as the reduction of the workweek and the workday, mitigate the disadvantages of living in the country. Concurrently the expanding size of cities makes them more difficult to live in. The population classed as "rural" is accordingly enlarged, both from cities and from true farms. For these reasons the "rural" population in some industrial countries never did fall in absolute size. In all the industrial countries, however, the population dependent on agriculture – which the reader will recognize as a more functional definition of the nonurban population than mere rural residence

– decreased in absolute as well as relative terms. In the U.S., for example, the net migration from farms totaled more than 27 million between 1920 and 1959 and thus averaged approximately 700,000 a year. As a result the farm population declined from 32.5 million in 1916 to 20.5 million in 1960, in spite of the large excess of births in farm families. In 1964, by a stricter American definition classifying as "farm families" only those families actually earning their living from agriculture, the farm population was down to 12.9 million. This number represented 6.8 percent of the nation's population; the comparable figure for 1880 was 44 percent. In Great Britain the number of males occupied in agriculture was, at its peak, 1.8 million, in 1851; by 1961 it had fallen to 0.5 million.

In the later stages of the cycle, then, urbanization in the industrial countries tends to cease. Hence the connection between economic development and the growth of cities also ceases. The change is explained by two circumstances. First, there is no longer enough farm population to furnish a significant migration to the cities. (What can 12.9 million American farmers contribute to the growth of the 100 million people already in urbanized areas?) Second, the rural nonfarm population, nourished by refugees from the expanding cities, begins to increase as fast as the city population. The effort of census bureaus to count fringe residents as urban simply pushes the definition of "urban" away from the notion of dense settlement and in the direction of the term "nonfarm." As the urban population becomes more "rural," which is to say less densely settled, the advanced industrial peoples are for a time able to enjoy the amenities of urban life without the excessive crowding of the past.

Here, however, one again encounters the fact that a cessation of urbanization does not necessarily mean a cessation of city growth. An example is provided by New Zealand. Between 1945 and 1961 the proportion of New Zealand's population classed as urban – that is, the ratio between urban and rural residents – changed hardly at all (from 61.3 percent to 63.6 percent) but the urban population increased by 50 percent. In Japan between 1940 and 1950 urbanization actually decreased slightly, but the urban population increased by 13 percent.

The point to be kept in mind is that once urbanization ceases, city growth becomes a function of general population growth. Enough farm-to-city migration may still occur to redress the difference in natural increase. The reproductive rate of urbanites tends, however, to increase when they live at lower densities, and the reproductive rate of "urbanized" farmers tends to decrease; hence little migration is required to make the urban increase equal the national increase.

I now turn to the currently underdeveloped countries. With the advanced nations having slackened their rate of urbanization, it is the others – representing three-fourths of humanity – that are mainly responsible for the rapid urbanization now characterizing the world as a whole. In fact, between 1950 and 1960 the proportion of the population in cities of 100,000 or more rose about a third faster in the underdeveloped regions than in the developed ones. Among the underdeveloped regions the pace was slow in eastern and southern Europe but in the rest of the under-developed world the proportion in cities rose twice as fast as it did in the industrialized countries, even though the latter countries in many cases broadened their definitions of urban places to include more suburban and fringe residents.

Because of the characteristic pattern of urbanization, the current rates of urbanization in underdeveloped countries could be expected to exceed those now existing in countries far advanced in the cycle. On discovering that this is the case one is tempted to say that the underdeveloped regions are now in the typical stage of urbanization associated with early economic development. This notion, however, is erroneous. In their urbanization the underdeveloped countries are definitely not recreating past history. Indeed, the best grasp of their present situation comes from analyzing how their course differs from the previous pattern of development.

The first thing to note is that today's underdeveloped countries are urbanizing not only more rapidly than the industrial nations are now but also more rapidly than the industrial nations did in the heyday of their urban growth. The difference, however, is not large. In 40 underdeveloped countries for which we have data in recent decades, the average gain in the proportion of the population urban was 20 percent per decade; in 16 industrial countries, during the decades of their most rapid urbanization (mainly in the nineteenth century), the average gain per decade was 15 percent.

This finding that urbanization is proceeding only a little faster in underdeveloped countries than it did historically in the advanced nations may be questioned by the reader. It seemingly belies the widespread impression that cities throughout the nonindustrial parts of the world are bursting with people. There is, however, no contradiction. One must recall the basic distinction between a change in the proportion of the population urban, which is a ratio, and the absolute growth of cities. The popular impression is correct: the cities in underdeveloped areas are growing at a disconcerting rate. They are far outstripping the city boom of the industrializing era in the nineteenth century. If they continue their recent rate of growth, they will double their population every 15 years.

In 34 underdeveloped countries for which we have data relating to the 1940s and 1950s, the average annual gain in the urban population was 4.5 percent. The figure is remarkably similar for the various regions: 4.7 percent in seven countries of Africa, 4.7 percent in 15 countries of Asia and 4.3 percent in 12 countries of Latin America. In contrast, in nine European countries during their period of fastest urban population growth (mostly in the latter half of the nineteenth century) the average gain per year was 2.1 percent. Even the frontier industrial countries – the U.S., Australia–New Zealand, Canada and Argentina – which received huge numbers of immigrants had a smaller population growth in towns and cities: 4.2 percent per year. In Japan and the U.S.S.R. the rate was respectively 5.4 and 4.3 percent per year, but their economic growth began only recently.

How is it possible that the contrast in growth between today's underdeveloped countries and yesterday's industrializing countries is sharper with respect to the absolute urban population than with respect to the urban share of the total population? The answer lies in another profound difference between the two sets of countries – a difference in total population growth, rural as well as urban. Contemporary underdeveloped populations have been growing since 1940 more than twice as fast as industrialized populations, and their increase far exceeds the growth of the latter at the peak of their expansion. The only rivals in an earlier day were the frontier nations, which had the help of great streams of immigrants. Today the underdeveloped nations – already densely settled, tragically impoverished and with gloomy economic prospects – are multiplying their people by sheer biological increase at a rate that is unprecedented. It is this population boom that is overwhelmingly responsible for the rapid inflation of city populations in such countries. Contrary to popular opinion both inside and outside those countries, the main factor is not rural–urban migration.

This point can be demonstrated easily by a calculation that has the effect of eliminating the influence of general population growth on urban growth. The calculation involves assuming that the total population of a given country remained constant over a period of time but that the percentage urban changed as it did historically. In this manner one obtains the growth of the absolute urban population that would have occurred if rural–urban migration were the only factor affecting it. As an example, Costa Rica had in 1927 a total population of 471,500, of which 88,600, or 18.8 percent, was urban. By 1963 the country's total population was 1,325,200 and the urban population was 456,600, or 34.5 percent. If the total population had remained at 471,500 but the percentage urban had still risen from 18.8 to 34.5, the absolute urban population in 1963 would have been only 162,700. That is the growth that would have occurred in the urban population if rural–urban migration had been the only factor. In actuality the urban population rose to 456,600. In other words, only 20 percent of the rapid growth of Costa Rica's towns and cities was attributable to urbanization per se; 44 percent was attributable solely to the country's general population increase, the remainder to the joint operation of both factors. Similarly, in Mexico between 1940 and 1960, 50 percent of the urban population increase was attributable to national multiplication alone and only 22 percent to urbanization alone.

The past performance of the advanced countries presents a sharp contrast. In Switzerland between 1850 and 1888, when the proportion urban resembled that in Costa Rica recently, general population growth alone accounted for only 19 percent of the increase of town and city people, and rural–urban migration alone accounted for 69 percent. In France between 1846 and 1911 only 21 percent of the growth in the absolute urban population was due to general growth alone.

The conclusion to which this contrast points is that one anxiety of governments in the underdeveloped nations is misplaced. Impressed by the mushrooming in their cities of shanty-towns filled with ragged peasants, they attribute the fantastically fast city growth to rural–urban migration. Actually this migration now does little more than make up for the small difference in the birth rate between city and countryside. In the history of the industrial nations, as we have seen, the sizable difference between urban and rural birth rates and death rates required that cities, if they were to grow, had to have an enormous influx of people from farms and villages. Today in the underdeveloped countries the towns and cities have only a slight disadvantage in fertility, and their old disadvantage in mortality not only has been wiped out but also in many cases has been reversed. During the nineteenth century the urbanizing nations were learning how to keep crowded populations in cities from dying like flies. Now the lesson has been learned, and it is being applied to cities even in countries just emerging from tribalism. In fact, a disproportionate share of public health funds goes into cities. As a result, throughout the nonindustrial world people in cities are multiplying as never before, and rural–urban migration is playing a much lesser role.

The trends just described have an important implication for the rural population. Given the explosive overall population growth in underdeveloped countries, it follows that if the rural population is not to pile up on the land and reach an economically absurd density, a high rate of rural–urban migration must be maintained. Indeed, the exodus from rural areas should be higher than in the past. But this high rate of internal movement is not taking place, and there is some doubt that it could conceivably do so.

To elaborate, I shall return to my earlier point that in the evolution of industrialized countries the rural citizenry often declined in absolute as well as relative terms. The rural population of France – 26.8 million in 1846 – was down to 20.8 million by 1926 and 17.2 million by 1962, notwithstanding a gain in the nation's total population during this period. Sweden's rural population dropped from 4.3 million in 1910 to 3.5 million in 1960. Since the category "rural" includes an increasing portion of urbanites living in fringe areas, the historical drop was more drastic and consistent specifically in the farm population. In the U.S., although the "rural" population never quite ceased to grow, the farm contingent began its long descent shortly after the turn of the century; today it is less than two-fifths of what it was in 1910.

This transformation is not occurring in contemporary underdeveloped countries. In spite of the enormous growth of their cities, their rural populations – and their more narrowly defined agricultural populations – are growing at a rate that in many cases exceeds the rise of even the urban population during the evolution of the now advanced countries. The poor countries thus confront a grave dilemma. If they do not substantially step up the exodus from rural areas, these areas will be swamped with underemployed farmers. If they do step up the exodus, the cities will grow at a disastrous rate.

The rapid growth of cities in the advanced countries, painful though it was, had the effect of solving a problem – the problem of the rural population. The growth of cities enabled agricultural holdings to be consolidated, allowed increased capitalization and in general resulted in greater efficiency. Now, however, the underdeveloped countries are experiencing an even more rapid urban growth – and are suffering from urban problems – but urbanization is not solving their rural ills.

A case in point is Venezuela. Its capital, Caracas, jumped from a population of 359,000 in 1941 to 1,507,000 in 1963; other Venezuelan towns and cities equaled or exceeded this growth. Is this rapid rise denuding the countryside of people? No, the Venezuelan farm population increased in the decade 1951–1961 by 11 percent. The only thing that declined was the amount of cultivated land. As a result the agricultural population density became worse. In 1950 there were some 64 males engaged in agriculture per square mile of cultivated land; in 1961 there were 78. (Compare this with 4.8 males occupied in agriculture per square mile of cultivated land in Canada, 6.8 in the U.S. and 15.6 in Argentina.) With each male occupied in agriculture there are of course dependants. Approximately 225 persons in Venezuela are trying to live from each square mile of cultivated land. Most of the growth of cities in Venezuela is attributable to overall population growth. If the general population had not grown at all, and internal migration had been large enough to produce the actual shift in the proportion in cities, the increase in urban population would have been only 28 percent of what it was and the rural population would have been reduced by 57 percent.

The story of Venezuela is being repeated virtually everywhere in the underdeveloped world. It is not only Caracas that has thousands of squatters living in self-constructed junk houses on land that does not belong to them. By whatever name they are called, the squatters are to be found in all major cities in the poorer countries. They live in broad gullies beneath the main plain in San Salvador and on the hillsides of Rio de Janeiro and Bogotá. They tend to occupy with implacable determination parks, school grounds and vacant lots. Amman, the capital of Jordan, grew from 12,000 in 1958 to 247,000 in 1961. A good part of

it is slums, and urban amenities are lacking most of the time for most of the people. Greater Baghdad now has an estimated 850,000 people; its slums, like those in many other underdeveloped countries, are in two zones: the central part of the city and the outlying areas. Here are the *sarifa* areas, characterized by self-built reed huts; these areas account for about 45 percent of the housing in the entire city and are devoid of amenities, including even latrines. In addition to such urban problems, all the countries struggling for higher living levels find their rural population growing too and piling up on already crowded land. I have characterized urbanization as a transformation that, unlike economic development, is finally accomplished and comes to an end. At the 1950–1960 rate the term "urbanized world" will be applicable well before the end of the century. One should scarcely expect, however, that mankind will complete its urbanization without major complications. One sign of trouble ahead turns on the distinction I made at the start between urbanization and city growth *per se*. Around the globe today city growth is disproportionate to urbanization. The discrepancy is paradoxical in the industrial nations and worse than paradoxical in the nonindustrial.

It is in this respect that the nonindustrial nations, which still make up the great majority of nations, are far from repeating past history. In the nineteenth and early twentieth centuries the growth of cities arose from and contributed to economic advancement. Cities took surplus manpower from the countryside and put it to work producing goods and services that in turn helped to modernize agriculture. But today in underdeveloped countries, as in present-day advanced nations, city growth has become increasingly unhinged from economic development and hence from rural–urban migration. It derives in greater degree from overall population growth, and this growth in nonindustrial lands has become unprecedented because of modern health techniques combined with high birth rates.

The speed of world population growth is twice what it was before 1940, and the swiftest increase has shifted from the advanced to the backward nations. In the latter countries, consequently, it is virtually impossible to create city services fast enough to take care of the huge, never-ending cohorts of babies and peasants swelling the urban masses. It is even harder to expand agricultural land and capital fast enough to accommodate the enormous natural increase on farms.

The problem is not urbanization, not rural–urban migration, but human multiplication. It is a problem that is new in both its scale and its setting, and runaway city growth is only one of its painful expressions.

As long as the human population expands, cities will expand too, regardless of whether urbanization increases or declines. This means that some individual cities will reach a size that will make nineteenth-century metropolises look like small towns. If the New York urbanized area should continue to grow only as fast as the nation's population (according to medium projections of the latter by the Bureau of the Census), it would reach 21 million by 1985 and 30 million by 2010. I have calculated that if India's population should grow as the U.N. projections indicate it will, the largest city in India in the year 2000 will have between 36 and 66 million inhabitants.

What is the implication of such giant agglomerations for human density? In 1950 the New York–Northeastern New Jersey urbanized area had an average density of 9,810 persons per square mile. With 30 million people in the year 2010, the density would be 24,000 per square mile. Although this level is exceeded now in parts of New York City (which averages about 25,000 per square mile) and many other cities, it is a high density to be spread over such a big area; it would cover, remember, the suburban areas to which people moved to escape high density. Actually, however, the density of the New York urbanized region is dropping, not increasing, as the population grows. The reason is that the territory covered by the urban agglomeration is growing faster than the population: it grew by 51 percent from 1950 to 1960, whereas the population rose by 15 percent.

If, then, one projects the rise in population and the rise in territory for the New York urbanized region one finds the density problem solved. It is not solved for long, though, because New York is not the only city in the region that is expanding. So are Philadelphia, Trenton, Hartford, New Haven and so on. By 1960 a huge stretch of territory about 600 miles long and 30 to 100 miles wide along the eastern seaboard contained some 37 million people. (I am speaking of a longer section of the seaboard than the Boston-to-Washington conurbation referred to by some other authors.) Since the whole area is becoming one big polynucleated city, its population cannot long expand without a rise in density. Thus persistent human multiplication promises to frustrate the ceaseless search for space – for ample residential lots, wide-open suburban school grounds, sprawling shopping centers, one-floor factories, broad freeways.

How people feel about giant agglomerations is best indicated by their headlong effort to escape them. The bigger the city, the higher the cost of space; yet the more the level of living rises, the more people are willing to pay for low-density living. Nevertheless, as urbanized areas expand and collide, it seems probable that life in low-density surroundings will become too dear for the great majority.

One can of course imagine that cities may cease to grow and may even shrink in size while the population in general continues to multiply. Even this dream, however, would not permanently solve the problem of space. It would eventually obliterate the distinction between urban and rural, but at the expense of the rural.

It seems plain that the only way to stop urban crowding and to solve most of the urban problems besetting both the developed and the underdeveloped nations is to reduce the overall rate of population growth. Policies designed to do this have as yet little intelligence and power behind them. Urban planners continue to treat population growth as something to be planned for, not something to be itself planned. Any talk about applying brakes to city growth is therefore purely speculative, overshadowed as it is by the reality of uncontrolled population increase.

"The Urban Revolution"

Town Planning Review (1950)

V. Gordon Childe

Editors' Introduction

V. Gordon Childe (1892–1957) is arguably the single most influential archaeologist of the twentieth century. Born in Australia, Childe won a scholarship to Queen's College, Oxford, returned to Australia where he briefly pursued a career in left-wing politics, then returned to the UK as Professor of Archaeology at the University of Edinburgh and, later, Director of the Institute of Archaeology at the University of London.

Childe's most important book, the one that revolutionized the world of archaeological research by laying out an entirely new theoretical framework for understanding the phases of human development throughout history and pre-history, was *Man Makes Himself* (1936). In that pioneering work, Childe threw out the "three age system" (Stone Age, Bronze Age, Iron Age) that had been left over from nineteenth-century conceptions of human historical development. In its place he proposed a series of four stages (paleolithic, neolithic, urban, industrial) punctuated by three "revolutions" (or, as we might term them today, "paradigm shifts"). According to Childe, the first revolution – from old Stone Age hunter–gatherer cultures to settled agriculture – was the Neolithic Revolution. The second – the movement from neolithic agriculture to complex, hierarchical systems of manufacturing and trade that began during the fourth and third millennia BCE – was the Urban Revolution. And the third major shift in the record of human cultural and historical development – the only truly new development since the rise of cities – was the Industrial Revolution of the eighteenth and nineteenth centuries.

Childe is best known for his writings on the first cities, which arose in Mesopotamia (present-day Iraq) beginning about 4000 BCE. These cities sprang up in the area bounded by the Tigris and Euphrates Rivers – often referred to as "the Fertile Crescent." Plate 1, "The Palace of Sargon II of Khorsabad," illustrates the form these first cities took. Monumental gates, massive mud-brick walls, courtyards, residences for priest-kings, and a ziggurat.

Childe's work continues to figure prominently in ongoing debates about when, where, and why the first cities arose and in the antecedent debate about what a city is. Not everyone has accepted Childe's notion that the shift from neolithic to urban was a total break with the past. Evidence of ancient earthworks, wells, irrigation systems, and even continental trade networks have been traced back as far as 10,000 years in a number of areas in both the Old World and the New. Archaeologist James Mellaart has argued that evidence from the great neolithic communities of Catal Hüyük and Hacilar in ancient Turkey, which predate the earliest Mesopotamian cities by some thousands of years, calls the entire Childe theory into question. Still, it is clear that in most locations agriculture generally predated the rise of the first cities by not just thousands, but tens of thousands of years, and that the full elaboration of those cultural institutions we associate with urban life only emerged with the rise of the Mesopotamian cities.

Not everyone agrees with Childe's definition of a city. Archaeologists excavating older, smaller, less culturally advanced settlements than the Mesopotamian cities Childe studied often argue that these settlements were urban enough to qualify as true cities. Scholars working in South and Central America point out that many of the cultural features Childe believed essential to the definition of a city (including the wheel, writing, and the plow) did not exist in large and culturally advanced Amerindian settlements that appear truly urban in other respects.

In the selection from *Town Planning Review* reprinted here, Childe details the constituent elements of the Urban Revolution that accompanied the initial rise of complex civilizations in Mesopotamia and elsewhere in the ancient Near East. Childe felt that the major factors motivating the transformation were rooted in the material base of the society: its means of production and its available physical and technological resources. Thus, the economic division of labor, the elaboration of socio-political hierarchies, and even the emergence of basic religious and intellectual patterns of thought characteristic of urban civilizations all rested on the underlying need to increase food production through massive irrigation systems and to protect the communities themselves through the erection of massive walls and fortifications.

Many modern scholars question the deterministic Marxist categories Childe employed. Although he stresses the importance of writing as an element of any truly urban society, Childe has been faulted for his apparent disregard of the primacy of non-material aspects of culture. His system has very little room for what Lewis Mumford (p. 92) called "the urban drama" or what Jane Jacobs (p. 114) called the "street ballet." Still, no one has ever called Childe's vision limited or ideologically cramped. On the contrary, he provided an expansive macro-historical foundation upon which generations of others have built.

A tireless researcher and writer, Childe produced a veritable stream of books, many of which are still classics. Among the most notable are: *The Dawn of European Civilization* (London: Routledge and Kegan Paul, 1925), *The Most Ancient East* (New York: Grove Press, 1928), *What Happened in History* (Harmondsworth: Penguin, 1942), *Social Evolution* (London: Watts, 1951).

Other books on Mesopotamian cities include Nicholas Postgate and J.N. Postgate, *Early Mesopotamia: Society and Economy at the Dawn of History* (London and New York: Routledge, 1994), Georges Roux, *Ancient Iraq* (New York: Penguin, 1993), and C. Leonard Wooley's classics: *The Sumerians* (Oxford: Oxford University Press, 1928) and *Ur of the Chaldees* (Oxford: Oxford University Press, 1929).

Books on the rise of the earliest cities elsewhere in the world include Mortimer Wheeler, *Civilizations of the Indus Valley and Beyond* (London: Thames and Hudson, 1966), Karl Wittfogel, *Oriental Despotism* (New Haven, CT: Yale University Press, 1957), Basil Davidson, *The Lost Cities of Africa* (Boston: Little, Brown, 1959), Richard E.W. Adams, *Prehistoric Mesoamerica* (Norman: University of Oklahoma Press, 1991), Sylvanus G. Morely and George W. Brainerd, *The Ancient Maya* (Stanford, CA: Stanford University Press, 1956), Jacques Soustelle, *The Daily Life of the Aztecs* (New York: Macmillan, 1962), James Mellaart, *Earliest Civilizations of the Near East* (New York: McGraw-Hill, 1965) and *Catal Hüyük* (New York: McGraw-Hill, 1967), and Paul Wheatley, *The Pivot of the Four Quarters: A Preliminary Inquiry into the Origins and Character of the Ancient Chinese City* (Chicago: Aldine, 1971).

The concept of 'city' is notoriously hard to define. The aim of the present essay is to present the city historically – or rather prehistorically – as the resultant and symbol of a 'revolution' that initiated a new economic stage in the evolution of society. The word 'revolution' must not of course be taken as denoting a sudden violent catastrophe; it is here used for the culmination of a progressive change in the economic structure and social organization of communities that caused, or was accompanied by, a dramatic increase in the population affected – an increase that would appear as an obvious bend in the population graph were vital statistics available. Just such a bend is observable at the time of the Industrial Revolution in England. Though not demonstrable statistically, comparable changes of direction must have occurred at two earlier points in the demographic history of Britain and other regions. Though perhaps less sharp and less durable, these too should indicate equally revolutionary changes in economy. They may then be regarded likewise as marking transitions between stages in economic and social development.

Sociologists and ethnographers last century classified existing pre-industrial societies in a hierarchy of three evolutionary stages, denominated respectively 'savagery,' 'barbarism' and 'civilization.' If they be defined by suitably selected criteria, the logical hierarchy of stages can be transformed into a temporal sequence of ages, proved archaeologically to follow one another in the same order wherever they occur. Savagery and barbarism are conveniently recognized and appropriately defined by the methods adopted for procuring food. Savages live exclusively on wild food obtained by collecting, hunting or fishing. Barbarians on the contrary at least supplement these natural resources by cultivating edible plants and – in the Old World north of the Tropics – also by breeding animals for food.

Throughout the Pleistocene Period – the Palaeolithic Age of archaeologists – all known human societies were savage in the foregoing sense, and a few savage tribes have survived in out of the way parts to the present day. In the archaeological record barbarism began less than ten thousand years ago with the Neolithic Age of archaeologists. It thus represents a later, as well as a higher stage, than savagery. Civilization cannot be defined in quite such simple terms. Etymologically the word is connected with 'city,' and sure enough life in cities begins with this stage. But 'city' is itself ambiguous so archaeologists like to use 'writing' as a criterion of civilization; it should be easily recognizable and proves to be a reliable index to more profound characters. Note, however, that, because a people is said to be civilized or literate, it does not follow that all its members can read and write, nor that they all lived in cities. Now there is no recorded instance of a community of savages civilizing themselves, adopting urban life or inventing a script. Wherever cities have been built, villages of preliterate farmers existed previously (save perhaps where an already civilized people have colonized uninhabited tracts). So civilization, wherever and whenever it arose, succeeded barbarism.

We have seen that a revolution as here defined should be reflected in the population statistics. In the case of the Urban Revolution the increase was mainly accounted for by the multiplication of the numbers of persons living together, i.e., in a single built-up area. The first cities represented settlement units of hitherto unprecedented size. Of course it was not just their size that constituted their distinctive character. We shall find that by modern standards they appeared ridiculously small and we might meet agglomerations of population today to which the name city would have to be refused. Yet a certain size of settlement and density of population is an essential feature of civilization.

Now the density of population is determined by the food supply which in turn is limited by natural resources, the techniques for their exploitation and the means of transport and food-preservation available. The last factors have proved to be variables in the course of human history, and the technique of obtaining food has already been used to distinguish the consecutive stages termed savagery and barbarism. Under the gathering economy of savagery population was always exceedingly sparse. In aboriginal America the carrying capacity of normal unimproved land seems to have been from .05 to .10 per square mile. Only under exceptionally favourable conditions did the fishing tribes of the Northwest Pacific coast attain densities of over one human to the square mile. As far as we can guess from the extant remains, population densities in Palaeolithic and pre-neolithic Europe were less than the normal American. Moreover such hunters and collectors usually live in small roving bands. At best several bands may come together for quite brief periods on ceremonial occasions such as the Australian corroborees. Only in exceptionally favoured regions can fishing tribes establish anything like villages. Some settlements on the Pacific coasts comprised thirty or so substantial and durable houses, accommodating groups of several hundred persons. But even these villages were only occupied during the winter; for the rest of the year their inhabitants disposed in smaller groups. Nothing comparable has been found in pre-neolithic times in the Old World.

The Neolithic Revolution certainly allowed an expansion of population and enormously increased the carrying capacity of suitable land. On the Pacific Islands neolithic societies today attain a density of 30 or more persons to the square mile. In pre-Columbian North America, however, where the land is not obviously restricted by surrounding seas, the maximum density recorded is just under 2 to the square mile.

Neolithic farmers could of course, and certainly did, live together in permanent villages, though, owing to the extravagant rural economy generally practised, unless the crops were watered by irrigation, the villages had to be shifted at least every twenty years. But on the whole the growth of population was not reflected so much in the enlargement of the settlement unit as in a multiplication of settlements. In ethnography neolithic villages can boast only a few hundred inhabitants (a couple of 'pueblos' in New Mexico house over a thousand, but perhaps they cannot be regarded as neolithic). In prehistoric Europe the largest neolithic village yet known, Barkaer in Jutland, comprised 52 small, one-roomed dwellings, but 16 to 30 houses was a more normal figure; so the average, local group in neolithic times would average 200 to 400 members.

These low figures are of course the result of technical limitations. In the absence of wheeled vehicles and roads for the transport of bulky crops men had to live within easy walking distance of their cultivations. At the same time the normal rural economy of the Neolithic Age, what is now termed slash-and-burn or *jhumming*, condemns much more than half the arable land to lie fallow so that large areas were required. As soon as the population of a settlement rose above the numbers that could be supported from the accessible land, the excess had to hive off and found a new settlement.

The Neolithic Revolution had other consequences beside increasing the population, and their exploitation might in the end help to provide for the surplus increase. The new economy allowed, and indeed required, the farmer to produce every year more food than was needed to keep him and his family alive. In other words it made possible the regular production of a social surplus. Owing to the low efficiency of neolithic technique, the surplus produced was insignificant at first, but it could be increased till it demanded a reorganization of society.

Now in any Stone Age society, palaeolithic or neolithic, savage or barbarian, everybody can at least in theory make at home the few indispensable tools, the modest cloths and the simple ornaments everyone requires. But every member of the local community, not disqualified by age, must contribute actively to the communal food supply by personally collecting, hunting, fishing, gardening or herding. As long as this holds good, there can be no full-time specialists, no persons nor class of persons who depend for their livelihood on food produced by others and secured in exchange for material or immaterial goods or services.

We find indeed today among Stone Age barbarians and even savages expert craftsmen (for instance flint-knappers among the Ona of Tierra del Fuego), men who claim to be experts in magic, and even chiefs. In Palaeolithic Europe too there is some evidence for magicians and indications of chieftainship in pre-neolithic times. But on closer observation we discover that today these experts are not full-time specialists. The Ona flintworker must spend most of his time hunting; he only adds to his diet and his prestige by making arrowheads for clients who reward him with

presents. Similarly a pre-Columbian chief, though entitled to customary gifts and services from his followers, must still personally lead hunting and fishing expeditions and indeed could only maintain his authority by his industry and prowess in these pursuits. The same holds good of barbarian societies that are still in the neolithic stage, like the Polynesians where industry in gardening takes the place of prowess in hunting. The reason is that there simply will not be enough food to go round unless every member of the group contributes to the supply. The social surplus is not big enough to feed idle mouths.

Social division of labour, save those rudiments imposed by age and sex, is thus impossible. On the contrary community of employment, the common absorption in obtaining food by similar devices guarantees a certain solidarity to the group. For co-operation is essential to secure food and shelter and for defence against foes, human and subhuman. This identity of economic interests and pursuits is echoed and magnified by identity of language, custom and belief; rigid conformity is enforced as effectively as industry in the common quest for food. But conformity and industrious co-operation need no State organization to maintain them. The local group usually consists either of a single clan (persons who believe themselves descended from a common ancestor or who have earned a mystical claim to such descent by ceremonial adoption) or a group of clans related by habitual inter-marriage. And the sentiment of kinship is reinforced or supplemented by common rites focused on some ancestral shrine or sacred place. Archaeology can provide no evidence for kinship organization, but shrines occupied the central place in preliterate villages in Mesopotamia, and the long barrow, a collective tomb that overlooks the presumed site of most neolithic villages in Britain, may well have been also the ancestral shrine on which converged the emotions and ceremonial activities of the villagers below. However, the solidarity thus idealized and concretely symbolized, is really based on the same principles as that of a pack of wolves or a herd of sheep; Durkheim has called it 'mechanical.'

Now among some advanced barbarians (for instance tattooers or woodcarvers among the Maori) still technologically neolithic we find expert craftsmen tending towards the status of full-time professionals, but only at the cost of breaking away from the local community. If no single village can produce a surplus large enough to feed a full-time specialist all the year round, each should produce enough to keep him a week or so. By going round from village to village an expert might thus live entirely from his craft. Such itinerants will lose their membership of the sedentary kinship group. They may in the end form an analogous organization of their own – a craft clan, which, if it remain hereditary, may become a caste, or, if it recruit its members mainly by adoption (apprenticeship throughout Antiquity and the Middle Ages was just temporary adoption), may turn into a guild. But such specialists by emancipation from kinship ties, have also forfeited the protection of the kinship organization which alone under barbarism, guaranteed to its members security of person and property. Society must be reorganized to accommodate and protect them.

In pre-history specialization of labour presumably began with similar itinerant experts. Archaeological proof is hardly to be expected, but in ethnography metal-workers are nearly always full-time specialists. And in Europe at the beginning of the Bronze Age metal seems to have been worked and purveyed by perambulating smiths who seem to have functioned like tinkers and other itinerants of much more recent times. Though there is no such positive evidence, the same probably happened in Asia at the beginning of metallurgy. There must of course have been in addition other specialist craftsmen whom, as the Polynesian example warns us, archaeologists could not recognize because they worked in perishable materials. One result of the Urban Revolution will be to rescue such specialists from nomadism and to guarantee them security in a new social organization.

About 5,000 years ago irrigation cultivation (combined with stockbreeding and fishing) in the valleys of the Nile, the Tigris-Euphrates and the Indus had begun to yield a social surplus, large enough to support a number of resident specialists who were themselves released from food-production. Water-transport, supplemented in Mesopotamia and the Indus valley by wheeled vehicles and even in Egypt by pack animals, made it easy to gather food stuffs at a few centres. At the same time dependence on river water for the irrigation of the crops restricted the cultivable areas while the necessity of canalizing the waters and protecting habitations against annual floods encouraged the aggregation of population. Thus arose the first cities – units of settlement ten times as great as any known neolithic village. It can be argued that all cities in the old world are offshoots of those of Egypt, Mesopotamia and the Indus basin. So the latter need not be taken into account if a minimum definition of civilization is to be inferred from a comparison of its independent manifestations.

But some three millennia later cities arose in Central America, and it is impossible to prove that the Mayas owed anything directly to the urban civilizations of the Old World. Their achievements must therefore be taken into account in our comparison, and their inclusion seriously complicates the task of defining the essential preconditions for the Urban Revolution. In the Old World the rural economy which yielded the surplus was based on the cultivation of cereals combined with stock-breeding. But this economy had been made more efficient as a result of the adoption of irrigation (allowing cultivation without prolonged fallow periods) and of important inventions and discoveries – metallurgy, the plough, the sailing boat and the wheel. None of these devices was known to the Maya; they bred no animals for milk or meat; though they cultivated the cereal maize, they used the same sort of slash-and-burn method as neolithic farmers in prehistoric Europe or in the Pacific Islands today. Hence the minimum definition of a city, the greatest factor common to the Old World and the New will be substantially reduced and impoverished by the inclusion of the Maya. Nevertheless ten rather abstract criteria, all deducible from archaeological data, serve to distinguish even the earliest cities from any older or contemporary village.

(1) In point of size the first cities must have been more extensive and more densely populated than any previous settlements, although considerably smaller than many villages today. It is indeed only in Mesopotamia and India that the first urban populations can be estimated with any confidence or precision. There excavation has been sufficiently extensive and intensive to reveal both the total area and the density of building in sample quarters and in both respects has disclosed significant agreement with the less industrialized Oriental cities today. The population of Sumerian cities, thus calculated, ranged between 7,000 and 20,000; Harappa and Mohenjo-daro in the Indus valley must have approximated to the higher figure. We can only infer that Egyptian and Maya cities were of comparable magnitude from the scale of public works, presumably executed by urban populations.

(2) In composition and function the urban population already differed from that of any village. Very likely indeed most citizens were still also peasants, harvesting the lands and waters adjacent to the city.

Figure 1 Plan of the city of Erek (Uruk)

But all cities must have accommodated in addition classes who did not themselves procure their own food by agriculture, stock-breeding, fishing or collecting – full-time specialist craftsmen, transport workers, merchants, officials and priests. All these were of course supported by the surplus produced by the peasants living in the city and in dependent villages, but they did not secure their share directly by exchanging their products or services for grains or fish with individual peasants.

(3) Each primary producer paid over the tiny surplus he could wring from the soil with his still very limited technical equipment as tithe or tax to an imaginary deity or a divine king who thus concentrated the surplus. Without this concentration, owing to the low productivity of the rural economy, no effective capital would have been available.

(4) Truly monumental public buildings not only distinguish each known city from any village but also symbolize the concentration of the social surplus. Every Sumerian city was from the first dominated by one or more stately temples, centrally situated on a brick platform raised above the surrounding dwellings and usually connected with an artificial mountain, the staged tower or ziggurat. But attached to the temples were workshops and magazines, and an important appurtenance of each principal temple was a great granary. Harappa, in the Indus basin, was dominated by an artificial citadel, girt with a massive rampart of kiln-baked bricks, containing presumably a palace

and immediately overlooking an enormous granary and the barracks of artisans. No early temples nor palaces have been excavated in Egypt, but the whole Nile valley was dominated by the gigantic tombs of the divine pharaohs while royal granaries are attested from the literary record. Finally the Maya cities are known almost exclusively from the temples and pyramids of sculptured stone round which they grew up.

Hence in Sumer the social surplus was first effectively concentrated in the hands of a god and stored in his granary. That was probably true in Central America while in Egypt the pharaoh (king) was himself a god. But of course the imaginary deities were served by quite real priests who, besides celebrating elaborate and often sanguinary rites in their honour, administered their divine masters' earthly estates. In Sumer indeed the god very soon, if not even before the revolution, shared his wealth and power with a mortal viceregent, the 'City-King,' who acted as civil ruler and leader in war. The divine pharaoh was naturally assisted by a whole hierarchy of officials.

(5) All those not engaged in food-production were of course supported in the first instance by the surplus accumulated in temple or royal granaries and were thus dependent on temple or court. But naturally priests, civil and military leaders and officials absorbed a major share of the concentrated surplus and thus formed a 'ruling class.' Unlike a palaeolithic magician or a neolithic chief, they were, as an Egyptian scribe actually put it, 'exempt from all manual tasks.' On the other hand, the lower classes were not only guaranteed peace and security, but were relieved from intellectual tasks which many find more irksome than any physical labour. Besides reassuring the masses that the sun was going to rise next day and the river would flood again next year (people who have not five thousand years of recorded experience of natural uniformities behind them are really worried about such matters!), the ruling classes did confer substantial benefits upon their subjects in the way of planning and organization.

(6) They were in fact compelled to invent systems of recording and exact, but practically useful, sciences. The mere administration of the vast revenues of a Sumerian temple or an Egyptian pharaoh by a perpetual corporation of priests or officials obliged its members to devise conventional methods of recording that should be intelligible to all their colleagues and successors, that is, to invent systems of writing and numeral notation. Writing is thus a significant, as well as a convenient, mark of civilization. But while writing is a trait common to Egypt, Mesopotamia, the Indus valley and Central America, the characters themselves were different in each region and so were the normal writing materials – papyrus in Egypt, clay in Mesopotamia. The engraved seals or stelae that provide the sole extant evidence for early Indus and Maya writing no more represent the normal vehicles for the scripts than do the comparable documents from Egypt and Sumer.

(7) The invention of writing – or shall we say the inventions of scripts – enabled the leisured clerks to proceed to the elaboration of exact and predictive sciences – arithmetic, geometry and astronomy. Obviously beneficial and explicitly attested by the Egyptian and Maya documents was the correct determination of the tropic year and the creation of a calendar. For it enabled the rulers to regulate successfully the cycle of agricultural operations. But once more the Egyptian, Maya and Babylonian calendars were as different as any systems based on a single natural unit could be. Calendrical and mathematical sciences are common features of the earliest civilizations and they too are corollaries of the archaeologists' criterion, writing.

(8) Other specialists, supported by the concentrated social surplus, gave a new direction to artistic expression. Savages even in palaeolithic times had tried, sometimes with astonishing success, to depict animals and even men as they saw them – concretely and naturalistically. Neolithic peasants never did that; they hardly ever tried to represent natural objects, but preferred to symbolize them by abstract geometrical patterns which at most may suggest by a few traits a fantastical man or beast or plant. But Egyptian, Sumerian, Indus and Maya artist-craftsmen – full-time sculptors, painters, or seal-engravers – began once more to carve, model or draw likenesses of persons or things, but no longer with the naive naturalism of the hunter, but according to conceptualized and sophisticated styles which differ in each of the four urban centres.

(9) A further part of the concentrated social surplus was used to pay for the importation of raw materials, needed for industry or cult and not available locally. Regular 'foreign' trade over quite long distances was a feature of all early civilizations and, though common enough among barbarians later, is not certainly attested in the Old World before 3000 B.C. nor in the New before the Maya 'empire.' Thereafter regular trade

extended from Egypt at least as far as Byblos on the Syrian coast while Mesopotamia was related by commerce with the Indus valley. While the objects of international trade were at first mainly 'luxuries,' they already included industrial materials, in the Old World notably metal the place of which in the New was perhaps taken by obsidian. To this extent the first cities were dependent for vital materials on long distance trade as no neolithic village ever was.

(10) So in the city, specialist craftsmen were both provided with raw materials needed for the employment of their skill and also guaranteed security in a State organization based now on residence rather than kinship. Itinerancy was no longer obligatory. The city was a community to which a craftsman could belong politically as well as economically.

Yet in return for security they became dependent on temple or court and were relegated to the lower classes. The peasant masses gained even less material advantages; in Egypt for instance metal did not replace the old stone and wood tools for agricultural work. Yet, however imperfectly, even the earliest urban communities must have been held together by a sort of solidarity missing from any neolithic village. Peasants, craftsmen, priests and rulers form a community, not only by reason of identity of language and belief, but also because each performs mutually complementary functions, needed for the well-being (as redefined under civilization) of the whole. In fact the earliest cities illustrate a first approximation to an organic solidarity based upon a functional complementarity and interdependence between all its members such as subsist between the constituent cells of an organism. Of course this was only a very distant approximation. However necessary the concentration of the surplus really were with the existing forces of production, there seemed a glaring conflict on economic interests between the tiny ruling class, who annexed the bulk of the social surplus, and the vast majority who were left with a bare subsistence and effectively excluded from the spiritual benefits of civilization. So solidarity had still to be maintained by the ideological devices appropriate to the mechanical solidarity of barbarism as expressed in the pre-eminence of the temple or the sepulchral shrine, and now supplemented by the force of the new State organization. There could be no room for sceptics or sectaries in the oldest cities.

These ten traits exhaust the factors common to the oldest cities that archaeology, at best helped out with fragmentary and often ambiguous written sources, can detect. No specific elements of town planning for example can be proved characteristic of all such cities; for on the one hand the Egyptian and Maya cities have not yet been excavated; on the other neolithic villages were often walled, an elaborate system of sewers drained the Orcadian hamlet of Skara Brae; two-storeyed houses were built in pre-Columbian pueblos, and so on.

The common factors are quite abstract. Concretely Egyptian, Sumerian, Indus and Maya civilizations were as different as the plans of their temples, the signs of their scripts and their artistic conventions. In view of this divergence and because there is so far no evidence for a temporal priority of one Old World centre (for instance, Egypt) over the rest nor yet for contact between Central America and any other urban centre, the four revolutions just considered may be regarded as mutually independent. On the contrary, all later civilizations in the Old World may in a sense be regarded as lineal descendants of those of Egypt, Mesopotamia or the Indus.

But this was not a case of like producing like. The maritime civilizations of Bronze Age Crete or classical Greece for example, to say nothing of our own, differ more from their reputed ancestors than these did among themselves. But the urban revolutions that gave them birth did not start from scratch. They could and probably did draw upon the capital accumulated in the three allegedly primary centres. That is most obvious in the case of cultural capital. Even today we use the Egyptians' calendar and the Sumerians' divisions of the day and the hour. Our European ancestors did not have to invent for themselves these divisions of time nor repeat the observations on which they are based; they took over – and very slightly improved – systems elaborated 5,000 years ago! But the same is in a sense true of material capital as well. The Egyptians, the Sumerians and the Indus people had accumulated vast reserves of surplus food. At the same time they had to import from abroad necessary raw materials like metals and building timber as well as 'luxuries.' Communities controlling these natural resources could in exchange claim a slice of the urban surplus. They could use it as capital to support full-time specialists – craftsmen or rulers – until the latters' achievement in technique and organization had so enriched barbarian economics that they too could produce a substantial surplus in their turn.

"The Polis"

from *The Greeks* (1951)

H.D.F. Kitto

Editors' Introduction

At its peak ancient Athens had only about as many residents as Peoria, Illinois (1990 population 113,504) – not a city that leaps out as a great center of world civilization. But British classicist Humphrey Davy Findley Kitto (1897–1982) reminds us not to commit the vulgar error of confusing size with significance. During its golden age, Athens and the 700 or so other tiny settlements of ancient Greece made a monumental contribution to human culture. What the Greeks achieved in philosophy, literature, drama, poetry, art, logic, mathematics, sculpture, and architecture has exercised a profound influence on Western civilization.

A Greek invention of enduring interest to urbanists is the polis. Since we have not got the thing, which the Greeks called "the polis," Kitto notes, we do not possess an equivalent word. "City-state" or, perhaps, "self-governing community" come closest.

The classical Greek polis came of age in the fifth century BCE, about halfway between the emergence of the great Mesopotamian cities Childe describes and the present time. The physical form of the polis stressed public space. Private houses were low and turned away from the street. In contrast the Greeks emphasized public temples, stadiums, the agora (a combined market place and public forum), and theaters like Athens' magnificent Theater of Dionysus illustrated in Plate 2. In the larger polei, like Athens, these public buildings were spacious and often beautifully constructed of marble. Even in the smaller ones the community devoted many of its resources to them.

If the physical form of the polis was often stunning, it was the social organization of the polis that remains of particular fascination. The polis represents a form of community, which has exerted a powerful fascination for more than two millennia.

In the following selection Kitto describes how the polis made it possible for each citizen to realize his spiritual, moral, and intellectual capacities. The polis was a living community; almost an extended family. While the Greeks were very private in many ways, Kitto notes that their public life was essentially communistic. The polis as a social institution defined the very nature of being human for its citizens.

Not that the polis supported development of every resident: women and slaves were not citizens and did not participate in much of the life of the polis. Foreigners could attend plays in the Greek theater, but were barred from many institutions reserved to the (free, non-foreign, male) citizens. Sir Peter Hall in *Cities in Civilization* further questions the extent to which many citizens actually participated in public affairs. He hypothesizes that only a small percentage of those eligible to participate in public decision-making actually did so. He also notes that while farmers and other of the least educated and least articulate citizens of the Greek polis may have been physically present and possessed the same voting rights as educated upper-class Athenians it is unlikely that they participated very effectively compared to the higher classes. For the most part, Hall believes, they were passive spectators rather than active participants in the public affairs.

While a balanced view of the polis must acknowledge the existence of slavery, exclusion of women from civic life, limitations on the rights of foreigners, and the influence of education and class on social relations, Athens and the other Greek poleis were astonishingly democratic compared to any other urban civilization that preceded them. It is easy to dismiss Kitto as a hopeless romantic and his depiction of the Greek polis as an ivory tower depiction of a Camelot that never was. But that may be too harsh. The Greek polis as a social institution did represent a remarkable advance over social relations in any previous society. And the values it represented for its citizens are of enduring importance in an imperfect world.

In the debate about why the polis arose in Greece when it did, Kitto rejects deterministic answers. Geographical and economic determinists argue that the mountainous terrain required little, separate city-states. Kitto attributes the rise of the polis to the *character of the Greeks* rather than these external factors.

Kitto expresses nostalgia for human qualities of life in the polis that appear threatened today. Compare the vision of polis as a supportive, humanistic, structure for human fulfillment with the vision of large modern cities as centers of alienation and anomie depicted by Louis Wirth (p. 97), or ghettos housing the Black underclass as described by William Julius Wilson (p. 126). Note the connections between humanistic values Kitto felt that the polis nurtured and Robert Putnam's concept of "social capital" growing out of civic engagement (p. 105), idealized life in J.B. Jackson's "almost perfect town" (p. 164), and the return to human-scale community Peter Calthorpe and William Fulton advocate in *The Regional City* (p. 331). Compare the positive values Kitto describes in the polis as well as the status and gender barriers with Dolores Hayden's vision of a non-sexist city (p. 448).

Other books helpful in understanding the polis and its significance are Christian Meier, *Athens: A Portrait of the City in its Golden Age* (New York: Metropolitan Books, 1998), Cecil Maurice Bowra, *The Greek Experience* (London: Weidenfeld and Nicolson, 1957), and a new edition of classic writings by a great classicist – Jacob Burckhardt, *The Greeks and Greek Civilization* (New York: St Martin's, 1998). Also of interest are Lisa Nevett, *House and Society in the Ancient Greek World* (Cambridge: Cambridge University Press, 1999) and Nicholas Cahill, *Household and City Organization at Olynthus* (New Haven, CT: Yale University Press, 2002). Two recent and excellent studies of Greek democracy are James O'Neil, *The Origins and Development of Ancient Greek Democracy* (Lanham, MD: Rowman and Littlefield, 1995) and Josiah Ober, *Political Dissent in Democratic Athens* (Princeton, NJ: Princeton University Press, 1998).

For accounts of Greek city planning see Richard Ernest Wycherley, *How the Greeks Built Cities*, 2nd edn (London: Macmillan, 1963) and *The Stones of Athens* (Princeton, NJ: Princeton University Press, I978). Spiro Kostof, "Polis and Akropolis," Chapter 7 of *A History of Architecture* (New York: Oxford University Press, 1980) is also helpful.

Two masterful accounts of the role of cities in civilization give particular emphasis to the contribution of the Greek polis. See Lewis Mumford's chapter on "The Emergence of the Polis" and "Citizen Versus Ideal City" in *The City in History* (New York: Harcourt Brace Jovanovich, 1961), and "The Fountainhead," the second chapter of Sir Peter Hall's *Cities in Civilization* (New York: Pantheon Books, 1998).

"Polis" is the Greek word which we translate "city-state". It is a bad translation, because the normal polis was not much like a city, and was very much more than a state. But translation, like politics, is the art of the possible; since we have not got the thing which the Greeks called "the polis", we do not possess an equivalent word. From now on, we will avoid the misleading term "city-state", and use the Greek word instead . . . We will first inquire how this political system arose, then we will try to reconstitute the word "polis" and recover its real meaning by watching it in action. It may be a long task, but all the time we shall be improving our acquaintance with the Greeks. Without a clear conception what the polis was, and what it meant to the Greek, it is quite impossible to understand properly Greek history, the Greek mind, or the Greek achievement.

First then, what was the polis? . . .

. . . In Crete . . . we find over fifty quite independent poleis, fifty small "states" . . . What is true of Crete is true of Greece in general, or at least of those parts which play any considerable part in Greek history . . .

It is important to realize their size. The modern reader picks up a translation of Plato's *Republic* or Aristotle's *Politics*; he finds Plato ordaining that his ideal city shall have 5,000 citizens, and Aristotle that each citizen should be able to know all the others by sight; and he smiles, perhaps, at such philosophic fantasies. But Plato and Aristotle are not fantasts. Plato is imagining a polis on the normal Hellenic scale; indeed he implies that many existing Greek poleis are too small – for many had less than 5,000 citizens. Aristotle says, in his amusing way . . . that a polis of ten citizens would be impossible, because it could not be self-sufficient, and that a polis of a hundred thousand would be absurd, because it could not govern itself properly . . . Aristotle speaks of a hundred thousand citizens; if we allow each to have a wife and four children, and then add a liberal number of slaves and resident aliens, we shall arrive at something like a million – the population of Birmingham, and to Aristotle an independent "state" as populous as Birmingham is a lecture-room joke . . .

In fact, only three poleis had more than 20,000 citizens: Syracuse and Acragas (Girgenti) in Sicily, and Athens. At the outbreak of the Peloponnesian War the population of Attica was probably about 350,000, half Athenian (men, women and children), about a tenth resident aliens, and the rest slaves. Sparta, or Lacedaemon, had a much smaller citizen-body, though it was larger in area. The Spartans had conquered and annexed Messenia, and possessed 3,200 square miles of territory. By Greek standards this was an enormous area: it would take a good walker two days to cross it. The important commercial city of Corinth had a territory of 330 square miles . . . The island of Ceos, which is about as big as Bute, was divided into four poleis. It had therefore four armies, four governments, possibly four different calendars, and, it may be, four different currencies and systems of measures – though this is less likely. Mycenae was in historical times a shrunken relic of Agamemnon's capital, but still independent. She sent an army to help the Greek cause against Persia at the battle of Plataea; the army consisted of eighty men. Even by Greek standards this was small, but we do not hear that any jokes were made about an army sharing a cab.

To think on this scale is difficult for us, who regard a state of ten million as small, and are accustomed to states which, like the U.S.A. and the U.S.S.R., are so big that they have to be referred to by their initials; but when the adjustable reader has become accustomed to the scale, he will not commit the vulgar error of confusing size with significance . . .

But before we deal with the nature of the polis, the reader might like to know how it happened that the relatively spacious pattern of pre-Dorian Greece became such a mosaic of small fragments. The Classical scholar too would like to know; there are no records, so that all we can do is to suggest plausible reasons. There are historical, geographical and economic reasons; and when these have been duly set forth, we may conclude perhaps that the most important reason of all is simply that this is the way in which the Greeks preferred to live.

[Here Kitto describes the evolution of the Greek acropolis from a fortified hilltop strong-point built for protection against Dorian invaders to a place of assembly, religion, and commerce.]

At this point we may invoke the very sociable habits of the Greeks, ancient or modern. The English farmer likes to build his house on his land, and to come into town when he has to. What little leisure he has he likes to spend on the very satisfying occupation of looking over a gate. The Greek prefers to live in the town or village, to walk out to his work, and to spend his rather ampler leisure talking in the town or village square. Therefore the market becomes a market-town, naturally beneath the acropolis. This became the center of the communal life of the people – and we shall see presently how important that was.

But why did not such towns form larger units? This is the important question.

There is an economic point. The physical barriers which Greece has so abundantly made the transport of goods difficult, except by sea, and the sea was not yet used with any confidence. Moreover, the variety of which we spoke earlier enabled quite a small area to be reasonably self-sufficient for a people who made such small material demands on life as the Greek. Both of these facts tend in the same direction; there was in Greece no great economic interdependence, no reciprocal pull between the different parts of the country, strong enough to counteract the desire of the Greek to live in small communities.

There is a geographical point. It is sometimes asserted that this system of independent poleis was imposed on Greece by the physical character of the country. The theory is attractive, especially to those who like to have one majestic explanation of any phenomenon, but it does not seem to be true. It is of course obvious that the physical subdivision of the country helped; the system could not have existed, for example, in Egypt, a country which depends entirely on the proper management of the Nile flood, and therefore must have a central government. But there are countries cut up quite as much as Greece – Scotland, for instance – which have never developed the polis-system; and conversely there were in Greece many neighbouring poleis, such as Corinth and Sicyon, which remained independent of each other although between them there was no physical barrier that would seriously incommode a modern cyclist. Moreover, it was precisely the most mountainous parts of Greece that never developed poleis, or not until later days – Arcadia and Aetolia, for example, which had something like a canton-system. The polis flourished in those parts where communications were relatively easy. So that we are still looking for our explanation.

Economics and geography helped, but the real explanation is the character of the Greeks . . . As it will take some time to deal with this, we may first clear out of the way an important historical point. How did it come about that so preposterous a system was able to last for more than twenty minutes?

The ironies of history are many and bitter, but at least this must be put to the credit of the gods, that they arranged for the Greeks to have the Eastern Mediterranean almost to themselves long enough to work out what was almost a laboratory-experiment to test how far, and in what conditions, human nature is capable of creating and sustaining a civilization . . . this lively and intelligent Greek people was for some centuries allowed to live under the apparently absurd system which suited and developed its genius instead of becoming absorbed in the dull mass of a large empire, which would have smothered its spiritual growth . . . no history of Greece can be intelligible until one has understood what the polis meant to the Greek; and when we have understood that, we shall also understand why the Greeks developed it, and so obstinately tried to maintain it. Let us then examine the word in action.

It meant at first that which was later called the Acropolis, the stronghold of the whole community and the centre of its public life . . . "polis" very soon meant either the citadel or the whole people which, as it were, "used" this citadel. So we read in Thucydides, "Epidamnus is a polis on the right as you sail into the Ionian gulf." This is not like saying "Bristol is a city on the right as you sail up the Bristol Channel", for Bristol is not an independent state which might be at war with Gloucester, but only an urban area with a purely local administration. Thucydides' words imply that there is a town – though possibly a very small one – called Epidamnus, which is the political centre of the Epidamnians, who live in the territory of which the town is the centre – not the "capital" – and are Epidamnians whether they live in the town or in one of the villages in this territory.

Sometimes the territory and the town have different names. Thus, Attica is the territory occupied by the Athenian people; it comprised Athens – the "polis" in the narrower sense – the Piraeus, and many villages; but the people collectively were Athenians, not Attics, and a citizen was an Athenian in whatever part of Attica he might live.

In this sense "polis" is our "state" . . . The actual business of governing might be entrusted to a monarch, acting in the name of all according to traditional usages, or to the heads of certain noble families, or to a council of citizens owning so much property, or to all the citizens. All these and many modifications of them, were natural forms of "polity"; all were sharply distinguished by the Greek from Oriental monarchy, in which the monarch is irresponsible, not holding his powers in trust by the grace of god, but being himself a god. If there were irresponsible government there was no polis . . .

. . . [T]he size of the polis made it possible for a member to appeal to all his fellow citizens in person, and this he naturally did if he thought that another member of the polis had injured him. It was the common assumption of the Greeks that the polis took its origin in the desire for Justice. Individuals are lawless, but the polis will see to it that wrongs are redressed. But not by an elaborate machinery of state-justice, for such a machine could not be operated except by individuals, who may be as unjust as the original wrongdoer. The injured party will be sure of obtaining Justice only if he can declare his wrongs to the whole polis. The word therefore now means "people" in actual distinction from state.

[. . .]

. . . Demosthenes the orator talks of a man who, literally, "avoids the city" – a translation which might lead the unwary to suppose that he lived in something corresponding to the Lake District, or Purley. But the phrase "avoids the polis" tells us nothing about his domicile; it means that he took no part in public life – and was therefore something of an oddity. The affairs of the community did not interest him.

We have now learned enough about the word polis to realize that there is no possible English rendering of such a common phrase as, "It is everyone's duty to help the polis." We cannot say "help the state", for that arouses no enthusiasm; it is "the state" that takes half our incomes from us. Not "the community", for with us "the community" is too big and too various to be grasped except theoretically. One's village, one's trade union, one's class, are entities that mean something to us at once, but "work for the community", though an admirable sentiment, is to most of us vague and flabby. In the years before the war, what did most parts of Great Britain know about the depressed areas? How much do bankers, miners and farmworkers understand each other? But the "polis" every Greek knew; there it was, complete, before his eyes. He could see the fields which gave it its sustenance – or did not, if the crops failed; he could see how agriculture, trade and industry dovetailed into one another; he knew the frontiers, where they were strong and where weak; if any malcontents were planning a *coup*, it was difficult for them to conceal the fact. The entire life of the polis, and the relation between its parts, were much easier to grasp, because of the small scale of things. Therefore to say "It is everyone's duty to help the polis" was not to express a fine sentiment but to speak the plainest and most urgent common sense. Public affairs had an immediacy and a concreteness which they cannot possibly have for us.

[. . .]

Pericles' Funeral Speech, recorded or recreated by Thucydides, will illustrate this immediacy, and will also take our conception of the polis a little further. Each year, Thucydides tells us, if citizens had died in war – and they had, more often than not – a funeral oration was delivered by "a man chosen by the polis". Today, that would be someone nominated by the Prime Minister, or the British Academy, or the BBC [British Broadcasting Corporation]. In Athens it meant that someone was chosen by the Assembly who had often spoken to that Assembly; and on this occasion Pericles spoke from a specially high platform, that his voice might reach as many as possible. Let us consider two phrases that Pericles used in that speech.

He is comparing the Athenian polis with the Spartan, and makes the point that the Spartans admit foreign visitors only grudgingly, and from time to time expel all strangers, "while we make our polis common to all". "Polis" here is not the political unit; there is no question of naturalizing foreigners – which the Greeks

did rarely, simply because the polis was so intimate a union. Pericles means here: "We throw open to all our common cultural life", as is shown by the words that follow, difficult though they are to translate: "nor do we deny them any instruction or spectacle" – words that are almost meaningless until we realize that the drama, tragic and comic, the performance of choral hymns, public recitals of Homer, games, were all necessary and normal parts of "political" life. This is the sort of thing Pericles has in mind when he speaks of "instruction and spectacle", and of "making the polis open to all".

But we must go further than this. A perusal of the speech will show that in praising the Athenian polis Pericles is praising more than a state, a nation, or a people: he is praising a way of life; he means no less when, a little later, he calls Athens the "school of Hellas". – And what of that? Do not we praise "the English way of life"? The difference is this; we expect our State to be quite indifferent to "the English way of life" – indeed, the idea that the State should actively try to promote it would fill most of us with alarm. The Greeks thought of the polis as an active, formative thing, training the minds and characters of the citizens; we think of it as a piece of machinery for the production of safety and convenience. The training in virtue, which the medieval state left to the Church, and the polis made its own concern, the modern state leaves to God knows what.

"Polis", then, originally "citadel", may mean as much as "the whole communal life of the people, political, cultural, moral" – even "economic", for how else are we to understand another phrase in this same speech, "the produce of the whole world comes to us, because of the magnitude of our polis"? This must mean "our national wealth".

Religion too was bound up with the polis – though not every form of religion. The Olympian gods were indeed worshipped by Greeks everywhere, but each polis had, if not its own gods, at least its own particular cults of these gods . . . But beyond these Olympians, each polis had its minor local deities, "heroes" and nymphs, each worshipped with his immemorial rite, and scarcely imagined to exist outside the particular locality where the rite was performed. So . . . there is a sense in which it is true to say that the polis is an independent religious, as well as political, unit . . .

[. . .]

. . . Aristotle made a remark which we most inadequately translate "Man is a political animal." What

Aristotle really said is "Man is a creature who lives in a polis"; and what he goes on to demonstrate, in his *Politics*, is that the polis is the only framework within which man can fully realize his spiritual, moral and intellectual capacities.

Such are some of the implications of this word . . . The polis was a living community, based on kinship, real or assumed – a kind of extended family, turning as much as possible of life into family life, and of course having its family quarrels, which were the more bitter because they were family quarrels.

This it is that explains not only the polis but also much of what the Greek made and thought, that he was essentially social. In the winning of his livelihood he was essentially individualist: in the filling of his life he was essentially "communist". Religion, art, games, the discussion of things – all these were needs of life that could be fully satisfied only through the polis – not, as with us, through voluntary associations of like-minded people, or through entrepreneurs appealing to individuals. (This partly explains the difference between Greek drama and the modern cinema.) Moreover, he wanted to play his own part in running the affairs of the community. When we realize how many of the necessary, interesting and exciting activities of life the Greek enjoyed through the polis, all of them in the open air, within sight of the same acropolis, with the same ring of mountains or of sea visibly enclosing the life of every member of the state – then it becomes possible to understand Greek history, to understand that in spite of the promptings of common sense the Greek could not bring himself to sacrifice the polis, with its vivid and comprehensive life, to a wider but less interesting unity . . .

[. . .]

First-Person Accounts of Great Cities of the Medieval and Early Modern World

Marco Polo, Ibn Battuta, Bernal Diaz, and Albrecht Dürer

Editors' Introduction

The period between the fall of the Roman Empire in Western Europe and the rise of the modern nation-states during the Renaissance was once called the medieval age, or even the "dark ages," but is now more often referred to as the pre-modern or early pre-modern period. Whatever the terminology, these centuries represented a profound transition in the history of technology, economy, religion, and social life . . . and in the emergence of modern urban civilization.

The early modern period was also an extraordinary age of travel and discovery, of cultures intersecting in ways that would determine the course of world history. The Venetian merchant Marco Polo (1254–1324) was not the first European to encounter China, but the account of his travels to the Mongol Empire of Kublai Khan and of his participation in the imperial Chinese government, first published as *The Book of Marco Polo* in 1299, was nothing less than a revelation. Accompanied by his father and his uncle, Polo left for the East in 1275 and did not return until 1295. During the intervening years, he experienced the wonders of a vast and advanced civilization that knew printing, used paper money, invented gunpowder, and had developed an intricate and efficient form of political and social organization that Europeans could only wonder at.

In the passage here reprinted – "Of the Noble and Magnificent City of Kin-sai" – Polo may be criticized for possible exaggerations and a self-aggrandizing habit of referring to himself in the third person. But all in all, his observations on the present-day city of Hangzhou are dazzling, full of wonder, and imbued with a quickening sense of possibility.

An even more extraordinary traveler of this period was Ibn Battuta (1307–1377). Born in Morocco, Ibn Battuta spent the better part of his life exploring and writing about the full extent of the Muslim world. In 1326, he traveled across North Africa to Cairo, Palestine, Syria, and Mecca. In the years following, up until 1354 when he wrote *The Rihla* (My Travels), he visited Persia and Iraq, Arabia and East Africa, Anatolia, the Central Asian steppes and the Empire of the Golden Horde, India and Ceylon, Malaysia and China, and finally Spain and West Africa.

In this passage, Ibn Battuta describes the city of Constantinople, the capital of the Eastern Roman Empire. A century later, in 1453, Constantinople would fall to the Ottoman Turks and be reconstituted as the Muslim city of Istanbul. But when Ibn Battuta encounters the city, the Byzantine emperor still rules (although Battuta refers to him as a "sultan"), and the court and the city are described as foreign, even exotic.

Bernal Diaz del Castillo (1492–1581?) came to the Aztec city of Tenochtitlán not as a traveler but as a conqueror, a conquistador. Born in Spain in the year of the first Columbus voyage, Diaz pursued gold and glory in Panama and the Yucatán before joining Hernán Cortés on the historic expedition to the Valley of Mexico (1519–1521) that brought the Aztec Empire under Spanish control.

As with Marco Polo's description of Kin-sai, Diaz's account of Mexico, especially the city of Tenochtitlán (now Mexico City), is full of astonishment and wonder. He and the other soldiers, some of whom had seen Constantinople and other European capitals, had never seen anything to compare to the magnificence and splendor of the Aztec city. As Diaz notes in his *Historia Verdadera de la Conquista de la Nueva España* (first published in 1636), the market-places, the temples, the court of Montezuma – all leave the Spanish awestruck. The biggest wonder of all, as Diaz freely admits, is that such an empire could fall to a desperate band of less than 400 soldiers!

Our final traveler is not an adventurer or a conqueror but a German artist. Albrecht Dürer (1471–1528) was one of the great painters and engravers of the Renaissance. Born in Nuremberg, Dürer studied art in Venice and became the court painter for two Holy Roman Emperors, Maximilian I and Charles V. He also traveled throughout Western Europe at a time when travel was still difficult and indulged in mostly by itinerant merchants. During his travels to the Low Countries, Dürer, something of an art merchant himself, visited other artists and sought out commissions for portraits from prosperous burghers.

Dürer also observed the cities and social customs of the people he visited, and his description of the Great Procession before the cathedral in Antwerp, Belgium, impressed Lewis Mumford (p. 92) as a notable example of the "urban drama." In *The City in History* (1960), Mumford writes, "Note the vast number of people arrayed in this procession. As in the church itself, the spectators were also communicants and participants: they engaged in the spectacle, watching it from within, not just from without . . . Prayer, mass, pageant, life-ceremony . . . the city itself was stage for these separate scenes of the drama . . ." Each in his own way, all of the travelers represented in this modest selection experienced a heightened sense of "urban drama" as diverse urban civilizations, formerly kept apart by history and geography, begin to interact. In the record of these contacts, we see the beginning of a process that today reaches an unprecedented level of intensity in the instantaneous communications between all parts of a globalized economy dominated by global urban command and control centers.

THE ITALIAN MERCHANT MARCO POLO DESCRIBES THE MAGNIFICENT CHINESE CITY OF KIN-SAI (1299)

Marco Polo, "Of the Noble and Magnificent City of Kin-sai," *The Travels of Marco Polo* (1299), translated by William Marsden and Manuel Komroff

Upon leaving Va-giu you pass, in the course of three days' journey, many towns, castles, and villages, all of them well inhabited and opulent. The people have abundance of provisions. At the end of three days you reach the noble and magnificent city of Kin-sai capital; Hang-chau, a name that signifies 'The Celestial City', and which it merits from its pre-eminence to all others in the world, in point of grandeur and beauty, as well as from its abundant delights, which might lead an inhabitant to imagine himself in paradise.

This city was frequently visited by Marco Polo, who carefully and diligently observed and inquired into every circumstance respecting it, all of which he recorded in his notes, from whence the following particulars are briefly stated. According to common estimation, this city is a hundred miles in circuit. Its streets and canals are extensive, and there are squares, or market-places, which being necessarily proportioned in size to the prodigious concourse of people by whom they are frequented, are exceedingly spacious. It is situated between a lake of fresh and very clear water on the one side, and a river of great magnitude on the other, the waters of which, by a number of canals, large and small, are made to run through every quarter of the city, carrying with them all the filth into the lake, and ultimately to the sea. This furnishes a communication by water, in addition to that by land, to all parts of the town. The canals and the streets being of sufficient width to allow boats on the one, and carriages on the other, to pass easily with articles necessary for the inhabitants.

It is commonly said that the number of bridges, of all sizes, amounts to twelve thousand. Those which are thrown over the principal canals and are connected with the main streets, have arches so high, and built with so much skill, that vessels with their masts can pass under them. At the same time, carts and horses can pass over, so well is the slope from the street graded to the height of the arch. If they were not so numerous, there would be no way of crossing from one place to another.

[. . .]

There are within the city ten principal squares or market-places, besides innumerable shops along the streets. Each side of these squares is half a mile in length, and in front of them is the main street, forty paces in width, and running in a direct line from one extremity of the city to the other. It is crossed by many low and convenient bridges. These market-squares are at the distance of four miles from each other. In a direction parallel to that of the main street, but on the opposite side of the squares runs a very large canal, on the nearer bank of which capacious warehouses are built of stone, for the accommodation of the merchants who arrive from India and other parts with their goods and effects. They are thus conveniently situated with respect to the market-places. In each of these, upon three days in every week, there is an assemblage of from forty to fifty thousand persons, who attend the markets and supply them with every article of provision that can be desired.

[. . .]

At all seasons there is in the markets a great variety of herbs and fruits, and especially pears of an extraordinary size, weighing ten pounds each, that are white in the inside, like paste, and have a very fragrant smell. There are peaches also, in their season, both of the yellow and white kind, and of a delicious flavour. Grapes are not produced there, but are brought in a dried state, and very good, from other parts. This applies also to wine, which the natives do not hold in estimation, being accustomed to their own liquor prepared from rice and spices. From the sea, which is fifteen miles distant, there is daily brought up the river, to the city, a vast quantity of fish; and in the lake also there is abundance, which gives employment at all times to persons whose sole occupation it is to catch them. The sorts are various according to the season of the year. At the sight of such an importation of fish, you would think it impossible that it could be sold; and yet, in the course of a few hours, it is all taken off, so great is the number of inhabitants, even of those classes which can afford to indulge in such luxuries, for fish and flesh are eaten at the same meal.

Each of the ten market-squares is surrounded with high dwelling-houses, in the lower part of which are shops, where every kind of manufacture is carried on, and every article of trade is sold; such, amongst others, as spices, drugs, trinkets, and pearls. In certain shops nothing is vended but the wine of the country, which they are continually brewing, and serve out fresh to their customers at a moderate price. The streets connected with the market-squares are numerous, and in some of them are many cold baths, attended by servants of both sexes. The men and women who frequent them have from their childhood been accustomed at all times to wash in cold water, which they reckon highly conducive to health. At these bathing places, however, they have apartments provided with warm water, for the use of strangers, who cannot bear the shock of the cold. All are in the daily practice of washing their persons, and especially before their meals.

In other streets are the quarters of the courtesans who are here in such numbers as I dare not venture to report. Not only near the squares, which is the situation usually appropriated for their residence, but in every part of the city they are to be found, adorned with much finery, highly perfumed, occupying well-furnished houses, and attended by many female domestics. These women are accomplished, and are perfect in the arts of caressing and fondling which they accompany with expressions adapted to every description of person. Strangers who have once tasted of their charms, remain in a state of fascination, and become so enchanted by their wanton arts, that they can never forget the impression. Thus intoxicated with sensual pleasures, when they return to their homes they report that they have been in Kin-sai, or The Celestial City, and look forward to the time when they may be enabled to revisit this paradise.

In other streets are the dwellings of the physicians and the astrologers, who also give instructions in reading and writing, as well as in many other arts. They have apartments also amongst those which surround the market-squares. On opposite sides of each of these squares there are two large edifices, where officers appointed by the Great Khan are stationed, to take immediate notice of any differences that may happen to arise between the foreign merchants, or amongst the inhabitants of the place. It is their duty likewise to see that the guards upon the several bridges in their respective vicinities are duly placed, and in cases of neglect, to punish the offenders at their discretion.

[. . .]

The inhabitants of the city are idolaters, and they use paper money as currency. The men as well as the women have fair complexions, and are handsome. The greater part of them are always clothed in silk, in consequence of the vast quantity of that material produced in the territory of Kin-sai, exclusively of what the merchants import from other provinces.

[. . .]

The natural disposition of the native inhabitants of Kin-sai is peaceful, and by the example of their former kings, who were themselves unwarlike, they have been accustomed to habits of tranquility. The management of arms is unknown to them, nor do they keep any in their houses. They conduct their mercantile and manufacturing concerns with perfect candour and honesty. They are friendly towards each other, and persons who inhabit the same street, both men and women, from the mere circumstance of neighbourhood, appear like one family.

In their domestic manners they are free from jealousy or suspicion of their wives, to whom great respect is shown, and any man would be accounted infamous who should presume to use indecent expressions to a married woman. To strangers also, who visit their city in the way of commerce, they give proofs of cordiality, inviting them freely to their houses, showing them friendly attentions, and furnishing them with the best advice and assistance in their mercantile transactions. On the other hand, they dislike the sight of soldiery, not excepting the guards of the Great Khan, for they remind them that they were deprived of the government of their native kings and rulers.

[. . .]

By a regulation which his Majesty has established, there is a guard of ten watchmen stationed, under cover, upon all the principal bridges, of whom five do duty by day and five by night. Each of these guards is provided with a sonorous wooden instrument as well as one of metal, together with a water device, by means of which the hours of the day and night are ascertained. As soon as the first hour of the night is expired, one of the watchmen gives a single stroke upon the wooden instrument, and also upon the metal gong, which announces to the people of the neighbouring streets that it is the first hour. At the expiration of the second, two strokes are given; and so on progressively, increasing the number of strokes as the hours advance. The guard is not allowed to sleep, and must be always on the alert. In the morning, as soon as the sun begins to appear, a single stroke is again struck, as in the evening, and so onwards from hour to hour.

Some of these watchmen patrol the streets, to observe whether any person has a light or fire burning after the hour appointed for extinguishing them. Upon making the discovery, they affix a mark to the door, and in the morning the owner of the house is taken before the magistrates, by whom, if he cannot assign a legitimate excuse for his offence, he is punished. Should they find any person abroad at an unseasonable hour, they arrest and confine him, and in the morning he is carried before the same tribunal. If they notice any person who from lameness or other infirmity is unable to work, they place him in one of the hospitals, of which there are several in every part of the city, founded by the ancient kings, and liberally endowed. When cured, he is obliged to work at some trade.

Immediately upon the appearance of fire breaking out in a house, they give the alarm by beating on the wooden machine, when the watchmen from all the bridges within a certain distance assemble to extinguish it, as well as to save the effects of the merchants and others, by removing them to the stone towers that have been mentioned. The goods are also sometimes put into boats, and conveyed to the islands in the lake. Even on such occasions the inhabitants dare not stir out of their houses, when the fire happens in the night, and only those can be present whose goods are actually being removed, together with the guard collected to assist, which seldom amounts to a smaller number than from one to two thousand men.

In cases also of tumult or insurrection amongst the citizens, the services of this police guard are necessary; but, independently of them, his Majesty always keeps on foot a large body of troops, both infantry and cavalry, in the city and its vicinity, the command of which he gives to his ablest officers.

[. . .]

At the distance of twenty-five miles from this city, in a direction to the northward of east, lies the sea, where there is an extremely fine port, frequented by all the ships that bring merchandise from India.

Marco Polo, happening to be in the city of Kin-sai at the time of making the annual report to his Majesty's commissioners of the amount of revenue and the number of inhabitants, had an opportunity of observing that the latter were registered at one hundred and sixty tomans of fire-places, that is to say, of families dwelling under the same roof; and as a toman is ten thousand, it follows that the whole city must have contained one million six hundred thousand families.

THE NORTH AFRICAN MUSLIM TRAVELLER IBN BATTUTA DESCRIBES BYZANTINE CONSTANTINOPLE (1354)

Ibn Battuta, "Constantinople the Great," from *The Travels of Ibn Battuta, AD 1325–1354*, H.A.R. Gibb (ed.), Hakluyt Society, 1962

Our entry into Constantinople was made about noon or a little later, and they beat their church gongs until the very skies shook with the mingling of their sounds. When we reached the first of the gates of the king's palace, we found it guarded by about a hundred men, who had an officer of theirs with them on top of a platform, and I heard them saying 'Sarakinu, Sarakinu' which means 'Muslims'. They would not let us enter, and when the members of the khatun's [the Byzantine Emperor's daughter] party told them that we had come in her suite they answered, 'They cannot enter except by permission', so we stayed by the gate. One of the khatun's party sent a messenger to tell her of this while she was still with her father. She told him about us, whereupon he gave orders to admit us and assigned us a house near the residence of the khatun. He wrote also on our behalf an order that we should not be molested wheresoever we might go in the city, and this order was proclaimed in the bazaars. We remained indoors for three nights, during which hospitality-gifts were sent us of flour, bread, sheep, fowls, ghee, fruit, fish, money and rugs, and on the fourth day we had an audience of the sultan [Emperor Andronicus III].

Account of the sultan of Constantinople

His name is Takfur, son of the sultan Jirjis. [Here Ibn Battuta uses Arabic names and Muslim titles to refer to the Eastern Roman Emperor.] His father . . . was still in the bond of life, but had renounced the world and become a monk, devoting himself to religious exercises in the churches, and had resigned the kingship to his son. . . . On the fourth day from our arrival at Constantinople, the khatun sent her page Sumbul the Indian to me, and he took my hand and led me into the palace. We passed through four gateways, each of which had porticoes in which were footsoldiers with their weapons, their officer being on a carpeted platform. When we reached the fifth gateway, the page Sumbul left me, and going inside returned with four

Greek pages, who searched me to see that I had no knife on my person. The officer said to me, 'This is a custom of theirs. Every person who enters the king's presence, be he noble or commoner, foreigner or native, must be searched'. The same practice is observed in the land of India.

Then, after they had searched me, the man in charge of the gate rose, took me by the hand, and opened the door. Four of the men surrounded me, two holding my sleeves and two behind me, and brought me into a large audience-hall, whose walls were of mosaic work, in which were pictured figures of creatures, both animate and inanimate. In the centre of it was a water-channel with trees on either side of it, and men were standing to right and left, silent, not one of them speaking. In the midst of the hall were three men standing, to whom those four men delivered me. These took hold of my garments as the others had done and so on a signal from another man led me forward. One of them was a Jew and he said to me in Arabic, 'Don't be afraid for this is their custom that they use with every visitor. I am the interpreter and I am originally from Sytia'. So I asked him how I should salute, and he told me to say 'al-salamu alaikum'.

I came then to a great pavilion; the sultan was there on his throne, with his wife, the mother of the khatun. Before him and at the foot of the throne were the khatun and her brothers. To the right of him were six men, to his left four and behind him four, every one of them armed. He signed to me, before I had saluted and reached him, to sit down for a moment, so that my apprehension might be calmed, and I did so. Then I approached him and saluted him, and he signed me to sit down, but I did not do so. He questioned me about Jerusalem, the Sacred Rock, the Church called al Qumama the cradle of Jesus [the Church of the Holy Sepulchre] and Bethlehem, and about the city of al-Khalil (peace be upon him) [Hebron], then about Damascus, Cairo, al-Iraq and the land of al-Rum, and I answered him on all his questions, the Jew interpreting between us. He was pleased with my replies and said to his sons, 'Honour this man and ensure his safety'. He then bestowed on me a robe of honour and ordered for me a horse with saddle and bridle, and a parasol of the kind that the king has carried above his head, that being a sign of protection. I asked him to designate someone to ride about the city with me every day, that I might see its wonders and curious sights and tell of them in my own country, and he designated such a guide for me. It is one of the customs among them that anyone who wears the king's robe of honour and rides on his horse is paraded through the city bazaars with trumpets, fifes and drums, so that the people may see him . . .

Account of the city

It is enormous in magnitude and divided into two parts, between which there is a great river in which there is a flow and ebb of tide [the Golden Horn] just as in the wadi of Sala in the country of the Maghrib [North Africa]. In the former time there was a bridge over it, built of stone, but the bridge has fallen into ruin and nowadays it is crossed in boats. The name of the river is Absumi. One of the two parts of the city is called Astanbiil; it is on the eastern bank of the river and includes the places of residence of the sultan, his officers of state, and the rest of the population. Its bazaars and streets are spacious and paved with flagstones, and the members of each craft have a separate place, no others sharing it with them. Each bazaar has gates which are closed upon it at night, and the majority of the artisans and sellers in them are women. The city is at the foot of a hill that projects about nine miles into the sea, and its breadth is the same or more. On top of the hill is a small citadel and the sultan's palace. This hill is surrounded by the city wall, which is a formidable one and cannot be taken by assault on the side of the sea. Within the wall are about thirteen inhabited villages. The principal church too is in the midst of this section of the city.

As for the other section of it, it is called al-Ghalata, and lies on the western bank of the river, somewhat like Ribat al-Fath in its proximity to the river. This section is reserved for the Christians of the Franks dwelling there. They are of different kinds, including Genoese, Venetians, men of Rome and people of France, and they are under the government of the King of Constantinople, who appoints over them one of their number whom they approve, and him they call the Qums. They are required to pay a tax every year to the King of Constantinople, but they often rebel against his authority and then he makes war on them until the Pope restores peace between them. They are all men of commerce, and their port is one of the greatest of ports; I saw in it about a hundred galleys, such as merchant vessels and other large ships, and as for the small ships they were too numerous to be counted. The bazaars in this section are good, but

overlaid with all kinds of filth, and traversed by a small, dirty and filth-laden stream. Their churches too are dirty and mean.

Account of the great church

I can describe only its exterior; as for its interior I did not see it. It is called in their language Aya Sufiya, and the story goes that it was an erection of Asaf the son of Barakhya, who was the son of the maternal aunt of Solomon (on whom be peace). It is one of the greatest churches of the Greeks; around it is a wall which encircles it so that it looks like a city in itself. Its gates are thirteen in number, and it has a sacred enclosure, which is about a mile long and closed by a great gate. No one is prevented from entering the enclosure, and in fact I went into it with the king's father, who will be mentioned later; it is like an audience-hall, paved with marble and traversed by a water-channel which issues from the church. This flows between two walls about a cubit high, constructed in marble inlaid with pieces of different colours and cut with the most skilful art, and trees are planted in rows on both sides of the channel. From the gate of the church to the gate of this hall there is a lofty pergola made of wood, covered with grape-vines and at the foot with jasmine and scented herbs. Outside the gate of this hall is a large wooden pavilion containing platforms, on which the guardians of this gate sit, and to the right of the 435 pavilions are benches and booths, mostly of wood, in which sit their qadis and the recorders of their bureaux. In the middle of the booths is a wooden pavilion, to which one ascends by a flight of wooden steps; in this pavilion is a great chair swathed in woollen cloth on which their qadi sits. We shall speak of him later. To the left of the pavilion which is at the gate of this hall is the bazaar of the druggists. The canal that we have described divides into two branches, one of which passes through the bazaar of the druggists and the other through the bazaar where the judges and the scribes sit.

At the door of the church there are porticoes where the attendants sit who sweep its paths, light its lamps and close its doors. They allow no person to enter it until he prostrates himself to the huge cross at their place, which they claim to be a relic of the wood on which the double of Jesus (on whom be peace) was crucified. This is over the door of the church, set in a golden frame about ten cubits in height, across which they have placed a similar golden frame so that it forms a cross. This door is covered with plaques of silver and gold, and its two rings are of pure gold. I was told that the number of monks and priests in this church runs into thousands, and that some of them are descendants of the Apostles, also that inside it is another church exclusively for women, containing more than a thousand virgins consecrated to religious devotions, and a still greater number of aged and widowed women. It is the custom of the king, his officers of state, and the rest of the inhabitants to come to visit this church every morning, and the Pope comes to it once in the year. When he is at a distance of four nights' journey from the town the king goes out to meet him and dismounts before him; when he enters the city, the king walks on foot in front of him, and comes to salute him every morning and evening during the whole period of his stay in Constantinople until he departs.

THE SPANISH CONQUISTADOR BERNAL DIAZ DESCRIBES TENOCHTITLÁN, CAPITAL OF AZTEC MEXICO (1521)

Bernal Diaz del Castillo, "About the Great and Solemn Reception which the Great Montezuma Gave Cortés upon Entering the Great City of Mexico," from *The Conquest of Mexico* (1521)

Gazing on such wonderful sights, we did not know what to say, or whether what appeared before us was real, for on one side, on the land, there were great cities, and in the lake ever so many more, and the lake itself was crowded with canoes, and in the Causeway were many bridges at intervals, and in front of us stood the great City of Mexico, and we – we did not even number four hundred soldiers! . . .

Let the curious readers consider whether there is not much to ponder over in this that I am writing. What men have there been in the world who have shown such daring? But let us get on, and march along the Causeway. When we arrived where another small causeway branches off . . . where there were some buildings like towers, which are their oratories, many more chieftains and Caciques approached clad in very rich mantles, the brilliant liveries of one chieftain differing from those of another, and the causeways were crowded with them. The Great Montezuma had sent these great Caciques in advance to receive us, and when they came before Cortés they bade us welcome

in their language, and as a sign of peace, they touched their hands against the ground, and kissed the ground with the hand.

[. . .]

As soon as we arrived and entered into the great court, the Great Montezuma took our Captain by the hand, for he was there awaiting him, and led him to the apartment and saloon where he was to lodge, which was very richly adorned according to their usage . . . A sumptuous dinner was provided for us according to their use and custom, and we ate it at once. So this was our lucky and daring entry into the great city of Tenochtitlán Mexico on the 8th day of November the year of our Saviour Jesus Christ 1519.

[. . .]

As I am almost tired of writing about this subject and my interested readers will be even more so, I will stop talking about it and tell how our Cortés in company with our captains and soldiers went to see Tlaltelolco, which is the great market-place of Mexico, and how we ascended the great Cue [temple pyramid] where stand the Idols Tezcatepuca and Huichilobos. This was the first time that our Captain went out to see the City, and I will relate what else happened.

As we had already been four days in Mexico and neither the Captain nor any of us had left our lodgings except to go to the houses and gardens, Cortés said to us that it would be well to go to the great Plaza and see the great Temple of Huichilobos, and that he wished to consult the Great Montezuma and have his approval . . . When Montezuma knew his wishes he sent to say that we were welcome to go; on the other hand, as he was afraid that we might do some dishonour to his idols, he determined to go with us himself with many of his chieftains . . . So he went on, and ascended the great Cue accompanied by many priests, and he began to burn incense and perform other ceremonies to Huichilobos . . .

When we arrived at the market-place, called Tlaltelolco, we were astounded at the number of people and the quantity of merchandise that it contained, and at the good order and control that was maintained, for we had never seen such a thing before. The chieftains who accompanied us acted as guides. Each kind of merchandise was kept by itself and had its fixed place marked out. Let us begin with the dealers in gold, silver, and precious stones, feathers, mantles, and embroidered goods. Then there were other wares consisting of Indian slaves both men and women; . . . and they brought them along tied to long poles, with collars round their necks so that they could not escape, and others they left free. Next there were other traders who sold great pieces of cloth and cotton, and articles of twisted thread, and there were cacahuateros who sold cacao . . . There were those who sold cloths of henequen and ropes and cotaras with which they are shod, which are made from the same plant, and sweet cooked roots, and other tubers which they get from this plant, all were kept in one part of the market in the place assigned to them. In another part there were skins of tigers and lions, of otters and jackals, deer and other animals and badgers and mountain cats, some tanned and others untanned, and classes of merchandise.

Let us go on and speak of those who sold beans and sage and other vegetables and herbs in another part, and to those who sold fowls, cocks with wattles, rabbits, hares, deer, mallards, young dogs and other things of that sort in their part of the market, and let us also mention the fruiterers, and the women who sold cooked food, dough and tripe in their own part of the market; then every sort of pottery made in a thousand different forms from great water jars to little jugs, these also had a place to themselves; then those who sold honey and honey paste and other dainties like nut paste, and those who sold lumber, boards, cradles, beams, blocks and benches, each article by itself, and the vendors of ocote firewood, and other things of a similar nature . . . But why do I waste so many words in recounting what they sell in that great market, for I shall never finish if I tell it all in detail. Paper, which in this country is called Amal, and reeds scented with liquidambar, and full of tobacco, and yellow ointments and things of that sort are sold by themselves, and much cochineal is sold under the arcades which are in that great market place, and there are many vendors of herbs and other sorts of trades. There are also buildings where three magistrates sit in judgement and there are executive officers like Alguacils who inspect the merchandise. I am forgetting those who sell salt, and those who make the stone knives, and how they split them off the stone itself; and the fisherwomen and others who sell some small cakes made from a sort of ooze which they get out of the great lake, which curdles, and from this they make a bread having a flavour something like cheese. There are for sale axes of brass and copper and tin, and gourds and gaily painted jars made of wood. I could wish that I had finished telling of all the things which are sold there, but they are so numerous and of such different quality and the great

market-place with its surrounding arcades was so crowded with people, that one would not have been able to see and inquire about it all in two days.

[. . .]

Now let us leave the great market-place . . . and arrive at the great courts and walls where the great Cue stands. Before reaching the great Cue there is a great enclosure of courts, it seems to me larger than the plaza of Salamanca, with two walls of masonry surrounding it and the court itself all paved with very smooth great white flagstones. And where there were not these stones it was cemented and burnished and all very clean, so that one could not find any dust or a straw in the whole place.

When we arrived near the great Cue and before we had ascended a single step of it, the Great Montezuma sent down from above, where he was making his sacrifices, six priests and two chieftains to accompany our Captain. On ascending the steps, which are one hundred and fourteen in number, they attempted to take him by the arms so as to help him to ascend (thinking that he would get tired), as they were accustomed to assist their lord Montezuma, but Cortés would not allow them to come near him. When we got to the top of the great Cue, on a small plaza which has been made on the top where there was a space like a platform with some large stones placed on it, on which they put the poor Indians for sacrifice, there was a bulky image like a dragon and other evil figures and much blood shed that very day.

When we arrived there Montezuma came out of an oratory where his cursed idols were, at the summit of the great Cue, and two priests came with him, and after paying great reverence to Cortés and to all of us he said: "You must be tired . . . from ascending this our great Cue," and Cortés replied through our interpreters who were with us that he and his companions were never tired by anything. Then Montezuma took him by the hand and told him to look at his great city and all the other cities that were standing in the water, and the many other towns on the land round the lake . . .

So we stood looking about us, for that huge and cursed temple stood so high that from it one could see over everything very well, and we saw the three causeways which led into Mexico . . . and we saw the fresh water that comes from Chapultepec which supplies the city, and we saw the bridges on the three causeways which were built at certain distances apart through which the water of the lake flowed in and out from one side to the other, and we beheld on that great lake a great multitude of canoes, some coming with supplies of food and others returning loaded with cargoes of merchandise; and we saw that from every house of that great city and of all the other cities that were built in the water it was impossible to pass from house to house except by drawbridges which were made of wood or in canoes; and we saw in those cities Cues and oratories like towers and fortresses and all gleaming white, and it was a wonderful thing to behold . . .

After having examined and considered all that we had seen we turned to look at the great market-place and the crowds of people that were in it, some buying and others selling, so that the murmur and hum of their voices and words that they used could be heard more than a league off. Some of the soldiers among us who had been in many parts of the world, in Constantinople, and all over Italy, and in Rome, said that so large a market-place and so full of people, and so well regulated and arranged, they had never beheld before.

THE GERMAN ARTIST ALBRECHT DÜRER DESCRIBES A COMMUNAL PROCESSION IN ANTWERP, BELGIUM (1519)

Albrecht Dürer, "Visit to Antwerp," from *Journal of a Voyage to Belgium* (1527)

The Church of our Lady (the Cathedral) at Antwerp is so very large that many masses are sung in it at one time without interfering with each other. The altars have wealthy endowments, and the best musicians are employed that can be had. The church has many devout services, much stone-work, and in particular a beautiful tower. I have also been into the rich Abbey of St Michael. There are, in the choir there, splendid stalls of sculptured stone-work. But at Antwerp they spare no cost on such things, for there is money enough . . .

On the Sunday after our dear Lady's Assumption I saw the great Procession from the Church of our Lady at Antwerp, when the whole town of every craft and rank was assembled, each dressed in his best according to his rank. And all ranks and guilds had their signs, by which they might be known. In the intervals great costly pole-candles were borne, and their long old Frankish trumpets of silver. There were also in the German fashion many pipers and drummers. All the instruments were loudly and noisily blown and beaten.

I saw the Procession pass along the street, the people being arranged in rows, each man some distance from his neighbour, but the rows close one behind another. There were the Goldsmiths, the Painters, the Masons, the Broderers, the Sculptors, the Joiners, the Carpenters, the Sailors, the Fishermen, the Butchers, the Leatherers, the Clothmakers, the Bakers, the Tailors, the Cordwainers – indeed workmen of all kinds, and many craftsmen and dealers who work for their livelihood. Likewise the shop-keepers and merchants and their assistants of all kinds were there. After these came the shooters with guns, bows, and crossbows and the horsemen and foot-soldiers also. Then followed the watch of the Lords Magistrates. Then came a fine troop all in red, nobly and splendidly clad. Before them however went all the religious Orders and the members of some Foundations very devoutly, all in their different robes. A very large company of widows also took part in this procession. They support themselves with their own hands and observe a special rule. They were all dressed from head to foot in white linen garments, made expressly for the occasion, very sorrowful to see. Among them I saw some very stately persons. Last of all came the Chapter of our Lady's Church with all their clergy, scholars, and treasurers. Twenty persons bore the image of the Virgin Mary with the Lord Jesus, adorned in the costliest manner, to the honour of the Lord God.

In this Procession very many delightful things were shown, most splendidly got up. Wagons were drawn along with masques upon ships and other structures. Behind them came the company of the Prophets in their order and scenes from the New Testament, such as the Annunciation, the Three Holy Kings riding on great camels and on other rare beasts, very well arranged; also how our Lady fled to Egypt – very devout – and many other things, which for shortness I omit. At the end came a great Dragon which St Margaret and her maidens led by a girdle; she was especially beautiful. Behind her came St George with his squire, a very goodly knight in armour. In this host also rode boys and maidens most finely and splendidly dressed in the costumes of many lands, representing various Saints. From beginning to end the Procession lasted more than two hours before it was gone past our house. And so many things were there that I could never write them all in a book, so I let it well alone.

"The Great Towns"
from *The Condition of the Working Class in England in 1844* (1845)

Friedrich Engels

Editors' Introduction

It was the peculiar fate of Friedrich Engels (1820–1895) to live most of his adult life in the shadow of his better-known friend and partner Karl Marx and to be remembered as a fiercely bearded icon of International Communism. It was, however, a more humanly accessible Engels who, full of youthful idealism at the age of only 24, came face to face with the social horrors of the Industrial Revolution. Young Engels was sent by his industrialist father to learn business management in the factories of Manchester in the English North West. The unintended consequences of that particular paternal decision was *The Condition of the Working Class in England in 1844* (1845), a book that ranks as one of the earliest masterpieces of urban sociology.

By 1844 the Industrial Revolution had transformed conditions in many English cities. Manchester, England, which Engels observed in detail, was emblematic of what the new industrial cities were like. Plate 4 from Augustus Welby Pugin's *Contrasts* (1841) contrasts the skyline of a fifteenth-century city dominated by church steeples to the same town in 1840. In the second view mills, factories, and a huge prison dominate the scene.

In the selection "The Great Towns" reprinted here, Engels employs a peripatetic method. Although he summarizes the socialist theory of the origin and historic role of the industrial working class, and although he quotes from many contemporaneous sources to bolster his analysis, Engels constructs the bulk of his argument by merely walking around the city and reporting what he sees. Quickly growing impatient with *telling* his readers about the social misery of working-class life, Engels begins *showing* them the horrors of industrial urbanism by conducting them on a tour of Manchester's working-class districts. As in Dante's *Inferno*, the tour descends deeper and deeper into the filth, misery, and despair that constitute the greater part of the Manchester conurbation.

Engels wrote just as the first daguerreotypes of cities were being produced, but unfortunately he did not illustrate his book with actual pictures of the conditions that he described. Later in the nineteenth century Jacob Riis, Lewis Hine, and other photographers were to document slum conditions. The response of these and other photographers to the new reality of the nineteenth-century city is discussed in the selection by Frederic Stout, "Visions of a New Reality" (p. 147).

No one can read *The Condition of the Working Class* without acknowledging that Engels had come to know the various neighborhoods of proletarian Manchester – Old Town, Irish Town, Long Millgate, and Salford – intimately and that his observations were acute and objective. Of particular interest are his descriptions of the public health consequences (in terms of air and water pollution) of unrestrained overbuilding. Engels also observed that the façades of the main thoroughfares mask the horrors that lie beyond from the eyes of the factory owners and the middle-class managers who commute into the city from outlying suburbs. Plate 5, from *Frank Leslie's Illustrated Newspaper* in 1887, contrasts the belching factories and slums that comprise the bulk of the city with a spacious main thoroughfare in the city.

The entire tradition of twentieth-century urban planning, capitalist and socialist alike, owes an enormous debt to Engels. The connection he draws between the physical decrepitude of the urban infrastructure and the alienation and despair of the urban poor remains valid to the present day. The urban parks movement and the construction of ideal company towns – Saltaire and Port Sunlight in the UK, Lowell and Pullman in the United States – as well as more recent attempts at inner-city redevelopment, all address issues first identified by Engels.

The conditions described by Engels form the basis for the social realist tradition in literature, a tradition that begins with Charles Dickens and Mrs Gaskell in England and is continued in the works of Upton Sinclair and Theodore Dreiser in the United States.

See Chapter 10 of Sir Peter Hall's *Cities in Civilization* (New York: Pantheon, 1998) "Manchester 1760–1840" for more on early Manchester. Hall's chapter contains a bibliography of source material on economic and social conditions in early Manchester.

Other significant investigations of urban poverty in England include Henry Mayhew's *London Labour and the London Poor* (four volumes, 1851–1862), Charles Booth's *Conditions and Occupation of the People in East London and Hackney* (*Journal of the Royal Historical Society*, 1887), Jack London's *People of the Abyss* (New York: Macmillan, 1903), and George Orwell's *Down and Out in Paris and London* (London: Secker and Warburg, 1933).

In America, key works include Jacob Riis, *How the Other Half Lives* (New York: Scribners, 1903) and a whole series of reports on conditions in the African-American ghettos such as W.E.B. Du Bois's *The Philadelphia Negro* (p. 119), St. Clair Drake and Horace Cayton's *Black Metropolis* (Chicago: University of Chicago Press, 1945), and William Julius Wilson's *When Work Disappears* (p. 126).

For a sampling of Engels's most important writings as well as those of his lifelong friend Karl Marx, see Robert C. Tucker, *The Marx–Engels Reader* (New York: W.W. Norton, 1978). For an excellent summary of nineteenth-century urban poverty conditions, consult "The City of Dreadful Night" in Peter Hall's *Cities of Tomorrow* (London: Basil Blackwell, 1988). For further information on Engels's Manchester, and its connections to an emerging social realism in fiction, consult literary historian Steven Marcus's magisterial *Engels, Manchester, and the Working Class* (New York: Random House, 1974). Also of interest is Robert Roberts's extraordinary *The Classic Slum* (Manchester: University of Manchester Press, 1971), a first-person account of growing up in Salford during the early years of the twentieth century.

A town, such as London, where a man may wander for hours together without reaching the beginning of the end, without meeting the slightest hint which could lead to the inference that there is open country within reach, is a strange thing. This colossal centralization, this heaping together of two and a half millions of human beings at one point, has multiplied the power of this two and a half millions a hundred-fold; has raised London to the commercial capital of the world, created the giant docks and assembled the thousand vessels that continually cover the Thames. I know nothing more imposing than the view which the Thames offers during the ascent from the sea to London Bridge. The masses of buildings, the wharves on both sides, especially from Woolwich upwards, the countless ships along both shores, crowding ever closer and closer together, until, at last, only a narrow passage remains in the middle of the river, a passage through which hundreds of steamers shoot by one another; all this is so vast, so impressive, that a man cannot collect himself, but is lost in the marvel of England's greatness before he sets foot upon English soil.

But the sacrifices which all this has cost become apparent later. After roaming the streets of the capital a day or two, making headway with difficulty through the human turmoil and the endless lines of vehicles, after visiting the slums of the metropolis, one realizes for the first time that these Londoners have been forced to sacrifice the best qualities of their human nature, to bring to pass all the marvels of civilization which crowd their city; that a hundred powers which slumbered within them have remained inactive, have been suppressed in order that a few might be developed more fully and multiply through union with those of others. The very turmoil of the streets has something repulsive, something against which human nature rebels. The hundreds of thousands of all classes and ranks crowding past each other, are they not all human beings with the same qualities and powers, and with the same interest in being happy? And have they not, in the end, to seek happiness in the same way, by the same means? And still they crowd by one another as though they had nothing in common, nothing to do with one another, and their only agreement is the tacit one, that each keep to his own side of the pavement, so as not to delay the opposing streams of the crowd, while it occurs to no man to honour another with so much as a glance. The brutal indifference, the unfeeling isolation of each in his private interest becomes the more repellent and offensive, the more these individuals are crowded together, within a limited space. And, however much one may be aware that this isolation of the individual, this narrow self-seeking is the fundamental principle of our society everywhere, it is nowhere so shamelessly barefaced, so self-conscious as just here in the crowding of the great city. The dissolution of mankind into monads, of which each one has a separate principle and a separate purpose, the world of atoms, is here carried out to its utmost extreme.

Hence it comes, too, that the social war, the war of each against all, is here openly declared. . . . , people regard each other only as useful objects; each exploits the other, and the end of it all is, that the stronger treads the weaker under foot, and that the powerful few, the capitalists, seize everything for themselves, while to the weak many, the poor, scarcely a bare existence remains.

What is true of London, is true of Manchester, Birmingham, Leeds, is true of all great towns. Everywhere barbarous indifference, hard egotism on one hand, and nameless misery on the other, everywhere social warfare, every man's house in a state of siege, everywhere reciprocal plundering under the protection of the law, and all so shameless, so openly avowed that one shrinks before the consequences of our social state as they manifest themselves here undisguised, and can only wonder that the whole crazy fabric still hangs together.

Since capital, the direct or indirect control of the means of subsistence and production, is the weapon with which this social warfare is carried on, it is clear that all the disadvantages of such a state must fall upon the poor. For him no man has the slightest concern. Cast into the whirlpool, he must struggle through as well as he can. If he is so happy as to find work, i.e. if the bourgeoisie does him the favour to enrich itself by means of him, wages await him which scarcely suffice to keep body and soul together; if he can get no work he may steal, if he is not afraid of the police, or

starve, in which case the police will take care that he does so in a quiet and inoffensive manner. During my residence in England, at least twenty or thirty persons have died of simple starvation under the most revolting circumstances, and a jury has rarely been found possessed of the courage to speak the plain truth in the matter. Let the testimony of the witnesses be never so clear and unequivocal, the bourgeoisie, from which the jury is selected, always finds some backdoor through which to escape the frightful verdict, death from starvation. The bourgeoisie dare not speak the truth in these cases, for it would speak its own condemnation. But indirectly, far more than directly, many have died of starvation, where long continued want of proper nourishment has called forth fatal illness, when it has produced such debility that causes which might otherwise have remained inoperative, brought on severe illness and death. The English working-men call this social murder, and accuse our whole society of perpetrating this crime perpetually. Are they wrong?

True, it is only individuals who starve, but what security has the working-man that it may not be his turn tomorrow? Who assures him employment, who vouches for it that, if for any reason or no reason his lord and master discharges him tomorrow, he can struggle along with those dependent upon him, until he may find some one else "to give him bread"? Who guarantees that willingness to work shall suffice to obtain work, that uprightness, industry, thrift, and the rest of the virtues recommended by the bourgeoisie, are really his road to happiness? No one. He knows that he has something today, and that it does not depend upon himself whether he shall have something tomorrow. He knows that every breeze that blows, every whim of his employer, every bad turn of trade may hurl him back into the fierce whirlpool from which he has temporarily saved himself, and in which it is hard and often impossible to keep his head above water. He knows that, though he may have the means of living today, it is very uncertain whether he shall tomorrow.

[. . .]

Manchester lies at the foot of the southern slope of a range of hills, which stretch hither from Oldham, their last peak, Kersallmoor, being at once the racecourse and the Mons Sacer of Manchester. Manchester proper lies on the left bank of the Irwell, between that stream and the two smaller ones, the Irk and the Medlock, which here empty into the Irwell. On the right bank of the Irwell, bounded by a sharp curve of the river, lies Salford, and farther westward Pendleton; northward from the Irwell lie Upper and Lower Broughton; northward of the Irk, Cheetham Hill; south of the Medlock lies Hulme; farther east Chorlton on Medlock; still farther, pretty well to the east of Manchester, Ardwick. The whole assemblage of buildings is commonly called Manchester, and contains about four hundred thousand inhabitants, rather more than less. The town itself is peculiarly built, so that a person may live in it for years, and go in and out daily without coming into contact with a working-people's quarter or even with workers; that is, so long as he confines himself to his business or to pleasure walks. This arises chiefly from the fact, that by unconscious tacit agreement, as well as with outspoken conscious determination, the working-people's quarters are sharply separated from the sections of the city reserved for the middle class; or, if this does not succeed, they are concealed with the cloak of charity. Manchester contains, at its heart, a rather extended commercial district, perhaps half a mile long and about as broad, and consisting almost wholly of offices and ware houses. Nearly the whole district is abandoned by dwellers, and is lonely and deserted at night; only watchmen and policemen traverse its narrow lanes with their dark lanterns. This district is cut through by certain main thoroughfares upon which the vast traffic concentrates, and in which the ground level is lined with brilliant shops. In these streets the upper floors are occupied, here and there, and there is a good deal of life upon them until late at night. With the exception of this commercial district, all Manchester proper, all Salford and Hulme, a great part of Pendleton and Chorlton, two-thirds of Ardwick, and single stretches of Cheetham Hill and Broughton are all unmixed working-people's quarters, stretching like a girdle, averaging a mile and a half in breadth, around the commercial district. Outside, beyond this girdle, lives the upper and middle bourgeoisie, the middle bour-geoisie in regularly laid out streets in the vicinity of the working quarters, especially in Chorlton and the lower-lying portions of Cheetham Hill; the upper bourgeoisie in remoter villas with gardens in Chorlton and Ardwick or on the breezy heights of Cheetham Hill, Broughton and Pendleton, in free, wholesome country air, in fine, comfortable homes, passed once every half or quarter hour by omnibuses going into the city. And the finest part of the arrangement is this, that the members of this money aristocracy can take the shortest road through the middle of all the labouring districts to their places of business, without ever seeing that they are in the midst of the grimy misery that lurks to the right and the left. For the thoroughfares leading from the Exchange in all directions out of the city are lined, on both sides, with an almost unbroken series of shops, and are so kept in the hands of the middle and lower bourgeoisie, which, out of self-interest, cares for a decent and cleanly external appearance and *can* care for it. True, these shops bear some relation to the districts which lie behind them, and are more elegant in the commercial and residential quarters than when they hide grimy working-men's dwellings; but they suffice to conceal from the eyes of the wealthy men and women of strong stomachs and weak nerves the misery and grime which form the complement of their wealth. So, for instance, Deansgate, which leads from the Old Church directly southward, is lined first with mills and warehouses, then with second-rate shops and alehouses; farther south, when it leaves the com-mercial district, with less inviting shops, which grow dirtier and more interrupted by beerhouses and gin palaces the farther one goes, until at the southern end the appearance of the shops leaves no doubt that workers and workers only are their customers. So Market Street running south east from the Exchange; at first brilliant shops of the best sort, with counting-houses or warehouses above; in the continuation, Piccadilly, immense hotels and warehouses; in the farther continuation, London Road, in the neighbour-hood of the Medlock, factories, beerhouses, shops for the humbler bourgeoisie and the working population; and from this point onward, large gardens and villas of the wealthier merchants and manufacturers. In this way any one who knows Manchester can infer the adjoining districts, from the appearance of the thoroughfare, but one is seldom in a position to catch from the street a glimpse of the real labouring districts. I know very well that this hypocritical plan is more or less common to all great cities; I know, too, that the retail dealers are forced by the nature of their business to take possession of the great highways; I know that there are more good buildings than bad ones upon such streets everywhere, and that the value of land is greater near them than in remoter districts; but at the same time I have never seen so systematic a shutting out of the working-class from the thoroughfares, so tender a concealment of everything which might affront the eye and the nerves of the bourgeoisie, as in Manchester. And yet, in other respects, Manchester is less built according to a plan, after official regulations, is more an outgrowth of accident, than any other city; and when I consider in this connection the eager assurances of the middle-class, that the working-class is doing famously, I cannot help feeling that the liberal manufacturers, the "Big Wigs" of Manchester, are not so innocent after all, in the matter of this sensitive method of construction.

I may mention just here that the mills almost all adjoin the rivers or the different canals that ramify throughout the city, before I proceed at once to describe the labouring quarters. First of all, there is the Old Town of Manchester [Figure 1], which lies between the northern boundary of the commercial district and the Irk. Here the streets, even the better ones, are narrow and winding, as Todd Street, Long Millgate, Withy Grove, and Shude Hill, the houses dirty, old, and tumble-down, and the construction of the side streets utterly horrible. Going from the Old Church to Long Millgate, the stroller has at once a row of old-fashioned houses at the right, of which not one has kept its original level; these are remnants of the old pre-manufacturing Manchester, whose former inhabitants have removed with their descendants into better-built districts, and have left the houses, which were not good enough for them, to a working-class population strongly mixed with Irish blood. Here one is in an almost undisguised working-men's quarter, for even the shops and beer-houses hardly take the trouble to exhibit a trifling degree of cleanliness. But all this is nothing in comparison with the courts and lanes which lie behind, to which access can be gained only through covered passages in which no two human beings can pass at the same time. Of the irregular cramming together of dwellings in ways which defy all rational plan, of the tangle in which they are crowded literally one upon the other, it is impossible to convey an idea. And it is not the buildings surviving from the old times of Manchester which are to blame for this; the confusion has only recently reached its height when every scrap of space left by the old way of building has been filled up and patched over until not a foot of land is left to be further occupied.

[. . .]

The south bank of the Irk is here very steep and between fifteen and thirty feet high. On this declivitous hillside there are planted three rows of houses, of which the lowest rise directly out of the river, while the front walls of the highest stand on the crest of the hill in Long Millgate. Among them are mills on the river; in short, the method of construction is as crowded and disorderly here as in the lower part of Long Millgate. Right and left a multitude of covered passages lead from the main street into numerous courts, and he who turns in thither gets into a filth and disgusting grime the equal of which is not to be found – especially in the courts which lead down to the Irk and which contain unqualifiedly the most horrible dwellings which I have yet beheld. In one of these courts there stands directly at the entrance, at the end of the covered passage, a privy without a door, so dirty that the inhabitants can pass into and out of the court only by passing through foul pools of stagnant urine and excrement. This is the first court on the Irk above Ducie Bridge – in case any one should care to look into it. Below it on the river there are several tanneries which fill the whole neighbourhood with the stench of animal putrefaction. Below Ducie Bridge the only entrance to most of the houses is by means of narrow dirty stairs and over heaps of refuse and filth. The first court below Ducie Bridge, known as Allen's Court, was in such a state at the time of the cholera that the sanitary police ordered it evacuated, swept, and disinfected with chloride of lime. Dr. Kay gives a terrible description of the state of this court at that time. Since then it seems to have been partially torn away and rebuilt; at least looking down from Ducie Bridge, the passer-by sees several ruined walls and heaps of debris with some newer houses. The view from this bridge,

Figure 1

mercifully concealed from mortals of small stature by a parapet as high as a man, is characteristic for the whole district. At the bottom flows, or rather stagnates, the Irk, a narrow, coal-black, foul-smelling stream full of debris and refuse, which it deposits on the shallower right bank. In dry weather, a long string of the most disgusting, blackish-green, slime pools are left standing on this bank, from the depths of which bubbles of miasmatic gas constantly arise and give forth a stench unendurable even on the bridge forty or fifty feet above the surface of the stream. But besides this, the stream itself is checked every few paces by high weirs, behind which slime and refuse accumulate and rot in thick masses. Above the bridge are tanneries, bonemills, and gasworks, from which all drains and refuse find their way into the Irk, which receives further the contents of all the neighbouring sewers and privies. It may be easily imagined, therefore, what sort of residue the stream deposits. Below the bridge you look upon the piles of debris, the refuse, filth, and offal from the courts on the steep left bank; here each house is packed close behind its neighbour and a piece of each is visible, all black, smoky, crumbling, ancient, with broken panes and window-frames. The background is furnished by old barrack-like factory buildings. On the lower right bank stands a long row of houses and mills; the second house being a ruin without a roof, piled with debris; the third stands so low that the lowest floor is uninhabitable, and therefore without windows or doors. Here the background embraces the pauper burial-ground, the station of the Liverpool and Leeds railway, and, in the rear of this, the Workhouse, the "Poor-Law-Bastille" of Manchester, which, like a citadel, looks threateningly down from behind its high walls and parapets on the hilltop, upon the working-people's quarter below.

Above Ducie Bridge, the left bank grows more flat and the right bank steeper, but the condition of the dwellings on both banks grows worse rather than better. He who turns to the left here from the main street, Long Millgate, is lost; he wanders from one court to another, turns countless corners, passes nothing but narrow, filthy nooks and alleys, until after a few minutes he has lost all clue, and knows not whither to turn. Everywhere half or wholly ruined buildings, some of them actually uninhabited, which means a great deal here; rarely a wooden or stone floor to be seen in the houses, almost uniformly broken, ill-fitting windows and doors, and a state of filth! Everywhere heaps of debris, refuse, and offal; standing pools for gutters, and a stench which alone would make it impossible for a human being in any degree civilized to live in such a district. The newly-built extension of the Leeds railway, which crosses the Irk here, has swept away some of these courts and lanes, laying others completely open to view. Immediately under the railway bridge there stands a court, the filth and horrors of which surpass all the others by far, just because it was hitherto so shut off, so secluded that the way to it could not be found without a good deal of trouble. I should never have discovered it myself, without the breaks made by the railway, though I thought I knew this whole region thoroughly. Passing along a rough bank, among stakes and washing-lines, one penetrates into this chaos of small one-storeyed, one-roomed huts, in most of which there is no artificial floor; kitchen, living and sleeping room all in one. In such a hole, scarcely five feet long by six broad, I found two beds – and such bedsteads and beds! – which, with a staircase and chimney-place, exactly filled the room. In several others I found absolutely nothing, while the door stood open, and the inhabitants leaned against it. Everywhere before the doors refuse and offal; that any sort of pavement lay underneath could not be seen but only felt, here and there, with the feet. This whole collection of cattle-sheds for human beings was surrounded on two sides by houses and a factory, and on the third by the river, and besides the narrow stair up the bank, a narrow doorway alone led out into another almost equally ill-built, ill-kept labyrinth of dwellings.

Enough! The whole side of the Irk is built in this way, a planless, knotted chaos of houses, more or less on the verge of uninhabitableness, whose unclean interiors fully correspond with their filthy external surroundings. And how could the people be clean with no proper opportunity for satisfying the most natural and ordinary wants? Privies are so rare here that they are either filled up every day, or are too remote for most of the inhabitants to use. How can people wash when they have only the dirty Irk water at hand, while pumps and water pipes can be found in decent parts of the city alone? In truth, it cannot be charged to the account of these helots of modern society if their dwellings are not more clean than the pig-sties which are here and there to be seen among them. The landlords are not ashamed to let dwellings like the six or seven cellars on the quay directly below Scotland Bridge, the floors of which stand at least two feet below the low-water level of the Irk that flows not six feet away from them; or like the upper floor of the corner-house on the opposite shore directly above the bridge, where the ground-floor, utterly uninhabitable, stands deprived of all fittings for doors and windows, a case by no means rare in this region, when this open ground-floor is used as a privy by the whole neighbourhood for want of other facilities!

If we leave the Irk and penetrate once more on the opposite side from Long Millgate into the midst of the working-men's dwellings, we shall come into a somewhat newer quarter, which stretches from St. Michael's Church to Withy Grove and Shude Hill. Here there is somewhat better order. In place of the chaos of buildings, we find at least long straight lanes and alleys or courts, built according to a plan and usually square. But if, in the former case, every house was built according to caprice, here each lane and court is so built, without reference to the situation of the adjoining ones. The lanes run now in this direction, now in that, while every two minutes the wanderer gets into a blind alley, or, on turning a corner, finds himself back where he started from; certainly no one who has not lived a considerable time in this labyrinth can find his way through it.

If I may use the word at all in speaking of this district, the ventilation of these streets and courts is, in consequence of this confusion, quite as imperfect as in the Irk region; and if this quarter may, nevertheless, be said to have some advantage over that of the Irk, the houses being newer and the streets occasionally having gutters, nearly every house has, on the other hand, a cellar dwelling, which is rarely found in the Irk district, by reason of the greater age and more careless construction of the houses. As for the rest the filth, debris, and offal heaps, and the pools in the streets are common to both quarters, and in the district now under discussion, another feature most injurious to the cleanliness of the inhabitants, is the multitude of pigs walking about in all the alleys, rooting into the offal heaps, or kept imprisoned in small pens. Here, as in most of the working-men's quarters of Manchester, the pork-raisers rent the courts and build pigpens in them. In almost every court one or even several such pens may be found into which the inhabitants of the court throw all refuse and offal, whence the swine grow fat; and the atmosphere, confined on all four sides, is utterly corrupted by putrefying animal and vegetable substances. Through this quarter, a broad and measurably decent street has been cut, Millers Street, and the background has been pretty successfully concealed. But if any one should be led by curiosity to pass through one of the numerous passages which lead into the courts, he will find this piggery repeated at every twenty paces.

Such is the Old Town of Manchester, and on re-reading my description, I am forced to admit that instead of being exaggerated, it is far from black enough to convey a true impression of the filth, ruin, and uninhabitableness, the defiance of all considerations of cleanliness, ventilation, and health which characterize the construction of this single district, containing at least twenty to thirty thousand inhabitants. And such a district exists in the heart of the second city of England, the first manufacturing city of the world. If any one wishes to see in how little space a human being can move, how little air – and *such* air! – he can breathe, how little of civilization he may share and yet live, it is only necessary to travel hither. True, this is the *Old* Town, and the people of Manchester emphasize the fact whenever any one mentions to them the frightful condition of this Hell upon Earth; but what does that prove? Everything which here arouses horror and indignation is of recent origin, belongs to the *industrial* epoch. The couple of hundred houses, which belong to old Manchester, have been long since abandoned by their original inhabitants; the industrial epoch alone has crammed into them the swarms of workers whom they now shelter; the industrial epoch alone has built up every spot between these old houses to win a covering for the masses whom it has conjured hither from the agricultural districts and from Ireland; the industrial epoch alone enables the owners of these cattlesheds to rent them for high prices to human beings, to plunder the poverty of the workers, to undermine the health of thousands, in order that they *alone*, the owners, may grow rich. In the industrial epoch alone has it become possible that the worker scarcely freed from feudal servitude could be used as mere material, a mere chattel; that he must let himself be crowded into a dwelling too bad for every other, which he for his hard-earned wages buys the right to let go utterly to ruin. This manufacture has achieved, which, without these workers, this poverty, this slavery could not have lived. True, the original construction of this quarter was bad, little good could have been made out of it; but, have the landowners, has the municipality done anything to improve it when rebuilding? On the contrary, wherever a nook or corner was free, a house has been run up; where a superfluous passage remained, it has been built up; the value of land rose with the blossoming out of manufacture, and the more it rose, the more madly was the work of building up carried on, without reference to the health or comfort of the inhabitants, with sole reference to the highest possible profit on the principle that *no hole is so bad but that some poor creature must take it who can pay for nothing better.*

[. . .]

It may not be out of place to make some general observations just here as to the customary construction of working-men's quarters in Manchester. We have seen how in the Old Town pure accident determined the grouping of the houses in general. Every house is built without reference to any other, and the scraps of space between them are called courts for want of another name. In the somewhat newer portions of the same quarter, and in other working-men's quarters, dating from the early days of industrial activity, a somewhat more orderly arrangement may be found. The space between two streets is divided into more regular, usually square courts.

These courts were built in this way from the beginning, and communicate with the streets by means of covered passages. If the totally planless construction is injurious to the health of the workers by preventing ventilation, this method of shutting them up in courts surrounded on all sides by buildings is far more so. The air simply cannot escape; the chimneys of the houses are the sole drains for the imprisoned atmosphere of the courts, and they serve the purpose only so long as fire is kept burning. Moreover, the houses surrounding such courts are usually built back to back, having the rear wall in common; and this alone suffices to prevent any sufficient through ventilation. And, as the police charged with care of the streets does not trouble itself about the condition of these courts, as everything quietly lies where it is thrown, there is no cause for wonder at the filth and heaps of ashes and offal to be found here. I have been in courts, in Millers Street, at least half a foot below the level of the thoroughfare, and without the slightest drainage for the water that accumulates in them in rainy weather! More recently another different method of building was adopted, and has now become general. Working-men's cottages are almost never built singly, but always by the dozen or score; a single contractor building up one or two streets at a time. These are then arranged as follows: One front is formed of cottages of the best class, so fortunate as to possess a back door and small court, and these command the highest rent. In the rear of these cottages runs a narrow alley, the back street, built up at both ends, into which either a narrow roadway or a covered

passage leads from one side. The cottages which face this back street command least rent, and are most neglected. These have their rear walls in common with the third row of cottages which face a second street, and command less rent than the first row and more than the second. The streets are laid out somewhat as in [Figure 2].

Figure 2

By this method of construction, comparatively good ventilation can be obtained for the first row of cottages, and the third row is no worse off than in the former method. The middle row, on the other hand, is at least as badly ventilated as the houses in the courts, and the back street is always in the same filthy, disgusting condition as they. The contractors prefer this method because it saves them space, and furnishes the means of fleecing better-paid workers through the higher rents of the cottages in the first and third rows. These three different forms of cottage building are found all over Manchester and throughout Lancashire and Yorkshire, often mixed up together, but usually separate enough to indicate the relative age of parts of towns. The third system, that of the back alleys, prevails largely in the great working-men's district east of St. George's Road and Ancoats Street, and is the one most often found in the other working-men's quarters of Manchester and its suburbs.

[. . .]

Such are the various working-people's quarters of Manchester as I had occasion to observe them personally during twenty months. If we briefly formulate the result of our wanderings, we must admit that 350,000 working-people of Manchester and its environs live, almost all of them, in wretched, damp, filthy cottages, that the streets which surround them are usually in the most miserable and filthy condition, laid out without the slightest reference to ventilation, with reference solely to the profit secured by the contractor. In a word, we must confess that in the working-men's dwellings of Manchester, no cleanliness, no convenience, and consequently no comfortable family life is possible; that in such dwellings only a physically degenerate race, robbed of all humanity, degraded, reduced morally and physically to bestiality, could feel comfortable and at home.

[. . .]

To sum up briefly the facts thus far cited. The great towns are chiefly inhabited by working-people, since in the best case there is one bourgeois for two workers, often for three, here and there for four; these workers have no property whatsoever of their own, and live wholly upon wages, which usually go from hand to mouth. Society, composed wholly of atoms, does not trouble itself about them; leaves them to care for themselves and their families, yet supplies them no means of doing this in an efficient and permanent manner. Every working-man, even the best, is therefore constantly exposed to loss of work and food, that is, to death by starvation, and many perish in this way. The dwellings of the workers are everywhere badly planned, badly built, and kept in the worst condition, badly ventilated, damp, and unwholesome. The inhabitants are confined to the smallest possible space, and at least one family usually sleeps in each room. The interior arrangement of the dwellings is poverty-stricken in various degrees, down to the utter absence of even the most necessary furniture. The clothing of the workers, too, is generally scanty, and that of great multitudes is in rags. The food is, in general, bad; often almost unfit for use, and in many cases, at least at times, insufficient in quantity, so that, in extreme cases, death by starvation results. Thus the working-class of the great cities offers a graduated scale of conditions in life, in the best cases a temporarily endurable existence for hard work and good wages, good and endurable, that is, from the worker's standpoint; in the worst cases, bitter want, reaching even homelessness and death by starvation. The average is much nearer the worst case than the best. And this series does not fall into fixed classes, so that one can say, this fraction of the working-class is well off, has always been so, and remains so. If that is the case here and there, if single branches of work have in general an advantage over others, yet the condition of the workers in each branch is subject to such great fluctuations that a single working-man may be so placed as to pass through the whole range from comparative comfort to the extremest need, even to death by starvation, while almost every English working-man can tell a tale of marked changes of fortune.

"The Drive-in Culture of Contemporary America"

from *The Crabgrass Frontier: The Suburbanization of the United States* (1985)

Kenneth T. Jackson

Editors' Introduction

V. Gordon Childe (p. 35) argues that there have only been three great revolutions in the history of humankind: the Neolithic Revolution, the Urban Revolution, and the Industrial Revolution. Today, others suggest that the post-Industrial Revolution, propelled by computer-based telecommunications technologies, is just as significant. Is it also possible to talk about a Suburban Revolution?

As Friedrich Engels (p. 58), Robert Fishman (p. 77), and others have shown, suburbia has a long history, extending back at least as far as the European and American railway suburbs that arose as retreats from the polluted industrial cities for the comfortable middle class. But the suburban project took on new form and historical significance in the 1920s and in the years following World War II. The initial locus was America, and the catalyzing technology was the automobile.

In *The Crabgrass Frontier*, Kenneth T. Jackson, the dean of American urban historians, provides a sweeping overview of the "suburban revolution" in the United States, and in the chapter entitled "The Drive-in Culture of Contemporary America" he lays out a devastating critique of the mostly negative social and cultural effects that the private automobile has had on urban society.

Kenneth Jackson did not originate the critique of suburbia. Indeed, the suburban developments of the 1940s, 1950s, and 1960s in America and elsewhere gave birth to a massive literature, most of it highly critical. Damned as culturally dead and socially/racially segregated, the post-World War II suburbs were called "sprawl" and stigmatized as "anti-cities" (to use Lewis Mumford's term to describe Los Angeles). Titles such as John Keats's *The Crack in the Picture Window* (1956), Richard Gordon's *The Split-level Trap* (1961), and Mark Baldassare's *Trouble in Paradise* (1986) capture the tone of much of the commentary. Indeed, James Howard Kunstler in *The Geography of Nowhere* (1993) calls the automobile suburbs "the evil empire," and another radical analysis screams *Bomb the Suburbs* (2001)! Nevertheless, Jackson's measured and well-documented analysis of suburban culture – everything from three-car garages to drive-in churches – stands as the definitive statement on how the automobile transformed both the structure and social life of modern cities.

Kenneth Jackson is the Jacques Barzun Professor of History and the Social Sciences at Columbia University. He earned his Ph.D. at the University of Chicago and is a past president of the Urban History Association and the Organization of American Historians. He is the editor of the *Encyclopedia of New York* (1995) and the author of several influential books including *The Ku Klux Klan in the City, 1915–1930* (1967) and *Cities in American History* (1972). Jackson has been called an "urban pessimist" because of his dark view of suburbanization – a current work in progress on American transportation policy is entitled *The Road To Hell* – but in a recent interview he noted that he now sees "a ray of hope" for cities "after a long decline."

Classic works on suburbia include Herbert Gans's *The Levittowners* (New York: Pantheon, 1967), remarkable for its overall positive view of middle-class tract-home life, Sam Bass Warner, Jr.'s *Streetcar Suburbs: The Process of Growth in Boston, 1870–1900* (Cambridge, MA: Harvard University Press, 1962), and Robert Fishman's *Bourgeois Utopias: The Rise and Fall of Suburbia* (New York: Basic Books, 1987).

Other recent studies of interest include J. Eric Oliver's *Democracy in Suburbia* (Princeton, NJ: Princeton University Press, 2001), Mark Salzman's *Lost in Place: Growing Up Absurd in Suburbia* (New York: Vintage, 1995), Valerie C. Johnson's *Black Power in the Suburbs* (New York: SUNY Press, 2002), Becky Nocolaides's *My Blue Heaven: Life and Politics in the Working-class Suburbs of Los Angeles* (Chicago: University of Chicago Press, 2002), and Adam Rome's *The Bulldozer and the Countryside: Suburban Sprawl and the Rise of American Environmentalism* (Cambridge: Cambridge University Press, 2001).

■ ■ ■ ■ ■ ■

The postwar years brought unprecedented prosperity to the United States, as color televisions, stereo systems, frost-free freezers, electric blenders, and automatic garbage disposals became basic equipment in the middle-class American home. But the best symbol of individual success and identity was a sleek, air-conditioned, high-powered, personal statement on wheels. Between 1950 and 1980, when the American population increased by 50 percent, the number of their automobiles increased by 200 percent. In high school the most important rite of passage came to be the earning of a driver's license and the freedom to press an accelerator to the floor. Educational administrators across the country had to make parking space for hundreds of student vehicles. A car became one's identity, and the important question was: "What does he drive?" Not only teenagers, but also millions of older persons literally defined themselves in terms of the number, cost, style, and horsepower of their vehicles. "Escape," thinks a character in a novel by Joyce Carol Oates. "As long as he had his own car he was an American and could not die."

Unfortunately, Americans did die, often behind the wheel. On September 9, 1899, as he was stepping off a streetcar at 74th Street and Central Park West in New York, Henry H. Bliss was struck and killed by a motor vehicle, thus becoming the first fatality in the long war between flesh and steel. Thereafter, the carnage increased almost annually until Americans were sustaining about 50,000 traffic deaths and about 2 million nonfatal injuries per year. Automobility proved to be far more deadly than war for the United States. It was as if a Pearl Harbor attack took place on the highways every two weeks, with crashes becoming so commonplace that an entire industry sprang up to provide medical, legal, and insurance services for the victims.

The environmental cost was almost as high as the human toll. In 1984 the 159 million cars, trucks, and buses on the nation's roads were guzzling millions of barrels of oil every day, causing traffic jams that shattered nerves and clogged the cities they were supposed to open up and turning much of the countryside to pavement. Not surprisingly, when gasoline shortages created long lines at the pumps in 1974 and 1979, behavioral scientists noted that many people experienced anger, depression, frustration, and insecurity, as well as a formidable sense of loss.

Such reactions were possible because the automobile and the suburb have combined to create a drive-in culture that is part of the daily experience of most Americans . . . Moreover, the American people have proven to be no more prone to motor vehicle purchases than the citizens of other lands. After World War II, the Europeans and the Japanese began to catch up, and by 1980 both had achieved the same level of automobile ownership that the United States had reached in 1950. In automotive technology, American dominance slipped away in the postwar years as German, Swedish, and Japanese engineers pioneered the development of diesel engines, front-wheel drives, disc brakes, fuel-injection, and rotary engines.

Although it is not accurate to speak of a uniquely American love affair with the automobile, and although John B. Rae claimed too much when he wrote in 1971 that "modern suburbia is a creature of the automobile and could not exist without it," the motor vehicle has fundamentally restructured the pattern of everyday life in the United States. As a young man, Lewis Mumford advised his countrymen to "forget the damned motor car and build cities for lovers and friends." As it was, of course, the nation followed a different pattern. Writing in the *American Builder* in 1929, the critic Willard Morgan noted that the building of drive-in structures to serve a motor-driven population had ushered in "a completely new architectural form."

THE INTERSTATE HIGHWAY

The most popular exhibit at the New York World's Fair in 1939 was General Motors' "Futurama." Looking twenty-five years ahead, it offered a "magic Aladdin-like flight through time and space." Fair-goers stood in hour-long lines, waiting to travel on a moving sidewalk above a huge model created by designer Norman Bel Geddes. Miniature superhighways with 50,000 automated cars wove past model farms en route to model cities . . . The message of "Futurama" was as impressive as its millions of model parts: "The job of building the future is one which will demand our best energies, our most fruitful imagination; and that with it will come greater opportunities for all."

The promise of a national system of impressive roadways attracted a diverse group of lobbyists, including the Automobile Manufacturers Association, state-highway administrators, motor-bus operators, the American Trucking Association, and even the American Parking Association – for the more cars on the road, the more cars would be parked at the end of the journey. Truck companies, for example, promoted legislation to spend state gasoline taxes on highways, rather than on schools, hospitals, welfare, or public transit. In 1943 these groups came together as the American Road Builders Association, with General Motors as the largest contributor, to form a lobbying enterprise second only to that of the munitions industry. By the mid-1950s, it had become one of the most broad-based of all pressure groups, consisting of the oil, rubber, asphalt, and construction industries; the car dealers and renters; the trucking and bus concerns; the banks and advertising agencies that depended upon the companies involved; and the labor unions. On the local level, professional real estate groups and home-builders associations joined the movement in the hope that highways would cause a spurt in housing turnover and a jump in prices. They envisaged no mere widening of existing roads, but the creation of an entirely new superhighway system and the initiation of the largest peacetime construction project in history.

[. . .]

Sensitive to mounting political pressure, President Dwight Eisenhower appointed a committee in 1954 to "study" the nation's highway requirements. Its conclusions were foregone, in part because the chairman was Lucius D. Clay, a member of the board of directors of General Motors. The committee considered no alternative to a massive highway system, and it suggested a major redirection of national policy to benefit the car and the truck. The Interstate Highway Act became law in 1956, when the Congress provided for a 41,000-mile (eventually expanded to a 42,500-mile) system, with the federal government paying 90 percent of the cost. President Eisenhower gave four reasons for signing the measure: current highways were unsafe; cars too often became snarled in traffic jams; poor roads saddled business with high costs for transportation; and modern highways were needed because "in case of atomic attack on our key cities, the road net must permit quick evacuation of target areas." Not a single word was said about the impact of highways on cities and suburbs, although the concrete thoroughfares and the thirty-five-ton tractor-trailers which used them encouraged the continued outward movement of industries toward the beltways and interchanges. Moreover, the interstate system helped continue the downward spiral of public transportation and virtually guaranteed that future urban growth would perpetuate a centerless sprawl . . .

[. . .]

The inevitable result of the bias in American transport funding, a bias that existed for a generation before the Interstate Highway program was initiated, is that the United States now has the world's best road system and very nearly its worst public transit offerings. Los Angeles, in particular, provides the nation's most dramatic example of urban sprawl tailored to the mobility of the automobile. Its vast, amorphous conglomeration of housing tracts, shopping centers, industrial parks, freeways, and independent towns blend into each other in a seamless fabric of concrete and asphalt, and nothing over the years has succeeded in gluing this automobile-oriented civilization into any kind of cohesion – save that of individual routine. Los Angeles's basic shape comes from three factors, all of which long preceded the freeway system. The first was cheap land (in the 1920s rather than 1970s) and the desire for single-family houses. In 1950, for example, nearly two-thirds of all the dwelling units in the Los Angeles area were fully detached, a much higher percentage than in Chicago (28 percent), New

York City (20 percent), or Philadelphia (15 percent), and its residential density was the lowest of major cities. The second was the dispersed-location of its oil fields and refineries, which led to the creation of industrial suburbs like Whittier and Fullerton and of residential suburbs like La Habra, which housed oil workers and their families. The third was its once excellent mass transit system, which at its peak included more than 1,100 miles of track and constituted the largest electric interurban railway in the world.

[. . .]

THE GARAGE

The drive-in structure that is closest to the hearts, bodies, and cars of the American family is the garage. It is the link between the home and the outside world. The word is French, meaning storage space, but its transformation into a multipurpose enclosure internally integrated with the dwelling is distinctively American.

[. .]

After World War I, house plans of the expensive variety began to include garages, and by the mid-1920s driveways were commonplace and garages had become important selling points. The popular 1928 *Home Builders* pattern book offered designs for fifty garages in wood, Tudor, and brick varieties. In affluent sections, such large and efficiently planned structures included housing above for the family chauffeur. In less pretentious neighborhoods, the small, single-purpose garages were scarcely larger than the vehicles themselves . . . Although there was a tendency to move garages closer to the house, they typically remained at the rear of the property before 1925, often with access via an alley which ran parallel to the street. The car was still thought of as something similar to a horse – dependable and important, but not something that one needed to be close to in the evening.

By 1935, however, the garage was beginning to merge into the house itself, and in 1937 the *Architectural Record* noted that "the garage has become a very essential part of the residence." The tendency accelerated after World War II, as alleys went the way of the horse-drawn wagon, as property widths more often exceeded fifty feet, and as the car became not only a status symbol, but almost a member of the family, to be cared for and sheltered. The introduction

of a canopied and unenclosed structure called a "car port" represented an inexpensive solution to the problem, particularly in mild climates, but in the 1950s the enclosed garage was back in favor and a necessity even in a tract house. Easy access to the automobile became a key aspect of residential design, and not only for the well-to-do. By the 1960s garages often occupied about 400 square feet (about one-third that of the house itself) and usually contained space for two automobiles and a variety of lawn and woodworking tools. Offering direct access to the house (a conveniently placed door usually led directly into the kitchen), the garage had become an integrated part of the dwelling, and it dominated the front facades of new houses. In California garages and driveways were often so prominent that the house could almost be described as accessory to the garage. Few people, however, went to the extremes common in England, where the automobile was often so precious that living rooms were often converted to garages.

THE MOTEL

As the United States became a rubber-tire civilization, a new kind of roadside architecture was created to convey an instantly recognizable image to the fast-moving traveler. Criticized as tasteless, cheap, forgettable, and flimsy by most commentators, drive-in structures did attract the attention of some talented architects, most notably Los Angeles's Richard Neutra. For him, the automobile symbolized modernity, and its design paralleled his own ideals of precision and efficiency. This correlation between the structure and the car began to be celebrated in the late 1960s and 1970s when architects Robert Venturi, Denise Scott Brown, and Steven Izenour developed such concepts as "architecture as symbol" and the "architecture of communication." Their book, *Learning From Las Vegas*, was instrumental in encouraging a shift in taste from general condemnation to appreciation of the commercial strip and especially of the huge and garish signs which were easily recognized by passing motorists.

A ubiquitous example of the drive-in culture is the motel. In the middle of the nineteenth century, every city, every county seat, every aspiring mining town, every wide place in the road with aspirations to larger size, had to have a hotel. Whether such structures were grand palaces on the order of Boston's Tremont House or New York's Fifth Avenue Hotel, or whether they were jerry-built shacks, they were typically located at the center of the business district, at the focal point of community activities. To a considerable extent, the hotel was the place for informal social interaction and business, and the very heart and soul of the city.

Between 1910 and 1920, however, increasing numbers of traveling motorists created a market for overnight accommodation along the highways. The first tourists simply camped wherever they chose along the road. By 1924, several thousand municipal camp-grounds were opened which offered cold water spigots and outdoor privies. Next came the "cabin camps," which consisted of tiny, white clapboard cottages arranged in a semicircle and often set in a grove of trees. Initially called "tourist courts," these estab-lishments were cheap, convenient, and informal, and by 1926 there were an estimated two thousand of them, mostly in the West and in Florida.

[. . .]

It was not until 1952 that Kemmons Wilson and Wallace E. Johnson opened their first "Holiday Inn" on Summer Avenue in Memphis. But long before that, in 1926, a San Luis Obispo, California, proprietor had coined a new word, "motel," to describe an establishment that allowed a guest to park his car just outside his room . . .

Motels began to thrive after World War II, when the typical establishment was larger and more expensive than the earlier cabins. Major chains set standards for prices, services, and respectability that the travel-ing public could depend on. As early as 1948, there were 26,000 self-styled motels in the United States. Hard-won respectability attracted more middle-class families, and by 1960 there were 60,000 such places, a figure that doubled again by 1972. By that time an old hotel was closing somewhere in downtown America every thirty hours. And somewhere in sub-urban America, a plastic and glass Shangri-La was rising to take its place.

[. . .]

THE DRIVE-IN THEATER

The downtown movie theaters and old vaudeville houses faced a similar challenge from the automobile. In 1933 Richard M. Hollinshead set up a 16-mm projector in front of his garage in Riverton, New Jersey, and then settled down to watch a movie. Recognizing

a nation addicted to the motorcar when he saw one, Hollinshead and Willis Smith opened the world's first drive-in movie in a forty-car parking lot in Camden on June 6, 1933. Hollinshead profited only slightly from his brainchild, however, because in 1938 the United States Supreme Court refused to hear his appeal against Loew's Theaters, thus accepting the argument that the drive-in movie was not a patentable item. The idea never caught on in Europe, but by 1958 more than four thousand outdoor screens dotted the American landscape. Because drive-ins offered bargain-base-ment prices and double or triple bills, the theaters tended to favor movies that were either second-run or second-rate. Horror films and teenage romance were the order of the night . . . Pundits often commented that there was a better show in the cars than on the screen.

In the 1960s and 1970s the drive-in movie began to slip in popularity. Rising fuel costs and a season that lasted only six months contributed to the problem, but skyrocketing land values were the main factor. When drive-ins were originally opened, they were typically out in the hinterlands. When subdivisions and shop-ping malls came closer, the drive-ins could not match the potential returns from other forms of investments. According to the National Association of Theater Owners, only 2,935 open-air theaters still operated in the United States in 1983, even though the total number of commercial movie screens in the nation, 18,772, was at a thirty-five-year high. The increase picked up not by the downtown and the neighborhood theaters, but by new multiscreen cinemas in shopping centers. Realizing that the large parking lots of indoor malls were relatively empty in the evening, shopping center moguls came to regard theaters as an important part of a successful retailing mix.

THE GASOLINE SERVICE STATION

The purchase of gasoline in the United States has thus far passed through five distinct epochs. The first stage was clearly the worst for the motorist, who had to buy fuel by the bucketful at a livery stable, repair shop, or dry goods store. Occasionally, vendors sold gasoline from small tank cars which they pushed up and down the streets. In any event, the automobile owner had to pour gasoline from a bucket through a funnel into his tank. The entire procedure was inefficient, smelly, wasteful, and occasionally dangerous.

The second stage began about 1905, when C.H. Laessig of St. Louis equipped a hot-water heater with a glass gauge and a garden hose and turned the whole thing on its end. With this simple maneuver, he invented an easy way to transfer gasoline from a storage tank to an automobile without using a bucket. Later in the same year, Sylvanus F. Bowser invented a gasoline pump which automatically measured the outflow. The entire assembly was labeled a "filling station." At this stage, which lasted until about 1920, such an apparatus consisted of a single pump outside a retail store which was primarily engaged in other businesses and which provided precious few services for the motorist . . .

Between 1920 and 1950, service stations entered into a third phase and became, as a group, one of the most widespread kinds of commercial buildings in the United States. Providing under one roof all the functions of gasoline distribution and normal automotive maintenance, these full-service structures were often built in the form of little colonial houses, Greek temples, Chinese pagodas, and Art Deco palaces. Many were local landmarks and a source of community pride . . .

After 1935 the gasoline station evolved again, this time into a more homogeneous entity that was standardized across the entire country and that reflected the mass-marketing techniques of billion-dollar oil companies. Some of the more familiar designs were innovative or memorable, such as the drum-like Mobile station by New York architect Frederick Frost, which featured a dramatically curving facade while conveying the corporate identity. Another popular service station style was the Texaco design of Walter Dorwin Teaguea, smooth white exterior with elegant trim and the familiar red star and bold red lettering. Whatever the product or design, the stations tended to be operated by a single entrepreneur and represented an important part of small business in American life.

The fifth stage of gasoline-station development began in the 1970s, with the slow demise of the traditional service-station businessman. New gasoline outlets were of two types. The first was the super station, often owned and operated by the oil companies themselves. Most featured a combination of self-service and full-service pumping consoles, as well as fully equipped "car care centers." Service areas were separated from the pumping sections so that the two functions would not interfere with each other. Mechanics never broke off work to sell gas.

The more pervasive second type might be termed the "mini-mart station." The operators of such establishments have now gone full circle since the early twentieth century. Typically, they know nothing about automobiles and expect the customers themselves to pump the gasoline. Thus, "the man who wears the star" has given way to the teenager who sells six-packs, bags of ice, and pre-prepared sandwiches.

THE SHOPPING CENTER

Large-scale retailing, long associated with central business districts, began moving away from the urban cores between the world wars. The first experiments to capture the growing suburban retail markets were made by major department stores in New York and Chicago in the 1920s . . .

Another threat to the primacy of the central business district was the "string street" or "shopping strip," which emerged in the 1920s and which was designed to serve vehicular rather than pedestrian traffic. These bypass roads encouraged city dwellers with cars to patronize businesses on the outskirts of town. Short parades of shops could already have been found near the streetcar and rapid transit stops, but, as has been noted, these new retailing thoroughfares generally radiated out from the city business district toward low-density, residential areas, functionally dominating the urban street system. They were the prototypes for the familiar highway strips of the 1980s which stretch far into the countryside.

[. . .]

The concept of the enclosed, climate-controlled mall, first introduced at the Southdale Shopping Center near Minneapolis in 1956, added to the suburban advantage . . .

During the 1970s, a new phenomenon – the super regional mall – added a more elaborate twist to suburban shopping. Prototypical of the new breed was Tyson's Corner, on the Washington Beltway in Fairfax County, Virginia. Anchored by Bloomingdale's, it did over $165 million in business in 1983 and provided employment to more than 14,000 persons. Even larger was Long Island's Roosevelt Field, a 180-store, 2.2 million square foot megamall that attracted 275,000 visitors a week and did $230 million in business in 1980. Most elaborate of all was Houston's Galleria, a world-famed setting for 240 prestigious boutiques, a quartet of cinemas, 26 restaurants, an Olympic-sized ice-skating pavilion, and two luxury hotels. There were few windows in these mausoleums of merchandising, and clocks were rarely seen – just as in gambling casinos.

Boosters of such megamalls argue that they are taking the place of the old central business districts and becoming the identifiable collecting points for the rootless families of the newer areas. As weekend and afternoon attractions, they have a special lure for teenagers, who often go there on shopping dates or to see the opposite sex. As one official noted in 1971: "These malls are now their street corners. The new shopping centers have killed the little merchant, closed most movies, and are now supplanting the older shopping centers in the suburbs." They are also especially attractive to mothers with young children and to the elderly, many of whom visit regularly to get out of the house without having to worry about crime or inclement weather.

[. . .]

THE HOUSE TRAILER AND MOBILE HOME

The phenomenon of a nation on wheels is perhaps best symbolized by the uniquely American development of the mobile home. "Trailers are here to stay," predicted the writer Howard O'Brien in 1936. Although in its infancy at that time, the mobile-home industry has flourished in the United States. The house trailer itself came into existence in the teens of this century as an individually designed variation on a truck or a car, and it began to be produced commercially in the 1920s. Originally, trailers were designed to travel, and they were used primarily for vacation purposes. During the Great Depression of the 1930s, however, many people, especially salesmen, entertainers, construction workers, and farm laborers, were forced into a nomadic way of life as they searched for work, any work. They found that these temporary trailers on rubber tires provided the necessary shelter while also meeting their economic and migratory requirements. Meanwhile, Wally Byam and other designers were streamlining the mobile home into the classic tear-drop form made famous by Airstream.

During World War II, the United States government got into the act by purchasing tens of thousands of trailers for war workers and by forbidding their sale to the general public. By 1943 the National Housing Agency alone owned 35,000 of the aluminum boxes,

and more than 60 percent of the nation's 200,000 mobile homes were in defense areas . . .

Not until the mid-1950s did the term "mobile home" begin to refer to a place where respectable people could marry, mature, and die. By then it was less a "mobile" than a "manufactured" home. No longer a trailer, it became a modern industrialized residence with almost all the accoutrements of a normal house. By the late 1950s, widths were increased to ten feet, the Federal Housing Administration (FHA) began to recognize the mobile home as a type of housing suitable for mortgage insurance, and the maturities on sales contracts were increased from three to five years.

In the 1960s, twelve-foot widths were introduced, and then fourteen, and manufacturers began to add fireplaces, skylights, and cathedral ceilings. In 1967 two trailers were attached side by side to form the first "double wide." These new dimensions allowed for a greater variety of room arrangement and became particularly attractive to retired persons with fixed incomes. They also made the homes less mobile. By 1979 even the single-width "trailer" could be seventeen feet wide (by about sixty feet long), and according to the Manufactured Housing Institute, fewer than 2 percent were ever being moved from their original site. Partly as a result of this increasing permanence, individual communities and the courts began to define the structures as real property and thus subject to real estate taxes rather than as motor vehicles subject only to license fees.

Although it continued to be popularly perceived as a shabby substitute for "stick" housing (a derogatory word used to describe the ordinary American balloon-frame dwelling), the residence on wheels reflected American values and industrial practices. Built with easily machined and processed materials, such as sheet metal and plastic, it represented a total consumer package, complete with interior furnishings, carpets, and appliances. More importantly, it provided a suburban-type alternative to the inner-city housing that would otherwise have been available to blue-collar workers, newly married couples, and retired persons . . .

A DRIVE-IN SOCIETY

Drive-in motels, drive-in movies, and drive-in shopping facilities were only a few of the many new institutions that followed in the exhaust of the internal-combustion engine. By 1984 mom-and-pop grocery stores had given way almost everywhere to supermarkets, most banks had drive-in windows, and a few funeral homes were making it possible for mourners to view the deceased, sign the register, and pay their respects without emerging from their cars. Odessa Community College in Texas even opened a drive-through registration window.

Particularly pervasive were fast-food franchises, which not only decimated the family-style restaurants but cut deeply into grocery store sales. In 1915, James G. Huneker, a raconteur whose tales of early twentieth-century American life were compiled as *New Cosmopolis*, complained of the infusion of cheap, quick-fire "food hells," and of the replacement of relaxed dining with "canned music and automatic lunch taverns." With the automobile came the notion of "grabbing" something to eat. The first drive-in restaurant, Royce Hailey's Pig Stand, opened in Dallas in 1921, and later in the decade, the first fast-food franchise, "White Tower," decided that families touring in motorcars needed convenient meals along the way. The places had to look clean, so they were painted white. They had to be familiar, so a minimal menu was standardized at every outlet. To catch the eye, they were built like little castles, replete with fake ramparts and turrets. And to forestall any problem with a land lease, the little white castles were built to be moveable.

The biggest restaurant operation of all began in 1954, when Ray A. Kroc, a Chicago area milkshake-machine salesman, joined forces with Richard and Maurice McDonald, the owners of a fast-food emporium in San Bernardino, California. In 1955 the first of Mr. Kroc's "McDonald's" outlets was opened in Des Plaines, a Chicago suburb long famous as the site of an annual Methodist encampment. The second and third, both in California, opened later in 1955 . . . [T]he McDonald's enterprise is based on free parking and drive-in access, and its methods have been copied by dozens of imitators. Late in 1984, on an interstate highway north of Minneapolis, McDonald's began construction of the most complete drive-in complex in the world. To be called McStop, it will feature a motel, gas station, convenience store, and, of course, a McDonald's restaurant.

[. . .]

THE CENTERLESS CITY

More than anyplace else, California became the symbol of the postwar suburban culture. It pioneered the booms in sports cars, foreign cars, vans, and motor homes, and by 1984 its 26 million citizens owned almost 19 million motor vehicles and had access to the world's most extensive freeway system. The result has been a new type of centerless city, best exemplified by once sleepy and out-of-the-way Orange County, just south and east of Los Angeles. After Walt Disney came down from Hollywood, bought out the ranchers, and opened Disneyland in 1955, Orange County began to evolve from a rural backwater into a suburb and then into a collection of medium and small towns. It had never had a true urban focus, in large part because its oil-producing sections each spawned independent suburban centers, none of which was particularly dominant over the others. The tradition continued when the area became a subdivider's dream in the 1960s and 1970s. By 1980 there were twenty-six Orange County cities, none with more than 225,000 residents. Like the begats of the Book of Genesis, they merged and multiplied into a huge agglomeration of two million people with its own Census Bureau metropolitan area designation – Anaheim, Santa Ana, Garden Grove. Unlike the traditional American metropolitan region, however, Orange County lacked a commutation focus, a place that could obviously be accepted as the center of local life. Instead, the experience of a local resident was typical: "I live in Garden Grove, work in Irvine, shop in Santa Ana, go to the dentist in Anaheim, my husband works in Long Beach, and I used to be the president of the League of Women Voters in Fullerton."

A centerless city also developed in Santa Clara County, which lies forty-five miles south of San Francisco and which is best known as the home of Silicon Valley. Stretching from Palo Alto on the north to the garlic and lettuce fields of Gilroy to the south, Santa Clara County has the world's most extensive concentration of electronics concerns. In 1940, however, it was best known for prunes and apricots, and it was not until after World War II that its largest city, San Jose, also became the nation's largest suburb. With fewer than 70,000 residents in 1940, San Jose exploded to 636,000 by 1980, superseding San Francisco as the region's largest municipality . . .

The numbers were larger in California, but the pattern was the same on the edges of every American city, from Buffalo Grove and Schaumburg near Chicago, to Germantown and Collierville near Memphis, to Creve Coeur and Ladue near St. Louis. And perhaps more important than the growing number of people living outside of city boundaries was the sheer physical sprawl of metropolitan areas. Between 1950 and 1970, the urbanized area of Washington, DC, grew from 181 to 523 square miles, of Miami from 116 to 429, while in the larger megalopolises of New York, Chicago, and Los Angeles, the region of settlement was measured in the thousands of square miles.

THE DECENTRALIZATION OF FACTORIES AND OFFICES

The deconcentration of post-World War II American cities was not simply a matter of split-level homes and neighborhood schools. It involved almost every facet of national life, from manufacturing to shopping to professional services. Most importantly, it involved the location of the workplace, and the erosion of the concept of suburb as a place from which wage-earners commuted daily to jobs in the center. So far had the trend progressed by 1970 that in nine of the fifteen largest metropolitan areas suburbs were the principal sources of employment, and in some cities, like San Francisco, almost three-fourths of all work trips were by people who neither lived nor worked in the core city. In Wilmington, Delaware, 66 percent of area jobs in 1940 were in the core city; by 1970, the figure had fallen below one-quarter. And despite the fact that Manhattan contained the world's highest concentration of office space and business activity, in 1970, about 78 percent of the residents in the New York suburbs also worked in the suburbs. Many outlying communities thus achieved a kind of autonomy from the older downtown areas . . .

Manufacturing is now among the most dispersed of nonresidential activities. As the proportion of industrial jobs in the United States work force fell from 29 percent to 23 percent of the total in the 1970s, those manufacturing enterprises that survived often relocated either to the suburbs or to the lower-cost South and West . . .

Office functions, once thought to be securely anchored to the streets of big cities, have followed the suburban trend. In the nineteenth century, businesses tried to keep all their operations under one centralized roof. It was the most efficient way to run a company

when the mails were slow and uncertain and communication among employees was limited to the distance that a human voice could carry. More recently, the economics of real estate and a revolution in communications have changed these circumstances, and many companies are now balkanizing their accounting departments, data-processing divisions, and billing departments. Just as insurance companies, branch banks, regional sales staffs, and doctors' offices have reduced their costs and presumably increased their accessibility by moving to suburban locations, so also have back-office functions been splitting away from front offices and moving away from central business districts.

[. . .]

Since World War II, the American people have experienced a transformation of the manmade environment around them. Commercial, residential, and industrial structures have been redesigned to fit the needs of the motorist rather than the pedestrian. Garish signs, large parking lots, one-way streets, drive-in windows, and throw-away fast-food buildings – all associated with the world of suburbia – have replaced the slower-paced, neighborhood-oriented institutions of an earlier generation. Some observers of the automobile revolution have argued that the car has created a new and better urban environment and that the change in spatial scale, based upon swift transportation, has formed a new kind of organic entity, speeding up personal communication and rendering obsolete the older urban settings. Lewis Mumford, writing from his small-town retreat in Amenia, New York, has emphatically disagreed. His prize-winning book, *The City in History*, was a celebration of the medieval community and an excoriation of "the formless urban exudation" that he saw American cities becoming. He noted that the automobile megalopolis was not a final stage in city development but an anti-city which "annihilates the city whenever it collides with it."

[. . .]

There are some signs that the halcyon days of the drive-in culture and automobile are behind us. More than one hundred thousand gasoline stations, or about one-third of the American total, have been eliminated in the last decade. Empty tourist courts and boarded-up motels are reminders that the fast pace of change can make commercial structures obsolete within a quarter-century of their erection. Even that suburban bellwether, the shopping center, which revolutionized merchandising after World War II, has come to seem small and out-of-date as newer covered malls attract both the trendy and the family trade. Some older centers have been recycled as bowling alleys or industrial buildings, and some have been remodeled to appeal to larger tenants and better-heeled customers. But others stand forlorn and boarded up. Similarly, the characteristic fast-food emporiums of the 1950s, with uniformed "car hops" who took orders at the automobile window, are now relics of the past. One of the survivors, Delores Drive-in, which opened in Beverly Hills in 1946, was recently proposed as an historic landmark, a sure sign that the species is in danger.

"Beyond Suburbia: The Rise of the Technoburb"

from *Bourgeois Utopias: The Rise and Fall of Suburbia* (1987)

Robert Fishman

Editors' Introduction

Robert Fishman is a professor of history at the University of Michigan who established his academic reputation with his first book, the magisterial *Urban Utopias in the Twentieth Century* (1977), a study of the work of Ebenezer Howard, Le Corbusier, and Frank Lloyd Wright. For his second book, he decided to address a totally prosaic, non-visionary subject – the history of suburbia – only to discover that "the suburban ideal" was, in the final analysis, yet another form of utopia, the utopia of the middle class.

The real focus of *Bourgeois Utopias* is the suburban ideal, more than suburbia itself and the logic of Fishman's analysis leads him to many surprising insights and conclusions. In the medieval period and up through the eighteenth century, suburbs were clusters of houses inhabited by poor and/or disreputable people on the outskirts of towns. When suburbs were first established for the upper and middle classes, the ideal was to create a perfect synthesis of urban sophistication and rural virtue. Here was a conception as utopian as that of any visionary social reformer but with an important difference: "Where other modern utopias have been collectivist," writes Fishman, "suburbia has built its vision of community on the primacy of private property and the individual family."

What suburbia has evolved into today is "technoburbia," a dominant new urban reality that can no longer be considered suburbia in the traditional sense. In Redmond, Washington (Plate 8), Microsoft Corporation's corporate headquarters mix with residential neighborhoods, retail centers, and even bands of open space to make up a new urban form where city and suburb; urbanized and unurbanized areas; high-tech and conventional development flow together.

To describe this new reality Fishman has coined two new terms "technoburb" and "techno-city." Fishman defines technoburbs as peripheral zones, perhaps as large as a county, that have emerged as viable socioeconomic units. The new technoburbs are spread out along highway growth corridors. Along the highways of metropolitan regions shopping malls, industrial parks, campus-like office complexes, hospitals, schools, and a whole range of housing types succeed each other. Silicon Valley in the San Francisco Bay Area is an example of such a technoburb.

By "techno-city" Fishman means the whole metropolitan region, which has been transformed by the coming of the technoburb. In Fishman's view we may still refer to the New York Metropolitan region as "New York City," but increasingly by "New York City" we mean the entire New York City region. And much of the economic and cultural life of the region no longer resides just in the core city. The old central cities have become increasingly marginal, while the technoburb has emerged as the focus of American life. In Fishman's view, the new technoburbs surrounding the old urban cores do not represent "the suburbanization of the United States," as Kenneth Jackson (p. 67) would have it, but "the end of suburbia in its traditional sense and the creation of a new kind of decentralized city." That suburbia has become the city itself is, perhaps, the final irony of modern urbanism.

Fishman lays out a strong indictment of what is wrong with technoburbs. They consist of an unplanned jumble of discordant elements – housing, industry, commerce, even agriculture – with little coherent pattern or structure.

They waste land. Technoburbs are dependent on highway systems, yet their highway systems are in a state of chronic chaos. They have no proper boundaries, but consist of a crazy quilt of separate and overlapping political jurisdictions that make meaningful region-wide planning virtually impossible.

Yet Fishman notes that all new urban forms appeared chaotic in their early stages. Even the most "organic" cityscapes of the past evolved slowly after much chaos and trial and error. For example, it took planners of genius like Frederick Law Olmsted (p. 307) and Ebenezer Howard (p. 309) to create orderly parks and garden suburbs out of the chaos of the nineteenth-century city or to imagine and actually build Garden Cities.

Fishman acknowledges that there is a functional logic to sprawl. Perhaps, he speculates, if sprawl is better understood and better managed it might prove to be a positive rather than a negative development. Fishman looks to Frank Lloyd Wrights's Broadacre City vision (p. 325) as an example of how inspired planners may yet devise an aesthetic to tame techoburbia.

The technocity, Fishman concludes, is still under construction both physically and culturally. How technocities will evolve is unclear. The jury is still out on whether technoburbia will ultimately be judged as an advance over earlier urban forms.

This selection is from Robert Fishman, *Bourgeois Utopias: The Rise and Fall of Surburbia* (New York: Basic Books, 1987. Fishman's other major book on cities is *Urban Utopias in the Twentieth Century* (New York: Basic Books, 1977).

For other views of emerging postmodern suburbia, see journalist Joel Garreau's *Edge City* (New York: Anchor, 1992), Manuel Castells, *The Rise of the Network Society* (Cambridge, MA: Blackwell, 1996), Manuel Castells and Peter Hall, *Technopoles of the World* (London and New York: Routledge, 1994), Edward Soja's studies of the Los Angeles region cited in the selection on "Taking Los Angeles Apart" (p. 189), and Peter Calthorpe's and William Fulton's description of the emerging "Regional City" (p. 331).

■ ■ ■ ■ ■ ■

If the nineteenth century could be called the Age of Great Cities, post-1945 America would appear to be the Age of Great Suburbs. As central cities stagnated or declined in both population and industry, growth was channeled almost exclusively to the peripheries. Between 1950 and 1970 American central cities grew by 10 million people, their suburbs by 85 million. Suburbs, moreover, accounted for at least three-quarters of all new manufacturing and retail jobs generated during that period. By 1970 the percentage of Americans living in suburbs was almost exactly double what it had been in 1940, and more Americans lived in suburban areas (37.6 percent) than in central cities (31.4 percent) or in rural areas (31 percent). In the 1970s central cities experienced a net out-migration of 13 million people, combined with an unprecedented deindustrialization, increasing poverty levels, and housing decay.

[. . .]

From its origins in eighteenth-century London, suburbia has served as a specialized portion of the expanding metropolis. Whether it was inside or outside the political borders of the central city, it was always functionally dependent on the urban core. Conversely, the growth of suburbia always meant a strengthening of the specialized services at the core.

In my view, the most important feature of postwar American development has been the almost simultaneous decentralization of housing, industry, specialized services, and office jobs; the consequent breakaway of the urban periphery from a central city it no longer needs; and the creation of a decentralized environment that nevertheless possesses all the economic and technological dynamism we associate with the city. This phenomenon, as remarkable as it is unique, is not suburbanization but a new city.

Unfortunately, we lack a convenient name for this new city, which has taken shape on the outskirts of all our major urban centers. Some have used the terms "exurbia" or "outer city." I suggest (with apologies) two neologisms: the "technoburb" and the "techno-city." By "technoburb" I mean a peripheral zone, perhaps as large as a county, that has emerged as a viable socioeconomic unit. Spread out along its highway growth corridors are shopping malls, industrial parks, campuslike office complexes, hospitals, schools, and a full range of housing types. Its residents look to their immediate surroundings rather than to the city for their jobs and other needs; and its industries find not only the employees they need but also the specialized services.

The new city is a technoburb not only because high tech industries have found their most congenial homes in such archetypal technoburbs as Silicon Valley in northern California and Route 128 in Massachusetts. In most technoburbs such industries make up only a small minority of jobs, but the very existence of the decentralized city is made possible only through the advanced communications technology which has so completely superseded the face-to-face contact of the traditional city. The technoburb has generated urban diversity without traditional urban concentration.

By "techno-city" I mean the whole metropolitan region that has been transformed by the coming of the technoburb. The techno-city usually still bears the name of its principal city, for example, "the New York metropolitan area"; its sports teams bear that city's name (even if they no longer play within the boundaries of the central city); and its television stations appear to broadcast from the central city. But the economic and social life of the region increasingly bypasses its supposed core. The techno-city is truly multicentered, along the pattern that Los Angeles first created. The technoburbs, which might stretch over seventy miles from the core in all directions, are often in more direct communication with one another – or with other techno-cities across the country – than they are with the core. The techno-city's real structure is aptly expressed by the circular superhighways or beltways that serve so well to define the perimeters of the new city. The beltways put every part of the urban periphery in contact with every other part without passing through the central city at all.

[. . .]

The old central cities have become increasingly marginal, while the technoburb has emerged as the focus of American life. The traditional suburbanite – commuting at ever-increasing cost to a center where the available resources barely duplicate those available much closer to home – becomes increasingly rare. In this transformed urban ecology the history of suburbia comes to an end.

PROPHETS OF THE TECHNO-CITY

Like all new urban forms, the techno-city and its technoburbs emerged not only unpredicted but unobserved. We are still seeing this new city through the intellectual categories of the old metropolis. Only two prophets, I believe, perceived the underlying forces that would lead to the techno-city at the time of their first emergence. Their thoughts are therefore particularly valuable in understanding the new city.

At the turn of the twentieth century, when the power and attraction of the great city was at its peak, H.G. Wells daringly asserted that the technological forces that had created the industrial metropolis were now moving to destroy it. In his 1900 essay "The Probable Diffusion of Great Cities," Wells argued that the seemingly inexorable concentration of people and resources in the largest cities would soon be reversed. In the course of the twentieth century, he prophesied, the metropolis would see its own resources drain away to decentralized "urban regions" so vast that the very concept of "the city" would become, in his phrase, "as obsolete as 'mailcoach.'"

Wells based his prediction on a penetrating analysis of the emerging networks of transportation and communication. Throughout the nineteenth century, rail transportation had been a relatively simple system favoring direct access to large centers. With the spread of branchlines and electric tramways, however, a complex rail network had been created that could serve as the basis for a decentralized region. (As Wells wrote, Henry E. Huntington was proving the truth of his propositions for the Los Angeles region.)

Wells pictured the "urban region" of the year 2000 as a series of villages with small homes and factories set in the open fields, yet connected by high speed rail transportation to any other point in the region. (It was a vision not very different from those who saw Los Angeles developing into just such a network of villages.) The old cities would not completely disappear, but they would lose both their financial and their industrial functions, surviving simply because of an inherent human love of crowds. The "post-urban" city, Wells predicted, will be "essentially a bazaar, a great gallery of shops and places of concourse and rendezvous, a pedestrian place, its pathways reinforced by lifts and moving platforms, and shielded from the weather, and altogether a very spacious, brilliant, and entertaining agglomeration." In short, the great metropolis will dwindle to what we would today call a massive shopping mall, while the productive life of the society would take place in the decentralized urban region.

Wells's prediction was taken up in the late 1920s and early 1930s by Frank Lloyd Wright, who moved from similar assumptions to an even more radical view. Wright had actually seen the beginnings of the automobile and truck era; he was, perhaps not coincidentally, living mostly in Los Angeles in the late 1910s and early 1920s. Wright, like Wells, argued that "the great city was no longer modern" and that it was destined to be replaced by a decentralized society.

He called this new society Broadacre City. It has often been confused with a kind of universal suburbanization, but for Wright "Broadacres" was the exact opposite of the suburbia he despised. He saw correctly that suburbia represented the essential extension of the city into the countryside, whereas Broadacres represented the disappearance of all previously existing cities.

As Wright envisioned it, Broadacres was based on universal automobile ownership combined with a network of superhighways, which removed the need for population to cluster in a particular spot. Indeed, any such clustering was necessarily inefficient, a point of congestion rather than of communication. The city would thus spread out over the countryside at densities low enough to permit each family to have its own homestead and even to engage in part-time agriculture. Yet these homesteads would not be isolated; their access to the superhighway grid would put them within easy reach of as many jobs and specialized services as any nineteenth-century urbanite. Traveling at more than sixty miles an hour, each citizen would create his own city within the hundreds of square miles he could reach in an hour's drive.

Like Wells, Wright saw industrial production inevitably leaving the cities for the space and convenience of rural sites. But Wright went one step further in his attempt to envision the way that a radically decentralized environment could generate that diversity and excitement which only cities had possessed.

He saw that even in the most scattered environment, the crossing of major highways would possess a certain special status. These intersections would be the natural sites of what he called the roadside market, a remarkable anticipation of the shopping center: "great spacious roadside pleasure places these markets, rising high and handsome like some flexible form of pavilion – designed as places of cooperative exchange, not only of commodities but of cultural facilities." To the roadside markets he added a range of highly civilized yet small scale institutions: schools, a modern cathedral, a center for festivities, and the like. In such an environment, even the entertainment functions of the city would disappear. Soon, Wright devoutly wished, the centralized city itself would disappear.

Taken together, Wells's and Wright's prophecies constitute a remarkable insight into the decentralizing tendencies of modern technology and society. Both were presented in utopian form, an image of the future presented as somehow "inevitable" yet without any sustained attention to how it would actually be achieved. Nevertheless, something like the transformation that Wells and Wright foresaw has taken place in the United States, a transformation all the more remarkable in that it occurred without a clear recognition that it was happening. While diverse groups were engaged in what they believed was "the suburbanization" of America, they were in fact creating a new city.

[. . .]

TECHNOBURB/TECHNO-CITY: THE STRUCTURE OF THE NEW METROPOLIS

To claim that there is a pattern or structure in the new American city is to contradict what appears to be overwhelming evidence. One might sum up the structure of the technoburb by saying that it goes against every rule of planning. It is based on two extravagances that have always aroused the ire of planners: the waste of land inherent in a single family house with its own yard, and the waste of energy inherent in the use of the personal automobile. The new city is absolutely dependent on its road system, yet that system is almost always in a state of chaos and congestion. The landscape of the technoburb is a hopeless jumble of housing, industry, commerce, and even agricultural uses. Finally, the technoburb has no proper boundaries; however defined, it is divided into a crazy quilt of separate and overlapping political jurisdictions, which make any kind of coordinated planning virtually impossible.

Yet the technoburb has become the real locus of growth and innovation in our society. And there is a real structure in what appears to be wasteful sprawl, which provides enough logic and efficiency for the technoburb to fulfill at least some of its promises.

If there is a single basic principle in the structure of the technoburb, it is the renewed linkage of work and residence. The suburb had separated the two into distinct environments; its logic was that of the massive commute, in which workers from the periphery traveled each morning to a single core and then dispersed each evening. The technoburb, however, contains both work and residence within a single decentralized environment.

By the standards of a preindustrial city where people often lived and worked under the same roof, or even of the turn of the century industrial zones where factories were an integral part of working class neighborhoods, the linkage between work and residence in the technoburb is hardly close. A recent study of New Jersey shows that most workers along the state's growth corridors now live in the same county in which they work. But this relative dispersion must be contrasted to the former pattern of commuting into urban cores like Newark or New York. In most cases traveling time to work diminishes, even when the distances traveled are still substantial; as the 1980 census indicates, the average journey to work appears to be diminishing both in distance and, more importantly, in time.

For commuting within the technoburb is multi-directional, following the great grid of highways and secondary roads that, as Frank Lloyd Wright understood, defines the community. This multiplicity of destinations makes public transportation highly inefficient, but it does remove that terrible bottleneck which necessarily occurred when work was concentrated at a single core within the region. Each house in a technoburb is within a reasonable driving time of a truly "urban" array of jobs and services, just as each workplace along the highways can draw upon an "urban" pool of workers.

Those who believed that the energy crisis of the 1970s would cripple the technoburb failed to realize that the new city had evolved its own pattern of transportation in which a multitude of relatively short automobile journeys in a multitude of different directions substitutes for that great tidal wash in and out of a single urban core which had previously defined commuting. With housing, jobs, and services all on the periphery, this sprawl develops its own form of relative efficiency. The truly inefficient form would be any attempted revival of the former pattern of long distance mass transit commuting into a core area. To account for the new linkage of work and residence in the technoburb, we must first confront this paradox: the new city required a massive and coordinated relocation of housing, industry, and other "core" functions to the periphery; yet there were no coordinators directing the process. Indeed, the technoburb emerged in spite of, not because of, the conscious purposes

motivating the main actors. The postwar housing boom was an attempt to escape from urban conditions; the new highways sought to channel traffic into the cities; planners attempted to limit peripheral growth; the government programs that did the most to destroy the hegemony of the old industrial metropolis were precisely those designed to save it.

This paradox can be seen clearly in the area of transportation policy. Wright had grasped the basic point in his Broadacre City plan: a fully developed highway grid eliminates the primacy of a central business district. It creates a whole series of highway crossings, which can serve as business centers while promoting the multidirectional travel that prevents any single center from attaining unique importance. Yet, from the time of Robert Moses to the present, highway planners have imagined that the new roads, like the older rail transportation, would enhance the importance of the old centers by funneling cars and trucks into the downtown area and the surrounding industrial belt. At most, the highways were to serve traditional suburbanization; in other words, the movement from the periphery to the core during morning rush hours and the reverse movement in the afternoon. The beltways, those crucial "Main Streets" of the technoburb, were designed simply to allow interstate traffic to avoid going through the central cities.

The history of the technoburb, therefore, is the history of those deeper structural features of modern society first described by Wells and Wright taking precedence over conscious intentions. For purposes of clarity I shall now divide this discussion of the making of the techno-city into two interrelated topics: housing and job location.

Housing

The great American postwar housing boom was perhaps the purest example of the suburban dream in action, yet its ultimate consequence was to render suburbia obsolete. Between 1950 and 1970, on the average, 1.2 million housing units were built each year, the vast majority as suburban single family dwellings; the nation's housing stock increased by 21 million units or over 50 percent. In the 1970s the boom continued even more strongly: twenty million more new units were added, almost as many as in the previous two decades. It was precisely this vast production of new residences that shifted the center of gravity in the

United States from the urban core to the periphery and thus ensured that these vital and expanding areas could no longer remain simply bedroom communities.

This great building boom, which seems so characteristic of post-1945 conditions, in fact had its origins early in the twentieth century in the first attempts to universalize suburbia throughout the United States. It can be seen essentially as a continuation of the 1920s building boom, which had been cut off for two decades by the Depression and the war. As George Sternlieb reminds us, the American automobile industry in 1929 was producing as many cars per capita as it did in the 1980s, and real estate developers had already plotted out subdivisions in out-lying areas that were only built up in the 1960s and 1970s.

[. . .]

Even the late 1970s combination of stagnant real income with high interest rates, gasoline prices, and land values did not diminish the desirability of the new single family house. In 1981 a median American family earned only 70 percent of what was needed to make the payments on the median priced house; by 1986, the median family could once again afford the median house. Single family houses still constitute 67 percent of all occupied units, down only 2 percent since 1970 despite the increase in costs; moreover, a survey of potential home buyers in 1986 showed that 85 percent intended to purchase a detached, single family suburban house, while only 15 percent were looking at condominium apartments or townhouses. The "single," as builders call it, is still alive and well on the urban periphery.

This continuing appeal of the single should not, however, obscure the crucial changes that have transformed the meaning and context of the house. The new suburban house of the 1950s, like its predecessors for more than a century, existed precisely to isolate women and the family from urban economic life; it defined an exclusive zone of residence between city and country. Now a new house might adjoin a landscaped office park with more square feet of new office space than in a downtown building, or might be just down the highway from an enclosed shopping mall with a sales volume that exceeds those of the downtown department stores, or might overlook a high tech research laboratory making products that are exported around the world. No longer a refuge, the single family detached house on the periphery is preferred as a convenient base from which both spouses can rapidly reach their jobs.

Without the simultaneous movement of jobs along with housing, the great "suburban" boom would surely have exhausted itself in ever longer journeys to workplaces in a crowded core on overburdened highways and mass transit facilities. And the new peripheral communities would have been in reality the "isolation wards" for women that critics have called them, instead of becoming the setting for the reintegration of middle-class women into the work force as they have. The unchanging image of the suburban house and the suburban bedroom community has obscured the crucial importance of this transformation in work location, the subject of the next section.

Job location

As those who have tried to plan the process have painfully learned, job location has its own autonomous rules. The movement of factories away from the urban core after 1945 took place independently of the housing boom and probably would have occurred without it. Nevertheless, the simultaneous movement of housing and jobs in the 1950s and 1960s created an unforeseen "critical mass" of entrepreneurship and expertise on the perimeters, which allowed the technoburb to challenge successfully the two century long economic dominance of the central city.

[. . .]

At the same time, the growing importance of trucking meant that factories were no longer as dependent on the confluence of rail lines which existed only in the old factory zones. Workers had their automobiles, so factories could scatter along the periphery without concern about the absence of mass transit. (The scattering of aircraft plants and other factories in Los Angeles in the 1930s prefigured this trend.) The process gained momentum as a result of thousands of uncoordinated decisions in which managers allowed their inner city plants to run down and directed new investment toward the outskirts . . .

These changes in job location during the 1950s and 1960s were, however, only a prelude to the real triumph of the technoburb: the luring of both managerial office employment and advanced technological laboratories and production facilities from the core to the peripheries. This process may be divided into three parts. First came the establishment of "high tech" growth corridors in such diverse locations as Silicon Valley, California; Silicon Prairie, between Dallas and

Fort Worth; the Atlanta Beltway; Route 1 between Princeton and New Brunswick, New Jersey; Westchester County, New York; Route 202 near Valley Forge, Pennsylvania; and Route 128 outside Boston. The second step was the movement of office bureaucracies, especially the so-called back office, from center city high-rises to technoburb office parks; and the final phase was the movement of production-service employment – banks, accountants, lawyers, advertising agencies, skilled technicians, and the like – to locations within the technoburb, thus creating that vital base of support personnel for larger firms.

Indeed, this dramatic surge toward the technoburb has been so sweeping that we must now ask whether Wright's ultimate prophecy will be fulfilled: the disappearance of the old urban centers. Is the present-day boom in downtown office construction and inner city gentrification simply a last hurrah for the old city before deeper trends in decentralization lead to its ultimate decay?

In my view, the final diffusion that Wells and Wright predicted is unlikely, if only because both underestimated the forces of economic and political centralization that continue to exist in the late twentieth century. If physical decentralization had indeed meant economic decentralization, then the urban cores would by now be ghost towns. But large and powerful organizations still seek out a central location that validates their importance, and the historic core of great cities still meets that need better than the office complexes on the outskirts. Moreover, the corporate and government headquarters in the core still attract a wide variety of specialized support services – law firms, advertising, publishing, media, restaurants, entertainment centers, museums, and more – that continue to make the center cities viable.

The old factory zones around the core have also survived, but only in the painfully anomalous sense of housing those too poor to earn admission to the new city of prosperity at the periphery. The big city, therefore, will not disappear in the foreseeable future, and residents of the technoburbs will continue to confront uneasily both the economic power and elite culture of the urban core and its poverty. Nevertheless, the technoburb has become the true center of American society.

THE MEANING OF THE NEW CITY

Beyond the structure of the techno-city and its technoburbs, there is the larger question: what is the impact of this decentralized environment on our culture? Can anyone say of the technoburb, as Olmsted said of the suburb a century ago, that it represents "the most attractive, the most refined, and the most soundly wholesome forms of domestic life, and the best application of the arts of civilization to which mankind has yet attained"? Most planners in fact say the exact opposite. Their indictment can be divided into two parts. First, decentralization has been a social and economic disaster for the old city and for the poor, who have been increasingly relegated to its crowded, decayed zones. It has resegregated American society into an affluent outer city and an indigent inner city, while erecting ever higher barriers that prevent the poor from sharing in the jobs and housing of the technoburbs.

Second, decentralization has been seen as a cultural disaster. While the rich and diverse architectural heritage of the cities decays, the technoburb has been built up as a standardized and simplified sprawl, consuming time and space, destroying the natural landscape. The wealth that postindustrial America has generated has been used to create an ugly and wasteful pseudocity, too spread out to be efficient, too superficial to create a true culture. The truth of both indictments is impossible to deny, yet it must be rescued from the polemical overstatements that seem to afflict anyone who deals with these topics. The first charge is the more fundamental, for it points to a genuine structural discontinuity in post-1945 decentralization. By detaching itself physically, socially, and economically from the city, the technoburb is profoundly antiurban as suburbia never had been. Suburbanization strengthened the central core as the cultural and economic heart of an expanding region; by excluding industry, suburbia left intact and even augmented the urban factory districts.

Technoburb development, however, completely undermines the factory district and potentially threatens even the commercial core. The competition from new sites on the outskirts renders obsolete the whole complex of housing and factory sites that had been built up in the years 1890 to 1930 and provides alternatives to the core for even the most specialized shopping and administrative services.

This competition, moreover, has occurred in the context of a massive migration of southern blacks to northern cities. Blacks, Hispanics, and other recent migrants could afford housing only in the old factory districts, which were being abandoned by both employers and the white working class. The result was a twentieth century version of Disraeli's "two nations." Now, however, the outer reaches of affluence include both the middle class and the better-off working class – a majority of the population; while the largely black and Hispanic minority are forced into decaying neighborhoods, which lack not only decent housing but jobs.

This bleak picture has been modified somewhat by the continued ability of the traditional urban cores to retain certain key areas of white collar and professional employment; and by the choice of some highly paid core workers to live in high-rise or recently renovated housing around the core. Compared both to the decaying factory zones and to peripheral expansion, the "gentrification" phenomenon has been highly visible yet statistically insignificant. It has done as much to displace low income city dwellers as to benefit them. The late twentieth century American environment thus shows all the signs of the two nations syndrome: one caught in an environment of poverty, cut off from the majority culture, speaking its own languages and dialects; the other an increasingly homogenized culture of affluence, more and more remote from an urban environment it finds dangerous.

[. . .]

The case against the technoburb can easily be summarized. Compared even to the traditional suburb, it at first appears impossible to comprehend. It has no clear boundaries; it includes discordant rural, urban, and suburban elements; and it can best be measured in counties rather than in city blocks. Consequently the new city lacks any recognizable center to give meaning to the whole. Major civic institutions seem scattered at random over an undifferentiated landscape.

Even planned developments – however harmonious they might appear from the inside – can be no more than fragments in a fragmented environment. A single house, a single street, even a cluster of streets and houses can be and frequently are well designed. But true public space is lacking or totally commercialized. Only the remaining pockets of undeveloped farmland maintain real openness, and these pockets are inevitably developed, precipitating further flight and further sprawl.

The case for the techno-city can only be made hesitantly and conditionally. Nevertheless, we can hope that its deficiencies are in large part the early awkwardness of a new urban type. All new city forms appear in their early stages to be chaotic. "There were a hundred thousand shapes and substances of incompleteness, wildly mingled out of their places, upside down, burrowing in the earth, aspiring in the earth, moldering in the water, and unintelligible as any dream." This was Charles Dickens describing London in 1848, in his novel *Dombey and Son* (Chapter 6). As I have indicated, sprawl has a functional logic that may not be apparent to those accustomed to more traditional cities. If that logic is understood imaginatively, as Wells and especially Wright attempted to do, then perhaps a matching aesthetic can be devised.

We must remember that even the most "organic" cityscapes of the past evolved slowly after much chaos and trial and error. The classic late nineteenth century railroad suburb – the standard against which critics judge today's sprawl – evolved out of the disorder of nineteenth century metropolitan growth. First, planners of genius like John Nash and Frederick Law Olmsted comprehended the process and devised aesthetic formulas to guide it. These formulas were then communicated – slowly and incompletely – to speculative builders, who nevertheless managed to capture the basic idea. Finally, individual property owners constantly upgraded their holdings to eliminate discordant elements and bring their community closer to the ideal.

We might hope that a similar process is now at work in the postsuburban outer city. As a starting point for a technoburb aesthetic, there are Wright's Broadacre City plans and drawings, which still repay study for anyone seeking a vision of a modern yet organic American landscape. More useful still is the American New Town tradition, starting from Radburn, New Jersey, with its careful designs intended to reconcile decentralization with older ideas of community. Already, New Town designs have been adopted by speculative builders, not only in a highly publicized project like James Rouse's Columbia, Maryland, but in hundreds of smaller planned communities, which are beginning to leave their mark on the landscape.

At the level of civic architecture there is Wright's Marin County Civic Center to serve as a model for public monuments in a decentralized environment. The multilevel, enclosed shopping mall has attained a spaciousness not unworthy of the great urban shopping districts of the past, while newly built college campuses and campuslike office complexes and research centers contribute significantly to the environment. Some commercial highway strips have been rescued from cacophony and have managed to achieve a liveliness that is not tawdry. (This evolution parallels the evolution of the nineteenth-century urban core, originally a remarkably ugly cluster of small buildings and large signs, which was transformed into a reasonably dignified center for commerce by the turn of the century.)

Most importantly, there is a growing sense that open land must be preserved as an integral part of the landscape, through regional land use plans, purchases for parklands, and tax abatements for working farms. These governmental measures, combined with thousands of small scale efforts by individuals, could create a fitting environment for the new city. These efforts, moreover, could provide the starting point for a more profound diversification of the outer city. An increased understanding and respect for the landscape of each region could lead to a growing rejection of a mass culture that erases all such distinctions.

The techno-city, therefore, is still under construction, both physically and culturally. Its economic and social successes are undeniable, as are its costs. Most importantly, the new pattern of decentralization has fundamentally altered the urban form on which suburbia had depended for its function and meaning. Whatever the fate of the new city, suburbia in its traditional sense now belongs to the past.

ONE

Plate 1 Palace of Sargon II of Khorsabad. The first cities arose about 4000 BCE in Mesopotamia between the Tigris and Euphrates Rivers during what V. Gordon Childe termed the "Urban Revolution." Note the mud brick walls, monumental gate, and ziggurat in this very substantial Mesopotamian palace.

Plate 2 Theater of Dionysus, Athens, Greece, fifth century BCE. The Greek polis stressed public over private life. Citizens conversed, shopped, and settled disputes in public agora, exercised and competed in public stadia and gymnasia, and shared in the political and cultural life of the community in large open-air structures like the magnificent Theater of Dionysus.

Plate 3 Medieval market-place, Montpazier, France. As Europe began to revive after the period of disorganization and strife that followed the fall of the Roman Empire, cities like Montpazier, France, fostered trade, economic expansion, and free institutions. (Photograph by Jay Vance.)

Plate 4 The nineteenth-century industrial city. During the first part of the nineteenth century new industrial cities based on steam-powered machinery sprang up in Europe. In these prints titled "A Catholic town in 1440" and "The same town in 1840," Augustus Welby Pugin, a contemporary observer, contrasts the same city before and after the Industrial Revolution. The first print shows a city where church spires are the dominant architectural element, the land surrounding the medieval city walls is largely empty, and the air and water are clean. In the second print, factory smokestacks have largely replaced steeples, the air is filled with smoke, development has sprawled to the once-empty land, and the foreground is dominated by a massive panopticon prison. Pugin subtitled his work: "A Parallel Between the Noble Edifices of the Middle Ages and the Corresponding Buildings of the Present Day Shewing the Present Decay of Taste."

Plate 5 Bird's-eye view of La Crosse, Wisconsin, showing the principal business street. One of the features which Friedrich Engels noted in mid-nineteenth-century Manchester, England, was the appearance of large streets where the upper classes could travel without coming into contact with the squalid living conditions of the residential slums which comprised the greater part of the city. This plate from a popular American illustrated weekly depicted with pride the principal business street in a Midwestern industrial city. The belching factories and packed residential slums, which then comprised much of La Crosse, are not mentioned in the caption.

Plate 6 **Levittown, New York, 1947.** While suburbs have a long and varied history, it was during the period after World War II that many of the suburbs surrounding US cities arose. Levittown, New York (and its counterpart Levittown, Pennsylvania), provided entirely new communities of affordable, cute, single-family houses on individual lots to returning GIs and other first-time (white) homebuyers.

Plate 7 **The auto-centered metropolis.** Developers show off the location of a proposed new "50 Foot Boulevard" in the Westwood Village area of Santa Monica in 1922. Capitalizing on the mass production of Henry Ford's Model-T, they and thousands of their counterparts built communities in Southern California that required a car. (Security Pacific Collection. Los Angeles Public Library.)

Plate 8 Technoburbia. Robert Fishman uses the term technoburbia to describe the form of urban development that jumbles business, residential, commercial, and other uses together in the area surrounding older core cities. Here Microsoft Corporate Headquarters in Redmond, Washington, forms part of a pattern that includes residential areas, open space, shopping centers, and other uses.

PART TWO

Urban culture and society

INTRODUCTION TO PART TWO

As Shakespeare wrote, and as urbanists ever since have never tired of quoting, "the people are the city." Urban history expresses the progressive evolution of the city as an institution. Urban form and design describe the physical appearance and infrastructural layout of cities. But it is the people of the city – their individual aspirations and collective struggles, their day-to-day lives and their moments of heightened awareness – that constitute the core subject of urban studies and the final purpose of city planning.

In turning to the people of the cities themselves, we move to a consideration of the subtle and ever-shifting interplay between society, community, and culture. This section addresses how urban society affects urban culture and how culture affects the daily lives and the life prospects of city dwellers. It asks what culture is in an urban context and how it expresses itself in different social contexts – either as high culture or popular culture. Finally, it analyzes what community is in an urban context and speculates about what it could be.

In studying the people of the city, the key methodology is sociology, that "science of society" that arose alongside the emergence of the modern industrial city itself. Urban sociology has always been allied with urban anthropology, and in recent years new formulations – social studies, social theory, social relations, culture studies – have joined the discipline either as adjuncts or rivals. But the basic sociological vision remains central to all investigations of people in cities.

There is no better person with whom to begin a discussion of "Urban Culture and Society" than Lewis Mumford, one of the great public intellectuals of the twentieth century. Mumford never lost sight of the human dimension of cities. For over sixty years he sparred with those who argued that cities arose and prospered for purely economic reasons or that cities were best defined in terms of size and density. Not so! thundered Mumford: cities are expressions of the human spirit and cities exist to contribute to the ever-evolving human personality. This perspective comes through loud and clear in "What Is a City?" To Mumford, defining a city in terms of population size, or density, or attributes of the built environment is inadequate. Rather, the human side of cities is their very essence, and city streets are a stage on which life's drama is played out. Like William Whyte (p. 429), Jane Jacobs (p. 114), and Sharon Zukin (p. 136), Mumford takes real delight in city life. For him, cities reflect and enlarge the human spirit, and he argues that the project of creating better, more human cities will enrich civilization itself.

Mumford, of course, was not alone in focusing on the connections between urban life and the human personality. In "Urbanism as a Way of Life," Louis Wirth asked the fundamental question "What does it mean to be urban?" and concluded that an urban "way of life" resulted in an "urban type" of character and personality. Wirth was one of a gifted group of sociologists at the University of Chicago who, in the 1920s and 1930s, developed a pioneering body of urban sociological theory that still shapes the field of urban sociology today. Studying rural migrants to Chicago from the peasant societies of Southern and Eastern Europe, Wirth perceived that the whole way of life in modern cities was fundamentally different from the way of life in rural cultures. In "Urbanism as a Way of Life," he attempts to abstract the essential characteristics of urban as opposed to rural life and to find the sources of the widely perceived urban characteristics of brusqueness and impersonality. As the face-to-face transactions of static rural village life are replaced by the distanced and mediated transactions of a large city, human personalities are transformed, and the new urbanites respond to each other and to society as a whole in entirely different ways than they did in their rural folk communities.

While Wirth's is a *theoretical* study, it is important to remember that his theories were generated by empirical observations that he and his colleagues conducted in Chicago during the early decades of the twentieth century. Architectural critic and urban community activist Jane Jacobs did her own kind of street-level social observation in writing *The Death and Life of Great American Cities* (1961), a book that shook the complacent world of establishment planning by re-introducing the values of community to the design of urban spaces. Architecture and urban design may not *determine* human behavior, but bad design can numb the human spirit and good design can have powerful, positive influences on human beings. Of the many values designers seek to build into their designs perhaps none is more important than fostering community and human interaction. To Jacobs, traffic engineering should be only one consideration in designing a street. In "The Uses of Sidewalks: Safety," she argues that a street designed so that people can see their children from house windows and will want to congregate on the front door stoops – one very much like her own Hudson Street in Greenwich Village – will be much more user-friendly. It will also be much safer than one, which moves traffic efficiently, but is inhospitable to neighborhood life and insensitive to the potential for street life to reduce urban crime. Jacobs stresses the importance of designing streets to promote safety, particularly for women. A safer environment, she argues, is essential to the creation and preservation of community.

The Black ghettoes of the United States have been the subject of an enormous body of sociological research that both celebrates the distinctive culture and analyzes the social pathology of segregated, poverty-ridden inner-city communities. In multi-ethnic societies, racial divisions compounded class distinctions to create an even further crisis of community in the form of racially segregated neighborhoods that have remained as symbols of inequality and oppression. *The Philadelphia Negro* (1899) by W.E.B. Du Bois specifically describes the African-American district of Philadelphia, Pennsylvania, as it developed in the years following the American Civil War, but the social and cultural dynamic of housing segregation and racial discrimination in the workplace that Du Bois describes can be applied to ghetto and barrio experiences throughout the United States and to Third-World immigrant communities worldwide. In the years since Du Bois first surveyed the life of the racially segregated ghetto community, conditions have in some ways grown worse: so much so that the persistence of racial segregation, and the emergence of an "underclass" population radically disconnected from the rest of the urban community, threaten the social stability of some of the largest, wealthiest cities in the world.

Beginning with W.E.B. Du Bois's study of *The Philadelphia Negro* (1899) (p . 119) and continuing through the work of E. Franklin Frazier (*The Negro Family in the United States*, 1939), St. Clair Drake and Horace Cayton (*Black Metropolis*, 1945), and Kenneth B. Clark (*Dark Ghetto*, 1965), African-American scholars have taken the lead in examining the social and cultural dynamics of ghetto communities in America's Northern cities. More recently, an important debate, called the "underclass" debate, arose concerning the plight of the mostly Black residents of American inner-city ghettos. One of the principals in that debate was William Julius Wilson, an African-American sociologist.

In "From Institutional to Jobless Ghettos" from *When Work Disappears* (1996), Wilson illustrates the role of ideology in shaping urban theory. As a leading liberal voice, he argues that the situation of the poorest urban Blacks in the United States has grown worse during the last generation and that poor ghetto Blacks today, especially youth, are in deep trouble. But why is this so . . . and what to do about it? Wilson stresses the loss of jobs accessible to unskilled Black youth. It is this loss of jobs, Wilson argues, that has destroyed ghetto family structure and lies at the root of crime, substance abuse, and other ghetto ills. Conservatives, of course, look at many of the same facts and reach different conclusions. As they see it, many young Black males are not "marriageable" because they lack minimum job skills, have substance abuse problems, or are in prison. Without "marriageable" males, many young Black women cannot form two-parent nuclear families, and teen pregnancies, out-of-wedlock births, and female-headed, welfare-dependent, single-parent households result.

Perhaps the biggest difference between liberals and conservatives in the underclass debate lies in their views on the role of government in solving social problems. Wilson would like to see government intervene with a universal, not race-specific, full employment program. If the poor of all races are employed, he reasons, family stability will return, and substance abuse and criminal behavior will drop. Conservatives argue that the last thing the poor need is more government assistance. For example, Charles Murray, author of *Losing Ground* (1984) and *The Bell Curve* (1994), argues that patronizing government programs sapped initiative from the poor and created perverse incentives to stay out of the labor market. Murray's remedy: cut government welfare and social service programs. When the previously unemployed know they are making it on their own by working long hours at hard jobs, without paternalistic government programs, Murray argues, they will regain self-respect and cut down on drugs and criminality.

The "underclass debate" changed direction when, in the mid-1990s, the American Congress and President Clinton changed "welfare as we know it" by radically cutting back on many welfare support programs in favor of decentralized "workfare." On balance, the results have seemed promising, with welfare rolls declining and employment trending upward. At the same time, however, another concern emerged – or rather re-emerged – about the fundamental quality of civic culture in contemporary urban society. The leading voice raising this concern was that of Robert Putnam, author of *Bowling Alone* (2000). Drawing on evidence from both Europe and the United States, and without down-playing the importance of the issues of race and poverty, Putnam asked urban leaders and members of the urban middle class to confront the evident social reality that people are no longer as connected to the basic institutions of their communities – the neighborhood groups, the fraternal organizations, even the political parties – as they once were.

Putnam attributes the growing lack of community participation – what he calls the decline in "social capital" – to many factors: the movement of women into the workforce, increased social and geographical mobility, and "the technological transformation of leisure." Another explanation might be that contemporary urban society is deeply divided and that every major city has now become culturally contested terrain. Culture used to mean "high culture" – the world of the symphony, opera, and ballet. But culture, of course, is not just the product of an identifiable artistic or intellectual class. Working-class neighborhoods and inner-city ghettos also produce formal poetry and the linguistic inventiveness of street talk, rhapsodies and jazz music, paintings and graffiti. These are surely elements of "the urban drama," but who will dominate a city's culture? Who will be hegemonic in the socially "contested city"? In "Whose Culture? Whose City?" from *The Cultures of Cities* (1995), sociologist and cultural analyst Sharon Zukin explores the ways that authentic popular culture is being increasingly privatized and commercialized in a contested urban environment. As private corporations take over public parks in the name of civility and as the fashion industry appropriates the attitudes and demeanors of the street culture of urban youth, an authentic and democratic public culture may, she fears, be irrevocably lost. In the final analysis, the urban cultural drama may not be just about the abstractions of class and socioeconomic dynamics. Instead, it must encompass the real, everyday experience of all citizens, rich and poor alike, in the public places of their cities. Respecting and making provision for these experiences will be the ultimate building blocks of the urban community.

Finally, in exploring urban society and culture, academic analysts of the urban condition have been trained to investigate a range of measurable conditions and observable social behaviors and to analyze their findings in terms of class, race, ethnicity, and socioeconomic status. But urban culture has other dimensions that do not lend themselves so easily to quantitative analysis or narrative description. Among these are the expressions of the creative arts and the other cultural productions that emerge from urban communities everywhere, indeed from the very conditions of modern urban life. In "Visions of a New Reality: The City and the Emergence of Modern Visual Culture," the image-essay that ends Part Two, Frederic Stout argues that popular illustrated journalism, photography, and cinema are among the most characteristic of urban cultural genres and that visual culture generally is an artifact of modern urban society.

"What Is a City?"

Architectural Record (1937)

Lewis Mumford

Editors' Introduction

Lewis Mumford (1895–1990) has been called America's last great public intellectual. Beginning with his first book in 1922 and continuing throughout a career that saw the publication of some twenty-five influential volumes, Mumford made signal contributions to social philosophy, American literary and cultural history, the history of technology and, preeminently, the history of cities and urban planning practice.

Mumford saw the urban experience as an integral component in the development of human culture and the human personality. He consistently argued that the physical design of cities and their economic functions were secondary to their relationship to the natural environment and to the spiritual values of human community. Mumford applied these principles to his architectural criticism for *The New Yorker* magazine in the 1920s, his work with the Regional Planning Association of America, his campaign against plans to build a highway through Washington Square in New York's Greenwich Village in the 1950s, and his life-long championing of the Garden City ideals of Ebenezer Howard.

In "What Is a City?" Mumford lays out his fundamental propositions about city planning and the human potential, both individual and social, of urban life. The city, he writes, is "a theater of social action," and everything else – art, politics, education, commerce – only serve to make the "social drama . . . more richly significant, as a stage-set, well-designed, intensifies and underlines the gestures of the actors and the action of the play." It was a theme and an image to which Mumford would return over and over again. In his chapter on "The Nature of the Ancient City" in *The City in History* (1961), he wrote that the city is "above all things a theater" and, as if commenting on the cultural conformity of the 1950s, warned that an urban civilization that has lost its sense of dramatic dialogue "is bound to have a fatal last act."

Mumford's influence on modern urban planning theory can hardly be overstated. His "urban drama" idea clearly resonates with an entire line of urban cultural analysts. Jane Jacobs, for example, talks about "street ballet" (p. 114). William Whyte (p. 429) says that a good urban plaza should function like a stage. Allan Jacobs and Donald Appleyard (p. 437) urge planners to fulfill human needs for "fantasy and exoticism." The city, they write, "has always been a place of excitement; it is a theater, a stage upon which citizens can display themselves and be seen by others."

As a historian, Mumford's emphasis on community values and the city's role in enlarging the potential of the human personality connects him with a long line of urban theorists that includes Louis Wirth (p. 97) and many others. *The City in History* (New York: Harcourt Brace, 1961) is undoubtedly Mumford's masterpiece, but an earlier version of the same material, *The Culture of Cities* (New York: Harcourt Brace, 1938), is still of interest. *The Urban Prospect* (New York: Harcourt Brace, 1968) is an outstanding collection of his essays on urban planning and culture, and *The Myth of the Machine* (New York; Harcourt Brace, 1967) and *The Pentagon of Power* (New York: Harcourt Brace, 1970) are excellent analyses of the influence of technology on human culture. The magisterial *The Transformations of Man* (New York: Harper, 1956) invites comparison with V. Gordon Childe's theory of the urban revolution (p 35). A sampling of Mumford's writings are included in Donald L. Miller (ed.), *The Lewis Mumford Reader* (Athens: University of Georgia Press, 1995).

Mumford's illuminating correspondence with Patrick Geddes is contained in Frank G. Novak, *Lewis Mumford and Patrick Geddes: The Correspondence* (London: Routledge, 1995). His correspondence with Frank Lloyd Wright *is contained in Frank Lloyd Wright and Lewis Mumford: Thirty Years of Correspondence* (New York: Princeton Architectural Press, 2001), and his writings for *The New Yorker* are contained in Robert Wojtowicz (ed.), *Sidewalk Critic: Lewis Mumford's Writings on New York* (New York: Princeton Architectural Press, 1998).

Mumford is being rediscovered by the current generation of environmental planners. Examples of recent books applying his perspective to current ecological issues are Mark Luccarelli Lewis, *Mumford and the Ecological Region: The Politics of Planning* (New York: Guilford Press, 1997) and Robert Wojtowicz, *Lewis Mumford and American Modernism: Eutopian Theories for Architecture and Urban Planning* (Cambridge: Cambridge University Press, 1998).

Biographies of Lewis Mumford are Donald L. Miller's *Lewis Mumford: A Life* (New York: Weidenfeld & Nicolson, 1989), Thomas P. Hughes and Agatha C. Hughes (eds), *Lewis Mumford: Public Intellectual* (Oxford: Oxford University Press, 1990), and Frank G. Novak, *Lewis Mumford* (New York: Twayne Publishers, 1998). An excellent bibliography of Mumford's writings is Elmer S. Newman, *Lewis Mumford: A Bibliography, 1914–1970* (New York: Harcourt Brace Jovanovich, 1971).

Most of our housing and city planning has been handicapped because those who have undertaken the work have had no clear notion of the social functions of the city. They sought to derive these functions from a cursory survey of the activities and interests of the contemporary urban scene. And they did not, apparently, suspect that there might be gross deficiencies, misdirected efforts, mistaken expenditures here that would not be set straight by merely building sanitary tenements or straightening out and widening irregular streets.

The city as a purely physical fact has been subject to numerous investigations. But what is the city as a social institution? The earlier answers to these questions, in Aristotle, Plato, and the Utopian writers from Sir Thomas More to Robert Owen, have been on the whole more satisfactory than those of the more systematic sociologists: most contemporary treatises on "urban sociology" in America throw no important light upon the problem. One of the soundest definitions of the city was that framed by John Stow, an honest observer of Elizabethan London, who said:

Men are congregated into cities and commonwealths for honesty and utility's sake, these shortly be the commodities that do come by cities, commonalties and corporations. First, men by this nearness of conversation are withdrawn from barbarous fixity and force, to certain mildness of manners, and to humanity and justice . . . Good behavior is yet called urbanitas because it is rather found in cities than elsewhere. In sum, by often hearing, men be better persuaded in religion, and for that they live in the eyes of others, they be by example the more easily trained to justice, and by shamefastness restrained from injury.

And whereas commonwealths and kingdoms cannot have, next after God, any surer foundation than the love and good will of one man towards another, that also is closely bred and maintained in cities, where men by mutual society and companying together, do grow to alliances, commonalties, and corporations.

It is with no hope of adding much to the essential insight of this description of the urban process that I would sum up the sociological concept of the city in the following terms:

The city is a related collection of primary groups and purposive associations: the first, like family and neighborhood, are common to all communities, while the second are especially characteristic of city life. These varied groups support themselves through economic organizations that are likewise of a more or less corporate, or at least publicly regulated, character; and they are all housed in permanent structures, within a relatively limited area. The essential physical means of a city's existence are the fixed site, the durable shelter, the permanent facilities for assembly, inter-change, and storage; the essential social means are the social division of labor, which serves not merely the economic life but the cultural processes. The city in its complete sense, then, is a geographic plexus, an economic organization, an institutional process, a theater of social action, and an aesthetic symbol of collective unity. The city fosters art and is art; the city creates the theater and is the theater. It is in the city, the city as theater, that man's more purposive activities are focused, and work out, through conflicting and cooperating personalities, events, groups, into more significant culminations.

Without the social drama that comes into existence through the focusing and intensification of group activity there is not a single function performed in the city that could not be performed – and has not in fact been performed – in the open country. The physical organization of the city may deflate this drama or make it frustrate; or it may, through the deliberate efforts of art, politics, and education, make the drama more richly significant, as a stage-set, well-designed, inten-sifies and underlines the gestures of the actors and the action of the play. It is not for nothing that men have dwelt so often on the beauty or the ugliness of cities: these attributes qualify men's social activities. And if there is a deep reluctance on the part of the true city dweller to leave his cramped quarters for the physically more benign environment of a suburb – even a model garden suburb! – his instincts are usually justified: in its various and many-sided life, in its very opportunities for social disharmony and conflict, the city creates drama; the suburb lacks it.

One may describe the city, in its social aspect, as a special framework directed toward the creation of differentiated opportunities for a common life and a significant collective drama. As indirect forms of association, with the aid of signs and symbols and specialized organizations, supplement direct face-to-face intercourse, the personalities of the citizens themselves become many-faceted: they reflect their specialized interests, their more intensively trained aptitudes, their finer discriminations and selections: the personality no longer presents a more or less unbroken traditional face to reality as a whole. Here lies the possibility of personal disintegration; and here lies the need for reintegration through wider participation in a concrete and visible collective whole. What men cannot imagine as a vague formless society, they can live through and experience as citizens in a city.

Their unified plans and buildings become a symbol of their social relatedness; and when the physical environment itself becomes disordered and incoherent, the social functions that it harbors become more difficult to express.

One further conclusion follows from this concept of the city: social facts are primary, and the physical organization of a city, its industries and its markets, its lines of communication and traffic, must be subservient to its social needs. Whereas in the development of the city during the last century we expanded the physical plant recklessly and treated the essential social nucleus, the organs of government and education and social service, as mere afterthought, today we must treat the social nucleus as the essential element in every valid city plan: the spotting and inter-relationship of schools, libraries, theaters, community centers is the first task in defining the urban neighborhood and laying down the outlines of an integrated city.

In giving this sociological answer to the question: What is a City? one has likewise provided the clue to a number of important other questions. Above all, one has the criterion for a clear decision as to what is the desirable size of a city – or may a city perhaps continue to grow until a single continuous urban area might cover half the American continent, with the rest of the world tributary to this mass? From the standpoint of the purely physical organization of urban utilities – which is almost the only matter upon which metropolitan planners in the past have concentrated – this latter process might indeed go on indefinitely. But if the city is a theater of social activity, and if its needs are defined by the opportunities it offers to differentiated social groups, acting through a specific nucleus of civic institutes and associations, definite limitations on size follow from this fact.

In one of Le Corbusier's early schemes for an ideal city, he chose three million as the number to be accommodated: the number was roughly the size of the urban aggregate of Paris, but that hardly explains why it should have been taken as a norm for a more rational type of city development. If the size of an urban unit, however, is a function of its productive organization and its opportunities for active social intercourse and culture, certain definite facts emerge as to adequate ratio of population to the process to be served. Thus, at the present level of culture in America, a million people are needed to support a university. Many factors may enter which will change the size of both the university and the population base;

nevertheless one can say provisionally that if a million people are needed to provide a sufficient number of students for a university, then two million people should have two universities. One can also say that, other things being equal, five million people will not provide a more effective university than one million people would. The alternative to recognizing these ratios is to keep on overcrowding and overbuilding a few existing institutions, thereby limiting, rather than expanding, their genuine educational facilities.

What is important is not an absolute figure as to population or area: although in certain aspects of life, such as the size of city that is capable of reproducing itself through natural fertility, one can already lay down such figures. What is more important is to express size *always as a function of the social relationships to be served* . . . There is an optimum numerical size, beyond which each further increment of inhabitants creates difficulties out of all proportion to the benefits. There is also an optimum area of expansion, beyond which further urban growth tends to paralyze rather than to further important social relationships. Rapid means of transportation have given a regional area with a radius of from forty to a hundred miles, the unity that London and Hampstead had before the coming of the underground railroad. But the activities of small children are still bounded by a walking distance of about a quarter of a mile; and for men to congregate freely and frequently in neighborhoods the maximum distance means nothing, although it may properly define the area served for a selective minority by a university, a central reference library, or a completely equipped hospital. The area of potential urban settlement has been vastly increased by the motor car and the airplane; but the necessity for solid contiguous growth, for the purposes of intercourse, has in turn been lessened by the telephone and the radio. In the Middle Ages a distance of less than a half a mile from the city's center usually defined its utmost limits. The block-by-block accretion of the big city, along its corridor avenues, is in all important respects a denial of the vastly improved type of urban grouping that our fresh inventions have brought in. For all occasional types of intercourse, the region is the unit of social life but the region cannot function effectively, as a well-knit unit, if the entire area is densely filled with people – since their very presence will clog its arteries of traffic and congest its social facilities.

Limitations on size, density, and area are absolutely necessary to effective social intercourse; and they are therefore the most important instruments of rational economic and civic planning. The unwillingness in the past to establish such limits has been due mainly to two facts: the assumption that all upward changes in magnitude were signs of progress and automatically "good for business," and the belief that such limitations were essentially arbitrary, in that they proposed to "decrease economic opportunity" – that is, opportunity for profiting by congestion – and to halt the inevitable course of change. Both these objections are superstitious.

Limitations on height are now common in American cities; drastic limitations on density are the rule in all municipal housing estates in England: that which could not be done has been done. Such limitations do not obviously limit the population itself: they merely give the planner and administrator the opportunity to multiply the number of centers in which the population is housed, instead of permitting a few existing centers to aggrandize themselves on a monopolistic pattern. These limitations are necessary to break up the functionless, hypertrophied urban masses of the past. Under this mode of planning, the planner proposes to replace the "mononucleated city," as Professor Warren Thompson has called it, with a new type of "polynucleated city," in which a cluster of communities, adequately spaced and bounded, shall do duty for the badly organized mass city. Twenty such cities, in a region whose environment and whose resources were adequately planned, would have all the benefits of a metropolis that held a million people, without its ponderous disabilities: its capital frozen into unprofitable utilities, and its land values congealed at levels that stand in the way of effective adaptation to new needs.

Mark the change that is in process today. The emerging sources of power, transport, and communication do not follow the old highway network at all. Giant power strides over the hills, ignoring the limitations of wheeled vehicles; the airplane, even more liberated, flies over swamps and mountains, and terminates its journey, not on an avenue, but in a field. Even the highway for fast motor transportation abandons the pattern of the horse-and-buggy era. The new highways, like those of New Jersey and Westchester, to mention only examples drawn locally, are based more or less on a system definitively formulated by Benton MacKaye in his various papers on the Townless Highway. The most complete plans form an independent highway network, isolated both from the adjacent countryside and the towns that they by-pass: as free from communal encroachments as the railroad system. In such a network no single center will, like the metropolis of old, become the focal point of all regional advantages: on the contrary, the "whole region" becomes open for settlement.

Even without intelligent public control, the likelihood is that within the next generation this dissociation and decentralization of urban facilities will go even farther. The Townless Highway begets the Highwayless Town in which the needs of close and continuous human association on all levels will be uppermost. This is just the opposite of the earlier mechanocentric picture of Roadtown, as pictured by Edgar Chambless and the Spanish projectors of the Linear City. For the highwayless town is based upon the notion of effective zoning of functions through initial public design, rather than by blind legal ordinances. It is a town in which the various functional parts of the structure are isolated topographically as urban islands, appropriately designed for their specific use with no attempt to provide a uniform plan of the same general pattern for the industrial, the commercial, the domestic, and the civic parts.

The first systematic sketch of this type of town was made by Messrs. Wright and Stein in their design for Radburn in 1929; a new type of plan that was repeated on a limited scale – and apparently in complete independence – by planners in Köln and Hamburg at about the same time. Because of restrictions on design that favored a conventional type of suburban house and stale architectural forms, the implications of this new type of planning were not carried very far in Radburn. But in outline the main relationships are clear: the differentiation of foot traffic from wheeled traffic in independent systems, the insulation of residence quarters from through roads; the discontinuous street pattern; the polarization of social life in specially spotted civic nuclei, beginning in the neighborhood with the school and the playground and the swimming pool. This type of planning was carried to a logical conclusion in perhaps the most functional and most socially intelligent of all Le Corbusier's many urban plans: that for Nemours in North Africa, in 1934.

Through these convergent efforts, the principles of the polynucleated city have been well established. Such plans must result in a fuller opportunity for the primary group, with all its habits of frequent direct meeting and face-to-face intercourse: they must also result in a more complicated pattern and a more comprehensive life for the region, for this geographic area can only now, for the first time, be treated as an instantaneous whole for all the functions of social existence. Instead of trusting to the mere massing of population to produce the necessary social concentration and social drama, we must now seek these results through deliberate local nucleation and a finer regional articulation. The words are jargon; but the importance of their meaning should not be missed. To embody these new possibilities in city life, which come to us not merely through better technical organization but through acuter sociological understanding, and to dramatize the activities themselves in appropriate individual and urban structures, forms the task of the coming generation.

"Urbanism as a Way of Life"

American Journal of Sociology (1938)

Louis Wirth

Editors' Introduction

Louis Wirth (1897–1952) was a member of the famed "Chicago School" of urban sociology that included such academic luminaries as Ernest W. Burgess (author of "The Growth of the City," p. 158), Robert E. Park, St Clair Drake, and Horace Cayton. Together, these scholars at the University of Chicago set out virtually to reinvent modern sociology by taking academic research to the streets and by using the city of Chicago itself as a "living laboratory" for the study of urban problems and social processes.

Wirth's major contribution to urban sociology was the formulation of nothing less fundamental than a meaningful and logically coherent "sociological definition" of urban life. As he lays it out in the magnificent synthesis that is his 1938 essay "Urbanism as a Way of Life," a "sociologically significant definition of the city" looks beyond the mere physical structure of the city, or its economic product, or its characteristic cultural institutions – however important all these may be – to discover those underlying "elements of urbanism which mark it as a distinctive mode of human group life."

Wirth argues that three key characteristics of cities – large population size, social heterogeneity, and population density – contribute to the development of a peculiarly "urban way of life" and, indeed, a distinct "urban personality." For centuries, at least as far back as Aesop's fable of the city mouse and the country mouse, casual observers have noted sharp personality differences between urban and rural people and between nature-based and machine-based styles of living. Wirth attempts to explain those differences in terms of the functional responses of urban dwellers to the characteristic environmental conditions of modern urban society. If, for example, city people are regarded as rather more socially tolerant that rural people – and, at the same time, more impersonal and seemingly less friendly – these are merely adaptations to the experience of living in large, dense, socially diverse urban environments.

Although some see Wirth's explanation of the sociology of urban life as nothing more than the social scientific verification of the obvious, others have argued that there is actually no such thing as an "urban personality" or an "urban way of life." Sociologist Herbert Gans, for example, argues in *The Levittowners* (1967) that both inner-city "urban villagers" and suburbanites tend to maintain their preexisting cultures and personalities, and Oscar Lewis's work on "the culture of poverty" suggests that culture and personality types differ widely with socioeconomic class. Wirth's work, however, led to the development of a whole school of urban social ecology, and Wirth's basic ideas about personality and adaptation to urban conditions inform the full range of more recent urban planning theories and the planning practitioners who attempt to create and nurture a sense of community in the urban environment.

Other books by Louis Wirth include *Contemporary Social Problems* (Chicago: University of Chicago Press, 1940), *The Effect of War on American Minorities* (New York: Social Science Research Council, 1943), *Community Life and Social Policy* (Chicago: University of Chicago Press, 1956), and *The Ghetto* (Chicago: University of Chicago Press, 1956). *Louis Wirth on Cities and Social Life: Selected Papers* (Chicago: University of Chicago

Press, 1964) is a useful collection. Also of interest is Roger A. Salerno, *Louis Wirth: A Bio-Bibliography* (Westport, CT: Greenwood Publishing, 1987).

For other important analyses of the relationship between urban life and the human personality, see Erving Goffman, *The Presentation of Self in Everyday Life* (Garden City, NY: Doubleday, 1959) and Richard Sennett, *The Uses of Disorder: Personal Identity and City Life* (New York: W.W. Norton, 1970).

▓ ▓ ▓ ▓ ▓ ▓

THE CITY AND CONTEMPORARY CIVILIZATION

Just as the beginning of Western civilization is marked by the permanent settlement of formerly nomadic peoples in the Mediterranean basin, so the beginning of what is distinctively modern in our civilization is best signalized by the growth of great cities. Nowhere has mankind been farther removed from organic nature than under the conditions of life characteristic of great cities . . . The city and the country may be regarded as two poles in reference to one or the other of which all human settlements tend to arrange themselves. In viewing urban-industrial and rural-folk society as ideal types of communities, we may obtain a perspective for the analysis of the basic models of human association as they appear in contemporary civilization.

A SOCIOLOGICAL DEFINITION OF THE CITY

Despite the preponderant significance of the city in our civilization, however, our knowledge of the nature of urbanism and the process of urbanization is meager. Many attempts have indeed been made to isolate the distinguishing characteristics of urban life. Geographers, historians, economists, and political scientists have incorporated the points of view of their respective disciplines into diverse definitions of the city. While it is in no sense intended to supersede these, the formulation of a sociological approach to the city may incidentally serve to call attention to the interrelations between them by emphasizing the peculiar characteristics of the city as a particular form of human association. A sociologically significant definition of the city seeks to select those elements of urbanism which mark it as a distinctive mode of human group life.

[. . .]

While urbanism, or that complex of traits which makes up the characteristic mode of life in cities, and

urbanization, which denotes the development and extensions of these factors, are thus not exclusively found in settlements which are cities in the physical and demographic sense, they do, nevertheless, find their most pronounced expression in such areas, especially in metropolitan cities. In formulating a definition of the city it is necessary to exercise caution in order to avoid identifying urbanism as a way of life with any specific locally or historically conditioned cultural influences which, while they may significantly affect the specific character of the community, are not the essential determinants of its character as a city.

It is particularly important to call attention to the danger of confusing urbanism with industrialism and modern capitalism. The rise of cities in the modern world is undoubtedly not independent of the emergence of modern power-driven machine technology, mass production, and capitalistic enterprise. But different as the cities of earlier epochs may have been by virtue of their development in a preindustrial and precapitalistic order from the great cities of today, they were, nevertheless, cities.

For sociological purposes a city may be defined as a relatively large, dense, and permanent settlement of socially heterogeneous individuals. On the basis of the postulates which this minimal definition suggests, a theory of urbanism may be formulated in the light of existing knowledge concerning social groups.

A THEORY OF URBANISM

In the rich literature on the city we look in vain for a theory of urbanism presenting in a systematic fashion the available knowledge concerning the city as a social entity. We do indeed have excellent formulations of theories on such special problems as the growth of the city viewed as a historical trend and as a recurrent process, and we have a wealth of literature presenting insights of sociological relevance and empirical studies offering detailed information on a variety of particular aspects of urban life. But despite the multi-plication of research and textbooks on the city, we do not as yet have a comprehensive body of competent hypotheses which may be derived from a set of pos-tulates implicitly contained in a sociological definition of the city, and from our general sociological know-ledge which may be substantiated through empirical research. The closest approximations to a systematic theory of urbanism that we have are to be found in a penetrating essay, "Die Stadt," by Max Weber, and a memorable paper by Robert E. Park titled "The City: Suggestions for the Investigation of Human Behavior in the Urban Environment." But even these excellent contributions are far from constituting an ordered and coherent framework of theory upon which research might profitably proceed.

In the pages that follow, we shall seek to set forth a limited number of identifying characteristics of the city. Given these characteristics we shall then indicate what consequences or further characteristics follow from them in the light of general sociological theory and empirical research. We hope in this manner to arrive at the essential propositions comprising a theory of urbanism. Some of these propositions can be sup-ported by a considerable body of already available research materials; others may be accepted as hypo-theses for which a certain amount of presumptive evidence exists, but for which more ample and exact verification would be required. At least such a pro-cedure will, it is hoped, show what in the way of systematic knowledge of the city we now have and what are the crucial and fruitful hypotheses for future research.

[. . .]

There are a number of sociological propositions concerning the relationship between (a) numbers of population, (b) density of settlement, (c) heterogeneity of inhabitants and group life, which can be formulated on the basis of observation and research.

SIZE OF THE POPULATION AGGREGATE

Ever since Aristotle's *Politics*, it has been recognized that increasing the number of inhabitants in a settle-ment beyond a certain limit will affect the relationships between them and the character of the city. Large numbers involve, as has been pointed out, a greater range of individual variation. Furthermore, the greater the number of individuals participating in a process of interaction, the greater is the potential differentiation between them. The personal traits, the occupations, the cultural life, and the ideas of the members of an urban community may, therefore, be expected to range between more widely separated poles than those of rural inhabitants.

That such variations should give rise to the spatial segregation of individuals according to color, ethnic heritage, economic and social status, tastes and prefer-ences, may readily be inferred. The bonds of kinship, of neighborliness, and the sentiments arising out of living together for generations under a common folk tradition are likely to be absent or, at best, relatively weak in an aggregate the members of which have such diverse origins and backgrounds. Under such circumstances competition and formal control mecha-nisms furnish the substitutes for the bonds of solidarity that are relied upon to hold a folk society together.

[. . .]

The multiplication of persons in a state of inter-action under conditions which make their contact as full personalities impossible produces that segmental-ization of human relationships which has sometimes been seized upon by students of the mental life of the cities as an explanation for the "schizoid" character of urban personality. This is not to say that the urban inhabitants have fewer acquaintances than rural inhabitants, for the reverse may actually be true; it means rather that in relation to the number of people whom they see and with whom they rub elbows in the course of daily life, they know a smaller proportion, and of these they have less intensive knowledge.

Characteristically, urbanites meet one another in highly segmental roles. They are, to be sure, dependent upon more people for the satisfactions of their life-needs than are rural people and thus are associated with a greater number of organized groups, but they are less dependent upon particular persons, and their dependence upon others is confined to a highly frac-tionalized aspect of the other's round of activity. This is essentially what is meant by saying that the city is characterized by secondary rather than primary contacts. The contacts of the city may indeed be face to face, but they are nevertheless impersonal, superficial, transitory, and segmental. The reserve, the indifference, and the blasé outlook which urbanites manifest in their relationships may thus be regarded as devices for immunizing themselves against the personal claims and expectations of others.

The superficiality, the anonymity, and the transitory character of urban social relations make intelligible, also, the sophistication and the rationality generally ascribed to city-dwellers. Our acquaintances tend to stand in a relationship of utility to us in the sense that the role which each one plays in our life is over-whelmingly regarded as a means for the achievement of our own ends. Whereas, therefore, the individual gains, on the one hand, a certain degree of eman-cipation or freedom from the personal and emotional controls of intimate groups, he loses, on the other hand, the spontaneous self-expression, the morale, and the sense of participation that comes with living in an integrated society. This constitutes essentially the state of anomie or the social void to which Durkheim alludes in attempting to account for the various forms of social disorganization in technological society.

The segmental character and utilitarian accent of interpersonal relations in the city find their institutional expression in the proliferation of specialized tasks which we see in their most developed form in the professions. The operations of the pecuniary nexus lead to predatory relationships, which tend to obstruct the efficient functioning of the social order unless checked by professional codes and occupational eti-quette. The premium put upon utility and efficiency suggests the adaptability of the corporate device for the organization of enterprises in which individuals can engage only in groups. The advantage that the corporation has over the individual entrepreneur and the partnership in the urban-industrial world derives not only from the possibility it affords of centralizing the resources of thousands of individuals or from the legal privilege of limited liability and perpetual succession, but from the fact that the corporation has no soul.

[. . .]

DENSITY

As in the case of numbers, so in the case of concen-tration in limited space certain consequences of relevance in sociological analysis of the city emerge. Of these only a few can be indicated.

As Darwin pointed out for flora and fauna and as Durkheim noted in the case of human societies, an increase in numbers when area is held constant (i.e. an increase in density) tends to produce differentiation and specialization, since only in this way can the area support increased numbers. Density thus reinforces the effect of numbers in diversifying men and their activities and in increasing the complexity of the social structure.

On the subjective side, as Simmel has suggested, the close physical contact of numerous individuals necessarily produces a shift in the mediums through which we orient ourselves to the urban milieu, espe-cially to our fellow-men. Typically, our physical contacts are close but our social contacts are distant. The urban world puts a premium on visual recognition. We see the uniform which denotes the role of the functionaries and are oblivious to the personal eccen-tricities that are hidden behind the uniform. We tend to acquire and develop a sensitivity to a world of artifacts and become progressively farther removed from the world of nature.

We are exposed to glaring contrasts between splendor and squalor, between riches and poverty, intelligence and ignorance, order and chaos. The com-petition for space is great, so that each area generally tends to be put to the use which yields the greatest economic return. Place of work tends to become dissociated from place of residence, for the proximity of industrial and commercial establishments makes an area both economically and socially undesirable for residential purposes.

Density, land values, rentals, accessibility, health-fulness, prestige, aesthetic consideration, absence of nuisances such as noise, smoke, and dirt determine the desirability of various areas of the city as places of settlement for different sections of the population . . . The different parts of the city thus acquire specialized functions. The city consequently tends to resemble a mosaic of social worlds in which the transition from one to the other is abrupt. The juxtaposition of divergent personalities and modes of life tends to produce a relativistic perspective and a sense of toleration of differences which may be regarded as prerequisites for rationality and which lead toward the secularization of life.

The close living together and working together of individuals who have no sentimental and emotional ties foster a spirit of competition, aggrandizement, and mutual exploitation. To counteract irresponsibility and potential disorder, formal controls tend to be resorted to. Without rigid adherence to predictable routines a large, compact society would scarcely be able to maintain itself. The clock and the traffic signal are symbolic of the basis of our social order in the urban world. Frequent close physical contact, coupled with great social distance, accentuates the reserve of unattached individuals toward one another and, unless compensated for by other opportunities for response, gives rise to loneliness. The necessary frequent movement of great numbers of individuals in a congested habitat gives occasion to friction and irritation. Nervous tensions which derive from such personal frustrations are accentuated by the rapid tempo and the complicated technology under which life in dense areas must be lived.

HETEROGENEITY

The social interaction among such a variety of personality types in the urban milieu tends to break down the rigidity of caste lines and to complicate the class structure, and thus induces a more ramified and differentiated framework of social stratification than is found in more integrated societies. The heightened mobility of the individual, which brings him within the range of stimulation by a great number of diverse individuals and subjects him to fluctuating status in the differentiated social groups that compose the social structure of the city, tends toward the acceptance of instability and insecurity in the world at large as a norm. This fact helps to account, too, for the sophistication and cosmopolitanism of the urbanite. No single group has the undivided allegiance of the individual. The groups with which he is affiliated do not lend themselves readily to a simple hierarchical arrangement. By virtue of his different interests arising out of different aspects of social life, the individual acquires membership in widely divergent groups, each of which functions only with reference to a single segment of his personality. Nor do these groups easily permit of a concentric arrangement so that the narrower ones fall within the circumference of the more inclusive ones, as is more likely to be the case in the rural community or in primitive societies. Rather the groups with which the person typically is affiliated are tangential to each other or intersect in highly variable fashion.

Partly as a result of the physical footlooseness of the population and partly as a result of their social mobility, the turnover in group membership generally is rapid. Place of residence, place and character of employment, income and interests fluctuate, and the task of holding organizations together and maintaining and promoting intimate and lasting acquaintanceship

between the members is difficult. This applies strikingly to the local areas within the city into which persons become segregated more by virtue of differences in race, language, income, and social status, than through choice or positive attraction to people like themselves. Overwhelmingly the city-dweller is not a home-owner, and since a transitory habitat does not generate binding traditions and sentiments, only rarely is he truly a neighbor. There is little opportunity for the individual to obtain a conception of the city as a whole or to survey his place in the total scheme. Consequently he finds it difficult to determine what is to his own "best interests" and to decide between the issues and leaders presented to him by the agencies of mass suggestion. Individuals who are thus detached from the organized bodies which integrate society comprise the fluid masses that make collective behavior in the urban community so unpredictable and hence so problematical.

Although the city, through the recruitment of variant types to perform its diverse tasks and the accentuation of their uniqueness through competition and the premium upon eccentricity, novelty, efficient performance, and inventiveness, produces a highly differentiated population, it also exercises a leveling influence. Wherever large numbers of differently constituted individuals congregate, the process of depersonalization also enters . . . Individuality under these circumstances must be replaced by categories. When large numbers have to make common use of facilities and institutions, an arrangement must be made to adjust the facilities and institutions to the needs of the average person rather than to those of particular individuals. The services of the public utilities, of the recreational, educational, and cultural institutions, must be adjusted to mass requirements. Similarly, the cultural institutions, such as the schools, the movies, the radio, and the newspapers, by virtue of their mass clientele, must necessarily operate as leveling influences. The political process as it appears in urban life could not be understood without taking account of the mass appeals made through modern propaganda techniques. If the individual would participate at all in the social, political, and economic life of the city, he must subordinate some of his individuality to the demands of the larger community and in that measure immerse himself in mass movements.

THE RELATION BETWEEN A THEORY OF URBANISM AND SOCIOLOGICAL RESEARCH

By means of a body of theory such as that illustratively sketched above, the complicated and many-sided phenomena of urbanism may be analyzed in terms of a limited number of basic categories. The sociological approach to the city thus acquires an essential unity and coherence enabling the empirical investigator not merely to focus more distinctly upon the problems and processes that properly fall in his province but also to treat his subject matter in a more integrated and systematic fashion. A few typical findings of empirical research in the field of urbanism, with special reference to the United States, may be indicated to substantiate the theoretical propositions set forth in the preceding pages, and some of the crucial problems for further study may be outlined.

On the basis of the three variables, number, density of settlement, and degree of heterogeneity, of the urban population, it appears possible to explain the characteristics of urban life and to account for the differences between cities of various sizes and types.

Urbanism as a characteristic mode of life may be approached empirically from three interrelated perspectives: (1) as a physical structure comprising a population base, a technology, and an ecological order; (2) as a system of social organization involving a characteristic social structure, a series of social institutions, and a typical pattern of social relationships; and (3) as a set of attitudes and ideas, and a constellation of personalities engaging in typical forms of collective behavior and subject to characteristic mechanisms of social control.

URBANISM IN ECOLOGICAL PERSPECTIVE

Since in the case of physical structure and ecological processes we are able to operate with fairly objective indices, it becomes possible to arrive at quite precise and generally quantitative results. The dominance of the city over its hinterland becomes explicable through the functional characteristics of the city which derive in large measure from the effect of numbers and density. Many of the technical facilities and the skills and organizations to which urban life gives rise can grow and prosper only in cities where the demand is sufficiently great. The nature and scope of the services rendered by these organizations and institutions and the advantage which they enjoy over the less developed facilities of smaller towns enhances the dominance of the city and the dependence of ever wider regions upon the central metropolis.

The urban population composition shows the operation of selective and differentiating factors. Cities contain a larger proportion of persons in the prime of life than rural areas which contain more old and very young people. In this, as in so many other respects, the larger the city the more this specific characteristic of urbanism is apparent. With the exception of the largest cities, which have attracted the bulk of the foreign-born males, and a few other special types of cities, women predominate numerically over men. The heterogeneity of the urban population is further indicated along racial and ethnic lines. The foreign born and their children constitute nearly two-thirds of all the inhabitants of cities of one million and over. Their proportion in the urban population declines as the size of the city decreases, until in the rural areas they comprise only about one-sixth of the total population. The larger cities similarly have attracted more Negroes and other racial groups than have the smaller communities. Considering that age, sex, race, and ethnic origin are associated with other factors such as occupation and interest, it becomes clear that one major characteristic of the urban-dweller is his dissimilarity from his fellows. Never before have such large masses of people of diverse traits as we find in our cities been thrown together into such close physical contact as in the great cities of America. Cities generally, and American cities in particular, comprise a motley of peoples and cultures, of highly differentiated modes of life between which there often is only the faintest communication, the greatest indifference and the broadest tolerance, occasionally bitter strife, but always the sharpest contrast.

The failure of the urban population to reproduce itself appears to be a biological consequence of a combination of factors in the complex of urban life, and the decline in the birth-rate generally may be regarded as one of the most significant signs of the urbanization of the Western world. While the proportion of deaths in cities is slightly greater than in the country, the outstanding difference between the failure of present-day cities to maintain their population and that of cities of the past is that in former times it was due to the exceedingly high death-rates in cities, whereas today, since cities have become more livable from a health standpoint, it is due to low birth-rates. These biological characteristics of the urban population are significant sociologically, not merely because they reflect the urban mode of existence but also because they condition the growth and future dominance of cities and their basic social organization. Since cities are the consumers rather than the producers of men, the value of human life and the social estimation of the personality will not be unaffected by the balance between births and deaths. The pattern of land use, of land values, rentals, and ownership, the nature and functioning of the physical structures, of housing, of transportation and communication facilities, of public utilities – these and many other phases of the physical mechanism of the city are not isolated phenomena unrelated to the city as a social entity, but are affected by and affect the urban mode of life.

URBANISM AS A FORM OF SOCIAL ORGANIZATION

The distinctive features of the urban mode of life have often been described sociologically as consisting of the substitution of secondary for primary contacts, the weakening of bonds of kinship, and the declining social significance of the family, the disappearance of the neighborhood, and the undermining of the traditional basis of social solidarity. All these phenomena can be substantially verified through objective indices. Thus, for instance, the low and declining urban reproduction rates suggest that the city is not conducive to the traditional type of family life, including the rearing of children and the maintenance of the home as the locus of a whole round of vital activities. The transfer of industrial, educational, and recreational activities to specialized institutions outside the home has deprived the family of some of its most characteristic historical functions. In cities mothers are more likely to be employed, lodgers are more frequently part of the household, marriage tends to be postponed, and the proportion of single and unattached people is greater. Families are smaller and more frequently without children than in the country. The family as a unit of social life is emancipated from the larger kinship group characteristic of the country, and the individual members pursue their own diverging interests in their vocational, educational, religious, recreational, and political life.

[. . .]

On the whole, the city discourages an economic life in which the individual in time of crisis has a basis of subsistence to fall back upon, and it discourages self-employment. While incomes of city people are on the average higher than those of country people, the cost of living seems to be higher in the larger cities. Home ownership involves greater burdens and is rarer. Rents are higher and absorb a large proportion of the income. Although the urban-dweller has the benefit of many communal services, he spends a large proportion of his income for such items as recreation and advancement and a smaller proportion for food. What the communal services do not furnish the urbanite must purchase, and there is virtually no human need which has remained unexploited by commercialism. Catering to thrills and furnishing means of escape from drudgery, monotony, and routine thus become one of the major functions of urban recreation, which at its best furnishes means for creative self-expression and spontaneous group association, but which more typically in the urban world results in passive spectatorism on the one hand, or sensational record-smashing feats on the other.

Being reduced to a stage of virtual impotence as an individual, the urbanite is bound to exert himself by joining with others of similar interest into organized groups to obtain his ends. This results in the enormous multiplication of voluntary organizations directed toward as great a variety of objectives as there are human needs and interests. While on the one hand the traditional ties of human association are weakened, urban existence involves a much greater degree of interdependence between man and man and a more complicated, fragile, and volatile form of mutual inter-relations over many phases of which the individual as such can exert scarcely any control. Frequently there is only the most tenuous relationship between the economic position or other basic factors that determine the individual's existence in the urban world and the voluntary groups with which he is affiliated. While in a primitive and in a rural society it is generally possible to predict on the basis of a few known factors who will belong to what and who will associate with whom in almost every relationship of life, in the city we can only project the general pattern of group formation and affiliation, and this pattern will display many incongruities and contradictions.

URBAN PERSONALITY AND COLLECTIVE BEHAVIOR

It is largely through the activities of the voluntary groups, be their objectives economic, political, educational, religious, recreational, or cultural, that the urbanite expresses and develops his personality, acquires status, and is able to carry on the round of activities that constitute his life-career. It may easily be inferred, however, that the organizational framework which these highly differentiated functions call into being does not of itself insure the consistency and integrity of the personalities whose interests it enlists. Personal disorganization, mental breakdown, suicide, delinquency, crime, corruption, and disorder might be expected under these circumstances to be more prevalent in the urban than in the rural community. This has been confirmed insofar as comparable indices are available; but the mechanisms underlying these phenomena require further analysis.

Since for most group purposes it is impossible in the city to appeal individually to the large number of discrete and differentiated individuals, and since it is only through the organizations to which men belong that their interests and resources can be enlisted for a collective cause, it may be inferred that social control in the city should typically proceed through formally organized groups. It follows, too, that the masses of men in the city are subject to manipulation by symbols and stereotypes managed by individuals working from afar or operating invisibly behind the scenes through their control of the instruments of communication. Self-government either in the economic, the political, or the cultural realm is under these circumstances reduced to a mere figure of speech or, at best, is subject to the unstable equilibrium of pressure groups. In view of the ineffectiveness of actual kinship ties we create fictional kinship groups. In the face of the disappearance of the territorial unit as a basis of social solidarity we create interest units. Meanwhile the city as a community resolves itself into a series of tenuous segmental relationships superimposed upon a territorial base with a definite center but without a definite periphery and upon a division of labor which far transcends the immediate locality and is world-wide

in scope. The larger the number of persons in a state of interaction with one another the lower is the level of communication and the greater is the tendency for communication to proceed on an elementary level, i.e. on the basis of those things which are assumed to be common or to be of interest to all.

It is obviously, therefore, to the emerging trends in the communication system and to the production and distribution technology that has come into existence with modern civilization that we must look for the symptoms which will indicate the probable future development of urbanism as a mode of social life. The direction of the ongoing changes in urbanism will for good or ill transform not only the city but the world. Some of the more basic of these factors and processes and the possibilities of their direction and control invite further detailed study.

It is only insofar as the sociologist has a clear conception of the city as a social entity and a workable theory of urbanism that he can hope to develop a unified body of reliable knowledge, which what passes as "urban sociology" is certainly not at the present time. By taking his point of departure from a theory of urbanism such as that sketched in the foregoing pages to be elaborated, tested, and revised in the light of further analysis and empirical research, it is to be hoped that the criteria of relevance and validity of factual data can be determined. The miscellaneous assortment of disconnected information which has hitherto found its way into sociological treatises on the city may thus be sifted and incorporated into a coherent body of knowledge. Incidentally, only by means of some such theory will the sociologists escape the futile practice of voicing in the name of sociological science a variety of often unsupportable judgments concerning such problems as poverty, housing, city-planning, sanitation, municipal administration, policing, marketing, transportation, and other technical issues. While the sociologist cannot solve any of these practical problems – at least not by himself – he may, if he discovers his proper function, have an important contribution to make to their comprehension and solution. The prospects for doing this are brightest through a general, theoretical, rather than through an *ad hoc* approach.

"Bowling Alone: America's Declining Social Capital"

Journal of Democracy (1995)

Robert D. Putnam

Editors' Introduction

Robert Putnam has been called "the most influential academic in the world today," and his work has been praised by political leaders Bill Clinton, Tony Blair, and George W. Bush. As a social critic, he stands in a tradition that includes Alexis de Tocqueville (*Democracy in America*, 1835), Paul Goodman (*Growing Up Absurd*, 1960), and Philip Slater (*The Pursuit of Loneliness: American Culture at the Breaking Point*, 1970). In this article from the *Journal of Democracy*, and in a subsequent book of the same title, Putnam asserts the "bowling alone" phenomenon as a metaphor for what urban life has become in contemporary middle-class America and for millions of others worldwide living in an increasingly materialistic and solipsistic culture of corporate work, obsessive consumption, and over-determined leisure.

Whereas cities once held out the promise of a wider, higher form of human community, Putnam argues that contemporary urbanites now follow a path of less, not more, civic engagement and that our collective stock of "social capital" – the meaningful human contacts of all kinds that characterize true communities – is so dangerously eroded that it verges on depletion.

In some ways, Putnam's critique is an updated version of Louis Wirth's 1938 essay "Urbanism as a Way of Life" (p. 97). Urban dwellers are not connected to one another through collective action as they once were. Instead, virtually all measures of social engagement – voting, participation in social organizations, active church membership, even friendships and family ties – seem to grow weaker every year. In part, this phenomenon arises from the well-known causes: social and geographic mobility, the decreasing importance of families as women join the corporate workforce, and the technological transformation of leisure. However understandable, Putnam argues, these forces have now risen to the level of social crisis and must be addressed by conscious policies to increase civic engagement of all sorts and to strengthen the connection between people in their roles as neighbors, co-workers, and fellow citizens.

Putnam's ideas about urban civic engagement emerge from a deep philosophical tradition that includes Peter Kropotkin's *Mutual Aid* (New York: Knopf, 1922, originally published in 1902) and John Gardner's "Building Community" (*Independent Sector*, 1991), and they raise questions as to whether any modern urban society can hope to regain the intimacy of the ancient Greek polis as described by H.D.F. Kitto (p. 43) or if the planning practices of New Urbanists such as Peter Calthorpe and William Fulton (p. 331) or Andres Duany and Elizabeth Plater-Zyberk (p. 207) can truly lead to the heightened degrees of "community" that Jane Jacobs describes in *The Death and Life of Great American Cities* (p. 114).

Robert Putnam is the Peter and Isabel Malkin Professor of Public Policy at Harvard University and has written widely in the fields of politics, comparative politics, international relations, and public policy. He attended Swarthmore and Balliol College, Oxford, before taking his Ph.D. at Yale in 1970. He is the author of *The Beliefs of Politicians: Ideology, Conflict, and Democracy in Britain and Italy* (New Haven, CT: Yale University Press, 1973) and, with Robert Leonardi and Rafaella Nanetti, *Making Democracy Work: Civic Traditions in Modern Italy*

(Princeton, NJ: Princeton University Press, 1993). Following the wide popular success of *Bowling Alone* (New York: Simon and Schuster, 2000), Putnam founded the Saguaro Seminar on Civic Engagement in America where he brings together community leaders and policy makers to develop working plans for increasing civic connectedness in specific urban contexts. He is also the editor of *Democracies in Flux: The Evolution of Social Capital in Contemporary Society* (New York: Oxford University Press, 2002).

Other sources of information on the idea of urban social capital include James S. Coleman's "Social Capital in the Creation of Human Capital" (*American Journal of Sociology*, 1988) and Robert Wuthnow's *Sharing the Journey: Support Groups and America's New Quest for Community* (New York: Free Press, 1994).

Many students of the new democracies that have emerged over the past decade and a half have emphasized the importance of a strong and active civil society to the consolidation of democracy. Especially with regard to the postcommunist countries, scholars and democratic activists alike have lamented the absence or obliteration of traditions of independent civic engagement and a widespread tendency toward passive reliance on the state. To those concerned with the weakness of civil societies in the developing or postcommunist world, the advanced Western democracies and above all the United States have typically been taken as models to be emulated. There is striking evidence, however, that the vibrancy of American civil society has notably declined over the past several decades.

Ever since the publication of Alexis de Tocqueville's *Democracy in America* [1835], the United States has played a central role in systematic studies of the links between democracy and civil society. Although this is in part because trends in American life are often regarded as harbingers of social modernization, it is also because America has traditionally been considered unusually "civic" (a reputation that, as we shall later see, has not been entirely unjustified).

When Tocqueville visited the United States in the 1830s, it was the Americans' propensity for civic association that most impressed him as the key to their unprecedented ability to make democracy work. "Americans of all ages, all stations in life, and all types of disposition," he observed, "are forever forming associations. There are not only commercial and industrial associations in which all take part, but others of a thousand different types – religious, moral, serious, futile, very general and very limited, immensely large and very minute. Nothing, in my view, deserves more attention than the intellectual and moral associations in America."

Recently, American social scientists of a neo-Tocquevillean bent have unearthed a wide range of empirical evidence that the quality of public life and the performance of social institutions (and not only in America) are indeed powerfully influenced by norms and networks of civic engagement. Researchers in such fields as education, urban poverty, unemployment, the control of crime and drug abuse, and even health have discovered that successful outcomes are more likely in civically engaged communities. Similarly, research on the varying economic attainments of different ethnic groups in the United States has demonstrated the importance of social bonds within each group. These results are consistent with research in a wide range of settings that demonstrates the vital importance of social networks for job placement and many other economic outcomes.

Meanwhile, a seemingly unrelated body of research on the sociology of economic development has also focused attention on the role of social networks. Some of this work is situated in the developing countries, and some of it elucidates the peculiarly successful "network capitalism" of East Asia. Even in less exotic Western economies, however, researchers have discovered highly efficient, highly flexible "industrial districts" based on networks of collaboration among workers and small entrepreneurs. Far from being paleo-industrial anachronisms, these dense interpersonal and interorganizational networks undergird ultramodern industries, from the high tech of Silicon Valley to the high fashion of Benetton.

The norms and networks of civic engagement also powerfully affect the performance of representative government. That, at least, was the central conclusion of my own 20-year, quasi-experimental study of subnational governments in different regions of Italy. Although all these regional governments seemed identical on paper, their levels of effectiveness varied dramatically. Systematic inquiry showed that the quality of governance was determined by longstanding traditions of civic engagement (or its absence). Voter turnout, newspaper readership, membership in choral societies and football clubs – these were the hallmarks of a successful region. In fact, historical analysis suggested that these networks of organized reciprocity and civic solidarity, far from being an epiphenomenon of socioeconomic modernization, were a precondition for it.

No doubt the mechanisms through which civic engagement and social connectedness produce such results – better schools, faster economic development, lower crime, and more effective government – are multiple and complex. While these briefly recounted findings require further confirmation and perhaps qualification, the parallels across hundreds of empirical studies in a dozen disparate disciplines and subfields are striking. Social scientists in several fields have recently suggested a common framework for understanding these phenomena, a framework that rests on the concept of social capital. By analogy with notions of physical capital and human capital – tools and training that enhance individual productivity – "social capital" refers to features of social organization such as networks, norms, and social trust that facilitate coordination and cooperation for mutual benefit.

For a variety of reasons, life is easier in a community blessed with a substantial stock of social capital. In the first place, networks of civic engagement foster sturdy norms of generalized reciprocity and encourage the emergence of social trust. Such networks facilitate coordination and communication, amplify reputations, and thus allow dilemmas of collective action to be resolved. When economic and political negotiation is embedded in dense networks of social interaction, incentives for opportunism are reduced. At the same time, networks of civic engagement embody past success at collaboration, which can serve as a cultural template for future collaboration. Finally, dense networks of interaction probably broaden the participants' sense of self, developing the "I" into the "we," or (in the language of rational-choice theorists) enhancing the participants' "taste" for collective benefits.

I do not intend here to survey (much less contribute to) the development of the theory of social capital. Instead, I use the central premise of that rapidly growing body of work – that social connections and civic engagement pervasively influence our public life, as well as our private prospects – as the starting point

for an empirical survey of trends in social capital in contemporary America. I concentrate here entirely on the American case, although the developments I portray may in some measure characterize many contemporary societies.

WHATEVER HAPPENED TO CIVIC ENGAGEMENT?

We begin with familiar evidence on changing patterns of political participation, not least because it is immediately relevant to issues of democracy in the narrow sense. Consider the well-known decline in turnout in national elections over the last three decades. From a relative high point in the early 1960s, voter turnout had by 1990 declined by nearly a quarter; tens of millions of Americans had forsaken their parents' habitual readiness to engage in the simplest act of citizenship. Broadly similar trends also characterize participation in state and local elections.

It is not just the voting booth that has been increasingly deserted by Americans. A series of identical questions posed by the Roper Organization to national samples ten times each year over the last two decades reveals that since 1973 the number of Americans who report that "in the past year" they have "attended a public meeting on town or school affairs" has fallen by more than a third (from 22 percent in 1973 to 13 percent in 1993). Similar (or even greater) relative declines are evident in responses to questions about attending a political rally or speech, serving on a committee of some local organization, and working for a political party. By almost every measure, Americans' direct engagement in politics and government has fallen steadily and sharply over the last generation, despite the fact that average levels of education – the best individual-level predictor of political participation – have risen sharply throughout this period. Every year over the last decade or two, millions more have withdrawn from the affairs of their communities.

Not coincidentally, Americans have also disengaged psychologically from politics and government over this era. The proportion of Americans who reply that they "trust the government in Washington" only "some of the time" or "almost never" has risen steadily from 30 percent in 1966 to 75 percent in 1992.

These trends are well known, of course, and taken by themselves would seem amenable to a strictly political explanation. Perhaps the long litany of political tragedies and scandals since the 1960s (assassinations, Vietnam, Watergate, Irangate, and so on) has triggered an understandable disgust for politics and government among Americans, and that in turn has motivated their withdrawal. I do not doubt that this common interpretation has some merit, but its limitations become plain when we examine trends in civic engagement of a wider sort.

Our survey of organizational membership among Americans can usefully begin with a glance at the aggregate results of the General Social Survey, a scientifically conducted, national-sample survey that has been repeated 14 times over the last two decades. Church-related groups constitute the most common type of organization joined by Americans; they are especially popular with women. Other types of organizations frequently joined by women include school-service groups (mostly parent–teacher associations), sports groups, professional societies, and literary societies. Among men, sports clubs, labor unions, professional societies, fraternal groups, veterans' groups, and service clubs are all relatively popular.

Religious affiliation is by far the most common associational membership among Americans. Indeed, by many measures America continues to be (even more than in Tocqueville's time) an astonishingly "churched" society. For example, the United States has more houses of worship per capita than any other nation on Earth. Yet religious sentiment in America seems to be becoming somewhat less tied to institutions and more self-defined.

How have these complex crosscurrents played out over the last three or four decades in terms of Americans' engagement with organized religion? The general pattern is clear: the 1960s witnessed a significant drop in reported weekly churchgoing – from roughly 48 percent in the late 1950s to roughly 41 percent in the early 1970s. Since then, it has stagnated or (according to some surveys) declined still further. Meanwhile, data from the General Social Survey show a modest decline in membership in all "church-related groups" over the last 20 years. It would seem, then, that net participation by Americans, both in religious services and in church-related groups, has declined modestly (by perhaps a sixth) since the 1960s.

For many years, labor unions provided one of the most common organizational affiliations among American workers. Yet union membership has been falling for nearly four decades, with the steepest decline occurring between 1975 and 1985. Since the mid-1950s, when union membership peaked, the unionized portion of the nonagricultural workforce in America has dropped by more than half, falling from 32.5 percent in 1953 to 15.8 percent in 1992. By now, virtually all of the explosive growth in union membership that was associated with the New Deal has been erased. The solidarity of union halls is now mostly a fading memory of aging men.

The parent–teacher association (PTA) has been an especially important form of civic engagement in twentieth-century America because parental involvement in the educational process represents a particularly productive form of social capital. It is, therefore, dismaying to discover that participation in parent–teacher organizations has dropped drastically over the last generation, from more than 12 million in 1964 to barely 5 million in 1982 before recovering to approximately 7 million now.

Next, we turn to evidence on membership in (and volunteering for) civic and fraternal organizations. These data show some striking patterns. First, membership in traditional women's groups has declined more or less steadily since the mid-1960s. For example, membership in the national Federation of Women's Clubs is down by more than half (59 percent) since 1964, while membership in the League of Women Voters (LWV) is off 42 percent since 1969.

Similar reductions are apparent in the numbers of volunteers for mainline civic organizations, such as the Boy Scouts (off by 26 percent since 1970) and the Red Cross (off by 61 percent since 1970). But what about the possibility that volunteers have simply switched their loyalties to other organizations? Evidence on "regular" (as opposed to occasional or "drop-by") volunteering is available from the Labor Department's Current Population Surveys of 1974 and 1989. These estimates suggest that serious volunteering declined by roughly one-sixth over these 15 years, from 24 percent of adults in 1974 to 20 percent in 1989. The multitudes of Red Cross aides and Boy Scout troop leaders now missing in action have apparently not been offset by equal numbers of new recruits elsewhere.

Fraternal organizations have also witnessed a substantial drop in membership during the 1980s and 1990s. Membership is down significantly in such groups as the Lions (off 12 percent since 1983), the Elks (off 18 percent since 1979), the Shriners (off 27 percent since 1979), the Jaycees (off 44 percent since 1979), and the Masons (down 39 percent since 1959). In sum, after expanding steadily throughout most of this century, many major civic organizations have experienced a sudden, substantial, and nearly simultaneous decline in membership over the last decade or two.

The most whimsical yet discomfiting bit of evidence of social disengagement in contemporary America that I have discovered is this: more Americans are bowling today than ever before, but bowling in organized leagues has plummeted in the last decade or so. Between 1980 and 1993 the total number of bowlers in America increased by 10 percent, while league bowling decreased by 40 percent. (Lest this be thought a wholly trivial example, I should note that nearly 80 million Americans went bowling at least once during 1993, nearly a third more than voted in the 1994 congressional elections and roughly the same number as claim to attend church regularly. Even after the 1980s' plunge in league bowling, nearly 3 percent of American adults regularly bowl in leagues.) The rise of solo bowling threatens the livelihood of bowling-lane proprietors because those who bowl as members of leagues consume three times as much beer and pizza as solo bowlers, and the money in bowling is in the beer and pizza, not the balls and shoes. The broader social significance, however, lies in the social interaction and even occasionally civic conversations over beer and pizza that solo bowlers forgo. Whether or not bowling beats balloting in the eyes of most Americans, bowling teams illustrate yet another vanishing form of social capital.

COUNTERTRENDS

At this point, however, we must confront a serious counterargument. Perhaps the traditional forms of civic organization whose decay we have been tracing have been replaced by vibrant new organizations. For example, national environmental organizations (like the Sierra Club) and feminist groups (like the National Organization for Women) grew rapidly during the 1970s and 1980s and now count hundreds of thousands of dues-paying members. An even more dramatic example is the American Association of Retired Persons (AARP), which grew exponentially from 400,000 card-carrying members in 1960 to 33 million in 1993, becoming (after the Catholic Church) the largest private organization in the world. The national administrators of these organizations are among the most feared lobbyists in Washington, in large part because of their massive mailing lists of presumably loyal members.

These new mass-membership organizations are plainly of great political importance. From the point of view of social connectedness, however, they are sufficiently different from classic "secondary associations" that we need to invent a new label – perhaps "tertiary associations." For the vast majority of their members, the only act of membership consists in writing a check for dues or perhaps occasionally reading a newsletter. Few ever attend any meetings of such organizations, and most are unlikely ever (knowingly) to encounter any other member. The bond between any two members of the Sierra Club is less like the bond between any two members of a gardening club and more like the bond between any two Red Sox fans (or perhaps any two devoted Honda owners): they root for the same team and they share some of the same interests, but they are unaware of each other's existence. Their ties, in short, are to common symbols, common leaders, and perhaps common ideals, but not to one another. The theory of social capital argues that associational membership should, for example, increase social trust, but this prediction is much less straightforward with regard to membership in tertiary associations. From the point of view of social connectedness, the Environmental Defense Fund and a bowling league are just not in the same category.

If the growth of tertiary organizations represents one potential (but probably not real) counterexample to my thesis, a second countertrend is represented by the growing prominence of nonprofit organizations, especially nonprofit service agencies. This so-called third sector includes everything from Oxfam and the Metropolitan Museum of Art to the Ford Foundation and the Mayo Clinic. In other words, although most secondary associations are nonprofits, most nonprofit agencies are not secondary associations. To identify trends in the size of the nonprofit sector with trends in social connectedness would be another fundamental conceptual mistake.

A third potential countertrend is much more relevant to an assessment of social capital and civic engagement. Some able researchers have argued that the last few decades have witnessed a rapid expansion in "support groups" of various sorts. Robert Wuthnow reports that fully 40 percent of all Americans claim to be "currently involved in [a] small group that meets regularly and provides support or caring for those who participate in it." Many of these groups are religiously affiliated, but many others are not. For example, nearly 5 percent of Wuthnow's national sample claim to participate regularly in a "self-help" group, such as Alcoholics Anonymous, and nearly as many say they belong to book-discussion groups and hobby clubs.

The groups described by Wuthnow's respondents unquestionably represent an important form of social capital, and they need to be accounted for in any serious reckoning of trends in social connectedness. On the other hand, they do not typically play the same role as traditional civic associations. As Wuthnow emphasizes,

> Small groups may not be fostering community as effectively as many of their proponents would like. Some small groups merely provide occasions for individuals to focus on themselves in the presence of others. The social contract binding members together asserts only the weakest of obligations. Come if you have time. Talk if you feel like it. Respect everyone's opinion. Never criticize. Leave quietly if you become dissatisfied . . . We can imagine that [these small groups] really substitute for families, neighborhoods, and broader community attachments that may demand lifelong commitments, when, in fact, they do not.

All three of these potential countertrends – tertiary organizations, nonprofit organizations, and support groups – need somehow to be weighed against the erosion of conventional civic organizations. One way of doing so is to consult the General Social Survey.

Within all educational categories, total associational membership declined significantly between 1967 and 1993. Among the college-educated, the average number of group memberships per person fell from 2.8 to 2.0 (a 26 percent decline); among high-school graduates, the number fell from 1.8 to 1.2 (32 percent); and among those with fewer than 12 years of education, the number fell from 1.4 to 1.1 (25 percent). In other words, at all educational (and hence social) levels of American society, and counting all sorts of group memberships, the average number of associational memberships has fallen by about a fourth over the last quarter-century. Without controls for educational levels, the trend is not nearly so clear, but the central point is this: more Americans than ever before are in social circumstances that foster associational involvement (higher education, middle age, and so on), but nevertheless aggregate associational membership appears to be stagnant or declining.

Broken down by type of group, the downward trend is most marked for church-related groups, for labor unions, for fraternal and veterans' organizations, and for school-service groups. Conversely, membership in professional associations has risen over these years, although less than might have been predicted, given sharply rising educational and occupational levels. Essentially the same trends are evident for both men and women in the sample. In short, the available survey evidence confirms our earlier conclusion: American social capital in the form of civic associations has significantly eroded over the last generation.

GOOD NEIGHBORLINESS AND SOCIAL TRUST

I noted earlier that most readily available quantitative evidence on trends in social connectedness involves formal settings, such as the voting booth, the union hall, or the PTA. One glaring exception is so widely discussed as to require little comment here: the most fundamental form of social capital is the family, and the massive evidence of the loosening of bonds within the family (both extended and nuclear) is well known. This trend, of course, is quite consistent with – and may help to explain – our theme of social decapitalization.

A second aspect of informal social capital on which we happen to have reasonably reliable time-series data involves neighborliness. In each General Social Survey since 1974 respondents have been asked, "How often do you spend a social evening with a neighbor?" The proportion of Americans who socialize with their neighbors more than once a year has slowly but steadily declined over the last two decades, from 72 percent in 1974 to 61 percent in 1993. (On the other hand, socializing with "friends who do not live in your neighborhood" appears to be on the increase, a trend that may reflect the growth of workplace-based social connections.)

Americans are also less trusting. The proportion of Americans saying that most people can be trusted fell by more than a third between 1960, when 58 percent chose that alternative, and 1993, when only 37 percent did. The same trend is apparent in all educational groups; indeed, because social trust is also correlated with education and because educational levels have risen sharply, the overall decrease in social trust is even more apparent if we control for education.

Our discussion of trends in social connectedness and civic engagement has tacitly assumed that all the forms of social capital that we have discussed are themselves coherently correlated across individuals. This is in fact true. Members of associations are much more likely than nonmembers to participate in politics, to spend time with neighbors, to express social trust, and so on.

The close correlation between social trust and associational membership is true not only across time and across individuals, but also across countries. Evidence from the 1991 World Values Survey demonstrates the following: across the 35 countries in this survey, social trust and civic engagement are strongly correlated; the greater the density of associational membership in a society, the more trusting its citizens. Trust and engagement are two facets of the same underlying factor – social capital.

America still ranks relatively high by cross-national standards on both these dimensions of social capital. Even in the 1990s, after several decades' erosion, Americans are more trusting and more engaged than people in most other countries of the world. The trends of the past quarter-century, however, have apparently moved the United States significantly lower in the international rankings of social capital. The recent deterioration in American social capital has been sufficiently great that (if no other country changed its position in the meantime) another quarter-century of change at the same rate would bring the United States, roughly speaking, to the midpoint among all these countries, roughly equivalent to South Korea, Belgium, or Estonia today. Two generations' decline at the same rate would leave the United States at the level of today's Chile, Portugal, and Slovenia.

WHY IS US SOCIAL CAPITAL ERODING?

As we have seen, something has happened in America in the last two or three decades to diminish civic engagement and social connectedness. What could that "something" be? Here are several possible explanations, along with some initial evidence on each.

The movement of women into the labor force. Over these same two or three decades, many millions of American women have moved out of the home into paid employment. This is the primary, though not the sole, reason why the weekly working hours of the average American have increased significantly during these years. It seems highly plausible that this social revolution should have reduced the time and energy available for building social capital. For certain organizations, such as the PTA, the League of Women Voters, the Federation of Women's Clubs, and the Red Cross, this is almost certainly an important part of the story. The sharpest decline in women's civic participation seems to have come in the 1970s; membership in such "women's" organizations as these has been virtually halved since the late 1960s. By contrast, most of the decline in participation in men's organizations occurred about 10 years later; the total decline to date has been approximately 25 percent for the typical organization. On the other hand, the survey data imply that the aggregate declines for men are virtually as great as those for women. It is logically possible, of course, that the male declines might represent the knock-on effect of women's liberation, as dishwashing crowded out the lodge, but time-budget studies suggest that most husbands of working wives have assumed only a minor part of the housework. In short, something besides the women's revolution seems to lie behind the erosion of social capital.

Mobility: The "re-potting" hypothesis

Numerous studies of organizational involvement have shown that residential stability and such related phenomena as homeownership are clearly associated with greater civic engagement. Mobility, like frequent re-potting of plants, tends to disrupt root systems, and it takes time for an uprooted individual to put down new roots. It seems plausible that the automobile, suburbanization, and the movement to the Sun Belt have reduced the social rootedness of the average American, but one fundamental difficulty with this hypothesis is apparent: the best evidence shows that residential stability and homeownership in America have risen modestly since 1965, and are surely higher now than during the 1950s, when civic engagement and social connectedness by our measures was definitely higher.

Other demographic transformations

A range of additional changes have transformed the American family since the 1960s – fewer marriages, more divorces, fewer children, lower real wages, and so on. Each of these changes might account for some of the slackening of civic engagement, since married, middle-class parents are generally more socially involved than other people. Moreover, the changes in scale that have swept over the American economy in these years – illustrated by the replacement of the corner grocery by the supermarket and now perhaps of the supermarket by electronic shopping at home, or the replacement of community-based enterprises by outposts of distant multinational firms – may perhaps have undermined the material and even physical basis for civic engagement.

The technological transformation of leisure

There is reason to believe that deep-seated techno-logical trends are radically "privatizing" or "individual-izing" our use of leisure time and thus disrupting many opportunities for social-capital formation. The most obvious and probably the most powerful instrument of this revolution is television. Time-budget studies in the 1960s showed that the growth in time spent watching television dwarfed all other changes in the way Americans passed their days and nights. Tele-vision has made our communities (or, rather, what we experience as our communities) wider and shallower. In the language of economics, electronic technology enables individual tastes to be satisfied more fully, but at the cost of the positive social externalities associated with more primitive forms of entertainment. The same logic applies to the replacement of vaudeville by the movies and now of movies by the VCR. The new "virtual reality" helmets that we will soon don to be entertained in total isolation are merely the latest extension of this trend. Is technology thus driving a wedge between our individual interests and our collective interests? It is a question that seems worth exploring more systematically.

WHAT IS TO BE DONE?

The last refuge of a social-scientific scoundrel is to call for more research. Nevertheless, I cannot forbear from suggesting some further lines of inquiry.

We must sort out the dimensions of social capital, which clearly is not a unidimensional concept, despite language (even in this essay) that implies the contrary.

What types of organizations and networks most effectively embody – or generate – social capital, in the sense of mutual reciprocity, the resolution of dilemmas of collective action, and the broadening of social identities? In this essay I have emphasized the density of associational life. In earlier work I stressed the structure of networks, arguing that "horizontal" ties represented more productive social capital than vertical ties.

Another set of important issues involves macro-sociological crosscurrents that might intersect with the trends described here. What will be the impact, for example, of electronic networks on social capital? My hunch is that meeting in an electronic forum is not the equivalent of meeting in a bowling alley – or even in a saloon – but hard empirical research is needed. What about the development of social capital in the workplace? Is it growing in counterpoint to the decline of civic engagement, reflecting some social analogue of the first law of thermodynamics – social capital is neither created nor destroyed, merely redistributed? Or do the trends described in this essay represent a deadweight loss?

A rounded assessment of changes in American social capital over the last quarter-century needs to count the costs as well as the benefits of community engagement. We must not romanticize small-town, middle-class civic life in the America of the 1950s. In addition to the deleterious trends emphasized in this essay, recent decades have witnessed a substantial decline in intolerance and probably also in overt discrimination, and those beneficent trends may be related in complex ways to the erosion of traditional social capital. Moreover, a balanced accounting of the social-capital books would need to reconcile the insights of this approach with the undoubted insights offered by Mancur Olson and others who stress that closely knit social, economic, and political organ-izations are prone to inefficient cartelization and to what political economists term "rent seeking" and ordinary men and women call corruption.

Finally, and perhaps most urgently, we need to explore creatively how public policy impinges on (or might impinge on) social-capital formation. In some well-known instances, public policy has destroyed highly effective social networks and norms. American slum-clearance policy of the 1950s and 1960s, for example, renovated physical capital, but at a very high cost to existing social capital. The consolidation of country post offices and small school districts has promised administrative and financial efficiencies, but full-cost accounting for the effects of these policies on social capital might produce a more negative verdict. On the other hand, such past initiatives as the county agricultural-agent system, community colleges, and tax deductions for charitable contributions illustrate that government can encourage social-capital formation. Even a recent proposal in San Luis Obispo, California, to require that all new houses have front porches illustrates the power of government to influence where and how networks are formed.

The concept of "civil society" has played a central role in the recent global debate about the preconditions for democracy and democratization. In the newer democracies this phrase has properly focused attention on the need to foster a vibrant civic life in soils traditionally inhospitable to self-government. In the established democracies, ironically, growing numbers of citizens are questioning the effectiveness of their public institutions at the very moment when liberal democracy has swept the battlefield, both ideologically and geopolitically. In America, at least, there is reason to suspect that this democratic disarray may be linked to a broad and continuing erosion of civic engagement that began a quarter-century ago. High on our scholarly agenda should be the question of whether a comparable erosion of social capital may be under way in other advanced democracies, perhaps in different institutional and behavioral guises. High on America's agenda should be the question of how to reverse these adverse trends in social connectedness, thus restoring civic engagement and civic trust.

"The Uses of Sidewalks: Safety"

from *The Death and Life of Great American Cities* (1961)

Jane Jacobs

Editors' Introduction

Jane Jacobs started writing about city life and urban planning as a neighborhood activist, not as a trained professional. Dismissed as the original "little old lady in tennis shoes" and derided as a political amateur more concerned about personal safety issues than state-of-the-art planning techniques, she nonetheless struck a responsive chord with a 1960s public eager to believe the worst about arrogant city planning technocrats and just as eager to rally behind movements for neighborhood control and community resistance to bulldozer redevelopment.

The Death and Life of Great American Cities hit the world of city planning like an earthquake when it appeared in 1961. The book was a frontal attack on the planning establishment, especially on the massive urban renewal projects that were being carried out by powerful redevelopment bureaucrats like Robert Moses in New York. Jacobs derided urban renewal as a process that only served to create instant slums. She questioned universally accepted articles of faith – for example that parks were good and that crowding was bad. Indeed she suggested that parks were often dangerous and that crowded neighborhood sidewalks were the safest places for children to play. Jacobs ridiculed the planning establishment's most revered historical traditions as "the Radiant Garden City Beautiful" – an artful phrase that not only airily dismissed the contributions of Le Corbusier (p. 317), Ebenezer Howard (p. 309), and Daniel Burnham but lumped them together as well!

The selection from *The Death and Life of Great American Cities* reprinted here presents Jane Jacobs at her very best. In "The Uses of Sidewalks: Safety," she outlines her basic notions of what makes a neighborhood a community and what makes a city livable. Safety – particularly for women and children – comes from "eyes on the street," the kind of involved neighborhood surveillance of public space that modern planning practice in the Corbusian tradition had destroyed with its insistence on superblocks and skyscraper developments. A sense of personal belonging and social cohesiveness comes from well-defined neighborhoods and narrow, crowded, multi-use streets. And, finally, basic urban vitality comes from residents' participation in an intricate "street ballet," a diurnal pattern of observable and comprehensible human activity that is possible only in places like Jacobs' own Hudson Street in her beloved Greenwich Village.

It was this last quality, her unabashed love of cities and urban life, that is Jane Jacobs' most obvious and enduring characteristic. *The Death and Life of Great American Cities* was a scathing attack on the planning establishment – and, in many ways, it was a grassroots political call to arms – but it was also a loving invitation to experience the joys of city living that led many young, college-educated people to seek out neighborhoods like Greenwich Village as places to live, struggle, and raise families. In one sense, the book encouraged and justified middle-class gentrification of formerly working-class neighborhoods. In another, it found itself oddly reflected in the fantasy-nostalgia of "Sesame Street." But in all ways it was committedly urban, never sub-urban, at a time when inner-city communities were being increasingly abandoned to the forces of poverty, decay, and neglect.

Contrast Louis Wirth's theory of how population size, density, and heterogeneity in cities create a distinct urban personality (p. 97) with Jacobs argument that these very same city characteristics may create neighborhood vitality, social cohesion, and the perception and reality of safety. Jacobs' notion of the "street ballet" invites comparison with Lewis Mumford's idea of the "urban drama" (p. 92), William Whyte's emphasis on the importance of public plazas (p. 429), and Robert Putnam's emphasis on "social capital" (p. 105). Jacobs' community activism in resistance to urban renewal places her within a long tradition that includes Paul Davidoff's "Advocacy and Pluralism in Planning" (p. 388) and Sherry Arnstein's "A Ladder of Citizen Participation" (p. 244).

Other important works by Jane Jacobs include *The Economy of Cities* (New York: Random House, 1969) and *Systems of Survival* (New York: Random House, 1992). In the former book Jacobs again turns conventional explanation on its head by arguing that the rise of cities may have proceeded, and even accounted for, rural agricultural development. The latter is a Platonic dialogue on "the moral foundations of commerce and politics." Max Allen, *Ideas that Matter: The Worlds of Jane Jacobs* (New York: Ginger Press, 1997) is of interest, and Robert A. Caro's masterful study of Jacobs's archenemy, *The Power Broker: Robert Moses and the Fall of New York* (New York: Random House, 1975), is essential.

Streets in cities serve many purposes besides carrying vehicles, and city sidewalks – the pedestrian parts of the streets – serve many purposes besides carrying pedestrians. These uses are bound up with circulation but are not identical with it and in their own right they are at least as basic as circulation to the proper workings of cities.

A city sidewalk by itself is nothing. It is an abstraction. It means something only in conjunction with the buildings and other uses that border it, or border other sidewalks very near it. The same might be said of streets, in the sense that they serve other purposes besides carrying wheeled traffic in their middles. Streets and their sidewalks, the main public places of a city, are its most vital organs. Think of a city and what comes to mind? Its streets. If a city's streets look interesting, the city looks interesting; if they look dull, the city looks dull.

More than that, and here we get down to the first problem, if a city's streets are safe from barbarism and fear, the city is thereby tolerably safe from barbarism and fear. When people say that a city, or a part of it, is dangerous or is a jungle what they mean primarily is that they do not feel safe on the sidewalks. But sidewalks and those who use them are not passive beneficiaries of safety or helpless victims of danger. Sidewalks, their bordering uses, and their users, are active participants in the drama of civilization versus barbarism in cities. To keep the city safe is a fundamental task of a city's streets and its sidewalks.

This task is totally unlike any service that sidewalks and streets in little towns or true suburbs are called upon to do. Great cities are not like towns, only larger. They are not like suburbs, only denser. They differ from towns and suburbs in basic ways, and one of these is that cities are, by definition, full of strangers. To any one person, strangers are far more common in big cities than acquaintances. More common not just in places of public assembly, but more common at a man's own doorstep. Even residents who live near each other are strangers, and must be, because of the sheer number of people in small geographical compass.

The bedrock attribute of a successful city district is that a person must feel personally safe and secure on the street among all these strangers. He must not feel automatically menaced by them. A city district that fails in this respect also does badly in other ways and lays up for itself, and for its city at large, mountain on mountain of trouble.

Today barbarism has taken over many city streets, or people fear it has, which comes to much the same thing in the end. "I live in a lovely, quiet residential area," says a friend of mine who is hunting another place to live. "The only disturbing sound at night is the occasional scream of someone being mugged." It does not take many incidents of violence on a city street, or in a city district, to make people fear the streets . . . And as they fear them, they use them less, which makes the streets still more unsafe.

To be sure, there are people with hobgoblins in their heads, and such people will never feel safe no matter what the objective circumstances are. But this is a different matter from the fear that besets normally prudent, tolerant and cheerful people who show nothing more than common sense in refusing to venture after dark – or in a few places, by day – into streets where they may well be assaulted, unseen or unrescued until too late. The barbarism and the real, not imagined, insecurity that gives rise to such fears cannot be tagged a problem of the slums. The problem is most serious, in fact, in genteel-looking "quiet residential areas" like that my friend was leaving.

It cannot be tagged as a problem of older parts of cities. The problem reaches its most baffling dimensions in some examples of rebuilt parts of cities, including supposedly the best examples of rebuilding, such as middle-income projects. The police precinct captain of a nationally admired project of this kind (admired by planners and lenders) has recently admonished residents not only about hanging around outdoors after dark but has urged them never to answer their doors without knowing the caller. Life here has much in common with life for the three little pigs or the seven little kids of the nursery thrillers. The problem of sidewalk and doorstep insecurity is as serious in cities which have made conscientious efforts at rebuilding as it is in those cities that have lagged. Nor is it illuminating to tag minority groups, or the poor, or the outcast with responsibility for city danger. There are immense variations in the degree of civilization and safety found among such groups and among the city areas where they live. Some of the safest sidewalks in New York City, for example, at any time of day or night, are those along which poor people or minority groups live. And some of the most dangerous are in streets occupied by the same kinds of people. All this can also be said of other cities.

[. . .]

The first thing to understand is that the public peace – the sidewalk and street peace – of cities is not kept primarily by the police, necessary as police are. It is kept primarily by an intricate, almost unconscious, network of voluntary controls and standards among the people themselves, and enforced by the people themselves. In some city areas – older public housing projects and streets with very high population turnover are often conspicuous examples – the keeping of public sidewalk law and order is left almost entirely to the police and special guards. Such places are jungles. No amount of police can enforce civilization where the normal, casual enforcement of it has broken down.

The second thing to understand is that the problem of insecurity cannot be solved by spreading people out more thinly, trading the characteristics of cities for the characteristics of suburbs. If this could solve danger on the city streets, then Los Angeles should be a safe city because superficially Los Angeles is almost all suburban. It has virtually no districts compact enough to qualify as dense city areas. Yet Los Angeles cannot, any more than any other great city, evade the truth that, being a city, it is composed of strangers not all of whom are nice. Los Angeles' crime figures are flabbergasting. Among the seventeen standard metropolitan areas with populations over a million, Los Angeles stands so pre-eminent in crime that it is in a category by itself. And this is markedly true of crimes associated with personal attack, the crimes that make people fear the streets.

[. . .]

This is something everyone already knows: A well-used city street is apt to be a safe street. A deserted city street is apt to be unsafe. But how does this work, really? And what makes a city street well used or shunned? . . . What about streets that are busy part of the time and then empty abruptly?

A city street equipped to handle strangers, and to make a safety asset, in itself, out of the presence of strangers, as the streets of successful city neighborhoods always do, must have three main qualities:

First, there must be a clear demarcation between what is public space and what is private space. Public and private spaces cannot ooze into each other as they do typically in suburban settings or in projects.

Second, there must be eyes upon the street, eyes belonging to those we might call the natural proprietors of the street. The buildings on a street equipped to handle strangers and to insure the safety of both residents and strangers must be oriented to the street. They cannot turn their backs or blank sides on it and leave it blind.

And third, the sidewalk must have users on it fairly continuously, both to add to the number of effective eyes on the street and to induce the people in buildings along the street to watch the sidewalks in sufficient numbers. Nobody enjoys sitting on a stoop or looking out a window at an empty street. Almost nobody does such a thing. Large numbers of people entertain themselves, off and on, by watching street activity.

In settlements that are smaller and simpler than big cities, controls on acceptable public behavior, if not on crime, seem to operate with greater or lesser success through a web of reputation, gossip, approval, disapproval and sanctions, all of which are powerful if people know each other and word travels. But a city's streets, which must control the behavior not only of the people of the city but also of visitors from suburbs and towns who want to have a big time away from the gossip and sanctions at home, have to operate by more direct, straightforward methods. It is a wonder cities have solved such an inherently difficult problem at all. And yet in many streets they do it magnificently.

It is futile to try to evade the issue of unsafe city streets by attempting to make some other features of a locality, say interior courtyards, or sheltered play spaces, safe instead. By definition again, the streets of a city must do most of the job of handling strangers, for this is where strangers come and go. The streets must not only defend the city against predatory strangers, they must protect the many, many peaceable and well-meaning strangers who use them, insuring their safety too as they pass through. Moreover, no normal person can spend his life in some artificial haven, and this includes children. Everyone must use the streets.

On the surface, we seem to have here some simple aims: to try to secure streets where the public space is unequivocally public, physically unmixed with private or with nothing-at-all space, so that the area needing surveillance has clear and practicable limits; and to see that these public street spaces have eyes on them as continuously as possible.

But it is not so simple to achieve these objects, especially the latter. You can't make people use streets they have no reason to use. You can't make people watch streets they do not want to watch. Safety on the streets by surveillance and mutual policing of one another sounds grim, but in real life it is not grim. The safety of the street works best, most casually, and with least frequent taint of hostility or suspicion precisely where people are using and most enjoying the city streets voluntarily and are least conscious, normally, that they are policing.

The basic requisite for such surveillance is a substantial quantity of stores and other public places sprinkled along the sidewalks of a district; enterprises and public places that are used by evening and night must be among them especially. Stores, bars and restaurants, as the chief examples, work in several different and complex ways to abet sidewalk safety.

First, they give people – both residents and strangers – concrete reasons for using the sidewalks on which these enterprises face.

Second, they draw people along the sidewalks past places which have no attractions to public use in themselves but which become traveled and peopled as routes to somewhere else; this influence does not carry very far geographically, so enterprises must be frequent in a city district if they are to populate with walkers those other stretches of street that lack public places along the sidewalk. Moreover, there should be many different kinds of enterprises, to give people reasons for crisscrossing paths.

Third, storekeepers and other small businessmen are typically strong proponents of peace and order themselves; they hate broken windows and hold-ups; they hate having customers made nervous about safety. They are great street watchers and sidewalk guardians if present in sufficient numbers.

Fourth, the activity generated by people on errands, or people aiming for food or drink, is itself an attraction to still other people.

This last point, that the sight of people attracts still other people, is something that city planners and city architectural designers seem to find incomprehensible. They operate on the premise that city people seek the sight of emptiness, obvious order and quiet. Nothing could be less true. People's love of watching activity and other people is constantly evident in cities everywhere.

[. . .]

Under the seeming disorder of the old city, wherever the old city is working successfully, is a marvelous order for maintaining the safety of the streets and the freedom of the city. It is a complex order. Its essence is intricacy of sidewalk use, bringing with it a constant succession of eyes. This order is all composed of movement and change, and although it is life, not art, we may fancifully call it the art form of the city and liken it to the dance – not to a simple-minded precision dance with everyone kicking up at the same time, twirling in unison and bowing off en masse, but to an intricate ballet in which the individual dancers and ensembles all have distinctive parts which miraculously reinforce each other and compose an orderly whole. The ballet of the good city sidewalk never repeats itself from place to place, and in any one place is always replete with new improvisations.

The stretch of Hudson Street where I live is each day the scene of an intricate sidewalk ballet. I make my own first entrance into it a little after eight when I put out the garbage can, surely a prosaic occupation, but I enjoy my part, my little clang, as the droves of junior high school students walk by the center of the stage dropping candy wrappers. (How do they eat so much candy so early in the morning?)

While I sweep up the wrappers I watch the other rituals of morning: Mr. Halpert unlocking the laundry's handcart from its mooring to a cellar door, Joe Cornacchia's son-in-law stacking out the empty crates from the delicatessen, the barber bringing out his sidewalk folding chair, Mr. Goldstein arranging the coils of wire which proclaim the hardware store is open, the wife of the tenement's superintendent depositing her chunky 3-year-old with a toy mandolin on the stoop, the vantage point from which he is learning the English his mother cannot speak. Now the primary children, heading for St. Luke's, dribble through to the south; the children for St. Veronica's cross, heading to the west, and the children for P.S. 41, heading toward the east. Two new entrances are being made from the wings: well-dressed and even elegant women and men with briefcases emerge from doorways and side streets . . . Most of these are heading for the bus and subways, but some hover on the curbs, stopping taxis which have miraculously appeared at the right moment, for the taxis are part of a wider morning ritual: having dropped passengers from midtown in the downtown financial district, they are now bringing downtowners up to midtown. Simultaneously, numbers of women in housedresses have emerged and as they crisscross with one another they pause for quick conversations that sound with either laughter or joint indignation; never, it seems, anything between. It is time for me to hurry to work too, and I exchange my ritual farewell with Mr. Lofaro, the short, thick-bodied, white-aproned fruit man who stands outside his doorway a little up the street, his arms folded, his feet planted, looking solid as earth itself. We nod; we each glance quickly up and down the street then look back to each other and smile. We have done this many a morning for more than ten years, and we both know what it means: All is well.

[. . .]

I know the deep night ballet and its seasons best from waking; long after midnight to tend a baby and,

sitting in the dark, seeing the shadows and hearing the sounds of the sidewalk. Mostly it is a sound like infinitely pattering snatches of party conversation and, about three in the morning, singing, very good singing. Sometimes there is sharpness and anger or sad, sad weeping, or a flurry of search for a string of beads broken. One night, a young man came roaring along, bellowing terrible language at two girls whom he had apparently picked up and who were disappointing him. Doors opened; a wary semicircle formed around him, not too close, until the police came. Out came the heads, too, along Hudson Street, offering opinion, "Drunk . . . Crazy . . . A wild kid from the suburbs." (He turned out to be a wild kid from the suburbs. Sometimes, on Hudson Street, we are tempted to believe the suburbs must be a difficult place to bring up children.)

I have made the daily ballet of Hudson Street sound more frenetic than it is, because writing it telescopes it. In real life, it is not that way. In real life, to be sure, something is always going on, the ballet is never at a halt, but the general effect is peaceful and the general tenor even leisurely. People who know well such animated city streets will know how it is. I am afraid people who do not will always have it a little wrong in their heads like the old prints of rhinoceroses made from travelers' descriptions of rhinoceroses. On Hudson Street, the same as in the North End of Boston or in any other animated neighborhoods of great cities, we are not innately more competent at keeping the sidewalks safe than are the people who try to live off the hostile truce of Turf in a blind-eyed city. We are the lucky possessors of a city order that makes it relatively simple to keep the peace because there are plenty of eyes on the street. But there is nothing simple about that order itself, or the bewildering number of components that go into it. Most of those components are specialized in one way or another. They unite in their joint effect upon the sidewalk, which is not specialized in the least. That is its strength.

"The Negro Problems of Philadelphia," "The Question of Earning a Living," and "Color Prejudice"

from *The Philadelphia Negro* (1899)

W.E.B. Du Bois

Editors' Introduction

William Edward Burghardt Du Bois (1868–1963) was one of the preeminent intellectuals of his generation. As a professor, editor, author, novelist, playwright, and politician he made notable contributions in history, sociology, ethnic studies, literature, politics, and other fields. A brilliant student, Du Bois excelled at Fisk University in Nashville, Tennessee, the University of Berlin where he studied with the great sociologist Max Weber, and at Harvard University, where in 1895 he obtained the first Ph.D. degree Harvard awarded to an African-American.

Du Bois defies easy classification. He was always an independent and critical thinker. During his long and varied career he was a pan-Africanist who advocated solidarity among Black Africans and Blacks elsewhere in the world; a radical pacifist who was indicted, tried, and acquitted as an unregistered foreign agent during the McCarthy era for circulating the Stockholm peace plan; a humanist who wrote novels and plays and published many of the writers of the "Harlem Renaissance"; a civil rights leader who founded the NAACP's publication *Crisis* in 1910 and served as its influential editor until 1934; a writer of children's books which taught Black pride; and a world political figure who urged United Nations protection for Black Americans as a nation within a nation. Du Bois joined the Communist Party at age 93 and became a Ghanaian citizen just before his death in 1963.

At the time that Du Bois completed his education, Philadelphia had the largest and oldest settlement of African-Americans in the Northern United States. The settlement house movement was under way, and some well-intentioned Philadelphians were concerned to understand "the Negro problem" and to help the many poor Blacks in the city. Two wealthy leaders of Philadelphia society suggested a study of Negroes in the Seventh Ward, the city's Black ghetto.

Du Bois was given a one-year appointment as an assistant instructor in the Sociology Department at the University of Pennsylvania. Living with his bride of three months in one room over a cafeteria in the worst part of Philadelphia's worst Black ghetto, with no contact with students and little with faculty, Du Bois wrote *The Philadelphia Negro* from which the following selection is taken. He was only 31 when his monumental study was published.

While Du Bois found many problems in Philadelphia's African-American community in the 1890s, there was work available for able-bodied laborers, no evidence of drug use, substantial homeownership, middle- and upper-income craftspeople, businessmen, and professionals to serve the community and act as role models, and little Black-on-Black violent crime. This is in marked contrast to William Julius Wilson's description of poor Black ghetto areas of Chicago in the 1980s (p. 126). Wilson describes "underclass" ghettos in Chicago consisting almost entirely of renters (many in public housing), with very few employed residents, extremely high concentrations of single-parent families, welfare dependency, drug use, and violent crime.

Ethnographic studies by sociologists and anthropologists often shed light on variations within communities, which are viewed as homogenous by outsiders. While white Philadelphians who never visited the Seventh Ward tended to view the area as homogenous and all African-Americans as similar, Du Bois found a physical and social structure within the neighborhood – alleys peopled by criminals, loafers, and prostitutes separate from streets of the working poor and still other streets where an established group of Black middle-class homeowners lived.

In addition to *The Philadelphia Negro* (Philadelphia: University of Pennsylvania Press, 1899) from which the following selection is taken, Du Bois's writings include *Suppression of the Slave Trade to the United States of America* (New York: Longmans, Green and Co., 1896), *Souls of Black Folk* (Chicago: A.C. McClurg and Co., 1903), *The Negro* (New York: Henry Holt and Co., 1915), *Black Reconstruction* (New York: Harcourt, Brace, 1935), and *The World and Africa* (New York: The Viking Press, 1947). There are many anthologies of Du Bois's writings and speeches. Perhaps the best is David Levering Lewis (ed.), *W.E.B. Du Bois: A Reader* (New York: Henry Holt, 1995). Also by Lewis, and of great interest, is *W.E.B. Du Bois: The Fight for Equality and the American Century, 1919–1963* (New York: Henry Holt, 1995).

For more by and about W.E.B. Du Bois see *The Autobiography of W.E.B. Du Bois* (New York: International Publishers, 1968), Francis L. Broderick, *W.E.B. Du Bois* (Palo Alto, CA: Stanford University Press, 1959), Walter Wilson (ed.), *The Selected Writings of W.E.B. Du Bois* (New York: New American Library, 1970), Henry Lee Moon, *The Emerging Thought of W.E.B Du Bois* (New York: Simon and Schuster, 1972), Marable Manning, *W.E.B. DuBois, Black Radical Democrat* (Boston: Twayne Publishers, 1986), and Patricia and Fredrick McKissack, *W.E.B. Du Bois* (New York: Franklin Watts, 1990).

4. THE NEGRO PROBLEMS OF PHILADELPHIA

In Philadelphia, as elsewhere in the United States, the existence of certain peculiar social problems affecting the Negro people are plainly manifest. Here is a large group of people – perhaps forty-five thousand, a city within a city – who do not form an integral part of the larger social group. This in itself is not altogether unusual; there are other unassimilated groups: Jews, Italians, even Americans; and yet in the case of the Negroes the segregation is more conspicuous, more patent to the eye, and so intertwined with a long historic evolution, with peculiarly pressing social problems of poverty, ignorance, crime and labor, that the Negro problem far surpasses in scientific interest and social gravity most of the other race or class questions.

The student of these questions must first ask, What is the real condition of this group of human beings? Of whom is it composed, what sub-groups and classes exist, what sort of individuals are being considered? Further, the student must clearly recognize that a complete study must not confine itself to the group, but must specially notice the environment; the physical environment of city, sections and houses, the far mightier social environment – the surrounding world of custom, wish, whim and thought which envelops this group and powerfully influences its social development.

[. . .]

The Seventh Ward starts from the historic center of Negro settlement in the city, South Seventh street and Lombard, and includes the long narrow strip, beginning at South Seventh and extending west, with South and Spruce streets as boundaries, as far as the Schuylkill River. The colored population of this ward numbered 3,621 in 1860, 4,616 in 1870, and 8,861 in 1890. It is a thickly populated district of varying character; north of it is the residence and business section of the city; south of it a middle class and workingmen's residence section; at the east end it joins Negro, Italian and Jewish slums; at the west end, the wharves of the river and an industrial section separating it from the grounds of the University of Pennsylvania and the residence section of West Philadelphia.

Starting at Seventh street and walking along Lombard, let us glance at the general character of the ward. Pausing a moment at the corner of Seventh and Lombard, we can at a glance view the worst Negro slums of the city. The houses are mostly brick, some wood, not very old, and in general uncared for rather than dilapidated. The blocks between Eighth, Pine, Sixth, and South have for many decades been the center of Negro population. Here the riots of the thirties took place, and here once was a depth of poverty and degradation almost unbelievable. Even today there are many evidences of degradation . . . The alleys near, as Ratcliffe street, Middle alley, Brown's court, Barclay street, etc., are haunts of noted criminals, male and female, of gamblers and prostitutes, and at the same time of many poverty-stricken people, decent but not energetic. There is an abundance of political clubs, and nearly all the houses are practically lodging houses, with a miscellaneous and shifting population. The corners, night and day, are filled with Negro loafers – able-bodied young men and women, all cheerful, some with good natured, open faces, some with traces of crime and excess, a few pinched with poverty. They are mostly gamblers, thieves and prostitutes, and few have fixed and steady occupation of any kind. Some are stevedores, porters, laborers and laundresses. On its face this slum is noisy and dissipated, but not brutal, although now and then highway robberies and murderous assaults in other parts of the city are traced to its denizens. Nevertheless a stranger can usually walk about here day and night with little fear of being molested if he be not too inquisitive.

Passing up Lombard, beyond Eighth, the atmosphere suddenly changes, because these next two blocks have few alleys and the residences are good-sized and pleasant. Here some of the best Negro families of the ward live. Some are wealthy in a small way, nearly all are Philadelphia born, and they represent an early wave of emigration from the old slum section . . .

[. . .]

21. THE QUESTION OF EARNING A LIVING

For a group of freedmen the question of economic survival is the most pressing of all questions; the problem as to how, under the circumstances of modern life, any group of people can earn a decent living, so as to maintain their standard of life, is not always easy to answer. But when the question is complicated by the fact that the group has a low degree of efficiency on account of previous training; is in competition with well-trained, eager and often ruthless competitors; is more or less handicapped by a somewhat wide-reaching discrimination; and finally is seeking not merely to maintain a standard of living but steadily to raise it to a higher plane – such a situation presents baffling problems to the sociologist and philanthropist.

Of the men 21 years of age and over, there were in gainful occupations, the following:

In the learned professions		61	2.0 per cent
Conducting business on their own account		207	6.5
In the skilled trades		236	7.0
Clerks, etc.		159	5.0
Laborers, better class	602		
Laborers, common class	852	1454	45.0
Servants		1079	34.0
Miscellaneous		11	0.5
		3207	100 per cent
Total male population 21 and over		3850	

Taking the occupations of women 21 years of age and over, we have:

Domestic servants	1262	37.0 per cent
Housewives and day laborers	937	27.0
Housewives	568	17.0
Day laborers, maids, etc.	297	9.0
In skilled trades	221	6.0
Conducting businesses	63	2.0
Clerks, etc.	40	1.0
Learned professions	37	1.0
	3425	100 per cent
Total female population 21 and over	3740	

47. COLOR PREJUDICE

Incidentally throughout this study the prejudice against the Negro has been again and again mentioned. It is time now to reduce this somewhat indefinite term to something tangible. Everybody speaks of the matter, everybody knows that it exists, but in just what form it shows itself or how influential it is few agree. In the Negro's mind, color prejudice in Philadelphia is that widespread feeling of dislike for his blood, which keeps him and his children out of decent employment, from certain public conveniences and amusements, from hiring houses in many sections, and in general, from being recognized as a man. Negroes regard this prejudice as the chief cause of their present unfortunate condition. On the other hand most white people are quite unconscious of any such powerful and vindictive feeling; they regard color prejudice as the easily explicable feeling that intimate social intercourse with a lower race is not only undesirable but impractical if our present standards of culture are to be maintained, and although they are aware that some people feel the aversion more intensely than others, they cannot see how such a feeling has much influence on the real situation or alters the social condition of the mass of Negroes.

As a matter of fact, color prejudice in this city is something between these two extreme views: it is not today responsible for all, or perhaps the greater part of the Negro problems, or of the disabilities under which the race labors; on the other hand it is a far more powerful social force than most Philadelphians realize. The practical results of the attitude of most of the inhabitants of Philadelphia towards persons of Negro descent are as follows:

1. As to getting work:
    No matter how well trained a Negro may be, or how fitted for work of any kind, he cannot in the ordinary course of competition hope to be much more than a menial servant.

    He cannot get clerical or supervisory work to do save in exceptional cases.

    He cannot teach save in a few of the remaining Negro schools.

    He cannot become a mechanic except for small transient jobs, and cannot join a trades union.

    A Negro woman has but three careers open to her in this city: domestic service, sewing, or married life.

2. As to keeping work:
    The Negro suffers in competition more severely than white men.

    Change in fashion is causing him to be replaced by whites in the better-paid positions of domestic service.

    Whim and accident will cause him to lose a hard-earned place more quickly than the same things would affect a white man.

    Being few in number compared with the whites the crime or carelessness of a few of his race is easily imputed to all, and the reputation of the good, industrious, and reliable suffer thereby.

    Because Negro workmen may not often work side by side with white workmen, the individual black workman is rated not only by his own efficiency, but by the efficiency of a whole group of black fellow workmen which may often be low.

    Because of these difficulties which virtually increase competition in his case, he is forced to take lower wages for the same work than white workmen.

3. As to entering new lines of work:
    Men are used to seeing Negroes in inferior positions; when, therefore, by any chance a Negro gets in a better position, most men immediately conclude that he is not fitted for it, even before he has a chance to show his fitness.

    If, therefore, he set up a store, men will not patronize him.

    If he is put into public position men will complain.

    If he gain a position in the commercial world, men will quietly secure his dismissal or see that a white man succeeds him.

4. As to his expenditure:
    The comparative smallness of the patronage of the Negro, and the dislike of other customers, makes it usual to increase the charges or difficulties in certain directions in which a Negro must spend money.

    He must pay more house-rent for worse houses than most white people pay.

    He is sometimes liable to insult or reluctant service in some restaurants, hotels and stores, at public resorts, theaters and places of recreation; and at nearly all barber shops.

5. As to his children:
    The Negro finds it extremely difficult to rear children in such an atmosphere and not have them

either cringing or impudent: if he impresses upon them patience with their lot, they may grow up satisfied with their condition; if he inspires them with ambition to rise, they may grow to despise their own people, hate the whites, and become embittered with the world.

His children are discriminated against, often in public schools.

They are advised when seeking employment to become waiters and maids.

They are liable to species of insult and temptation peculiarly trying to children.

6. As to social intercourse:

In all walks of life the Negro is liable to meet some objection to his presence or some discourteous treatment; and the ties of friendship or memory seldom are strong enough to hold across the color line.

If an invitation is issued to the public for any occasion, the Negro can never know whether he would be welcomed or not; if he goes he is liable to have his feelings hurt and get into unpleasant altercation; if he stays away, he is blamed for indifference.

If he meet a lifelong white friend on the street, he is in a dilemma; if he does not greet the friend he is put down as boorish and impolite; if he does greet the friend he is liable to be flatly snubbed.

If by chance he is introduced to a white woman or man, he expects to be ignored on the next meeting, and usually is.

White friends may call on him, but he is scarcely expected to call on them, save for strictly business matters.

If he gain the affections of a white woman and marry her he may invariably expect that slurs will be thrown on her reputation and on his, and that both his and her race will shun their company. When he dies he cannot be buried beside white corpses.

7. The result:

Any one of these things happening now and then would not be remarkable or call for especial comment; but when one group of people suffer all these little differences of treatment and discriminations and insults continually, the result is either discouragement, or bitterness, or over-sensitiveness, or recklessness. And a people feeling thus cannot do their best.

Presumably the first impulse of the average Philadelphian would be emphatically to deny any such marked and blighting discrimination as the above against a group of citizens in this metropolis. Every one knows that in the past color prejudice in the city was deep and passionate; living men can remember when a Negro could not sit in a street car or walk many streets in peace. These times have passed, however, and many imagine discrimination against the Negro has passed with them. Careful inquiry will convince any such one of his error. To be sure a colored man to-day can walk the streets of Philadelphia without personal insult; he can go to theaters, parks and some places of amusement without meeting more than stares and discourtesy; he can be accommodated at most hotels and restaurants, although his treatment in some would not be pleasant. All this is a vast advance and augurs much for the future. And yet all that has been said of the remaining discrimination is but too true.

During the investigation of 1896 there was collected a number of actual cases, which may illustrate the discriminations spoken of. So far as possible these have been sifted and only those which seem undoubtedly true have been selected:

I. As to getting work

It is hardly necessary to dwell upon the situation of the Negro in regard to work in the higher walks of life: the white boy may start in the lawyer's office and work himself into a lucrative practice; he may serve a physician as office boy or enter a hospital in a minor position, and have his talent alone between him and affluence and fame; if he is bright in school, he may make his mark in a university, become a tutor with some time and much inspiration for study, and eventually fill a professor's chair. All these careers are at the very outset closed to the Negro on account of his color; what lawyer would give even a minor case to a Negro assistant? What university would appoint a promising young Negro as tutor? Thus the young white man starts in life knowing that within some limits and barring accidents, talent and application will tell. The young Negro starts knowing that on all sides his advance is made difficult if not wholly shut off by his color. Let us come, however, to ordinary occupations which concern more nearly the mass of Negroes. Philadelphia is a great industrial and business center with thousands of foremen, managers and clerks – the lieutenants of industry who direct its progress. They are paid for thinking and for skill to direct, and naturally such positions are coveted because they are well paid, well thought-of and carry some authority. To such positions Negro boys and girls may not aspire no matter what their qualifications. Even as teachers and ordinary clerks and stenographers they find almost no openings. Let us note some actual instances:

A young woman who graduated with credit from the Girls Normal School in 1892 has taught in the kindergarten, acted as substitute, and waited in vain for a permanent position. Once she was allowed to substitute in a school with white teachers; the principal commended her work, but when the permanent appointment was made a white woman got it.

A girl who graduated from a Pennsylvania high school and from a business college sought work in the city as a stenographer and typewriter. A prominent lawyer undertook to find her a position; he went to friends and said, "Here is a girl that does excellent work and is of good character; can you not give her work?" Several immediately answered yes. "But," said the lawyer, "I will be perfectly frank with you and tell you she is colored"; and not in the whole city could he find a man willing to employ her. It happened, however, that the girl was so light in complexion that few not knowing would have suspected her descent. The lawyer therefore gave her temporary work in his own office until she found a position outside the city. "But," said he, "to this day I have not dared to tell my clerks that they worked beside a Negress." Another woman graduated from the high school and the Palmer College of Shorthand, but all over the city has met with nothing but refusal of work.

Several graduates in pharmacy have sought three years' required apprenticeship in the city and in only one case did one succeed, although they offered to work for nothing. One young pharmacist came from Massachusetts and for weeks sought in vain for work here at any price; "I wouldn't have a darky to clean out my store, much less to stand behind the counter," answered one druggist.

A colored man answered an advertisement for a clerk in the suburbs. "What do you suppose we'd want of a nigger?" was the plain answer. A graduate of the University of Pennsylvania in mechanical engineering, well recommended, obtained work in the city, through an advertisement, on account of his excellent record. He worked a few hours and then was discharged because he was found to be colored. He is now a waiter at the University Club, where his white fellow graduates dine. Another young man attended Spring Garden Institute and studied drawing for lithography. He had good references from the institute and elsewhere, but application at the five largest establishments in the city could secure him no work. A telegraph operator has hunted in vain for an opening, and two graduates of the Central High School have sunk to menial labor. "What's the use of an education?" asked one. Mr. A— has elsewhere been employed as a traveling salesman. He applied for a position here by letter and was told he could have one. When they saw him they had no work for him.

Such cases could be multiplied indefinitely. But that is not necessary; one has but to note that, notwithstanding the acknowledged ability of many colored men, the Negro is conspicuously absent from all places of honor, trust, emolument, as well as from those of respectable grade in commerce and industry.

Even in the world of skilled labor the Negro is largely excluded. Many would explain the absence of Negroes from higher vocations by saying that while a few may now and then be found competent, the great mass are not fitted for that sort of work and are destined for some time to form a laboring class. In the matter of the trades, however, there can be raised no serious question of ability; for years the Negroes filled satisfactorily the trades of the city, and to-day in many parts of the South they are still prominent. And yet in Philadelphia a determined prejudice, aided by public opinion, has succeeded nearly in driving them from the field:

A——, who works at a bookbinding establishment on Front street, has learned to bind books and often does so for his friends. He is not allowed to work at the trade in the shop, however, but must remain a porter at a porter's wages.

B——is a brushmaker; he has applied at several establishments, but they would not even examine his testimonials. They simply said: "We do not employ colored people."

C——is a shoemaker; he tried to get work in some of the large department stores. They "had no place" for him.

D——was a bricklayer, but experienced so much trouble in getting work that he is now a messenger.

E——is a painter, but has found it impossible to get work because he is colored.

F——is a telegraph line man, who formerly worked in Richmond, Va. When he applied here he was told that Negroes were not employed.

G——is an iron puddler, who belonged to a Pittsburgh union. Here he was not recognized as a union man and could not get work except as a stevedore.

H——was a cooper, but could get no work trials, and is now a common laborer.

I——is a candy-maker, but has never been able to find employment in the city; he was always told the white help would not work with him.

J——is a carpenter; he can only secure odd jobs or work where only Negroes are employed.

K——was an upholsterer, but could get no work save in the few colored shops which had workmen; he is now a waiter on a dining car.

L——was a first-class baker; he applied for work some time ago near Green street and was told shortly, "We don't work no niggers here."

[. . .]

"From Institutional to Jobless Ghettos"

from *When Work Disappears: The World of the New Urban Poor* (1996)

William Julius Wilson

Editors' Introduction

Harvard sociologist William Julius Wilson spent much of his career at the University of Chicago. Like the earlier Chicago School sociologists Ernest W. Burgess (p. 156) and Louis Wirth (p. 97) writing in the 1920s and 1930s, and St. Clair Drake and Horace Cayton writing in the 1940s, Wilson uses careful empirical studies of Chicago to generate important urban theory. An African-American, Wilson has been particularly concerned about the situation of poor Blacks in America's decaying central cities.

Wilson is critical of timid liberals who avoid confronting tough questions about race and poverty because they are afraid anything negative they say about Blacks will appear racist. He argues that there is an *urban underclass* (a term many liberals will not use) and that residents of poor Black ghettos today are socially isolated and caught in a tangle of pathology characterized by unemployment, crime (including violent Black-on-Black crime), teenage pregnancy, out-of-wedlock births, welfare dependency, and drug use. Wilson feels that the situation of Blacks, particularly poor urban Blacks, requires objective research and honest reportage. He is unwilling to let conservatives dominate theoretical discourse about the causes of and cures for Black poverty.

Wilson's research has convinced him that conditions for poor Black ghetto residents are far worse in many ways than a century ago when W.E.B. Du Bois studied *The Philadelphia Negro* (p. 119) or in 1945 when St. Clair Drake and Horace Cayton published *Black Metropolis*, their study of the African-American community of Chicago at that time. Both Du Bois's and Drake and Cayton's studies found more Blacks employed and role models of upward mobility present, a higher proportion of nuclear families, less Black-on-Black violence, and much less drug use than exist in the poorest Black ghettos today.

Wilson argues that the changed structure of the US economy is more responsible for the plight of poor Blacks today than is racism. A generation ago, Wilson argues, an able-bodied unskilled Black man could readily find work sufficient to support himself and a family – albeit often physically hard, racially segregated, and dirty work. But in the last generation unskilled manual urban jobs have largely disappeared. Without work, Black males cannot support a family. Hence, Wilson argues, Black ghettos today contain few "marriageable" Black males capable of supporting a family. A high incidence of out-of-wedlock births, family dissolution, and welfare-dependent female-headed households follow directly from that fact. With little sense of self-worth, Wilson argues, unemployed Blacks naturally turn to drug dependence and crime.

One of Wilson's most controversial contentions is that race and racism are declining in importance as causes of Black distress. Paradoxically, he argues, less racial discrimination has made matters in Black ghettos worse. According to Wilson, as upwardly mobile Blacks move out of Black ghetto areas, community leadership and positive role models disappear and pathology is concentrated.

Wilson is skeptical that *race-specific* policies like affirmative action will address problems as pervasive and profound as he describes. Rather he favors universal social policies aimed at improving the lot of all poor people regardless of race: new education, training, and particularly full employment policy.

This selection is from *When Work Disappears: The World of the New Urban Poor* (New York: Knopf, 1996). Wilson's two most influential prior books are *The Declining Significance of Race: Blacks and Changing American Institutions*, 2nd edn (Chicago: University of Chicago Press, 1980) and *The Truly Disadvantaged* (Chicago: University of Chicago Press, 1987). Also of interest is *The Bridge Over the Racial Divide: Rising Inequality and Coalition Politics* (Los Angeles: University of California Press, 2001).

The literature on urban social inequality, welfarism, and the "underclass debate" is vast. Classic studies include Richard A. Cloward and Frances Fox Piven, *Regulating the Poor: The Functions of Public Welfare*, 2nd edn (New York: Vintage, 1993), Michael B. Katz, *In the Shadow of the Poorhouse: A Social History of Welfare in America* (New York: Basic Books, 1987), *The Undeserving Poor: From the War on Poverty to the War on Welfare* (New York: Random House, 1990), and Christopher Jencks, *Rethinking Social Policy: Race, Poverty, and the Underclass* (New York: Harpers, 1993). More recent studies of interest include Benjamin I. Page and James Roy Simmons, *What Government Can Do: Dealing with Poverty and Inequality* (Chicago: University of Chicago Press, 2000), R. Kent Weaver and Michael H. Armacost, *Ending Welfare as we Know it* (Washington, DC: Brookings, 2000), and Theda Skocpol and Richard C. Leone, *The Missing Middle: Working Families and the Future of American Social Policy* (New York: W.W. Norton, 2001). Interesting conservative explanations of and suggested public policy regarding poverty and race include Edward Banfield, *The Unheavenly City Revisited* (Boston: Little Brown, 1974) and Charles Murray, *Losing Ground* (New York: Basic Books, 1984).

For a recent overview of the literature on the underclass debate and writings by Latino scholars exploring the relevance and limitations of underclass theory for America's varied Latino communities, see Joan Moore and Raquel Pinderhughes, *In the Barrios: Latinos and the Underclass Debate* (New York: Russell Sage Foundation, 1993).

▪ ▪ ▪ ▪ ▪ ▪

An elderly woman who has lived in one inner-city neighborhood on the South Side of Chicago for more than forty years reflected:

> I've been here since March 21, 1953. When I moved in, the neighborhood was intact. It was intact with homes, beautiful homes, mini mansions, with stores, laundromats, with cleaners, with Chinese [cleaners]. We had drugstores. We had hotels. We had doctors over on Thirty-ninth Street. We had doctors' offices in the neighborhood. We had the middle class and upper middle class. It has gone from affluent to where it is today. And I would like to see it come back, that we can have some of the things we had. Since I came in young, and I'm a senior citizen now, I would like to see some of the things come back so I can enjoy them like we did when we first came in.

[. . .]

A 91-year-old woman spoke of safety concerns: "It's not safe anymore because the streets aren't. When all the black businesses and shows closed down, the economy went to the dogs. The stores, the businesses, the shows, everywhere was lighted, the stores and businesses have disappeared."

The negative social forces triggered a decision by a concerned mother to send her son away.

> I have a 13-year-old. I sent him away when he was nine because the gangs was at him so tough, because he wouldn't join – he's a basketball player. That's all he ever cared about. They took his gym shoes off his feet. They took his clothes. Made him walk home from school. Jumped on him every day. Took his jacket off his back in subzero weather. You know, and we only live two blocks from the school . . . A boy pulled a gun to his head and told him, "If you don't join, next week you won't be here." I had to send him out of town. His father stayed out of town. He came here last week for a week. He said, "Mom, I want to come home so bad," I said no!

The social deterioration of ghetto neighborhoods is the central concern expressed in the testimony of these residents. As a representative from the media put it, the ghetto has gone "from bad to worse." Few observers of the urban scene in the late 1960s anticipated the extensive breakdown of social institutions and the sharp rise in rates of social dislocation that have since swept the ghettos and spread to other neighborhoods that were once stable. For example, in the neighborhood of Woodlawn, located on the South Side of Chicago, there were over eight hundred commercial and industrial establishments in 1950. Today, it is estimated that only about a hundred are left, many of them represented by "tiny catering places, barber shops, and thrift stores with no more than one or two employees." As Löic Wacquant, a member of the Urban Poverty and Family Life Study research team, put it:

> The once-lively streets – residents remember a time, not so long ago, when crowds were so dense at rush hour that one had to elbow one's way to the train station – now have the appearance of an empty, bombed-out war zone. The commercial strip has been reduced to a long tunnel of charred stores, vacant lots littered with broken glass and garbage, and dilapidated buildings left to rot in the shadow of the elevated train line. At the corner of Sixty-third Street and Cottage Grove Avenue, the handful of remaining establishments that struggle to survive are huddled behind wrought iron bars . . . The only enterprises that seem to be thriving are liquor stores and currency exchanges, these "banks of the poor" where one can cash checks, pay bills and buy money orders for a fee.

[. . .]

In 1950, almost two-thirds of Woodlawn's population was white; by 1960 the white population had declined to just 10 percent. Despite the sudden white exodus, the number of residents in the neighborhood increased slightly during this period. After 1960, however, a sizable exodus of black residents followed, including a significant number of working- and middle-class families. The population of the neighborhood declined from over 80,000 in 1960 to 53,814 in 1970; it further slipped to 36,323 in 1980 and finally to 24,473 in 1990. The loss of residents was accompanied by a substantial reduction in the economic, social, and political resources that make a community vibrant. Woodlawn is only one of a growing number of poor black neighborhoods in Chicago plagued by depopulation and social and economic deterioration.

When the black respondents in our large . . . survey were asked to rate their neighborhood as a place to live, only a third said that their area was a good or very good place to live and only 18 percent of those in the ghetto poverty census tracts felt that their neighborhood was a desirable place to live. (The Bureau of the Census defines a census tract as "a relatively homogeneous area with respect to population characteristics, economic status, and living conditions with an average population of 4,000." Poverty tracts are those in which at least 20 percent of the residents are poor, and ghetto poverty tracts are those in which at least 40 percent are poor.)

[. . .]

Many of the respondents described the negative effects of their neighborhood on their own personal outlook. An unmarried, employed clerical worker from a ghetto poverty census tract on the West Side stated:

> There is a more positive outlook if you come from an upwardly mobile neighborhood than you would here. In this type of neighborhood, all you hear is negative [things] and that can kind of bring you down when you're trying to make it. So your neighborhood definitely has something to do with it.

This view was shared by a 17-year-old college student and part-time worker from an impoverished West Side neighborhood.

> I'd say about 40 percent in my neighborhood . . . I'd say 40 percent are alcoholics . . . And . . . only 5 percent of the alcoholics have homes. Then you got the other 35 percent who are in the street . . . They probably live somewhere, but they in street, on the corner every day, same old thing, because they don't have no chance in life. They live based on today. [They say] "Oh, we gonna get high today." "Oh, whoopee!" "What you gonna do tomorrow, man?" "I don't know, man, I don't know." You can ask any of 'em: "What you gonna do tomorrow?" "I don't know, man. I know when it gets here." And I can really understand, you know, being in that state. If you around totally negative people, people who are not doing anything, that's the way you gonna be regardless.

The state of the inner-city public schools was another major concern expressed by our respondents. The complaints ranged from overcrowded conditions to unqualified and uncaring teachers. Sharply voicing her views on these subjects, a 25-year-old married mother of two children from a South Side census tract that just recently became poor stated: "My daughter ain't going to school here, she was going to a nursery school where I paid and of course they took the time and spent it with her, 'cause they was getting the money. But the public schools, no! They are over-crowded and the teachers don't care."

[. . .]

The respondents were also asked whether their neighborhoods had changed as a place to live over the years. Seventy-one percent of the African-American respondents felt that their neighborhoods had either stayed the same or had gotten worse.

An unemployed black man from a West Side housing project felt that the only thing that had changed in his neighborhood was that it was "going down instead of going back up." He further stated, "It ain't like it used to be. They laid off a lot of people. There used to be a time when you got a broken window, you call up housing and they send someone over to fix it, but it ain't like that no more."

Respondents frequently made statements about the increase in drug trafficking and drug consumption when discussing how their neighborhood had changed. "Well, OK, I realize there was drugs when I was growing up but they weren't as open as they are now," stated a divorced telephone dispatcher and mother of five children from a neighborhood that recently changed from a nonpoverty to a poverty area. "It's nothing to see a 10-year-old kid strung out or a 10-year-old kid selling drugs. I mean, when they were doing it back then they were sneaking around doing it. It's like an open thing now."

[. . .]

The feelings of many of the respondents in our study were summed up by a 33-year-old married mother of three from a very poor West Side neighborhood:

If you live in an area in your neighborhood where you have people that don't work, don't have no means of support, you know, don't have no jobs, who're gonna break into your house to steal what you have, to sell to get them some money, then you can't live in a neighborhood and try to concentrate on tryin' to get ahead, then you get to work and you have to worry if somebody's breakin' into your house or not. So, you know, it's best to try to move in a decent area, to live in a community with people that works.

In 1959, less than one-third of the poverty population in the United States lived in metropolitan central cities. By 1991, the central cities included close to half of the nation's poor. Many of the most rapid increases in concentrated poverty have occurred in African-American neighborhoods. For example, in the ten community areas that represent the historic core of Chicago's Black Belt (see Figure 1), eight had rates of poverty in 1990 that exceeded 45 percent, including three with rates higher than 50 percent and three that surpassed 60 percent. Twenty-five years earlier, in 1970, only two of these neighborhoods had poverty rates above 40 percent.

In recent years, social scientists have paid particular attention to the increases in urban neighborhood poverty. "Defining an urban neighborhood for ana-lytical purposes is no easy task." The community areas of Chicago referred to in Figure 1 include a number of adjacent census tracts. The seventy-seven community areas within the city of Chicago represent statistical units derived by urban sociologists at the University of Chicago for the 1930 census in their effort to analyze varying conditions within the city. These delineations were originally drawn up on the basis of settlement and history of the area, local identification and trade patterns, local institutions, and natural and artificial barriers. There have been major shifts in population and land use since then. But these units remain useful in tracing changes over time, and they continue to capture much of the contemporary reality of Chicago neighborhoods.

Other cities, however, do not have such convenient classifications of neighborhoods, which means that comparison across cities cannot be drawn using community areas. The measurable unit considered most appropriate to represent urban neighborhoods is the census tract. In attempts to examine this problem of ghetto poverty across the nation empirically, social scientists have tended to define ghetto neighborhoods as those located in the *ghetto poverty* census tracts. As indicated earlier, ghetto poverty census tracts are those in which at least 40 percent of the residents are poor. For example, Paul Jargowsky and Mary Jo Bane state: "Visits to various cities confirmed that the 40 percent criterion came very close to identifying areas that looked like ghettos in terms of their housing condi-tions. Moreover, the areas selected by the 40 percent

1. West Garfield Park
2. East Garfield Park
3. North Lawndale
4. Near West Side
5. Near South Side

6. Douglas
7. Oakland
8. Grand Boulevard
9. Washington Park
10. Englewood

Figure 1 Community areas in Chicago's Black Belt

criterion corresponded closely with the neighborhoods that city officials and local Census Bureau officials considered ghettos." The ghetto poor in Jargowsky and Bane's study are therefore designated as those among the poor who live in these ghetto poverty areas. Three-quarters of all the ghetto poor in metropolitan areas reside in one hundred of the nation's largest central cities; however, it is important to remember that the ghetto areas in these central cities also include a good many families and individuals who are not poor.

In the nation's one hundred largest central cities, nearly one in seven census tracts is at least 40 percent poor. The number of such tracts has more than doubled since 1970 – indeed, it is alarming that 579 tracts fell to ghetto poverty level in these cities between 1970 and 1980, and 624 additional tracts joined these ranks in the following decade.

Paul Jargowsky's research reveals that a vast majority of people (almost seven out of eight) living in metropolitan-area ghettos in 1990 were minority group members. The number of African-Americans in these ghettos grew by more than one-third from 1980 to 1990, reaching nearly 6 million. Most of this growth involved poor people. The proportion of metropolitan blacks who live in ghetto areas climbed from more than a third (37 percent) to almost half (45 percent). Indeed, the metropolitan black poor are becoming increasingly isolated. The poverty rate among metropolitan blacks who reside in ghettos increased while the rate among those who live in nonghettos decreased.

The increase in the *number* of ghetto blacks is related to the *geographical spread* of the ghetto. Jargowsky and Bane found that in the cities they studied (Philadelphia, Cleveland, Milwaukee, and Memphis) areas that had become ghettos by 1980 had been mixed-income tracts in 1970 – but tracts that were contiguous to areas identified as ghettos. The exodus of the nonpoor from mixed-income areas was a major factor in the spread of ghettos in these cities in the 1970s. Since 1980, ghetto census tracts have increased in a substantial majority of the metropolitan areas in the country, including those with fewer people living in them. Nine new ghetto census tracts were added in Philadelphia, even though it experienced one of the largest declines in the proportion of people living in ghetto tracts. In a number of other cities, including Baltimore, Boston, and Washington, D.C., a smaller percentage of poor blacks live in a larger number of ghetto census tracts. Chicago had a 61.5 percent increase in the number of ghetto census tracts from 1980 to 1990, even though the number of poor residing in those areas increased only slightly.

Jargowsky reflects on the significance of the substantial spread of ghetto areas:

The geographic size of a city's ghetto has a large effect on the perception of the magnitude of the problem associated with ghetto poverty. How big an area of the city do you consider off limits? How far out of your way will you drive not to go through a dangerous area? Indeed, the lower density exacerbated the problem. More abandoned buildings mean more places for crack dens and criminal enterprises. Police trying to protect a given number of citizens have to be stretched over a wider number of square miles, making it less likely that criminals will be caught. Lower density also makes it harder for a sense of community to develop, or for people to feel that they can find safety in

numbers. From the point of view of local political officials, the increase in the size of the ghetto is a disaster. Many of those leaving the ghetto settle in non-ghetto areas outside the political jurisdiction of the central city. Thus, geographic size of the ghetto is expanding, cutting a wider swath through the hearts of our metropolitan areas.

In sum, the 1970s and 1980s witnessed a sharp growth in the number of census tracts classified as ghetto poverty areas, an increased concentration of the poor in these areas, and sharply divergent patterns of poverty concentration between racial minorities and whites. One of the legacies of historic racial and class subjugation in America is a unique and growing concentration of minority residents in the most impoverished areas of the nation's metropolises.

Some have argued that this concentration of poverty is not new but mirrors conditions prevalent in the 1930s. According to Douglas Massey and Nancy Denton, during the Depression poverty was just as concentrated in the ghettos of the 1930s as in those of the 1970s. The black communities of the 1930s and those of the 1970s shared a common experience: a high degree of racial segregation from the larger society. Massey and Denton argue that "concentrated poverty is created by a pernicious interaction between a group's overall rate of poverty and its degree of segregation in society. When a highly segregated group experiences a high or rising rate of poverty, geographically concentrated poverty is the inevitable result." However convincing the logic of that argument, it does not explain the following: In the ten neighborhoods that make up Chicago's Black Belt, the poverty rate increased almost 20 percent between 1970 and 1980 (from 32.5 to 50.4 percent) despite the fact that the overall black poverty rate for the city of Chicago increased only 7.5 percent during this same period (from 25.1 to 32.6 percent).

Concentrated poverty may be the inevitable result when a highly segregated group experiences an increase in its overall rate of poverty. But segregation does not explain why the concentration of poverty in *certain* neighborhoods of this segregated group should increase to nearly three times the group's *overall* rate of poverty increase. There is no doubt that the disproportionate concentration of poverty among African-Americans is one of the legacies of historic racial segregation. It is also true that segregation often compounds black vulnerability in the face of other changes in the society, including, as we shall soon see, economic changes. Nonetheless, to focus mainly on segregation to account for the growth of concentrated poverty is to overlook some of the dynamic aspects of the social and demographic changes occurring in cities like Chicago. Given the existence of segregation, we must consider the way in which other changes in society have interacted with segregation to produce the dramatic social transformation of inner-city neighborhoods, especially since 1970.

For example, the communities that make up the Black Belt in Chicago have been overwhelmingly black for the last four decades, yet they lost almost half their residents between 1970 and 1980. This rapid depopulation has had profound consequences for the social and economic deterioration of segregated Black Belt neighborhoods, including increases in concentrated poverty and joblessness. If comparisons are made strictly between the Depression years of the 1930s and the 1980s, rates of ghetto poverty and joblessness in these neighborhoods will indeed be similar. But such a comparison obscures significant changes that have occurred in these neighborhoods across the fifty-year span between those two points.

. . . Many of the gains made in inner-city neighborhoods following the Depression were wiped out after 1970. To maintain that concentrated black poverty in the 1970s or in the 1980s is equivalent in severity and pervasiveness to that which occurred during the Depression does not explain its dramatic rise since 1970; nor does it address a far more fundamental problem that is at the heart of the extraordinary increases in and spread of concentrated poverty – namely, the rapid growth of joblessness, which accelerated through these two decades. The problems reported by the residents of poor Chicago neighborhoods are not a consequence of poverty alone. Something far more devastating has happened that can only be attributed to the emergence of concentrated and persistent joblessness and its crippling effects on neighborhoods, families, and individuals. The city of Chicago epitomizes these changes.

[. . .]

The most fundamental difference between today's inner-city neighborhoods and those studied by Drake and Cayton [in the 1940s] is the much higher levels of joblessness. Indeed, there is a new poverty in our nation's metropolises that has consequences for a range of issues relating to the quality of life in urban areas, including race relations.

By "the new urban poverty," I mean poor, segregated neighborhoods in which a substantial majority of individual adults are either unemployed or have dropped out of the labor force altogether. For example, in 1990 only one in three adults aged 16 and over in the twelve Chicago community areas with ghetto poverty rates held a job in a typical week of the year. Each of these community areas, located on the South and West Sides of the city, is overwhelmingly black. We can add to these twelve high-jobless areas three additional predominantly black community areas, with rates approaching ghetto poverty, in which only 42 percent of the adult population were working in a typical week in 1990. Thus, in these fifteen black community areas – comprising a total population of 425,125 – only 37 percent of all the adults were gainfully employed in a typical week in 1990. By contrast, 54 percent of the adults in the seventeen other predominantly black community areas in Chicago – a total population of 545,408 – worked in a typical week in 1990. This was close to the citywide employment figure of 57 percent for all adults. Finally, except for one Asian community area with an employment rate of 46 percent, and one Latino community area with an employment rate of 49 percent, a majority of the adults held a job in a typical week in each of the remaining forty-five community areas of Chicago.

But Chicago is by no means the only city that features new poverty neighborhoods. In the ghetto census tracts of the nation's one hundred largest central cities, there were only 65.5 employed persons for every hundred adults who did not hold a job in a typical week in 1990. In contrast, the nonpoverty areas contained 182.3 employed persons for every hundred of those not working. In other words, the ratio of employed to jobless persons was three times greater in census tracts not marked by poverty.

Looking at Drake and Cayton's Bronzeville, I can illustrate the magnitude of the changes that have occurred in many, inner-city ghetto neighborhoods in recent years. A majority of adults held jobs in the three Bronzeville areas in 1950, but by 1990 only four in ten in Douglas worked in a typical week, one in three in Washington Park, and one in four in Grand Boulevard. In 1950, 69 percent of all males 14 and over who lived in the Bronzeville neighborhoods worked in a typical week, and in 1960, 64 percent of this group were so employed. However, by 1990 only 37 percent of all males 16 and over held jobs in a typical week in these three neighborhoods.

Upon the publication of the first edition of *Black Metropolis* in 1945, there was much greater class integration within the black community. As Drake and Cayton pointed out, Bronzeville residents had limited success in "sorting themselves out into broad community areas designated as 'lower class' and 'middle class' . . . Instead of middle-class areas, Bronzeville tends to have middle-class buildings in all areas, or a few middle-class blocks here and there." Though they may have lived on different streets, blacks of all classes in inner-city areas such as Bronzeville lived in the same community and shopped at the same stores. Their children went to the same schools and played in the same parks. Although there was some class antagonism, their neighborhoods were more stable than the inner-city neighborhoods of today; in short, they featured higher levels of what social scientists call "social organization."

When I speak of social organization I am referring to the extent to which the residents of a neighborhood are able to maintain effective social control and realize their common goals. There are three major dimensions of neighborhood social organization: (1) the prevalence, strength, and interdependence of social networks; (2) the extent of collective supervision that the residents exercise and the degree of personal responsibility they assume in addressing neighborhood problems; and (3) the rate of resident participation in voluntary and formal organizations. Formal institutions (e.g., churches and political party organizations), voluntary associations (e.g., block clubs and parent–teacher organizations), and informal networks (e.g., neighborhood friends and acquaintances, coworkers, marital and parental ties) all reflect social organization.

Neighborhood social organization depends on the extent of local friendship ties, the degree of social cohesion, the level of resident participation in formal and informal voluntary associations, the density and stability of formal organizations, and the nature of informal social controls. Neighborhoods in which adults are able to interact in terms of obligations, expectations, and relationships are in a better position to supervise and control the activities and behavior of children. In neighborhoods with high levels of social organization, adults are empowered to act to improve the quality of neighborhood life – for example, by breaking up congregations of youths on street corners and by supervising the leisure activities of youngsters.

Neighborhoods plagued by high levels of joblessness are more likely to experience low levels of social organization: the two go hand in hand. High rates of joblessness trigger other neighborhood problems that undermine social organization, ranging from crime, gang violence, and drug trafficking to family breakups and problems in the organization of family life.

Consider, for example, the problems of drug trafficking and violent crime. As many studies have revealed, the decline in legitimate employment opportunities among inner-city residents has increased incentives to sell drugs. The distribution of crack in a neighborhood attracts individuals involved in violence and lawlessness. Between 1985 and 1992, there was a sharp increase in the murder rate among men under the age of 24; for men 18 years old and younger, murder rates doubled. Black males in particular have been involved in this upsurge in violence. For example, whereas the homicide rate for white males between 14 and 17 increased from 8 per 100,000 in 1984 to 14 in 1991, the rate for black males tripled during that time (from 32 per 100,000 to 112). This sharp rise in violent crime among younger males has accompanied the widespread outbreak of addiction to crack cocaine. The association is especially strong in inner-city ghetto neighborhoods plagued by joblessness and weak social organization.

Violent persons in the crack-cocaine marketplace have a powerful impact on the social organization of a neighborhood. Neighborhoods plagued by high levels of joblessness, insufficient economic opportunities, and high residential mobility are unable to control the volatile drug market and the violent crimes related to it. As informal controls weaken, the social processes that regulate behavior change.

As a result, the behavior and norms in the drug market are more likely to influence the action of others in the neighborhood, even those who are not involved in drug activity. Drug dealers cause the use and spread of guns in the neighborhood to escalate, which in turn raises the likelihood that others, particularly the youngsters, will come to view the possession of weapons as necessary or desirable for self-protection, settling disputes, and gaining respect from peers and other individuals.

Moreover, as Alfred Blumstein pointed out, the drug industry actively recruits teenagers in the neighborhood "partly because they will work more cheaply than adults, partly because they may be less vulnerable to the punishments imposed by the adult criminal justice system, partly because they tend to be daring and willing to take risks that more mature adults would eschew." Inner-city black youths with limited prospects for stable or attractive employment are easily lured into drug trafficking and therefore increasingly find themselves involved in the violent behavior that accompanies it.

A more direct relationship between joblessness and violent crime is revealed in recent research by Delbert Elliott of the University of Colorado, a study based on National Longitudinal Youth Survey data collected from 1976 to 1989, covering ages 11 to 30. As Elliott points out, the transition from adolescence to adulthood usually results in a sharp drop in most crimes as individuals take on new adult roles and responsibilities. "Participation in serious violent offending behavior (aggravated assault, forcible rape, and robbery) increases [for all males] from ages 11 and 12 to ages 15 and 16, then declines dramatically with advancing age." Although black and white males reveal similar age curves, "the negative slope of the age curve for blacks after age 20 is substantially less than that of whites."

The black–white differential in the proportion of males involved in serious violent crime, although almost even at age 11, increases to 3 : 2 over the remaining years of adolescence, and reaches a differential of nearly 4 : 1 during the late twenties. However, when Elliott compared only *employed* black and white males, he found no significant differences in violent behavior patterns among the two groups by age 21. Employed black males, like white males, experienced a precipitous decline in serious violent behavior following their adolescent period. Accordingly, a major reason for the racial gap in violent behavior after adolescence is joblessness; a large proportion of jobless black males do not assume adult roles and responsibilities, and their serious violent behavior is therefore more likely to extend into adulthood. The new poverty neighborhoods feature a high concentration of jobless males and, as a result, suffer rates of violent criminal behavior that exceed those in other urban neighborhoods.

. . . In 1990, 37 percent of Woodlawn's 27,473 adults were employed and only 23 percent of Oakland's 4,935 adults were working. When asked how much of a problem unemployment was in their neighborhood, 73 percent of the residents in Woodlawn and 76 percent in Oakland identified it as a major problem. The responses to the survey also revealed the residents' concerns about a series of related problems, such as crime and drug abuse, that are symptomatic of severe problems of social organization. Indeed, crime was identified as a major problem by 66 percent of the residents in each neighborhood. Drug abuse was cited as a major problem by as many as 86 percent of the adult residents in Oakland and 79 percent of those in Woodlawn.

Although high-jobless neighborhoods also feature concentrated poverty, high rates of neighborhood poverty are less likely to trigger problems of social organization if the residents are working. This was the case in previous years when the working poor stood out in areas like Bronzeville. Today, the nonworking poor predominate in the highly segregated and impoverished neighborhoods.

The rise of new poverty neighborhoods represents a movement away from what the historian Allan Spear has called an institutional ghetto – whose structure and activities parallel those of the larger society, as portrayed in Drake and Cayton's description of Bronzeville – toward a jobless ghetto, which features a severe lack of basic opportunities and resources, and inadequate social controls.

What can account for the growing proportion of jobless adults and the corresponding increase in problems of social organization in innercity communities such as Bronzeville? An easy answer is racial segregation. However, a race-specific argument is not sufficient to explain recent changes in neighborhoods like Bronzeville. After all, Bronzeville was just as *segregated by skin color in 1950* as it is today, yet the level of employment was much higher then.

Nonetheless, racial segregation does matter. If large segments of the African-American population had not been historically segregated in inner-city ghettos, we would not be talking about the new urban poverty. The segregated ghetto is not the result of voluntary or positive decisions on the part of the residents who live there. As Massey and Denton have carefully documented, the segregated ghetto is the product of systematic racial practices such as restrictive covenants, redlining by banks and insurance companies, zoning, panic peddling by real estate agents, and the creation of massive public housing projects in low-income areas.

Segregated ghettos are less conducive to employment and employment preparation than are other areas of the city. Segregation in ghettos exacerbates employment problems because it leads to weak informal employment networks and contributes to the social isolation of individuals and families, thereby reducing their chances of acquiring the human capital skills, including adequate educational training, that facilitate mobility in a society. Since no other group in society experiences the degree of segregation, isolation, and poverty concentration as do African-Americans, they are far more likely to be disadvantaged when they have to compete with other groups in society, including other despised groups, for resources and privileges.

To understand the new urban poverty, one has to account for the ways in which segregation interacts with other changes in society to produce the recent escalating rates of joblessness and problems of social organization in inner-city ghetto neighborhoods.

"Whose Culture? Whose City?"

from *The Cultures of Cities* (1995)

Sharon Zukin

Editors' Introduction

One way to look at urban society is through the prism of the arts and popular culture. Novels and short stories – as well as paintings, sculptures, musical compositions, and architectural designs – are all created out of specific urban contexts and all reflect, to one degree or another, the social, political, and economic conditions of the urban cultures that gave birth to them. And increasingly, movies, advertising, fashion, and expressions of the "street culture" of urban youth have supplanted high culture as the subjects of postmodern urban cultural analysis.

In "Whose Culture? Whose City?" sociologist Sharon Zukin looks at the way urban public space in New York is increasingly being appropriated and privatized by corporate and commercial forces, thus making a study of urban popular culture one part of a larger analysis of an on-going cultural war on the streets of the city. "In recent years," she writes, "culture has . . . become a more explicit site of conflicts over social differences and urban fears." Many of these conflicts are between middle-class, mostly white citizens on the one hand and homeless, poor, and ethnic minority citizens on the other. In the case of Bryant Park (next to the New York Public Library and directly across the street from Zukin's office at the Graduate Center of the City University of New York), a private nonprofit corporation dominated by middle-class and corporate interests and inspired by William Whyte's ideas (p. 429) took over management of the space to rescue it from homeless panhandlers and drug dealers. Although many of the results seem quite positive – lively cafes and fashion shows brought many more users into the park than ever before – the underlying reality that disturbs Zukin is the gradual loss of meaningful *public* life under the control of inclusive, democratic forces that creeping privatization implies.

Like Jane Jacobs (p. 114), Zukin has a real love for what she proudly calls "my city," and, like Lewis Mumford (p. 92), she has a well-developed sensitivity to city life as a form of social theater.

This selection is excerpted from *The Cultures of Cities* (London: Blackwell, 1995). Other books by Zukin on urban culture are *Landscapes of Power: From Detroit to Disney World* (Berkeley: University of California Press, 1991), *Loft Living: Culture and Capital in Urban* Change (Baltimore, MD: Johns Hopkins University Press, 1982), and, with Michael Sorkin, *After the World Trade Center* (New York: Routledge, 2002) (p. 290).

A consideration of urban culture opens the door to an entire body of analytical literature on the relationship of the arts – both high and popular – to society as a whole. For literature, Richard Lehan's *The City in Literature: An Intellectual and Cultural History* (Berkeley: University of California Press, 1998) and Burton Pike's *The Image of the City in Modern Literature* (Princeton, NJ: Princeton University Press, 1981) are excellent overviews. Raymond Williams's *The Country and the City* (London: Oxford University Press, 1973) is especially insightful on the relationship of Charles Dickens to nineteenth-century London.

For painting, consult Arnold Hauser's chapter on "Impressionism" in *The Social History of Art* (New York: Knopf, 1952) and T.J. Clark's *The Painting of Modern Life* (New York: Knopf, 1985). For an alternative view of privatization, Tyler Cowen's *In Praise of Commercial Culture* (Cambridge, MA: Harvard University Press, 1998) is a stimulating analysis that celebrates the role of the market in arts production and urban culture. For a specifically

American view of the image of the city as a cultural force, see Morton and Lucia White's classic *The Intellectual versus the City: From Thomas Jefferson to Frank Lloyd Wright* (Cambridge, MA: Harvard University Press and MIT Press, 1962).

For the relationship of architecture to urban life – an enormous field – one can begin with Spiro Kostof's *A History of Architecture: Settings and Rituals* (New York: Oxford University Press, 1985), Craig Whitaker's *Architecture and the American Dream* (New York: Three Rivers Press, 1996), and Mark Girouard's *Cities and People: A Social and Architectural History* (New Haven, CT: Yale University Press, 1985). Also of interest is Henri Lefebvre's *The Production of Space* (London: Blackwell, 1991).

Other interesting perspectives on the general relationship of the arts to urban history and culture are Roland Barthes's "Semiology and the Urban" from M. Gottdiener and Alexandros Lagopoulos (eds), *The City and the Sign* (New York: Columbia University Press, 1986), Vera Solberg's *Constructing a Sociology of the Arts* (Cambridge: Cambridge University Press, 1990), Richard Sennett's *Flesh and Stone: The Body and the City in Western Civilization* (New York: W.W. Norton, 1994), and Ray Oldenburg's *The Great Good Place: Cafes, Coffee Shops, Community Centers, Beauty Parlors, General Stores, Bars, Hangouts and How They Get You Through the Day* (New York: Paragon House, 1989).

Cities are often criticized because they represent the basest instincts of human society. They are built versions of Leviathan and Mammon, mapping the power of the bureaucratic machine or the social pressures of money. We who live in cities like to think of "culture" as the antidote to this crass vision. The Acropolis of the urban art museum or concert hall, the trendy art gallery and cafe, restaurants that fuse ethnic traditions into culinary logos – cultural activities are supposed to lift us out of the mire of our everyday lives and into the sacred spaces of ritualized pleasures.

Yet culture is also a powerful means of controlling cities. As a source of images and memories, it symbolizes "who belongs" in specific places. As a set of architectural themes, it plays a leading role in urban redevelopment strategies based on historic prese. vation or local "heritage." With the disappearance of local manufacturing industries and periodic crises in government and finance, culture is more and more the business of cities – the basis of their tourist attractions and their unique, competitive edge. The growth of cultural consumption (of art, food, fashion, music, tourism) and the industries that cater to it fuels the city's symbolic economy, its visible ability to produce both symbols and space.

In recent years, culture has also become a more explicit site of conflicts over social differences and urban fears. Large numbers of new immigrants and ethnic minorities have put pressure on public institutions, from schools to political parties, to deal with their individual demands. Such high culture institutions as art museums and symphony orchestras have been driven to expand and diversify their offerings to appeal to a broader public. These pressures, broadly speaking, are both ethnic and aesthetic. By creating policies and ideologies of "multiculturalism," they have forced public institutions to change.

On a different level, city boosters increasingly compete for tourist dollars and financial investments by bolstering the city's image as a center of cultural innovation, including restaurants, avant garde performances, and architectural design. These cultural strategies of redevelopment have fewer critics than multiculturalism. But they often pit the self-interest of real estate developers, politicians, and expansion-minded cultural institutions against grassroots pressures from local communities.

At the same time, strangers mingling in public space and fears of violent crime have inspired the growth of private police forces, gated and barred communities, and a movement to design public spaces for maximum surveillance. These, too, are a source of contemporary urban culture. If one way of dealing with the material inequalities of city life has been to aestheticize diversity, another way has been to aestheticize fear.

Controlling the various cultures of cities suggests the possibility of controlling all sorts of urban ills, from violence and hate crime to economic decline. That this is an illusion has been amply shown by battles over multiculturalism and its warring factions – ethnic politics and urban riots. Yet the cultural power to create an image, to frame a vision, of the city has become more important as publics have become more mobile and diverse, and traditional institutions – both social classes and political parties – have become less relevant mechanisms of expressing identity. Those who create images stamp a collective identity. Whether they are media corporations like the Disney Company, art museums, or politicians, they are developing new spaces for public cultures. Significant public spaces of the late 19th and early 20th century – such as Central Park, the Broadway theater district, and the top of the Empire State Building – have been joined by Disney World, Bryant Park, and the entertainment-based retail shops of Sony-Plaza. By accepting these spaces without questioning their representations of urban life, we risk succumbing to a visually seductive, privatized public culture.

THE SYMBOLIC ECONOMY

Anyone who walks through midtown Manhattan comes face to face with the symbolic economy. A significant number of new public spaces owe their particular shape and form to the intertwining of cultural symbols and entrepreneurial capital.

- The AT&T Building, whose Chippendale roof was a much criticized icon of postmodern architecture, has been sold to the Japanese entertainment giant Sony; the formerly open public areas at street level have been enclosed as retail stores and transformed into Sony-Plaza. Each store sells Sony products: video cameras in one shop, clothes and accessories related to performers under contract to Sony's music or film division in another . . .
- Two blocks away, Andre Emmerich, a leading contemporary art dealer, rented an empty storefront in a former bank branch to show three huge abstract canvases by the painter Al Held. Entitled *Harry, If I Told You, Would You Know?* the group of paintings was exhibited in raw space, amid falling plaster, peeling paint, exposed wires, and unfinished floors, and passersby viewed the exhibit from the street through large plate glass windows. The work of art was certainly for sale, yet it was displayed as if it were a free, public good; and it would never have been there had the storefront been rented by a more usual commercial tenant.
- On 42nd Street, across from my office, Bryant Park is considered one of the most successful public

spaces to be created in New York City in recent years. After a period of decline, disuse, and daily occupation by vagrants and drug dealers, the park was taken over by a not-for-profit business association of local property owners and their major corporate tenants, called the Bryant Park Restoration Corporation. This group redesigned the park and organized daylong programs of cultural events; they renovated the kiosks and installed new food services; they hired a phalanx of private security guards. All this attracted nearby office workers, both women and men, who make the park a lively midday gathering place, as it had been prior to the mid 1970s – a public park under private control.

Building a city depends on how people combine the traditional economic factors of land, labor, and capital. But it also depends on how they manipulate symbolic languages of exclusion and entitlement. The look and feel of cities reflect decisions about what – and who – should be visible and what should not, on concepts of order and disorder, and on uses of aesthetic power. In this primal sense, the city has always had a symbolic economy. Modern cities also owe their existence to a second, more abstract symbolic economy devised by "place entrepreneurs," officials and investors whose ability to deal with the symbols of growth yields "real" results in real estate development, new businesses, and jobs.

Related to this entrepreneurial activity is a third, traditional symbolic economy of city advocates and business elites who, through a combination of philanthropy, civic pride, and desire to establish their identity as a patrician class, build the majestic art museums, parks, and architectural complexes that represent a world-class city. What is new about the symbolic economy since the 1970s is its symbiosis of image and product, the scope and scale of selling images on a national and even a global level, and the role of the symbolic economy in speaking for, or representing, the city.

[. . .]

The growth of the symbolic economy in finance, media, and entertainment may not change the way entrepreneurs do business. But it has already forced the growth of towns and cities, created a vast new work force, and changed the way consumers and employees think. In the early 1990s, employment in "entertainment and recreation" in the United States grew slightly more than in health care and six times more than in the auto industry. The facilities where these employees work – hotels, restaurants, expanses of new construction and undeveloped land – are more than just workplaces. They reshape geography and ecology; they are places of creation and transformation.

The Disney Company, for example, makes films and distributes them from Hollywood. It runs a television channel and sells commercial spinoffs, such as toys, books, and videos, from a national network of stores. Disney is also a real estate developer in Anaheim, Orlando, France, and Japan and the proposed developer of a theme park in Virginia and a hotel and theme park in Times Square. Moreover, as an employer, Disney has redefined work roles. Proposing a model for change in the emerging service economy, Disney has shifted from the white-collar worker described by C. Wright Mills in the 1950s to a new chameleon of "flexible" tasks. The planners at its corporate headquarters are "imagineers"; the costumed crowd-handlers at its theme parks are "cast members." Disney suggests that the symbolic economy is more than just the sum of the services it provides. The symbolic economy unifies material practices of finance, labor, art, performance, and design.

The prominence of culture industries also inspires a new language dealing with difference. It offers a coded means of discrimination, an undertone to the dominant discourse of democratization. Styles that develop on the streets are cycled through mass media, especially fashion and "urban music" magazines and MTV, where, divorced from their social context, they become images of cool. On urban billboards advertising designer perfumes or jeans, they are recycled to the streets, where they become a provocation, breeding imitation and even violence. The beachheads of designer stores, from Armani to A/X, from Ralph Lauren to Polo, are fiercely parodied for the "props" of fashion-conscious teenagers in inner city ghettos. The cacophony of demands for justice is translated into a coherent demand for jeans. Claims for public space by culture industries inspire the counterpolitics of display in late 20th century urban riots.

The symbolic economy recycles real estate as it does designer clothes. Visual display matters in American and European cities today, because the identities of places are established by sites of delectation. The sensual display of fruit at an urban farmers' market or gourmet food store puts a neighborhood "on the map" of visual delights and reclaims it for gentrification. A sidewalk cafe takes back the street from casual workers and homeless people. In Bryant Park, enormous white tents and a canopied walkway set the scene for spring and fall showings of New York fashion designers. Twice a year, the park is filled by the fashion media, paparazzi, store buyers, and supermodels doing the business of culture and reclaiming Bryant Park as a vital, important place. We New Yorkers become willing participants in the drama of the fashion business. As cultural consumers, we are drawn into the interrelated production of symbols and space.

Mass suburbanization since the 1950s has made it unreasonable to expect that most middle-class men and women will want to live in cities. But developing small places within the city as sites of visual delectation creates urban oases where everyone *appears* to be middle class. In the fronts of the restaurants or stores, at least, consumers are strolling, looking, eating, drinking, sometimes speaking English and sometimes not. In the back regions, an ethnic division of labor guarantees that immigrant workers are preparing food and cleaning up. This is not just a game of representations: developing the city's symbolic economy involves recycling workers, sorting people in housing markets, luring investment, and negotiating political claims for public goods and ethnic promotion. Cities from New York to Los Angeles and Miami seem to thrive by developing small districts around specific themes. Whether it is Times Square or el Calle Ocho, a commercial or an "ethnic" district, the narrative web spun by the symbolic economy around a specific place relies on a vision of cultural consumption and a social and an ethnic division of labor.

[. . .]

I also see public culture as socially constructed on the micro-level. It is produced by the many social encounters that make up daily life in the streets, shops, and parks – the spaces in which we experience public life in cities. The right to be in these spaces, to use them in certain ways, to invest them with a sense of our selves and our communities – to claim them as ours and to be claimed in turn by them – make up a constantly changing public culture. People with economic and political power have the greatest opportunity to shape public culture by controlling the building of the city's public spaces in stone and concrete. Yet public space is inherently democratic. The question of who can occupy public space, and so define an image of the city, is open-ended.

Talking about the cultures of cities in purely visual terms does not do justice to the material practices of politics and economics that create a symbolic economy. But neither does a strictly political-economic approach suggest the subtle powers of visual and spatial strategies of social differentiation. As I suggested in *Landscapes of Power* (1991), the rise of the cities' symbolic economy is rooted in two long-term changes – the economic decline of cities compared to suburban and nonurban spaces and the expansion of abstract financial speculation – and in such short-term factors, dating from the 1970s and 1980s, as new mass immigration, the growth of cultural consumption, and the marketing of identity politics. This is an inclusive, structural, and materialist view. If I am right, we cannot speak about cities today without understanding:

- how cities use culture as an economic base,
- how capitalizing on culture spills over into the privatization and militarization of public space, and
- how the power of culture is related to the aesthetics of fear.

CULTURE AS AN ECONOMIC BASE

Suppose we turn the old Marxist relation between a society's base and its superstructure on its head and think of culture as a way of producing basic goods. In fact, culture supplies the basic information – including symbols, patterns, and meaning – for nearly all the service industries. In our debased contemporary vocabulary, the word *culture* has become an abstraction for any economic activity that does not create material products like steel, cars, or computers. Stretching the term is a legacy of the advertising revolution of the early 20th century and the more recent escalation in political image making. Because culture is a system for producing symbols, every attempt to get people to buy a product becomes a culture industry. The sociologist Daniel Bell used to tell a joke about a circus employee whose job it was to follow the elephant and clean up after it; when asked, she said her job was in "the entertainment business." Today, she might say she was in "the culture industry." Culture is intertwined with capital and identity in the city's production systems.

From one point of view, cultural institutions establish a competitive advantage over other cities for attracting new businesses and corporate elites. Culture suggests the coherence and consistency of a brand name product. Like any commodity, "cultural" landscape has the possibility of generating other commodities. Historically, of course, the arrow of causality goes the other way. Only an economic surplus – sufficient to fund sacrifices for the temple, Michelangelos for the chapel, and bequests to art museums in the wills of robber barons – generates culture. But in American and European cities during the 1970s, culture became more of an instrument in the entrepreneurial strategies of local governments and business alliances. In the shift to a post-postwar economy, who could build the biggest modern art museum suggested the vitality of the financial sector. Who could turn the waterfront from docklands rubble to parks and marinas suggested the possibilities for expansion of the managerial and professional corps. This was probably as rational a response as any to the unbeatable isolationist challenge of suburban industrial parks and office campuses. The city, such planners and developers as James Rouse believed, would counter the visual homogeneity of the suburbs by playing the card of aesthetic diversity.

Yet culture also suggests a labor force that is well suited to the revolution of diminished expectations that began in the 1960s. In contrast to high-rolling rappers and rockers, "high" cultural producers are supposed to live on the margins; and the incomes of most visual artists, art curators, actors, writers, and musicians suggest they must be used to deprivation. A widespread appreciation of culture does not really temper the work force's demands. But, in contrast to workers in other industries, artists are flexible on job tasks and work hours, do not always join labor unions, and present a docile or even "cultured" persona. These qualities make them, like immigrants, desirable employees in service industries. Dissatisfaction with menial and dead-end jobs does not boil over into protest because their "real" identity comes from an activity outside the job.

[. . .]

Whether there is a singular, coherent vision no longer depends on the power of a single elite group. Constant political pressures by interest groups and complex interwoven networks of community groups, corporations, and public officials signal multiple visions. The ability to arrange these visions artfully, to orchestrate and choreograph images of diversity to speak for a larger whole, has been claimed by major nonprofit cultural institutions. This is especially true of art museums.

Since the 1980s, museums have fallen victim to their own market pressures. Reduced government funding and cutbacks in corporate support have made them more dependent than ever on attracting paying visitors ("gate"). They rely on their gift shops to contribute a larger share of their operating expenses. They try out new display techniques and seek crowd-pleasing exhibit ideas. In an attempt to reach a broader public, the Metropolitan Museum of Art and the Museum of Modern Art in New York have upgraded their restaurants and offer jazz performances on week-end evenings. Yet financial pressures have also led museums to capitalize on their visual holdings. By their marketing of cultural consumption, great art has become a public treasure, a tourist attraction, and a representation – divorced from the social context in which the art was produced – of public culture. Like Calvin Klein jeans on a bus stop billboard, the work of art and the museum itself have become icons of the city's symbolic economy.

[. . .]

PUBLIC SPACE

The fastest growing kind of public space in America is prisons. More jails are being built than housing, hospitals, or schools. No matter how well designed or brightly painted they may be, prisons are still closely guarded, built as cheaply as possible, and designed for surveillance. I can think of more pleasant public spaces, especially parks that I use in New York City. But is the Hudson River Park, near Battery Park City, or Bryant Park, on 42nd Street, less secure or exclusive than a prison? They share with the new wave of prison building several characteristics symptomatic of the times. Built or rebuilt as the city is in severe financial distress, they confirm the withdrawal of the public sector, and its replacement by the private sector, in defining public space. Reacting to previous failures of public space – due to crime, a perceived lower-class and minority-group presence, and disrepair – the new parks use design as an implicit code of inclusion and exclusion. Explicit rules of park use are posted in the parks and enforced by large numbers of sanitation workers and security guards, both public and private. By cleaning up public space, nearby property owners restore the attractiveness of their holdings and reconstruct the image of the city as well.

It is important to understand the histories of these symbolically central public spaces. The history of Central Park, for example, shows how, as definitions of who should have access to public space have changed, public cultures have steadily become more inclusive and democratic. From 1860 to 1880, the first uses of the park – for horseback riders and carriages – rapidly yielded to sports activities and promenades for the mainly immigrant working class. Over the next 100 years, continued democratization of access to the park developed together with a language of political equality. In the whole country, it became more difficult to enforce outright segregation by race, sex, or age.

[. . .]

In 1989, a private organization that manages Central Park, the Central Park Conservancy, demanded demolition of the Naumberg Bandshell, site of popular concerts from the 1930s to the 1950s, where homeless people gathered. Similarly, the Bryant Park Restoration Corporation started cleaning up the midtown business district by adopting the social design principles developed by William H. Whyte. Whyte's basic idea is that public spaces are made safe by attracting lots of "normal" users. The more normal users there are, the less space there will be for vagrants and criminals to maneuver. The Bryant Park Restoration Corporation intended their work to set a prototype for urban public space. They completely reorganized the landscape design of the park, opening it up to women, who tended to avoid the park even during daylight, and selling certain kinds of buffet food. They established a model of pacification by cappuccino.

Central Park, Bryant Park, and the Hudson River Park show how public spaces are becoming progressively less public: they are, in certain ways, more exclusive than at any time in the past 100 years. Each of these areas is governed, and largely or entirely financed, by a private organization, often working as a quasi-public authority. These private groups are much better funded than the corresponding public organization. Design in each park features a purposeful vision of urban leisure. A heightened concern for security inspires the most remarkable visible features: gates, private security guards, and eyes keeping the space under surveillance. The underlying assumption is that of a paying public, a public that values public space as an object of visual consumption. Yet it has become inconceivable in public discussions that control of the parks be left in public hands. When the *New York Times* praised plans to require developers to provide public access to the city's extensive water-front, the newspaper said that only a public–private partnership could raise the funds to maintain such a significant public space.

A major reason for privatization of some public parks is that city governments cannot pay for taking care of them. Since the 1960s, while groups of all sorts have requested more use of the parks, the New York City Parks Department has been starved of government funds. Half the funding for Central Park is now raised privately by the Central Park Conservancy, which enjoys a corresponding influence on parks policy. Founded by private donors in 1980, the conservancy's original mission was to raise funds in the private sector to offset the park's physical deterioration. But it soon developed an authoritative cultural voice. The conservancy publicly defends the intentions of Olmsted and Vaux, the park's original designers, to create a "natural" landscape for contemplation. Most often, they beautify the park by restoring its 19th century buildings and bridges or setting up a nature program or skating facilities on one of its landscaped ponds. The conservancy has also become an arbiter between groups that want to use the park for sports or demonstrations, thus mediating between the homeless and the joggers, between athletes who come to the park from all over the city and those who come from low-income neighborhoods on the park's northern borders. The conservancy, moreover, has spoken loudly and often in favor of hiring nonunion labor . . .

In midtown, Bryant Park is an even more aggressive example of privatization. Declared a New York City landmark in 1975, the nine-acre park is essentially run by the Bryant Park Restoration Corporation, whose biggest corporate members are Home Box Office (HBO), a cable television network, and NYNEX, a regional telecommunications company. Like the Central Park Conservancy, the Bryant Park Restoration Corporation raises most of the park's budget, supervises maintenance, and decides on design and amenities.

The design of Bryant Park, in 1934, was based on an Olmstedian separation of a rural space of contemplation from the noisy city. By the late 1970s, this was determined to have the effect of walling off the park's intended public of office workers outside from drug dealers and loiterers inside. When the restoration corporation was formed, it took as its major challenge the development of a new design that would visually and spatially ensure security. The wall around the park was lowered, and the ground leveled to bring it closer to the surrounding streets. The restoration corporation bought movable chairs and painted them green, as in Parisian parks, responding to William H. Whyte's suggestion that park users like to create their own small spaces. Whyte recommended keeping "the undesirables" out by making a park *attractive*. Victorian kiosks selling cappuccino and sandwiches were built and painted, paths were repaved and covered with pebbles, a central lawn was opened up, and performers were enlisted to offer free entertainment in the afternoons. The restoration corporation hired its own security guards and pressured the New York City Police Department to supply uniformed officers. Four uniformed New York City police officers and four uniformed private security guards are on duty all day.

Plainclothes private security guards are also on patrol. A list posted at all entrances prohibits drug use, picking flowers, and drinking alcohol except for beverages bought at park concessions, which are limited to certain seating areas. It states the park's hours, 9 a.m. to 7 p.m., coinciding roughly with the business day. The rules specify that only homeless people connected to a particular shelter in the neighborhood have the right to rummage through the garbage cans for returnable bottles and cans. Unlike Parks Department workers, Bryant Park maintenance workers do not belong to a labor union. Starting salary for a maintenance worker is $6 an hour, half the starting rate of unionized workers in other city parks.

On a sunny summer day at noon, Bryant Park is full of office workers out to lunch – between 1,500 and 6,000 of them. The movable chairs and benches are filled; many people are sitting on the grass, on the edge of the fountain, even on the pebbled paths. Men and women eat picnic lunches singly, in couples, and in groups. Some traditional social hierarchies are subverted. Women feel free to glance at men passing by. Most men do not ogle the women. The dominant complexion of park users is white, with minority group members clustered outside the central green. Few people listen to the subsidized entertainment, an HBO comedian shouting into a microphone; no one notices when she finishes the show. A large sculpture by Alexander Calder stands in the middle of the lawn, on loan from an art gallery, both an icon and a benediction on the space. At sunset in the summer, HBO shows free movies from their stock of old films, a "take back the night" activity similar to those now being tried in other cities. This is a very deliberate exception to the rule of closing the park at night. During lunchtime, at least, the park visually represents an urban middle class: men and women who work in offices, jackets off, sleeves rolled up, mainly white. On the same day, at the same hour, another public space a block away – the tellers' line at Citibank – attracts a group that is not so well dressed, with more minority group members. The cultural strategies that have been chosen to revitalize Bryant Park carry with them the implication of controlling diversity while re-creating a consumable vision of civility.

[. . .]

Since its renovation, Bryant Park has changed character. It has become a place for people to be with others, to see others, a place of public sociability. John Berger once criticized New Yorkers for eating while walking alone on the street, alienating a social ritual from its proper context. Yet now, in the park, eating becomes a public ritual, a way of trusting strangers while maintaining private identities. Because of the police and security guards, the design, and the food, the park has become a visual and spatial representation of a middle-class public culture. The finishing touch will be a privately owned, expensive restaurant, whose rent payments will help finance the park's maintenance. This, however, is a degree of privatization that has stirred prolonged controversy. First envisioned in the 1980s, the restaurant remained the subject of public approvals processes until 1994.

The disadvantage of creating public space this way is that it owes so much to private-sector elites, both individual philanthropists and big corporations. This is especially the case for centrally located public spaces, the ones with the most potential for raising property values and with the greatest claim to be symbolic spaces for the city as a whole. Handing such spaces over to corporate executives and private investors means giving them carte blanche to remake public culture. It marks the erosion of public space in terms of its two basic principles: public stewardship and open access.

The Central Park Conservancy, a group of 30 private citizens who choose their own replacements, represents large corporations with headquarters in the city, major financial institutions, and public officials. The membership echoes both the new (the nonelected, tripartite Emergency Financial Control Board that has overseen New York City's budget since the fiscal crisis of 1975) and the old (the board of "gentlemen" trustees that originally guided the planning of Central Park in the 1860s) . . .

The Bryant Park Restoration Corporation, a subsidiary of the Bryant Park Business Improvement District, follows a fairly new model in New York State, and in smaller cities around the United States, that allows business and property owners in commercial districts to tax themselves voluntarily for maintenance and improvement of public areas and take these areas under their control. The concept originated in the 1970s as special assessment districts; in the 1980s, the name was changed to a more upbeat acronym, business improvement districts (BIDs). A BID can be incorporated in any commercial area. Because the city government has steadily reduced street cleaning and trash pickups in commercial streets since the fiscal crisis of 1975, there is a real incentive for business and property owners to take up the slack . . .

What kind of public culture is created under these conditions? Do urban BIDs create a Disney World in the streets, take the law into their own hands, and reward their entrepreneurial managers as richly as property values will allow? If elected public officials continue to urge the destruction of corrupt and bankrupt public institutions, I imagine a scenario of drastic privatization, with BIDs replacing the city government. As Republican Mayor Rudolph Giuliani said enthusiastically at the second annual NYC BIDs Association Conference in 1994, "This is a difficult time for the city and the country as we redefine ourselves. BIDs are one of the true success stories in the city. It's a tailormade form of local government."

[. . .]

In their own way, under the guise of improving public spaces, BIDs nurture a visible social stratification. Like the Central Park Conservancy, they channel investment into a central space, a space with both real and symbolic meaning for elites as well as other groups. Like the Central Park Conservancy, the resources of the rich Manhattan BIDs far outstrip those even potentially available in other areas of the city, even if those areas set up BIDS. The rich BIDs' opportunity to exceed the constraints of the city's financial system confirms the fear that the prosperity of a few central spaces will stand in contrast to the impoverishment of the entire city.

BIDs can be equated with a return to civility, "an attempt to reclaim public space from the sense of menace that drives shoppers, and eventually store owners and citizens, to the suburbs." But rich BIDs can be criticized on the grounds of control, accountability, and vision. Public space that is no longer controlled by public agencies must inspire a liminal public culture open to all but governed by the private sector. Private management of public space does create some savings: saving money by hiring nonunion workers, saving time by removing design questions from the public arena. Because they choose an abstract aesthetic with no pretense of populism, private organizations avoid conflicts over representations of ethnic groups that public agencies encounter when they subsidize public art, including murals and statues.

Each area of the city gets a different form of visual consumption catering to a different constituency: culture functions as a mechanism of stratification. The public culture of midtown public space diffuses down through the poorer BIDs. It focuses on clean design, visible security, historic architectural features, and the sociability among strangers achieved by suburban shopping malls. Motifs of local identity are chosen by merchants and commercial property owners. Since most commercial property owners and merchants do not live in the area of their business or even in New York City, the sources of their vision of public culture may be eclectic: the nostalgically remembered city, European piazzas, suburban shopping malls, Disney World. In general, however, their vision of public space derives from commercial culture.

[. . .]

SECURITY, ETHNICITY, AND CULTURE

One of the most tangible threats to public culture comes from the politics of everyday fear. Physical assaults, random violence, hate crimes that target specific groups: the dangers of being in public spaces utterly destroy the principle of open access. Elderly men and women who live in cities commonly experience fear as a steady erosion of spaces and times available to them. An elderly Jewish politician who in the 1950s lived in Brownsville, a working-class Jewish neighborhood in Brooklyn where blacks began to move in in greater numbers as whites moved out, told me, "My wife used to be able to come out to meet me at night, after a political meeting, and leave the kids in our apartment with the door unlocked." A Jewish woman remembers about that same era, "I used to go to concerts in Manhattan wearing a fur coat and come home on the subway at 1 a.m." There may be some exaggeration in these memories, but the point is clear.

And it is not altogether different from the message behind crimes against black men who venture into mainly white areas of the city at night or attacks on authority figures such as police officers and firefighters who try to exercise that authority against street gangs, drug dealers, and gun-toting kids. Cities are not safe enough for people to participate in a public culture.

"Getting tough" on crime by building more prisons and imposing the death penalty are all too common answers to the politics of fear. "Lock up the whole population," I heard a man say on the bus, at a stroke reducing the solution to its ridiculous extreme. Another answer is to privatize and militarize public space – making streets, parks, and even shops more secure but less free, or creating spaces, such as shopping malls and Disney World, that only *appear* to be public spaces because so many people use them for common purposes. It is not so easy, given a language of social equality, a tradition of civil rights, and a market economy, to enforce social distinctions in public space. The flight from "reality" that led to the privatization of public space in Disney World is an attempt to create a different, ultimately more menacing kind of public culture.

In *City of Quartz* (1990), Mike Davis describes the reshaping of public spaces in Los Angeles by surveillance and security procedures. Helicopters buzz the skies over ghetto neighborhoods, police hassle teenagers as putative gang members, homeowners buy into the type of armed defense they can afford . . . or have nerve enough to use. While Los Angeles may represent an extreme, high-tech example, I have also seen "Eyes on the Street" surveillance signs on lamp posts in small towns in Vermont and the design of Bryant Park gives evidence of a relatively low-tech but equally suggestive concern for public order. Indeed, Bryant Park may be a more typical public space than downtown Los Angeles because it has been "secured" within a democratic discourse of aestheticizing both cities and fear.

Gentrification, historic preservation, and other cultural strategies to enhance the visual appeal of urban spaces developed as major trends during the late 1960s and early 1970s. Yet these years were also a watershed in the institutionalization of urban fear. Voters and elites – a broadly conceived middle class in the United States – could have faced the choice of approving government policies to eliminate poverty, manage ethnic competition, and integrate everyone into common public institutions. Instead, they chose to buy protection, fueling the growth of the private security industry . . .

From the viewpoint of political economy, the withdrawal from public to private security employees is part of a general shift to privatization. Fiscal austerity limits government spending increases, even on the police. Yet private security cannot be free. The security costs borne by the private sector are passed on to the public by excluding potential criminal acts from segregated spaces, leaving the rest of the city to watch out for itself and be watched by the police. Crime, the criminal justice system, and private security forces absorb a high percentage of the unemployed, a "reserve army" in a more literal sense than Marx intended in his famous phrase about "the reserve army of the unemployed." While factory jobs disappear, urban workers, especially minority-group members, seek security jobs, and their mainly white colleagues in small, rural towns go to work in prisons.

[. . .]

In the past, those people who lived so close together they had to work out some etiquette for sharing, or dividing, public space were usually the poor. An exception that affected everyone was the system of racial segregation that worked by law in the south and by convention in many northern states until the 1960s, when – not surprisingly – perceptions of danger among whites increased. Like segregation, a traditional etiquette of public order of the urban poor involves dividing up territory by ethnic groups . . . Among city dwellers today, innumerable informal etiquettes for survival in public spaces flourish. The "streetwise" scrutiny of passersby . . . is one means for unarmed individuals to secure the streets. I think ethnicity – a cultural strategy for producing difference – is another, and it survives on the politics of fear by requiring people to keep their distance from certain aesthetic markers. These markers vary over time. Pants may be baggy or pegged, heads may be shaggy or shaved. Like fear itself, ethnicity becomes an aesthetic category.

[. . .]

In such a landscape, there are no safe places. The Los Angeles uprising of 1992 showed that, unlike in earlier riots, the powerless respect fewer geographical boundaries, except perhaps the neighborhoods where rich people live. Carjackings – the ultimate American violence – occur on the highway and in the parking lots of fast food restaurants . . . Patrons of 24-hour automatic teller machines are robbed so often that the NYCE Network, with 10,000 machines in New York

City, distributes a pamphlet of safety tips worthy of a military base: "As you approach an ATM, be aware of your surroundings . . . When using an ATM at night, be sure it is located in a well lit area. And consider having someone accompany you." Someone, that is, other than the homeless man who stood by the door with an empty paper cup in his hand, until the New York City Council passed a law that forbade panhandlers to stand within 15 feet of an ATM. Or, as a Spanish-language subway advertisement cautions, "Mantengase alerta. Sus ojos, oidos y instinto son sus recursos naturales de seguridad en la ATM." In Chicago and Los Angeles, ATMs have been installed in police stations, so residents with bank accounts in the poorest neighborhoods will have a safe place to get cash.

For a brief moment in the late 1940s and early 1950s, working-class urban neighborhoods held the possibility of integrating white Americans and African-Americans in roughly the same social classes. This dream was laid to rest by movement to the suburbs, continued ethnic bias in employment, the decline of public services in expanding racial ghettos, criticism of integration movements for being associated with the Communist party, and fear of crime. Over the next 15 years, enough for a generation to grow up separate, the inner city developed its stereotyped image of "Otherness." The reality of minority groups' working-class life was demonized by a cultural view of the inner city "made up of four ideological domains: a physical environment of dilapidated houses, disused factories, and general dereliction; a romanticized notion of white working-class life with particular emphasis on the centrality of family life; a pathological image of black culture; and a stereotypical view of street culture."

By the 1980s, the development of a large black middle class with incomes more or less equal to white households' and the increase in immigrant groups raised a new possibility of developing ethnically and racially integrated cities. This time, however, there is a more explicit struggle over who will occupy the image of the city. Despite the real impoverishment of most urban populations, the larger issue is whether cities can again create an inclusive public culture. The forces of order have retreated into "small urban spaces," like privately managed public parks that can be refashioned to project an image of civility. Guardians of public institutions – teachers, cops – lack the time or inclination to *understand* the generalized ethnic Other . . .

Yet the groups that have inherited the city have a claim on its central symbolic spaces. Not only to the streets that serve as major parade routes, not only to the central parks, but also to the monumental spaces that confirm identity by offering visual testimony to a group's presence in history.

[. . .]

Many social critics have begun to write about new public spaces formed by the "transactional space" of telecommunications and computer technology, but my interest . . . is in public spaces as places that are physically *there*, as geographical and symbolic centers, as points of assembly where strangers mingle. Many Americans, born and raised in the suburbs, accept shopping centers as the preeminent public spaces of our time. Yet while shopping centers are undoubtedly gathering places, their private ownership has always raised questions about whether all the public has access to them and under what conditions. In the 1980s and 1990s, shopping centers became sites for hotels, post offices, and even schools, suggesting that public institutions can indeed function on private property . . .

When Disneyland recruited teenagers in South Central Los Angeles for summer jobs following the riots of 1992, it thrust into prominence a new confluence between the sources of contemporary public culture: a confluence between commercial culture and ethnic identity. Defining public culture in these terms recasts the way we view and describe the cultures of cities. Real cities are both material constructions, with human strengths and weaknesses, and symbolic projects developed by social representations, including affluence and technology, ethnicity and civility, local shopping streets and television news. Real cities are also macro-level struggles between major sources of change – global and local cultures, public stewardship and privatization, social diversity and homogeneity – and mid-level negotiations of power. Real cultures, for their part, are not torn by conflict between commercialism and ethnicity; they are made up of one-part corporate image selling and two-parts claims of group identity, and get their power from joining autobiography to hegemony – a powerful aesthetic fit with a collective lifestyle. This is the landscape of a symbolic economy . . . on sites as geographically and socially diverse as Disney World, the Massachusetts Museum of Contemporary Art, New York art worlds, Times Square, New York City restaurants, and ghetto shopping streets like 125th Street in Harlem. These are my sources; this is my "city."

How do we connect what we experience in public space with ideologies and rhetorics of public culture?

On the streets, the vernacular culture of the powerless provides a currency of economic exchange and a language of social revival. In other public spaces – grand plazas, waterfronts, and shopping streets reorganized by business improvement districts – another landscape incorporates vernacular culture or opposes it with its own image of identity and desire. Fear of reducing the distance between "us" and "them," between security guards and criminals, between elites and ethnic groups, makes culture a crucial weapon in reasserting order. Militant rhetoric belongs to the forces of order. "We will fight for every house in the city," said the New York City Police Commissioner in his inaugural address in 1993. "We will fight for every street. We will fight for every borough. And we will win." This Churchillian call echoes the appeal of right-wing journalist Patrick Buchanan, who explicitly identified cities and culture when he addressed the 1992 Republican National Convention in Houston: "And as those boys [in the National Guard] took back the streets of Los Angeles, block by block, my friends, we must take back our cities, and take back our culture, and take back our country."

But whose city? I ask. And whose culture?

"Visions of a New Reality: The City and the Emergence of Modern Visual Culture"

(1999)

Frederic Stout

A fundamental precept of Marxist cultural analysis is that superstructures of thought and artistic expression rest upon and derive from a material base rooted in social and economic reality. Thus, each historical era creates characteristic forms of expression and explanatory discourse that reflect, indeed construct, the social reality of the period. Lewis Mumford spoke to this process when, in *The City in History* (1961), he wrote that in "the Book of Job, one beholds Jerusalem; in Plato, Sophocles, and Euripides, Athens; in Shakespeare and Marlowe . . . Elizabethan London."

For the cities of the Industrial Revolution, a number of forms of expression and modes of critical analysis arose to make sense of the dramatic and rapidly changing social reality. In literature, Balzac and Dickens pioneered a tradition of social realism particularly focused on the struggles of the urban poor which later attracted such followers as Emile Zola, Theodore Dreiser, Mrs. Gaskell, Upton Sinclair . . . and thousands more. In the realm of social and political analysis, the use of statistical evidence based on survey data compiled by government commissions and philanthropic organizations, the construction of great explanatory theories (such as those of Marx and Weber), and the development of social science methodologies helped observers of the urban milieu make sense of the revolutionary changes that were taking place around them. And in the realm of art, a whole new kind of visual culture emerged rooted in the observation of the new urban reality, both social and physical. It was a culture that began with nineteenth-century popular illustrated journalism and evolved to include main-stream traditions in the twentieth-century history of photography and cinema.

In the history of the fine arts, urban social realism plays a brief but highly visible role. Although much artistic social activism was largely confined to the satirical printmakers (except in those countries where state-sponsored socialist realism became the pre-scribed orthodoxy), works like William Hogarth's depictions of eighteenth-century London and Honoré Daumier's glimpses of nineteenth-century Paris are as influential as they are familiar. In painting, however, the realism of Courbet and Millet quickly give way to the impressionist celebration of light and the post-impressionist analysis of pure form. In America, the urban imagery of The Eight (dubbed "The Ashcan School" by a hostile critic) was revolutionary and shocking at their group show in 1907 but strictly old hat by the time modern abstraction was introduced at the Armory Show in 1912. One explanation for this is that The Eight stood not at the beginning but at the end of a fifty-year social realist tradition in the visual arts, a tradition that took place not in the salons of the easel painters but in the pages of the illustrated newspapers and magazines that flourished as a ubiquitous element of urban popular culture during the last half of the nineteenth century. Several of the Ashcan School painters – among them John Sloan, William Glackens, Everett Shinn, and Edward Hopper – had previously worked as journalistic illustrators, and it is to the pages of the popular newspapers that one must turn to see the earliest representations of the modern industrial city.

Illustrated journalism began in England as early as the 1820s, and in America in the 1840s, in response to a growing demand for information and entertainment on the part of a marginally literate but fully enfranchised working class and petit bourgeoisie. The first images of cities were almost always long-range or bird's-eye views (similar to the view of La Crosse, Wisconsin, from a 1887 issue of *Frank Leslie's Illustrated Newspaper* that appears as Plate 5). These static images were imitative of a received landscape tradition that had been commonly applied to rural and wild nature scenes, but they conveyed relatively little information about urban subject matter – note the inclusion of a street-scene cut-out in the La Crosse picture – and were soon replaced by more kinetic images and vignettes. "Contrast pictures" – both dual images such as "Music in the Street/Music in the Parlor" (Plate 9) from an 1868 issue of *The Illustrated News* and single images such as "The Hearth-stone of the Poor" (Plate 10) from an 1876 issue of *Harper's Weekly* by the exceptionally talented Sol Eytinge, Jr. – were particularly useful for comparing the lives of the urban rich and poor. Soon, a new kind of composite image emerged – "City Sketches" by the illustrator C.A. Barry from an 1855 issue of *Ballou's Pictorial Drawing-Room Companion* (Plate 11) is an interesting early example – that more accurately reflected the diverse, jumbled-together chaos that was the common experience of urban street life in the nineteenth century. Eventually, such composite pictures became a staple of popular illustrated journalism, and the step-by-step depiction of industrial processes became a favored topic, as in "Bicycles and Tricycles – How They Are Made" (Plate 12) from an 1887 issue of *Frank Leslie's Illustrated Newspaper.*

Handdrawn popular illustration died out around the turn of the century when a new technology of visual representation, photography, took over the journalistic duties of the newspaper illustrators. As a genre, popular illustration had accomplished much. It had created a visual culture embedded in the social reality of urban life and had urged visual art generally away from landscape toward cityscape, from stasis to kinesis. The contribution of photography would be even greater. Here was a technology of visual representation perfectly suited to its age. In the hands of journalists, it created a new and powerfully accurate kind of documentary record. In the hands of social activists, it created images of passion and outrage that could not be ignored. In the hands of the popular masses, it created snapshots and memories, both social and personal. And in the hands of artists – and who with a camera in his hands was not an artist? – it created art.

Photography is arguably the single most characteristic medium of expression and representation of the modern period, and a complete history of photography would include consideration of its use in science, medicine, criminology, education, and commerce as well as art. But as a medium of artistic expression, photography perfectly exemplified the spirit of the modern age, addressing every known genre and inventing new ones of its own. Nature and landscape, still life and the nude, staged heroic set pieces and candid genre scenes – all found new vitality in the hands of the masters of the new technology of visual representation. But it was urban subjects more than any others – the inescapable social and physical reality of the modern city – that captured photography's unique potential as an expressive medium and gave photography its historic and artistic *raison d'être.*

Although photography did not completely supplant handdrawn illustrations in journalism until the 1890s and early 1900s, the very first news photo may well have been "Burning Mills, Oswego, New York" (Plate 13), a daguerreotype by George N. Barnard that was printed in the *Oswego Daily Times* in 1853. While pictures of factory fires had been a staple of popular illustration, photography brought a new realism and intensity to the images presented to the public. Even today, dramatic pictures of big fires, even if they are from cities far away, have a special fascination and are often prominently placed in newspapers and the evening television news broadcasts.

The spectacle of the burning factory was a special case in a larger melodrama: the ongoing, day-to-day reality of modern urban life. Many of the very earliest photographs by Daguerre, Nadar, and Fox Talbot were simple street scenes made by the photographer simply turning his camera out the studio window. Like the "city sketches" type of popular illustration, these photographic street scenes conveyed a special sense of the vitality of urban life. Edward Anthony's "A Rainy Day on Broadway" (Plate 14) shows New York in 1859. The motion of those on the street and on the sidewalks is frozen in time – clearly Anthony's shutter speed was fast enough to eliminate most of the blurring that was so common in early photographs of moving subjects – but the image itself does not seem frozen. Rather, the image speaks directly to the viewer's own experience of street life, and the interaction of the viewed image and the viewer's imagination almost magically captures the hustle and bustle of New York street life.

Not all streets in the modern city were so pleasant. In "Bandits' Roost, 39½ Mulberry Street," of 1889 (Plate 15), Jacob Riis invites the viewer to look into one of New York's most notorious and dangerous back alleys. The author of *How the Other Half Lives* (1890), Riis was a crusading social reformer who had observed the dark side of the city as a police reporter for the local newspapers and he used his skills as a photographer to supplement his activism. Many middle-class viewers were undoubtedly repelled by images such as this one. It was the kind of view that one might catch fleetingly as one passed along a main thoroughfare, something to glance at quickly and then turn away. But Riis the reformer would have us not look away, and photography rivets the gaze on an unpleasant but inescapable social reality. In this, Riis's photographic work is directly linked to Friedrich Engels's descriptions of the Manchester slums in *The Condition of the Working Class in England in 1844*.

Jacob Riis was but one in a long and distinguished line of photographers who turned their art and their talents to the ends of political activism and social protest. One of the greatest of these visual documentors of the modern city was Lewis Hine. Hine's picture of a young girl tending a massive cloth-weaving machine in a North Carolina mill has become a nearly universal icon of industrial child labor, and his photographs of high-steel construction workers casually toiling hundreds of feet above the city are gut-wrenching images of working-class heroism. For a time, Hine worked at Ellis Island, the New York port of entry for immigrants to America. His portraits of the immigrants (Plate 16) are masterpieces of photographic humanism. Portraiture holds an important place in the history of photography. Whereas oil-painting portraiture had served the interests of the aristocracy and the haute bourgeoisie almost exclusively, the new medium allowed members of the broad urban middle class, and the new industrial working class as well, to record their images for posterity. And whereas the images of immigrants in popular illustration had tended toward racial and ethnic caricatures, Hine's portraits reveal the individuality and humanism behind the stereotypes. Another photograph depicting the immigrant condition – all the more powerful for its juxtaposition of social commentary and pure artistic composition – was Alfred Stieglitz's "The Steerage" of 1907 (Plate 17). Here, the interplay between the immigrants, mostly enveloped in shadows, and the physical structures of gangway and ladder combine simultaneously to fascinate the eye and touch the soul.

The Great Depression of the 1930s strengthened the documentary and social activist tendencies in photography that had been pioneered by Riis, Hine, and the photo-journalists, especially in work that had been commissioned by the Works Progress Administration and other New Deal agencies. The photographs of Southern sharecroppers by Walker Evans and the images of Harlem streetlife by Gordon Parkes are notable examples of the intersection of art and social protest. Dorothea Lange was another activist photographer who, like Evans, was drawn to the documentation of rural poverty. Her pictures of dispossessed dust bowl emigres are among the most famous and most powerful images of the Depression era. And in her picture of a roadside sign (Plate 18), published in a 1939 Farm Security Administration report on rural migration, Lange captured both the bleakness of the Depression and the kinds of fears and insecurities that underlay Frank Lloyd Wright's "Broadacre City" proposal – a homestead of at least one acre per person – and other back to the land schemes.

The people of the city and their problems were the dominant, but not the sole, subject of urban documentary photography. The very physical presence of the modern city was often evoked through images of architecture and infrastructure. Stieglitz's moody impression of the Flatiron Building cloaked in mist is one well-known example, and Margaret Bourke-White's formalist, monumental depiction of Hoover Dam that appeared on the cover of the first *Life* magazine is another. Charles Sheeler's "Ford Plant, Detroit" of 1927 (Plate 19) demonstrates the way in which pure architectural form – an assemblage of strong verticals and diagonals revealed through light and shadow – could be read both as pure cubist composition and powerful, if somewhat distanced, social documentation.

If the architecture of the city could present images of foreboding power or lyrical freedom, it was nonetheless the lives of the people of the city themselves that became the constantly recurring theme of photojournalists and artists alike. The work of Weegee (Arthur Fellig) is a case in point. Although most of his pictures were made in the 1940s and 1950s and featured startling images of auto wrecks, transvestite bars, and gangland killings, the individual faces communicate directly to the viewer with a sense of direct, immediate reality that is the transcendent essence of urban life. "The Critic" (Plate 20), for example, was taken in 1943, but the people and the streetlife situation – so like the contrast pictures in popular illustration – are instantly recognizable as big-city icons of the painful urban conflation of luxury and poverty, indifference and despair.

The project of comprehending the modern city visually played, and continues to play, a central role in the history of art and consciousness. The great themes of the city – its kinetic activity, its juxtapositions and ironies, its massive forms and tiny details – provided the artist with a subject matter that could not be ignored and pioneered modes of visual perception and communication that were to fundamentally transform the nature of social life. Whereas popular illustrated journalism liberated visual images from the control of privileged elites, photography completed the democratization of the visual by placing it in the hands of the masses.

Eventually, the new modes of visual perception would be connected to narrative – particularly to the "urban narratives" of young rural innocents encountering the experience of the city for the first time, a plot line as old as Gilgamesh and given renewed meaning as a result of the millions of immigration stories that fueled the urbanization process during the nineteenth and twentieth centuries. And out of the interconnection of urban narrative and visual representation would come cinema, the ultimate realization of the kinetic imagery that urban life characterized and photography mirrored. In King Vidor's *The Crowd* of 1926 (Plate 21), we see the once-optimistic individual lost in the enveloping urban mass. In Fritz Lang's *Metropolis* of 1929 (Plate 22), we see the modern citadels of faceless power looming over the dehumanizing structures of class segregation and oppression. In both cases, the viewer confronts and subconsciously confirms the artist's perception of the reality of modern urban life. The confrontation may be discomfiting, but the ubiquity of modern visual culture makes the confrontations inescapable and memorable, perhaps more memorable even than actual encounters on the streets of the city. Writing about the historic city, Lewis Mumford found that they were characterized by an "urban drama" and that the dramatic dialogue was "one of the ultimate expressions of life in the city." In the modern city, it is the image – sometimes celebratory, sometimes haunting, always definitive in its explanatory value – that is paramount both as spectacle and revelation.

MUSIC IN THE STREET

MUSIC IN THE PARLOR—p. 15

Plate 9   **"Music in the Street/Music in the Parlor"** (Unknown artist, 1868, *The Illustrated News*).

HARPER'S WEEKLY.

A JOURNAL OF CIVILIZATION.

Vol. XX.—No. 998.] NEW YORK, SATURDAY, FEBRUARY 12, 1876. [WITH A SUPPLEMENT. PRICE TEN CENTS.

Entered according to Act of Congress, in the Year 1876, by Harper & Brothers, in the Office of the Librarian of Congress, at Washington.

THE HEARTH-STONE OF THE POOR.—WASTE STEAM NOT WASTED.—DRAWN BY SOL EYTINGE, JR.—[SEE PAGE 133.]

Plate 10 **"The Hearth-stone of the Poor"** (Sol Eytinge, Jr., 1876, *Harper's Weekly*).

Plate 11 **"City Sketches"** (C.A. Barry, 1855, *Ballou's Pictorial Drawing-Room Companion*).

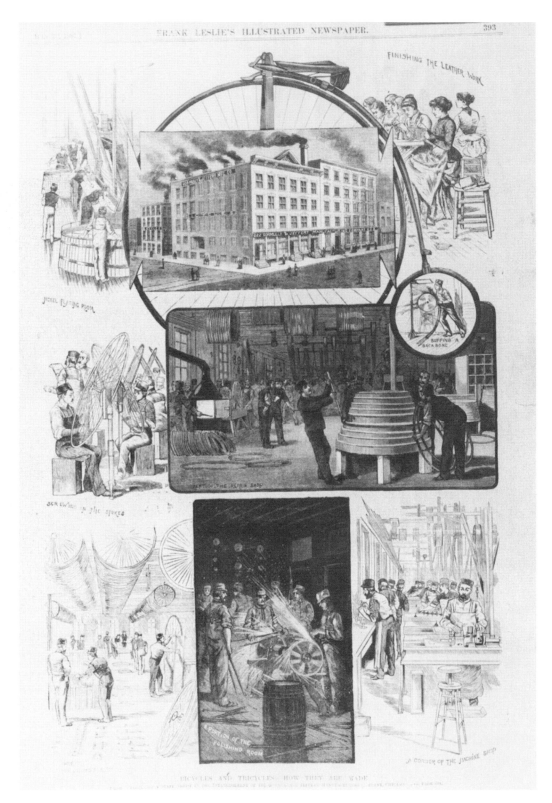

Plate 12 **"Bicycles and Tricycles – How They Are Made"** (Unknown artist, 1887, *Frank Leslie's Illustrated Newspaper*).

Plate 13 **"Burning Mills, Oswego, New York"** (George N. Barnard, 1853, *Oswego Daily Times*). (Courtesy of George Eastman House.)

Plate 14 **"A Rainy Day on Broadway"** (Edward Anthony, 1859). (Courtesy of George Eastman House.)

Plate 15 **"Bandits' Roost, 39½ Mulberry Street"** (Jacob Riis, New York, 1889. (The Jacob A. Riis collection # 101.) (Copyright © Museum of the City of New York.)

WELSH COAL MINER

SLAVIC STEEL-WORKER

Photos by Hine

ITALIAN LABORER

IRISH IRON WORKER

Plate 16 Ellis Island immigrant portraits (Lewis Hine, ca. 1910). In Grove S. Dow, *Social Problems of Today* (New York: Thomas Y. Crowell, 1925). Public domain.

Plate 17 **"The Steerage"** (Alfred Stieglitz, 1907). (Courtesy of the Museum of Modern Art, New York.)

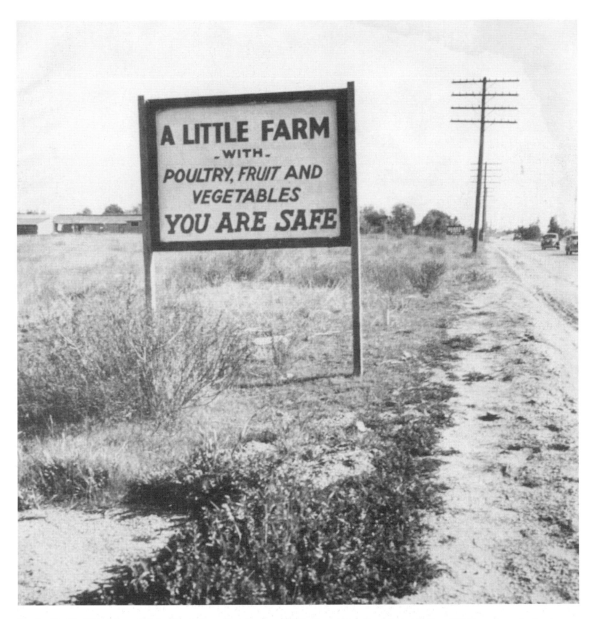

Plate 18 **Untitled (roadside sign)** (Dorothea Lange, 1939). Farm Security Administration, US Department of Agriculture. Printed in C.E. Lively and Conrad Tauber, *Rural Migration in the United States* (Washington, DC: Works Progress Administration, 1939).

Plate 19 **"Ford Plant, Detroit"** (Charles Sheeler, 1927). (Courtesy of the Museum of Modern Art, New York.)

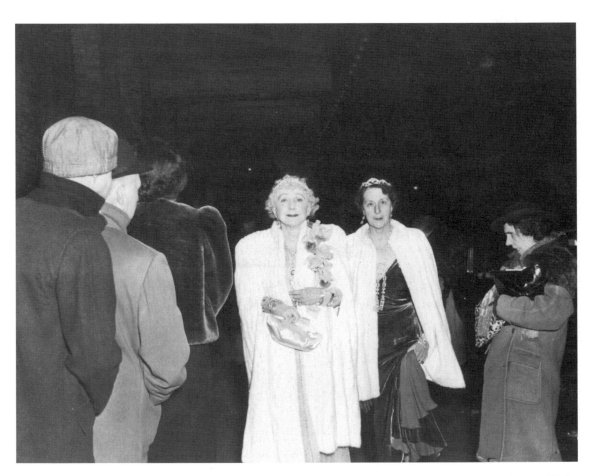

Plate 20 **"The Critic"** (Weegee (Arthur Fellig), 1943). (Courtesy of the International Center of Photography, New York. Bequest of Wilma Wilcox.)

Plate 21 **The Crowd** (King Vidor, still from 1926). (Courtesy of the Museum of Modern Art, New York, Film Stills Archive.)

Plate 22 **Still from *Metropolis*** (Fritz Lang, 1929). (Courtesy of the Museum of Modern Art, New York, Film Stills Archive.)

PART THREE

Urban space

INTRODUCTION TO PART THREE

The physical form of cities described in Part One on "The Evolution of Cities" ranged from Ur's mudbrick walls and ziggurat to the marble agora of the Greek polis; from the polluted slums of nineteenth-century Manchester to the high-tech research "campuses," malls, and residential areas of today's technoburbs. The size, density, spatial distribution of functions, physical and cultural characteristics of cities described in that section display an astonishing variety over time.

This section is concerned with *urban geography*. It contains classic and contemporary writings on the physical and social structure of cities. Parts Five, Six, and Seven on "Urban Planning History and Visions", "Urban Planning Theory and Practice" and "Perspectives on Urban Design" return to this material, but extend the discussion to interventions to shape the form and design of cities and regions.

For many years geography focused on the study of physical space and the mapping of topgraphical and political features. Today academic geography has been reinvigorated by a new generation of scholars who define the discipline broadly and include analysis of social and cultural reality. Scholars from many disciplines have contributed to our understanding of urban space. In addition to geographers such as Edward Soja and David Harvey, this selection includes writings from the disciplines of sociology (Janet Abu-Lughod and Ernest Burgess), landscape architecture (J.B. Jackson), and urban planning (Andres Duany and Elizabeth Plater-Zyberk). Writers from these and other disciplines throughout this book inform our understanding of urban geography. Many cities have grown organically with little or no explicit overall plan or centrally controlled regulation. They contain vernacular architecture created by ordinary people without specialized training in the design professions. Other cities display a mix of organic growth and development planned at different periods in their history. The old crooked streets and irregular lots that constitute most of central Boston, Massachusetts, contrast with the regularity of Back Bay. There a swampy area was developed during the latter part of the nineteenth century following a plan by Frederick Law Olmsted. Disorderly districts of London that have evolved since the Middle Ages contrast with the area rebuilt after the great London fire of 1666. Only a few entire cities like Australia's capital city Canberra and Brazil's capital city of Brasilia have been built according to a consistent master plan and even they show vernacular building that does not fit the plan.

It was sociologists, not geographers, who pioneered the modern field of urban geography. In addition to path-breaking work on the social structure of cities by Louis Wirth (p. 97), Robert Park, Ernest W. Burgess and others members of the Chicago school of urban sociology developed the first systematic theory about the physical form of cities. The first selection in this section, Ernest W. Burgess's "The Growth of the City" (p. 156), advances many provocative hypotheses. Burgess argued that there was an underlying logic to the physical form of cities. In his view they were organized in a series of concentric rings moving out from a central business district. Each ring had a distinct set of residents and functions. Physical form and human life were intimately linked. According to Burgess cities were not static. They were dynamic organisms with a constant flow of new residents coming into the inner rings and a flow from these rings outward over time. Burgess and other members of the Chicago School thought of cities as similar to natural ecological systems. Each part of an urban system had distinct characteristics and played a unique role in the total system. But each was related to the others and the whole system was in a state of constant flux.

Sociologists inspired by Burgess, land economists, and many geographers have studied city form during the last seventy-five years. Burgess's "concentric zone" theory is just one of a number of theories which seek to theoretically describe the internal structure of cities. It is often juxtaposed with theories developed by real estate economist Homer Hoyt in the 1930s, postulating an organization of cities in *sectors* radiating out along transportation corridors, and with an essay by geographers Chauncy Harris and Edward Ullman in the 1940s concluding that most cities have multiple nuclei rather than either concentric zone or sectoral organization. Even in the large postmodern Los Angeles region, geographer Edward Soja (p. 189) draws on this early theory and discerns a concentric and sectoral logic.

Burgess wrote about the way in which a *single city* is organized. A related topic in the study of urban form concerns *systems of cities*. Central place theory, developed by German geographer Walter Christaller, focuses on how cities within an entire region relate to each other. In this tradition Saskia Sassen (p. 212) argues that a world system of cities is emerging. As computerized statistical analysis and geographical information systems for analyzing and mapping data rapidly increase our ability to understand urban form, the debates about the internal structure of cities and systems of cities continue.

The selection by landscape architect J.B. Jackson (p. 164) also notes an underlying logic to the physical form, social structure, and function of cities. According to Jackson even small cities that evolved without city planners and architects display an underlying regularity and logic. For Jackson the built environment of every place expresses the belief system and values of the humans who created it.

Despite the fact that until recently most cities have grown with little or no formal planning or professional design, little scholarly attention was devoted to studying vernacular urban form and popular design until J.B. Jackson pioneered the study of *vernacular* landscapes, the environments built by common people without trained architects, planners, or other professionals. Jackson was fascinated with the physical form of barns, fences, billboards, and grain silos; pioneer settlements with stumps in the field and muddy roads; dying former railroad towns where the train no longer stops; humble mobile homes in rural New Mexico; and gas stations and frozen custard stands along the principal artery across the American Southwest, Highway 66. Jackson was one of the first to argue that a study of vernacular architecture provides important understanding of the culture and values of the people who have built the built environment. According to Jackson, a close look at Highway 66 in the 1950s reveals a great deal about the worldview of the highway builders and residents along the highway. Today, largely as a result of Jackson's influence, there is a large literature describing and analyzing the function and cultural meaning of Las Vegas casinos, White Tower hamburger stands, suburban tract homes, billboards, and other vernacular architecture.

Sociologist Janet Abu-Lughod is sensitive to the culture that produced Islamic cities. Her effort to define the essence of Islamic cities (p. 172) is an excellent lesson in comparative urbanism that can sensitize readers not familiar with Islamic cities to what is unique about them. It is also intended as a resource for urban planners in the Middle East, North Africa, Pakistan, Indonesia, and other countries with large Muslim populations who seek to plan new cities for these fast-growing countries that preserve what is best about historic Islamic city form but are appropriate for modern Islamic societies.

Ali Madanipour shares Abu-Lughod's interest in the relationship between society and the built environment. In his essay in this section (p. 181) Madanipour is particularly concerned with the way in which Europeans may knowingly or unwittingly exclude people from other cultures from the full benefits of their cities. Madanipour distinguishes between *economic discrimination*, in which members of a group are excluded from access to employment, *political discrimination*, in which they are excluded from political power, and *cultural exclusion* in which the group members are marginalized from the symbols, meanings, rituals, and discourses of the dominant culture. Since exclusion often has a spatial dimension he suggests a number of strategies to break down spatial exclusion and increase inclusion. Subsidized housing, for example, may permit low-income foreign immigrants to live in parts of a city they could not otherwise afford, giving them access to job opportunities and better education for their children.

University of California, Los Angeles, geography professor Edward Soja uses the Los Angeles metropolitan region as the basis for a discussion of the postmodern metropolis (p. 189). He is bemused by the bewildering contrasts to be found in the Sixty-Mile Circle surrounding downtown Los Angeles. The pieces of the Los Angeles landscape Soja takes apart display astonishing contrasts: huge military bases and tiny eighteenth-century missions; gated communities for the rich and squalid ghettos for the poor;

buildings and whole regions engaged solely in post-industrial research and managerial activities; and a thriving industrial sector employing low-wage immigrant labor. Soja emphasizes the fantastical imagery and theme park quality of the Los Angeles region. But underlying his irony is a serious message. Beneath the glitz and glitter, Los Angeles has a dark side. According to Soja, police stations, courts, and prisons are central features to downtown. The region's industry could not exist without poorly paid immigrant labor. Violent and depressing ghettos are threaded through the Los Angeles region. The natural ecology of this congested and smog-blanketed metropolis is in deep trouble.

In "Fortress L.A." (p. 201) social critic Mike Davis addresses the same issue – the dark side of the postmodern metropolis – and he also uses Los Angeles as his exemplar city. Davis describes a built environment complete with surveillance cameras, barrel-shaped park benches designed to keep people from sleeping on them, overhead sprinklers to douse the homeless, windowless concrete hotel walls facing streets, entrances to public buildings reminiscent of the fortifications in front of medieval European castles, and gated communities where the rich can be (or at least feel) safe from outsiders. All these artifacts provide a disturbing glimpse of Los Angeles and, by implication, postmodern cities emerging around the world.

Andres Duany and Elizabeth Plater-Zyberk are interested in urban space because as architects and urban planners space is critical to their work. Their selection is an excellent illustration of how professionals who actually create the built environment can use insights from the academic and theoretical writings of geographers, sociologists, and anthropologists to create humane and interesting cities. They provide an antidote to Mike Davis's cynicism (p. 201) and practical suggestions to build places that do not have the worst features Davis and Soja deplore in twenty-first-century Los Angeles. Duany and Plater-Zyberk anticipate material on urban planning and urban design in the following three sections of *The City Reader*. Duany and Plater-Zyberk are a husband and wife team who run an architectural and urban planning practice in Florida. They are founders of a movement in architecture and planning called the New Urbanism. There are strong neo-traditionalist elements in their design thinking. Like J.B. Jackson (p. 164) they see positive values in small American towns of the past. They have carefully studied vernacular architecture and incorporate insights from the way in which people built small towns, without architectural training, into their urban plans. Their selection has some parallels with Janet Abu-Lughod's selection. They are concerned with the physical form produced by a distinct culture and want to help professionals build relevant twenty-first-century cities that capture the essence of a tradition, but in ways that are relevant to twenty-first-century realities. Abu-Lughod focuses on Islamic cities; Duany and Plater-Zyberk on small American towns and suburbs.

The final selection in this section, "The Impact of the New Technologies and Globalization on Cities," by urban planner Saskia Sassen (p. 212), moves from Soja and Davis's discussion of the contradictions in postmodern metropoles like Los Angeles to an analysis of the underlying economic characteristics of global cities. Based on her research on the economy of world cities, Sassen concludes that a truly global system of cities has emerged. New York, London, and Tokyo are the main bastions of international finance and control. Decisions made in these three cities affect the residents of cities around the world. Sassen concludes that globalization and information technology are the driving forces behind the new world spatial order, but rejects some facile stereotypes. Some writers argue that because phones, fax machines, e-mail, and other more advanced communication technologies make it possible to communicate easily and instantaneously anywhere in the world, place no longer matters and large cities will wither away. Sassen finds no evidence for this view. Rather she notes further growth of population and power in the commanding global cities. She believes that information technology is already restructuring the physical space of global city regions. According to Sassen, huge new office complexes like La Défense in Paris and nodes of activity following "information highways" are two of the emerging models by which global city metropolitan regions are being restructured. While Sassen agrees that the new world economic order and the rise of global cities are creating enormous inequality she dismisses facile divisions of the world into a developed center and an undeveloped margin. Rather, she argues, there are *de facto* Third World communities in developed global cities and outposts of global cities throughout the Third World.

"The Growth of the City: An Introduction to a Research Project"

from Robert Park *et al.*, *The City* (1925)

Ernest W. Burgess

Editors' Introduction

Ernest W. Burgess (1886–1966) was a member of the famed Sociology Department at the University of Chicago that included such luminaries as Louis Wirth, author of "Urbanism as a Way of Life" (p. 97), and former newspaper reporter and social reformer Robert E. Park. Together, these scholars set out to virtually reinvent modern sociology by taking academic research to the streets and by using the city of Chicago itself as a "living laboratory" for the study of urban problems and social dynamics.

Throughout a long and productive career, Burgess addressed a whole series of issues that connected the social dynamics of the city as a whole with the lives of its citizens. He wrote extensively on issues related to marriage and the family, the relation of personality to social groups, and, in the final decades of his life, problems of the elderly. However, his most famous contribution to the study of the city was the 1925 essay reprinted here: "The Growth of the City."

Subtitled "An Introduction to a Research Project," Burgess's seminal analysis of the inter-relation of the social growth and the physical expansion of modern cities served generations of other urban sociologists, geographers, and planners as a kind of prolegomena to all future study of the city. Seeking to describe what he called "the pulse of the community," Burgess devised a theory that was thoroughly organic, dynamic, and developmental. "In the expansion of the city," he wrote, "a process of distribution takes place, which sifts and sorts and relocates individuals and groups by residence and occupation." And it was this dynamic process – "process" was one of Burgess's favorite words – that "gives form and character to the city."

Central to Burgess's analysis of urban growth was his famous model based on a series of concentric circles that divided the city into five zones. On one level, the concentric zone model was merely a static map of contemporaneous Chicago. On another level, the model was a theoretical diagram of a dynamic process Burgess called "succession," a term he borrowed from the science of plant ecology to describe "urban metabolism and mobility." By borrowing terminology from the natural sciences, and by drawing analogies between the urban and the natural worlds, Burgess established the study of "social ecology" as a distinct approach to understanding the underlying patterns of urban growth and development.

Following the publication of Burgess's essay, a number of urban theorists offered modifications and even refutations of the simple elegance of the concentric zone model. In 1939, real estate economist Homer Hoyt proposed a "sectoral model" for modern capitalist cities based on "wedges of activity" extending outward from the city center along transportation corridors. In 1945, geographers Chauncy Harris and Edward Ullman suggested a "multiple nuclei model," arguing that cities developed around several, not just one, center of economic activity.

Geographer Edward Soja, discussing postmodern Los Angeles (p. 189), notes that even in the polycentric archipelago of modern Los Angeles "population densities do mound up around the center of cities" and that "there is an accompanying concentric residential rhythm" to the region. Soja also notes "the emanation of fortuitous wedges or sectors starting from the center," suggesting that Homer Hoyt's sectoral model still has some explanatory power even in postmodern Los Angeles. Soja notes that the Wilshire corridor, for example, "extends the citadels of the central city almost twenty miles westwards to the Pacific."

Burgess's influence was both widespread and long-standing, if not for the details of his model, then for the spirit of its basic insight. Burgess's view may be compared to J.B. Jackson's notion that there is an underlying logic to urban form that occurs even in the absence of formal planning (p. 164). The work of dozens of urban sociologists – for example William Julius Wilson's study of the urban underclass (p. 126) and Ali Madanipour's analysis of "Social Exclusion and Space" (p. 181) – owe a profound debt to Burgess. And just as importantly, Burgess's sense of urban dynamism and vitality captures the essence of those more lyric conceptions of the city such as Lewis Mumford's notion of "the urban drama" (p. 92) and Jane Jacobs's "ballet of the streets" (p. 114).

Computer technology, including statistical packages and geographic information systems (GIS) software, now makes it possible for present-day geographers and sociologists to summarize vast amounts of data and map the internal structure of cities in ever more sophisticated ways. Thus, understanding of the relationship between social groups and urban form pioneered by Burgess continues to advance by leaps and bounds.

Robert E. Park, Ernest W. Burgess, and Roderick D. McKenzie, *The City*, 3rd edn, revised (Chicago: University of Chicago Press, 1969), from which this selection is taken, is the best introductory collection of the works of the Chicago School of urban sociology. For the works of Burgess in particular, including a comprehensive bibliography, see Leonard S. Cottrell, Jr., *et al.* (eds), *On Community, Family and Delinquency* (Chicago: University of Chicago Press, 1973). Homer Hoyt's sectoral theory is described in *The Structure and Growth of Residential Neighborhoods in American Cities* (Washington, DC: Federal Housing Administration, 1939); Chauncy Harris and Edward Ullman's multiple nuclei model is presented in "The Nature of Cities," *Annals of the American Academy of Political and Social Science*, 242 (1945).

The outstanding fact of modern society is the growth of great cities. Nowhere else have the enormous changes which the machine industry has made in our social life registered themselves with such obviousness as in the cities. In the United States the transition from a rural to an urban civilization, though beginning later than in Europe, has taken place, if not more rapidly and completely, at any rate more logically in its most characteristic forms.

All the manifestations of modern life which are peculiarly urban – the skyscraper, the subway, the department store, the daily newspaper, and social work – are characteristically American. The more subtle changes in our social life, which in their cruder manifestations are termed "social problems," problems that alarm and bewilder us, such as divorce, delinquency, and social unrest, are to be found in their most acute forms in our largest American cities. The profound and "subversive" forces which have wrought these changes are measured in the physical growth and expansion of cities. That is the significance of the comparative statistics of Weber, Bucher, and other students.

These statistical studies, although dealing mainly with the effects of urban growth, brought out into clear relief certain distinctive characteristics of urban as compared with rural populations. The larger proportion of women to men in the cities than in the open country, the greater percentage of youth and middle-aged, the higher ratio of the foreign-born, the increased heterogeneity of occupation increase with the growth of the city and profoundly alter its social structure. These variations in the composition of population are indicative of all the changes going on in the social organization of the community. In fact, these changes are a part of the growth of the city and suggest the nature of the processes of growth.

The only aspect of growth adequately described by Bucher and Weber was the rather obvious process of the aggregation of urban population. Almost as overt a process, that of expansion, has been investigated from a different and very practical point of view by groups interested in city planning, zoning, and regional surveys. Even more significant than the increasing density of urban population is its correlative tendency to overflow, and so to extend over wider areas, and to incorporate these areas into a larger communal life. This paper, therefore, will treat first of the expansion of the city, and then of the less-known processes of urban metabolism and mobility which are closely related to expansion.

EXPANSION AS PHYSICAL GROWTH

The expansion of the city from the standpoint of the city plan, zoning, and regional surveys is thought of almost wholly in terms of its physical growth. Traction studies have dealt with the development of transportation in its relation to the distribution of population throughout the city. The surveys made by the Bell Telephone Company and other public utilities have attempted to forecast the direction and the rate of growth of the city in order to anticipate the future demands for the extension of their services. In the city plan the location of parks and boulevards, the widening of traffic streets, the provision for a civic center, are all in the interest of the future control of the physical development of the city.

This expansion in area of our largest cities is now being brought forcibly to our attention by the Plan for the Study of New York and Its Environs, and by the formation of the Chicago Regional Planning Association, which extends the metropolitan district of the city to a radius of 50 miles, embracing 4,000 square miles of territory. Both are attempting to measure expansion in order to deal with the changes that accompany city growth. In England, where more than one-half of the inhabitants live in cities having a population of 100,000 and over, the lively appreciation of the bearing of urban expansion on social organization is thus expressed by C.B. Fawcett:

> One of the most important and striking developments in the growth of the urban populations of the more advanced peoples of the world during the last few decades has been the appearance of a number of vast urban aggregates, or conurbations, far larger and more numerous than the great cities of any preceding age. These have usually been formed by the simultaneous expansion of a number of neighboring towns, which have grown out toward each other until they have reached a practical coalescence in one continuous urban area. Each such conurbation still has within it many nuclei of denser town growth, most of which represent the central areas of the various towns from which it has grown, and these nuclear patches are connected by the less densely urbanized areas which began as suburbs of these towns. The latter are still usually rather less continuously occupied by buildings, and often have many open spaces.

These great aggregates of town dwellers are a new feature in the distribution of man over the earth. At the present day there are from thirty to forty of them, each containing more than a million people, whereas only a hundred years ago there were, outside the great centers of population on the waterways of China, not more than two or three. Such aggregations of people are phenomena of great geographical and social importance; they give rise to new problems in the organization of the life and well-being of their inhabitants and in their varied activities. Few of them have yet developed a social consciousness at all proportionate to their magnitude, or fully realized themselves as definite groupings of people with many common interests, emotions and thoughts.

In Europe and America the tendency of the great city to expand has been recognized in the term "the metropolitan area of the city," which far overruns its political limits, and, in the case of New York and Chicago, even state lines. The metropolitan area may be taken to include urban territory that is physically contiguous, but it is coming to be defined by that facility of transportation that enables a business man to live in a suburb of Chicago and to work in the loop, and his wife to shop at Marshall Field's and attend grand opera in the Auditorium.

EXPANSION AS A PROCESS

No study of expansion as a process has yet been made, although the materials for such a study and intimations of different aspects of the process are contained in city planning, zoning, and regional surveys. The typical processes of the expansion of the city can best be illustrated, perhaps, by a series of concentric circles, which may be numbered to designate both the successive zones of urban extension and the types of areas differentiated in the process of expansion [Figure 1].

[Figure 1] represents an ideal construction of the tendencies of any town or city to expand radially from its central business district – on the map "the Loop" (I). Encircling the downtown area there is normally an area

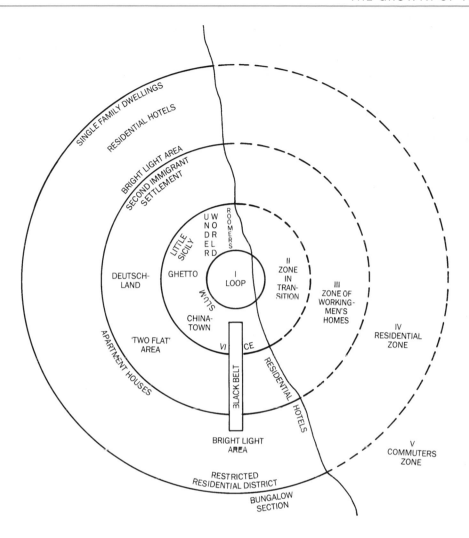

Figure 1

in transition, which is being invaded by business and light manufacture (II). A third area (III) is inhabited by the workers in industries who have escaped from the area of deterioration (II) but who desire to live within easy access of their work. Beyond this zone is the "residential area" (IV) of high-class apartment buildings or of exclusive "restricted" districts of single family dwellings. Still farther, out beyond the city limits, is the commuters' zone: suburban areas, or satellite cities, within a thirty- to sixty-minute ride of the central business district.

This [figure] brings out clearly the main fact of expansion, namely, the tendency of each inner zone to extend its area by the invasion of the next outer zone. This aspect of expansion may be called *succession*, a process which has been studied in detail in plant ecology. If this [figure] is applied to Chicago, all four of these zones were in its early history included in the circumference of the inner zone, the present business district. The present boundaries of the area of deterioration were not many years ago those of the zone now inhabited by independent wage-earners, and within the memories of thousands of Chicagoans contained the residences of the "best families." It hardly needs to be added that neither Chicago nor any other city fits perfectly into this ideal scheme. Complications are introduced by the lake front, the Chicago River, railroad lines, historical factors in the location of industry, the relative degree of the resistance of communities to invasion, etc.

Besides extension and succession, the general process of expansion in urban growth involves the antagonistic and yet complementary processes of concentration and decentralization. In all cities there is the natural tendency for local and outside transportation to converge in the central business district. In the downtown section of every large city we expect to find the department stores, the skyscraper office buildings, the railroad stations, the great hotels, the theaters, the art museum, and the city hall. Quite naturally, almost inevitably, the economic, cultural, and political life centers here. The relation of centralization to the other processes of city life may be roughly gauged by the fact that over half a million people daily enter and leave Chicago's "loop." More recently sub-business centers have grown up in outlying zones. These "satellite loops" do not, it seems, represent the "hoped for" revival of the neighborhood, but rather a telescoping of several local communities into a larger economic unity. The Chicago of yesterday, an agglomeration of country towns and immigrant colonies, is under-going a process of reorganization into a centralized decentralized system of local communities coalescing into sub-business areas visibly or invisibly dominated by the central business district. The actual processes of what may be called centralized decentralization are now being studied in the development of the chain store, which is only one illustration of the change in the basis of the urban organization.

Expansion, as we have seen, deals with the physical growth of the city, and with the extension of the technical services that have made city life not only livable, but comfortable, even luxurious. Certain of these basic necessities of urban life are possible only through tremendous development of communal existence. Three millions of people in Chicago are dependent upon one unified water system, one giant gas company, and one huge electric light plant. Yet, like most of the other aspects of our communal urban life, this economic co-operation is an example of co-operation without a shred of what the "spirit of co-operation" is commonly thought to signify. The great public utilities are a part of the mechanization of life in great cities, and have little or no other meaning for social organization.

Yet the processes of expansion, and especially the rate of expansion, may be studied not only in the physical growth and business development, but also in the consequent changes in the social organization and in personality types. How far is the growth of the city, in its physical and technical aspects, matched by a natural but adequate readjustment in the social organization? What, for a city, is a normal rate of expansion, a rate of expansion with which controlled changes in the social organization might successfully keep pace?

SOCIAL ORGANIZATION AND DISORGANIZATION AS PROCESSES OF METABOLISM

These questions may best be answered, perhaps, by thinking of urban growth as a resultant of organization and disorganization analogous to the anabolic and katabolic processes of metabolism in the body. In what way are individuals incorporated into the life of a city? By what process does a person become an organic part of his society? The natural process of acquiring culture is by birth. A person is born into a family already adjusted to a social environment – in this case the modern city. The natural rate of increase of population most favorable for assimilation may then be taken as the excess of the birth-rate over the death-rate, but is this the normal rate of city growth? Certainly, modern cities have increased and are increasing in population at a far higher rate. However, the natural rate of growth may be used to measure the disturbances of metabolism caused by any excessive increase, as those which followed the great influx of southern Negroes into northern cities since the war. In a similar way all cities show deviations in composition by age and sex from a standard population such as that of Sweden, unaffected in recent years by any great emigration or immigration. Here again, marked variations, as any great excess of males over females, or of females over males, or in the proportion of children, or of grown men or women, are symptomatic of abnormalities in social metabolism.

Normally the processes of disorganization and organization may be thought of as in reciprocal relationship to each other, and as co-operating in a moving equilibrium of social order toward an end vaguely or definitely regarded as progressive. So far as disorganization points to reorganization and makes for more efficient adjustment, disorganization must be conceived not as pathological, but as normal. Disorganization as preliminary to reorganization of attitudes and conduct is almost invariably the lot of the newcomer to the city, and the discarding of the habitual, and often of what has been to him the moral, is not infrequently accompanied by sharp mental conflict and sense of personal loss. Oftener, perhaps, the change gives sooner or later a feeling of emancipation and an urge toward new goals.

In the expansion of the city a process of distribution takes place which sifts and sorts and relocates individuals and groups by residence and occupation. The resulting differentiation of the cosmopolitan American city into areas is typically all from one pattern, with only interesting minor modifications. Within the central business district or on an adjoining street is the "main stem" of "hobohemia," the teeming Rialto of the homeless migratory man of the Middle West. In the zone of deterioration encircling the central business section are always to be found the so-called "slums" and "bad lands," with their submerged regions of poverty, degradation, and disease, and their underworlds of crime and vice. Within a deteriorating area are rooming-house districts, the purgatory of "lost souls." Nearby is the Latin Quarter, where creative and rebellious spirits resort. The slums are also crowded to over flowing with immigrant colonies – the Ghetto, Little Sicily, Greek town, Chinatown – fascinatingly combining old world heritages and American adaptations. Wedging out from here is the Black Belt with its free and disorderly life. The area of deterioration, while essentially one of decay, of stationary or declining population, is also one of regeneration, as witness the mission, the settlement, the artists' colony, radical centers – all obsessed with the vision of a new and better world.

The next zone is also inhabited predominatingly by factory and shop workers, but skilled and thrifty. This is an area of second immigrant settlement, generally of the second generation. It is the region of escape from the slum, the *Deutschland* of the aspiring Ghetto family. For *Deutschland* (literally "Germany") is the name given, half in envy, half in derision, to that region beyond the Ghetto where successful neighbors appear to be imitating German Jewish standards of living. But the inhabitant of this area in turn looks to the "Promised Land" beyond, to its residential hotels, its apartment-house region, its "satellite loops," and its "bright light" areas.

This differentiation into natural economic and cultural groupings gives form and character to the city. For segregation offers the group, and thereby the individuals who compose the group, a place and a role in the total organization of city life. Segregation limits development in certain directions, but releases it in others. These areas tend to accentuate certain traits, to attract and develop their kind of individuals, and so to become further differentiated.

The division of labor in the city likewise illustrates disorganization, reorganization and increasing differentiation. The immigrant from rural communities in Europe and America seldom brings with him economic skill of any great value in our industrial, commercial, or professional life. Yet interesting occupational selection has taken place by nationality, explainable more by racial temperament or circumstance than by old-world economic background as Irish policemen, Greek ice-cream parlors, Chinese laundries, Negro porters, Belgian janitors, etc.

The facts that in Chicago one million (996,589) individuals gainfully employed reported 509 occupations, and that over 1,000 men and women in *Who's Who* gave 116 different vocations give some notion of how in the city the minute differentiation of occupation "analyzes and sifts the population, separating and classifying the diverse elements." These figures also afford some intimation of the complexity and complication of the modern industrial mechanism and the intricate segregation and isolation of divergent economic groups. Interrelated with this economic division of labor is a corresponding division into social classes and into cultural and recreational groups. From this multiplicity of groups, with their different patterns of life, the person finds his congenial social world and – what is not feasible in the narrow confines of a village – may move and live in widely separated, and perchance conflicting, worlds. Personal disorganization may be but the failure to harmonize the canons of conduct of two divergent groups.

If the phenomena of expansion and metabolism indicate that a moderate degree of disorganization may and does facilitate social organization, they indicate as well that rapid urban expansion is accompanied by excessive increases in disease, crime, disorder, vice, insanity and suicide, rough indexes of social disorganization. But what are the indexes of the causes, rather than of the effects, of the disordered social metabolism of the city? The excess of the actual over the natural increase of population has already been suggested as a criterion. The significance of this increase consists in the immigration into a metropolitan city like New York and Chicago of tens of thousands of persons annually. Their invasion of the city has the effect of a tidal wave inundating first the immigrant colonies, the ports of first entry, dislodging thousands of inhabitants who overflow into the next zone, and so on and on until the momentum of the wave has spent its force on the last urban zone. The whole effect is to speed up expansion, to speed up industry, to speed up the "junking" process in the area of deterioration (II). These internal movements of the population become the more significant for study. What movement is going on in the city, and how may this movement be measured? It is easier, of course, to classify movement within the city than to measure it. There is the movement from residence to residence, change of occupation, labor turnover, movement to and from work, movement for recreation and adventure. This leads to the question: what is the significant aspect of movement for the study of the changes in city life? The answer to this question leads directly to the important distinction between movement and mobility.

MOBILITY AS THE PULSE OF THE COMMUNITY

Movement, per se, is not an evidence of change or of growth. In fact, movement may be a fixed and unchanging order of motion, designed to control a constant situation, as in routine movement. Movement that is significant for growth implies a change of movement in response to a new stimulus or situation. Change of movement of this type is called *mobility*. Movement of the nature of routine finds its typical expression in work. Change of movement, or mobility, is characteristically expressed in adventure. The great city, with its "bright lights," its emporiums of novelties and bargains, its palaces of amusement, its underworld of vice and crime, its risks of life and property from accident, robbery, and homicide, has become the region of the most intense degree of adventure and danger, excitement and thrill.

Mobility, it is evident, involves change, new experience, stimulation. Stimulation induces a response of the person to those objects in his environment which afford expression for his wishes. For the person, as for the physical organism, stimulation is essential to growth. Response to stimulation is wholesome so long as it is a correlated integral reaction of the entire personality. When the reaction is segmental, that is, detached from, and uncontrolled by, the organization of personality, it tends to become disorganizing or

pathological. That is why stimulation for the sake of stimulation, as in the restless pursuit of pleasure, partakes of the nature of vice.

The mobility of city life, with its increase in the number and intensity of stimulations, tends inevitably to confuse and to demoralize the person. For an essential element in the mores and in personal morality is consistency, consistency of the type that is natural in the social control of the primary group. Where mobility is the greatest, and where in consequence primary controls break down completely, as in the zone of deterioration in the modern city, there develop areas of demoralization, of promiscuity, and of vice.

In our studies of the city it is found that areas of mobility are also the regions in which are found juvenile delinquency, boys' gangs, crime, poverty, wife desertion, divorce, abandoned infants, vice.

These concrete situations show why mobility is perhaps the best index of the state of metabolism of the city. Mobility may be thought of, in more than a fanciful sense, as the "pulse of the community." Like the pulse of the human body, it is a process which reflects and is indicative of all the changes that are taking place in the community, and which is susceptible of analysis into elements which may be stated numerically.

The elements entering into mobility may be classified under two main heads: (1) the state of mutability of the person, and (2) the number and kind of contacts or stimulations in his environment. The mutability of city populations varies with sex and age composition, and the degree of detachment of the person from the family and from other groups. All these factors may be expressed numerically. The new stimulations to which a population responds can be measured in terms of change of movement or of increasing contacts. Statistics on the movement of urban population may only measure routine, but an increase at a higher ratio than the increase of population measures mobility. In 1860 the horse-car lines of New York City carried about 50,000,000 passengers; in 1890 the trolley cars (and a few surviving horse-cars) transported about 500,000,000; in 1921, the elevated, subway, surface, and electric and steam suburban lines carried a total of more than 2,500,000,000 passengers. In Chicago the total annual rides per capita on the surface and elevated lines were 164 in 1890; 215 in 1900; 320 in 1910; and 338 in 1921. In addition, the rides per capita on steam and electric suburban lines almost doubled between 1916

(23) and 1921 (41), and the increasing use of the automobile must not be overlooked. For example, the number of automobiles in Illinois increased from 131,140 in 1915 to 833,920 in 1923.

Mobility may be measured not only by these changes of movement, but also by increase of contacts. While the increase of population of Chicago in 1912–22 was less than 25 percent (23.6 percent), the increase of letters delivered to Chicagoans was double that (49.6 percent) – from 693,048,196 to 1,038,007,854. In 1912 New York had 8.8 telephones; in 1922, 16.9 per 100 inhabitants. Boston had, in 1912, 10.1 telephones; ten years later, 19.5 telephones per 100 inhabitants. In the same decade the figures for Chicago increased from 12.3 to 21.6 per 100 population. But increase of the use of the telephone is probably more significant than increase in the number of telephones. The number of telephone calls in Chicago increased from 606,131,928 in 1914 to 944,010,586 in 1922, an increase of 55.7 percent, while the population increased only 13.4 percent.

Land values, since they reflect movement, afford one of the most sensitive indexes of mobility. The highest land values in Chicago are at the point of greatest mobility in the city, at the corner of State and Madison streets, in the Loop. A traffic count showed that at the rush period 31,000 people an hour, or 210,000 men and women in sixteen and one-half hours, passed the southwest corner. For over ten years land values in the Loop have been stationary but in the same time they have doubled, quadrupled and even sextupled in the strategic corners of the "satellite loops," an accurate index of the changes which have occurred. Our investigations so far seem to indicate that variations in land values, especially where correlated with differences in rents, offer perhaps the best single measure of mobility, and so of all the changes taking place in the expansion and growth of the city.

In general outline, I have attempted to present the point of view and methods of investigation which the department of sociology is employing in its studies in the growth of the city, namely, to describe urban expansion in terms of extension, succession, and concentration; to determine how expansion disturbs metabolism when disorganization is in excess of organization; and, finally, to define mobility and to propose it as a measure both of expansion and metabolism, susceptible to precise quantitative formulation, so that it may be regarded almost literally as the pulse of the community. In a way, this statement might serve as an introduction to any one of five or six research projects under way in the department. The project, however, in which I am directly engaged is an attempt to apply these methods of investigation to a cross-section of the city – to put this area, as it were, under the microscope, and so to study in more detail and with greater control and precision the processes which have been described here in the large. For this purpose the West Side Jewish community has been selected. This community includes the so-called "Ghetto," or area of first settlement, and Lawndale, the so-called "Deutschland," or area of second settlement. This area has certain obvious advantages for this study, from the standpoint of expansion, metabolism, and mobility. It exemplifies the tendency to expansion radially from the business center of the city. It is now relatively a homogeneous cultural group. Lawndale is itself an area in flux, with the tide of migrants still flowing in from the Ghetto and a constant egress to more desirable regions of the residential zone. In this area, too, it is also possible to study how the expected outcome of this high rate of mobility in social and personal disorganization is counteracted in large measure by the efficient communal organization of the Jewish community.

"The Almost Perfect Town"

Landscape (1952)

John Brinckerhoff Jackson

Editors' Introduction

John Brinckerhoff Jackson (1909–1997) can best be described as a historian of landscapes as physical, cultural, and conceptual artifacts. To Jackson, the landscape was neither wilderness nor rural nature "out there," but the totality of natural and man-made environments that simultaneously surround and infuse all forms of human activity.

Jackson taught for many years in schools of landscape architecture and geography at Harvard and the University of California, Berkeley. He was the founder of the influential journal *Landscape* and its editor from 1951 to 1968. He may be regarded as an heir to the nineteenth-century French and English landscape gardening traditions and to the pioneering work of Frederick Law Olmsted (p. 302) and the American parks movement. What sets him clearly apart, however, from those traditional roots is that, for Jackson, the urban landscape is not just a city's parks, public gardens, official buildings, and tree-lined boulevards, but its highways, its shopping malls, its run-down warehouse districts, its standard-built two-bedroom houses, and its slums as well. Jackson sees these many elements of the man-made landscape not just as physical objects, but also as social constructs full of meaning and moral implication. It is hardly surprising, then, that J.B. Jackson's approximate version of urban utopia – his *almost* perfect place – is the commonplace, vernacular, Main Street environment of a typical American small town.

In "The Almost Perfect Town," written in the early 1950s, Jackson describes a semi-mythical Optimo City that is located in the American Southwest, but could just as easily have been found almost anywhere in North America, Europe, Australia, and even parts of Asia and Africa beyond the margins of the great metropolitan regions. It is important that Optimo is a small place. Jackson observes that the "world of Optimo City is still complete" precisely because "the ties between country and town have not yet been broken." It is also important that Optimo has a history, however slender, that can be read in its architecture, in the layout of its streets, and in its traditional rivalry with Apache Center twenty miles away.

Jackson uses his loving, elegiac description of Optimo City as a way of criticizing developments in city planning after World War II that threatened to destroy local communities in the name of economic progress. He compares Optimo's unexceptional Courthouse Square to the Spanish plazas and the great public squares of the Baroque era that serve as socially unifying communal centers. And he notes with dismay that some of Optimo's business leaders want to tear the courthouse down to build a parking lot and to replace it with the typical "bureaucrat modernism" of so many contemporary civic centers. Jackson would like to preserve positive features of small towns that Peter Calthorpe (p. 331) and other architects associated with the "New Urbanism" seek to recapture in pedestrian pockets and related neo-traditional designs.

Jackson's Optimo City provides a revealing snapshot of small-town America in the 1950s before freeway construction, mass auto-ownership, and the proliferation of highway-based shopping centers. It also represents a way of thinking about urban space that is timeless.

Among the best of J.B. Jackson's book-length studies and collections of essays are *Landscapes* (Amherst: University of Massachusetts Press, 1970), *American Space* (New York: Norton, 1972), *Discovering the Vernacular*

Landscape (New Haven, CT: Yale University Press, 1984), *The Necessity for Ruins, and Other Topics* (Amherst: University of Massachusetts Press, 1980), and *A Sense of Place, A Sense of Time* (New Haven, CT: Yale University Press, 1994). The journal *Landscape* that Jackson founded and edited for many years contains additional writings by J.B. Jackson himself and articles exploring the meaning of landscape by the many academic and applied landscape architects, urban planners, landscape designers, historians, and others Jackson inspired.

Books by Jackson's student and his successor at Harvard are John R. Stilgoe, *Common Landscape of America, 1580 to 1845* (New Haven, CT: Yale University Press, 1986) and *Outside Lies Magic: Regaining History and Awareness in Everyday Places* (New York: Walker and Co., 1998). Paul E. Groth and Todd Bressi, *Understanding Ordinary Landscapes* (New Haven, CT: Yale, 1997) contains essays in Jackson's tradition.

Other books on vernacular architecture include Robert Venturi, Denise Scott Brown, and Steven Izenour, *Learning from Las Vegas: The Forgotten Symbolism of Architectural Form* (Cambridge, MA: MIT Press, 1972), Dell Upton and John M. Vlach, *Common Places: Readings in American Vernacular* (Atlanta: University of Georgia Press, 1986), Ronald W. Haase, *Classic Cracker: Florida's Wood-frame Vernacular Architecture* (Sarasota, FL: Pineapple Press, 1992), John A. Kouwenhoven, *Made in America: The Arts in Modern Civilization* (Garden City, NY: Doubleday, 1948), Thomas Carter (ed.), *Images of an American Land: Vernacular Architecture in the Western United States* (Albuquerque: University of New Mexico Press, 1997), Jim Heimann, *California Crazy: Roadside Vernacular Architecture* (San Francisco: Chronicle Books, 1981), Colleen Josephine Sheehy, *The Flamingo in the Garden: American Yard Art and the Vernacular Landscape* (London: Garland, 1998), Fred E.H. Schroeder, *Front Yard America: The Evolution and Meanings of a Vernacular Domestic Landscape* (Bowling Green, OH: Bowling Green State University, 1993), John Chase, *Glitter Stucco and Dumpster Diving: Reflections on Building Production in the Vernacular City* (London: Verso, 1998), Bernard Rudofsky, *Architecture Without Architects: A Short Introduction to Non-pedigreed Architecture* (Albuquerque: University of New Mexico Press, 1987), and Paul Oliver (ed.), *Encyclopedia of Vernacular Architecture of the World* (Cambridge: Cambridge University Press, 1998).

Optimo City (pop. 10,783, alt. 2,100 ft.), situated on a small rise overlooking the N. branch of the Apache River, is a farm and ranch center served by a spur of the S.P. County seat of Sheridan Co. Optimo City (originally established in 1843 as Ft. Gaffney). It was the scene of a bloody encounter with a party of marauding Indians in 1857. (See marker on courthouse lawn.) It is the location of a state Insane Asylum, of a sorghum processing plant and an overall factory. Annual County Fair and Cowboy Roundup Sept. 4. The highway now passes through a rolling countryside devoted to grain crops and cattle raising.

Thus would the state guide dispose of Optimo City and hasten on to a more spirited topic if Optimo City as such existed. Optimo City, however, is not one town, it is a hundred or more towns, all very much alike, scattered across the United States from the Alleghenies to the Pacific, most numerous west of the Mississippi and south of the Platte. When, for instance, you travel through Texas and Oklahoma and New Mexico and even parts of Kansas and Missouri, Optimo City is the blur of filling stations and motels you occasionally pass;

the solitary traffic light, the glimpse up a side street of an elephantine courthouse surrounded by elms and sycamores, the brief congestion of mud-spattered pickup trucks that slows you down before you hit the open road once more. And fifty miles farther on Optimo City's identical twin appears on the horizon, and a half dozen more Optimos beyond that until at last, with some relief, you reach the metropolis with its new housing developments and factories and the cluster of downtown skyscrapers.

Optimo City, then, is actually a very familiar feature of the American landscape. But since you have never stopped there except to buy gas, it might be well to know it a little better. What is there to see? Not a great deal, yet more than you would at first suspect.

Optimo, being after all an imaginary average small town, has to have had an average small-town history, or at least a Western version of that average. The original Fort Gaffney (named after some inconspicuous worthy in the U.S. Army) was really little more than a stockade on a bluff overlooking a ford in the river; a few roads or trails in the old days straggled out into the plain (or desert as they called it then), lost heart and disappeared two or three miles from town. Occasionally even today someone digs up a fragment of the palisade or a bit of rust-eaten hardware in the backyards of the houses near the center of town, and the historical society possesses what it claims is the key to the principal gate. But on the whole, Optimo City is not much interested in its martial past. The fort as a military installation ceased to exist during the Civil War, and the last of the pioneers died a half century ago before anyone had the historical sense to take down his story. And when the county seat was located in the town the name was changed from Fort Gaffney with its frontier connotation to Optimo, which means (so the townspeople will tell you) "I hope for the best" in Latin.

What Optimo is really proud of even now is its identity as county seat. Sheridan County (and you will do well to remember that it was NOT named after the notorious Union general but after Horace Sheridan, an early member of the territorial legislature; Optimo still feels strongly about what it calls the War between the States) was organized in the 1870s and there ensued a brief but lively competition for the possession of the courthouse between Optimo and the next largest settlement, Apache Center, twenty miles away. Optimo City won, and Apache Center, a cowtown with one paved street, is not allowed to forget the fact. The football and basketball games between the Optimo Cougars and the Apache Braves are still characterized by a very special sort of rivalry. No matter how badly beaten Optimo City often is, it consoles itself by remembering that it is still the county seat, and that Apache Center, in spite of the brute cunning of its team, has still only one street paved. We shall presently come back to the meaning of that boast.

To get on with the history of Optimo.

THE INFLEXIBLE GRIDIRON

Aided by the state and Army engineers, the city fathers, back in the 1870s, surveyed and laid out the new metropolis. As a matter of course they located a square or public place in the center of the town and eventually they built their courthouse in the middle of the square; such having been the layout of every county seat these Western Americans had ever seen. Streets led from the center of each side of the square, being named Main Street North and South, and Sheridan Street East

and West. Eventually these four streets and the square were surrounded by a gridiron pattern of streets and avenues – all numbered or lettered, and all of them totally oblivious of the topography of the town. Some streets charge up impossibly steep slopes, straight as an arrow; others lead off into the tangle of alders and cottonwoods near the river and get lost.

Strangely enough, this inflexibility in the plan has had some very pleasant results. South Main Street, which leads from the square down to the river, was too steep in the old days for heavily laden wagons to climb in wet weather, so at the foot of it on the flats near the river those merchants who dealt in farm produce and farm equipment built their stores and warehouses. The blacksmith and welder, the hay and grain supply, and finally the auction ring and the farmers' market found South Main the best location in town for their purpose – which purpose being primarily dealing with out-of-town farmers and ranchers. And when, after considerable pressure on the legislature and much resistance from Apache Center (which already had a railroad), the Southern Pacific built a spur to Optimo, the depot was naturally built at the foot of South Main. And of course the grain elevator and the stockyards were built near the railroad. The railroad spur was intended to make Optimo into a manufacturing city, and never did; all that ever came was a small overall factory and a plant for processing sorghum with a combined payroll of about 150. Most of the workers in the two establishments are Mexicans from south of the border – locally referred to in polite circles as "Latinos" or "Hispanos." They have built for themselves flimsy little houses under the cottonwoods and next to the river. "If ever we have an epidemic in Optimo," the men at the courthouse remark, "it will break out first of all in those Latino shacks." But they have done nothing as yet about providing them with better houses, and probably never will.

DOWNTOWN AND UPTOWN

Depot, market, factories, warehouses, slum – these features, combined with the fascination of the river bank and stockyards and the assorted public of railroaders and Latinos and occasional ranch hands – have all given South Main a very definite character: easy-going, loud, colorful, and perhaps during fair week or at shipping time a little disreputable. Boys on the Cougar football squad have specific orders to stay away from South Main, but they don't. Actually the whole of Optimo looks on the section with indulgence and pride; it makes the townspeople feel that they understand metropolitan problems when they can compare South Main with the New York waterfront.

North Main, up on the heights beyond the Courthouse Square and past the two or three blocks of retail stores, is (on the other hand) the very finest part of Optimo. The northwestern section of town, with its tree-shaded streets, its view over the river and the prairie, its summer breezes, has always been identified with wealth and fashion as Optimo understands them. Colonel Ephraim Powell (Confederate Army, Ret., owner of some of the best ranch country in the region) built his bride a handsome limestone house with a slate roof and a tower, and Walter Slymaker, proprietor of Slymaker's Mercantile and of the grain elevator, not to be outdone, built an even larger house farther up Main; so did Hooperson, first president of the bank. There are a dozen such houses in all, stone or Milwaukee brick with piazzas (or galleries, as the old timers still call them) and large, untidy gardens around them. It is worth noting, by the way, that the brightest claim to aristocratic heritage is this: grandfather came out West for his health. New England may have its "Mayflower" and "Arabella," east Texas its Three Hundred Founding Families, New Mexico its Conquistadores; but Optimo is loyal to the image of the delicate young college graduate who arrived by train with his law books, his set of Dickens, his taste for wine, and the custom of dressing for dinner. This legendary figure has about seen his day in the small talk of Optimo society, and the younger generation frankly doubts his having ever existed; but he (or his ghost) had a definite effect on local manners and ways of living. At all events, because of this memory Optimo looks down on those Western mining towns where Sarah Bernhardt and de Reszke and Oscar Wilde seem to have played so many one-night stands in now-vanished opera houses.

A WORLD IN ITSELF

Wickedness – or the suggestion of wickedness – at one end of Main, affluence and respectability at the other. How about Sheridan Street running East and West? That is where you'll find most of the stores; in the first four or five blocks on either side of the Courthouse Square. They form a rampart: narrow brick houses, most of them two stories high with elaborate cornices and long narrow windows; all of them devoid of modern commercial graces of chromium and black glass and artful window display, all of them ugly but all of them pretty uniform; and so you have on Sheridan Street something rarely seen in urban America: a harmonious and restful and dignified business section. Only eight or ten blocks of it in all, to be sure; turn any corner and you are at once in a residential area.

Here there is block after block of one-story frame houses with trees in front and picket fences or hedges; no sidewalk after the first block or so; a hideous church (without a cemetery of course); a small-time auto repair shop in someone's back yard; dirt roadway; and if you follow the road a few blocks more – say to 10th Street (after that there are no more signs) – you are likely to see a tractor turn into someone's drive with wisps of freshly cut alfalfa clinging to the vertical sickle bar. The countryside is that close to the heart of Optimo City, farmers are that much part of the town. And the glimpse of the tractor (like the glimpse of a deer or a fox driven out of the hills by a heavy winter) restores for a moment a feeling for an old kinship that seemed to have been destroyed forever. But this is what makes Optimo, the hundreds of Optimos throughout America, so valuable; the ties between country and town have not yet been broken. Limited though it may well be in many ways, the world of Optimo City is still complete.

The center of this world is Courthouse Square, with the courthouse, ponderous, barbaric, and imposing, in the center of that. The building and its woebegone little park not only interrupts the vistas of Main and Sheridan – it was intended to do this – it also interrupts the flow of traffic in all four directions. A sluggish eddy of vehicles and pedestrians is the result, Optimo's animate existence slowed and intensified. The houses on the four sides of the square are of the same vintage (and the same general architecture) as the monument in their midst: mid-nineteenth-century brick or stone; cornices like the brims of hats, fancy dripstones over the arched windows like eyebrows; painted blood-red or mustard-yellow or white; identical except for the six-story Gaffney Hotel and the classicism of the First National Bank.

Every house has a tin roof porch extending over the sidewalk, a sort of permanent awning which protects passersby and incidentally conceals the motley of store windows and signs. To walk around the square and down Sheridan Street under a succession of these galleries or metal awnings, crossing the strips of bright sunlight between the roofs of different height, is one of the delights of Optimo – one of its amenities in the English use of that word. You begin to understand why the Courthouse Square is such a popular part of town.

SATURDAY NIGHTS – BRIGHT LIGHTS

Saturday, of course, is the best day for seeing the full tide of human existence in Sheridan County. The rows of parked pickups are like cattle in a feed lot; the sidewalks in front of Slymaker's Mercantile, the Ranch Cafe, Sears, the drugstore, resound to the mincing steps of cowboy boots; farmers and ranchers, thumbs in their pants pockets, gather in groups to lament the drought (there is always a drought) and those men in Washington, while their wives go from store to movie house to store. Radios, jukeboxes, the bell in the courthouse tower; the teenagers doing "shave-and-a-haircut; bay rum" on the horns of their parents' cars as they drive round and round the square. The smell of hot coffee, beer, popcorn, exhaust, alfalfa, cow manure. A man is trying to sell a truckload of grapefruit so that he can buy a truckload of cinderblocks to sell somewhere else. Dogs; 10-year-old cowboys firing cap pistols at each other. The air is full of pigeons, floating candy wrappers, the flat strong accent erroneously called Texan.

All these people are here in the center of Optimo for many reasons – for sociability first of all, for news, for the spending and making of money; for relaxation. "Jim Guthrie and wife were in town last week, visiting friends and transacting business," is the way the *Sheridan Sentinel* describes it; and almost all of Jim Guthrie's business takes place in the square. That is one of the peculiarities of Optimo and one of the reasons why the square as an institution is so important. For it is around the square that the oldest and most essential urban (or county) services are established. Here are the firms under local control and ownership, those devoted almost exclusively to the interest of the surrounding countryside. Upstairs are the lawyers, doctors, dentists, insurance firms, the public stenographer, the Farm Bureau. Downstairs are the bank, the prescription drugstore, the newspaper office, and of course Slymaker's Mercantile and the Ranch Cafe.

INFLUENCE OF THE COURTHOUSE

Why have the chain stores not invaded this part of town in greater force? Some have already got a foothold, but most of them are at the far end of Sheridan or even out on the Federal Highway. The presence of the courthouse is partly responsible. The traditional services want to be as near the courthouse as they can, and real-estate values are high. The courthouse itself attracts so many out-of-town visitors that the problem of parking is acute. The only solution that occurs to the enlightened minds of the Chamber of Commerce is to tear the courthouse down, use the place for parking, and build a new one somewhere else. They have already had an architect draw a sketch of a new courthouse to go at the far end of Main Street: a chaste concrete cube with vertical motifs between the windows – a fine specimen of bureaucrat modernism. But the trouble is, where to get the money for a new courthouse when the old one is still quite evidently adequate and in constant use?

If you enter the courthouse you will be amazed by two things: the horrifying architecture of the place, and the variety of functions it fills. Courthouse means of course courtrooms, and there are two of those. Then there is the office of the County Treasurer, the Road Commissioner, the School Board, the Agricultural Agent, the Extension Agent, Sanitary Inspector, and usually a group of Federal agencies as well – PMA, Soil Conservation, FHA and so on. Finally the Red Cross, the Boy Scouts, and the District Nurse. No doubt many of these offices are tiresome examples of government interference in private matters; just the same, they are for better or worse part of almost every farmer's and rancher's business, and the courthouse, in spite of all the talk about county consolidation, is a more important place than ever.

As it is, the ugly old building has conferred upon Optimo a blessing which many larger and richer American towns can envy: a center for civic activity and a symbol for civic pride – something as different from the modern "civic center" as day is from night. Contrast the array of classic edifices, lost in the midst of waste space, the meaningless pomp of flagpoles and war memorials and dribbling fountains of any American city from San Francisco to Washington with the animation and harmony and the almost domestic intimacy of Optimo Courthouse Square, and you have a pretty good measure of what is wrong with much American city planning: civic consciousness has been divorced from everyday life, put in a special zone all by itself. Optimo City has its zones; but they are organically related to one another.

Doubtless the time will never come that the square is studied as a work of art. Why should it be? The craftsmanship in the details, the architecture of the building, the notions of urbanism in the layout of the square itself are all on a very countrified level. Still, such a square as this has dignity and even charm. The charm is perhaps antiquarian – a bit of rural America of seventy-five years ago; the dignity is something else again. It derives from the function of the courthouse and the square, and from its peculiarly national character.

COMMUNAL CENTER

The practice of erecting a public building in the center of an open place is in fact pretty well confined to America – more specifically to nineteenth-century America. The vast open areas favored by eighteenth-century European planners were usually kept free of construction, and public buildings – churches and palaces and law-courts – were located to face these squares; to command them, as it were. But they were not allowed to interfere with the original open effect. Even the plans of eighteenth-century American cities, such as Philadelphia and Reading and Savannah and Washington, always left the square or public place intact. Spanish America, of course, provides the best illustrations of all; the plaza, nine times out of ten, is surrounded by public buildings, but it is left free. Yet almost every American town laid out after (say) 1820 deliberately planted a public building in the center of its square. Sometimes it was a school, sometimes a city hall, more often a courthouse, and it was always approachable from all four sides and always as conspicuous as possible.

Why? Why did these pioneer city fathers go counter to the taste of the past in this matter? One guess is as good as another. Perhaps they were so proud of their representative institutions that they wanted to give their public buildings the best location available. Perhaps frontier America was following an aesthetic movement, already at that date strong in Europe, that held that an open space was improved when it contained some prominent free-standing object – an obelisk or a statue or a triumphal arch. However that may have been, the pioneer Americans went Europe

one better, and put the largest building in town right in the center of the square.

Thus the square ceased to be thought of in nineteenth-century America as a vacant space; it became a container or (if you prefer) a frame. A frame, so it happened, not merely for the courthouse, but for all activity of a communal sort. Few aesthetic experiments have ever produced such brilliantly practical results. A society which had long since ceased to rally around the individual leader and his residence and which was rapidly tiring of rallying around the meetinghouse or church all at once found a new symbol: local representative government, or the courthouse. A good deal of flagwaving resulted – as European travelers have always told us – and a good deal of very poor "representational" architecture; but Optimo acquired something to be proud of, something to moderate that American tendency to think of every town as existing entirely for money-making purposes.

SYMBOL OF INDEPENDENCE

At this juncture the protesting voice of the Chamber of Commerce is heard. "One moment. Before you finish with our courthouse you had better hear the other side of the question. If the courthouse were torn down we would not only have more parking space – sorely needed in Optimo – we would also get funds for widening Main Street into a four-lane highway. If Main Street were widened Optimo could attract many new businesses catering to tourists and other transients – restaurants and motels and garages and all sorts of drive-in establishments. In the last ten years" (continues the Chamber of Commerce) "Optimo has grown by twelve hundred. Twelve hundred! At that rate we'll still be a small town of less than twenty thousand in 1999. But if we had new businesses we'd grow fast and have better schools and a new hospital, and the young people wouldn't move to the cities. Or do you expect Optimo to go on depending on a few hundred tight-fisted farmers and ranchers for its livelihood?" The voice, now shaking with emotion, adds something about "eliminating" South Main by means of an embankment and a clover leaf and picnic grounds for tourists under the cottonwoods where the Latinos still reside.

These suggestions are very sensible ones on the whole. Translate them into more general terms and what they amount to is this: if we want to get ahead, the best thing to do is break with our own past, become as independent as possible of our immediate environment and at the same time become almost completely dependent for our well-being on some remote outside resource. Whatever you may think of such a program, you cannot very well deny that it has been successful for a large number of American towns. Think of the hayseed communities which have suddenly found themselves next to an oil field or a large factory or an Army installation, and which have cashed in on their good fortune by transforming themselves overnight, turning their backs on their former sources of income, and tripling their population in a few years! It is true that these towns put all their eggs in one basket, that they are totally at the mercy of some large enterprise quite beyond their control. But think of the freedom from local environment; think of the excitement and the money! Given the same circumstances – and the Southwest is full of surprises still – why should Optimo not do the same?

A COMMON DESTINY

Because there are many different kinds of towns just as there are many different kinds of men, a development which is good for one kind can be death on another. Apache Center (to use that abject community as an example), with its stockyards and its one paved street and its very limited responsibility to the county, as a community might well become a boom-town and no one would be worse off. Optimo seems to have a different destiny. For almost a hundred years – a long time in this part of the world – it has been identified with the surrounding landscape and been an essential part of it. Whatever wealth it possesses has come from the farms and ranches, not from the overall factory or from tourists. The bankers and merchants will tell you, of course, that without their ceaseless efforts and their vision the countryside could never have existed; the farmers and ranchers consider Optimo's prosperity and importance entirely their own creation. Both parties are right to the extent that the town is part of the landscape – one might even say part of every farm, since much farm business takes place in the town itself.

Now if Optimo suddenly became a year-round tourist resort, or the overall capital of the Southwest; what would happen to that relationship, do you sup-

pose? It would vanish. The farmers and ranchers would soon find themselves crowded out, and would go elsewhere for those services and benefits which they now enjoy in Optimo. And as for Optimo itself, it would soon achieve the flow of traffic, the new store fronts, the housing developments, the payrolls and bank accounts it cannot help dreaming about; and in the same process achieve a total social and physical dislocation, and a loss of a sense of its own identity. County Seat of Sheridan County? Yes; but much more important: Southwestern branch of the "American Cloak and Garment Corporation"; or the LITTLE TOWN WITH THE BIG WELCOME – 300 tourist beds which, when empty for one night out of three, threaten bankruptcy to half the town.

As of the present, Optimo remains pretty much as it has been for the last generation. The Federal Highway still bypasses the center (what a roadblock, symbolical as well as actual, that courthouse is!); so if you want to see Optimo, you had better turn off at the top of the hill near the watertower of the lunatic asylum – now called Fairview State Rest Home, and with the hideous high fence around it torn down. The dirt road eventually becomes North Main. The old Slymaker place is still intact. The Powell mansion, galleries and all, belongs to the American Legion, and a funeral home has taken over the Hooperson house. Then comes downtown Optimo; and then the courthouse, huge and graceless, in detail and proportion more like a monstrous birdhouse than a monument. Stop here. You'll find nothing of interest in the stores, and no architectural gems down a side street. Even if there were, no one would be able to point them out. The historical society, largely in the hands of ladies, thinks of antiquity in terms of antiques, and art as anything that looks pretty on the mantelpiece.

The weather is likely to be scorching hot and dry, with a wild ineffectual breeze in the elms and sycamores. You'll find no restaurant in town with atmosphere – no chandeliers made out of wagon wheels, no wall decorations of famous brands, no bar disguised as the Hitching Rail or the Old Corral. Under a high ceiling with a two-bladed fan in the middle, you'll eat ham hock and beans, hot bread, iced tea without lemon, and like it or go without. But as compensation of sorts at the next table there will be two ranchers eating with their hats on, and discussing the affairs, public and private, of Optimo City. To hear them talk, you'd think they owned the town.

That's about all. There's the market at the foot

of South Main, the Latino shacks around the overall factory, a grove of cottonwoods, and the Apache River (North Branch) trickling down a bed ten times too big; and then the open country. You may be glad to have left Optimo behind.

Or you may have liked it, and found it pleasantly old-fashioned. Perhaps it is; but it is in no danger of dying out quite yet. As we said to begin with, there is another Optimo City fifty miles farther on. The country is covered with them. Indeed they are so numerous that it sometimes seems as if Optimo and rural America were one and indivisible.

THREE

"The Islamic City: Historic Myths, Islamic Essence, and Contemporary Relevance"

International Journal of Middle East Studies (1987)

Janet Abu-Lughod

Editors' Introduction

The physical form, social structure, politics, and economics of "non-Western" cities in the Middle East, Asia, Africa, Latin and Central America, and elsewhere in the world differ markedly from "Western" cities in North America and Europe. In an increasingly globally interconnected world it is important to understand the nature of non-Western cities.

In this selection sociologist Janet Abu-Lughod describes forces that shaped Islamic cities. She criticizes sloppy and overly-generalized thinking about city classification, and warns planners to be sensitive to the *processes* – good and bad – that shaped Islamic cities and not merely to try to replicate old physical forms today.

It is tempting for writers who perceive cities in some part of the world as like each other and unlike cities elsewhere to lump together all the cities in a region, country, or even a continent and apply broad generalizations to, for example, "Chinese cities" or "Asian cities," or "South American cities," or "African cities." But this is like grouping all North American or all European cities together as a single city type. Honolulu, East St. Louis, New Orleans, and Salt Lake City are all "North American cities." London, Pamplona, and Biarritz are all "European cities." But readers even remotely familiar with these North American and European cities will be aware of how different they are from each other and how misleading lumping them together as a single city type would be.

Sophisticated comparative urbanists like Abu-Lughod are sensitive to the variety of city types within a region, country, or continent. They are cautious about classifying cities into simple "ideal types." This selection is exemplary for the subtlety and detail of Abu-Lughod's description of what Islamic cities were and are like, the intellectual rigor she displays in sorting out what is truly distinctive about Islamic cities from what is not, her rejection of deterministic explanations, and her honesty critiquing religious, class, and gender discrimination that contributed the formation of Islamic cities.

Like Lewis Mumford (p. 92), Abu-Lughod is aware of how history and culture have shaped cities. She rejects simple explanations of how city characteristics were *determined* by the geography of other factors. Terrain (often hot and arid) and transportation technology (foot, animal, or sail powered) shaped Islamic cities in the Middle East, North Africa, and India, but Abu-Lughod points out that these factors are hardly unique to Islam and did not determine the form Islamic cities took. Nor does technology explain the essence of Islamic cities. There is nothing distinctively "Islamic" about pack animals, handicraft production, or small-scale markets. Islamic religion – with its emphasis on a significant gathering at a single site for Friday prayers – is an essentially "urban" religion. But Abu-Lughod notes that Christianity and Judaism are also urban religions. Religion was a crucial contributing factor in shaping Islamic cities, but Abu-Lughod argues that it was not *the* determining factor.

Rather, Abu-Lughod argues, distinctly Islamic cities are best defined by a combination of social structure, particularly distinctions between the community of believers and non-believers in Islam, the status of women, and Islamic law's decentralized decision-making about land use and social control rather than religion. It was social, political, and legal characteristics of Islam that shaped (but did not determine) the processes whereby Islamic cities were created, transformed, and transformed again.

Abu-Lughod hypothesizes that three factors were particularly important: the definition of legally distinct social classes in Islamic law based on religion and nationality, gender segregation, and decentralization of social and land use control to the neighborhood level.

Historically Islamic law created distinctions between classes based on religion and nationality. In most Islamic cities Jews, Christians, Zoroastrians, and other religious or nationality groups lived together in distinct city neighborhoods (quarters), though residential segregation in Islamic cities was not universal and was generally voluntary rather than required. The behavior of foreigners and non-believers was controlled. Islamic societies specified what non-believers could wear, what trades they could engage in, and sometimes even consumption practices such as the right to use horses rather than mules. While the extent of discrimination may not have been as great as W.E.B. DuBois describes in the Black wards of nineteenth-century Philadelphia (p. 119), not all scholars would agree with Abu-Lughod's assessment that the regulation of non-believers in Islamic cities was seldom viewed as oppressive.

Gender segregation also shaped Islamic cities. Abu-Lughod considers the division between male and female turf to be perhaps the most important defining characteristic of the Islamic city. She describes a range of features designed to *physically* shelter women from the unwanted attention of men and architectural features – crooked entryways, lattice-wood screens, doors that do not face doors on the opposite side of the street, windows set so women could view men, but not the other way around – that create *visual* segregation of the sexes. Public, semiprivate, and private spaces in Islamic cities were all defined largely around gender discrimination.

A final hypothesis Abu-Lughod advances is that the way in which neighborhoods functioned – particularly to assure proper behavior between men and women and to defend the neighborhood from crime, drug dealing, and other outside threats. Long before Jane Jacobs urged modern city planners to design cities so that residents could keep "eyes on the street" (p. 114), neighborhoods in Islamic cities were organized to observe and protect their turf.

While Abu-Lughod is sensitive to Moslem planners who want to plan cities consistent with the best cultural traditions of Islam, she argues that attempts to replicate historic Islamic city physical *forms* without an understanding of the *processes* that created them is terribly inappropriate. Looking for techniques and tricks in past Islamic city planning and architecture misses the point. She argues that the beauty, privacy, and community that represent the best in Islamic cities of the past can be created in modern Islamic cities without replicating the gender, class, and religious discrimination that was present. She argues that modern Islamic cities can have better planning and government oversight than historic Islamic cities without destroying their best aspects.

Janet Abu-Lughod is Professor Emerita of Sociology and Historical Studies from the graduate faculty of the New School for Social Research in New York City and the department of sociology of Northwestern University in Chicago. Her books include *New York, Chicago, Los Angeles: America's Global Cities* (Minneapolis: University of Minnesota Press, 1999), *From Urban Village to East Village: The Battle for New York's Lower East Side* (Oxford: Blackwell, 1994), *Changing Cities* (New York: Addison, 1991), and other works on urbanism, sociology, and Islamic history and culture. Her *Before European Hegemony: The World System AD 1250–1350* (London and New York: Oxford University Press, 1991) won an American Sociological Association award for distinguished scholarship.

Other books on Islamic cities past and present include Stefano Bianco, *Urban Form in the Arab World* (London and New York: Thames and Hudson, 2000), Paul Wheatley, *Cities Where Men Pray: Cities in Islamic Lands, Seventh Through the Tenth Centuries* (Chicago: University of Chicago Press, 2001), Suquitaki Sato (ed.), *Islamic Urbanism in Human History* (New York: Columbia University Press, 1996), Hooshang Amirahmadi and Salah S. El-Shakhs (eds), *Urban Development in the Muslim World* (New Brunswick, NJ: Center for Urban Policy Research Press, 1993) and Besim Selim Habim, *Arabic-Islamic Cities* (New York: Columbia University Press, 1986).

At the present time of a resurgence in Islamic beliefs, the question of the Islamic city has once again come to the fore. In many parts of the Arab world, and especially in Saudi Arabia and the Persian Gulf, urban planners with a newfound respect for the great achievements of the past are searching for ways to reproduce in today's cities some of the patterns of city building that have been identified as Islamic. They have been influenced, whether wittingly or not, by a body of literature produced by Western Orientalists purporting to describe the essence of the Islamic city . . .

THE *ISNAD* OF THE ISLAMIC CITY

In some ways, historiography takes the same form as the traditions of the Prophet. The authenticity of any proposition is judged by the *isnad* or "chain" by which it descended from the past. Certain chains are deemed more trustworthy than others. One makes reference to an earlier authority in order to substantiate a statement's authenticity or truth. The truth, therefore, is only as good as the *isnad* (chain) of its "construction."

[. . .]

One of the earliest codifications of the characteristics of the Islamic city, at least the earliest generally cited in subsequent literature, was William Marçais's 1928 article, "L'Islamisme et la vie urbaine." The article introduces several themes that appear over and over again in subsequent discussions of the Islamic city.

The first is that Islam is essentially an urban religion. In support of this contention, Marçais notes that Muhammad himself was an urbanite suspicious of nomads, that the leadership cadres of the early Islamic proselytizers were members of the urban bourgeoise of the Arabian peninsula, and that the requirement that the Friday communal prayer be solemnized at a congregational mosque made urban living necessary for the full Muslim life. Marçais uses an earlier link in the chain of Orientalism when he cites Joseph Ernest Renan, the French philologist and historian, to legitimate his view. The phrase he quotes is a simple allegation: "The mosque, like the synagogue and the Church, is a thing essentially urban (*citadine*). Islamism [sic] is a religion of cities." This quotation is particularly intriguing because it undermines the whole enterprise of defining the unique character of the Islamic city; it suggests that Islam shares with Judaism and Christianity the same quality of urbanity.

However, Marçais makes a second point as well. He notes that new cities were often founded by new powers/dynasties in Islamdom, thus acknowledging that Islamic civilization was not merely a set of religious beliefs and laws but also a functioning society that was Islamic in the sense that it organized the life of Muslims into a community not just of believers but of doers.

Finally, Marçais introduces several characteristic elements of the physical city. Citing the North African historian Ibn-Khaldun, Arab geographers, and legal doctrines, he reaches a definition of the Islamic city that he contends is quintessential: a city must have a congregational Friday mosque and it must have a market/chief bazaar nearby. Associated with the *jami suq* (mosque-market) complex was a third physical feature of Islamic cities, the public bath (*hammam*), of functional significance to prepare believers for the Friday prayer. Paraphrasing Renan, though, we might note that when the church was also the temporal power, medieval European cities were also defined by the presence of the cathedral and the marketplace in front of it. Thus far, therefore, we have only a very modestly etched idea of the Islamic city, one that poorly distinguishes it from cities in other religious/cultural contexts and one that has as yet no topography.

[. . .]

It is very important to note here that virtually all of the cases cited by Marçais in his article are North African . . .

[. . .]

The fact is that most studies still focus on a single case and try to generalize rather than start with the more fundamental question: Why would one expect Islamic cities to be similar and in what ways? . . .

Before this, however, a brief biographical note. I went to India in the late seventies, looking for (and hoping not to find) the Islamic city. To my surprise, it was easy to distinguish Muslim from Hindu urban quarters. Several cues seemed to trigger a subliminal response. First, the ratio of males to females on the street and in public places was higher in Muslim than in Hindu areas. Second, butcher shops and trades related to hides and animal products were located almost exclusively in Muslim areas. Third, it seemed that the decibel level of sounds was higher and more animated in Muslim quarters, of which the call to prayer was only one of the added elements.

These semiotics were chiefly in public space; it was harder to penetrate semiprivate and private space, except in several Hindu areas to which I had personal entrée. The quarter, in the sense of a semiprivate lane or courtyard apartment house, seemed as typical of poor Hindu areas as of Muslim areas, but circumspect behavior outside the private living quarters seemed less in Hindu than in Muslim areas. One explanation for this might be differences in the rules of veiling. Hindu women veil primarily before close relatives (especially fathers-in-law) whereas Muslim women veil chiefly from outsiders. Given this, one would expect Muslims to make a greater distinction between in and out of the house. In brief, it seemed that social patterns of gender segregation and social patterns of proscribed foods were the chief religion-linked variables distinguishing Muslim from Hindu quarters in mixed cities.

[. . .]

What, then, can it be that is distinctive?

It is easy to say what it is not. It is not a form per se. The careful reader will have noted that the Orientalists' discussions of the Islamic city focus on a unique conjunction of forces that created a few cities they take to be prototypical. The forms of these cities at certain points in time are taken as ideal types and are further abstracted to obtain a final ideal that is created out of congruent forms. This, they suggest, was (is?) the Islamic city. Not only scholars but present practicing Muslims looking to build new Islamic cities have accepted this approach and are trying to find in the planning and architectural repertoire of the past the tricks and techniques that will reproduce it. However, these approaches – scholarly or practical – miss the point. Cities are the products of many forces, and the forms that evolve in response to these forces are unique to the combination of those forces. A city at one point in time is a still photograph of a complex system of building and destroying, of organizing and reorganizing, and so on. In short, the intellectual question we need to ask ourselves is: Out of what forces were the prototypical Islamic cities created?

WHAT CREATED THE "TRADITIONAL" CITY IN THE ISLAMIC WORLD

A modest list of the forces that created the traditional Islamic city would include: a terrain/climate; a technology of production, distribution, and transportation; a system of social organization; and a legal/political system that, in Islamic places and times, could vary considerably.

It is exceedingly hard to unpack this complex bundle to determine the extent to which Islam influenced any one of them at any point in time. We must dismiss terrain entirely as being Islamic, even though the Arab region in which this genre of city building was developed had a characteristic climate/terrain as well as a historic inheritance that encouraged common solutions. We must also dismiss technology as being Islamic, for there is nothing religious about pack animals, handicraft production, small-scale market arrangements, and so forth, any more than there is something religious about terrain and climate. Once one eliminates the influence of these factors, one is left with exploring the social, political, and legal characteristics of Islam that shaped, but did not determine, the processes whereby Islamic cities were formed, transformed, and transformed again . . .

Let us begin with an extreme statement: namely, that the division of the Middle Eastern–North African city into a nested set of territories with clear markers and defended borders was not Islamic per se but was reflective of a social order that had much in common with other societies based upon the family writ large (tribalism, clans, and ethnicity). Fluctuations in the strength of the markers and the degree to which boundaries were defended were contingent more upon the state of law and order than on shifts in religious ideology. This is a clear indication that religion was not the determining variable.

However, to say that Islam was not the only cause of urban form is not to say that it was unimportant. On the contrary, it was a crucial contributing factor in shaping cities within its realms. It contributed in several important ways.

First, it made rough juridical distinctions among population classes on the basis of their relation to the *Umma* (community of believers) and thus the state. These distinctions were available in the repertoire of territoriality and could be translated into spatial segregation under certain conditions. Furthermore, the frequent inability of the state to transcend communal organizations and the laissez-faire attitude of the state toward civil society left important functions to other units of social organization that strengthened them. Since many of these functions were vicinal ones (such as maintaining streets and utilities, guarding turf, providing lighting, and supervising and sanctioning behavior), and since many vicinal units were composed of socially related people, what we would call the neighborhood became a crucial building block of cities in the Arab world during medieval and even later times . . .

Second, by encouraging gender segregation, Islam created a set of architectural and spatial imperatives that had only a limited range of solutions. What Islam required was some way of dividing functions and places on the basis of gender and then of creating a visual screen between them. This structuring of space was different from what would have prevailed had freer mixing of males and females been the pattern . . .

Finally, one returns to the system of property laws that governed rights and obligations vis-à-vis both other property owners and the state. Such customary laws and precedents set in motion a process whereby a pattern of space was continually reproduced. Of primary importance were the preexisting rights of individual or collective users of land and immovable property. Of secondary importance were the rights and responsibilities of proximate neighbors, followed by those of more distant ones. Then, at last, as a residual, there was the right of the collectivity or larger administrative unit. Under such circumstances, access to entrances took priority over major thoroughfares or the reservation of land for public purposes.

I would like here to speculate on these three themes, drawing upon evidence from both medieval and modern time periods . . .

TERRITORIALITY IN THE ARABO-ISLAMIC CITY

One of the most striking features of the cities of the Middle East and North Africa, certainly during medieval times but to some extent persisting feebly to this day in the older residential quarters, is its subdivision into smaller quarters whose approximate boundaries remain relatively constant over time and whose names continue to be employed as important referential terms, even when they do not appear on modern markers of street names.

[. . .]

The design of the Arabo-Islamic city, with is convoluted paths, was intended to subdivide space into relatively permanent quarters, but a recognition of this fact simply begs the question. Medieval European towns were equally devoid of right angles and through streets; they also were subdivided into potentially organizable subpockets. Nevertheless, they are quite different in physical pattern and were quite different in social organization from the medieval Arab City.

One must look for more than design to explain the signification of turf in the Arabo-Islamic city. One must look for a common creator of boundaries. There may be an overarching Islamic reason for the latter. Islamic property laws about differential responsibility to neighbors and control over accessways to dwellings may have been partially responsible for the typical cellular pattern found in residential quarters of medieval Islamic cities. However, I see Islamic law as an adaptive mechanism for helping the society to achieve its goals rather than as a deus ex machina determining them. Therefore, we must go behind the issue of how imageable cells of residence were created and maintained to explore why they were so typically the building block of urban society. This brings us, then, to the three hypotheses I would like to advance.

TURF AND JURIDICAL CLASSES

States that make juridical distinctions among residents lay the foundation for what can evolve into a system of spatial segregation. It is a necessary but, however, not a sufficient cause of residential apartheid. That is because social distance and physical distance are not necessarily the same thing. By social distance, we mean the degree to which open egalitarian interaction is blocked or, rather, the amount of work that must be done by two parties to overcome social barriers to intimacy. By physical distance, we mean the degree to which physical contact is blocked by space or, rather, the amount of work that must be done to overcome spatial barriers to face-to-face contact. Clearly, we can all think of cases in which maximum social distance can coexist with minimal spatial segregation (master–valet, master–house slave) and, conversely, other cases where minimum social distance can be sustained over great physical distance (loved ones in other countries). Indeed, it is generally when lines of social distance become less marked that physical distance is intensified.

Medieval Islamic cities certainly did maintain the distinction between juridical classes through social distance (as evidenced by sumptuary regulations, the semiotics of clothes, body postures, and so on), but spatial distance was not always a mechanism for maintaining social distance. On occasion, though, particularly during periods of tension, physical segregation was employed to intensify the social boundary markers.

For example, a juridically different category in most Middle Eastern theocracies consisted of the *dhimmis* (Christians, Jews, and Zoroastrians enjoying a special status of protection), who, where relevant, might be subclassified by specific faith or place of origin. Always there were rules governing their behavior and regulations concerning their collective responsibilities to the state. Often there were specific restrictions on the occupations open to them. Occasionally, there were rules specifying consumption patterns (whether they could ride horses or were confined to mules, and the like) and even permitted dress. For the most part, these regulations do not seem to have been viewed as oppressive. (When they became so, as, for instance, during the reign of al-Hakim in Cairo, they occasioned great comment about the sanity of the ruler, indicating how great a deviation from normal they were.)

Residential segregation, however, was not invariable and was seldom involuntary. Voluntary concentrations are noted over and over again in urban histories, either in relation to certain economic functions (Coptic quarters near ports in Cairo) or to certain political advantages (Jewish quarters near the palace of the ruler). Such concentrations facilitated the exercise of self-rule in matters of personal status and helped, in the proximity-based city of the time, to gather the density required to support common special services and institutions. These common services and institutions, in turn, created markers for quarters that indicated to outsiders who was supposed to live there and indicated to insiders that they belonged there.

[. . .]

GENDER SEGREGATION AND THE ARABO-ISLAMIC CITY FORM

The creation of male and female turf is perhaps the most important element of the structure of the city contributed by Islam. It is important to remember, however, that the rules of turf were not only to establish physically distinctive regions; more important, they were to establish visually distinctive or insulated regions. The object was not only to prevent physical contact but to protect visual privacy. Line-of-sight distance, rather than physical distance, was the object of urban design. Thus, Islamic law regulated the placement of windows, the heights of adjacent buildings, and the mutual responsibilities of neighbors toward one another so as to guard visual privacy. Architecture assisted this process. Not only the devices of mashribiyya (lattice-wood) screening but the layout of houses and even of quarters created the strangely asymmetrical reality that women could see men but men could not see women, except those in certain relationships with them. Here is Stanley Lane-Poole's description of an Egyptian upper-class house in the nineteenth century:

> [As one enters the house there] is a passage, which bends sharply after (the first yard or two, and bars any view into the interior from the open door. At the end of this passage we emerge into an open court . . . Here is no sign of life; the doors are jealously closed, the windows shrouded . . . We shall see nothing of the domestic life of the inhabitants; for the women's apartments are carefully shut off from the court . . . The lower rooms, opening directly off the court are those into which a man may walk with impunity and no risk of meeting any of the women . . . [another] door opens out of the court into the staircase leading to the harim rooms, and here no man but the master of the house dare penetrate . . . When a man returns there he is in the bosom of his family, and it would need a very urgent affair to induce the doorkeeper to summon him down to anyone who called to see him.

One need not take the testimony of a foreign observer to substantiate the universality of some of the principles. In the *aqsar* (castles) of southern Morocco, one can find the same bent entrances designed to create a visual blindspot. Urban building regulations were replete with requirements that the doors of buildings occupying opposite sides of the street must not face one another, another mechanism of visual control.

Similarly, within the dwelling, the ideal was to segregate public from private space so that males could circulate without interfering with the movement and activities of females. Obviously, this was possible only, if at all, for the very wealthy, such as the household home that was described above . . . For the poor, no such absolute segregation was possible. Rather, signs and codes helped regulate spatial symbiosis, often by rules that governed timing.

The most obvious semiotic of sexual segregation in the Islamic city was the sign used in front of the public bath to indicate ladies' day. Subtler signs governed other divisions of time and/or space, however. Take, for example, the zone just outside the houses that share a common accessway in a dead-end quarter. These are found in most parts of the Arab world, from the Fertile Crescent to Morocco. I have elsewhere termed this space semiprivate space, a third category between public and private that is found infrequently in sex-integrated societies but is often found in sex-segregated societies.

The fact is that the ideal of separation between the sexes is best achieved by the wealthy, who can afford to duplicate space and can afford the servant or slave girls who were never guarded from male sight or contact. Most poor women were less able to meet the ideal. For them, the family writ large permitted the doing of tasks as well as the protection from strange males because the local neighborhood was an extension of the home and therefore the family. The blind alley or dead-end court street was such a device for achieving this compromise between the exigencies of life and the directives of female seclusion . . .

[. . .]

It is clear that when densities are high and houses too small to contain the manifold activities women are supposed to do in them, the spillover space becomes appropriate as semiprivate space and coresidents who might inadvertently have visual access are appropriated into a fictive kinship relationship to neutralize danger. Dress is an important part of the semiotics of space. As Nadim notes, "clothing which is acceptable for a woman within the lodging is also acceptable in the harah. Nor is it only in Egypt that such adaptations take place. Elizabeth W. Fernea's descriptions of her *Street in Marrakech* indicate that when women ran next door within the enclosed portion of their street, they did not cover themselves as fully as they would have had they been going into public space. It was all in the family. Clearly, then, one of the reasons why the older pattern of city building has been maintained in many sections of Arabo–Islamic cities, even today, is that it

is still well adapted to the complex demands for visual privacy for females.

I am often struck, as I wander around Arab cities, with how easy it is to tell whether I am in public space or have blundered into semiprivate space. I have often tried to identify the markers that indicate this. A sudden narrowing of the path, particularly if that narrowing has been exaggerated by the implanting of low stone posts or even a pile of bricks, is a sign of the shift, especially when the load widens again soon afterward. Even when the spatial semiotics are absent, however, the personal ones are present. There is the questioning look or the approach of someone wanting to help but clearly also wanting to know.

Institutions have been retained from earlier periods, one of which is the *nadorgi* (from *nazara*, "to sight"), who, in Nadim's Cairo hara, was responsible not only for overseeing proper behavior between male and female hara residents but also for spotting strangers. As she describes it, the

> *nadorgi* . . . is responsible for keeping an eye on those entering the harah and detecting their movements. He is usually someone whose shop or house is close to the entrance to the harah where he remains most of the time. Besides being a source of information concerning external movement into the harah, the *nadorgi* can provide equally valuable insights into the internal movements of the various harah residents . . . Whenever illegal activities occur in the harah, and in most cases this is the smoking or trading of hashish, the role of the *nadorgi* becomes vital since he quickly warns of the entrance of outsiders into the harah. The *nadorgi* will approach the outsider under the pretense of wishing to help him find whomever he wants. This tactic serves two purposes: first, it detains the intruder and secondly, it provides the *nadorgi* with information about the outsider's destination and contact.

I shall return to this point when I investigate the other function of the neighborhood in the Arabo-Islamic city, namely, defensive space.

THE NEIGHBORHOOD AS A KEY ELEMENT IN CIVIL SOCIETY AND THE STATE

The final way in which Islam shaped the traditional Arabo-Islamic city was through neglect, ironic as that may seem. By failing to concern itself with matters of day-to-day maintenance, Islamic states often encouraged the vitality of other substate functional units. One of these was definitely the residential neighborhood.

The rather more rigid segregation between commercial and residential quarters in the classic Islamic city has been attributed to the need to separate private (that is, female) from public (that is, male) space. Whatever the cause, such segregation certainly did have important effects. It left to the residential areas a large measure of autonomy since many of the public functionaries (the supervisors of the marketplaces or the supervisors of public morals) operated largely in the commercial sections of the city. Neighborhoods handled many of their internal functions on a more ad hoc basis, being unable to afford more commercialized services . . .

A second factor that strengthened the neighborhood was its role as protector. I would like to explore the issue of turf and defended neighborhoods because I find the literature produced by Orientalists on the role of *dhu'ar* (militant), *futuwwa* (chivalrous society), and so forth, in the medieval Islamic city highly deficient in sociological sense. I have been struck over and over again with the fact that the traditional Arabo–Islamic city was designed to maximize what Oscar Newman has termed defensible space.

In the introduction to his book, *Defensible Space: Crime Prevention Through Urban Design*, Newman writes that he is trying to find an architectural solution to the rising disorder in American cities. He claims:

Architectural design can make evident by the physical layout that an area is the shared extension of the private realms of a group of individuals. For one group to be able to set the norms of behavior and the nature of activity possible within a particular space, it is necessary that it have clear, unquestionable control over what can occur there. Design can make it possible for both inhabitant and stranger to perceive that an area is under the undisputed influence of a particular group, that they dictate the activity taking place within it, and who its users are to be. This can be made so clearly evident that residents will not only feel confident, but that it is incumbent upon them to question the comings and goings of people to ensure the continued safety of the defined areas. Any intruder will be made to anticipate that his presence will be under question and open to challenge; so much so that a criminal can be deterred from even contemplating entry. Defensible space is a model for residential environments which inhibit crime by creating the physical expression of a social fabric that defends itself.

Certainly, what Newman has just described is the Arabo-Islamic semiprivate quarter *par excellence*, but is the picture as benign as he has drawn it? Yes, the neighborhood defends itself – but perhaps it is defending its criminal activities or its warfare with the rest of society. Two nonbenign parallels present themselves. These are boys' gangs and the militia turf in embattled Beirut. Both offer a seamier view of the defensible space advocated by Newman.

Boys' gangs were certainly a feature of Arabo-Islamic cities in the past and continue to be present today . . .

[.]

Historically, in Arabo-Islamic cities, the neighborhood has been in dialectical process with the external society. When central power was strong and when the citywide hierarchical structure was working smoothly, agents of the central administration operated within the neighborhoods to provide information to the center and ensure conformance with central directives. This was certainly the case with the sheikh of the hara in Cairo at certain points, when he was essentially an informer for the police as well as for the *muhtasib* (inspector of morals). He acted, in his capacity as real estate expert, to "steer" or supervise who should have access to vacant dwelling units in the quarter. Sometimes, the neighborhood was an administrative subset of the state.

More often, however, the quarter played the opposite role, that of a defended neighborhood, particularly when chaos reigned. One reads, in the historical accounts, of civil strife/invasions/street battles and the recurring phrase, "and people closed the gates to their harat." Alternatively, to gain control over the city, conquerors always had to destroy the gates to the harat, as Napoléon's forces did when they invaded Cairo.

[. . .]

CONCLUSIONS AND A NOTE OF WARNING

In the first part of this chapter, I attempted to deconstruct Orientalist thinking about the Islamic city by showing not only that the idea itself was "created" on the basis of two few cases but, even worse, was a model of outcomes rather than one of processes. By that I mean that the goal was to generalize about a specific form of city at one long historic moment without unpacking the various causes of that particular outcome. That form was then equated with the Islamic city, regardless of whether there was anything especially Islamic about the causes.

The reason it is important to criticize this approach is that in a number of Arab countries today planners are trying to re-create Islamic cities – but by means that are terribly inappropriate because they focus on outcomes rather than processes. Such planners hope, by edict and ordinance, to preserve and to build anew cities on an Islamic pattern. It should be clear by now, though, why this approach is likely to fail.

Cities are processes, not products. The three Islamic elements that set in motion the processes that give rise to Islamic cities were: a distinction between the members of the Umma and outsiders, which led to juridical and spatial distinction by neighborhoods; the segregation of the sexes, which gave rise to a particular solution to the question of spatial organization; and a legal system, which, rather than imposing general regulations over land uses of various types in various places, left to the litigation of neighborhoods the detailed adjudication of mutual rights over space and use. These three factors were Islamic *per se*.

However, in addition, the historic cities that developed in Arabo-Islamic lands in premodern times were deeply influenced by such non-Islamic factors as climate, terrain, technologies of construction, circulation, and production, as well as political variables, such as the relation between rulers and the ruled, the general level of intercommunal strife, and fluctuations in the degree of internal and external security. Furthermore, the nature of any Islamic city at any point in time was the result not only of the contemporaneous nature of these variables but the inherited forms that took shape under earlier and different circumstances.

It is clear, then, that one does not have the capacity to re-create Islamic cities by edict. One has only the capacity to create conditions that might set in motion processes that, in the past, generated the forms of the traditional city in the Arabo-Islamic world. It must be recognized that one cannot do that without being willing to live with the three conditions mentioned above, namely: (1) juridical distinctions between Muslims and/or citizens and outsiders; (2) segregation by gender and a virtually complete division of labor according to it; and (3) a fully decentralized and ex post facto system of land use and governmental regulation over space. In today's world, these three are considered retrogressive.

First, modern states accord basic rights and responsibilities in an egalitarian manner – at least in theory. Where distinctions are made on the basis of ascribed status, they have attracted the criticism of the world . . .

Second, throughout the world, there has been a trend toward increased equality between the sexes. Integration, not segregation, has been the ideal toward which many cultures are moving.

Finally, modern municipal governments stress the provision of community facilities through a centralized system and stress the establishment of laws that apply to whole classes of places and uses, that is, zoning laws, building codes, street alignments, subdivision regulations, and so on. Such regulations, as I have tried to show, are the antithesis of the assumptions and mechanisms of property law under Islamic legal approaches.

Therefore, none of the conditions still exist that would permit us to reconstruct Islamic cities by design. Only a view of the Islamic city such as that held by earlier Orientalists would allow one to even entertain such a notion.

That is not to say, however, that we could not build better cities in the contemporary Arabo–Islamic world if we paid closer attention to some of the true achievements of the past and if we learned from them. The historic Islamic city often achieved community, privacy, and beauty. It would be wise to seek these same goals, even though the old means are no longer available. Since cities are living processes rather than formalistic shells for living, they cannot be built by us. We can only encourage them to grow in the desired direction. Can we nurture neighborhoods that are supportive but not defensive? Can we foster privacy not for women alone but for households? Can we guard the rights of neighbors while still applying laws consistently? That is the task Arab city planners must set for themselves.

"Social Exclusion and Space"

from Ali Madanipour, Goran Cars, and Judith Allen (eds), Social Exclusion in European Cities: Processes, Experiences, and Responses (1998)

Ali Madanipour

Editors' Introduction

Exclusion of groups of city residents from access to all that the city has to offer on the basis of race, religion, income, or national origin has been and continues to be a pressing issue in cities throughout the world. While sociologist Ali Madanipour analyzes spatial aspects of social exclusion in contemporary European cities in this selection, his analysis is very relevant to understanding cities everywhere in the world both today and in the past.

Some of the most dynamic urban societies have welcomed foreigners and included them in the life of the city. H.D.F. Kitto notes that twenty-five centuries ago foreigners (metics) participated in most aspects of the life of the Greek polis (p. 43). They lived throughout the polis rather in geographically segregated foreigners' neighborhoods, worked as merchants and tradesmen on an equal footing with Athenian citizens, and contributed significantly to philosophical, scientific, literary, and artistic achievements of the polis. But they were not Athenian citizens and were excluded from participation in Athens' otherwise extraordinarily inclusive and democratic political institutions. In other cities law and/or cultural norms have excluded social groups. Janet Abu-Lughod (p. 172) notes that in medieval Islamic cities Christians and Jews were sometimes excluded from specified occupations, forbidden to wear the same clothes as members of the Islamic community of believers, and forced to live in designated quarters of the city. W.E.B. Du Bois describes in painful detail how Blacks in late nineteenth-century Philadelphia were spatially isolated in just a few wards of the city and systematically barred from white schools, most public facilities, and good-paying jobs. Members of the Chicago School of sociology such as Louis Wirth (p. 97) and Ernest W. Burgess (p. 156) described the spatial separation and social exclusion of early twentieth-century Chicago as waves of immigrants poured into the city.

As globalization continues to bring people from throughout the world into closer contact and as the pace of immigration increases the issue of exclusion becomes ever more pressing. All the major cities of the European Union have significant numbers of immigrants from Third World countries – often with skin color, religion, educational backgrounds, and cultures very different from the host countries. In what different ways are these people "excluded" from participation in the life of the cities where they live? How is exclusion expressed in urban space?

Madanipour distinguishes between *economic discrimination*, in which members of a group are excluded from access to employment resulting in poverty and economic marginalization, *political discrimination*, in which they are excluded from political power by being denied voting rights or full political representation, and *cultural exclusion* in which the group members are marginalized from the symbols, meanings, rituals, and discourses of the dominant culture. He sees exclusion as a continuum from full integration into society at one end of the spectrum to nearly total lack of integration at the other.

While some societal rules about exclusion are benign – the right of strangers to enter a person's home at will is unacceptable in almost all cultures – exclusion of groups from the opportunities and advantages that cities possess is painful to members of the group and damaging to the society at large, as evidenced by the racial and religious conflict in cities around the world.

Exclusion frequently has a *spatial* dimension. Members of a group are sometimes excluded from areas of a city by law as when medieval Venetian Jews were restricted to the city's ghetto or Chinese (along with dogs!) prohibited from entering parks in nineteenth-century Shanghai. Even when people are legally free to enter areas of the city, as Sharon Zukin (p. 136) and Mike Davis (p. 201) point out, subtle and not-so-subtle signs and cues may signal that members of a particular group are not welcome.

Madanipour suggests two potentially more promising approaches – "decommodifying" space so that the private real estate market plays a less decisive role in where different groups are located within the city and deliberate city planning to de-spatialize social exclusion. Building "inclusionary" housing units for low- and moderate-income households in neighborhoods they could otherwise not afford is an example of the first strategy. Mixed-use zoning to promote social diversity is an example of the second.

Madanipour concludes his analysis by advocating inclusionary practices – to assure that "outsiders" are more fully included in urban societies. He wants to break the trap of socio-spatial exclusion and provide more possibilities for inclusion.

Ali Madanipour teaches at the School of Architecture, Planning, and Landscape at the University of Newcastle in the UK. His special area of interest is the relationship between society and urban space. His books include *Public and Private Spaces of the City* (London: Routledge, 2003), *Terhan: The Making of a Metropolis* (New York: John Wiley and Sons, 1998), *Design of Urban Space: An Inquiry into a Socio-spatial Process* (New York: John Wiley, 1996) and three co-edited anthologies: *The Governance of Place* (Aldershot, Hampshire: Ashgate, 1996) co-edited with Angela Hull and Patsy Healey; *Urban Governance, Institutional Capacity, and Social Milieux* (Aldershot, Hampshire: Ashgate, 2002) co-edited with Goran Cars and Patsy Healey; and *Social Exclusion in European Cities: Processes, Experiences, and Responses* (London: Jessica Kingsley, 1996) co-edited with Goran Cars and Judith Allen, from which this selection is taken.

■ ■ ■ ■ ■ ■

This chapter concentrates on the relationship between social exclusion and space, exploring some of the frameworks which institute barriers to spatial practices. Its particular emphasis is on the way these barriers to movement are intertwined with social exclusionary processes. This shows that exclusion should be regarded as a socio-spatial phenomenon.

[. . .]

DIMENSIONS OF SOCIAL EXCLUSION

There is little disagreement on some of the major problems facing European cities. Challenges of competition from a global economy marked by a multiplicity of competitors and the European response in the form of moving into an integrative partnership are both aspects of globalization which have reshaped the social and spatial geography of cities. The re-structuring of cities and societies, however, has been a costly exercise, as it has been parallel with a growing social divide, long-term unemployment and joblessness, especially for men, and casualization of work, undermining the quality of life for large groups of the population. These symptoms have led to concerns for the fragmentation of the social world, where some members of society are excluded in the 'mainstream' and where this exclusion is painful for the excluded and harmful for society as a whole.

Yet the concept of social exclusion still appears to be in need of clarification due to the variety of the cultural and political contexts in which it has been used. For some it is the question of poverty which should remain the focus of attention, while for others social exclusion makes sense in the broader perspective of citizenship and integration into the social context. Social exclusion, therefore, is not necessarily equated with economic exclusion, although this form of exclusion is often the cause of a wider suffering and deprivation.

As a concept, social exclusion still suffers from a lack of clarity, as it is interpreted and analysed differently. We come across a degree of ambiguity especially between poverty and social exclusion. Some researchers, who have concentrated on the problems of poverty, find social exclusion a vague concept which, for whatever reason, takes attention away from poverty and deprivation. Furthermore, it is argued that the concept of social exclusion is rooted in a certain intellectual and cultural tradition (Catholic, solidarity) and a particular welfare regime (corporatist) and as such is not shared by other (especially liberal) cultures and welfare regimes. On the other hand, those who find social exclusion a useful concept criticize an emphasis on poverty as too narrow. They seek to open the discussion to accommodate the general issues of social integration and citizenship. To confront this ambiguity and contradiction, we need to clarify the concept of social exclusion first.

The overall constitution of the social world is such that different forms of exclusion are fundamental to any social relationship. For example, the division of social life into public and private spheres means drawing boundaries round some spatial and temporal domains and excluding others from these domains. In this way, exclusion becomes an operating mechanism, an institutionalized form of controlling access: to places, to activities, to resources and information. Individual actions as well as legal, political and cultural structures rely heavily upon this operating mechanism and reproduce it constantly. Institutionally organized or individually improvised, it appears that we are all engaged in exclusionary processes that are essential for our social life.

Yet we know that, whatever their importance, these exclusionary processes work in close relationship with inclusionary activities to maintain a social fabric. Maintaining the continuity of the social world is only possible through a combination of and a fine balance between these two processes. At the individual level, seeking privacy without seeking social interaction would lead to isolation. At the social level, exclusion without inclusion would lead to a collapse of social structures. What is a negative state of affairs, therefore, is not exclusion in all its forms but an absence of inclusionary processes, a lack of a balance between exclusion and inclusion.

But what are the dimensions of the social world in which inclusion and exclusion take place? It is often mentioned that social exclusion is multidimensional.

To be able to identify and analyse these dimensions, we should look at the dimensions of the social world in which exclusion and inclusion take place. We can identify economic, political and cultural arenas as the three broad spheres of social life in which social inclusion and exclusion are manifested and, therefore, can be analysed and understood.

In the economic arena, the main form of inclusion is access to resources, which is normally secured through employment. The main form of exclusion, therefore, is a lack of access to employment. Marginalization and long-term exclusion from the labour market lead to an absence of opportunity for production and consumption, which can in turn lead to acute forms of social exclusion.

Exclusion from the economic arena is often considered to be a crucial and painful form of exclusion. Poverty and unemployment are therefore frequently at the heart of most discussions of social exclusion, to the extent that poverty and economic exclusion are equated with social exclusion. There is a tendency in the literature to use these terms interchangeably. It is true that long-term economic exclusion can break down the political and cultural ties of the affected individuals and social groups. It is important, however, to note that there are other forms of social exclusion in political and cultural spheres.

In the political arena, the main form of inclusion is to have a stake in power, to participate in decision making. In European liberal democracies, inclusion is often ensured through voting and other processes associated with it. The most obvious form of social exclusion, therefore, is lack of political representation. This may take various forms: from the underrepresentation of women in parliaments and governments, to the complete exclusion of immigrant groups from political decision making; from the argument by smaller political parties for a new system of representation which would allow them a fairer share of power, to a withdrawal from political participation by those excluded in the economic and cultural arenas.

In the cultural arena, the main form of inclusion is to share a set of symbols and meanings. The most powerful of these have historically been language, religion and nationality. Some of the new sets of symbolic relationships include the way individual and group identities are formed through association with patterns of consumption, from necessities of daily life to cultural products. For example, in what has been termed a visual culture, aesthetics of social behaviour has become an essential part of social life. The main form of exclusion in the cultural arena, therefore, becomes a marginalization from these symbols, meanings, rituals and discourses. The forms of cultural exclusion vary widely, as experienced by minorities whose language, race, religion and lifestyle are different from those of the larger society.

Different social groups may experience varying degrees of these different but highly interrelated forms of social exclusion. The most acute forms of social exclusion, however, are those that simultaneously include elements of economic, political and cultural exclusion. The other end of the spectrum is occupied by citizens who are fully integrated in the mainstream of society through these three dimensions. Between these two extremes, there is a wide range of variations in which individuals and groups are included in some areas but excluded in others. A major trend is that more and more people suffer from anxiety and uncertainty, as there are ever larger numbers in transition from inclusion to exclusion.

SPATIALITY OF SOCIAL EXCLUSION

Social exclusion, therefore, should be understood in its political, economic and cultural dimensions. Exclusion from the political arena, i.e. the denial of participation in decision making, can alienate individuals and social groups. In the cultural arena, exclusion from common channels of cultural communication and integration can have similar effects. The exclusion from work and its impacts are widely known as undermining the ability of individuals and households to participate actively in social processes. When combined, these forms of exclusion can create an acute form of social exclusion which keeps the excluded at the very margins of the society, a phenomenon all too often marked by a clear spatial manifestation in deprived inner city or peripheral areas . . .

[. . .]

In the past, this spatiality of social exclusion had led to attempts to dismantle such pockets of deprivation without necessarily dismantling the causes of deprivation or the forces bringing them together in particular enclaves. The dismantling of spatial concentrations of deprivation has been a continuous trend: from Baron Haussmann's wide boulevards in the middle of poor neighbourhoods in the nineteenth century, to the slum

clearance programmes and more subtle forms of housing management in the twentieth century. These have been attempts to despatialize social exclusion, which is evidence of its inherent and re-emerging spatiality. The latest form of despatialization and re-spatialization of social exclusion is homelessness, a process in which some groups are cut off from their previous socio-spatial contexts and are apparently without a home base. They, however, have clustered in particular parts of cities, spatializing again what was thought to be despatialized.

SPATIALITY AND DIFFERENCE

The absence of homogeneity is most apparent in cities, as they are sites of difference. Large cities have often grown by attracting people from around the country in which they are located or even from around the world. Cities have always been known as the meeting places of different people. As Aristotle noted: 'A city is composed of different kinds of men; similar people cannot bring a city into existence.' The unprecedented growth of cities since the nineteenth century has permanently brought forward the issue of difference in the city as a feature of urban life. Wirth, in his celebrated theory of urbanism, saw heterogeneity as a determining feature of the city, along with population size and density. For him, the city was a 'melting-pot of races, peoples, and cultures, and a most favourable breeding-ground of new biological and cultural hybrids'. In the city, individual differences have 'not only [been] tolerated but rewarded'. Such emphasis on the heterogeneity of cities has led to conceiving it as a world of strangers.

Two sets of reactions to the diversity in the city can be identified: there are those who have tried to impose an order onto it so that it becomes understandable and manageable and those who promote a celebration of diversity. However, both these reactions, which indeed represent modernist and postmodernist thinking, have been unable to deal with the issue of social marginalization and exclusion. Concentrations of disadvantage have remained in cities, despite the large-scale redevelopment schemes of the rationalist tendency and the more sensitive spatial transformations which followed. On the one hand, emphasis on the eradication of difference and seeing the city as a melting pot has led to undermining sensitivities and to disruption of lives. On the other hand, the emphasis on difference has led to social fragmentation and tribalism. Both have failed to cure the wounds of those living on the edge of the society.

BARRIERS TO SPATIAL PRACTICES

But how do we analyse space? There are many gaps and dilemmas associated with understanding space. From the centuries-old philosophical divide between absolute and relational space, to the gap between mental and real space, between physical and social space, between abstract and differential space, to the relationship between space and mass, space and time, and the variety of perspectives from which space can be studied, all bear the possibility of confusion and collision. It is possible to show, however, that to avoid the gaps and dilemmas associated with understanding space, we need to concentrate on the processes which produce the built environment. By analysing the inter-section between space production and everyday life practices, we will be able to arrive at a dynamic understanding of space. We will then be able to understand and explain material space and its social and psychological contexts and attributes.

The question of social exclusion and integration, it can be argued, largely revolves around access. It is access to decision making, access to resources, and access to common narratives, which enable social integration. Many of these forms of access have clear spatial manifestations, as space is the site in which these different forms of access are made possible or denied. There is a direct relationship between our general sense of freedom and well-being with the choices open to us in our spatial practices. The more restricted our social options, the more restricted will be our spatial options, and the more excluded we feel or become. On the other hand, if we have a wide range of social options, we would have a wide range of places to go to, places for living, working and entertainment. Two extreme cases of the existence or absence of spatial freedom may be jetsetting executives versus prisoners. Whereas for one, the world may be shrinking to seem like a global village open for communication and interaction, for the other the world outside is large and out of reach. For most of us, however, our spaces are a continuum from accessible to non-accessible places. The space around us is a collection of open, closed or controlled places.

But how is the urban space organized and how are spatial practices controlled and regulated? We all have an understanding of the places where we can or cannot go, as over the years through our spatial practices, we have accumulated a knowledge about places and their patterns of accessibility. The physical organization of space, using elements from the natural or the built environment, has been socially and symbolically employed to put visible and strict limits on our spatial practices. For example, topography has always been used to institute difference and segregation, from ancient times when the hilltops were the place of gods for Greeks and Mesopotamians, to our own time when they are the living places of the rich and powerful. There is also a mental space, our perceptions of space. This may be regulated through codes and signs, preventing us from entering some spaces through outright warning or more subtle deterrents. Mental space may also be controlled through our fears and perceptions of activities in places. For example, we may be hesitant to enter an expensive-looking shopping centre if we do not have access to the resources needed for the activities there, even though there may not be any physical barriers which would prevent us from going there. A third form of barrier to our spatial behaviour is social control, which can range from legal prohibitions on entering places to constructing formal barriers along publicly recognized borders. National borders and public–private boundaries are examples of this form. A combination of formalized rules and regulations, informal codes and signs, and fears and desires control our spatial behaviour and alert us to the limitations on our access. Through these, we have come to know whether we can enter a place, are welcomed in another and excluded from others. More restrictions on our access to our surroundings would bring about the feeling of being trapped, alienated and excluded from our social space.

Space has, therefore, a major role in the integration or segregation of urban society. It is a manifestation of social relationships while affecting and shaping the geometries of these relationships. This leads us to the argument that social exclusion cannot be studied without also looking at spatial segregation and exclusion. Social cohesion or exclusion, therefore, are indeed socio-spatial phenomena . . . We know that all human societies have their own forms of social and spatial exclusion. So exclusionary processes per se are not the source of social fragmentation and disintegration. It is the absence of social integration which causes social exclusion, as individuals do not find the possibility and channels of participating in the main-stream society.

GLOBAL AND NATIONAL SPACE

National borders are the largest means of socio-spatial exclusion. The modern nation state exerts an exclusionary process along its boundaries, from lines on maps to barbed wires on the landscape. Those who are left outside need to go through special checks and controls to be allowed in. The same applies to those who are in and want to go out. The control of cross-border movement by the nation state, or by blocks of nation states as in the European Union, is a form of exclusion legitimated openly through political processes. A national territory, therefore, is a spatial manifestation of an institutionalized exclusionary process.

Other administrative boundaries, although potentially exclusionary, do not have such a forceful character, nor are they associated with such a degree of public awareness, such historical significance, or guarded by military might. No other form of exclusion has been associated with such high costs in human life, sacrifice and misery. Attempts to change or to protect national borders have inflicted the highest cost in human lives in the twentieth century, as experienced by two world wars and many regional conflicts. The birth of a nation state, when the multi-ethnic empires and states break up, can be a bloody process in which every means is used to exclude others. The surgical subdivision of national space, whether through external forces as in postwar Germany or by exploding internal forces as in the former Yugoslavia, has been equally difficult for those excluded from what they have regarded as their home.

In the national space enclosed within these boundaries, narratives of nationalism have been employed to legitimize the exclusion of others beyond these boundaries. Indeed, exclusionary narratives, which determine how 'we' are different from others, are often essential in binding individuals together as a group. The most dangerous of these narratives has been the rhetoric of hatred against other nations, races and groups. But there are many such exclusionary narratives which do not necessarily promote violence and hatred and still have a binding power. With these narratives, which often rely on a common historical experience, large groups of people have been associated with each other. The focal point of this association has been the nation state, which holds the power of controlling the national borders.

The narratives of nationalism attempt to create homogeneity out of an enormous diversity. As individuals have come together to create a democratic civil society, such narratives have helped the organization of modern democratic states . . .

[. . .]

NEIGHBOURHOODS, MARKETS AND REGULATION

At the local level, by following two processes, land and property development on the one hand and spatial planning on the other hand, we can see how a socio-spatial geometry of difference and segregation, which is the foundation of exclusion, emerges. We come across the term neighbourhood in a variety of distinct but interrelated usages. In one sense, the term is used loosely to address a locality. This daily usage is based on the images and understandings by individuals and groups of their surroundings. This is a view from below and, as such, can lead us to see a city as a collection of overlapping neighbourhoods. Research on people's perception of neighbourhood shows major differences according to age, gender, class and ethnicity. At the other end of the scale, there is a concept of neighbourhood from above, from the viewpoint of such experts as managers, planners and designers. Here neighbourhood refers to a particular part of a town and is used to understand urban structure and change in urban society. It is also used as a tool for management. From this viewpoint, the city is seen as a collection of segregated neighbourhoods.

Neighbourhoods as constituent parts of cities have long been the focus of attention by urban designers and planners. Drawing upon historic precedents and for practical reasons, neighbourhood has provided them with an intimate scale of the urban whole to understand and to deal with. Historically, neighbourhoods have been the sites and physical manifestations of close social relationships and so have been praised by town planners, especially those who have looked nostalgically to the feudal bonds of the medieval towns and the communal bonds of working-class neighbourhoods in the industrial city. A dichotomy emerged as a result of the unprecedented growth of the cities: between *gesellschaft* and *gemeinschaft*, between the alienation of the big city and the romanticized, small communities of towns and villages. To recreate the social cohesion of these small communities, it was thought, cities should be broken into smaller parts, into neighbourhoods. On the other hand, it was thought that the communitarianism of small neighbourhoods could overcome the individualism of the suburbs, those bourgeois utopias.

It is this association of neighbourhood as a physical entity with neighbourhood as a cohesive social unit that led to a series of reformist ideas throughout the twentieth century. From the widely used, and discredited, concept of neighbourhood unit, to Lynch's districts, which are still promoted to make cities legible, and today's urban villages and new urbanist neighbourhoods, there has been a long line of managerial attempts to promote social cohesion by spatial organization.

Along with this promotion of spatial subdivision by town planning, there has been a promotion of socio-spatial segregation by market forces through the ways in which space is produced, exchanged and used. The producers of space, such as volume housebuilders, tend to build in large-scale housing estates, creating an urban fabric which is a collection of different subdivisions. The land and property markets have operated so as to ensure the segregation of income groups and social classes. Commodification of space has led to different patterns of access to space and hence a differential spatial organization and townscape. Wherever there has been a tendency to decommodify space, as in the postwar social housing schemes, town planners and designers have ensured that a degree of spatial subdivision still prevailed.

We can therefore identify two processes: a land and property market which sees space as a commodity and tends to create socio-spatial segregation through differential access to this commodity, and a town planning and design tendency to regulate and rationalize space production by the imposition of some form of order. When we look at these two processes together, the picture which emerges is a collectivization of difference, of exclusion, which can lead to enclaves for the rich and the creation of new ghettos for the poor.

[. . .]

PUBLIC AND PRIVATE SPACE

Another form of socio-spatial exclusion, which is enforced with a rigour somewhat similar to the protection of national borders, is the separation between public and private territories. We guard our private spheres from intruders by whatever means, in some countries even legitimately by firearms. Privacy, private property and private space are intertwined, demarcated through a variety of objects and signs: from subtle variations of colour and texture to fences and high walls. Those who are in are entitled to be, excluding those who are not. This is an exclusionary process legitimized through public discourse, through custom or law. Violation of this exclusionary process is regarded as, at best, inconvenience and, at worst, crime. Public space, which is one of the manifestations of society's public sphere, is maintained by public agencies in the public interest and is accessible to the public. Access to public space, however, is subject to exclusionary processes. Public space is guarded from intrusion by private interests, a process which is regarded as essential for the health of the society. Some of the main currents in social and political thought that offer concepts of public space appear to stress the need to keep the public and private spheres distinctive and apart, despite the criticism that this idealizes the distinction.

[. . .]

The changing nature of development companies and the entry of the finance industry into built environment production and management has partly led to what is widely known as the privatization of space. Large-scale developers and financiers expect their commodities to be safe for investment and maintenance, hence their inclination to reduce as much as possible all the levels of uncertainty which could threaten their interests. This trend is parallel with the increasing fear of crime, rising competition from similar developments, and the rising expectations of the consumers, all encouraging the development of totally managed environments. What has emerged is an urban space where increasingly large sections are managed by private companies, as distinctive from those controlled by public authorities. Examples of these fragmented and privatized spaces are gated neighbourhoods, shopping malls and city centre walkways, under heavy private surveillance and separated from the public realm by controlled access and clear boundaries. This total management of parts of the city is in part an attempt to control crime. Crime acts as a counter-claim to space and as such is itself an exclusionary force, keeping many groups vulnerable and marginalized.

CONCLUSION: SOCIAL INTEGRATION AND SPATIAL FREEDOM

Social exclusion combines lack of access to resources, to decision making, and to common narratives. The multidimensional phenomenon of social exclusion finds spatial manifestation, in its acute forms, in deprived inner or peripheral urban areas. This spatiality of social exclusion is constructed through the physical organization of space as well as through the social control of space, as ensured by informal codes and signs and formal rules and regulations. These formal channels act at all scales of space. Global space is fragmented by national spaces, which have a tendency to deny difference and homogenize social groups. At the scale of local space, spatialization of social exclusion takes place through land and property markets. These markets tend to fragment, differentiate and commodify space through town planning mechanisms which tend to fragment, rationalize and manage

space, and also through the legal and customary distinctions between the public and private spheres, with a constant tension between the two and a tendency for the privatization of space.

To break the trap of socio-spatial exclusion, one strategy could be to challenge these deep-seated forms of differentiation. We know, however, that wholesale challenges can be problematic themselves, as exemplified by attempts to redefine the public–private relationship in Eastern Europe. Furthermore, we know that any human society is likely to have some form of exclusionary process in its constitution. Nevertheless, it is true that the form of these exclusionary processes changes over time. A reflexive revisiting of the processes of differentiation is therefore a constantly necessary task. At the same time what is necessary and urgent is to institute and promote inclusionary processes, to strike a balance between exclusion and integration, to provide the possibility of integration and to break the trap of socio-spatial exclusion. We have seen that space is a major component part of social exclusion. Revisiting spatial barriers and promoting accessibility and more spatial freedom can therefore be the way spatial planning can contribute to promoting social integration.

"Taking Los Angeles Apart: Towards A Postmodern Geography"

from *Postmodern Geographies: The Reassertion of Space in Critical Social Theory* (1989)

Edward Soja

Editors' Introduction

Because of its size, fragmentation, diversity, and dynamism and its role as the epicenter of global image and fantasy, Los Angeles is often held out as the quintessential postmodern city. In this selection University of California geographer and city planning professor Edward Soja "deconstructs" the greater metropolitan Los Angeles region. He takes as his area of enquiry the Sixty-Mile Circle around downtown Los Angeles – a five-county area with twelve million people, one hundred and thirty-two cities, and a $250 billion economy.

Soja's description of Los Angeles' Sixty-Mile Circle is rich in complexity and paradox. Freeway offramps lead to huge military establishments like Edwards Air Force Base, to an Indian reservation nearly devoid of Indians, and a condor refuge where the last condor has been transplanted to a local zoo to avoid lead poisoning. Within the Sixty-Mile Circle are ski slopes and orange groves; Spanish missions and Latino barrios; leisure towns and prisons; planned communities and unplanned sprawl; all-white stealth communities and ethnic neighborhoods; multiple-center cities and Edge Cities; science parks, and technopoles and technoburbs employing well-paid, largely white knowledge workers and industrial zones relying almost exclusively on Hispanic and Asian workers.

Among the many contradictions and paradoxes of the greater Los Angeles metropolitan area Soja notes is the area's extraordinary *industrial* dynamism. While William Julius Wilson (p. 126), Saskia Sassen (p. 212), Michael Porter (p. 277) and many other writers emphasize the shift of manufacturing jobs away from developed countries to the Third World, Soja characterizes greater Los Angeles as the premier industrial growth pole of the twentieth century. According to Soja Los Angeles may perhaps be accurately described as *post-Fordist*, but hardly *post-industrial*.

Contrary to appearances, in this bastion of private capitalism, government investment is important to the economic health of the Los Angeles region. Soja emphasizes the role of the state, particularly the military, in the creation of this "federated metro-sea of state-rescued capitalism." Defense and aerospace contracts for huge projects like the space shuttle and high-tech weapons systems fuel the growth of the region.

Despite all this growth and change, Soja sees an underlying logic to the region that reflects classic urban geographers' models. There is still a core central business district (CBD) in Los Angeles displaying some of the attributes of a concentric zone internal structure as Ernest Burgess hypothesized (p. 156). According to Soja there is still a radial logic to the region too. For example, Wilshire Boulevard is a twenty-mile "sector" radiating out from downtown Los Angeles in the form that real estate economist Homer Hoyt described seventy years ago as underlying city development.

The Los Angeles CBD contains a huge police headquarters building, courts, city hall, the largest women's prison in America, and the central organs of press and pulpit such as the Times/Mirror Building and St Vibiana's cathedral. In the newly emerging "left bank" of Los Angeles, yuppies sip cappuccinos just a short distance from ethnic neighborhoods populated by Vietnamese, Latino, and Chinese immigrant families. Conventioneers in this "dual" city enjoy luxury hotels walled off from panhandlers just outside. Downtown Los Angeles exhibits the cultural contradictions that David Harvey (p. 227) and Sharon Zukin (p. 136) describe in postmodern New York City.

Following an approach favored in postmodern critical theory, Soja seeks to "deconstruct" this dazzling array of elements. He seeks to get beneath the symbolism of the area – the semiotic blanket of words and artifacts created by the culture(s) that make up Los Angeles. In many ways the Los Angeles metropolitan region resembles a theme park. Amidst the whimsy and pastiche, Soja notes that it is tempting to become bemused and detached. But underneath the glitz and glitter, Soja points to what he sees as the hard edge of a capitalist, racist, patriarchal landscape.

This selection is from Edward Soja, *Postmodern Geographies: The Reassertion of Space in Critical Social Theory* (London and New York: Verso, 1989). Other books by Soja include *Postmetropolis: Critical Studies of Cities and Regions* (London: Blackwell, 2000), *The City: Los Angeles and Urban Theory at the End of the Twentieth Century* (Berkeley: University of California Press, 1996) co-edited with Allen J. Scott, and *Thirdspace: Journeys to Los Angeles and Other Real-and-imagined Places* (Cambridge, MA: Blackwell, 1996).

Books on Los Angeles include William Fulton, *The Reluctant Metropolis* (Point Arena, CA: Solano Press, 1998), Norman Smith, *The History of Forgetting: Los Angeles and the Erasure of Memory* (London and New York: Verso, 1997), Dolores Hayden, *The Power of Place: Urban Landscapes and Public History* (Cambridge: MIT Press, 1997) and Kevin Starr's series on California and the American Dream: *Americans and the California Dream, 1850–1915* (New York: Oxford University Press, 1986), *Inventing the Dream: California Through the Progressive Era* (New York: Oxford University Press, 1985), *Material Dreams: California Through the 1920s* (New York: Oxford University Press, 1990), *Endangered Dreams: The Great Depression in California* (New York: Oxford University Press, 1986), and *The Dream Endures: California Enters the 1940s* (New York: Oxford University Press, 1996).

Other books which explore postmodern landscapes in southern California and elsewhere include Michael Sorkin (ed.), *Variations on a Theme Park: The New American City and the End of Public Space* (New York: Hill and Wang, 1992), Sharon Zukin, *Landscapes of Power: From Detroit to Disney World* (Berkeley: University of California Press, 1993), and Mark Gottdiener, *The Theming of America: Dreams, Visions, and Commercial Spaces* (New York: HarperCollins, 1997).

■ ■ ■ ■ ■ ■

'The Aleph?' I repeated.

'Yes, the only place on earth where all places are – seen from every angle, each standing clear, without any confusion or blending' (10–11)

. . . Then I saw the Aleph . . . And here begins my despair as a writer. All language is a set of symbols whose use among its speakers assumes a shared past. How, then, can I translate into words the limitless Aleph, which my floundering mind can scarcely encompass? (12–13)

(Jorge Luis Borges, 'The Aleph')

Los Angeles, like Borges's Aleph, is exceedingly tough-to-track, peculiarly resistant to conventional description. It is difficult to grasp persuasively in a temporal narrative for it generates too many conflicting images, confounding historicization, always seeming to stretch laterally instead of unfolding sequentially. At the same time, its spatiality challenges orthodox analysis and interpretation, for it too seems limitless and constantly in motion, never still enough to encompass, too filled with 'other spaces' to be informatively described. Looking at Los Angeles from the inside, introspectively, one tends to see only fragments and immediacies, fixed sites of myopic understanding impulsively generalized to represent the whole. To the more far-sighted outsider, the visible aggregate of the whole of Los Angeles churns so confusingly that it induces little more than illusionary stereotypes or self-serving caricatures – if its reality is ever seen at all.

What is this place? Even knowing where to focus, to find a starting point, is not easy, for, perhaps more than any other place, Los Angeles is everywhere. It is global in the fullest sense of the word. Nowhere is this more evident than in its cultural projection and ideological reach, its almost ubiquitous screening of itself as a rectangular dream machine for the world. Los Angeles broadcasts its self-imagery so widely that probably more people have seen this place – or at least fragments of it – than any other on the planet . . .

[. . .]

. . . What follows then is a succession of fragmentary glimpses, a freed association of reflective and interpretive field notes which aim to construct a critical human geography of the Los Angeles urban region. My observations are necessarily and contingently incomplete and ambiguous, but the target I hope will remain clear: to appreciate the specificity and uniqueness of a particularly restless geographical landscape while simultaneously seeking to extract insights at higher levels of abstraction, to explore through Los Angeles glimmers of the fundamental spatiality of social life, the adhesive relations between society and space, history and geography, the splendidly idiographic and the enticingly generalizable features of a postmodern urban geography.

A ROUND AROUND LOS ANGELES

[. . .]

We must have a place to start, to begin reading the context . . . just such a reductionist mapping has popularly presented itself. It is defined by an embracing circle drawn sixty miles (about a hundred kilometres) out from a central point located in the downtown core of the City of Los Angeles.

The Sixty-Mile Circle, so inscribed, covers the thinly sprawling 'built-up' area of five counties, a population of more than 12 million individuals, at least 132 incorporated cities and, it is claimed, the greatest concentration of technocratic expertise and militaristic imagination in the USA. Its workers produce, when last estimated, a gross annual output worth nearly $250 billion, more than the 800 million people of India produce each year. This is certainly Greater Los Angeles, a dizzying world.

CIRCUMSPECTION

Securing the Pacific rim has been the manifest destiny of Los Angeles, a theme which defines its sprawling urbanization perhaps more than any other analytical construct . . . It is not always easy to see the imprint of this imperial history on the material landscape, but an imaginative cruise directly above the contemporary

circumference of the Sixty-Mile Circle can be unusually revealing. Figure 1 will help to find the way.

The Circle cuts the south coast at the border between Orange and San Diego Counties, near one of the key checkpoints regularly set up to intercept the northward flow of undocumented migrants, and not far from the San Clemente 'White House' of Richard Nixon and the fitful SONGS of the San Onofre Nuclear Generating Station. The first rampart to watch, however, is Camp Pendleton Marine Corps Base, the largest military base in California in terms of personnel, the freed spouses of whom have helped to build a growing hightechnology complex in northern San Diego County . . .

Another quick hop over Sunnymead, the Box Spring Mountains, and Redlands takes us to Rampart #3, Norton Air Force Base, next to the city of San Bernardino and just south of the almost empty San Manuel Indian Reservation. The guide books tell us that the primary mission of Norton is military airlifts, just in case. To move on we must rise still higher to pass over the skisloped peaks of the San Bernardino Mountains and National Forest, through Cajon Pass and passing the old Santa Fe Trail, into the picturesque Mojave Desert . . .

The next leg is longer and more serene: over the Antelope Valley and the Los Angeles Aqueduct (tapping the Los Angeles-owned segments of the lifegiving but rapidly dying Owens River Valley two hundred miles further away); across Interstate 5 (the main freeway corridor to the north), a long stretch of Los Padres National Forest and the Wild Condor Refuge – the last remaining condors were recently removed to zoos after leadpoisoning threatened their extinction in the 'wild' – to the idyllized town of Ojai (site for the filming of 'Lost Horizon'), and then to the Pacific again at the Mission of San Buenaventura, in Ventura County . . .

It is startling how much of the circumference is owned and preserved by the Federal Government in one way or another. Premeditation may be impossible to ascribe, but postmeditation on the circumscriptive federal presence is certainly in order.

ENCLOSURES

[. . .]

If there has emerged a compelling focus to the recent academic literature on Los Angeles, it is the

1	Los Angeles	3 144 800	6	Huntington Beach	179 900	11	Pasadena	126 600
2	Long Beach	381 900	7	Glendale	148 800	12	Oxnard	124 300
3	Amajeo	234 706	8	Torrance	135 100	13	Ontario	108 100
4	Santa Ana	221 600	9	San Bernardmo	134 700	14	Fullerton	107 700
5	Riverside	183 400	10	Garden Grove	130 300	15	Pomona	106 300

Figure 1 A view of the outer spaces of Los Angeles

The urban core is outlined in the shape of a pentagon, with the Central City denoted by the black triangle. The major military bases on the perimeter of the Sixty-Mile Circle are identified and the black squares are the sites of the largest defence contractors in the region. Also shown are county boundaries, the freeway system outside the central pentagon, and the location of all cities with more than 100,000 inhabitants (small open circles)

discovery of extraordinary industrial production . . . Yet it is no exaggeration to claim that the Sixty-Mile Circle contains the premier industrial growth pole of the twentieth century, at least within the advanced capitalist countries. Oil, orange groves, films and flying set the scene at the beginning of the century and tend to remain fixed in many contemporary images of industrious, but not industrial, Los Angeles. Since 1930, however, Los Angeles has probably led all other major

metropolitan areas in the USA, decade by decade, in the accumulation of new manufacturing employment.

[. . .]

. . . In the past half century, no other area has been so pumped with federal money as Los Angeles, via the Department of Defense to be sure, but also through numerous federal programmes subsidizing suburban consumption (suburbsidizing?) and the development of housing, transportation and water delivery systems.

From the last Great Depression to the present, Los Angeles has been the prototypical Keynesian state-city, a federalized metro-sea of state-rescued capitalism enjoying its place in the sunbelt, demonstrating decade by decade its redoubtable ability to go first and multiply the public seed money invested in its promising economic landscape. No wonder it remains so protected. In it are embedded many of the crown jewels of advanced industrial capitalism.

OUTERSPACES

The effulgent Star Wars colony currently blooming around Los Angeles International Airport (LAX) is part of a much larger outer city which has taken shape along the Pacific slope of Los Angeles County. In the context of this landscape, through the story line of the aerospace industry, can be read the explosive history and geography of the National Security State and what Mike Davis (1984) has called the 'Californianization of Late-Imperial America'.

If there is a single birthplace for this Californianization, it can be found at old Douglas Field in Santa Monica . . . Over half a million people now live in this 'Aerospace Alley', as it has come to be called. During working hours, perhaps 800,000 are present to sustain its global preeminence. Untold millions more lie within its extended orbit.

Attached around the axes of production are the representative locales of the industrialized outer city: the busy international airport; corridors filled with new office buildings, hotels, and global shopping malls; neatly packaged playgrounds and leisure villages; specialized and masterplanned residential communities for the high technocracy; armed and guarded housing estates for top professionals and executives; residual communities of low-pay service workers living in overpriced homes; and the accessible enclaves and ghettoes which provide dependable flows of the cheapest labour power to the bottom bulge of the bimodal local labour market. The LAX–City compage reproduces the segmentation and segregation of the inner city based on race, class, and ethnicity, but manages to break it down still further to fragment residential communities according to specific occupational categories, household composition, and a broad range of individual attributes, affinities, desired lifestyles and moods.

This extraordinary differentiation, fragmentation, and social control over specialized pools of labour is expensive. Housing prices and rental costs in the outer city are easily among the highest in the country and the provision of appropriate housing increasingly absorbs the energy not only of the army of real estate agents but of local corporate and community planners as well, often at the expense of longtime residents fighting to maintain their foothold in 'preferred' locations. From the give and take of this competition have emerged peculiarly intensified urban landscapes. Along the shores of the South Bay, for example, part of what Rayner Banham once called 'Surfurbia', there has developed the largest and most homogenous residential enclave of scientists and engineers in the world. Coincidentally, this beachhead of the high technocracy is also one of the most formidable racial redoubts in the region. Although just a few miles away, across the fortifying boundary of the San Diego freeway, is the edge of the largest and most tightly segregated concentration of Blacks west of Chicago, the sunbelted beach communities stretching south from the airport have remained almost 100 per cent white. The worldview from this highly engineered environment is stereotypically caricatured in Figure 2.

The Sixty-Mile Circle is ringed with a series of these outer cities at varying stages of development, each a laboratory for exploring the contemporaneity of capitalist urbanization . . .

[. . .]

. . . these new territorial complexes seem to be turning the industrial city inside-out, recentring the urban to transform the metropolitan periphery into the core region of advanced industrial production. Decentralization from the inner city has been taking place selectively for at least a century all over the world, but only recently has the peripheral condensation become sufficiently dense to challenge the older urban cores as centres of industrial production, employment nodality, and urbanism. This restructuring process is far from being completed but it is beginning to have some profound repercussions on the way we think about the city, on the words we use to describe urban forms and functions, and on the language of urban theory and analysis.

BACK TO THE CENTRE

[. . .]

To see more of Los Angeles, it is necessary to move away from the riveting periphery and return,

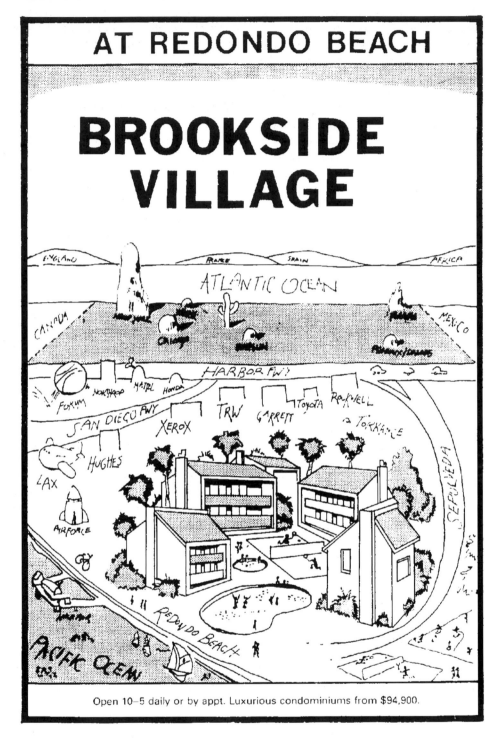

Figure 2 Worldview from Redondo Beach

literally and figuratively, to the centre of things to the still adhesive core of the urbanized landscape. In Los Angeles as in every city, the nodality of the centre defines and gives substance to the specificity of the urban, its distinctive social and spatial meaning. Urbanization and the spatial divisions of labour associated with it revolve around a socially constructed pattern of nodality and the power of the occupied centres both to cluster and disperse, to centralize and decentralize, to structure spatially all that is social and socially produced. Nodality situates and contextualizes urban society by giving material form to essential social relations. Only with a persistent centrality can there be outer cities and peripheral urbanization. Otherwise, there is no urban at all.

It is easy to overlook the tendential processes of urban structuration that emanate from the centre, especially in the postmodern capitalist landscape. Indeed, in contemporary societies the authoritative and allocative power of the urban centre is purposefully obscured or, alternatively, detached from place, ripped out of context, and given the stretched-out appearance of democratic ubiquity. In addition, as we have seen, the historical development of urbanization over the past century has been marked by a selective dispersal and decentralization, emptying the centre of many of the activities and populations which once aggregated densely around it. For some, this has signalled a negation of nodality, a submergence of the power of central places, perhaps even a Derridean deconstruction of all differences between the 'central' and the 'marginal'.

Yet the centres hold. Even as some things fall apart, dissipate, new nodalities form and old ones are reinforced. The specifying centrifuge is always spinning but the centripetal force of nodality never disappears. And it is the persistent residual of political power which continues to precipitate, specify, and contextualize the urban, making it all stick together.

[. . .]

To maintain adhesiveness, the civic centre has always served as a key surveillant node of the state, supervising locales of production, consumption and exchange. It still continues to do so, even after centuries of urban recomposition and restructuring, after waves of reagglomerative industrialization. It is not production or consumption or exchange in themselves that specifies the urban, but rather their collective surveillance, supervision and anticipated control within the power-filled context of nodality. In Foucauldian terms, cities are the convergent sites of (social) space,

knowledge, and power, the headquarters of societal modes of regulation (from *regula* and *regere*, to rule; the root of our keyword: region).

[. . .]

SIGNIFYING DOWNTOWN

The downtown core of the City of Los Angeles, which the signs call 'Central City', is the agglomerative and symbolic nucleus of the Sixty-Mile Circle, certainly the oldest but also the newest major node in the region. Given what is contained within the Circle, the physical size and appearance of downtown Los Angeles seem almost modest, even today after a period of enormous expansion. As usual, however, appearances can be deceptive.

Perhaps more than ever before, downtown serves in ways no other place can as a strategic vantage point, an urban panopticon counterposed to the encirclement of watchful military ramparts and defensive outer cities. Like the central well in Bentham's eminently utilitarian design for a circular prison, the original panopticon, downtown can be seen (when visibility permits) by each separate individual, from each territorial cell, within its orbit. Only from the advantageous outlook of the centre, however, can the surveillant eye see everyone collectively, disembedded but interconnected. Not surprisingly, from its origin, the central city has been an aggregation of overseers, a primary locale for social control, political administration, cultural codification, ideological surveillance, and the incumbent regionalization of its adherent hinterland.

Looking down and out from City Hall, the site is especially impressive to the observer. Immediately below and around is the largest concentration of government offices and bureaucracy in the country outside the federal capital district. To the east, over a pedestrian skyway, are City Hall East and City Hall South, relatively new civic additions enclosing a shopping mall, some murals, a children's museum, and the Triforium, a splashy sixty-foot fountain of water, light, and music entertaining the lunchtime masses. Just beyond is the imposing police administration building, Parker Center, hallowing the name of a former police chief of note. Looking further outside the central well of downtown but within its eastern salient, one can see an area which houses 25 per cent of California's prison population, at least 12,000 inmates held in four jails designed to hold half that number. Included within this carceral wedge are the largest women's prison in the country (Sybil Brand) and the seventh largest men's prison (Men's Central). More enclosures are being insistently planned by the state to meet the rising demand.

On the south, along First Street, are the State Department of Transportation (CALTRANS) with its electronic wall maps monitoring the arterial freeways of the region, the California State Office Building, and the headquarters of the fourth estate, the monumental Times-Mirror building complex, which many have claimed houses the unofficial governing power of Los Angeles, the source of many stories that mirror the times and spaces of the city. Near the spatial sanctum of the *Los Angeles Times* is also St Vibiana's Cathedral, mother church to one of the largest Catholic arch-dioceses in the world (nearly four million strong) and controller of another estate of significant proportions. The Pope slept here, across the street from Skid Row missions temporarily closed so that he could not see all his adherents.

Looking westward now, toward the Pacific and the smog-hued sunsets which brilliantly paint the night-falls of Los Angeles, is first the Criminal Courts Building, then the Hall of Records and Law Library, and next the huge Los Angeles County Courthouse and Hall of Administration, major seats of power for what is by far the country's largest county in total population (now over eight million). Standing across Grand Avenue is the most prominent cultural centre of Los Angeles, described by Unique Media Incorporated in their pictorial booster maps of downtown as 'the cultural crown of Southern California, reigning over orchestral music, vocal performance, opera, theatre and dance'. They add that the Music Center 'tops Bunker Hill like a contemporary Acropolis', one which has dominated civil cultural life since it was inaugurated in 1964. Just beyond this cultural crown is the Depart-ment of Water and Power (surrounded by usually waterless fountains) and a multi-level extravaganza of freeway interchanges connecting with every corner of the Sixty-Mile Circle, a peak point of accessibility within the regional transportation network. On its edge, one of Japan's greatest architects has designed a Gateway Building to punctuate the teeming sea.

Along the northern flank is the Hall of Justice, the US Federal Courthouse, and the Federal Building, completing the ring of local, city, state and federal government authority which comprises the potent civic centre. Sitting more tranquilly just beyond, cut off by a swathe of freeway, is the preserved remains of the old civic centre, now part of El Pueblo de Los Angeles State Historical Park, additional testimony to the lasting power of the central place. Since the origins of Los Angeles the sites described have served as the political citadel, designed with other citadels to command, protect, socialize and dominate the surrounding urban population.

There is still another segment of the citadel–panopticon which cannot be overlooked. Its form and function may be more specific to the contemporary capitalist city but its mercantile roots entwine his-torically with the citadels of all urbanized societies. Today, it has become the acknowledged symbol of the urbanity of Los Angeles, the visual evidence of the successful 'search for a city' by the surrounding sea of suburbs. This skylined sight contains the bunched castles and cathedrals of corporate power, the gleam-ing new 'central business district' of the 'central city', pinned next to its ageing predecessor just to the east. Here too the LA-leph's unending eyes are kept open and reflective, reaching out to and mirroring global spheres of influence, localizing the world that is within its reach.

Nearly all the landmarks of the new LA CBD have been built over the past fifteen years and flashily signify the consolidation of Los Angeles as a world city. Now more than half the major properties are in part or wholly foreign owned, although much of this landed presence is shielded from view. The most visible wardens are the banks which light up their logos atop the highest towers: Security Pacific (there again), First Interstate, Bank of America (co-owner of the sleek-black Arco Towers before their recent purchase by the Japanese), Crocker, Union, Wells Fargo, Citicorp (billing itself as 'the newest city in town'). Reading the skyline one sees the usual corpo-rate panorama: large insurance companies (Manulife, Transamerica, Prudential), IBM and major oil com-panies, the real estate giant Coldwell Banker, the new offices of the Pacific Stock Exchange, all serving as attachment points for silvery webs of financial and commercial transactions extending practically every-where on earth.

. . . Contrary to popular opinion, Los Angeles is a tightly planned and plotted urban environment, especially with regard to the social and spatial divisions of labour necessary to sustain its pre-eminent indus-trialization and consumerism. Planning choreographs Los Angeles through the fungible movements of the zoning game and the flexible staging of supportive community participation (when there are communities to be found), a dance filled with honourable intent, dedicated expertise, and selective beneficence. It has excelled, however, as an ambivalent but nonetheless enriching pipeline and place-maker to the domestic and foreign developers of Los Angeles, using its influential reach to prepare the groundwork and facilitate the selling of specialized locations and populations to suit the needs of the most powerful organizers of the urban space economy.

. . . Through a historic act of preservation and renewal, there now exists around downtown a deceptively harmonized showcase of ethni-cities and specialized economic enclaves which play key roles, albeit somewhat noisily at times, in the contemporary redevelopment and internationalization of Los Angeles. Primarily responsible for this packaged and planned production of the inner city is the Community Redevelopment Agency, probably the leading public entrepreneur of the Sixty-Mile Circle.

There is a dazzling array of sites in this compartmentalized corona of the inner city: the Vietnamese shops and Hong Kong housing of a redeveloping Chinatown; the Big Tokyo financed modernization of old Little Tokyo's still resisting remains; the induced pseudo-SoHo of artists' lofts and galleries hovering near the exhibitions of the 'Temporary Contemporary' art warehouse; the protected remains of El Pueblo along Calmexified Olvera Street and in the renewed Old Plaza; the strangely anachronistic wholesale markets for produce and flowers and jewellery growing bigger while other downtowns displace their equivalents; the foetid sweatshops and bustling merchandise marts of the booming garment district; the Latino retail festival along pedestrian-packed Broadway (another preserved zone and inch-for-inch probably the most profitable shopping street in the region); the capital site of urban homelessness in the CRA-gilded skid row district; the enormous muralled *barrio* stretching eastward to the still unincorporated East Los Angeles; the de-industrializing and virtually resident-less wholesaling City of Vernon to the south filled with chickens and pigs awaiting their slaughter; the Central American and Mexican communities of Pico-Union and Alvarado abutting the high-rises on the west; the obtrusive oil wells and aggressive graffiti in the backyards of predominantly immigrant Temple-Beaudry progressively being eaten away by the spread of Central City West (now being called 'The Left Bank' of downtown); the

intentionally yuppifying South Park redevelopment zone hard by the slightly seedy Convention Center; the revenue-milked towers and fortresses of Bunker Hill; the resplendently gentrified pocket of 'Victorian' homes in old Angelino Heights overlooking the citadel; the massive new Koreatown pushing out west and south against the edge of Black Los Angeles; the Filipino pockets to the north-west still uncoalesced into a 'town' of their own; and so much more: a constellation of Foucauldian heterotopias 'capable of juxtaposing in a single real place several spaces, several sites that are in themselves incompatible' but 'function in relation to all the space that remains'.

What stands out from a hard look at the inner city seems almost like an obverse (and perverse) reflection of the outer city, an agglomerative complex of dilapidated and overcrowded housing, low technology workshops, relics and residuals of an older urbanization, a sprinkling of niches for recentred professionals and supervisors, and, above all, the largest concentration of cheap, culturally splintered/occupationally manipulable Third World immigrant labour to be found so tangibly available in any First World urban region. Here in this colonial corona is another of the crown jewels of Los Angeles, carefully watched over, artfully maintained and reproduced to service the continued development of the manufactured region.

The extent and persistence of agglomerated power and ever-watchful eyes in downtown Los Angeles cannot be ignored by either captive participants or outside observers. The industrialization of the urban periphery may be turning the space economy of the region inside-out, but the old centre is more than holding its own as the preeminent political and economic citadel . . .

[. . .]

LATERAL EXTENSIONS

Radiating from the specifying nodality of the central city are the hypothesized pathways of traditional urban theory . . . Formal models of urban morphology have conventionally begun with the assumption of a structuring central place organizing an adherent landscape into discoverable patterns of hinterland development and regionalization . . .

Population densities do mound up around the centres of cities, even in the polycentric archipelago of Los Angeles (where there may be several dozen such mounds, although the most pronounced still falls off from the central city). There is also an accompanying concentric residential rhythm associated with the family life cycle and the relative premiums placed on access to the dense peaks versus the availability of living space in the sparseness of the valleys (at least for those who can afford such freedoms of choice). Land values (when they can be accurately calculated) and some job densities also tend to follow in diminishing peaks outwards from the centre, bringing back to mind those tented webs of the urban geography textbooks.

Adding direction to the decadence of distance reduces the Euclidean elegance of concentric gradations, and many of the most mathematical of urban geometricians have accordingly refused to follow this slightly unsettling path. But direction does indeed induce another fit by pointing out the emanation of fortuitous wedges or sectors from the centre. The sectoral wedges of Los Angeles are especially pronounced once you leave the inner circle around downtown.

The Wilshire Corridor, for example, extends the citadels of the central city almost twenty miles westwards to the Pacific, picking up several other prominent but smaller downtowns en route (the Miracle Mile that initiated this extension, Beverly Hills, Century City, Westwood, Brentwood, Santa Monica). Watching above it is an even lengthier wedge of the wealthiest residences, running with almost staggering homogeneities to the Pacific Palisades and the privatized beaches of Malibu, sprinkled with announcements of armed responsiveness and signs which say that 'trespassers will be shot'. Here are the hearths of the most vocal homeowners' movements, arms raised to slow growth and preserve their putative neighbourhoods in the face of the encroaching, view-blocking, street-clogging, and declassé downtowns.

As if in counterbalance, on the other side of the tracks east of downtown is the salient containing the largest Latino *barrio* in Anglo-America, where many of those who might be shot are carefully barricaded in poverty and there is at least one more prominent wedge, stretching southward from downtown to the twin ports of Los Angeles–Long Beach, still reputed to be one of the largest consistently industrial urban sectors in the world. This is the primary axis of Ruhral Los Angeles.

A third ecological order perturbs the geometrical neatness still further, punching holes into the mono-

centric gradients and wedges as a result of the territorial segregation of races and ethnicities. Segregation is so noisy that it overloads the conventional statistical methods of urban factorial ecology with scores of tiny but 'significant' eco-components. In Los Angeles, arguably the most segregated city in the country, these components are so numerous that they operate statistically to obscure the spatiality of social class relations deeply embedded in the zones and wedges of the urban landscape, as if they needed to be obscured any further.

These broad social geometries provide an attractive model of the urban geography of Los Angeles, but like most of the inherited overviews of formal urban theory they are seriously diverting and illusory. They mislead not because there is disagreement over their degree of fit – such regular empiricist arguments merely induce a temporary insensibility by forcing debate on to the usually sterile grounds of technical discourse. Instead, they deceive by involuting explanation, by the legerdemain of making the nodality of the urban explain itself through its mere existence, one outcome explaining another. Geographical covariance in the form of empirico-statistical regularity is elevated to causation and frozen into place without a history – and without a human geography which recognizes that the organization of space is a social product filled with politics and ideology, contradiction and struggle, comparable to the making of history. Empirical regularities are there to be found in the surface geometry of any city, including Los Angeles, but they are not explained in the discovery, as is so often assumed.

Different routes and different roots must be explored to achieve a practical understanding and critical reading of urban landscapes. The illusions of empirical opaqueness must be shattered, along with the other disciplining effects of Modern Geography.

DECONSTRUCTION

Back in the centre, shining from its circular turrets of bronzed glass, stands the Bonaventure Hotel, an amazingly storeyed architectural symbol of the splintered labyrinth that stretches sixty miles around it. Like many other Portmanteaus which dot the eyes of urban citadels in New York and San Francisco, Atlanta and Detroit, the Bonaventure has become a concentrated representation of the restructured spatiality of the late capitalist city: fragmented and fragmenting,

homogeneous and homogenizing, divertingly packaged yet curiously incomprehensible, seemingly open
in presenting itself to view but constantly pressing
to enclose, to compartmentalize, to circumscribe, to
incarcerate. Everything imaginable appears to be
available in this micro-urb, but real places are difficult
to find, its spaces confuse an effective cognitive
mapping, its pastiche of superficial reflections bewilder
co-ordination and encourage submission instead.
Entry by land is forbidding to those who carelessly walk
but entrance is nevertheless encouraged at many
different levels, from the truly pedestrian skyways
above to the bunkered inlets below. Once inside,
however, it becomes daunting to get out again without
bureaucratic assistance. In so many ways, its architecture recapitulates and reflects the sprawling
manufactured spaces of Los Angeles.

[. . .]

From the centre to the periphery, in both inner
and outer cities, the Sixty-Mile Circle today encloses a
shattered metro-sea of fragmented yet homogenized
communities, cultures, and economies confusingly
arranged to contingently ordered spatial division of
labour and power . . .

Over 130 other municipalities and scores of county-
administered areas adhere loosely around the irregular
City of Los Angeles in a dazzling, sprawling patchwork mosaic. Some have names which are startlingly
self-explanatory. Where else can there be a City of
Industry and a City of Commerce, so flagrantly com-
memorating the fractions of capital which guaranteed
their incorporation. In other places, names casually
try to recapture a romanticized history (as in the many
new communities called Rancho something-or-other)
or to ensconce the memory of alternative geographies
(as in Venice, Naples, Hawaiian Gardens, Ontario,
Manhattan Beach, Westminster). In naming, as in so
many other contemporary urban processes, time and
space, the 'once' and the 'there', are being increasingly
played with and packaged to serve the needs of the
here and the now, making the lived experience of
the urban increasingly vicarious, screened through
simulacra, those exact copies for which the real
originals have been lost.

A recent clipping from the *Los Angeles Times* tells of
the 433 signs which bestow identity within the hyper-
space of the City of Los Angeles, described as 'A City
Divided and Proud of It.' . . . One of the founders of
the programme pondered its development: 'At first,
in the early 1960s, the Traffic Department took the

position that all the communities were part of Los
Angeles and we didn't want cities within cities . . . but
we finally gave in. Philosophically it made sense. Los
Angeles is huge. The city had to recognize that there
were communities that needed identification . . .'

For at least fifty years, Los Angeles has been defy-
ing conventional categorical description of the urban,
of what is city and what is suburb, of what can be
identified as community or neighbourhood, of what co-
presence means in the elastic urban context. It has in
effect been deconstructing the urban into a confusing
collage of signs which advertise what are often little
more than imaginary communities and outlandish
representations of urban locality. I do not mean to
say that there are no genuine neighbourhoods to be
found in Los Angeles. Indeed, finding them through
car-voyages of exploration has become a popular
local pastime, especially for those who have become
so isolated from propinquitous community in the
repetitive sprawl of truly ordinary-looking landscapes
that make up most of the region. But again the urban
experience becomes increasingly vicarious, adding
more layers of opaqueness to *l'espace vécu*.

Underneath this semiotic blanket there remains
an economic order, an instrumental nodal structure, an
essentially exploitative spatial division of labour,
and this spatially organized urban system has for the
past half century been more continuously productive
than almost any other in the world. But it has also
been increasingly obscured from view, imaginatively
mystified in an environment more specialized in the
production of encompassing mystifications than
practically any other you can name. As has so often
been the case in the United States, this conservative
deconstruction is accompanied by a numbing depoliti-
cization of fundamental class and gender relations
and conflicts. When all that is seen is so fragmented
and filled with whimsy and pastiche, the hard edges of
the capitalist, racist and patriarchal landscape seem
to disappear, melt into air.

With exquisite irony, contemporary Los Angeles
has come to resemble more than ever before a gigantic
agglomeration of theme parks, a lifespace comprised
of Disneyworlds. It is a realm divided into showcases
of global village cultures and mimetic American landscapes, all-embracing shopping malls and crafty
Main Streets, corporation-sponsored magic kingdoms,
high-technology-based experimental prototype communities of tomorrow, attractively packaged places
for rest and recreation all cleverly hiding the buzzing

workstations and labour processes which help to keep it together: like the original 'Happiest Place on Earth', the enclosed spaces are subtly but tightly controlled by invisible overseers despite the open appearance of fantastic freedoms of choice . . .

AFTERWORDS

[. . .]

I have been looking at Los Angeles from many different points of view and each way of seeing assists in sorting out the interjacent medley of the subject landscape. The perspectives explored are purposeful, eclectic, fragmentary, incomplete, and frequently contradictory, but so too is Los Angeles and, indeed, the experienced historical geography of every urban landscape . . . The task of comprehensive, holistic regional description may therefore be impossible, as may be the construction of a complete historico-geographical materialism.

There is hope nonetheless. The critical and theoretical interpretation of geographical landscapes has recently expanded into realms that functionally had been spatially illiterate for most of the twentieth century. New and avid readers abound as never before, many are directly attuned to the specificity of the urban, and several have significantly turned their eyes to Los Angeles. Moreover, many practised readers of surface geographies have begun to see through the alternatively myopic and hypermetropic distortions of past perspectives to bring new insight to spatial analysis and social theory. Here too Los Angeles has attracted observant readers after a history of neglect and misapprehension, for it insistently presents itself as one of the most informative palimpsests and paradigms of twentieth-century urban-industrial development and popular consciousness.

As I have seen and said in various ways, everything seems to come together in Los Angeles, the totalizing Aleph. Its representations of spatiality and historicity are archetypes of vividness, simultaneity, and interconnection. They beckon inquiry at once into their telling uniqueness and, at the same time, into their assertive but cautionary generalizability. Not all can be understood, appearances as well as essences persistently deceive, and what is real cannot always be captured even in extraordinary language. But this makes the challenge more compelling, especially if once in a while one has the opportunity to take it all apart and reconstruct the context. The reassertion of space in critical social theory – and in critical political praxis – will depend upon a continued deconstruction of a still occlusive historicism and many additional voyages of exploration into the heterotopias of contemporary postmodern geographies.

"Fortress L.A."

from *City of Quartz: Excavating the Future in Los Angeles* (1990)

Mike Davis

Editors' Introduction

Mike Davis is to contemporary Los Angeles what Friedrich Engels was to mid-nineteenth-century Manchester, England. Engels was a kind of explorer, reporting to an educated, middle-class audience about the horrors of the industrial city and the miserable lives of the new industrial proletarian class of modern capitalism. Similarly, Davis, 150 years later, explores the dark side of the postmodern metropolis and reports on the hopelessness and despair of the *post-industrial* "underclass" (largely defined by race, gender, and ethnicity) to an audience largely comprised of young, disaffected intellectuals and academics. The parallels are striking. Davis may be the heir to Engels simply because the contemporary metropolis – characterized by wealth and homelessness and divided against itself along the fault lines that separate suburban enclaves from inner-city slums – is the heir to the geographical and social-class divisions of the cities of the Industrial Revolution.

Mike Davis is an acerbic Southern California social critic who makes his social-class sympathies and anti-establishment bias perfectly clear on every page of *City of Quartz*, which one admirer praised as a "visionary rant."

In *The Condition of the Working Class in England in 1844* (p. 58), Engels noted the boulevards that intersected the city of Manchester and how the façades of those broad thoroughfares served to mask and disguise the hovels of the poor that lay beyond the view of the middle-class commuter. Davis, employing an eclectic "culture studies" methodology, makes a similar but even subtler point about contemporary Los Angeles. The freeways allow middle-class suburbanites to navigate the city as a whole without encountering the lives of the residents of the inner-city neighborhoods. But beyond this, as Madanipour describes in less vitriolic prose (p. 181), the city itself has become, in the semiotic culture of postmodernism, a vast and continuous system of signs that we read and obey mostly on a subconscious level. "Today's upscale, pseudo-public spaces," Davis writes, " . . . are full of invisible signs warning off the underclass 'Other.' Although architectural critics are usually oblivious to how the built environment contributes to segregation, pariah groups – whether poor Latino families, young Black men, or elderly homeless white females – read the meaning immediately."

Another similarity between Engels and Davis is that both men, as they proceeded to "read the streets" of their respective paradigm cities, reach the limits of language's ability to describe the physical and psychological conditions being reported. Engels compiled a mountain of personal observations, journalistic reports, and official survey data to create a virtual catalog of social horror. Davis, on the other hand, relies on a more emotional strategy. His over-heated rhetorical excesses often overwhelm rational discourse. Davis turned out to be a prophet: the ghettos and barrios of South Central Los Angeles erupted into open rebellion in the Rodney King riots two years after the publication of *City of Quartz*.

Whereas Engels saw massive social dislocations and systematically set about to fashion a theory of revolutionary socialism in response to the observed reality, Davis offers no similarly optimistic solution. In a sense, Davis *personalizes* the horrors of the contemporary city, and his voice is one of desperation and despair offering no obvious way out of the current impasse. He leaves it to others, like William Julius Wilson (p. 126), Ali Madanipour

(p. 181), and Paul Davidoff (p. 388) to propose how government officials, urban designers, and city planners might advocate on behalf of the poor and oppressed to make cities better.

Davis's insights regarding the underclass "other" suggest comparisons with such classic analysts of ghettoization as W.E.B. Du Bois (p. 119), and William Julius Wilson (p. 126). But it is difficult to compare Davis with social scientists and social policy and planning professionals whose focus is on objectively analyzing reality and influencing the existing city systems by working within them.

This selection is from Mike Davis, *City of Quartz: Excavating the Future in Los Angeles* (London and New York: Verso, 1990). *The Grit Beneath the Glitter: Tales from the Real Las Vegas* (Berkeley: University of California Press, 2002) co-edited with Hal Rothman contains a mixture of personal narratives and scholarly articles on the seamier sides of Las Vegas. In addition to social criticism Davis has written interesting and controversial books on the abuse of the natural environment by humans: *Dead Cities: A Natural History* (New York: New Press, 2002), *Late Victorian Holocausts: El Niño Families and the Making of the Third World* (London: Verso, 2001), and *The Ecology of Fear: Los Angeles and the Imagination of Disaster* (New York: Metropolitan Books, 1998) and on oppression of immigrants: *Prisoners of the American Dream* (London: Verso, 1986), *Operation Gatekeeper: The Rise of the 'Illegal Alien' and the Reinvention of the US–Mexico Border* (London: Routledge, 2001). A final book by Davis – *Magical Urbanism: Latinos Reinvent the U.S. Big City* (London: Verso, 2001) – in addition to characteristic "slash and burn" prose excoriating oppression of Hispanic immigrants contains some uncharacteristically positive comments about how large US cities are being improved by immigration.

For more on Los Angeles, consult William Fulton, *The Reluctant Metropolis: The Politics of Urban Growth in Los Angeles* (Baltimore, MD: Johns Hopkins University Press, 2001), R.A. Nadeau's *Los Angeles: From Mission to Modern City* (New York: Longman, 1960) and R. Marchand's *The Emergence of Los Angeles: Population and Housing in the City of Dreams* (London: Pion, 1986). Interesting architectural and cultural analyses of Los Angeles can be found in Rayner Banham's *Los Angeles: The Architecture of Four Ecologies* (London: Penguin, 1971) and Dolores Hayden's *The Power of Place: Urban Landscapes as Public History* (Cambridge, MA: MIT Press, 1995).

For analyses of the troubled race relations of Los Angeles, see Mauricio Mazon, *The Zoot-suit Riots* (Austin: University of Texas Press, 1984), Raphael J. Sonenshein, *Politics in Black and White* (Princeton, NJ: Princeton University Press, 1993), and Mark Baldassare (ed.), *The Los Angeles Riots: Lessons for the Urban Future* (Boulder, CO: Westview Press, 1994).

The carefully manicured lawns of Los Angeles' West-side sprout forests of ominous little signs warning: "Armed Response!" Even richer neighborhoods in the canyons and hillsides isolate themselves behind walls guarded by gun-toting private police and state-of-the-art electronic surveillance. Downtown, a publicly subsidized "urban renaissance" has raised the nation's largest corporate citadel, segregated from the poor neighborhoods around it by a monumental archi-tectural glacis. In Hollywood, celebrity architect Frank Gehry, renowned for his "humanism," apotheosizes the siege look in a library designed to resemble a foreign-legion fort. In the Westlake district and the San Fernando Valley the Los Angeles Police barricade streets and seal off poor neighborhoods as part of their "war on drugs." In Watts, developer Alexander Haagen demonstrates his strategy for recolonizing inner-city retail markets: a panopticon shopping mall surrounded

by staked metal fences and a substation of the LAPD in a central surveillance tower. Finally, on the horizon of the next millennium, an ex-chief of police crusades for an anti-crime "giant eye" – a geo-synchronous law enforcement satellite – while other cops discreetly tend versions of "Garden Plot," a hoary but still viable 1960s plan for a law-and-order armageddon.

Welcome to post-liberal Los Angeles, where the defense of luxury lifestyles is translated into a prolif-eration of new repressions in space and movement, undergirded by the ubiquitous "armed response." This obsession with physical security systems, and, collaterally, with the architectural policing of social boundaries, has become a zeitgeist of urban restruc-turing, a master narrative in the emerging built environ-ment of the 1990s. Yet contemporary urban theory, whether debating the role of electronic technologies in precipitating "postmodern space," or discussing the dispersion of urban functions across poly-centered metropolitan "galaxies," has been strangely silent about the militarization of city life so grimly visible at the street level. Hollywood's pop apocalypses and pulp science fiction have been more realistic, and politically perceptive, in representing the programmed hardening of the urban surface in the wake of the social polarizations of the Reagan era. Images of carceral inner cities (*Escape from New York, Running Man*), high-tech police death squads (*Blade Runner*), sentient buildings (*Die Hard*), urban bantustans (*They Live!*), Vietnam-like street wars (*Colors*), and so on, only extrapolate from actually existing trends.

Such dystopian visions grasp the extent to which today's pharaonic scales of residential and commercial security supplant residual hopes for urban reform and social integration. The dire predictions of Richard Nixon's 1969 National Commission on the Causes and Prevention of Violence have been tragically fulfilled: we live in "fortress cities" brutally divided between "fortified cells" of affluent society and "places of terror" where the police battle the criminalized poor. The "Second Civil War" that began in the long hot summers of the 1960s has been institutionalized into the very structure of urban space. The old liberal paradigm of social control, attempting to balance repression with reform, has long been superseded by a rhetoric of social warfare that calculates the interests of the urban poor and the middle classes as a zero-sum game. In cities like Los Angeles, on the bad edge of postmodernity, one observes an unprecedented tendency to merge urban design, architecture and the police apparatus into a single, comprehensive security effort.

This epochal coalescence has far-reaching consequences for the social relations of the built environment. In the first place, the market provision of "security" generates its own paranoid demand. "Security" becomes a positional good defined by income access to private "protective services" and membership in some hardened residential enclave or restricted suburb. As a prestige symbol – and sometimes as the decisive borderline between the merely well-off and the "truly rich" – "security" has less to do with personal safety than with the degree of personal insulation, in residential, work, consumption and travel environments, from "unsavory" groups and individuals, even crowds in general.

Secondly, as William Whyte has observed of social intercourse in New York, "fear proves itself." The social perception of threat becomes a function of the security mobilization itself, not crime rates. Where there is an actual rising arc of street violence, as in Southcentral Los Angeles or Downtown Washington D.C., most of the carnage is self-contained within ethnic or class boundaries. Yet white middle-class imagination, absent from any firsthand knowledge of inner-city conditions, magnifies the perceived threat through a demonological lens. Surveys show that Milwaukee suburbanites are just as worried about violent crime as inner-city Washingtonians, despite a twentyfold difference in relative levels of mayhem. The media, whose function in this arena is to bury and obscure the daily economic violence of the city, ceaselessly throw up spectres of criminal underclasses and psychotic stalkers. Sensationalized accounts of killer youth gangs high on crack and shrilly racist evocations of marauding Willie Hortons foment the moral panics that reinforce and justify urban apartheid.

Moreover, the neo-military syntax of contemporary architecture insinuates violence and conjures imaginary dangers. In many instances the semiotics of so-called "defensible space" are just about as subtle as a swaggering white cop. Today's upscale, pseudo-public spaces – sumptuary malls, office centers, culture acropolises, and so on – are full of invisible signs warning off the underclass "Other." Although architectural critics are usually oblivious to how the built environment contributes to segregation, pariah groups – whether poor Latino families, young Black men, or elderly homeless white females – read the meaning immediately.

THE DESTRUCTION OF PUBLIC SPACE

The universal and ineluctable consequence of this crusade to secure the city is the destruction of accessible public space. The contemporary opprobrium attached to the term "street person" is in itself a harrowing index of the devaluation of public spaces. To reduce contact with untouchables, urban redevelopment has converted once vital pedestrian streets into traffic sewers and transformed public parks into temporary receptacles for the homeless and wretched. The American city, as many critics have recognized, is being systematically turned inside out – or, rather, outside in. The valorized spaces of the new megastructures and super-malls are concentrated in the center, street frontage is denuded, public activity is sorted into strictly functional compartments, and circulation is internalized in corridors under the gaze of private police.

The privatization of the architectural public realm, moreover, is shadowed by parallel restructurings of electronic space, as heavily policed, pay-access "information orders," elite databases and subscription cable services appropriate parts of the invisible agora. Both processes, of course, mirror the deregulation of the economy and the recession of non-market entitlements. The decline of urban liberalism has been accompanied by the death of what might be called the "Olmstedian vision" of public space. Frederick Law Olmsted, it will be recalled, was North America's Haussmann, as well as the Father of Central Park. In the wake of Manhattan's "Commune" of 1863, the great Draft Riot, he conceived public landscapes and parks as social safety-valves, mixing classes and ethnicities in common (bourgeois) recreations and enjoyments. As Manfredo Tafuri has shown in his well-known study of Rockefeller Center, the same principle animated the construction of the canonical urban spaces of the La Guardia–Roosevelt era.

This reformist vision of public space – as the emollient of class struggle, if not the bedrock of the American *polis* – is now as obsolete as Keynesian nostrums of full employment. In regard to the "mixing" of classes, contemporary urban America is more like Victorian England than Walt Whitman's or La Guardia's New York. In Los Angeles, once-upon-a-time a demi-paradise of free beaches, luxurious parks, and "cruising strips," genuinely democratic space is all but extinct. The Oz-like archipelago of Westside pleasure domes – a continuum of tony malls, arts centers and gourmet strips – is reciprocally dependent upon the social imprisonment of the third-world service proletariat who live in increasingly repressive ghettoes and barrios. In a city of several million yearning immigrants, public amenities are radically shrinking, parks are becoming derelict and beaches more segregated, libraries and playgrounds are closing, youth congregations of ordinary kinds are banned, and the streets are becoming more desolate and dangerous.

Unsurprisingly, as in other American cities, municipal policy has taken its lead from the security offensive and the middle-class demand for increased spatial and social insulation. De facto disinvestment in traditional public space and recreation has supported the shift of fiscal resources to corporate-defined redevelopment priorities. A pliant city government – in this case ironically professing to represent a bi-racial coalition of liberal whites and Blacks – has collaborated in the massive privatization of public space and the subsidization of new, racist enclaves (benignly described as "urban villages"). Yet most current, giddy discussions of the "postmodern" scene in Los Angeles neglect entirely these overbearing aspects of counter-urbanization and counter-insurgency. A triumphal gloss – "urban renaissance," "city of the future," and so on – is laid over the brutalization of inner-city neighborhoods and the increasing South Africanization of its spatial relations. Even as the walls have come down in Eastern Europe, they are being erected all over Los Angeles.

The observations that follow take as their thesis the existence of this new class war (sometimes a continuation of the race war of the 1960s) at the level of the built environment. Although this is not a comprehensive account, which would require a thorough analysis of economic and political dynamics, these images and instances are meant to convince the reader that urban form is indeed following a repressive function in the political furrows of the Reagan–Bush era. Los Angeles, in its usual prefigurative mode, offers an especially disquieting catalogue of the emergent liaisons between architecture and the American police state.

THE FORBIDDEN CITY

The first militarist of space in Los Angeles was General Otis of the *Times*. Declaring himself at war with labor, he infused his surroundings with an unrelentingly bellicose air:

> He called his home in Los Angeles the Bivouac. Another house was known as the Outpost. The *Times* was known as the Fortress. The staff of the paper was the Phalanx. The *Times* building itself was more fortress than newspaper plant, there were turrets, battlements, sentry boxes. Inside he stored fifty rifles.

A great, menacing bronze eagle was the *Times*'s crown; a small, functional cannon was installed on the hood of Otis's touring car to intimidate onlookers. Not surprisingly, this overwrought display of aggression produced a response in kind. On 1 October 1910 the heavily fortified *Times* headquarters – citadel of the open shop on the West Coast – was destroyed in a catastrophic explosion blamed on union saboteurs.

Eighty years later, the spirit of General Otis has returned to subtly pervade Los Angeles' new "post-modern" Downtown: the emerging Pacific Rim financial complex which cascades, in rows of skyscrapers, from Bunker Hill southward along the Figueroa corridor. Redeveloped with public tax increments under the aegis of the powerful and largely unaccountable Community Redevelopment Agency (CRA), the Downtown project is one of the largest postwar urban designs in North America. Site assemblage and clearing on a vast scale, with little mobilized opposition, have resurrected land values, upon which big developers and off-shore capital (increasingly Japanese) have planted a series of billion-dollar, block-square megastructures: Crocker Center, the Bonaventure Hotel and Shopping Mall, the World Trade Center, the Broadway Plaza, Arco Center, CitiCorp Plaza, California Plaza, and so on. With historical landscapes erased, with megastructures and superblocks as primary components, and with an increasingly dense and self-contained circulation system, the new financial district is best conceived as a single, demonically self-referential hyperstructure, a Miesian skyscape raised to dementia.

Like similar megalomaniac complexes, tethered to fragmented and desolated Downtowns (for instance, the Renaissance Center in Detroit, the Peachtree and Omni Centers in Atlanta, and so on), Bunker Hill and the Figueroa corridor have provoked a storm of liberal objections against their abuse of scale and composition, their denigration of street landscape, and their confiscation of so much of the vital life activity of the center, now sequestered within subterranean concourses or privatized malls. Sam Hall Kaplan, the crusty urban critic of the *Times*, has been indefatigable in denouncing the anti-pedestrian bias of the new corporate citadel, with its fascist obliteration of street frontage. In his view the superimposition of "hermetically sealed fortresses" and air-dropped "pieces of suburbia" has "dammed the rivers of life" Downtown.

Yet Kaplan's vigorous defense of pedestrian democracy remains grounded in hackneyed liberal complaints about "bland design" and "elitist planning practices." Like most architectural critics, he rails against the oversights of urban design without recognizing the dimension of foresight, of explicit repressive intention, which has its roots in Los Angeles' ancient history of class and race warfare. Indeed, when Downtown's new "Gold Coast" is viewed en bloc from the standpoint of its interactions with other social areas and landscapes in the central city, the "fortress effect"

emerges, not as an inadvertent failure of design, but as deliberate socio-spatial strategy.

The goals of this strategy may be summarized as a double repression: to raze all association with Downtown's past and to prevent any articulation with the non-Anglo urbanity of its future. Everywhere on the perimeter of redevelopment this strategy takes the form of a brutal architectural edge or glacis that defines the new Downtown as a citadel vis-à-vis the rest of the central city. Los Angeles is unusual amongst major urban renewal centers in preserving, however negligently, most of its circa 1900–30 Beaux Arts commercial core. At immense public cost, the corporate headquarters and financial district was shifted from the old Broadway–Spring corridor six blocks west to the greenfield site created by destroying the Bunker Hill residential neighborhood. To emphasize the "security" of the new Downtown, virtually all the traditional pedestrian links to the old center, including the famous Angels' Flight funicular railroad, were removed.

The logic of this entire operation is revealing. In other cities developers might have attempted to articulate the new skyscape and the old, exploiting the latter's extraordinary inventory of theaters and historic buildings to create a gentrified history – a gaslight district, Faneuil Market or Ghirardelli Square – as a support to middle-class residential colonization. But Los Angeles' redevelopers viewed property values in the old Broadway core as irreversibly eroded by the area's very centrality to public transport, and especially by its heavy use by Black and Mexican poor. In the wake of the Watts rebellion, and the perceived Black threat to crucial nodes of white power (spelled out in lurid detail in the McCone Commission Report), resegregated spatial security became the paramount concern. The Los Angeles Police Department abetted the flight of business from Broadway to the fortified redoubts of Bunker Hill by spreading scare literature typifying Black teenagers as dangerous gang members.

As a result, redevelopment massively reproduced spatial apartheid. The moat of the Harbor Freeway and the regraded palisades of Bunker Hill cut off the new financial core from the poor immigrant neighborhoods that surround it on every side. Along the base of California Plaza, Hill Street became a local Berlin Wall separating the publicly subsidized luxury of Bunker Hill from the lifeworld of Broadway, now reclaimed by Latino immigrants as their primary shopping and entertainment street. Because politically connected speculators are now redeveloping the northern end of the Broadway corridor (sometimes known as "Bunker Hill East"), the CRA is promising to restore pedestrian linkages to the Hill in the 1990s, including the Angels' Flight incline railroad. This, of course, only dramatizes the current bias against accessibility – that is to say, against any spatial interaction between old and new, poor and rich, except in the framework of gentrification or recolonization. Although a few white-collars venture into the Grand Central Market – a popular emporium of tropical produce and fresh foods – Latino shoppers or Saturday strollers never circulate in the Gucci precincts above Hill Street. The occasional appearance of a destitute street nomad in Broadway Plaza or in front of the Museum of Contemporary Art sets off a quiet panic; video cameras turn on their mounts and security guards adjust their belts.

Photographs of the old Downtown in its prime show mixed crowds of Anglo, Black and Latino pedestrians of different ages and classes. The contemporary Downtown "renaissance" is designed to make such heterogeneity virtually impossible. It is intended not just to "kill the street" as Kaplan fears, but to "kill the crowd," to eliminate that democratic admixture on the pavements and in the parks that Olmsted believed was America's antidote to European class polarizations. The Downtown hyperstructure – like some Buckminster Fuller post-Holocaust fantasy – is programmed to ensure a seamless continuum of middle-class work, consumption and recreation, without unwonted exposure to Downtown's working-class street environments. Indeed the totalitarian semiotics of ramparts and battlements, reflective glass and elevated pedways, rebukes any affinity or sympathy between different architectural or human orders. As in Otis's fortress *Times* building, this is the archisemiotics of class war.

"The Neighborhood, the District, and the Corridor"

from Peter Katz and Vincent Scully, Jr. (eds), *The New Urbanism: Toward an Architecture of Community* (1993)

Andres Duany and Elizabeth Plater-Zyberk

Editors' Introduction

The husband and wife team of Andres Duany and Elizabeth Plater-Zyberk are leaders in a current intellectual movement in architecture and urban planning called the *New Urbanism*. They are interested in urban space because they want to reshape urban and suburban patterns to create places where people can interact with their neighbors, enjoy public space, and get from home to work, school, and stores by foot and bicycle as well as by car. Duany and Plater-Zyberk are notable not only for their theorizing about better urban spaces, but also because they have designed and built successful new communities such as Seaside, Florida that actually implement their ideas.

Duany and Plater-Zyberk are neo-traditionalists. They find much of value in the traditional vernacular architecture and people-oriented layout of the kind of small American towns J.B. Jackson lovingly describes (p. 164). Like Jane Jacobs (p. 114) they feel that promoting human interaction and community are as important values as designing for efficiency and are skeptical of the way in which urban planners have segregated land uses and imposed artificial and de-humanizing order on urban spaces. Duany and Plater-Zyberk are often described as postmodernists. They reject the emphasis on machine-age efficiency, large scale, and speed that dominated the thinking of modernist architects and planners like Le Corbusier (p. 317) in the middle of the twentieth century.

In this essay Duany and Plater-Zyberk describe how architects and planners can design three urban elements to create urban spaces much better than the isolating, auto-dependent suburbs common today. Their approach of breaking down urban space into distinct elements – neighborhoods, districts, edges, and corridors – is influenced by Kevin Lynch (p. 424) who identified key "elements" that make up the image of the city and urged design professionals to follow principles of good design for each element.

Duany and Plater-Zyberk are particularly interested in *neighborhoods* – urbanized areas with a balanced mix of uses. Their ideas about the ideal neighborhood are very concrete and elements of their ideal model fit together into a unified vision for an alternative to the pattern of development that evolved after World War II. The ideal neighborhood they envisage is *small* – a five-minute walk from center to edge. It is *diverse*, containing a balanced mix of dwellings, workplaces, shops, parks, and civic institutions such as schools and churches. Their ideal neighborhood has a *center* dominated by civic buildings and public space, and an *edge* of some kind (perhaps defined by fields and forest; perhaps by a highway or rail line). Land use in their ideal neighborhood articulates with a system of pedestrian-friendly and transit-oriented transportation corridors that offer residents opportunities to walk, bicycle, drive, or take public transit to work, school, shopping, and entertainment. This vision stands in sharp contrast to the predominant suburban pattern today: sprawling, auto-dependent, homogenous areas lacking in public space and without clear centers or edges.

New Urbanists like Duany and Plater-Zyberk believe the modernist values that produced modern suburbia are no longer relevant. While the segregation of housing from the smoke and stench of nineteenth-century factories may have been an improvement over the conditions which Friedrich Engels described in nineteenth-century Manchester (p. 58), they argue that such a segregation of land uses is no longer necessary in the post-industrial information age. While the street pattern built to accommodate mass auto use allowed people to move efficiently from suburban homes to city-centered work locations, this is no longer the pattern in the technoburbia Fishman (p. 77) describes. Duany, Plater-Zyberk, and other New Urbanists feel the auto-oriented corridor system has had a devastating impact on community and the quality of life in suburbia that needs to be changed.

Duany and Plater-Zyberk have a private architectural/urban planning practice in Florida specializing in developments consistent with the design principles in this selection. They are founders of the Congress for New Urbanism – a loose association of like-minded architects, planners, and other professionals. Their most notable accomplishment is the design of Seaside, Florida, an entire new town built on New Urbanist principles that have attracted worldwide attention.

This reading is from Peter Katz and Vincent Scully, Jr. (eds), *The New Urbanism: Toward an Architecture of Community* (New York: McGraw-Hill, 1993), a collection of essays and illustrations of New Urbanist developments. Duany and Plater-Zyberk's most recent statement of their critique of suburbia and alternative vision is *Suburban Nation: The Rise of Sprawl and the Decline of the American Dream* (New York: North Point Press, 2001) co-authored with Jeff Speck. Duany co-authored *The Smart Growth Manual* (New York: McGraw-Hill, 2002) with Jeff Speck and edited *New Civic Art: Elements of Town Planning* (London: Rizzoli, 2002).

For a description of how Seaside, Florida was designed and built see David Mohoney and Keasterling Keller (eds), *Seaside: Making a Town in America* (Princeton, NJ: Princeton Architectural Press, 1991). Critiques of Seaside – both positive and negative – are contained in Tod Bressi (ed.), *The Seaside Debates: A Critique of the New Urbanism* (London: Rizzoli, 2002).

Other writings on the New Urbanism include Michael Lecesse and Kathleen McCormick (eds), *Charter of the New Urbanism* (New York: McGraw-Hill, 1999), John A. Dutton, *New American Urbanism: Re-forming the Suburban Metropolis* (New York: Skira, 2001), Peter Calthorpe, *The Next American Metropolis: Ecology, Community, and the American Dream* (Princeton, NJ: Princeton Architectural Press, 1993), and Philip Langdon, *A Better Place to Live: Reshaping the American Suburb* (Amherst: University of Massachusetts Press, 2001). James Howard Kunstler has popularized critiques of suburbia and New Urbanist ideas in *The Geography of Nowhere: The Rise and Decline of America's Man-made Landscape* (New York: Touchstone Books, 1994) and *Home from Nowhere: Remaking Our Everyday World for the Twenty-first Century* (New York: Touchstone Books, 1998).

The fundamental organizing elements of the New Urbanism are the neighborhood, the district and the corridor. Neighborhoods are urbanized areas with a balanced mix of human activity; districts are areas dominated by a single activity; corridors are connectors and separators of neighborhoods and districts.

A single neighborhood standing free in the landscape is a village. Cities and towns are made up of multiple neighborhoods and districts, organized by corridors of transportation or open space. Neighborhoods, districts and corridors are urban elements. By contrast, suburbia, which is the result of zoning laws that separate uses, is composed of pods, highways and interstitial spaces.

THE NEIGHBORHOOD

The nomenclature may vary, but there is general agreement regarding the physical composition of the neighborhood. The "neighborhood unit" of the 1929 New York Regional Plan, the "quartier" identified by Leon Krier, the "traditional neighborhood development" (TND) and "transit oriented development" (TOD) share similar attributes. They all propose a model of urbanism that is limited in area and structured around a defined center. While the population density may vary, depending on its context, each model offers a balanced mix of dwellings, workplaces, shops, civic buildings and parks.

Like the habitat of any species, the neighborhood possesses a natural logic that can be described in physical terms. The following are the principles of an ideal neighborhood design: (1) the neighborhood has a center and an edge; (2) the optimal size of a neighborhood is a quarter mile from center to edge; (3) the neighborhood has a balanced mix of activities – dwelling, shopping, working, schooling, worshipping and recreating; (4) the neighborhood structures building sites and traffic on a fine network of interconnecting streets; (5) the neighborhood gives priority to public space and to the appropriate location of civic buildings.

The neighborhood has a center and an edge

The combination of a focus and a limit contribute to the social identity of the community. The center is a necessity, the edge not always so. The center is always a public space, which may be a square, a green or an important street intersection. It is near the center of the urban area unless compelled by some geographic circumstance to be elsewhere. Eccentric locations are justified if there is a shoreline, a transportation corridor or a place with an engaging view.

The center is the locus of the neighborhood's public buildings, ideally a post office, a meeting hall, a day care center, and sometimes religious and cultural institutions. Shops and workplaces are usually associated with the center, especially in a village. In the aggregations of multiple neighborhoods which occur in a town or city, retail buildings and workplaces may be at the edge of the neighborhood, where they can combine with others and intensify commercial and community activity.

Neighborhood edges may vary in character: they can be natural, such as a forest, or man-made, such as infrastructure. In villages, the edge is usually defined by land designated for cultivation such as farms, orchards and nurseries or for conservation in a natural state as woodland, desert, wetland or escarpment. The edge may also be assigned to very low-density residential use with lots of at least 10 acres. When a community cannot afford to sustain large tracts of public open land, such large private ownerships are a way to maintain a green edge.

In cities and towns, edges can be formed by the systematic accretion between the neighborhoods of recreational open spaces, such as parks, schoolyards and golf courses. It is important that golf courses be confined to the edge of neighborhoods, because fairways obstruct direct pedestrian ways to the neighborhood center. These continuous green edges can be part of a larger network of corridors, connecting urban open space with rural surroundings, as described in the 1920s by Benton McKaye.

In high-density urban areas, the neighborhood edge is often defined by infrastructure, such as rail lines and high traffic thoroughfares that best remain outside the neighborhood. The latter, if generously lined with trees, become parkways that reinforce the legibility of the edge and, over a long distance, form the corridors connecting urban neighborhoods.

The optimal size of a neighborhood is a quarter mile from center to edge

This distance is the equivalent of a five-minute walk at an easy pace. The area thus circumscribed is the neighborhood proper, to differentiate it from the green edge, which extends beyond the discipline of the quarter mile. The limited area gathers the population of a neighborhood within walking distance of many of their daily needs, such as a convenience store, post office, community police post, automatic bank teller, school, daycare center and transit stop.

The stop's location among other neighborhood services and within walking distance of home or work makes the transit system convenient. When an automobile trip is necessary to arrive at a transit stop, most potential users will simply continue driving to their destinations. But the neighborhood, which focuses the required user population within walking distance of the stop, makes transit viable at densities that a suburban pattern cannot sustain.

Pedestrian-friendly and transit-oriented neighborhoods permit a region of cities, towns and villages to be accessible without singular reliance on cars. Such a system gives access to the major cultural and social institutions, the variety of shopping and the broad job base that can only be supported by the larger population of an aggregation of neighborhoods.

The neighborhood has a balanced mix of activities – dwelling, shopping, working, schooling, worshipping and recreating

This is particularly important for those who are unable to drive and thus depend on others for mobility. For instance, the young are able to walk or bicycle to school and other activities, freeing their parents from the responsibility and tedium of chauffeuring. The size of a school should be determined by the number of children who can walk or bicycle to it from adjacent neighborhoods.

And the elderly, who relinquish their willingness to drive before they lose their ability to walk, can age in place with dignity rather than being forced into specialized retirement communities, which are the attendant creations of the suburban pattern.

Even those for whom driving may not be a burden enjoy secondary advantages. The proximity of daily destinations and the convenience of transit reduces the number and length of trips, decreases the private stress of time in traffic and minimizes the public-borne expenses of road construction and atmospheric pollution.

The neighborhood's fine-grained mix of activities includes a range of housing types for a variety of incomes, from the wealthy business owner to the school teacher and the gardener. Suburban areas, which are most commonly segregated by income, do not provide for the full range of society. The true neighborhood, however, offers a variety of affordable housing choices: garage apartments in conjunction with single-family houses, apartments above shops and apartment buildings adjacent to shopping and workplaces. The latter's transitional sites are not provided within the suburban pattern whose rigorous, sanitized segregation of uses precludes them.

But the greatest contribution to affordable housing may be realized by the neighborhood's ability to reduce multiple automobile ownership and many of its associated costs. By enabling households to own one less vehicle, the average annual operating cost of $5,000 can be applied toward an additional $50,000 increment of mortgage financing at 10 percent. No other action of the designer can achieve an improvement in the availability of housing for the middle class comparable to the sensible organization of a good neighborhood plan.

The neighborhood structures building sites and traffic on a fine network of interconnecting streets

Neighborhood streets are configured to create blocks of appropriate building sites and to shorten pedestrian routes. They are designed to keep local traffic off regional roads and to keep through traffic off local streets. An interconnecting pattern of streets provides multiple routes that diffuse traffic congestion.

This contrasts to the easily congested single trajectories standard to the suburban pattern: culs-de-sac spill onto collector streets, which connect at single points to arterials, which in turn supply the highways. The suburban traffic model is more concerned with speeding traffic through a place than with the quality of the place itself; the pedestrian is assumed to be elsewhere on separate "walkways" or nonexistent.

Neighborhood streets of varying types are detailed to provide equitably for pedestrian comfort and for automobile movement. Slowing the automobile and increasing pedestrian activity encourages the casual meetings that form the bonds of community.

The neighborhood gives priority to public space and to the appropriate location of civic buildings

Public spaces and buildings represent community identity and foster civic pride. The neighborhood plan structures its streets and blocks to create a hierarchy of public spaces and locations for public buildings. Squares and streets have their size and geometry defined by the intention to create special places. Public buildings occupy important sites, overlooking a square or terminating a street vista.

The suburban practice of locating government buildings, places of worship, schools and even public art according to the expediencies of land cost is ineffective. The importance of these civic and community structures is enhanced by their suitable siting, without incurring additional costs to the infrastructure.

THE DISTRICT

The district is an urbanized area that is functionally specialized. Although districts preclude the full range of activities necessary for a complete neighborhood, they are not the rigorously single activity zones of suburbia: the office parks, housing subdivisions or shopping centers. The specialization of a district still allows multiple activities to support its primary identity. Typical are theater districts, which have restaurants and bars to support and intensify their nightlife; tourist districts, which concentrate hotels, retail activity and entertainment; and the capitol area and the college campus, which are dominated by a large institution. Others accommodate large-scale transportation or manufacturing, such as airports, container terminals and refineries.

Although a degree of specialization for certain urban areas enhances their character and efficiency, in reality, few pure districts are really justified. Thanks to industrial evolution and environmental regulation, the reasons for segregating uses recede with time. The modern North American workplace is no longer a bad neighbor to dwellings and shops.

The organizational structure of the district parallels that of the neighborhood and similarly, for a good fit within the greater region, relies on its relationship to transit. An identifiable focus encourages the formation of special communities: a park for workers at lunch, a square for theater-goers to meet, a mall for civic gatherings. Clear boundaries and dimensions facilitate the formation of special taxing or management organizations. Interconnected circulation supports the pedestrian, enhances transit viability and ensures security. And like the neighborhood, attention to the character of the public spaces creates a sense of place for its users, even if their home is elsewhere.

THE CORRIDOR

The corridor is at once the connector and the separator of neighborhoods and districts. Corridors include natural and man-made elements, ranging from wildlife trails to rail lines. The corridor is not the haphazardly residual space that remains outside subdivisions and shopping centers in suburbia. Rather, it is an urban element characterized by its visible continuity. It is defined by its adjacent districts and neighborhoods and provides entry to them.

The corridor's location and type is determined by its technological intensity and nearby densities. Heavy rail corridors are tangent to towns and traverse the industrial districts of cities. Light rail and trolleys may occur within a boulevard at the neighborhood edge. As such, they are detailed for pedestrian use and to accommodate the frontages of buildings. Bus corridors can pass through neighborhood centers on conventional streets. All of these should be landscaped to reinforce their continuity.

In low-density areas, the corridor may be the continuous green edge between neighborhoods, providing long-distance walking and bicycle trails, other recreational amenities and a continuous natural habitat.

The corridor is a significant element of the New Urbanism because of its inherently civic nature. In the age of the metropolis, with villages, towns, neighborhoods and districts aggregated in unprecedented quantity, the most universally used public spaces are the corridors that serve connection and mobility. Of the three elements – the neighborhood, the district and the corridor – the latter, in its optimum form, is the most difficult to implement because it requires regional coordination.

CONCLUSION

The conventional suburban practice of segregating uses by zones is the legacy of the "dark satanic mills," which were once genuine hazards to public welfare. The separation of dwelling from workplace in the course of the last century was the great achievement of the nascent planning profession and remains institutionalized in zoning ordinances. The suburbs and cities of today continue to separate the naturally integrated human activities of dwelling, working, shopping, schooling, worshipping and recreating.

The hardship caused by this separation has been mitigated by widespread automobile ownership and use, which in turn has increased the demand for vehicular mobility. The priority given to road building at the expense of other civic programs during the last four decades has brought our country to the multiple crises of environmental degradation, economic bankruptcy and social disintegration.

The New Urbanism offers an alternative future for the building and re-building of regions. Neighborhoods that are compact, mixed-use and pedestrian friendly; districts of appropriate location and character; and corridors that are functional and beautiful can integrate natural environments and man-made communities into a sustainable whole.

"The Impact of the New Technologies and Globalization on Cities"

from Arie Graafland and Deborah Hauptmann (eds),
Cities in Transition (2001)

Saskia Sassen

Editors' Introduction

Saskia Sassen, a professor of Sociology at the University of Chicago, has carefully examined data on information technology, the economies, and the organization of physical space in the most advanced cities and metropolitan regions in the world. In the following selection she describes how globalization and information technology appear to be creating a new global system of cities, changing relationships among cities, and reconfiguring the physical arrangement of activities within metropolitan space.

Sassen is particularly interested in global *cities* – where international financial functions are disproportionately concentrated and whose economies are most closely integrated with the world economy. Sassen argues that globalization is both concentrating and simultaneously dispersing activity at the global, national, and metropolitan levels. At the global scale, economic power is increasingly *concentrated* in New York, London, and Tokyo. But cities as dispersed as Mexico City, Taipei, Bangkok, Buenos Aires, São Paulo, Frankfurt, Zurich, and Sydney are also global cities.

As economic activity is increasingly globalized and as industries need more specialized services, Sassen believes a new world "system of cities" is emerging unlike anything that has previously existed. What Sassen terms "corporate service complexes" – sophisticated networks of high-level financiers, lawyers, accountants, advertising professionals, and other skilled professionals serving international corporations – are clustered in global cities. Decisions made by joint headquarters/corporate services complexes in London, New York, Tokyo, and other global cities affect not only the neighborhoods in these cities, but jobs, wages, and the economic health of cities as dispersed around the globe as Kuala Lumpur, Malaysia, Ho Chi Minh City, Vietnam, and Santiago, Chile. If financial analysts advise their corporate clients that Argentina's economy is weakening and lawyers inform them that legal reforms in China present new opportunities for super profits, the corporations may pull billions of dollars out of Argentina and channel the funds to Shanghai – assisted by a small army of advertising executives touting their products in China's vast emerging markets.

A traditional reason why businesses cluster in large cities has been to be in touch with other businesses and with lawyers, accountants, bankers, advertising firms and other specialized service providers that help them do business. It was efficient to walk next door to a lawyer's office and down the block to do business with a major business partner. Cities had what economists call "agglomeration economies." Now in the information age as Manuel Castells (p. 475) and William Mitchell (p. 497) discuss, information can travel almost instantly to anywhere in the world. Businessmen no longer need to walk next door to communicate with their lawyers or down the street to meet with a business partner: they can just phone, fax, e-mail, or video conference next door or to a remote location anywhere in the world, so long as the telecommunications infrastructure permits. Highly developed telecommunications infrastructure in global cities facilitate transmission of information in staggering quantities at lightning speed. International banks in Rio de Janeiro can bounce a year's worth of financial records to a New York bank via satellite in seconds.

Anticipating these trends Melvin Webber (p. 470) more than three decades ago hypothesized that information technology would make space increasingly irrelevant and as a consequence cities would diminish in importance. But according to Sassen that has not happened. Sassen has found that global cities such as New York, London, and Tokyo have grown in population, wealth, and power since the information revolution. Their status and importance continues to grow, not decline. On the other hand, many cities which historically served as secondary corporate command and control centers are in economic decline as corporate power continues to concentrate in the most advanced global cities. Paris is growing in economic power and wealth; Marseilles is declining.

Sassen points out that production and retailing are becoming more dispersed. Many corporations design products at a headquarters location – perhaps (but not necessarily) in a global city like London. They contract with firms in a developing country like Malaysia to produce the products they have designed. And then they market the finished products in Germany. This kind of production process requires sophisticated support to manage dispersed and rapidly changing operations all over the world – lawyers familiar with British law, accountants who understand Malaysian accounting practices, and advertising executives sensitive to the cultural preferences of German consumers. Corporations rarely have the internal capacity to do all that. Instead they turn to networks of specialized firms located in global cities to provide the services they need.

Sassen questions the whole notion of "rich" countries and "rich" cities; places central to the world economy, and those that exist at the margin. She argues that economic inequality is sharply increasing – particularly in global cities at the center of the world economy. The opportunity for hyper-profits in international finance is creating extraordinary wealth for Wall Street bankers. But many of the low-paid janitors and file clerks who work on Wall Street were born in Third World countries and live in ethnic New York neighborhoods just a short distance away. In São Paulo, wealthy Brazilian nationals and expatriates earn salaries and participate in lifestyles more like those of wealthy New Yorkers than those of the people in São Paulo neighborhoods a few blocks from where they work. An important public policy question facing countries all over the world is how to promote economic equality and help more of their citizens benefit from the new world economic order. Michael Sorkin (p. 290) raises the important public policy question of whether investment to rebuild after the attack on the World Trade Center should be concentrated in the same area of New York City or dispersed to benefit more economically and culturally diverse areas of the city.

Regions are also changing as a result of globalization and information technology. Sassen argues that economic activity within the metropolitan areas of global cities manifests a dynamic of both concentration and dispersion just as is occurring in the world system of cities. At the time that Ernest W. Burgess developed his concentric zone theory (p. 156), Chicago and many other cities had a distinct central business district (CBD) where intense economic activity was concentrated. In advanced metropolitan areas today there are often no longer single, clearly demarcated CBDs. Sassen believes four different models are emerging. In some cities something close to a classic CBD still exists. New York City's Wall Street area is an example. In others there is a new pseudo-CBD just outside the historic city center. The massive planned office complex named La Défense just outside the center of Paris is the example Sassen cites. In other regions Sassen sees nodes of intense business activity emerging along "cyber routes" or "digital highways." Sassen points out that these spaces along which information flows often follow historic infrastructure for highways, high-speed rail lines, and airports. Twentieth-century infrastructure appears to be shaping the spatial organization of twenty-first-century cities. Sassen also discerns agglomeration and centralization across physical space and within cyber space – transterritorial centers of intense economic activity and centrality in electronically generated space.

The opportunity for what Sassen calls "super-profits" in global cities may change the entire pattern of land values in these areas most connected to the global economy. Boutiques and expensive restaurants may outbid community-serving local businesses.

Sassen's work is central to debates about the effects of global economic restructuring on cities. Some writers on information technology stress the decentralizing effects communications technology has had and will likely have in the future and disagree with Sassen's assessment that corporate command functions will be increasingly centralized. They believe that cities will decline in importance as a result of advances in information technology.

Saskia Sassen is the Ralph Lewis Professor of Sociology at the University of Chicago, and Centennial Visiting Professor at the London School of Economics. Her books include *Denationalization: Economy and Polity in a Global Digital Age* (Princeton, NJ: Princeton University Press, 2003), *Global Networks/Linked Cities* (New York and London: Routledge, 2002), *The Global City*, 2nd edn (Princeton, NJ: Princeton University Press, 2001), *Guests and Aliens* (New York: New Press, 1999), *Globalization and its Discontents* (New York: New Press, 1998) co-authored with Anthony Appiah, *Losing Control?: Sovereignty in an Age of Globalization* (New York: Columbia University Press, 1996), *Cities in a Global Economy* (Thousand Oaks, CA: Pine Forge Press, 1994), and *The Mobility of Labor and Capital: A Study in International Investment and Labor Flows* (New York: Cambridge University Press, 1988).

For an excellent set of readings on global cities see Neil Brenner and Roger Keil (eds), *The Global Cities Reader* (London and New York: Routledge, 2003), Stephen Graham (ed.), *The Cybercities Reader* (London and New York: Routledge, 2003), contains writings on the impact of information technology on cities.

Telecommunications and globalization have emerged as major forces shaping the organization of urban space. This reorganization ranges from the spatial virtualization of a growing number of social and economic activities to the reconfiguration of the geography of the built environment for these activities. Whether in electronic space or in the geography of the built environment, this reorganization involves a repositioning of the urban and of urban centrality in particular.

The growth of global markets for finance and specialized services, the need for transnational servicing networks due to sharp increases in international investment, the reduced role of the government in the regulation of international economic activity and the corresponding ascendance of other institutional arenas, notably global markets and corporate headquarters – all these point to the existence of a series of transnational networks of cities. We can see here the formation, at least incipient, of transnational urban systems. To a large extent it seems to me that the major business centers in the world today draw their importance from these transnational networks. There is no such thing as a single global city – and in this sense there is a sharp contrast with the erstwhile capitals of empires.

[. . .]

WORLDWIDE NETWORKS AND CENTRAL COMMAND FUNCTIONS

The geography of globalization contains both a dynamic of dispersal and of centralization, the latter a condition that began receiving recognition only recently. The massive trends towards the spatial dispersal of economic activities at the metropolitan, national and global level which we associate with globalization have contributed to a demand for new forms of territorial centralization of top-level management and control operations. The rapid growth of affiliates illustrates this dynamic. By 1998 firms had about half a million affiliates outside their home countries. The sheer number of dispersed factories and service outlets that are part of a firm's integrated operation creates massive new needs for central coordination and servicing. Thus the spatial dispersal of economic activity made possible by telecommunications and the new legal frameworks for globalization contributes to an expansion of central functions if this dispersal is to take place under the continuing concentration in control, ownership and profit appropriation that characterizes the current economic system.

Another instance today of this negotiation between a global cross-border dynamic and territorially specific sites is that of the global financial markets. The orders of magnitude in these transactions have risen sharply, as illustrated by the $75 U.S. trillion in turnover in the global capital market, a major component of the global economy. These transactions are partly embedded in telecommunications systems that make possible the instantaneous transmission of money/information around the globe. Much attention has gone to the capacity for instantaneous transmission of the new technologies. But the other half of the story is the extent to which the global financial markets are located in particular cities in the highly developed countries;

indeed, the degrees of concentration are unexpectedly high, a subject I discuss empirically in a later section.

Stock markets worldwide have become globally integrated. Besides deregulation in the 1980s in all the major European and North American markets, the late 1980s and early 1990s saw the addition of such markets as Buenos Aires, São Paulo, Mexico City, Bangkok, Taipei, etc. The integration of a growing number of stock markets has contributed to raise the capital that can be mobilized through stock markets. Worldwide market value reached over 20 trillion dollars in 1998 . . .

The specific forms assumed by globalization over the last decade have created particular organizational requirements. The emergence of global markets for finance and specialized services, the growth of investment as a major type of international transaction, all have contributed to the expansion in command functions and in the demand for specialized services for firms.

By central functions I do not only mean top-level headquarters; I am referring to all the top-level financial, legal, accounting, managerial, executive, planning functions necessary to run a corporate organization operating in more than one country, and increasingly in several countries. These central functions are partly embedded in headquarters, but also in good part in what has been called the corporate services complex, that is, the network of financial, legal, accounting, advertising firms that handle the complexities of operating in more than one national legal system, national accounting system, advertising culture, etc. and do so under conditions of rapid innovations in all these fields. Such services have become so specialized and complex, that headquarters increasingly buy them from specialized firms rather than producing them in-house. These agglomerations of firms producing central functions for the management and coordination of global economic systems, are disproportionately concentrated in the highly developed countries – particularly, though not exclusively, in the kinds of cities I call global cities . . . Such concentrations of functions represent a strategic factor in the organization of the global economy, and they are situated right here, in New York, in Paris, in Amsterdam.

[. . .]

NEW FORMS OF CENTRALITY

Today there is no longer a simple straightforward relation between centrality and such geographic entities as the downtown, or the central business district. In the past, and up to quite recently in fact, the center was synonymous with the downtown or the CBD. Today, the spatial correlate of the center can assume several geographic forms. It can be the CBD, as it still is largely in New York City, or it can extend into a metropolitan area in the form of a grid of nodes of intense business activity, as we see in Frankfurt and Zurich. The center has been profoundly altered by telecommunications and the growth of a global economy, both inextricably linked; they have contributed to a new geography of centrality (and marginality). Simplifying, one could identify four forms assumed by centrality today.

First, while there is no longer a simple straightforward relation between centrality and such geographic entities as the downtown, or the central business district as was the case in the past, the CBD remains a key form of centrality. But the CBD in major international business centers is one profoundly reconfigured by technological and economic change.

We may be seeing a difference in the pattern of global city formation in parts of the United States and in parts of Western Europe. In the United States, major cities such as New York and Chicago have large centers that have been rebuilt many times, given the brutal neglect suffered by much urban infrastructure and the imposed obsolescence so characteristic of US cities. This neglect and accelerated obsolescence produce vast spaces for rebuilding the center according to the requirements of whatever regime of urban accumulation or pattern of spatial organization of the urban economy prevails at a given time. In Europe, urban centers are far more protected and they rarely contain significant stretches of abandoned space; the expansion of workplaces and the need for intelligent buildings necessarily will have to take place partly outside the old centers. One of the most extreme cases is the complex of La Défense, the massive, state-of-the-art office complex developed right outside Paris to avoid harming the built environment inside the city. This is an explicit instance of government policy and planning aimed at addressing the growing demand for central office space of prime quality. Yet another variant of this expansion of the "center" onto hitherto peripheral land can be seen in London's Docklands.

Similar projects for recentralizing peripheral areas were launched in several major cities in Europe, North America, and Japan during the 1980s.

Second, the center can extend into a metropolitan area in the form of a grid of nodes of intense business activity. One might ask whether a spatial organization characterized by dense strategic nodes spread over a broader region does or does not constitute a new form of organizing the territory of the "center," rather than, as in the more conventional view, an instance of suburbanization or geographic dispersal. Insofar as these various nodes are articulated through cyber-routes or digital highways, they represent a new geographic correlate of the most advanced type of "center." The places that fall outside this new grid of digital highways, however, are peripheralized. This regional grid of nodes represents, in my analysis, a reconstitution of the concept of region. Far from neutralizing geography the regional grid is likely to be embedded in conventional forms of communications infrastructure, notably rapid rail and highways connecting to airports. Ironically perhaps, conventional infrastructure is likely to maximize the economic benefits derived from telematics. I think this is an important issue that has been lost somewhat in discussions about the neutralization of geography through telematics.

Third, we are seeing the formation of a transterritorial "center" constituted via telematics and intense economic transactions . . . The most powerful of these new geographies of centrality at the inter-urban level binds the major international financial and business centers: New York, London, Tokyo, Paris, Frankfurt, Zurich, Amsterdam, Los Angeles, Sydney, Hong Kong, among others. But this geography now also includes cities such as São Paulo and Mexico City. The intensity of transactions among these cities, particularly through the financial markets, trade in services, and investment, has increased sharply, and so have the orders of magnitude involved. At the same time, there has been a sharpening inequality in the concentration of strategic resources and activities between each of these cities and others in the same country. For instance, Paris now concentrates a larger share of leading economic sectors and wealth in France than it did fifteen years ago, while Marseilles, once a major economic hub, has lost its share and is suffering severe decline.

[. . .]

Fourth, new forms of centrality are being constituted in electronically generated spaces. Electronic space is often read as a purely technological event and in that sense a space of innocence. But if we consider for instance that strategic components of the financial industry operate in such space we can see that these are spaces where profits are produced and power is thereby constituted. Insofar as these technologies strengthen the profit-making capability of finance and make possible the hyper-mobility of finance capital, they also contribute to the often devastating impacts of the ascendance of finance on other industries, on particular sectors of the population, and on whole economies. Cyberspace, like any other space, can be inscribed in a multiplicity of ways, some benevolent or enlightening; others, not. My argument is that structures for economic power are being built in electronic space and that their highly complex configurations contain points of coordination and centralization.

[. . .]

A CONCENTRATION AND THE REDEFINITION OF THE CENTER: SOME EMPIRICAL REFERENTS

The trend towards concentration of top-level management, coordination and servicing functions is evident at the national and international scales in all highly developed countries. For instance, the Paris region accounts for over 40 percent of all producer services in France, and over 80 percent of the most advanced ones. New York City is estimated to account for between a fourth and a fifth of all US producer services exports though it has only 3 percent of the US population. London accounts for 40 percent of all exports of producer services in the UK. Similar trends are also evident in Zurich, Frankfurt, and Tokyo, all located in much smaller countries.

[. . .]

In the financial district in Manhattan, the use of advanced information and telecommunication technologies has had a strong impact on the spatial organization of the district because of the added spatial requirements of "intelligent" buildings. A ring of new office buildings meeting these requirements was built over the last decade immediately around the old Wall Street core, where the narrow streets and lots made this difficult; furthermore, renovating old buildings in the Wall Street core is extremely expensive and often not possible. The new buildings in the district were mostly corporate headquarters and financial services industry facilities. These firms tend to be extremely intensive users of telematics, and the availability of the most advanced forms typically is a major factor in their real estate and locational decisions. They need complete redundancy of telecommunications systems, high carrying capacity, often their own private branch exchange, etc. With this often goes a need for large spaces. For instance, the technical installations backing a firm's trading floor are likely to require additional space equivalent to the size of the trading floor itself.

The case of Sydney illuminates the interaction of a vast, continental economic scale and pressures towards spatial concentration. Rather than strengthening the multipolarity of the Australian urban system, the developments of the 1980s – increased internationalization of the Australian economy, sharp increases in foreign investment, a strong shift towards finance, real estate and producer services – contributed to a greater concentration of major economic activities and actors in Sydney. This included a loss of share of such activities and actors by Melbourne, long the center of commercial activity and wealth in Australia.

[. . .]

THE INTERSECTION OF SERVICE INTENSITY AND GLOBALIZATION

To understand the new or sharply expanded role of a particular kind of city in the world economy since the early 1980s, we need to focus on the intersection of two major processes. The first is the sharp growth in the globalization of economic activity; this has raised the scale and the complexity of transactions, thereby feeding the growth of top-level multinational headquarter functions and the growth of advanced corporate services. It is important to note that even though globalization raises the scale and complexity of these operations, they are also evident at smaller geographic scales and lower orders of complexity, as is the case with firms that operate regionally. Thus while regionally oriented firms need not negotiate the complexities of international borders and the regulations of different countries, they are still faced with a regionally dispersed network of operations that requires centralized control and servicing.

The second process we need to consider is the growing service intensity in the organization of all industries. This has contributed to a massive growth in the demand for services by firms in all industries, from mining and manufacturing to finance and consumer services. Cities are key sites for the production of services for firms. Hence the increase in service intensity in the organization of all industries has had a significant growth effect on cities in the 1980s. It is important to recognize that this growth in services for firms is evident in cities at different levels of a nation's urban system. Some of these cities cater to regional or subnational markets; others cater to national markets and yet others cater to global markets. In this context, globalization becomes a question of scale and added complexity.

The key process from the perspective of the urban economy is the growing demand for services by firms in all industries and the fact that cities are preferred production sites for such services, whether at the global, national, or regional level. As a result we see in cities the formation of a new urban economic core of banking and service activities that comes to replace the older typically manufacturing-oriented core.

In the case of cities that are major international business centers, the scale, power, and profit levels of this new core suggest that we are seeing the formation of a new urban economy. This is so in at least two regards. First, even though these cities have long been centers for business and finance, since the late 1970s there have been dramatic changes in the structure of the business and financial sectors, as well as sharp increases in the overall magnitude of these sectors and their weight in the urban economy. Second, the ascendance of the new finance and services complex, particularly international finance, engenders what may be regarded as a new economic regime, that is, although this sector may account for only a fraction of the economy of a city, it imposes itself on that larger economy. Most notably, the possibility for super-profits in finance has the effect of devalorizing manufacturing insofar as the latter cannot generate the super-profits typical in much financial activity. This is not to say that everything in the economy of these cities has changed. On the contrary, they still show a great deal of continuity and many similarities with cities that are not global nodes. Rather, the implantation of global processes and markets has meant that the internationalized sector of the economy has expanded sharply and has imposed a new valorization dynamic – that is, a new set of criteria for valuing or pricing various economic activities and outcomes. This has had devastating effects on large sectors of the urban economy. High prices and profit levels in the internationalized sector and its ancillary activities, such as top-of-the-line restaurants and hotels, have made it increasingly difficult for other sectors to compete for space and investments. Many of these other sectors have experienced considerable downgrading and/or displacement, as, for example, neighborhood shops tailored to local needs are replaced by upscale boutiques and restaurants catering to new high-income urban elites.

Though at a different order of magnitude, these trends also became evident during the late 1980s in a number of major cities in the developing world that have become integrated into various world markets: São Paulo, Buenos Aires, Bangkok, Taipei, and Mexico City are only a few examples. Also here the new urban core was fed by the deregulation of financial markets, ascendance of finance and specialized services, and integration into the world markets. The opening of stock markets to foreign investors and the privatization of what were once public sector firms have been crucial institutional arenas for this articulation. Given the vast size of some of these cities, the impact of this new core on the broader city is not always as evident as in central London or Frankfurt, but the transformation is still very real.

[. . .]

The formation of a new production complex

According to standard conceptions about information industries, the rapid growth and disproportionate concentration of producer services in central cities should not have happened. Because they are thoroughly embedded in the most advanced information technologies, producer services could be expected to have locational options that bypass the high costs and congestion typical of major cities. But cities offer agglomeration economies and highly innovative environments. The growing complexity, diversity, and specialization of the services required have contributed to the economic viability of a freestanding specialized service sector.

The production process in these services benefits from proximity to other specialized services. This is especially the case in the leading and most innovative sectors of these industries. Complexity and innovation often require multiple highly specialized inputs from several industries. The production of a financial instrument, for example, requires inputs from accounting, advertising, legal expertise, economic consulting, public relations, designers, and printers. The particular characteristics of production of these services, especially those involved in complex and innovative operations, explain their pronounced concentration in major cities. The commonly heard explanation that high-level professionals require face-to-face interactions needs to be refined in several ways. Producer services, unlike other types of services, are not necessarily dependent on spatial proximity to the consumers, i.e. firms, served. Rather, economies occur in such specialized firms when they locate close to others that produce key inputs or whose proximity makes possible joint production of certain service offerings. The accounting firm can service its clients at a distance, but the nature of its service depends on proximity to specialists, lawyers, programmers. Moreover, concentration arises out of the needs and expectations of the people likely to be employed in these new high-skill jobs, who tend to be attracted to the amenities and lifestyles that large urban centers can offer. Frequently, what is thought of as face-to-face communication is actually a production process that requires multiple simultaneous inputs and feedbacks. At the current stage of technical development, immediate and simultaneous access to the pertinent experts is still the most effective way, especially when dealing with a highly complex product. The concentration of the most advanced telecommunications and computer network facilities in major cities is a key factor in what I refer to as the production process of these industries.

[. . .]

This combination of constraints suggests that the agglomeration of producer services in major cities actually constitutes a production complex. This producer services complex is intimately connected to the world of corporate headquarters; they are often thought of as forming a joint headquarters–corporate services complex. But in my reading, we need to distinguish the two. Although it is true that headquarters still tend to be disproportionately concentrated in cities, over the last two decades many have moved out. Headquarters can indeed locate outside cities, but they need a producer services complex somewhere in order to buy or contract for the needed specialized services and financing. Further, headquarters of firms with very high overseas activity or in highly innovative and complex lines of business tend to locate in major cities. In brief, firms in more routinized lines of activity, with predominantly regional or national markets, appear to be increasingly free to move or install their headquarters outside cities. Firms in highly competitive and innovative lines of activity and/or with a strong world market orientation appear to benefit from being located at the center of major international business centers, no matter how high the costs.

[. . .]

The region in the global information age

The massive use of telematics in the economy and the corresponding possibility for geographic dispersal and mobility of firms suggest that the whole notion of regional specialization and of the region may become obsolete. But there are indications that, as is the case for large cities, so also for regions the hypermobility of information industries and the heightened capacity for geographic dispersal may be only part of the story. The evidence on regional specialization in the US and in other highly developed countries along with new insights into the actual work involved in producing these services point to a different set of outcomes.

What is important from the perspective of the region is that the existence of, for instance, a producer services complex in the major city or cities in a region creates a vast concentration of communications infrastructure which can be of great use to other economic nodes in that region. Such nodes can (and do) connect with the major city or cities in a region and thereby to a worldwide network of firms and markets. The issue from the regional perspective is, then, that somewhere in its territory the region connects with state-of-the-art communication facilities which connect it with the world and which brings foreign firms from all over the world to the region. Given a regional grid of economic nodes, the benefits of this concentration in the major city or cities are no longer confined only to firms located in those cities.

Secondly, given the nature of the production process in advanced information industries, as described in the preceding section, the geographic dispersal of activities has limits. The importance of actual face-to-face transactions means that a metropolitan or regional network of firms will need conventional communications infrastructure, e.g. highways or rapid rail, and locations not farther than something like two hours. One of the ironies of the new information technologies

is that to maximize their use we need access to conventional infrastructure. In the case of international networks it takes airports and planes; and in the case of metropolitan or regional networks, trains and cars.

The importance of conventional infrastructure in the operation of economic sectors that are heavy users of telematics has not received sufficient attention. The dominant notion seems to be that telematics obliterates the need for conventional infrastructure. But it is precisely the nature of the production process in advanced industries, whether they operate globally or nationally, which contributes to explain the immense rise in business travel we have seen in all advanced economies over the last decade, the new electronic era. The virtual office is a far more limited option than a purely technological analysis would suggest. Certain types of economic activities can be run from a virtual office located anywhere. But for work processes requiring multiple specialized inputs, considerable innovation and risk taking, the need for direct interaction with other firms and specialists remains a key locational factor. Hence the metropolitanization and regionalization of an economic sector has boundaries that are set by the time it takes for a reasonable commute to the major city or cities in the region. The irony of today's electronic era is that the older notion of the region and older forms of infrastructure re-emerge as critical for key economic sectors. This type of region in many ways diverges from older forms of region. It corresponds rather to the second form of centrality posited above in this paper – a metropolitan grid of nodes connected via telematics. But for this digital grid to work, conventional infrastructure – ideally of the most advanced kind – is also a necessity.

[. . .]

THE INTERSECTION BETWEEN ACTUAL AND DIGITAL SPACE

There is a new topography of economic activity sharply evident in this subeconomy. This topography weaves in and out between actual and digital space. There is today no fully virtualized firm or economic sector. Even finance, the most digitalized, dematerialized and globalized of all activities, has a topography that weaves back and forth between actual and digital space. To different extents in different types of sectors and different types of firms, a firm's tasks now are distributed across these two kinds of spaces; further the actual configurations are subject to considerable transformation as tasks are computerized or standardized, markets are further globalized, etc. More generally, telematics and globalization have emerged as fundamental forces reshaping the organization of economic space.

The question I have for architects here is whether the point of intersection between these two kinds of spaces in a firm's or a dynamic's topography of activity, is one worth thinking about, theorizing, exploring. This intersection is unwittingly, perhaps, thought of as a line that divides two mutually exclusive zones. I would propose, again, to open up this line into an "analytic borderland" which demands its own empirical specification and theorization, and contains its own possibilities for architecture. The space of the computer screen, which one might posit as one version of the intersection, will not do, or is at most a partial enactment of this intersection.

What does contextuality mean in this setting? A networked subeconomy that operates partly in actual space and partly in globe-spanning digital space cannot easily be contextualized in terms of its surroundings. Nor can the individual firms. The orientation is simultaneously towards itself and towards the global. The intensity of its internal transactions is such that it overrides all considerations of the broader locality or region within which it exists. On another, larger scale, in my research on global cities I found rather clearly that these cities develop a stronger orientation towards the global markets than to their hinterlands. Thereby they override a key proposition in the urban systems literature, to wit, that cities and urban systems integrate, articulate national territory. This may have been the case during the period when mass manufacturing and mass consumption were the dominant growth machines in developed economies and thrived on the possibility of a national scale. But it is not today with the ascendance of digitalized, globalized, dematerialized sectors such as finance. The connections with other zones and sectors in its "context" are of a special sort – one that connects worlds that we think of as radically distinct. For instance, the informal economy in several immigrant communities in New York provides some of the low-wage workers for the "other" jobs on Wall Street, the capital of global finance. The same is happening in Paris, London, Frankfurt, Zurich . . .

[. . .]

CONCLUSION

Economic globalization and telecommunications have contributed to produce a spatiality for the urban that pivots on cross-border networks and territorial locations with massive concentrations of resources. This is not a completely new feature. Over the centuries cities have been at the crossroads of major, often worldwide, processes. What is different today is the intensity, complexity and global span of these networks, the extent to which significant portions of economies are now dematerialized and digitalized and hence the extent to which they can travel at great speeds through some of these networks, and, thirdly, the numbers of cities that are part of cross-border networks operating at vast geographic scales.

The new urban spatiality thus produced is partial in a double sense: it accounts for only part of what happens in cities and what cities are about, and it inhabits only part of what we might think of as the space administrative boundaries or in the sense of a city's public imaginary. What stands out, however, is the extent to which the city remains an integral part of these new configurations.

PART FOUR

Urban politics, governance, and economics

"WHO STOLE THE PEOPLE'S MONEY?" — DO TELL . N.Y.TIMES.   'TWAS HIM.

INTRODUCTION TO PART FOUR

The material on urban space in Part Three and on urban society and culture in Part Two raises important issues about how cities should be governed and the economy of cities. The conflicts related to gated communities and sprinklers that douse the homeless that Mike Davis discusses (p. 201) pose political questions: what *should* government, particularly local government, do when different groups want to use urban space in different ways? The selection by William Julius Wilson (p. 126) in Part Two illustrates how sociological issues of race and class are intimately related to both economic questions and questions of urban *politics and governance*. Part Four focuses on these questions of urban politics, governance, and economics, beginning with geographer David Harvey's selection on contested cities, social process, and spatial form.

Harvey's selection nicely bridges the material in Part Three on urban space and the material in this section on urban politics, governance, and economics. While Harvey is a geographer the issues he discusses are essentially political ones. Harvey sees cities a centers of conflict based on ideology, race, gender, and interests. Civil rights groups, feminists, Republicans, environmentalists, businessmen, and punks have different interests and values. The land that businessmen want for a new office building may be the exact same land that environmentalists feel should be a park or permanent open space. This inevitably leads to conflict among different groups in cities. The processes used to resolve these kinds of conflicts are political processes. Harvey is sympathetic to the "militant particularism" of some groups who focus on their specific interest or locality, but feels narrow values are more effective if they are generalized.

Many of the best writings about urban politics are by political scientists like John Mollenkopf (p. 235). Urban politics is a distinct subfield within the social science discipline of political science. Political sociology is a subfield within sociology. In the 1950s and 1960s, seminal work by sociologist Floyd Hunter and political scientist Robert Dahl began a major debate between two schools of thought. In his book *Community Power Structure* (1953), based on research in Atlanta, Georgia, Hunter concluded that in Atlanta – and by implication other cities – a small, interlocking *elite* consisting of key businessmen and members of established and socially prominent families made all the really important decisions about Atlanta, including the governmental decisions. They sat on the boards of each other's corporations, intermarried, chatted at the same social clubs, and ran things in their interest. Because Hunter concluded that a small elite ruled Atlanta, his model of urban community power is referred to as the *elitist* model of community power. In *Who Governs?* (1961) Dahl reached almost diametrically opposite conclusions. Dahl concluded that local political power in New Haven, Connecticut – and by implication other cities – was fragmented. Dahl reported that many different people from a variety of walks of life were involved in decision-making by the New Haven city government and many people influenced the outcome of different political decisions. As a result, Dahl advanced a *pluralist model* of urban community power. Dahl and Hunter's competing models stimulated debate between elitists and pluralists and the further elaboration of theories of urban politics. Some theorists critical of pluralist interpretations of community power focused on structural features of global capitalism. Marxists in particular feel that the global capitalist system really determines what city governments can and cannot do. Structuralists argue that if bankers in New York and London decide to pull investment out of Buenos Aires, Argentina, the city government in Buenos Aires cannot stop them. And if billions of dollars flow out of Buenos Aires, the city government cannot build roads or schools or carry out other parts of their political agenda.

John Mollenkopf, a political scientist at the Graduate School of the City University of New York, reviews the current status of the debate between elitists and pluralists in the first selection in this section: "How to Study Urban Political Power." Mollenkopf summarizes theories of urban politics. He notes that pluralist explanations were widely accepted in the 1960s and early 1970s. Elitist and structuralist explanations of urban political power gained wider acceptance during the 1970s and 1980s – particularly among Marxists. Extreme structuralists even questioned whether the study of urban politics itself makes sense. They felt that urban outcomes are so largely determined by economic and social structures that urban politics is irrelevant. Many political scientists today draw on both elitist and pluralist theory. They accept some of what structualists say, but within bounds. While they see urban politics as influenced by economic and societal forces that cities cannot control, balanced theorists argue that local elected officials and political interest groups can shape their cities' futures.

Mollenkopf rejects the extreme structuralist interpretation that urban politics is irrelevant. He borrows from both elitist and pluralist theory. His own answer to the question of how to study urban political power is synthetic and subtle. It acknowledges the importance of structural constraints, but notes that the real world of urban politics involves many conflicting forces and leaves plenty of room for politics. Mollenkopf stresses the connection between formal political structures and processes, corporate decision-making, and the exercise of power by many other institutions such as churches, unions, professional associations, and neighborhood groups that may form alliances with elected officials in order to get things done. Mollenkopf marshals evidence that there is broad participation in local government, and he believes that many participants do influence decisions.

During the 1960s Sherry Arnstein was the chief advisor on citizen participation in the US Department of Housing and Urban Development's Model Cities Program – a federal government program that provided billions of dollars to help lower-income city neighborhoods develop and implement physical and social programs. Arnstein begins her selection by noting that "the idea of citizen participation is a little like eating spinach: no one is against it in principle because it is good for you." But whether or not city dwellers participate *effectively* in government programs that affect them and their neighborhoods has varied greatly.

A little background will help clarify Arnstein's selection. Many urban renewal programs in the United States during the 1950s and 1960s were intended to and did remove low-income and minority residents and replace their homes and community institutions with office buildings, luxury housing, garages, and other developments totally unrelated to the former residents. The funky, but communal, Italian neighborhood in Boston's West End that Jane Jacobs lovingly describes (p. 114) was leveled by Boston's West End urban renewal project. The "street ballet" she describes has been replaced by yuppies going to Starbucks coffee shops and smart boutiques. In the urban renewal program invitations to neighborhood residents to help decide what the urban renewal project should be like were usually a sham. In sharp contrast the US "War on Poverty" in the late 1960s emphasized "maximum feasible participation of the poor." Many decisions about how to use federal anti-poverty funds were made by residents themselves. Many local poverty programs lacked the capacity to manage programs or dissolved into internal feuding. Cutting local elected officials and established agencies out of the loop to design programs and manage anti-poverty funds created a major political backlash. The Model Cities program, where Arnstein worked, sought to strike a balance between the top-down urban renewal approach in which there was no real citizen participation in decisions and the War on Poverty model in which neighborhood groups often fought among themselves, wasted money, and accomplished little.

Arnstein asks readers to imagine a "ladder" with different "rungs" of citizen participation from lowest to highest. The rungs of the ladder range from non-participation (*manipulation* and *therapy*) at the bottom of the ladder to *citizen control* at the top. While Arnstein herself favors citizen control as the ultimate objective of urban programs, she acknowledges all but the bottom two rungs of her ladder as useful to varying degrees. In the decades since Arnstein's classic article was written there has been a succession of urban development programs in the United States and Western European countries, most recently the UK's Enterprize Zone program and the US Empowerment Zones/Enterprise Communities program. Each of these programs calls for some degree of citizen participation. Arnstein's ladder is useful in understanding how to create meaningful citizen participation in these urban programs. It also helps in understanding the theoretical concept of pluralism as Mollenkopf (p. 235) discusses it at the grassroots level.

Anthony Downs, a policy analyst at the Brookings Institution – a Washington, DC think-tank – focuses on a major problem with the way in which local government within metropolitan regions is organized. While Downs concentrates on local government in the United States, his critique and proposed solutions are relevant for metropolitan regions everywhere in the world. The problem is that government structures that were invented to govern small, separate places before the massive population increase of the recent past are not adequate spatial and political units to manage huge, interconnected metropolitan regions. Most of the 101 local governments in the San Francisco Bay Area, for example, were incorporated before World War II when the population of the region was less than half what it is today. Many of the local governments were no more than villages separated from each other. Little local governments like these are adequate to decide how to patch potholes in the local street or how much of the city budget should go to the local beauty queen pageant. They are poorly equipped to handle complex regional decisions about air and water quality, traffic congestion, and the preservation of endangered species. Dividing up the region into 101 little governments – each protective of its own interests – makes it hard to get low-income housing built or assure that immigrants and low-income people have access to housing, schools, and jobs throughout the region. The same kind of fragmentation of local government authority and obsolete spatial organization of local government is common in many metropolitan areas of Europe, Asia, Africa, and Latin America. Downs's selection argues strongly for *regional* solutions to metropolitan problems. It proposes a new vision that eliminates the legal/governmental distinctions between impacted inner-city ghettos and comfortable suburban communities. It argues for integrated metropolitan solutions to traffic congestion and other problems that cannot be solved on a city-by-city basis.

One of the most important – and certainly among the most sensitive – local government functions everywhere in the world is police work. In "Broken Windows" (p. 267), James Q. Wilson and George L. Kelling propose a theory about *how* crime comes to dominate declining neighborhoods and what to do about it. Wilson and Kelling emphasize the importance of citizen perceptions of crime. They argue that whether or not crime has actually increased in a neighborhood, if residents think crime has increased, they will become more reclusive and less involved in the community. And that in turn will open the door to real crime. A single broken window may be a trivial problem, Wilson and Kelling argue, but if it is not fixed the signal that no one cares enough to fix it will lead to more broken windows and then to drug dealing and violent crime. The remedy Wilson and Kelling suggest – which has been widely adopted since this article was written – is community policing where highly visible police regularly patrol marginal neighborhoods on foot. In community policing, officers are assigned to a neighborhood for a long enough time to learn about the neighborhood and its residents. They will come to understand community norms and values. Wilson and Kelling go on to advance the controversial view that these community police should informally enforce community norms of appropriate civil behavior as the neighborhood itself defines them, even if that calls for extralegal or perhaps illegal controls. Wilson and Kelling note that a transitional neighborhood may be inhabited both by "neighborhood regulars" and by strangers. The regulars in turn consist of what Wilson and Kelling term "decent folk" and derelicts and drunks who are not so decent but "know their place." So long as questionable street behavior stays within neighborhood-defined norms, community police will look the other way. Police will leave the well-known and harmless local drunk alone even if he is technically breaking the law. But if rowdy teenagers, prostitutes, dope dealers, or other strangers violate community norms – regardless of whether they violate the law – community police will intervene, perhaps by ordering the strangers to leave even if there is no legal basis for such an order.

The kinds of neighborhoods in transition that Wilson and Kelling discuss and the residents they contain are products of a changing world economy. The world economy has become global and the fate of individual cities is now linked to world trends as never before. In this new global economy there has been a massive economic restructuring that affects city neighborhoods.

As Saskia Sassen (p. 212) discusses, two of the most significant trends today are globalization and the growth of high-tech, information management-oriented advanced service jobs. These trends create problems of job loss and unemployment among some sectors in some cities and boomtowns elsewhere. Both types of cities imply dramatic changes in the wealth and class structure of cities. The selection in this section by Harvard business professor Michael Porter (p. 277) approaches the issue of urban inequality and what to do about it from a private sector perspective. Porter believes that the economic and social health of inner cities depends upon economic development by the private sector. To succeed, he argues, economic development must be based on the economic self-interest of private firms rather than phony businesses propped up by government subsidies and preference programs. The key to success, according to Porter, lies in capitalizing on the competitive advantage which inner-city neighborhoods possess: strategic location, local market demand, their capacity to integrate with regional business clusters, and human resources. Porter urges an economic, not a social, model for development. He counsels corporations to shift their philanthropic priorities away from providing social services, such as daycare or food for the homeless, to providing managerial expertise to develop neighborhood economies.

The final selection in this section by architect Michael Sorkin examines one of the most emotionally charged decisions facing any city in the world – how New York City should rebuild after the attack on the World Trade Center. Sorkin argues that rebuilding should not be directed just at the World Trade Center site itself, but should be dispersed to strengthen neighborhoods throughout the city. Sorkin's selection fits nicely with Harvey's admonition to consider process as well as what is built. Arnstein's ladder of citizen participation is highly relevant to understanding the process that will unfold. Decisions this important require a decision-making process that meaningfully involves citizens – placing them at the high rungs of her ladder. As reinvestment flows into New York City, the residents will have to grapple with whether to pursue a social model or Porter's hard-nosed private sector-oriented economic model. Whether reinvestment will be concentrated at the World Trade Center site itself, as some members of the business community advocate, dispersed to create a "village structure" with a "network of autonomous, comprehensible places," as Sorkin advocates, or used throughout the region to implement Downs's new vision for metropolis, raises fundamental questions of urban politics, governance, and economics.

The varied selections in this section demonstrate that both economics and politics matter a great deal to cities. Cities are affected by the forces of global economic restructuring which they often barely comprehend and cannot control. Globalization and information technology are massively restructuring cities and metropolitan regions. The role of local government in the new world order is changing. The writers in this section point towards a new global economy that requires new ways of thinking about urban politics, governance, and economics.

"Contested Cities: Social Process and Spatial Form"

from Nick Jewson and Susanne MacGregor (eds),
Transforming Cities (1997)

David Harvey

Editors' Introduction

David Harvey was trained as a geographer and teaches in the Geography Department of Johns Hopkins University, but his prolific and influential writings over the last three decades bridge the concerns of all of the proceeding sections of *The City Reader* on urban history, society, culture, and space as well as the selections in this section on urban politics. Harvey writes theoretically at a high level of abstraction. He skillfully blends Marxist and postmodernist approaches to understanding cities. Harvey challenges fundamental assumptions about history, epistemology, social justice, community, ecology, and the physical organization of cities.

Urban politics is a distinct subfield within the discipline of political science and much good writing on urban politics and governance comes from political scientists like John Mollenkopf (p. 235). But questions of politics and governance also interest geographers like David Harvey, policy analysts like Anthony Downs (p. 256), architects like Michael Sorkin (p. 290), social workers like Sherry Arnstein (p. 244), and people from other academic disciplines and professional fields.

Harvey starts this selection by reminding us of the massive urbanization of the human population which Kingsley Davis describes (p. 25). Cities are critical to understanding the current human condition. Yet, Harvey notes, cities as a category are strikingly absent from many discussions of modernization, modernity, post-modernity, and capitalist and industrial society. He would like to see cities included in discussions of these topics.

Harvey emphasizes the importance of thinking about cities in terms of *processes* rather than just *things*. For him cities are sites of conflict – based on race, ideology, gender, and other values. Harvey thinks dialectically. He argues that processes are both *shaped by* time and place and *shape* time and place. For example, when communist Chinese leader Mao Tse Tung decided to send twenty million urban Chinese intellectuals out of Shanghai, Beijing, Guangzhou, and other large Chinese cities into the countryside during the great proletarian cultural revolution, this was a social process shaped by a specific place (China) at a specific time (during the Chinese communist revolution). For a while this social process produced a distinct kind of chaotic and austere revolutionary city. Historical sites and buildings with European architecture were destroyed; massive public works projects, bomb shelters, and monuments to revolutionary heroes were built. The cultural revolution effectively ended with Mao's death in 1976. Some of the urban things built (and destroyed) during the great proletarian cultural revolution period are still evident in Chinese cities, but Chinese society and politics are remarkably fluid and much has changed.

Some processes lead to very durable outcomes. The decision to build the Three Mile Island nuclear power plant, dam the Yangtze River, or rebuild the World Trade Center site produce physical forms and social structures that last for a very long time. What looks like a good solution to urban development to one generation, such as a massive highway construction program to solve transportation problems, building identical suburban tract housing to deal with the housing crisis, or creating a nuclear power grid to meet urban energy needs, may not look so good to the next generation. This is why Harvey urges us to think how flexible or adjustable cities can be and to develop fluid social processes that can change over time.

Two additional themes in Harvey's selection deal with "community" and the relationship between the natural and built environment. Louis Wirth (p. 97), Robert Putnam (p. 105), Ali Madanipour (p. 181), Jane Jacobs (p. 114), Peter Calthorpe and William Fulton (p. 331), Andres Duany and Elizabeth Plater-Zyberk (p. 207), and many others are concerned that cities alienate people from each other and break down community. They all favor agendas to build greater community in cities. Harvey raises some important questions about community building. He argues that many community-building projects are recipes for isolation. The positive identification of some groups is often achieved by first defining other groups as the other, the devalued, semi-human. For example, people within a gated community in Los Angeles of the kind that Mike Davis describes (p. 201) may become a "community" that play golf together and have drinks at their country club safe within the gates. They will not be alienated and isolated from each other. But this kind of community is based on walling themselves off from lower-income people and people of color outside the gates of their artificial community.

Harvey criticizes the belief that the proper design of things will solve social process problems. For him, this is an example of a thing taking precedence over a process. Rather, Harvey argues, the social processes underlying even the best designed community-enhancing place need to be cultivated and sustained. Just building a remarkable *physical* community like Ebenezer Howard's garden city of Letchworth, England or Andres Duany and Elizabeth Plater-Zyberk's neo-traditionalist new town of Seaside, Florida, will not, Harvey argues, create community.

Harvey notes that community activism is often built around "militant particularisms" in which a group coheres around a value, like social justice or environmental conservation, related to a very narrow and very local concern. For example, groups may focus on getting higher wages for low-paid garment workers at the Ajax garment factory or saving an endangered salamander in the Bolinas lagoon. These militant particularisms may contribute to society, but Harvey argues that they will be much more useful if they spill over into wider, more universal constructions concerns; for example, if militant environmentalists generalize their concern for one breed of salamander into a more universal (and perhaps more balanced) environmental movement or the militant garment workers generalize their concerns into a broad movement for all workers to be paid a living wage.

Harvey critiques the artificial distinction some environmentalists draw between the natural and the built environments. He agrees with Stephen Wheeler (p. 486) and Timothy Beatley (p. 399) that the natural and built environments need to evolve together. The sustainable urban development, green urbanism, design with nature, and ecological design movements all follow this notion of harmony between the natural and built environments.

David Harvey is Professor of Geography and Environmental Engineering at Johns Hopkins University; Senior Research Fellow, St Peter's College, University of Oxford; and Miliband Visiting Fellow, London School of Economics and Political Science. From 1987 to 1993 he was Halford McKinder Professor of Geography at the University of Oxford. He has received a Guggenheim Fellowship (1976), the Gill Memorial Prize of the Royal Geographical Society in London (1972), and the Outstanding Contributor Award of the Association of American Geographers (1980).

This selection is from Nick Jewson and Susanne MacGregor (eds), *Transforming Cities* (London and New York: Routledge, 1997). Books by Harvey include *Spaces of Capital* (London and New York: Routledge, 2002), *Spaces of Hope* (Berkeley: University of California Press, 2000), *Limits to Capital* (London: Verso, 1999), *Justice, Nature, and the Geography of Difference* (Oxford: Blackwell, 1996), *The Condition of Postmodernity: An Enquiry into the Origins of Cultural Change* (Oxford: Blackwell, 1989), *The Urban Experience* (Malden, MA: Blackwell, 1990), *The Urbanization of Capital: Studies in the History and Theory of Capitalist Urbanization* (Baltimore, MD: Johns Hopkins University Press, 1985), *Consciousness and the Urban Experience: Studies in the History and Theory of Capitalist Urbanization* (Baltimore, MD: Johns Hopkins University Press, 1985), *The Limits to Capital* (Oxford: Blackwell, 1982), *Social Justice and the City* (Oxford: Basil Blackwell, 1973, new edition 1988), and *Explanation in Geography* (London: Edward Arnold, 1969).

Other postmodernist writings on social processes and spatial form include Henri Lefebvre, *Production of Space* (Oxford: Blackwell, 1991) and *Writings on Cities* (Oxford: Blackwell, 1995), Edward Soja, *Postmodern*

Geographies (London: Verso, 1997) and *Postmetropolis* (London: Blackwell, 2000), Mike Craig and Nigel Thrift (eds), *Thinking Space* (London and New York: Routledge, 2000), and Fredric Jameson, *Postmodernism or the Cultural Logic of Capitalism* (Durham, NC: Duke University Press, 1992).

■ ■ ■ ■ ■ ■

At the beginning of this century, there were little more than a dozen or so cities in the world with more than a million people. They were all in the advanced capitalist countries and London, by far the largest of them, had just under 7 million. At the beginning of this century too, no more than 7 per cent of the world's population could reasonably be classified as 'urban'. By the year 2000 there may be as many as 500 cities with more than a million inhabitants. The largest of them (like Tokyo, São Paulo, Bombay and possibly Shanghai) will boast populations of more than 20 million, trailed by a score of cities, mostly in the so-called developing countries, with upwards of 10 million. Sometime early next century, if present trends continue, more than half of the world's population will be classified as urban rather than rural.

The twentieth century has been the century of urbanization. There has been a massive reorganization of the world's population, of its political and its institutional structures and of the very ecology of the earth.

These observations immediately suggest some fundamental questions. First, given these transformations, why is it that the urban so frequently disappears from our discussions of broader political-economic processes and social trends? Most of the writing about our recent history has failed to take into account this massive reorganization and its consequences. The urban rarely appears as a salient category in our analyses. The crucial categories seem to be those of modernization, modernity, post-modernity, capitalist and industrial society. So what has happened to the category 'urban'? This question is important because the qualities of urban living in the next century will define the qualities of life for the mass of humanity. And all political-economic processes we observe are mediated through the filter of urban organization. Discussions of contemporary politics, for example, often proceed as if a concept like that of 'democracy' can remain unaffected by urban transformations when, plainly, there is a huge difference between democracy in ancient Athens and democracy in contemporary São Paulo.

If we think about the likely qualities of life in the next century by projecting forward current trends in our cities, most commentators would end up with a somewhat dystopian view. We are producing marginalization, disempowerment, alienation, pollution and degradation. It might be said that this is nothing new and that, in the nineteenth century, conditions were even worse. In the past, however, urbanization and the consequences of urbanization were taken rather more seriously than they are today. In the late nineteenth century, the bourgeoisie at least had some notion that cities were important places and, therefore, that urban reform was necessary. This generated a bourgeois reform movement – from Birmingham to Chicago – which included figures such as Jane Addams, Octavia Hill, Charles Booth, Patrick Geddes, Ebenezer Howard and many others. All of these had some vision for the future and a clear grasp of the need for reform. The nineteenth century faced the difficulties of the urban in a very positive and powerful way. It blended socialist sentiments, anarchist ideas, notions of bourgeois reformism and social responsibility into a programmatic attempt to clean up the cities. The 'gas and water socialism' of the late nineteenth and early twentieth centuries did a great deal to improve the conditions of urban life for the mass of the population. There are many contemporary analysts who, armed with the insights of Foucault, will assert that these innovations were merely about social control, which indeed in part they were. But having acknowledged this point, I think we have also to recognize that a significant proportion of the population found itself living in better circumstances as a result. Moreover, inherent in these interventions was a visionary notion of an alternative city – a city beautiful with facilities and services that would, indeed, pacify alienated populations.

Some of that concern would be helpful to have back in our cities right now. In the past, capital regarded cities as important places which had to be efficiently organized and where social controls needed to operate in some sort of meaningful way. We now find that capital is no longer concerned about cities. Capital needs fewer workers and much of it can move all over the world, deserting problematic places and popu-lations at will. As a result, the coalition between big capital and bourgeois reformism has disappeared. Moreover, the bourgeoisie itself seems to have lost much of its guilty conscience about cities. It has, I think, concluded there is little to fear from socialist revolution, and so has attenuated its engagement with reformism. Increasingly the wealthy seal themselves off in those fanciful, gated communities – which are being built all over the United States – that enable the bourgeoisie to cut themselves off from what their representatives call by the hateful term 'the underclass'. 'The under-class' is left inside the ghetto, along with drugs, AIDS, epidemics of tuberculosis and much else. In this new politics, the poor no longer matter. The marginalization of the poor is accompanied by a blasé indifference on the part of the rich and powerful.

This blasé indifference is a matter of great concern. Accordingly I would like to highlight some funda-mental questions and beliefs about the role of the city in political, economic, social and ecological life. In defining that role, we are also formulating a notion of the kind of cities we would like to construct into the next century.

I would like to begin with a fundamental method-ological question: what is the relationship between process and form? . . . In my own work – from the standpoint of historical, geographical materialism and very strongly in the dialectical tradition – one of the rules of engagement which I have always tried to follow is to say that process takes precedence over things. We should focus on processes rather than things and we should think of things as products of processes. From this standpoint, we have to ask some fundamental questions about the nature of the categories we use to describe the world. Most of the categories we use tend to be 'thing' categories. If instead we examine dynamics and processes, we may try to do so by con-ceiving them as relationships between pre-existing things. But if things too are not pre-existing, but are actually constituted in some way by a process, then you have to have a rather different vision. This transformation in our way of thought seems to me absolutely essential if we are going to get to the heart of what the city is about.

Tony Leeds, an urban anthropologist, towards the end of his life wrote this:

In earlier years I thought of society . . . as a structure of positions, roles, statuses, groups, institutions and

so on, all given shape . . . by the cultures on which they draw. Process I saw as 'forces', movement, connection, pressures, taking place in and among these loci or nodes of organization, peopled by individuals. Although this still seems largely true to me, it has also come to seem a static view – more societal order than societal becoming . . . Since it does not seem inherent in nature . . . that these loci exist, it seems unacceptable simply to take them as axiomatic; rather we must search for ways to account for their appearances and forms. More and more, the problems of becoming . . . have led me to look at society as continuous process out of which structure or order precipitates in the forms of the loci listed above.

This, then, is a conceptualization in which process takes priority over things and which focuses on the way in which things get precipitated out of process.

Two terms or words deserve closer examination in our discussions. One is 'urbanization' – which we can convert into the 'urban process' or the 'urbanizing process' or the 'urbanization process'. The other is a 'thing-type' word – 'the city'. It is important to consider the relationship between the urbanizing process and this thing called the city. Now, from a dialectical standpoint, the relationship between process and thing becomes complicated because things, once constituted, have the habit of affecting the very processes which constituted them. The ways that particular 'thing-like structures' (such as political-administrative territories, built environments, fixed networks of social relations) precipitate out of fluid social processes and the fixed forms these things then assume have a powerful influ-ence upon the way that social processes can operate. Moreover, different fixed forms have been precipitated out at different historical moments and assume qualities reflective of social processes at work in particular times and places. The result is an urban environment con-stituted as a palimpsest, a series of layers constituted and constructed at different historical moments all superimposed upon each other. The question then becomes how does the life process work in and around all of those things which have been constituted at different historical periods? How are new meanings given to them? How are new possibilities constructed? I suggest that attention to this relationship between process and form will help us understand why the urban has been neglected and, furthermore, will enable us to change completely the terms of the debate.

In this vein, I want to suggest that the reduction of the urban – or the portrayal of the city as a minor feature of social organization – can only occur when particular assumptions are made about the nature of space and time or space/time. There are three different ways of understanding spatiality or space/time that are worth noting here. The first way is the absolute notion of space/time – attributable to Newton, Descartes and Kant – in which space and time are mere containers of social action. They are passive, neutral containers. These passive, neutral containers simply allow us to locate the action which is occurring. I would like to suggest that there is a parallel here with thinking that conceives of cities as passive, neutral containers of processes and contests. These ways of thinking focus on contestations occurring within the city – the city happens to be the mere site of a process of contestation (over gender, race, class or whatever). A radically different approach is one which sees the city not so much as a site of contestation but as something to be constructed and in which the contestation is over the construction, or framing, of the city itself. What would that imply about notions of space and time?

There is a well-known alternative to the absolute view of space/time: that is, the relative view attributed mainly to Einstein and worked on by others since. In this view, space and time, although they are still containers, are not neutral with respect to the processes they contain. Metrics of space and time can and do vary depending upon the nature of the processes under consideration. In geography, this idea has been adapted to think of different ways of measuring and mapping distances. Physical distance is different from distance measured in terms of the cost or time taken to move between points, and in the last two cases the space described is not necessarily Euclidean. Different metrics yield different maps of the space/time coordinates within which social interaction occurs.

A third perspective on space/time that I have employed – indeed it was incorporated in *Social Justice in the City* more than twenty years ago – is a relational view . . . This view is that space and time do not exist outside of process: process defines space/time. Each particular kind of process will define its own distinctive spatio-temporality. Our studies should, therefore, aim to explain the way in which different processes define spatio-temporality, and then, having defined that spatio-temporality, find themselves bound by its rules in certain kinds of ways. Moreover, our cities are constituted not by one but by multiple spatio-temporalities, producing multiple frameworks within which conflictual social processes are worked out.

From this standpoint, we have to take very seriously the notion . . . that space and time are not simply constituted by but are also constitutive of social processes. This is also true for the urban. The urban and the city are not simply constituted by social processes, they are constitutive of them. We have to understand that dialectic in order to appreciate how urbanization is constructed and produces all of these thing-like configurations which we call cities – with political organization, social organization and physical structures. We have to appreciate better the centrality of that moment of urban construction, which is fundamental to how the social process operates. In exactly the same way, we have to take seriously the idea of that moment of construction of spatio-temporality, which then defines how the system itself will operate. From this standpoint, it is possible to reposition the urban as fundamental in contemporary debates. At the same time we transform our notion of urbanization. We would abandon the view of the urban as simply a site or a container of social action in favour of the idea that it is, in itself, a set of conflictual heterogeneous processes which are producing spatio-temporalities as well as producing things, structures and permanencies in ways which constrain the nature of the social process. Social processes, in giving rise to things, create the things which then enhance the nature of those particular social processes.

One outcome may be that we find ourselves stuck for a very long time with a particular kind of social process. An example would be nuclear power. Once nuclear power stations exist all sorts of things follow. If a nuclear power station goes on the blink, can you imagine calling a town meeting to discuss democratically what to do about it? The answer is no, you can't. In these circumstances, we are immediately driven back to the realms of expert knowledge and expert decision-making. So a thing has been created which for as long as it lasts – which is going to be a very long time – is by its very nature going to be basically undemocratic in terms of the sort of social process that supports it. Here is a social process that has defined a certain spatio-temporality for the next 10,000 years, which in turn implies perpetuation of a certain kind of social order if it is not to unravel in highly destructive ways.

We have to be thinking in these kinds of terms about the nature of cities. What kinds of cities we create, how we create them, how flexible they are, how adjustable they can be: these are the questions we need to ask in order to understand better the relationship between process and thing. Our aim and objective should be to liberate emancipatory processes of social change. In so doing, however, we must understand that liberatory impulses and politics are always going to be contained and constrained by the nature of things which have been produced in the past.

This, then, is my first major point. We have to reconceptualize the urban as the production of space and the production of spatio-temporality, understood as a dialectical relationship between process and thing.

The second major point I would like to make concerns the currently widespread invocation of the word 'community'. It too entails an exploration of process/thing relationships. One of the aspects of much contemporary debate about the urban which I find particularly striking is the tendency when faced with all sorts of difficulties again and again to reach into this bag called 'community', on the assumption that 'community' is going to save us all. Community, endowed with salving powers, is perceived as capable of redeeming the mess which we are creating in our cities. This mode of thinking goes all the way from Prince Charles and the construction of urban villages through to communitarian philosophies that, it is believed, will save us from crass individualistic materialism.

There is here too an issue about the relationship between the thing called community and the processes which constitute it. What kinds of processes constitute community? Is a community, once constituted, going to liberate or imprison further social processes? A lot of community construction projects are, in the end, a recipe for isolation. They isolate groups from the city as a whole. They move them towards a fragmented notion of what the urban process is about. Here I find myself in agreement with Iris Marion Young when she says:

> Racism, ethnic chauvinism and class devaluation, I suggest, grow partly from the desire for community . . . Practically speaking, such mutual understanding can be approximated only within a homogeneous group that defines itself by common attributes. Such common identification, however, entails reference also to those excluded. In the dynamics of racism and ethnic chauvinism in the United States today,

the positive identification of some groups is often achieved by first defining other groups as the other, the devalued, semihuman.

What, then, are the implications of current notions of community? In answer, I would like to propose a dialectical view of relationships between process and community.

I think it is important to acknowledge that a lot of community activism is absolutely fundamental to many forms of social struggle. As a form of mobilization of power of people in place it can sometimes be extremely important and extremely useful. Community activism can simply be a way of containing discontent but it can also be a very important moment in more general mobilization. In this context, we have to think about the construction of community not as an end in itself but as a moment in a process. Here I refer to critiques of the nineteenth-century thinking which I described earlier. There were two flaws in that thinking. The first was the belief that, somehow or other, the proper design of things would solve all of the problems in the social process. It was assumed that if you could just build your urban village, like Ebenezer Howard, or your Radiant City, like Le Corbusier, then the thing would have the power to keep the process forever in harmonious state. The problem of these thinkers was not that they had a totalizing vision or subscribed to master narratives or indulged in master planning. Their problem was not that they had a conception of the city or the social process as a whole. Their problem was that they took this notion of thing and gave it power over the process. Their second flaw was that they did much the same with community. Much of the ideology that came out of Geddes and Ebenezer Howard was precisely about the construction of community. In particular, the construction of communities which were fixed and had certain qualities with respect to class and gender relations. Once again, the domination of things seems to me to be the fundamental flaw.

What then is the significance of community mobilization? The concept I wish to use here is the one that Raymond Williams tentatively suggested, and which he then shrank away from, but which I want to resurrect. It is what Williams calls 'militant particularism'. This idea suggests that almost all radical movements have their origin in some place, with a particular set of issues which people are pursuing and following. The key issue is whether that militant particularism simply remains localized or whether, at some point or other, it spills over into some more universal construction. Williams suggested that the whole history of socialism had to be read as a series of militant particularisms which generated what he described as the extraordinary claim that there is an alternative kind of society, called socialist, which would be a universal kind of condition to which we could all reasonably aspire. In other words, in this view foundational values and beliefs were discovered in particular struggles and then translated onto a broader terrain of conflict. It seems to me that the notion of community, viewed in this way, can be a positive moment within a political process. However, it is only a positive moment if it ceases to be an end in itself, ceases to be a thing which is going to solve all of our problems, and starts to be a moment in this process of broader construction of a more universal set of values which are going to be about how the city is going to be as a whole.

The third major point I am going to make is this: until very recently there was almost no mention of cities in the ecological literature. Cities were always regarded as the high point of the pollution and plundering of planet Earth. The environment was equated with nature; it was certainly not the built environment of cities. There is something curious about ecological rhetoric here (although I am probably misrepresenting some of the current thinking because it is getting a bit more sophisticated). Ecological rhetoric is committed to a totalizing perspective in the sense that, quite rightly, it perceives that everything relates to everything else. However, it has also failed to address the environment of cities and the 50 per cent of the world's population that are living in urban circumstances.

Why is it that we tend to think of the built environment of cities as somehow or other not being the environment? Where did that separation come from? Again it comes back to the notion that there is a thing called a city, which has various qualities and attributes, that is not part of a process. It seems to me that we have to think of environment and environmental modification as a fundamental process which we have always been engaged in and will always continue to be engaged in. The environmental modification process then has to be understood as producing certain kinds of structures and things, such as fields, forests and cities. That environmental modification process cannot be separated from the whole question of urban living. There is, it seems to me, nothing particularly anti-ecological about cities. Why should we think of them

that way? When does the built, constructed environment end and 'the natural environment' begin? Where does society begin and nature end? Go and look in a field of wheat and say where nature begins and society ends. You can't do it.

Here too, then, there is a dichotomy which works its way through our thinking, which we have to challenge, in which the relationship between processes and things is fundamental. We have to pay serious attention to the nature of the ecological modification process, and understand it not as something which is simply resident in nature. For example, one of the major ecological variables at work in the world right now is money flow. Just think of what would happen to the ecosystems of the world if the money flow stopped. How many ecosystems of this world are actually supported by money flow? Vast areas of the world would undergo radical ecological change if the money supply or commodity exchange was suddenly cut off or stopped. Some radical ecologists appear to relish such an outcome, as a transformation back to some ecologically sustainable condition in which the alienation of self from nature can be overcome by human beings treading far more lightly on the surface of the earth. But I believe we must pursue much more positive ecological politics. Ecological transformations are an inevitable facet of how human beings live their lives and construct their historical geographies. Urbanization is an ecological process and we desperately need creative ways to think and act on that relation. Conversely, it is impossible to talk of ecological politics without concomitantly examining urban processes in all their complexity and fullness.

We have to move the urban, and the urbanizing process, into a more central position in our debates and discussions about ecological, social, political and economic change. From this standpoint, there are a number of myths that we have to confront and contest.

The first myth is the simple idea that when we have got the economy right then we can spend money to get our cities right. This sort of thinking takes the view that cities are relatively unimportant: when we have got enough money and we have organized ourselves right then we can spend a little time fixing them up. From my perspective that is entirely the wrong way round. Getting things economically right in our cities is the path towards economic change and economic development, even to economic growth. To treat the cities as the secondary feature of this whole dynamic is essentially wrong.

The same is true with respect to social relations. We should not wait upon some great political revolution to tell us how to reorganize our cities in a socialist or eco-feminist or some other way. No, what we have to do is to work on the nature of the social relations in the cities. If there is going to be a revolution, it is going to be a long revolution, located within the urban process. That long revolution of social relations is going to have to comprise a steady working out, over a long period of time, of transformations. Here, I think again, community mobilization and the transformation of militant particularism have a vital role to play, enabling us to find the universal concerns that exist within a realm of difference. There is a certain dialectic here of unity and difference, universal and particular, which has to be worked out. We should not retain the notion of community as particularity or difference. We have to transcend those particularities and look for a negotiation of universalities through which to talk about how the cities of the future should be.

The point is not to see cities as anti-ecological. Cities are fundamental ecological features in themselves and the processes that build cities are ecological processes. The world of ecology and that of cities are part and parcel of each other; what we have to do is link them together much more strongly, in a more programmatic way. It is only in those terms that we can really push towards a full understanding of . . . 'contested cities'. This issue is not simply about contestation inside cities but more importantly concerns contests over the construction and framing of cities – especially what they are going to be in the future.

"How to Study Urban Political Power"

from *A Phoenix in the Ashes: The Rise and Fall of the Koch Coalition in New York City Politics* (1992)

John Mollenkopf

Editors' Introduction

"Is urban politics worth studying at all, or is the urban political realm so subordinate to, dependent on, and constrained by its economic and social context that factors from this domain have little independent explanatory power?" In posing this question, John Mollenkopf raises the fundamental issue of whether urban residents are actually members of a self-governing political community – citizens of a modern polis – or merely the helpless pawns of larger, faceless forces.

John Mollenkopf is a professor of Political Science and Director of the Public Policy Program at the Graduate School of the City University of New York. He is the co-editor of *The Urban Politics Reader* (London and New York: Routledge, 2003) – a companion anthology to *The City Reader* that contains readings on urban politics. Professor Mollenkopf previously taught at Stanford University, where he was one of the founders of the Program on Urban Studies. In 1983, he published *The Contested City*, a pioneering study of postwar American urban politics. Based on the experience of Boston and San Francisco, Mollenkopf outlined the process by which federal development programs were employed to help forge powerful pro-growth political coalitions at the local level. In "How to Study Urban Political Power", he applies a similar analysis to the politics of the city of New York during the mayoralty of Edward Koch.

"How to study Urban Political Power" sets the stage for Mollenkopf's analysis of New York politics and the study of urban politics generally. Analysts of city politics, he explains, may be divided into two camps – *pluralists* and *structuralists* – and the recent history of urban political theory has been a back-and-forth struggle between the contending conceptual frameworks.

To begin with, writes Mollenkopf, an extensive body of theory by both sociologists and political scientists argued that local decision-making is dominated by entrenched elites. Marxist theorists in particular feel capitalist social structures really determine political outcomes. In opposition to this prevailing orthodoxy, a number of pluralist scholars argued that no elitist model of governance could easily explain what they actually observed as they studied urban politics in different cities. Noting that almost one-quarter of the gross national product of the United States "passes through the public sector . . . much of it through urban governments," pluralist political analysts tended to see urban politics "as an autonomous realm that possessed real authority and commanded important resources."

But in an intellectual counter-attack, a new body of structuralist theory arose out of the social upheavals of the 1960s and 1970s that argued that the pluralist dispersion of power was mostly illusion, especially to inner-city minority communities, and that the imperatives of capitalism, in both the economic and social structural realms, repeatedly and inevitably established the basic parameters of local development policy.

After exploring the two poles of urban political theory, and providing a catalog of their principal exponents, Mollenkopf proposes a synthesis that simultaneously avoids the pitfalls of one-dimensionality and recognizes the legitimate claims of each school. "How can we develop a vocabulary for analyzing politics and state action," he asks, "that reconciles the political system's independent impact on social outcomes with its observed systemic bias in favor of capital?" He returns to his initial question – whether urban politics is worth studying at all – and answers in the affirmative. Although the structuralists are correct in stressing the importance of underlying economic forces, Mollenkopf feels that they typically lack "a well developed theory of the state," and thus he concludes that "economy-centered theorizing" must always be tempered by "polity-centered" thinking.

This selection is from *A Phoenix in the Ashes: The Rise and Fall of the Koch Coalition in New York City Politics* (Princeton, NJ: Princeton University Press, 1992). Other books by John Mollenkopf include *New York City in the 1980s: A Social, Economic, and Political Atlas* (New York: Simon & Schuster, 1993), *Power, Culture, and Place: Essays on New York City* (New York: Russell Sage, 1988), *The Contested City* (Princeton, NJ: Princeton University Press, 1983), and *Dual City: Restructuring New York* (New York: Russell Sage, 1991) co-edited with Manuel Castells.

The classic elitist/structuralist study of urban community power is Floyd Hunter, *Community Power Structure* (New York: Anchor, 1953). The classic pluralist study of urban community power is Robert Dahl's *Who Governs?* (New Haven, CT: Yale University Press, 1961).

Other studies of urban politics include Douglas Yates, *The Ungovernable City* (Cambridge, MA: MIT Press, 1977), Paul E. Peterson, *City Limits* (Chicago: University of Chicago Press, 1981), Barbara Ferman, *Governing the Ungovernable City* (Philadelphia, PA: Temple University Press, 1985), John R. Logan and Harvey L. Molotch, *Urban Fortunes: The Political Economy of Place* (Berkeley and Los Angeles: University of California Press, 1987), Timothy Barnekov, Robin Boyle, and Daniel Rich, *Privatism and Urban Policy in Britain and the United States* (Oxford: Oxford University Press, 1989), John R. Logan and Todd Swandstrom, *Beyond the City Limits: Urban Policy and Economic Restructuring in Comparative Perspective* (Philadelphia, PA: Temple University Press, 1990).

■ ■ ■ ■ ■ ■

What is the appropriate way to conceptualize the organization of political power in New York City during the Koch era? The dialogue between the pluralist interpreters of urban power and their structuralist critics has produced a rich variety of answers to this question. In the early 1960s, pluralist political scientists launched an attack on the previously accepted view, established by sociologists, that socioeconomic elites dominated urban politics. The success of this assault enabled pluralists to establish their view as the norm in political science.

From the mid-1970s onward, however, a new generation of structurally oriented critics challenged the pluralist point of view. While they were able to undermine the prevailing wisdom, they did not manage to supplant it with a new one, in part because of defects in their arguments that pluralists were quick to point out. More recently, students of urban politics have attempted to synthesize the strengths of both approaches. With respect to framing the study of how the Koch administration amassed and exercised political power, the debate between pluralists and their critics focuses our attention on four interrelated questions:

1 Is urban politics worth studying at all, or is the urban political realm so subordinate to, dependent on, and constrained by its economic and social context that factors from this domain have little independent explanatory power?
2 If urban politics does have an independent impact, how should we conceptualize power relations among interests or actors?
3 In particular, what factors govern the construction of a dominant political coalition within a given set of structural constraints and opportunities?
4 In constructing such a coalition, how important is promoting private investment compared to other strategies, such as increasing social spending to incorporate potentially insurgent groups?

THE PLURALIST CONCEPTION OF THE URBAN POLITICAL ORDER

The classic pluralist studies of a generation ago, like Banfield's *Political Influence*, Dahl's *Who Governs?*, or Sayre and Kaufman's *Governing New York City*, made important theoretical and methodological advances over the so-called elitists they attacked. They did not deduce power relations from the interlocks between economic and political elites. Instead, they went into the field to examine the tangled complexity of interest alignments around actual policy decisions and disputes. Pluralist scholars showed that no model of direct control by a unified economic or status elite could easily explain what they saw.

While most pluralists did not dwell theoretically on the larger relationship between the state and the economy, they implicitly rejected the notion that some underlying structural logic subordinated local politics to the private economy. They saw politics as an autonomous realm that possessed real authority and commanded important resources. They explicitly rejected the notion that economic or social notables controlled the state in any instrumental sense. Since they argued that every "legitimate" group commanded some important resource (if only the capacity to resist) and no one group commanded sufficient resources to control all others, pluralists argued that the bargaining among a multiplicity of groups defined the urban power structure.

In this view, coalition building was central to the definition of power. Political leaders and private interests built coalitions around specific issues, the coalitions varied from issue to issue, and they tended to be short lived. By selecting a range of different policy decisions as case studies for research, pluralists seemed to imply that urban development and social service issues had an equal importance in organizing political competition.

In the face of examples where entrenched interest groups dominated their own particular, fragmented policy areas over time to the exclusion of the public interest, the pluralist approach developed a clearly critical strand of analysis. But these scholars simply saw the dark side of the pluralist worldview without fundamentally challenging its basic assumptions or deflating the optimistic claims about system openness or responsiveness prevailing among other pluralists.

[. . .]

While the pluralist studies may have been convincing and accurate portraits of urban politics in the 1950s and early 1960s, the eruption of turmoil and political mobilization in the 1960s and the fiscal crisis of the 1970s soon revealed basic flaws in the pluralist analysis. Except for Robert Dahl's work, *Who Governs?*, these studies lacked a context in economic and political development. Despite obligatory opening chapters covering economic, social, and political trends, pluralist studies such as Sayre and Kaufman's did not treat the changing structure of urban economies or racial succession as problematic for the urban political order. It would, they thought, simply absorb and adapt to these changes. While Dahl provided a fine treatment of the transition from patrician dominance to what he argued were the dispersed inequalities of pluralist democracy in New Haven, he also failed to see that blacks might be led to challenge the system, not just participate in it as a minor interest. Neither Dahl nor his colleagues foresaw how economic transformation and racial succession might fundamentally challenge the previously observed "normal" patterns.

[. . .]

Dahl explicitly denied that economic and social inequalities would overlap and reinforce each other in the political arena. Other pluralist scholars also did not recognize the possibility that nonelite elements of the urban population would feel systematically excluded from power and would react by pressing for greater representation and more vigorously redistributive policies. As a result, the urban battles that erupted in the latter 1960s in New York City and elsewhere made their relatively tranquil picture of urban politics as a kind of market equilibrium-reaching mechanism seem anachronistic.

STRUCTURALIST CRITIQUES

As the pluralist political equilibrium unraveled on the ground, it came under increasing challenge from structuralist critics. The broad outlines of their progress may be traced from Peter Bachrach and Morton Baratz's classic essay on the "two faces of power" to Clarence Stone's work on "systemic power" to John Manley's "class analysis of pluralism." Bachrach and Baratz attacked pluralists for focusing on the "first face" of power, namely its exercise, while ignoring the second, namely the way that the relationship between the state and the underlying socioeconomic system shapes the political agenda. "Power may be, and often is," they said, "exercised by confining the scope of decision-making to relatively 'safe' issues." But while making a case for analyzing how the values embedded in institutional practices bias the rules of the game, they do not specify the mechanisms that promote some interests and issues while dampening others.

Stone advanced this line of thought by shifting the locus of analysis from decisions ("market exchange") toward the mechanisms that create systemic or strategic advantages for some interests over others ("production"). The unequal distribution of private resources, he argues, creates a differential capacity among political actors to shape the flow of benefits from the basic rules of the game, the construction of particular agendas, and the making of specific decisions. Business, in particular, derives systemic power not only from its juridical status and economic resources but from its attractiveness as an ally for those who advance any policy change and from the shared subculture from which private and public officials both emerged.

Despite the structuralist leaning in his concept of "systemic power," Stone did not break decisively with the pluralist interplay of interests around decisions. Manley's Marxist critique does make this break. He embraced the argument that the legal and structural primacy enjoyed by private ownership of capital requires the state to reinforce the systemic inequalities that result from the drive for private profit. He attacked pluralists, even the later work of Dahl and Lindblom that concedes that business enjoys a privileged position in pluralist competition, for lacking a theory of exploitation and, hence, an objective standard of a just or equal distribution. In Manley's view, the juridical protection of private property inevitably commits the state to control workers and promote capital.

While neo-Marxist work similar to Manley's stressed the systematic subordination of the state and politics to capital accumulation and the private market, a parallel and quite nonradical strand of public choice analysis reached quite similar conclusions. Focusing on the notion that cities compete to attract well-off residents and private investment, this line of analysis . . . is logically quite similar to some neo-Marxist critiques of pluralism.

NEO-MARXIST CRITIQUES

Structuralists have decisively transcended the pluralist vocabulary. They provided the social and economic context missing from pluralism and highlighted the ways that private property, market competition, wealth and income inequality, the corporate system, and the stage of capitalist development pervasively shape the terrain on which political competition occurs. They underscored the need to analyze how basic patterns of the economic, political, or cultural rules of the game bias the capacity of different interests to realize their ends through politics and the state.

Most importantly, neo-Marxist structuralists were able to empirically investigate these mechanisms, refuting the pluralist retort that "nondecisions" either must be studied just like decisions or else are unobservable ideological constructs. They have shown cases in which the systemic and cumulative inequality of political capacity undergirded, and indeed was ideologically reinforced by, a superficial pluralism. Structuralist studies may be flawed by economic determinism, but they are factually on target in observing and describing mechanisms that generate systemic, cumulative, political inequality, which has a more profound impact on outcomes than the coalition patterns studied by pluralists. Such critiques won relatively broad support among the younger generation of scholars, if not their elders. They may be subclassified into theories that stress the political logic of capital accumulation, social control, or the interplay of accumulation and legitimation. Each offers a different perspective on the central mechanisms that generate cumulative political inequality.

Theorists influenced by Marx's economic works have tended to argue that the mode of production stamps its pattern more or less directly on the organization of the state and on the dynamics of political competition. Marxists as different as David Harvey and David Gordon have both argued that the stage of capitalist development and the circuits of capital have determined urban spatial patterns, the bureaucratic state, and for Harvey even urban consciousness. While this strand of Marxist thinking made a breakthrough in orienting analysts to the importance of the process of capital accumulation, it has generally lacked a well-developed theory of the state that either identifies the instrumental mechanisms that link state actions to the power of capital or grants the relative autonomy to the state.

This literature does stress one mechanism, however: the state's dependence on private investment for public revenues. If the mobility of capital can discipline the state and constrain political competition, then competition among polities (whether cities or nations) to attract investment leads them to grant systematic benefits for capital, a dynamic that Alford and Friedland have called "power without participation." As Harvey wrote,

The successful urban region is one that evolves the right mix of life-styles and cultural, social, and political forms to fit with the dynamics of capital accumulation . . . Urban regions racked by class struggle or ruled by class alliances that take paths antagonistic to accumulation . . . at some point have to face the realities of competition for jobs, trade, money, investments, services, and so forth.

Sooner or later, the state and political competition will be subordinated to the needs of capital.

Several analysts, including Friedland and Palmer as well as Molotch and Logan, abstracted this mechanism from the larger Marxian vocabulary and made it central to their analysis of urban power. Friedland and Palmer argued that, while businesses do directly influence policy-making, such intervention is logically secondary. "The growth of locales depends on the fortunes of their firms," according to Friedland and Palmer, thus "dominant and mobile [corporate] actors set the boundaries within which debate over public policy takes place." As capital has become more mobile and less tied to specific locations, the need for business to intervene directly in politics has waned, while the structural subordination of local government to the general interests of business has waxed.

Molotch and Logan took a different tack on the same course. While conceding that the mobility of capital gives local government a powerful incentive to defer to capitalists, they argued that certain classes of business are not mobile: real estate developers, utilities, newspapers, and others with a fixed relationship to a place. Large sunk costs give these interests a powerful incentive to intervene in and dominate local politics in order to get local government to promote new investment. They saw this "growth machine" as a ubiquitous, inevitable, and at best weakly challenged feature of American cities . . .

[. . .]

While this strand of thinking argued that the multiplicity of competing local governments forces the state to reproduce and protect basic features of the advanced capitalist economy, a second, equally important school of neo-Marxist thinking stressed the way urban politics serves to dampen and regulate the conflicts inevitably generated by capitalist urbanization. Castells's work on "collective consumption" and urban social movements, Piven and Cloward's studies of urban protest, and Katznelson's studies of the absorptive capacity of local bureaucracies and the bias against class issues in urban politics represent the best of this work.

While these analysts differed over how the state coopts movements that challenge urban governments, they share the idea that this process is a central feature of urban politics in advanced capitalist societies. Not everyone, even on the left, has agreed with these contentions. Theret, Mingione, and Gottdiener have criticized the explanatory power of the notion of collective consumption, while Ceccarelli has argued that urban social movements did not turn out to be the force in west European urban politics that Castells portrayed them to be. Whatever the situation in Europe, the civil rights movement, urban unrest, and community organization clearly had a profound impact on urban politics in the United States after the 1960s, particularly in the rise of programs designed to absorb and deflect these forces.

[. . .]

PUBLIC CHOICE CRITIQUES

Neo-Marxist thinking is not the only source of structural criticism of the pluralist paradigm, however. Microeconomics, in the form of public choice theory, has contributed its own critique. Tiebout's seminal work led to Forrester's simulation of urban systems and ultimately to Paul Peterson's sophisticated "unitary" theory of urban politics. This tradition, born of the economists' distrust of state allocation of resources, has sought a functional equivalent to the marketplace in the multiplicity of local governments. They would compete, Tiebout argued, for residents of different means and desires by providing different service packages at various tax costs. An equilibrium would thus be reached in the sorting of populations across urban and suburban jurisdictions within the metropolis. This equilibrium would represent an efficient production of public goods, matching the marginal prospective resident with the jurisdiction's need to add (or subtract) residents on its own margin to provide services at the most efficient scale.

Such thinking has undergirded much of the orthodox literature on urban economics and local public finance. Urban housing, for example, has been analyzed as a function of how consumers trade off housing and commuting costs, given various levels of residential amenities. Forrester built the underlying assumptions into a model, influential for a time, that implied that whatever cities do to provide housing or social services for the poor will attract more of them, drive out the better off, and erode the tax base.

[. . .]

STRUCTURALISM RECONSIDERED

By providing the missing economic and social-structural context, these structuralist critiques achieved a considerable advance over pluralist analysis. Cities can no longer be taken as independent entities isolated from the larger economic and social forces that operate on them. Analysts can no longer ignore the impact of global and national economic restructuring on large central cities. Since cities cannot retard these global economic trends (though New York and others may propagate them), nor remake their populations at will, they clearly navigate in a sea of externally generated constraints and imperatives.

The structuralist critiques also make it clear that urban politics can no longer be considered to be un-related to the cumulative pattern of inequality in the economy and society. They have focused attention on how the state's dependence on private investment fosters political outcomes that systematically favor business interests. Structuralists have explored specific mechanisms that produce this result, such as the invidious competition among fragmented, autonomous urban governments for investment, the segregation of local government functions into quasi-private agencies that promote investment and politically exposed agencies that absorb and deflect protest, and the organization of the channels of political representation so as to articulate interests in some ways but not others. By stressing that advanced capitalism characteristically generates urban social movements and political con-flicts, some structuralists have also implied that political action can alter some of the constraints capitalism imposes on democracy.

Despite these strengths, however, structuralist perspectives also have grave flaws. The assertion that the state "must" undertake activities that favor capital tends to be functionalist. Such a standpoint begs the question of how these "imperatives" are put in place and reproduced over time, which inevitably must be through the medium of politics. As a result, struc-turalists may not see that political actors can fail to fulfill or to maximize their supposed imperatives. Dominant urban political coalitions have certainly done things

that cost them elections and the ability to exercise power; they have persisted in increasing the tax burden on private capital and imposing exactions on private developers even after the point that they diminish further investment. Others have chosen to increase budget deficits and risk their bond ratings.

Given the right conditions, nothing is inevitable about an administration's pursuit of electoral success, private investment, well-managed social tensions, or even good bond ratings. Nothing guarantees that city government will be willing or able to fulfill the functions structuralists have assigned to it. As Piven and Friedland observed in rejecting a "smoothly functioning determinism," a structural analysis cannot be adequate until it specifies "the political processes through which . . . systematic imperatives are trans-lated into government policies."

Structural critiques also tend not to be discon-firmable. For example, if structuralists argue that the use of legal injunctions by the conservative Republican administrations before the New Deal illustrates how the state supports capitalism, while the New Deal's recognition and promotion of trade unions also illustrate state support for capitalism, then they are explaining everything and nothing. Put another way, structural theories tend to have a hard time explaining the real and important variation over time and across places. The basic features of capitalism are common across nations and evolve slowly, while the political outcomes that capitalism is supposed to drive are highly varied and change more quickly.

Finally, when structuralists appeal to an ultimate economic determinism, they aggravate these prob-lems. Agency fades out of the analytic picture. To be sure, the most attractive variants of neo-Marxism sought to avoid this trap by using the concepts of the "relative autonomy of the state" and "conflicting imperatives of accumulation and legitimation" to introduce a political dimension into an otherwise inadequate economic determinism. Yet in their dis-comfort with granting politics a co-equal causal role, even these variants ultimately retreat to the view that politics is subordinate to economics: autonomy is after all only relative. From Gottdiener at one end of the spectrum to Peterson on the other, structuralist ana-lysts have made no bones about calling politics analytically irrelevant in the face of the economy's ability to constrain and impel.

To summarize, for all their strengths, structuralists conceptualized the political system as ultimately subordinate to economic structure. They tended to reduce urban politics to the fulfillment of economic imperatives; even social control achieved through political means serves capitalist ends. The most promising threads of structuralist thinking examined how systemic imperatives might conflict with each other or generate system-threatening conflict, thus opening the way for political indeterminacy. Here, however, they risked moving outside and beyond a structuralist paradigm. Indeed, orthodox Marxism (or for that matter orthodox neoclassical economics) simply does not provide a good basis for building a theory of politics. To the extent that structuralist theorists held true to the logic of their argument, they underplayed the importance of politics. They did not appreciate that policies that promote private investment must be constructed in a political environment that may favor but by no means guarantees this outcome. Indeed, popular, social, and communal forces pressure the state and the political process just as strongly in different, and often opposed, directions.

This tendency to trivialize politics removes a way to explain why outcomes vary even though capitalism is constant. States may be constrained, but they are also sovereign. They exercise a monopoly on the legitimate use of force, establish the juridical basis for private property, and shape economic development in myriad ways. Economies are delicate. They depend on political order and have been deformed or smashed by political disorder. State actions may be conditioned by economic structure, but they cannot be reduced to it. Many substantially different capitalisms are possible, and politics determine which ones evolve. Just as the state is dependent on the economy, economic institutions depend on and are vulnerable to the state and its changing political circumstances.

[. . .]

"Polity-centered" thinking must thus augment the "economy-centered" theorizing of the structuralist critiques. This does not require an equally one-sided political determinism. Rather, it requires us to extend the lines of structuralist thinking that stress conflict among imperatives or developmental tendencies until we go beyond the limits of economic determinism. We must recognize that "state power is *sui generis*, not reducible to class power," as Block put it. Or as Manuel Castells recently reflected, "experience was right and Marxist theory was wrong" about the central theoretical importance of urban social movements and the impossibility of reducing them to a class basis.

But if we give politics an analytic weight equal to that of economic structure, how can we avoid returning to a voluntaristic pluralism? How can we develop a vocabulary for analyzing politics and state action that reconciles the political system's independent impact on social outcomes with its observed systemic bias in favor of capital? A satisfactory approach must operate at three interrelated levels: (1) how the local state's relationship to the economy and society conditions its capacity to act; (2) how the "rules of the game" of local politics shape the competition among interests and actors to construct a dominant political coalition able to exercise that capacity to act; and (3) how economic and social change and the organization of political competition shape the mobilization of these interests.

TOWARD A THEORETICAL SYNTHESIS

We can begin to build such an approach by recognizing that city government and its political leaders interact with the resident population and constituency interests in its political and electoral operating environment and with market forces and business interests in its economic operating environments. This approach emphasizes two primary interactions: first, between the leaders of city government and their political/electoral base; and second, between the leaders of city government and their economic environment. It also suggests that political entrepreneurs who seek to direct the actions of city government must contend with three distinct sets of interests: (1) public sector producer interests inside local government; (2) popular or constituency interests (which are also public sector consumer interests), especially as they are organized in the electoral system; and (3) private market interests, particularly corporations with discretion over capital investment, as they are organized in the local economy.

To be sure, these interests are highly complex in a city like New York and cannot be captured by simple dichotomies like black versus white or capitalist versus worker . . . The city's residential communities are highly heterogeneous. Terms like "minority" hide far more than they reveal; even "black" or "Latino" blur important distinctions regarding nativity and ethnicity. Business interests come in many sizes, industries, and competitive situations; even corporate elites vary greatly. Still, a focus on the relationships among state, citizenry, and marketplace provides an entry point for analyzing what determines the shape of the urban political arena.

The concept of a "dominant political coalition" gives us a focal point for this analysis. A dominant political coalition is a working alliance among different interests that can win elections for executive office and secure the cooperation it needs from other public and private power centers in order to govern. To have an opportunity to become dominant, it must first win election to the chief executive office. To remain dominant, it must use the powers of government to consolidate its electoral base, win subsequent elections, and gain support from those other wielders of public authority and private resources whose co-operation is necessary for state action to go forward. Put another way, a dominant coalition must organize working control over both its political and its private market operating environments.

This formulation improves on the pluralist approach by directing our attention toward how the relationship between politics and markets biases outcomes in favor of private market interests, as structuralist approaches have pointed out. The notion of a dominant political coalition would not sit well with pluralists, who have argued that coalitions are unstable, form or reform according to the issue, and may be stymied by the capacity of any sizable group to resist. We posit instead that coalitions can be stable, operate across issues, and create persistent winners and losers. Challenging and supplanting such coalitions have generally been difficult, particularly for constituencies that lack resources or are particularly vulnerable to sanction. Effective challenges generally arise only at moments of crisis in periods of rapid social and economic change.

This formulation also improves on the structuralist approach by according the political/electoral arena an influence equal to that of economic forces. It also points us toward how strategies to control the direction of city government are shaped by (and in turn shape) the political environment and by the public sector producer interests that have a permanent stake in its operation. It posits a scope for political choice and innovation that is lacking in the structuralist perspective.

This approach points us toward the following central questions: how do political entrepreneurs seek to organize such coalitions, what enables them to succeed in the first instance, and how do they sustain success over time? In what ways can such coalitions be bound together? What interests do dominant coalitions include and exclude and why? How do the economic and political contexts affect these binding relationships? And what tensions or conflicts undermine dominant coalitions, opening the way for power realignment?

As structuralists have shown, one part of the answer to these questions lies in the relationship of politics to the structure of economic interests. Efforts to explore this relationship may be found in Stone's studies of Atlanta, Shefter's study of New York, and my own work on pro-growth coalitions in Boston and San Francisco. Stone distinguished three levels at which to analyze the relationship between a dominant coalition and various urban interests. The least interesting is the pluralist domain of individual decisions or "command power" in which one actor induces or coerces others to follow his or her bidding. The two others are more relevant to this analysis.

Political actors wield "coalition power" when they join together to exercise the policy powers of the state to produce a steady flow of benefits to their allies, without the need for coercing or inducing specific actions. Stone showed how the Atlanta regime used public and private subcontracts to minority business enterprises to cement its political support, but he gave relatively little attention to other aspects of how the coalition tried to dominate its political operating environment.

Instead, he emphasized the "preemptive" or "systemic power" enjoyed by private interests whose command over private resources is so great as to make their support crucial to the dominant coalition. Among the mechanisms of preemptive power in Atlanta, Stone identified the unity of a well-organized downtown business community, newspaper and television support for policies that favored downtown development, the reliance of politicians on campaign contributions from developers, business control over the equity and credit that government needed to carry out its plans, and the business community's ability to provide or deny access to upward mobility for the black middle class. These systemic powers made corporate interests ideal allies for politicians seeking to achieve and sustain political dominance.

While the arrival of a black majority in the electorate, militant new black leadership, and neighborhood mobilization eventually destabilized and modified the tradition of white dominance that prevailed in Atlanta until Maynard Jackson was elected mayor in 1973,

Stone argued that they did not overturn the preemptive power of corporate interests. Jackson's successor, Andrew Young, chose to abandon the fragmented and undisciplined neighborhood movement in favor of pro-growth politics with a new face, consolidated by white business support for set-asides to minority entrepreneurs.

[. . .]

My own study of how political entrepreneurs constructed pro-growth coalitions in Boston, San Francisco, and other large cities in the late 1950s and 1960s also advanced reasons why politicians would want to forge alliances with private sector elites. Promoting private development was an obvious way to bring together such otherwise disparate elements as a Republican corporate elite, regular Democratic party organizations, and reform-oriented rising public sector and nonprofit professionals.

These perspectives on how dominant political coalitions shape development politics and budget policy to secure business support, while convincing, remain incomplete. The "preemptive power" of business interests only explains part of how political actors construct a coalition to direct city government in the exercise of its powers. As Lincoln Steffens long ago observed, dominant coalitions needed to develop a grassroots base of legitimacy as well as support from elite interests. However much they may need corporate support, dominant coalitions must also have support from popular constituencies organized by such organizations as political parties, labor unions, and community organizations.

Mayors can lead dominant political coalitions only when they win electoral majorities and keep potential sources of electoral challenge fragmented or de-mobilized. The mobilization of blacks and Latinos and the neighborhood organization that began in the mid-1960s and continues today have prompted many currently dominant political coalitions to adopt policies that do not follow from a devotion to private market interests or public sector producer interests. For example, dominant coalitions must respond to mobil-izations against the negative impacts of downtown growth and inadequate public services, or demands for government programs that provide upward mobility for excluded groups.

Rufus Browning, Dale Marshall, and David Tabb have argued, for example, that when insurgent liberal biracial coalitions came to power in a few of the northern California cities they studied, the coalitions shifted policies in favor of the formerly under-represented groups that helped to elect them. Albert Karnig and Susan Welch's and Peter Eisinger's studies of the impact of black mayors reach a similar conclusion. My work on Boston and San Francisco showed that, while neighborhood protest and greater mayoral sensitivity toward neighborhood concerns did not halt the transformation of these two cities, they did produce numerous specific policy changes. In a comparative context, studies by Michael Aiken and Guido Martinotti and Edmond Preteceille suggest that left-wing local governments in Europe produced more progressive policy outputs. Castells concludes that "grassroots mobilization has been a crucial factor in the shaping of the city, as well as the decisive element in urban innovation against prevailing social interests."

[. . .]

In sum, the analysis of how Edward Koch and his allies constructed a new dominant political coalition in New York City must be framed in terms of three broad sets of factors. Building on the structuralists, it must understand how the local political system's interaction with private interests creates constraints and imperatives for the local state but also oppor-tunities that astute political entrepreneurs can seize. Second, it must go beyond the structuralists by recognizing that how popular constituencies are organized in the city's political and electoral arena has an equally strong impact on the strategies pursued by coalition builders. Finally, a sound theory must be sensitive to how the organization of interests within the public sector, embodied in political practices as well as formal authority, also influenced their choices and actions . . .

FOUR

"A Ladder of Citizen Participation"

Journal of the American Institute of Planners (1969)

Sherry Arnstein

Editors' Introduction

The previous selections by Harvey (p. 227) and Mollenkopf (p. 235) argue that local government is important and plural actors can influence the outcome of policies and programs that affect their lives. It depicts local government as an increasingly important part of a new global order where public, private, and nonprofit sectors work at odds with each other or together in complex regimes. This new order raises local citizens' stakes in having their interests taken into account by local decision makers. But how, exactly, should citizens "participate" in local government decision-making? Guidance as to how this might best be done comes from a classic article by Sherry Arnstein titled "A Ladder of Citizen Participation."

Arnstein had firsthand experience with neighborhood organizations and the tumultuous urban politics of the 1960s. She uses the metaphor of a ladder to describe gradations of "citizen participation" in urban programs that affect their lives. She makes clear her own personal commitment to a redistribution of power from haves to have-nots by empowering the poor and powerless.

At the lowest level of Arnstein's ladder are two forms of nonparticipation, which she terms *manipulation* and *therapy*. According to Arnstein, some governmental organizations have contrived phony forms of participation, which are really aimed at educating citizens to accept a predetermined course of action. While gullible citizens may think they are participating in decision-making at these lowest levels of the ladder, Arnstein says they really are not. They are simply being used by decision makers. Almost at the bottom of the ladder is another form of nonparticipation, which Arnstein identifies as *therapy*. Arnstein brands this form of nonparticipation both dishonest and arrogant. Here the intent is to "cure" participants of attitudes and behaviors that local government officials do not like under the guise of seeking their advice.

Legitimate, but low, rungs of the ladder are *informing* and *consultation*. Informing citizens of the facts about a government program and their rights, responsibilities and options is a good first step, particularly if it is designed to go beyond a one-way flow of information. Consultation – getting citizens' opinions – is even better if the process is honest and their opinions are really considered. Surveys, for example, may provide real input from citizens to decision makers, but if that is the only form of participation they would not go far in assuring that citizen views really carry weight. *Placation* – in which government gives in to some citizen demands – goes a step further. But a model in which government throws citizens some crumbs to placate them is not really a satisfactory relationship.

The highest rungs on Arnstein's ladder are *partnership*, three rungs from the top, *delegated power*, one rung below the top, and *citizen control* at the very top of the ladder. During the "War on Poverty" in the 1960s, local government delegated power to run programs to some citizen groups or gave them full control over programs. Delegated power and citizens' control have been rare since that time. Opponents of citizen control advance many of the arguments that Arnstein identifies – that citizen control arguably balkanizes public services, may be costly and inefficient, can enable citizen hustlers to be opportunistic, and may be symbolic politics.

Today public–private–nonprofit partnerships are popular. Arnstein places true partnerships relatively high on her eight-rung ladder because partnerships represent a redistribution of power arrived at through negotiation. Where the odd bedfellows of local government, private corporations, and neighborhood nonprofit community-based organizations form joint planning and decision-making structures, citizen views can have real weight.

Both Sherry Arnstein and Paul Davidoff were engaged liberals who wrote their classic statements about citizen participation and advocacy planning in the late 1960s. Compare the approach of Davidoff, the lawyer, who argues in favor of skilled professionals advocating on behalf of powerless clients (p. 388) with the approach of Arnstein, the social work professional, who favors empowering individuals and communities by involving them directly in planning and decision-making.

Other books on citizen participation in urban planning and programs include Milan J. Dluhy, *Building Coalitions in the Human Services* (Thousand Oaks, CA: Sage, 1990), Michael Fagence, *Citizen Participation in Planning* (Oxford and New York: Pergamon Press, 1977), Ronald L. Thomas, Mary C. Means, Margaret A. Grieve, and Neil R. Peirce, *Taking Charge: How Communities are Planning their Futures: A Special Report on Long Range/Strategic Planning Trends and Innovations* (Washington, DC: International City/County Management Association, 1988).

Peter Marris and Martin Rein's *Dilemmas of Social Reform*, 2nd edn (Chicago: University of Chicago Press, 1982) describes community-based urban programs and articulates a philosophy of social change that influenced US urban policy in the 1960s. Two very different views on the US "War on Poverty" are Sar Levitan, *The Great Society's Poor Law* (Baltimore, MD: Johns Hopkins University Press, 1969), and Daniel Patrick Moynihan, *Maximum Feasible Misunderstanding* (New York: Free Press, 1969). The US Model Cities program, its antecedents, and the initial phase of the successor Community Development Block Grant program are discussed in Bernard J. Frieden and Marshal Kaplan, *The Politics of Neglect: Urban Aid from Model Cities to Revenue Sharing* (Cambridge, MA: MIT Press, 1975).

A practitioner's book on how to encourage community participation in urban planning is Henry Sanoff, *Community Participation Methods in Design and Planning* (New York: Wiley, 1999).

Books on public participation in urban planning and programs in Europe include James Barlow, *Public Participation in Urban Development: The European Experience* (Washington, DC: Brookings, 1995), Her Majesty's Stationery Office, *Community Involvement in Planning and Development Processes* (London: HMSO, 1995), and Albert Mabileau, *Local Politics and Participation in Britain and France* (Cambridge: Cambridge University Press, 1990).

The idea of citizen participation is a little like eating spinach: no one is against it in principle because it is good for you. Participation of the governed in their government is, in theory, the cornerstone of democracy – a revered idea that is vigorously applauded by virtually everyone. The applause is reduced to polite handclaps, however, when this principle is advocated by the have-not blacks, Mexican Americans, Puerto Ricans, Indians, Eskimos, and whites. And when the have-nots define participation as redistribution of power, the American consensus on the fundamental principle explodes into many shades of outright racial, ethnic, ideological, and political opposition.

There have been many recent speeches, articles, and books which explore in detail *who* are the have-nots of our time. There has been much recent documentation of *why* the have-nots have become so offended and embittered by their powerlessness to deal with the profound inequities and injustices pervading their daily lives. But there has been very little analysis of the content of the current controversial slogan: "citizen participation" or "maximum feasible participation." In short: *What* is citizen participation and what is its relationship to the social imperatives of our time?

Citizen participation is citizen power

Because the question has been a bone of political contention, most of the answers have been purposely buried in innocuous euphemisms like "self-help" or "citizen involvement." Still others have been embellished with misleading rhetoric like "absolute control" which is something no one – including the President of the United States – has or can have. Between understated euphemisms and exacerbated rhetoric, even scholars have found it difficult to follow the controversy. To the headline reading public, it is simply bewildering.

My answer to the critical *what* question is simply that citizen participation is a categorical term for citizen power. It is the redistribution of power that enables the have-not citizens, presently excluded from the political and economic processes, to be deliberately included in the future. It is the strategy by which the have-nots join in determining how information is shared, goals and policies are set, tax resources are allocated, programs are operated, and benefits like contracts and patronage are parceled out. In short, it is the means by which they can induce significant social reform which enables them to share in the benefits of the affluent society.

EMPTY REFUSAL VERSUS BENEFIT

There is a critical difference between going through the empty ritual of participation and having the real power needed to affect the outcome of the process. This difference is brilliantly capsulized in a poster painted last spring [1968] by the French students to explain the student-worker rebellion. (See Figure 1.) The poster highlights the fundamental point that participation without redistribution of power is an empty and frustrating process for the powerless. It allows the powerholders to claim that all sides were considered, but makes it possible for only some of those sides to benefit. It maintains the status quo. Essentially, it is what has been happening in most of the 1,000 Community Action Programs, and what promises to be repeated in the vast majority of the 150 Model Cities programs.

Types of participation and "nonparticipation"

A typology of eight *levels* of participation may help in analysis of this confused issue. For illustrative purposes the eight types are arranged in a ladder pattern with each rung corresponding to the extent of

Figure 1 French student poster. In English, "I participate, you participate, he participates, we participate, you participate . . . they profit"

citizens' power in determining the end product. (See Figure 2.)

The bottom rungs of the ladder are (1) *Manipulation* and (2) *Therapy*. These two rungs describe levels of "nonparticipation" that have been contrived by some to substitute for genuine participation. Their real objective is not to enable people to participate in planning or conducting programs, but to enable powerholders to "educate" or "cure" the participants. Rungs 3 and 4 progress to levels of "tokenism" that allow the have-nots to hear and to have a voice: (3) *Informing* and (4) *Consultation*. When they are proffered by power-holders as the total extent of participation, citizens may indeed hear and be heard. But under these conditions they lack the power to insure that their views will be *heeded* by the powerful. When participation is restricted to these levels, there is no follow-through, no "muscle," hence no assurance of changing the status quo. Rung (5) *Placation* is simply a higher level tokenism because the groundrules allow have-nots to advise, but retain for the powerholders the continued right to decide.

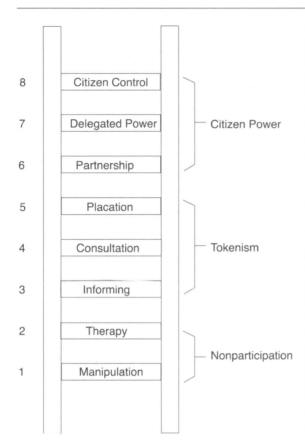

Figure 2 Eight rungs on the ladder of citizen participation

Further up the ladder are levels of citizen power with increasing degrees of decision-making clout. Citizens can enter into a (6) *Partnership* that enables them to negotiate and engage in trade-offs with traditional power holders. At the topmost rungs, (7) *Delegated Power* and (8) *Citizen Control*, have-not citizens obtain the majority of decision-making seats, or full managerial power.

Obviously, the eight-rung ladder is a simplification, but it helps to illustrate the point that so many have missed – that there are significant gradations of citizen participation. Knowing these gradations makes it possible to cut through the hyperbole to understand the increasingly strident demands for participation from the have-nots as well as the gamut of confusing responses from the powerholders.

Though the typology uses examples from federal programs such as urban renewal, anti-poverty, and Model Cities, it could just as easily be illustrated in the church, currently facing demands for power from priests and laymen who seek to change its mission;

colleges and universities which in some cases have become literal battlegrounds over the issue of student power; or public schools, city halls, and police departments (or big business which is likely to be next on the expanding list of targets). The underlying issues are essentially the same – "nobodies" in several arenas are trying to become "somebodies" with enough power to make the target institutions responsive to their views, aspirations, and needs.

LIMITATIONS OF THE TYPOLOGY

The ladder juxtaposes powerless citizens with the powerful in order to highlight the fundamental divisions between them. In actuality, neither the have-nots nor the powerholders are homogeneous blocs. Each group encompasses a host of divergent points of view, significant cleavages, competing vested interests, and splintered subgroups. The justification for using such simplistic abstractions is that in most cases the have-nots really do perceive the powerful as a monolithic "system," and powerholders actually do view the have-nots as a sea of "those people," with little comprehension of the class and caste differences among them.

It should be noted that the typology does not include an analysis of the most significant roadblocks to achieving genuine levels of participation. These roadblocks lie on both sides of the simplistic fence. On the powerholders' side, they include racism, paternalism, and resistance to power redistribution. On the have-nots' side, they include inadequacies of the poor community's political socioeconomic infrastructure and knowledge-base, plus difficulties of organizing a representative and accountable citizens' group in the face of futility, alienation, and distrust.

Another caution about the eight separate rungs on the ladder: In the real world of people and programs, there might be 150 rungs with less sharp and "pure" distinctions among them. Furthermore, some of the characteristics used to illustrate each of the eight types might be applicable to other rungs. For example, employment of the have-nots in a program or on a planning staff could occur at any of the eight rungs and could represent either a legitimate or illegitimate characteristic of citizen participation. Depending on their motives, powerholders can hire poor people to coopt them, to placate them, or to utilize the have-nots' special skills and insights. Some mayors, in private, actually boast of their strategy in hiring militant black leaders to muzzle them while destroying their credibility in the black community.

Characteristics and illustrations

It is in this context of power and powerlessness that the characteristics of the eight rungs are illustrated by examples from current federal social programs.

1. MANIPULATION

In the name of citizen participation, people are placed on rubberstamp advisory committees or advisory boards for the express purpose of "educating" them or engineering their support. Instead of genuine citizen participation, the bottom rung of the ladder signifies the distortion of participation into a public relations vehicle by powerholders.

This illusory form of "participation" initially came into vogue with urban renewal when the socially elite were invited by city housing officials to serve on Citizen Advisory Committees (CACs). Another target of manipulation were the CAC subcommittees on minority groups, which in theory were to protect the rights of Negroes in the renewal program. In practice, these subcommittees, like their parent CACs, functioned mostly as letterheads, trotted forward at appropriate times to promote urban renewal plans (in recent years known as Negro removal plans).

At meetings of the Citizen *Advisory* Committees, it was the officials who educated, persuaded, and advised the citizens, not the reverse. Federal guidelines for the renewal programs legitimized the manipulative agenda by emphasizing the terms "information-gathering," public relations," and "support" as the explicit functions of the committees.

This style of nonparticipation has since been applied to other programs encompassing the poor. Examples of this are seen in Community Action Agencies (CAAs) which have created structures called "neighborhood councils" or "neighborhood advisory groups." These bodies frequently have no legitimate function or power. The CAAs use them to "prove" that "grass-roots people" are involved in the program. But the program may not have been discussed with "the people." Or it may have been described at a meeting in the most general terms; "We need your signatures on this proposal for a multiservice center which will house, under one roof, doctors from the health department, workers from the welfare department, and specialists from the employment service."

The signatories are not informed that the $2 million-per-year center will only refer residents to the same old waiting lines at the same old agencies across town. No one is asked if such a referral center is really needed in his neighborhood. No one realizes that the contractor for the building is the mayor's brother-in-law, or that the new director of the center will be the same old community organization specialist from the urban renewal agency.

After signing their names, the proud grassrooters dutifully spread the word that they have "participated" in bringing a new and wonderful center to the neighborhood to provide people with drastically needed jobs and health and welfare services. Only after the ribbon-cutting ceremony do the members of the neighborhood council realize that they didn't ask the important questions, and that they had no technical advisors of their own to help them grasp the fine legal print. The new center, which is open 9 to 5 on weekdays only, actually adds to their problems. Now the old agencies across town won't talk with them unless they have a pink paper slip to prove that they have been referred by "their" shiny new neighborhood center.

Unfortunately, this chicanery is not a unique example. Instead it is almost typical of what has been perpetrated in the name of high-sounding rhetoric like "grassroots participation." This sham lies at the heart of the deep-seated exasperation and hostility of the have-nots toward the powerholders.

One hopeful note is that, having been so grossly affronted, some citizens have learned the Mickey Mouse game, and now they too know how to play. As a result of this knowledge, they are demanding genuine levels of participation to assure them that public programs are relevant to their needs and responsive to their priorities.

2. THERAPY

In some respects group therapy, masked as citizen participation, should be on the lowest rung of the ladder because it is both dishonest and arrogant. Its administrators – mental health experts from social workers to psychiatrists – assume that powerlessness is synonymous with mental illness. On this assumption, under a masquerade of involving citizens in planning, the experts subject the citizens to clinical group therapy. What makes this form of "participation" so invidious is that citizens are engaged in extensive activity, but the focus of it is on curing them of their "pathology" rather than changing the racism and victimization that create their "pathologies."

Consider an incident that occurred in Pennsylvania less than one year ago. When a father took his seriously ill baby to the emergency clinic of a local hospital, a young resident physician on duty instructed him to take the baby home and feed it sugar water. The baby died that afternoon of pneumonia and dehydration. The overwrought father complained to the board of the local Community Action Agency. Instead of launching an investigation of the hospital to determine what changes would prevent similar deaths or other forms of malpractice, the board invited the father to attend the CAA's (therapy) child-care sessions for parents, and promised him that someone would "telephone the hospital director to see that it never happens again."

Less dramatic, but more common examples of therapy, masquerading as citizen participation, may be seen in public housing programs where tenant groups are used as vehicles for promoting control-your-child or cleanup campaigns. The tenants are brought together to help them "adjust their values and attitudes to those of the larger society." Under these ground rules, they are diverted from dealing with such important matters as: arbitrary evictions; segregation of the housing project; or why there is a three-month time lapse to get a broken window replaced in winter.

The complexity of the concept of mental illness in our time can be seen in the experiences of student/civil rights workers facing guns, whips, and other for s of terror in the South. They needed the help of socially attuned psychiatrists to deal with their fears and to avoid paranoia.

3. INFORMING

Informing citizens of their rights, responsibilities, and options can be the most important first step toward legitimate citizen participation. However, too frequently the emphasis is placed on a one-way flow of information – from officials to citizens – with no channel provided for feedback and no power for negotiation. Under these conditions, particularly when information is provided at a late stage in planning,

people have little opportunity to influence the program designed "for their benefit." The most frequent tools used for such one-way communication are the news media, pamphlets, posters, and responses to inquiries.

Meetings can also be turned into vehicles for one-way communication by the simple device of providing superficial information, discouraging questions, or giving irrelevant answers. At a recent Model Cities citizen planning meeting in Providence, Rhode Island, the topic was "tot-lots." A group of elected citizen representatives, almost all of whom were attending three to five meetings a week, devoted an hour to a discussion of the placement of six tot-lots. The neighborhood is half black, half white. Several of the black representatives noted that four tot-lots were proposed for the white district and only two for the black. The city official responded with a lengthy, highly technical explanation about costs per square foot and available property. It was clear that most of the residents did not understand his explanation. And it was clear to observers from the Office of Economic Opportunity that other options did exist which, considering available funds, would have brought about a more equitable distribution of facilities. Intimidated by futility, legalistic jargon, and prestige of the official, the citizens accepted the "information" and endorsed the agency's proposal to place four lots in the white neighborhood.

4. CONSULTATION

Inviting citizens' opinions, like informing them, can be a legitimate step toward their full participation. But if consulting them is not combined with other modes of participation, this rung of the ladder is still a sham since it offers no assurance that citizen concerns and ideas will be taken into account. The most frequent methods used for consulting people are attitude surveys, neighborhood meetings, and public hearings.

When powerholders restrict the input of citizens' ideas solely to this level, participation remains just a window-dressing ritual. People are primarily perceived as statistical abstractions, and participation is measured by how many come to meetings, take brochures home, or answer a questionnaire. What citizens achieve in all this activity is that they have "participated in participation." And what powerholders achieve is the evidence that they have gone through the required motions of involving "those people."

Attitude surveys have become a particular bone of contention in ghetto neighborhoods. Residents are increasingly unhappy about the number of times per week they are surveyed about their problems and hopes. As one woman put it: "Nothing ever happens with those damned questions, except the surveyor gets $3 an hour, and my washing doesn't get done that day." In some communities, residents are so annoyed that they are demanding a fee for research interviews.

Attitude surveys are not very valid indicators of community opinion when used without other input from citizens. Survey after survey (paid for out of anti-poverty funds) has "documented" that poor housewives most want tot-lots in their neighborhood where young children can play safely. But most of the women answered these questionnaires without knowing what their options were. They assumed that if they asked for something small, they might just get something useful in the neighborhood. Had the mothers known that a free prepaid health insurance plan was a possible option, they might not have put tot-lots so high on their wish lists.

A classic misuse of the consultation rung occurred at a New Haven, Connecticut, community meeting held to consult citizens on a proposed Model Cities grant. James V. Cunningham, in an unpublished report to the Ford Foundation, described the crowd as large and mostly hostile:

> Members of The Hill Parents Association demanded to know why residents had not participated in drawing up the proposal. CAA director Spitz explained that it was merely a proposal for seeking Federal planning funds – that once funds were obtained, residents would be deeply involved in the planning. An outside observer who sat in the audience described the meeting this way:
> "Spitz and Mel Adams ran the meeting on their own. No representatives of a Hill group moderated or even sat on the stage. Spitz told the 300 residents that this huge meeting was an example of 'participation in planning.' To prove this, since there was a lot of dissatisfaction in the audience, he called for a 'vote' on each component of the proposal. The vote took this form: 'Can I see the hands of all those in favor of a health clinic? All those opposed?' It was a little like asking who favors motherhood."

It was a combination of the deep suspicion aroused at this meeting and a long history of similar forms of "window-dressing participation" that led New Haven residents to demand control of the program.

By way of contrast, it is useful to look at Denver where technicians learned that even the best intentioned among them are often unfamiliar with, and even insensitive to, the problems and aspirations of the poor. The technical director of the Model Cities program has described the way professional planners assumed that the residents, victimized by high-priced local storekeepers, "badly needed consumer education." The residents, on the other hand, pointed out that the local storekeepers performed a valuable function. Although they overcharged, they also gave credit, offered advice, and frequently were the only neighborhood place to cash welfare or salary checks. As a result of this consultation, technicians and residents agreed to substitute the creation of needed credit institutions in the neighborhood for a consumer education program.

5. PLACATION

It is at this level that citizens begin to have some degree of influence though tokenism is still apparent. An example of placation strategy is to place a few hand-picked "worthy" poor on boards of Community Action Agencies or on public bodies like the board of education, police commission, or housing authority. If they are not accountable to a constituency in the community and if the traditional power elite hold the majority of seats, the have-nots can be easily outvoted and outfoxed. Another example is the Model Cities advisory and planning committees. They allow citizens to advise or plan ad infinitum but retain for power-holders the right to judge the legitimacy or feasibility of the advice. The degree to which citizens are actually placated, of course, depends largely on two factors: the quality of technical assistance they have in articulating their priorities; and the extent to which the community has been organized to press for those priorities.

It is not surprising that the level of citizen participation in the vast majority of Model Cities programs is at the placation rung of the ladder or below. Policy-makers at the Department of Housing and Urban Development (HUD) were determined to return the genie of citizen power to the bottle from which it had escaped (in a few cities) as a result of the provision stipulating "maximum feasible participation" in poverty programs. Therefore, HUD channeled its physical-social-economic rejuvenation approach for blighted neighborhoods through city hall. It drafted legislation requiring that all Model Cities' money flow to a local City Demonstration Agency (CDA) through the elected city council. As enacted by Congress, this gave local city councils final veto power over planning and programming and ruled out any direct funding relationship between community groups and HUD.

HUD required the CDAs to create coalition, policy-making boards that would include necessary local powerholders to create a comprehensive physical-social plan during the first year. The plan was to be carried out in a subsequent five-year action phase. HUD, unlike OEO, did not require that have-not citizens be included on the CDA decision-making boards. HUD's Performance Standards for Citizen Participation only demanded that "citizens have clear and direct access to the decision-making process."

Accordingly, the CDAs structured their policy-making boards to include some combination of elected officials; school representatives; housing, health, and welfare officials; employment and police department representatives; and various civic, labor, and business leaders. Some CDAs included citizens from the neighborhood. Many mayors correctly interpreted the HUD provision for "access to the decision-making process" as the escape hatch they sought to relegate citizens to the traditional advisory role.

Most CDAs created residents' advisory committees. An alarmingly significant number created citizens' policy boards and citizens' policy committees which are totally misnamed as they have either no policy-making function or only a very limited authority. Almost every CDA created about a dozen planning committees or task forces on functional lines: health, welfare, education, housing, and unemployment. In most cases, have-not citizens were invited to serve on these committees along with technicians from relevant public agencies. Some CDAs, on the other hand, structured planning committees of technicians and parallel committees of citizens.

In most Model Cities programs, endless time has been spent fashioning complicated board, committee, and task force structures for the planning year. But the rights and responsibilities of the various elements of those structures are not defined and are ambiguous. Such ambiguity is likely to cause considerable conflict at the end of the one-year planning process. For at this point, citizens may realize that they have once again extensively "participated" but have not profited beyond the extent the powerholders decide to placate them.

Results of a staff study (conducted in the summer of 1968 before the second round of seventy-five planning grants were awarded) were released in a December 1968 HUD bulletin. Though this public document uses much more delicate and diplomatic language, it attests to the already cited criticisms of non-policy-making policy boards and ambiguous complicated structures, in addition to the following findings:

1. Most CDAs did not negotiate citizen participation requirements with residents.
2. Citizens, drawing on past negative experiences with local powerholders, were extremely suspicious of this new panacea program. They were legitimately distrustful of city hall's motives.
3. Most CDAs were not working with citizens' groups that were genuinely representative of model neighborhoods and accountable to neighborhood constituencies. As in so many of the poverty programs, those who were involved were more representative of the upwardly mobile working-class. Thus their acquiescence to plans prepared by city agencies was not likely to reflect the views of the unemployed, the young, the more militant residents, and the hard-core poor.
4. Residents who were participating in as many as three to five meetings per week were unaware of their minimum rights, responsibilities, and the options available to them under the program. For example, they did not realize that they were not required to accept technical help from city technicians they distrusted.
5. Most of the technical assistance provided by CDAs and city agencies was of third-rate quality, paternalistic, and condescending. Agency technicians did not suggest innovative options. They reacted bureaucratically when the residents pressed for innovative approaches. The vested interests of the old-line city agencies were a major – albeit hidden – agenda.
6. Most CDAs were not engaged in planning that was comprehensive enough to expose and deal with the roots of urban decay. They engaged in "meetingitis" and were supporting strategies that resulted in "projectitis," the outcome of which was a "laundry list" of traditional programs to be conducted by traditional agencies in the traditional manner under which slums emerged in the first place.

7. Residents were not getting enough information from CDAs to enable them to review CDA developed plans or to initiate plans of their own as required by HUD. At best, they were getting superficial information. At worst, they were not even getting copies of official HUD materials.

8. Most residents were unaware of their rights to be reimbursed for expenses incurred because of participation – babysitting, transportation costs, and so on. The training of residents, which would enable them to understand the labyrinth of the federal–state–city systems and networks of sub-systems, was an item that most CDAs did not even consider.

These findings led to a new public interpretation of HUD's approach to citizen participation. Though the requirements for the seventy-five "second-round" Model City grantees were not changed, HUD's twenty-seven-page technical bulletin on citizen participa-tion repeatedly advocated that cities share power with residents. It also urged CDAs to experiment with subcontracts under which the residents' groups could hire their own trusted technicians.

A more recent evaluation was circulated in February 1969 by OSTI, a private firm that entered into a contract with OEO to provide technical assis-tance and training to citizens involved in Model Cities programs in the north-east region of the country. OSTI's report to OEO corroborates the earlier study. In addition it states:

> In practically no Model Cities structure does citizen participation mean truly shared decision-making, such that citizens might view themselves as "the partners in this program . . ."
>
> In general, citizens are finding it impossible to have a significant impact on the comprehensive planning which is going on. In most cases the staff planners of the CDA and the planners of existing agencies are carrying out the actual planning with citizens having a peripheral role of watchdog and, ultimately, the "rubber stamp" of the plan generated. In cases where citizens have the direct responsibility for generating program plans, the time period allowed and the independent technical resources being made available to them are not adequate to allow them to do anything more than generate very traditional approaches to the problems they are attempting to solve.

In general, little or no thought has been given to the means of insuring continued citizen partici-pation during the stage of implementation. In most cases, traditional agencies are envisaged as the implementors of Model Cities programs and few mechanisms have been developed for encouraging organizational change or change in the method of program delivery within these agencies or for insuring that citizens will have some influence over these agencies as they implement Model Cities programs . . . By and large, people are once again being planned *for*. In most situations the major planning decisions are being made by CDA staff and approved in a formalistic way by policy boards.

6. PARTNERSHIP

At this rung of the ladder, power is in fact redistributed through negotiation between citizens and power-holders. They agree to share planning and decision-making responsibilities through such structures as joint policy boards, planning committees, and mechanisms for resolving impasses. After the groundrules have been established through some form of give-and-take, they are not subject to unilateral change.

Partnership can work most effectively when there is an organized power-base in the community to which the citizen leaders are accountable; when the citizens' group has the financial resources to pay its leaders reasonable honoraria for their time-consuming efforts; and when the group has the resources to hire (and fire) its own technicians, lawyers, and community organizers. With these ingredients, citizens have some genuine bargaining influence over the outcome of the plan (as long as both parties find it useful to maintain the partnership). One community leader described it "like coming to city hall with hat on head instead of in hand."

In the Model Cities program only about fifteen of the so-called first generation of seventy-five cities have reached some significant degree of power-sharing with residents. In all but one of those cities, it was angry citizen demands, rather than city initiative, that led to the negotiated sharing of power. The negotiations were triggered by citizens who had been enraged by previous forms of alleged participation. They were both angry and sophisticated enough to refuse to be "conned" again. They threatened to oppose the awarding of a planning grant to the city. They sent delegations to HUD in Washington. They used abrasive language. Negotiation took place under a cloud of suspicion and rancor.

In most cases where power has come to be shared it was *taken by the citizens*, not given by the city. There is nothing new about that process. Since those who have power normally want to hang onto it, historically it has had to be wrested by the powerless rather than proffered by the powerful.

Such a working partnership was negotiated by the residents in the Philadelphia model neighborhood. Like most applicants for a Model Cities grant, Philadelphia wrote its more than 400-page application and waved it at a hastily called meeting of community leaders. When those present were asked for an endorsement, they angrily protested the city's failure to consult them on preparation of the extensive application. A community spokesman threatened to mobilize a neighborhood protest *against* the application unless the city agreed to give the citizens a couple of weeks to review the application and recommend changes. The officials agreed.

At their next meeting, citizens handed the city officials a substitute citizen participation section that changed the groundrules from a weak citizens' advisory role to a strong shared power agreement. Philadelphia's application to HUD included the citizens' substitution word for word. (It also included a new citizen prepared introductory chapter that changed the city's description of the model neighborhood from a paternalistic description of problems to a realistic analysis of its strengths, weaknesses, and potentials.) Consequently, the proposed policy-making committee of the Philadelphia CDA was revamped to give five out of eleven seats to the residents' organization, which is called the Area Wide Council (AWC). The AWC obtained a subcontract from the CDA for more than $20,000 per month, which it used to maintain the neighborhood organization, to pay citizen leaders $7 per meeting for their planning services, and to pay the salaries of a staff of community organizers, planners, and other technicians. AWC has the power to initiate plans of its own, to engage in joint planning with CDA committees, and to review plans initiated by city agencies. It has a veto power in that no plans may be submitted by the CDA to the city council until they have been reviewed, and any differences of opinion have been successfully negotiated with the AWC. Representatives of the AWC (which is a federation of neighborhood organizations grouped

into sixteen neighborhood "hubs") may attend all meetings of CDA task forces, planning committees, or sub-committees.

Though the city council has final veto power over the plan (by federal law), the AWC believes it has a neighborhood constituency that is strong enough to negotiate any eleventh-hour objections the city council might raise when it considers such AWC proposed innovations as an AWC Land Bank, an AWC Economic Development Corporation, and an experimental income maintenance program for 900 poor families.

7. DELEGATED POWER

Negotiations between citizens and public officials can also result in citizens achieving dominant decision making authority over a particular plan or program. Model City policy boards or CAA delegate agencies on which citizens have a clear majority of seats and genuine specified powers are typical examples. At this level, the ladder has been scaled to the point where citizens hold the significant cards to assure accountability of the program to them. To resolve differences, powerholders need to start the bargaining process rather than respond to pressure from the other end.

Such a dominant decision-making role has been attained by residents in a handful of Model Cities including Cambridge, Massachusetts; Dayton and Columbus, Ohio; Minneapolis, Minnesota; St. Louis, Missouri; Hartford and New Haven, Connecticut; and Oakland, California.

In New Haven, residents of the Hill neighborhood have created a corporation that has been delegated the power to prepare the entire Model Cities plan. The city, which received a $117,000 planning grant from HUD, has subcontracted $110,000 of it to the neighborhood corporation to hire its own planning staff and consultants. The Hill Neighborhood Corporation has eleven representatives on the twenty-one-member CDA board which assures it a majority voice when its proposed plan is reviewed by the CDA.

Another model of delegated power is separate and parallel groups of citizens and powerholders, with provision for citizen veto if differences of opinion cannot be resolved through negotiation. This is a particularly interesting coexistence model for hostile citizen groups too embittered toward city hall – as a result of past "collaborative efforts" – to engage in joint planning.

Since all Model Cities programs require approval by the city council before HUD will fund them, city councils have final veto powers even when citizens have the majority of seats on the CDA Board. In Richmond, California, the city council agreed to a citizens' counter-veto, but the details of that agreement are ambiguous and have not been tested.

Various delegated power arrangements are also emerging in the Community Action Program as a result of demands from the neighborhoods and OEO's most recent instruction guidelines which urged CAAs "to exceed (the) basic requirements" for resident participation. In some cities, CAAs have issued subcontracts to resident dominated groups to plan and/or operate one or more decentralized neighborhood program components like a multipurpose service center or a Headstart program. These contracts usually include an agreed upon line-by-line budget and program specifications. They also usually include a specific statement of the significant powers that have been delegated, for example: policy-making; hiring and firing; issuing subcontracts for building, buying, or leasing. (Some of the subcontracts are so broad that they verge on models for citizen control.)

8. CITIZEN CONTROL

Demands for community controlled schools, black control, and neighborhood control are on the increase. Though no one in the nation has absolute control, it is very important that the rhetoric not be confused with intent. People are simply demanding that degree of power (or control) which guarantees that participants or residents can govern a program or an institution, be in full charge of policy and managerial aspects, and be able to negotiate the conditions under which "outsiders" may change them.

A neighborhood corporation with no intermediaries between it and the source of funds is the model most frequently advocated. A small number of such experimental corporations are already producing goods and/or social services. Several others are reportedly in the development stage, and new models for control will undoubtedly emerge as the have-nots continue to press for greater degrees of power over their lives.

Though the bitter struggle for community control of the Ocean Hill-Brownsville schools in New York City

has aroused great fears in the headline reading public, less publicized experiments are demonstrating that the have-nots can indeed improve their lot by handling the entire job of planning, policy-making, and managing a program. Some are even demonstrating that they can do all this with just one arm because they are forced to use their other one to deal with a continuing barrage of local opposition triggered by the announcement that a federal grant has been given to a community group or an all black group.

Most of these experimental programs have been capitalized with research and demonstration funds from the Office of Economic Opportunity in co-operation with other federal agencies. Examples include:

1. A $1.8 million grant was awarded to the Hough Area Development Corporation in Cleveland to plan economic development programs in the ghetto and to develop a series of economic enterprises ranging from a novel combination shopping-center-public-housing project to a loan guarantee program for local building contractors. The membership and board of the nonprofit corporation is composed of leaders of major community organizations in the black neighborhood.

2. Approximately $1 million ($595,751 for the second year) was awarded to the Southwest Alabama Farmers' Cooperative Association (SWAFCA) in Selma, Alabama, for a ten-county marketing co-operative for food and livestock. Despite local attempts to intimidate the coop (which included the use of force to stop trucks on the way to market) first year membership grew to 1,150 farmers who earned $52,000 on the sale of their new crops. The elected coop board is composed of two poor black farmers from each of the ten economically depressed counties.

3. Approximately $600,000 ($300,000 in a supplemental grant) was granted to the Albina Corporation and the Albina Investment Trust to create a black-operated, black-owned manufacturing concern using inexperienced management and unskilled minority group personnel from the Albina district. The profitmaking wool and metal fabrication plant will be owned by its employees through a deferred compensation trust plan.

4. Approximately $800,000 ($400,000 for the second year) was awarded to the Harlem Commonwealth Council to demonstrate that a community-based

development corporation can catalyze and implement an economic development program with broad community support and participation. After only eighteen months of program development and negotiation, the council will soon launch several large-scale ventures including operation of two supermarkets, an auto service and repair center (with built-in manpower training program), a finance company for families earning less than $4,000 per year, and a data processing company. The all black Harlem-based board is already managing a metal castings foundry.

Though several citizen groups (and their mayors) use the rhetoric of citizen control, no Model City can meet the criteria of citizen control since final approval power and accountability rest with the city council.

Daniel P. Moynihan argues that city councils are representative of the community, but Adam Walinsky illustrates the nonrepresentativeness of this kind of representation:

Who . . . exercises "control" through the representative process? In the Bedford-Stuyvesant ghetto of New York there are 450,000 people – as many as in the entire city of Cincinnati, more than in the entire state of Vermont. Yet the area has only one high school, and 80 per cent of its teenagers are dropouts; the infant mortality rate is twice the national average; there are over 8000 buildings abandoned by everyone but the rats, yet the area received not one dollar of urban renewal funds during the entire first 15 years of that program's operation; the unemployment rate is known only to God.

Clearly, Bedford-Stuyvesant has some special needs; yet it has always been lost in the midst of the city's eight million. In fact, it took a lawsuit to win for this vast area, in the year 1968, its first Congressman. In what sense can the representative system be said to have "spoken for" this community, during the long years of neglect and decay?

Walinsky's point on Bedford-Stuyvesant has general applicability to the ghettos from coast to coast.

It is therefore likely that in those ghettos where residents have achieved a significant degree of power in the Model Cities planning process, the first-year action plans will call for the creation of some new community institutions entirely governed by residents with a specified sum of money contracted to them. If the groundrules for these programs are clear and if citizens understand that achieving a genuine place in the pluralistic scene subjects them to its legitimate forms of give-and-take, then these kinds of programs might begin to demonstrate how to counteract the various corrosive political and socioeconomic forces that plague the poor.

In cities likely to become predominantly black through population growth, it is unlikely that strident citizens' groups like AWC of Philadelphia will eventually demand legal power for neighborhood self-government. Their grand design is more likely to call for a black city achieved by the elective process. In cities destined to remain predominantly white for the foreseeable future, it is quite likely that counterpart groups to AWC will press for separatist forms of neighborhood government that can create and control decentralized public services such as police protection, education systems, and health facilities. Much may depend on the willingness of city governments to entertain demands for resource allocation weighted in favor of the poor, reversing gross imbalances of the past.

Among the arguments against community control are: it supports separatism; it creates balkanization of public services; it is more costly and less efficient; it enables minority group "hustlers" to be just as opportunistic and disdainful of the have-nots as their white predecessors; it is incompatible with merit systems and professionalism; and ironically enough, it can turn out to be a new Mickey Mouse game for the have-nots by allowing them to gain control but not allowing them sufficient dollar resources to succeed. These arguments are not to be taken lightly. But neither can we take lightly the arguments of embittered advocates of community control – that every other means of trying to end their victimization has failed!

"The Need for a New Vision for the Development of Large U.S. Metropolitan Areas"

Saloman Brothers (1989)

Anthony Downs

Editors' Introduction

In the United States and throughout the world large, interconnected metropolitan regions have arisen. Often dozens of independent local governments exist in a single metropolitan region. Urban development in large metropolitan regions is often driven by the limited and self-serving plans of individual developers or the narrow self-interest of individual local governments without much concern for their neighbors or the public welfare. In a few rare instances "regional plans" with a clear vision have been prepared and implemented. Portland, Oregon, for example, has an elected "metro" government that is pursuing a regional vision to develop multiple city centers surrounded by open space. But in most cities "visions" of a desirable metropolitan future are sadly lacking. Toronto, Canada has incorporated surrounding suburban governments into a single metropolitan government. In addition to Downs, "new regionalist" urban planners and architects such as Peter Calthorpe and William Fulton (p. 331) and Andres Duany and Elizabeth Plater-Zyberk (p. 207) have worked out principles for regional plans that would address the problems Downs identifies.

In "The Need for a New Vision for the Development of Large U.S. Metropolitan Areas," Anthony Downs proposes a radical overhaul of metropolitan planning which would abandon the prevailing vision of single-family homes, an over-reliance on the automobile, and a governance structure that separates the needs of inner cities and suburban communities. In place of this outmoded vision, Downs argues for a more integrated system that provides suburban homes and apartments for low-wage workers and creates "governance structures that preserve substantial local authority – but within a framework that compels local governments to act responsibly to meet area-wide needs."

Downs is a senior fellow at the Brookings Institution, a Washington, DC, think-tank. He has written extensively over the past three decades on poverty, race, housing, urban sprawl, traffic congestion, metropolitan planning, and other urban issues. He often takes the logic of strategies that have been suggested or tried on a modest scale to their conceptual limit: What would it *really* take to completely disperse the population of every poor Black ghetto in America? How much would it really cost to completely eliminate traffic congestion? How much housing at what cost would really be required to provide every American household a decent, safe, and sanitary affordable housing unit?

Often, as in this selection, Downs concludes that strategies to achieve broad social goals would require massive government action and significantly increased levels of funding. This emphasis on government programs and spending differentiates Downs from conservatives like Michael Porter (p. 277) who oppose government spending to solve urban problems. But Downs is no bleeding-heart liberal. He advocates triage strategies for urban neighborhoods – leaving the worst neighborhoods to deteriorate while directing scarce resources to poor but salvageable communities. This puts him at odds with many liberals. So does the argument that Downs once put forward that the dispersal of Black ghettos would be accepted by the white majority in America only if Blacks were dispersed into white areas in such a way that the white majority would remain numerically and culturally dominant.

The selection reprinted here argues strongly for a sweeping planning strategy that eliminates the legal/governmental distinctions between impacted inner-city ghettos and comfortable suburban communities. It argues for integrated metropolitan solutions to traffic congestion and other problems that cannot be solved on a city-by-city basis. Originally published by the Saloman Brothers brokerage house, "The Need for a New Vision for the Development of Large U.S. Metropolitan Areas" is a visionary, yet hard-headed, approach to solving what has long been an intractable problem and one which threatens the stability of any urban future in a diverse society separated by race and class divisions.

Other books by Anthony Downs include *Urban Problems and Prospects* (Chicago: Markham, 1970), *Opening Up the Suburbs: An Urban Strategy for America* (New Haven, CT: Yale University Press, 1973), *Stuck in Traffic* (Washington, DC and Cambridge: Brookings Institution and Lincoln Institute of Land Policy, 1992), and *New Visions for Metropolitan America* (Washington, DC and Cambridge: Brookings Institution and Lincoln Institute of Land Policy, 1992).

Many of Downs's essays on urban issues are reprinted in *The Selected Essays of Anthony Downs*, vol. 1, *Political Theory and Public Choice* and vol. 2, *Urban Affairs and Urban Policy* (Cheltenham, UK and Northampton, MA: Edward Elgar, 1998).

For more on regional planning solutions to metropolitan problems see Eugenie Birch (ed.), *The City and Regional Planning Reader* (London and New York: Routledge, 2003) and Stephen M. Wheeler and Timothy Beatley (eds), *The Sustainable Urban Development Reader* (London and New York: Routledge, 2003).

INTRODUCTION

The very success of the long-dominant American ideal vision of how our metropolitan areas should develop now threatens its continuance, as that vision contains several major inner inconsistencies. This paradox is manifest in the rising protest across the nation against worsening suburban traffic congestion and declines in urban environmental quality. Up to now, these protests have been concentrated in our largest and most prosperous metropolitan areas, where the inherent inconsistencies of our dominant ideal vision have become most apparent. But some of that vision's failings will soon be visible in many more metropolitan areas.

Therefore, American society needs a new ideal vision to guide the future development of its large metropolitan areas. For us to formulate and accept such a new vision would be almost equivalent to a paradigm shift concerning urban development. Formulating such a new vision is vitally important to all participants in real estate in large metropolitan areas, including commercial developers, institutional investors, homebuilders, realtors, local officials, and homeowners. This article examines the nature and failings of the still-dominant ideal vision, and proposes an approach to a new alternative.

THE CURRENTLY DOMINANT IDEAL VISION OF DESIRABLE METROPOLITAN DEVELOPMENT

For the past few decades, one major vision about how US metropolitan areas ought to be developed has become totally dominant. This ideal vision reflects the views of the vast majority of American households – especially the more than 45% who live in the suburban portions of US metropolitan areas. The same vision is held by almost all suburban government officials. Hence, it has immensely influenced nearly all their policies concerning land use, transportation, etc.

This dominant ideal vision is built upon four pillars. Each is a key desire or aspiration shared by nearly all American households:

The first pillar is ownership of detached, single-family homes on spacious lots. Repeated polls show that over 90% of all American households would like to own their own homes, and the vast majority want single-family detached units. Hence, ownership of such a unit has become the heart of "the American dream" – the prevailing image of how a household "makes it" in contemporary society. Realization of this aspiration implies the dominance of very low-density settlement patterns.

The second pillar is ownership and use of a personal, private automotive vehicle. Every American wants to be able to leap into his or her own car and zoom off on an uncongested road, to wherever he or she wants to go, in total privacy and great comfort – and to arrive there in not more than 20 minutes. This factor has rapidly escalated total vehicle ownership and use. In 1983, over 87% of all U.S. households owned at least one car or truck, and 53% owned two or more such vehicles. Even among households with 1983 incomes under $10,000, over 60% owned at least one car or truck; among those with incomes of $40,000 and over, 98.9% owned at least one vehicle and 86.7% owned two or more. In 1987, there were 162 million licensed drivers in the United States, but 167 million cars and trucks in use.

During the five years 1983–1987, the United States added 9.2 million persons to its total human population, but more than twice as many cars and trucks – 20.1 million – to its total number of vehicles in use.

The third pillar of the dominant ideal vision involves the structure of suburban workplaces. They are visualized as consisting predominantly of low-rise office or industrial buildings or shopping centers, in attractively landscaped, park-like settings. Each such structure ought to be surrounded by a large supply of its own free parking so that those who work in it or patronize it can conveniently drive and park there without cost.

The fourth pillar for this ideal vision concerns governance. Most Americans want to live in small communities with strong local self-governments. They want those governments to control land use, public

schools, and other key elements affecting what they perceive as the quality of neighborhood life. This institutional structure permits existing residents to have a strong voice in controlling their local environments.

The above four elements define the prevailing ideal vision of "the American dream" as it is conceived of by the vast majority of suburbanites and by many city-dwellers. All four elements express what might be termed unconstrained individualism. They represent the pursuit of an environment that maximizes one's own well being, without regard to the collective results of such behavior. This ideal vision has been reinforced over the past few decades by promotional efforts made by parts of the real estate industry and by suburban communities. Such promoters include homebuilders selling new dwellings, realtors reselling them, advertisers highlighting suburban lifestyles, politicians running for election, and planning officials trying to carry out the policies of local politicians and voters. The dominant ideal vision has become so strongly entrenched that it has become almost political suicide to challenge openly any of these four pillars of "the American dream."

THE FIRST FLAW WITHIN THE IDEAL VISION: EXCESSIVE TRAVEL

Unfortunately, the prevailing ideal vision of how metropolitan areas ought to be developed contains four major flaws. They involve basic inconsistencies between its key elements and the real preferences of the citizenry concerning how suburban life should be lived.

The first flaw is that the low-density settlement patterns required by single-family housing and low-density workplaces generate immense travel requirements. Low density inherently spreads both homes and jobs widely across the landscape. That forces people to travel long distances between where they live and where they work, shop, or play. Moreover, low-density settlement patterns cannot efficiently support mass transit; hence they can only be served by private automotive vehicles. Mass transit, whether buses or fixed-rail, can be effective only if at least one end of most journeys is relatively concentrated in a few points.

The resulting requirement for massive daily use of private vehicles is consistent with the second element of the dominant vision but has several negative consequences. The most obvious is traffic congestion. The more fully Americans achieve their ideal vision of widespread low-density homeownership and private vehicle ownership, the more congested their commuting journeys become. In addition, these heavy travel requirements cause severe air pollution, costly demands for road and other infrastructure construction, and high-level consumption of energy.

Most people who believe in this ideal vision have not yet realized that their success in achieving it is causing these adverse outcomes. Moreover, they are reluctant to come to this conclusion. Doing so would force them to admit that their own ideals and behavior are the main causes of their principal problems.

Consequently, most suburbanites suffering from excessive traffic congestion, air pollution, and other results of low-density settlement patterns blame both real estate developers and the latest-arriving residents in their suburban communities and adopt antigrowth or growth-limiting policies. But these policies do not attack the fundamental causes of the problems concerned; hence, they cannot solve such problems.

THE SECOND FLAW: NO HOUSING PROVISION FOR LOW-WAGE WORKERS

The second flaw in the dominant ideal vision is that it contains only relatively high-cost housing, providing no residences where low- and moderate-income households can afford to live. Yet such households are an integral part of American life, because they provide workers for relatively low-wage jobs. Those jobs are vital to the efficient operation of every community, including the wealthiest exurban enclaves. Low-wage workers are essential to many service firms, including but not limited to fast-food establishments, gas stations, laundries, dry cleaners, hospitals, retail stores, shopping centers, construction firms and subcontractors, and gardening and lawn-care firms. And moderate-wage workers are often the backbone of local government services, including police, firefighters, teachers, and administrative personnel.

These households cannot afford to live in the spacious, single-family homes that populate the ideal vision. In fact, many cannot afford to live in any type of single-family home. Among the 88.4 million U.S. households in 1985, 62.2% lived in detached single-family units, and 4.6% lived in attached single-family units. But 5.4% lived in mobile homes, 11.6% lived in structures containing two and four units, and 16.1% lived in structures containing five or more units. Thus, about one-third of all American households did not live in single-family homes, and over one-third were not homeowners, but renters. The ideal vision of metropolitan development does not contain any dwelling units appropriate to this "forgotten fraction."

True, U.S. suburbs – including the newest ones – contain a lot of housing other than single-family homes. Such housing is in fact being built in our metropolitan areas. Nevertheless, it does not form part of the ideal vision of how those areas should be developed. Therefore, its construction is often strongly opposed by local residents whose views are dominated by that ideal vision.

Moreover, many low- and moderate-income households cannot afford to live in brand new housing units since such units have their highest relative prices when they are first built. At that moment, they contain the most modern and up-to-date amenities and design available, and they have not yet been subjected to wear and tear. Moreover, they are most often located at the edge of the built-up territory within their metropolitan area, which is then usually among the most fashionable neighborhoods in which to live.

As time passes, these units become relatively less desirable. They no longer contain the most up-to-date amenities; they have experienced some normal wear and tear and deterioration; and their neighborhoods are no longer at the edge of the built-up territory, which has now expanded beyond them.

Hence, their relative prices decline, even though their absolute prices may rise along with inflation. If this life cycle continues, these units eventually fall in both relative prices and absolute prices enough so that low- and moderate-income households can afford to live in them. This life-cycle movement of relative prices is part of the "filtering" or "trickle-down" process through which older housing is made available to those households that cannot afford brand new units.

Most American suburbs have been built since World War II. The nation's total suburban population rose from 40.9 million in 1950 to 108.6 million in 1986 – an increase of 165.5%, compared to a 58.7% rise in total population. By 1986, the fraction of all U.S. residents living in suburbs had risen to 44.9% from 27% in 1950. In very large metropolitan areas, the newest suburbs are now located quite far from the older portions of both the central city and older suburbs.

This is particularly true in those metropolitan areas that have experienced the fastest growth in total population, such as many in California and Florida.

Historically, employers in newer suburbs have relied upon the more affordable housing inventory in older neighborhoods to provide dwellings for their low- and moderate-income workers. This arrangement permitted the newest communities to avoid providing any housing directly affordable to all the persons working within their boundaries. Hence, these communities could develop themselves in full accord with the ideal vision of universal homeownership, involving mainly detached single-family units.

However, such reliance upon the "trickle-down" process to house low-wage workers has recently become much less effective, for several reasons. First, jobs have spread out into the suburbs much more than in the past. In fact, many more new jobs are being created in the suburbs than in central cities. They include jobs located in relatively far-out suburbs, as well as close in ones. Second, the sheer size of the suburban portions of large, fast-growing metropolitan areas means that much of the older, less costly housing is many miles from new-growth regions where most new jobs are located. Low- and moderate-income workers who live in older, more central neighborhoods are far from those new jobs. Hence, they have a hard time finding job openings, or commuting to jobs they find (often because of lack of public transportation). Third, soaring land prices in many such metropolitan areas have helped raise housing prices there to extremely high levels. Few middle-income households can afford most new single-family units there. Even the few multifamily units being built there have rents far above the payment ability of most low- or moderate-income households. Fourth, the long economic expansion that began in 1982 has soaked up most available workers. This has forced the unemployment rate well below 6% nationally, and below 2% in some booming suburban areas. In addition, the changes in U.S. age distribution are creating a nationwide shortage of entry-level workers, tightening the market further. The higher wages resulting from this prosperity have become embedded in housing costs, increasing prices beyond the reach of many households.

As a result, far-out suburban portions of many large, fast-growth metropolitan areas are experiencing acute shortages of low-wage and even moderate-wage labor. Workers willing to take such jobs cannot afford to live anywhere within reasonable commuting distance of those jobs. Thus, the ideal vision's exclusive focus upon relatively new single-family housing, which is quite expensive, is inconsistent with the actual need for low- and moderate-wage workers that arises in every community.

THE THIRD FLAW: NO CONSENSUS ON HOW TO FINANCE INFRASTRUCTURES FAIRLY

A third major flaw springs from the absence of any consensus in the dominant vision about how best to finance new infrastructures, more roads, sewage systems, water systems, and school systems. They also include costs of increasing the capacity of existing arterial facilities in established areas through which new residents will pass on their way to and from work or shopping. The absence of a widespread consensus about how to pay for such added facilities "fairly" has two negative consequences: It creates severe political conflicts in many suburban areas, and it often results in gross underfunding – and therefore under-provision – of needed facilities and services. Residents already living in fast-growth areas want all the added facilities but expect services required by newcomers to be paid for by the latter. Existing residents consider it justifiable to load those marginal costs entirely onto new developments through various impact fees, exactions, proffers, and permit fees. They regard this as merely requiring the newcomers to pay "their fair share" of such costs, since newcomers are the main beneficiaries of those facilities. Hence, they vote against any increases in general taxes or general bonding powers to pay for such facilities.

In contrast, developers and potential newcomers to growth areas believe the entire community should share in paying for these added infrastructures. They cite three reasons for such sharing: First, because many existing residents benefited from past general financing of similar infrastructures when their subdivisions were being built, it is unfair of them to change the rules of the game once they have received such benefits. Second, growth creates many gains for society in general – including existing residents. Its main benefit is greater economic prosperity, which aids everyone. Therefore, society should pay for some of the marginal costs of growth. Third, loading all the marginal costs of growth onto new developments raises housing costs. That unfairly reduces the homeownership and rental opportunities of households with low and moderate incomes. From this perspective, existing residents also ought to pay "a fair share" of the costs of added growth.

Political resolution of this controversy is inherently biased in favor of existing residents because potential newcomers do not yet reside in the areas concerned. Consequently, they cannot vote on local government policies that directly affect their welfare, whereas existing residents can. The deck is thus stacked in favor of loading most marginal costs of development onto newcomers, rather than sharing those costs throughout the community.

However, realistically it is quite difficult to force newcomers to pay all the marginal costs of growth through impact fees or other exactions. Some costs of growth spring from more intensive use of existing facilities, such as expressways, arterial streets, sewage-treatment plants, and school systems. Those facilities and services are shared between newcomers and previous residents. Moreover, many infrastructures can only be built in certain minimum sizes that have substantial capacity, thus presenting communities with sizable costs. For example, the addition of 200 households to an area where the sewage-treatment plant has reached its absolute capacity volume of 20,000 households may require construction of a whole new treatment plant with a minimum capacity of 10,000 households. It is not possible to load the entire cost of this new plant onto those 200 newcomers. Instead, existing residents must bear some of that initial cost.

Faced by this situation, existing residents often choose to provide inadequate facilities for newcomers to hold down taxes. They make such choices without fear of being outvoted by potential newcomers, as the latter are not yet present to vote on their own behalf.

As a result, new housing and commercial subdivisions often are built without adequate infrastructures and other supporting facilities. Existing roads, streets, schools, parks, and utility systems become grossly overloaded and congested. This imposes heavy costs not only upon newcomers, but also upon the existing residents themselves. Precisely this sequence of events has been happening in Florida for some time. The Florida State government has been unable to agree on some means of financing the infrastructure to accommodate Florida's rapid population growth.

Such developments aggravate inherent political tensions between residents trying to minimize their taxes and newcomers and developers trying to accom-

modate new growth. Many existing residents would like to prohibit all additional growth completely. But doing so is impossible in practice because it violates basic freedoms of movement, private property rights, and contract guaranteed by the U.S. Constitution. The result in many fast growth suburban areas is an undesirable combination of high political tensions and facility overload because of underfunding and congestion. The dominant vision of how metropolitan areas ought to develop contains no means or even guidance concerning how to resolve these issues.

THE FOURTH FLAW: INABILITY TO ACCOMMODATE LULUS

The dominant ideal vision's fourth major flaw is that it contains no effective political mechanisms for resolving inevitable conflicts between the welfare of society as a whole and the welfare of geographically small parts of society. This results from extreme spatial fragmentation of government decision-making powers. Yet such fragmentation is required by the governance arrangements that most Americans prefer. They want to live in geographically small communities, each of which fully controls the land use, public schools, and other environmental aspects within its own boundaries. Such "local sovereignty" is one of the pillars of the ideal vision. But it means that each government has a highly parochial viewpoint. Its officials are concerned only with the welfare of the local residents who elect them. As a result, no public officials are primarily concerned with the welfare of society as a whole, or even of the metropolitan area as a whole.

However, every modern society must contain those facilities that benefit society as a whole, but have negative "spillover effects" upon their immediate surroundings (e.g. airports, expressways, jails, garbage incinerators, and landfills). These essential facilities are known as Locally Undesirable Land Uses (LULUs). Because of their undesirable impacts upon their immediate neighbors, no residents want to permit a LULU to be located near them. This attitude has become known as the Not In My Back Yard (NIMBY) syndrome.

Whenever a LULU has to be built, there is no effective way to find a location for it in a society designed in accordance with the predominant ideal vision. Residents near every potential site pressure their local governments to oppose locating the LULU there.

Those governments have the power to reject the LULU because controls over land uses have been divided up among myriad local entities. And officials within all those entities are motivated to reject the LULU because they are politically responsible only to their own residents. Hence, it becomes almost impossible to find politically acceptable locations for facilities that every metropolitan area needs to operate efficiently. The resulting paralysis has virtually halted both major airport construction and prison expansion in most of the United States, and blocked creation of thousands of other badly needed LULUs.

Furthermore, there is a strong single-family-home bias built into the dominant ideal vision. As a result, nearly all new high-density real estate developments – especially those involving any multifamily housing – are usually considered LULUs, and are therefore extremely difficult to build. But low- and moderate-income households cannot afford to occupy new single-family homes. So the NIMBY syndrome reinforces the second flaw in the dominant ideal vision: Its failure to provide housing that low- and moderate-income households can afford.

In fact, low- and moderate-income households themselves are regarded as LULUs by many middle- and upper-income households. The latter consequently adopt a NIMBY attitude toward all housing affordable to low- and moderate-income households. They believe that such "undesirable" people near them will reduce the market values of their costly homes. About three-fourths of the financial net worth of the average American household consists of its equity in the home it owns and lives in. Hence, its members are quite sensitive about any changes in their neighborhoods that they believe would reduce the value of those equities, and very little such housing gets built in most U.S. suburbs.

SOCIAL INCONSISTENCIES RESULTING FROM THE IDEAL VISION'S FLAWS

The four major flaws in the ideal vision described above have caused the emergence and persistence of certain negative conditions in suburban America, the very heartland of that ideal vision. Hence, we must conclude that the dominant vision itself contains inherent characteristics that render it less ideal than we previously believed it to be – or than most Americans still believe it to be. These characteristics can be considered "social inconsistencies," as they are inconsistent with the high quality of life promised by the ideal vision. (Note: This situation illustrates a fundamental problem in democratic societies. They have great difficulty solving problems that arise as the long-run result of a majority's pursuit of policies that produce short-run benefits. Once a majority of citizens has begun receiving those short-run benefits, its members do not want to give them up. But they often fail to agree about how to distribute the long-run costs necessary to sustain those short-run benefits. Each subgroup among the beneficiaries tries to shift as much of its share of the costs as possible onto other subgroups. No consensus arises about how to pay for these costs. In some cases, the costs are not paid at all. That is precisely why the United States has been unable to reduce its huge recurrent Federal deficits.)

THE ALTERNATIVE VISION OFFERED BY THE URBAN PLANNING PROFESSION

The American urban planning profession has long been aware of some of the internal inconsistencies in the dominant ideal vision described above. In particular, urban planners have repeatedly attacked the heavy reliance upon automotive vehicle travel required by very low-density settlement patterns. In fact, many urban planners have advocated an alternative vision that rejects such reliance. But their alternative has never gained substantial public acceptance in America.

This alternative has been based upon the European pattern of urban development. It featured high-density residential settlements, high-density workplaces, tightly circumscribed land use patterns that prevent peripheral sprawl, and massive use of heavily subsidized public transit systems for movement. These traits are only possible under a governance system that centralizes power over the land use patterns in each metropolitan area in a single governing body with authority over the entire area. Accompanying such centralization in Europe is use of large amounts of publicly subsidized rental housing. Such housing accommodates a sizable fraction of the entire population in subsidized multi-family units scattered throughout each metropolitan area.

For decades, U.S. urban planners have been advocating an alternative vision of future metropolitan development based largely upon this pattern. But they have been unable to persuade either the public or local government officials to accept it because this alternative directly opposes the dominant vision described earlier that most Americans cherish. It rejects too many elements of that dominant vision, including primary reliance upon owner-occupied single-family housing, widespread use of private automotive vehicles, and decentralized local government.

Furthermore, some of the key elements in the European pattern do not appeal to Americans. For example, placing large clusters of high-rise apartments right around fixed-rail transit stops repels most suburban residents, who do not like high-rise housing. In addition, the immense cost of fixed-rail transit systems, and their lack of flexibility, are not acceptable to many Americans. The only large-scale fixed-rail transit system built in the U.S. since 1945 without major Federal subsidies is the Bay Area Rapid Transit System. All the others probably never would have been created if local residents had been compelled to pay their full costs without massive Federal subsidies.

Moreover, the urban planning profession has been guilty of largely ignoring the second major flaw in the dominant vision (its inability to provide housing for low- and moderate-income households). Few urban planners have ever developed comprehensive plans that specifically indicated where to accommodate poor households. Most comprehensive plans ignore the need for low-rent housing, because explicitly locating it within the plan would arouse immense opposition from nonpoor residents.

Yet the urban planning profession has been correct in claiming that the dominant vision would not work well in the long run. Most suburbanites in fast-growth metropolitan areas are now discovering this fact, to their own dismay.

CRITERIA FOR A NEW IDEAL VISION OF FUTURE METROPOLITAN AREA DEVELOPMENT

The preceding analysis shows that American society needs a new ideal vision of how future development ought to occur in our large metropolitan areas. The current dominant vision contains too many serious inconsistencies and undesirable outcomes. The new ideal vision should conform to several specific criteria. Such conformance would enable this vision both to (1) meet the future needs of American society; and

(2) avoid the problems generated by the currently dominant vision.

As noted earlier, development and widespread acceptance of a new vision would be similar to a paradigm shift in science – that is, the replacement of one fundamentally governing perception by another one. Such replacements may appear to occur suddenly, but they actually build up gradually over time.

The new ideal vision of future metropolitan growth will not require the complete rejection of the pillars on which the existing vision has been based. Rather, it will modify those pillars. This new perspective will be much more conscious of the collective impacts of individual decisions than the currently dominant perspective. Hence, it will embody individualism sensitive to collective behavior patterns, rather than unconstrained individualism. For example, we need not deny the desirability of homeownership or of single-family homes in order to recognize that a balanced society also needs rental housing and much higher average densities than we have considered ideal in the past. Similarly, we will certainly not abandon widespread use of individual vehicles. But those vehicles may be powered differently, and more people will be traveling together in shared rides or transit than at present.

Specific criteria of this new ideal vision are as follows:

First, the new vision must contain sizable areas of at least moderately high-density development, especially of housing, but also of workplaces. This is necessary to achieve three key goals: the creating of housing that is less costly than single-family homes; the reduction of the area required to accommodate both jobs and workers, thereby cutting travel requirements and reducing both congestion and air pollution; the development of more efficient spatial interrelations among buildings within suburban commercial and industrial developments. This would reduce travel requirements within such developments and promote more face-to-face contacts there. The third of these goals is far less important than the first two.

In order to make such higher-density areas politically acceptable to suburban residents, owners of single-family homes must be convinced that they can tolerate multifamily housing nearby. Such housing need not consist of big high-rise towers, but can be a mixture of low-rise and mid-rise structures, with an occasional high-rise. Owners of existing single-family homes must become confident that such housing will neither jeopardize the investments they have made in their own homes, nor expose their families to crime and violence. That will probably require careful studies of the impact of well-designed existing multifamily projects upon the market values of surrounding single-family homes. Owners of single-family homes must also be convinced that moderate-density housing can be designed to look good and therefore not reduce the aesthetic quality of their areas.

Second, the new ideal vision must encourage people to live nearer to where they work. This means that each subarea within a metropolitan area must contain a "balanced blend" of different types and prices of housing, all close to the jobs within the same subarea. Then it will be possible for workers holding jobs at all wage levels to live close to those jobs and still occupy housing they can afford.

True, people cannot be forced to live nearby their workplaces. Today, many people could live closer to where they work, but they prefer traveling longer distances in order to live in what they deem an ideal environment. But this preference is changing as congestion increases the time required to move any specific distance during peak commuting periods. Eventually, more people will decide that suffering the time losses and frustration of such worsened congestion is not worthwhile and will move either their jobs or their homes to reduce the distance between them. But this is possible only if they are compelled to live near their jobs by the appropriate cost of the housing close to those jobs. That, in turn, is possible only if each sizable subregion of the entire metropolitan area contains a mixture of different types of housing available at widely varying occupancy costs.

Achieving such "balanced blends" of housing in all parts of a large metropolitan area will not be easy. The logical spatial units within which such "balancing" should occur will rarely coincide with existing governmental boundaries. Moreover, at least one job cluster in each (i.e. downtown) contains so many jobs that not all of its workers could possibly live nearby. In order to reduce commuting travel, it may be necessary in the long run to "decant" some existing down-towns by shifting some of their existing jobs elsewhere. That is not a happy thought for owners and managers of downtown office buildings and other properties. But the goal of this article is not to determine how these criteria might be met, but rather to formulate them in the hope that others might determine how they might be implemented.

Third, the new ideal vision must contain governance structures that preserve substantial local authority – but within a framework that compels local governments to act responsibly to meet area-wide needs. That probably requires a comprehensive planning framework imposed by state governments. In such a framework, every local government would be responsible for drawing up comprehensive plans for its own future development. But all such plans would have to meet certain general criteria laid down by the state government. Such criteria might include provision of land zoned for low- and moderate-income housing, and arrangements that pushed the choosing of sites for certain region-affecting facilities up to regional or state levels to avoid complete paralysis. Moreover, the state government would have to create mechanisms to review and modify local government plans to insure that they met statewide criteria. Oregon has already put such arrangements into practice, and several other states are doing so. Thus, in considering possible future changes in existing governance within metropolitan areas, one should expect neither true metropolitan government, nor any total abandonment of local sovereignty. Simply, such sovereignty must be placed within a broader framework that enforces wider social responsibility. While this probably can occur only at the state level, where the constitutional power over local government resides, the Federal government might provide financial incentives to get more states to engage in this type of restructuring.

Fourth, the new vision should contain incentive arrangements that encourage individuals and households to take a more realistic account of the collective costs of their behavioral choices. For example, individual auto drivers are not required to pay more for taking trips alone during rush hours than during off-peak hours, because auto drivers are never charged directly for using roads. For decades, economists have pointed out that each driver who enters a well-traveled highway during peak hours creates a substantial social cost. By adding to the prevailing congestion, that driver slows down all other drivers to some degree, thereby imposing a time loss upon them. In contrast, a driver who takes the same trip during nonpeak hours does not generate any such social cost. Therefore, it would be economically efficient for society if drivers were charged a direct fee for driving during peak hours, but not during off-peak hours. Such charges would encourage more drivers to shift to off-peak periods, thereby reducing peak-hour congestion. And those charges could raise money useful in improving the nation's road system.

This type of incentive arrangement is especially important in view of the fact that most peak-hour auto trips made in the nation's largest metropolitan areas are nonwork trips, many of which could be shifted to other times relatively easily.

Unfortunately, politicians in most democracies have consistently refused to adopt differential road-use pricing. They argue that such pricing would penalize low-income drivers and favor higher-income ones, who could more easily afford to pay peak-hour charges. In a democracy where 87% of all households have cars, but most do not have high incomes, differential road pricing has always seemed to be a politically losing proposition.

This situation will change only when major freeways become so overcrowded that traffic literally slows to a crawl for several hours per day. Many southern California freeways are close to such prolonged gridlock today, and are likely to reach it soon. When they do, even low-income auto-driving voters will become frustrated enough to accept almost any proposed solution likely to be effective. Peak-hour road pricing will then be perceived in a much more favorable light because of the realistic incentives it presents to drivers.

So-called "exclusionary zoning" arrangements or "linkage fees" concerning housing could be another form of such incentives. Such exclusionary arrangements are created and sustained by political pressures on local governments from middle- and upper-income homeowners. They believe socioeconomic exclusion helps maximize the market values of their homes and improve the quality of their local environments.

In many suburban areas, exclusionary zoning has made the cost of housing units too great for most low- and moderate-income households. The typical application of this type of zoning has been the "down zoning" of lots where minimum sizes are often two acres and up. In the end, this imposes a cost upon society as a whole by requiring many such households to live far from their suburban jobs and commute long distances each day. The result is more overall traffic congestion and air pollution, plus the imposition of very high travel costs upon low- and moderate-income commuters.

Therefore, those who impose exclusionary arrangements on their suburban areas should be required to offset the social costs they are generating through various arrangements that charge them for engaging in exclusionary practices. Such charges would also discourage exclusion by raising the price of housing to those who engage in it.

Several forms of such incentive arrangements concerning housing have been tried in different areas. Local or state governments could add a fee onto all market-priced housing units built or sold in exclusionary regions, and use the proceeds to subsidize units available at costs affordable to low- and moderate-income households. Another tactic would be to add a transfer tax onto the sale of single-family homes and then put the proceeds into a housing trust fund used to help finance construction of low-rent units in otherwise exclusionary communities. Another approach would be to require each developer to allocate a certain percentage of the units he or she builds for low- and moderate-income occupancy. The resulting higher costs for market-priced units could be offset to some extent by permitting the developer to build at higher-than-usual average densities in the projects concerned. Yet another tactic would be to charge developers of commercial and industrial properties in exclusionary areas "linkage fees" that are used to subsidize housing for low- and moderate-income households.

Not all of these arrangements would have the same negative incentive impact upon exclusionary zoning practices that peak-hour-driving charges would have upon rush-hour trips. But they all would compel persons engaging in exclusionary zoning practices to bear some of the collective costs their individual behavioral choices are creating for society.

Fifth, the new vision should incorporate stable and predictable strategies that would adequately finance infrastructures built to accommodate growth. Such strategies need to be stable and predictable so that all concerned can efficiently plan how to carry out new developments, both public and private. These strategies also must be fair and politically acceptable both to existing residents on the one hand, and to developers and potential newcomers on the other. Financing strategies appropriate to slow-growth areas may be different from those appropriate to fast-growth areas. In all areas, such strategies will probably have to combine some impact fees and exactions upon newcomers, with some increases in general taxes upon all residents.

Another key criterion is that the new ideal vision should not depend upon construction of costly fixed-rail mass transit systems that are politically acceptable in each area only when heavily subsidized with funds from other parts of the nation. Many urban planners will reject this criterion. But this author believes the American people will accept it – as they have for many decades. Perhaps some forms of light rail transit can be created inexpensively enough to be incorporated within viable metropolitan area transportation systems. But most such systems must be based upon freely moving automotive vehicles, whether they are cars, trucks, vans, big buses, or mini-buses. Thus, designing urban transportation systems that will reduce congestion and pollution poses a tremendous challenge to transportation planners.

Further, the ideal vision should incorporate workplace designs that combine efficient interchange among individual workplaces with big enough critical masses of jobs to create productive interchange possibilities. These characteristics are now found within traditional downtowns and within large regional shopping malls. In both places, office buildings or retail shops are located right next to each other along attractive pedestrian walkways, thereby encouraging maximum efficiency of movement among different facilities. Parking is not placed around each structure, but in external zones that are nearby but do not interfere with movements among individual units. Yet enough facilities are massed together to permit efficient comparison shopping or interchanges of services and meetings among persons working in different enterprises.

These traits are now almost totally absent from all suburban office and industrial parks. Each such structure is separated from others nearby by a sea of parking or landscaping. Therefore, meeting this criterion might require forcing owners to build new structures between existing ones, as near Tyson's Corner, in Fairfax County, Virginia, so as to recreate the interchange efficiency of downtown streets. It cannot be determined how this could be done with present fragmented private site ownership. But considering the ingenuity with which Wall Street investment bankers have reorganized existing companies into multiple entities, or created new securities, we should be equally ingenious about land-use controls and forms of ownership and management.

Finally, the new ideal vision should be internally consistent, in the dual sense that (1) the amount of travel it requires does not lead to the levels of traffic congestion or air pollution we are now encountering; and (2) that some homes in the vision are affordable to the low- and moderate-income households needed to do much of the work. We need complex computerized transportation and land-use models to test such relationships. Designing both the models and the transportation elements of the new ideal vision will not be easy. But few aspects of urban life are more important.

CONCLUSION

Is it really possible to create an alternative ideal vision of future large metropolitan area development – a new paradigm – that meets all these criteria? Our society desperately needs to make an attempt to find out. We have overwhelming current evidence that the currently dominant ideal vision does not work effectively. But we cannot displace that vision unless we have some plausible, attractive, and persuasive alternative to offer. The option traditionally offered by urban planners is never going to win widespread acceptance among the American people.

"Broken Windows"

Atlantic Monthly (1982)

James Q. Wilson and George L. Kelling

Editors' Introduction

Why is urban crime a problem in inner-city neighborhoods? What can and should government do about it? That is the subject of the following selection by James Q. Wilson and George L. Kelling.

Wilson and Kelling studied police behavior in troubled neighborhoods – particularly in Newark, New Jersey. One of the authors (Kelling) spent many months walking Newark neighborhoods with local police officers, observing what was going on in the neighborhoods and how the police actually handled neighborhood problems. Harvard professor Wilson worked with Kelling's empirical observations to jointly develop this classic article. Wilson and Kelling were particularly interested in police discretion and how the police handled troublesome behavior at the borderline between acceptable and unacceptable; legal and illegal. Wilson and Kelling are interested as much in neighborhood residents' perceptions of crime as in crime itself. A particular focus of their research and theory building is to develop a model of policing that neighborhood residents (at least those residents the authors term "decent folk") will support. Based on this research, they advanced a controversial theory of neighborhood transition and an influential model for community policing which has been implemented in many communities.

The central metaphor of this selection is of a single broken window. Imagine a diverse city neighborhood like the Boston's West end or New York City's Greenwich Village as described by Jane Jacobs (p. 114) – a diverse, viable, exciting urban neighborhood, but with some crime. Most of the people in the neighborhood Jane Jacobs sees each morning as the "street ballet" begins are regulars – people who live or work in the neighborhood or are frequently in the area. Some are strangers. Most of the people in the neighborhood – the grocer putting out his vegtables, the school children on their way to school – are not criminals and do not engage in behavior which harms other people. The behavior of some people in the neighborhood might startle or even scare conventional Bostonians or New Yorkers, but it is understood and tolerated by neighborhood regulars. Some of the people in the area, however, *are* dangerous and do engage in criminal behavior: drug dealing, prostitution, theft, assault, and other serious crimes. Civil order exists, but it is fragile. Imagine that someone breaks a single window in the neighborhood. According to Wilson and Kelling, how the police respond to that trivial, but unacceptable, act is fraught with consequences. One response is to do nothing. This, Wilson and Kelling argue, will signal that no one cares about unacceptable conduct. It will be an invitation for people to break more windows. Neighborhood residents will start avoiding one another, stop participating in neighborhood block parties, and eventually cower in their own homes.

Wilson and Kelling argue that historically neighborhood residents judged the success of police activity by whether it succeeded in maintaining order. Governments often tolerated (or encouraged) police to keep order through *informal means* without much regard to legal niceties. A drunk might have a *legal* right to sit on a neighborhood park bench. But if his presence sufficiently disturbed "decent" neighborhood residents, the local cop was encouraged to make him move along elsewhere. Gang members in Chicago's crime-ridden Robert Taylor Homes might have a legal right to loiter by a playground, but the authors argue that project residents would want the local police to get rid of them ("kick ass" is the term Wilson and Kelling use). Wilson and Kelling take the controversial position that broad police discretion to enforce neighborhood standards is acceptable, even desirable.

Compare Wilson and Kelling's approach to Mike Davis's description of how he feels the Los Angeles police oppress low-income minorities (p. 201). How would you draw the line between "decent" and not-so-decent folk in New York neighborhoods as culturally heterogeneous as Sharon Zukin (p. 136) describes? Whose culture should define neighborhood norms?

The argument in this selection is further developed in George L. Kelling and Catherine M. Coles, *Fixing Broken Windows* (New York: Martin Kessler, 1996). Other books by James Q. Wilson related to crime prevention include *Crime*, co-edited with Joan Petersilia (San Francisco: ICS Press, 1995), *Crime and Public Policy* (San Francisco: ICS Press, 1983), *Drugs and Crime*, co-edited with Michael Tonry (Chicago: University of Chicago Press, 1990), *Families, Schools, and Delinquency Prevention*, co-edited with Glenn C. Loury (New York: Springer-Verlag, 1987), and *Understanding and Controlling Crime*, co-edited with David P. Farrington and Lloyd E. Ohlin (New York: Springer-Verlag, 1986).

Books on community policing include Elizabeth M. Watson, Alfred R. Stone, and Stuart M. DeLuca, *Strategies for Community Policing* (Upper Saddle River, NJ: Prentice-Hall, 1998), Wesley G. Skogan and Susan M. Hartnett, *Community Policing, Chicago Style* (New York: Oxford University Press, 1997), Kenneth J. Peak and Ronald W. Glensor, *Community Policing and Problem Solving: Strategies and Practices* (Upper Saddle River, NJ: Prentice-Hall, 1996), and Nigel Fielding, *Community Policing* (Oxford and New York: Oxford University Press, 1995).

■ ■ ■ ■ ■ ■

In the mid-1970s, the state of New Jersey announced a "Safe and Clean Neighborhoods Program," designed to improve the quality of community life in twenty-eight cities. As part of that program, the state provided money to help cities take police officers out of their patrol cars and assign them to walking beats. The governor and other state officials were enthusiastic about using foot patrol as a way of cutting crime, but many police chiefs were skeptical. Foot patrol, in their eyes, had been pretty much discredited. It reduced the mobility of the police, who thus had difficulty responding to citizen calls for service, and it weakened headquarters control over patrol officers.

Many police officers also disliked foot patrol, but for different reasons: it was hard work, it kept them outside on cold, rainy nights, and it reduced their chances for making a "good pinch." In some departments, assigning officers to foot patrol had been used as a form of punishment. And academic experts on policing doubted that foot patrol would have any impact on crime rates; it was, in the opinion of most, little more than a sop to public opinion. But since the state was paying for it, the local authorities were willing to go along.

Five years after the program started, the Police Foundation, in Washington, DC, published an evaluation of the foot-patrol project. Based on its analysis of a carefully controlled experiment carried out chiefly in

Newark, the foundation concluded, to the surprise of hardly anyone, that foot patrol had not reduced crime rates. But residents of the foot-patrolled neighborhoods seemed to feel more secure than persons in other areas, tended to believe that crime had been reduced, and seemed to take fewer steps to protect themselves from crime (staying at home with the doors locked, for example). Moreover, citizens in the foot-patrol areas had a more favorable opinion of the police than did those living elsewhere. And officers walking beats had higher morale, greater job satisfaction, and a more favorable attitude toward citizens in their neighborhoods than did officers assigned to patrol cars.

These findings may be taken as evidence that the skeptics were right – foot patrol has no effect on crime; it merely fools the citizens into thinking that they are safer. But in our view, and in the view of the authors of the Police Foundation study (of whom Kelling was one), the citizens of Newark were not fooled at all. They knew what the foot-patrol officers were doing, they knew it was different from what motorized officers do, and they knew that having officers walk beats did in fact make their neighborhoods safer.

But how can a neighborhood be "safer" when the crime rate has not gone down – in fact, may have gone up? Finding the answer requires first that we understand what most often frightens people in public places.

Many citizens, of course, are primarily frightened by crime, especially crime involving a sudden, violent attack by a stranger. This risk is very real, in Newark as in many large cities. But we tend to overlook or forget another source of fear – the fear of being bothered by disorderly people. Not violent people, nor, necessarily, criminals, but disreputable or obstreperous or unpredictable people: panhandlers, drunks, addicts, rowdy teenagers, prostitutes, loiterers, the mentally disturbed.

What foot-patrol officers did was to elevate, to the extent they could, the level of public order in these neighborhoods. Though the neighborhoods were predominantly black and the foot patrolmen were mostly white, this "order-maintenance" function of the police was performed to the general satisfaction of both parties.

One of us (Kelling) spent many hours walking with Newark foot-patrol officers to see how they defined "order" and what they did to maintain it. One beat was typical: a busy but dilapidated area in the heart of Newark, with many abandoned buildings, marginal shops (several of which prominently displayed knives and straight-edged razors in their windows), one large department store, and, most important, a train station and several major bus-stops. Though the area was run-down, its streets were filled with people, because it was a major transportation center. The good order of this area was important not only to those who lived and worked there but also to many others, who had to move through it on their way home, to supermarkets, or to factories.

The people on the street were primarily black; the officer who walked the street was white. The people were made up of "regulars" and "strangers." Regulars included both "decent folk" and some drunks and derelicts who were always there but who "knew their place." Strangers were, well, strangers, and viewed suspiciously, sometimes apprehensively. The officer – call him Kelly – knew who the regulars were, and they knew him. As he saw his job, he was to keep an eye on strangers, and make certain that the disreputable regulars observed some informal but widely understood rules. Drunks and addicts could sit on the stoops, but could not lie down. People could drink on side streets, but not at the main intersection. Bottles had to be in paper bags. Talking to, bothering, or begging from people waiting at the bus stop was strictly forbidden. If a dispute erupted between a businessman and a customer, the businessman was assumed to

be right, especially if the customer was a stranger. If a stranger loitered, Kelly would ask him if he had any means of support and what his business was; if he gave unsatisfactory answers, he was sent on his way. Persons who broke the informal rules, especially those who bothered people waiting at bus stops, were arrested for vagrancy. Noisy teenagers were told to keep quiet.

These rules were defined and enforced in collaboration with the "regulars" on the street. Another neighborhood might have different rules, but these, everybody understood, were the rules for this neighborhood. If someone violated them, the regulars not only turned to Kelly for help but also ridiculed the violator. Sometimes what Kelly did could be described as "enforcing the law," but just as often it involved taking informal or extralegal steps to help protect what the neighborhood had decided was the appropriate level of public order. Some of the things he did probably would not withstand a legal challenge.

A determined skeptic might acknowledge that a skilled foot-patrol officer can maintain order but still insist that this sort of "order" has little to do with the real sources of community fear – that is, with violent crime. To a degree, that is true. But two things must be borne in mind. First, outside observers should not assume that they know how much of the anxiety now endemic in many big-city neighborhoods stems from a fear of "real" crime and how much from a sense that the street is disorderly, a source of distasteful, worrisome encounters. The people of Newark, to judge from their behavior and their remarks to interviewers, apparently assign a high value to public order, and feel relieved and reassured when the police help them maintain that order.

Second, at the community level, disorder and crime are usually inextricably linked, in a kind of developmental sequence. Social psychologists and police officers tend to agree that if a window in a building is broken *and is left unrepaired*, all the rest of the windows will soon be broken. This is as true in nice neighborhoods as in run-down ones. Window-breaking does not necessarily occur on a large scale because some areas are inhabited by determined window-breakers whereas others are populated by window-lovers; rather, one unrepaired broken window is a signal that no one cares, and so breaking more windows costs nothing. (It has always been fun.)

Philip Zimbardo, a Stanford psychologist, reported in 1969 on some experiments testing the broken-window theory. He arranged to have an automobile without license plates parked with its hood up on a street in the Bronx and a comparable automobile on a street in Palo Alto, California. The car in the Bronx was attacked by "vandals" within ten minutes of its "abandonment." The first to arrive were a family – father, mother, and young son – who removed the radiator and battery. Within twenty-four hours, virtually everything of value had been removed. Then random destruction began – windows were smashed, parts torn off, upholstery ripped. Children began to use the car as a playground. Most of the adult "vandals" were well-dressed, apparently clean-cut whites. The car in Palo Alto sat untouched for more than a week. Then Zimbardo smashed part of it with a sledge-hammer. Soon, passersby were joining in. Within a few hours, the car had been turned upside down and utterly destroyed. Again, the "vandals" appeared to be primarily respectable whites.

Untended property becomes fair game for people out for fun or plunder, and even for people who ordinarily would not dream of doing such things and who probably consider themselves law-abiding. Because of the nature of community life in the Bronx – its anonymity, the frequency with which cars are abandoned and things are stolen or broken, the past experience of "no one caring" – vandalism begins much more quickly than it does in staid Palo Alto, where people have come to believe that private possessions are cared for, and that mischievous behavior is costly. But vandalism can occur anywhere once communal barriers – the sense of mutual regard and the obligations of civility – are lowered by actions that seem to signal that "no one cares."

We suggest that "untended" behavior also leads to the breakdown of community controls. A stable neighborhood of families who care for their homes, mind each other's children, and confidently frown on unwanted intruders can change, in a few years or even a few months, to an inhospitable and frightening jungle. A piece of property is abandoned, weeds grow up, a window is smashed. Adults stop scolding rowdy children; the children, emboldened, become more rowdy. Families move out, unattached adults move in. Teenagers gather in front of the corner store. The merchant asks them to move; they refuse. Fights occur. Litter accumulates. People start drinking in front of the grocery; in time, an inebriate slumps to the sidewalk and is allowed to sleep it off. Pedestrians are approached by panhandlers.

At this point it is not inevitable that serious crime will flourish or violent attacks on strangers will occur. But many residents will think that crime, especially violent crime, is on the rise, and they will modify their behavior accordingly. They will use the streets less often, and when on the streets will stay apart from their fellows, moving with averted eyes, silent lips, and hurried steps. "Don't get involved." For some residents, this growing atomization will matter little, because the neighborhood is not their "home" but "the place where they live." Their interests are elsewhere; they are cosmopolitans. But it will matter greatly to other people, whose lives derive meaning and satisfaction from local attachments rather than worldly involvement; for them, the neighborhood will cease to exist except for a few reliable friends whom they arrange to meet.

Such an area is vulnerable to criminal invasion. Though it is not inevitable, it is more likely that here, rather than in places where people are confident they can regulate public behavior by informal controls, drugs will change hands, prostitutes will solicit, and cars will be stripped. That the drunks will be robbed by boys who do it as a lark, and the prostitutes' customers will be robbed by men who do it purposefully and perhaps violently. That muggings will occur.

Among those who often find it difficult to move away from this are the elderly. Surveys of citizens suggest that the elderly are much less likely to be the victims of crime than younger persons, and some have inferred from this that the well-known fear of crime voiced by the elderly is an exaggeration: perhaps we ought not to design special programs to protect older persons; perhaps we should even try to talk them out of their mistaken fears. This argument misses the point. The prospect of a confrontation with an obstreperous teenager or a drunken panhandler can be as fear-inducing for defenseless persons as the prospect of meeting an actual robber; indeed, to a defenseless person, the two kinds of confrontation are often indistinguishable. Moreover, the lower rate at which the elderly are victimized is a measure of the steps they have already taken – chiefly, staying behind locked doors – to minimize the risks they face. Young men are more frequently attacked than older women, not because they are easier or more lucrative targets but because they are on the streets more.

Nor is the connection between disorderliness and fear made only by the elderly. Susan Estrich, of the Harvard Law School, has recently gathered together a number of surveys on the sources of public fear. One, done in Portland, Oregon, indicated that three-fourths of the adults interviewed cross to the other side of a street when they see a gang of teenagers; another survey, in Baltimore, discovered that nearly half would cross the street to avoid even a single strange youth. When an interviewer asked people in a housing project where the most dangerous spot was, they mentioned a place where young persons gathered to drink and play music, despite the fact that not a single crime had occurred there. In Boston public housing projects, the greatest fear was expressed by persons living in the buildings where disorderliness and incivility, not crime, were the greatest. Knowing this helps one understand the significance of such otherwise harmless displays as subway graffiti. As Nathan Glazer has written, the proliferation of graffiti, even when not obscene, confronts the subway rider with the "inescapable knowledge that the environment he must endure for an hour or more a day is uncontrolled and uncontrollable, and that anyone can invade it to do whatever damage and mischief the mind suggests."

In response to fear, people avoid one another, weakening controls. Sometimes they call the police. Patrol cars arrive, an occasional arrest occurs, but crime continues and disorder is not abated. Citizens complain to the police chief, but he explains that his department is low on personnel and that the courts do not punish petty or first-time offenders. To the residents, the police who arrive in squad cars are either ineffective or uncaring; to the police, the residents are animals who deserve each other. The citizens may soon stop calling the police, because "they can't do anything."

The process we call urban decay has occurred for centuries in every city. But what is happening today is different in at least two important respects. First, in the period before, say, World War II, city dwellers – because of money costs, transportation difficulties, familial and church connections – could rarely move away from neighborhood problems. When movement did occur, it tended to be along public-transit routes. Now mobility has become exceptionally easy for all but the poorest or those who are blocked by racial prejudice. Earlier crime waves had a kind of built-in self-correcting mechanism: the determination of a neighborhood or community to reassert control over its turf. Areas in Chicago, New York, and Boston would experience crime and gang wars, and then normalcy would return, as the families for whom no alternative residences were possible reclaimed their authority over the streets.

Second, the police in this earlier period assisted in that reassertion of authority by acting, sometimes violently, on behalf of the community. Young toughs were roughed up, people were arrested "on suspicion" or for vagrancy, and prostitutes and petty thieves were routed. "Rights" were something enjoyed by decent folk, and perhaps also by the serious professional criminal, who avoided violence and could afford a lawyer.

This pattern of policing was not an aberration or the result of occasional excess. From the earliest days of the nation, the police function was seen primarily as that of a night watchman: to maintain order against the chief threats to order – fire, wild animals, and disreputable behavior. Solving crimes was viewed not as a police responsibility but as a private one. In the March, 1969, *Atlantic,* one of us (Wilson) wrote a brief account of how the police role had slowly changed from maintaining order to fighting crimes. The change began with the creation of private detectives (often ex-criminals), who worked on a contingency-fee basis for individuals who had suffered losses. In time, the detectives were absorbed into municipal police agencies and paid a regular salary; simultaneously, the responsibility for prosecuting thieves was shifted from the aggrieved private citizen to the professional prosecutor. This process was not complete in most places until the twentieth century.

In the 1960s, when urban riots were a major problem, social scientists began to explore carefully the order-maintenance function of the police, and to suggest ways of improving it – not to make streets safer (its original function) but to reduce the incidence of mass violence. Order-maintenance became, to a degree, coterminous with "community relations." But, as the crime wave that began in the early 1960s continued without abatement throughout the decade and into the 1970s, attention shifted to the role of the police as crime-fighters. Studies of police behavior ceased, by and large, to be accounts of the order-maintenance function and became, instead, efforts to propose and test ways whereby the police could solve more crimes, make more arrests, and gather better evidence. If these things could be done, social scientists assumed, citizens would be less fearful.

A great deal was accomplished during this transition, as both police chiefs and outside experts emphasized the crime-fighting function in their plans, in the allocation of resources, and in deployment of personnel. The police may well have become better crime-fighters as a result. And doubtless they remained aware of their responsibility for order. But the link between order-maintenance and crime-prevention, so obvious to earlier generations, was forgotten.

That link is similar to the process whereby one broken window becomes many. The citizen who fears the ill-smelling drunk, the rowdy teenager, or the importuning beggar is not merely expressing his distaste for unseemly behavior; he is also giving voice to a bit of folk wisdom that happens to be a correct generalization – namely, that serious street crime flourishes in areas in which disorderly behavior goes unchecked. The unchecked panhandler is, in effect, the first broken window. Muggers and robbers, whether opportunistic or professional, believe they reduce their chances of being caught or even identified if they operate on streets where potential victims are already intimidated by prevailing conditions. If the neighborhood cannot keep a bothersome panhandler from annoying passersby, the thief may reason, it is even less likely to call the police to identify a potential mugger or to interfere if the mugging actually takes place.

Some police administrators concede that this process occurs, but argue that motorized-patrol officers can deal with it as effectively as foot-patrol officers. We are not so sure. In theory, an officer in a squad car can observe as much as an officer on foot; in theory, the former can talk to as many people as the latter. But the reality of police–citizen encounters is powerfully altered by the automobile. An officer on foot cannot separate himself from the street people; if he is approached, only his uniform and his personality can help him manage whatever is about to happen. And he can never be certain what that will be – a request for directions, a plea for help, an angry denunciation, a teasing remark, a confused babble, a threatening gesture.

In a car, an officer is more likely to deal with street people by rolling down the window and looking at them. The door and the window exclude the approaching citizen; they are a barrier. Some officers take advantage of this barrier, perhaps unconsciously, by acting differently if in the car than they would on foot. We have seen this countless times. The police car pulls up to a corner where teenagers are gathered. The window is rolled down. The officer stares at the youths. They stare back. The officer says to one, "C'mere." He saunters over, conveying to his friends by his elaborately casual style the idea that he is not

intimidated by authority "What's your name?" "Chuck." "Chuck who?" "Chuck Jones." "What'ya doing, Chuck?" "Nothin'." "Got a P.O. [parole officer]?" "Nah." "Sure?" "Yeah." "Stay out of trouble, Chuckie." Meanwhile, the other boys laugh and exchange comments among themselves, probably at the officer's expense. The officer stares harder. He cannot be certain what is being said, nor can he join in and, by displaying his own skill at street banter, prove that he cannot be "put down." In the process, the officer has learned almost nothing, and the boys have decided the officer is an alien force who can safely be disregarded, even mocked.

Our experience is that most citizens like to talk to a police officer. Such exchanges give them a sense of importance, provide them with the basis for gossip, and allow them to explain to the authorities what is worrying them (whereby they gain a modest but significant sense of having "done something" about the problem). You approach a person on foot more easily, and talk to him more readily, than you do a person in a car. Moreover, you can more easily retain some anonymity if you draw an officer aside for a private chat. Suppose you want to pass on a tip about who is stealing handbags, or who offered to sell you a stolen TV. In the inner city, the culprit, in all likelihood, lives nearby. To walk up to a marked patrol car and lean in the window is to convey a visible signal that you are a "fink."

The essence of the police role in maintaining order is to reinforce the informal control mechanisms of the community itself. The police cannot, without committing extraordinary resources, provide a substitute for that informal control. On the other hand, to reinforce those natural forces the police must accommodate them. And therein lies the problem.

Should police activity on the street be shaped, in important ways, by the standards of the neighborhood rather than by the rules of the state? Over the past two decades, the shift of police from order-maintenance to law-enforcement has brought them increasingly under the influence of legal restrictions, provoked by media complaints and enforced by court decisions and departmental orders. As a consequence, the order-maintenance functions of the police are now governed by rules developed to control police relations with suspected criminals. This is, we think, an entirely new development. For centuries, the role of the police as watchmen was judged primarily not in terms of its compliance with appropriate procedures but rather in terms of its attaining a desired objective. The objective was order, an inherently ambiguous term but a condition that people in a given community recognized when they saw it. The means were the same as those the community itself would employ, if its members were sufficiently determined, courageous, and authoritative. Detecting and apprehending criminals, by contrast, was a means to an end, not an end in itself; a judicial determination of guilt or innocence was the hoped-for result of the law-enforcement mode. From the first, the police were expected to follow rules defining that process, though states differed in how stringent the rules should be. The criminal-apprehension process was always understood to involve individual rights, the violation of which was unacceptable because it meant that the violating officer would be acting as a judge and jury – and that was not his job. Guilt or innocence was to be determined by universal standards under special procedures.

Ordinarily, no judge or jury ever sees the persons caught up in a dispute over the appropriate level of neighborhood order. That is true not only because most cases are handled informally on the street but also because no universal standards are available to settle arguments over disorder, and thus a judge may not be any wiser or more effective than a police officer. Until quite recently in many states, and even today in some places, the police make arrests on such charges as "suspicious person" or "vagrancy" or "public drunkenness" – charges with scarcely any legal meaning. These charges exist not because society wants judges to punish vagrants or drunks but because it wants an officer to have the legal tools to remove undesirable persons from a neighborhood when informal efforts to preserve order in the streets have failed.

Once we begin to think of all aspects of police work as involving the application of universal rules under special procedures, we inevitably ask what constitutes an "undesirable person" and why we should "criminalize" vagrancy or drunkenness. A strong and commendable desire to see that people are treated fairly makes us worry about allowing the police to rout persons who are undesirable by some vague or parochial standard. A growing and not-so-commendable utilitarianism leads us to doubt that any behavior that does not "hurt" another person should be made illegal. And thus many of us who watch over the police are reluctant to allow them to perform, in the only way they can, a function that every neighborhood desperately wants them to perform.

This wish to "decriminalize" disreputable behavior that "harms no one" – and thus remove the ultimate sanction the police can employ to maintain neighborhood order – is, we think, a mistake. Arresting a single drunk or a single vagrant who has harmed no identifiable person seems unjust, and in a sense it is. But failing to do anything about a score of drunks or a hundred vagrants may destroy an entire community. A particular rule that seems to make sense in the individual case makes no sense when it is made a universal rule and applied to all cases. It makes no sense because it fails to take into account the connection between one broken window left untended and a thousand broken windows. Of course, agencies other than the police could attend to the problems posed by drunks or the mentally ill, but in most communities – especially where the "deinstitutionalization" movement has been strong – they do not.

The concern about equity is more serious. We might agree that certain behavior makes one person more undesirable than another, but how do we ensure that age or skin color or national origin or harmless mannerisms will not also become the basis for distinguishing the undesirable from the desirable? How do we ensure, in short, that the police do not become the agents of neighborhood bigotry?

We can offer no wholly satisfactory answer to this important question. We are not confident that there is a satisfactory answer, except to hope that by their selection, training, and supervision, the police will be inculcated with a clear sense of the outer limit of their discretionary authority. That limit, roughly, is this – the police exist to help regulate behavior, not to maintain the racial or ethnic purity of a neighborhood.

Consider the case of the Robert Taylor Homes in Chicago, one of the largest public-housing projects in the country. It is home for nearly 20,000 people, all black, and extends over ninety-two acres along South State Street. It was named after a distinguished black who had been, during the 1940s, chairman of the Chicago Housing Authority. Not long after it opened, in 1962, relations between project residents and the police deteriorated badly. The citizens felt that the police were insensitive or brutal; the police, in turn, complained of unprovoked attacks on them. Some Chicago officers tell of times when they were afraid to enter the Homes. Crime rates soared.

Today, the atmosphere has changed. Police–citizen relations have improved – apparently, both sides learned something from the earlier experience.

Recently, a boy stole a purse and ran off. Several young persons who saw the theft voluntarily passed along to the police information on the identity and residence of the thief, and they did this publicly, with friends and neighbors looking on. But problems persist, chief among them the presence of youth gangs that terrorize residents and recruit members in the project. The people expect the police to "do something" about this, and the police are determined to do just that.

But do what? Though the police can obviously make arrests whenever a gang member breaks the law, a gang can form, recruit, and congregate without breaking the law. And only a tiny fraction of gang-related crimes can be solved by an arrest; thus, if an arrest is the only recourse for the police, the residents' fears will go unassuaged. The police will soon feel helpless, and the residents will again believe that the police "do nothing." What the police in fact do is to chase known gang members out of the project. In the words of one officer, "We kick ass."

Project residents both know and approve of this. The tacit police–citizen alliance in the project is reinforced by the police view that the cops and the gangs are the two rival sources of power in the area, and that the gangs are not going to win.

None of this is easily reconciled with any conception of due process or fair treatment. Since both residents and gang members are black, race is not a factor. But it could be. Suppose a white project confronted a black gang, or vice versa. We would be apprehensive about the police taking sides. But the substantive problem remains the same: how can the police strengthen the informal social-control mechanisms of natural communities in order to minimize fear in public places? Law enforcement, per se, is no answer. A gang can weaken or destroy a community by standing about in a menacing fashion and speaking rudely to passersby without breaking the law.

We have difficulty thinking about such matters, not simply because the ethical and legal issues are so complex but because we have become accustomed to thinking of the law in essentially individualistic terms. The law defines *my* rights, punishes *his* behavior, and is applied by *that* officer because of *this* harm. We assume, in thinking this way, that what is good for the individual will be good for the community, and what doesn't matter when it happens to one person won't matter if it happens to many. Ordinarily, those are plausible assumptions. But in cases where behavior that is tolerable to one person is intolerable to many others, the reactions of the others – fear, withdrawal, flight – may ultimately make matters worse for everyone, including the individual who first professed his indifference.

It may be their greater sensitivity to communal as opposed to individual needs that helps explain why the residents of small communities are more satisfied with their police than are the residents of similar neighborhoods in big cities. Elinor Ostrom and her co-workers at Indiana University compared the perception of police services in two poor, all-black Illinois towns – Phoenix and East Chicago Heights – with those of three comparable all-black neighborhoods in Chicago. The level of criminal victimization and the quality of police–community relations appeared to be about the same in the towns and the Chicago neighborhoods, but the citizens living in their own villages were much more likely than those living in the Chicago neighborhoods to say that they do not stay at home for fear of crime, to agree that the local police have "the right to take any action necessary" to deal with problems, and to agree that the police "look out for the needs of the average citizen." It is possible that the residents and the police of the small towns saw themselves as engaged in a collaborative effort to maintain a certain standard of communal life, whereas those of the big city felt themselves to be simply requesting and supplying particular services on an individual basis.

If this is true, how should a wise police chief deploy his meager forces? The first answer is that nobody knows for certain, and the most prudent course of action would be to try further variations on the Newark experiment, to see more precisely what works in what kinds of neighborhoods. The second answer is also a hedge – many aspects of order-maintenance in neighborhoods can probably best be handled in ways that involve the police minimally, if at all. A busy, bustling shopping center and a quiet, well-tended suburb may need almost no visible police presence. In both cases, the ratio of respectable to disreputable people is ordinarily so high as to make informal social control effective.

Even in areas that are in jeopardy from disorderly elements, citizen action without substantial police involvement may be sufficient. Meetings between teenagers who like to hang out on a particular corner and adults who want to use that corner might well lead to an amicable agreement on a set of rules about how many people can be allowed to congregate, where, and when.

Where no understanding is possible – or if possible, not observed – citizen patrols may be a sufficient response. There are two traditions of communal involvement in maintaining order. One, that of the "community watchmen," is as old as the first settlement of the New World. Until well into the nineteenth century, volunteer watchmen, not policemen, patrolled their communities to keep order. They did so, by and large, without taking the law into their own hands – without, that is, punishing persons or using force. Their presence deterred disorder or alerted the community to disorder that could not be deterred. There are hundreds of such efforts today in communities all across the nation. Perhaps the best known is that of the Guardian Angels, a group of unarmed young persons in distinctive berets and T-shirts, who first came to public attention when they began patrolling the New York City subways but who claim now to have chapters in more than thirty American cities. Unfortunately, we have little information about the effect of these groups on crime. It is possible, however, that whatever their effect on crime, citizens find their presence reassuring, and that they thus contribute to maintaining a sense of order and civility.

The second tradition is that of the "vigilante." Rarely a feature of the settled communities of the East, it was primarily to be found in those frontier towns that grew up in advance of the reach of government. More than 350 vigilante groups are known to have existed; their distinctive feature was that their members did take the law into their own hands, by acting as judge, jury, and often executioner as well as policeman. Today, the vigilante movement is conspicuous by its rarity, despite the great fear expressed by citizens that the older cities are becoming "urban frontiers." But some community-watchmen groups have skirted the line, and others may cross it in the future. An ambiguous case, reported in *The Wall Street Journal*, involved a citizens' patrol in the Silver Lake area of Belleville, New Jersey. A leader told the reporter, "We look for outsiders." If a few teenagers from outside the neighborhood enter it, "we ask them their business," he said. "If they say they're going down the street to see Mrs. Jones, fine, we let them pass. But then we follow them down the block to make sure they're really going to see Mrs. Jones."

Though citizens can do a great deal, the police are plainly the key to order-maintenance. For one thing, many communities, such as the Robert Taylor Homes, cannot do the job by themselves. For another, no citizen in a neighborhood, even an organized one, is likely to feel the sense of responsibility that wearing a badge confers. Psychologists have done many studies on why people fail to go to the aid of persons being attacked or seeking help, and they have learned that the cause is not "apathy" or "selfishness" but the absence of some plausible grounds for feeling that one must personally accept responsibility. Ironically, avoiding responsibility is easier when a lot of people are standing about. On streets and in public places, where order is so important, many people are likely to be "around," a fact that reduces the chance of any one person acting as the agent of the community. The police officer's uniform singles him out as a person who must accept responsibility if asked. In addition, officers, more easily than their fellow citizens, can be expected to distinguish between what is necessary to protect the safety of the street and what merely protects its ethnic purity.

But the police forces of America are losing, not gaining, members. Some cities have suffered substantial cuts in the number of officers available for duty. These cuts are not likely to be reversed in the near future. Therefore, each department must assign its existing officers with great care. Some neighborhoods are so demoralized and crime-ridden as to make foot patrol useless; the best the police can do with limited resources is respond to the enormous number of calls for service. Other neighborhoods are so stable and serene as to make foot patrol unnecessary. The key is to identify neighborhoods at the tipping point where the public order is deteriorating but not unreclaimable, where the streets are used frequently but by apprehensive people, where a window is likely to be broken at any time, and must quickly be fixed if all are not to be shattered.

Most police departments do not have ways of systematically identifying such areas and assigning officers to them. Officers are assigned on the basis of crime rates (meaning that marginally threatened areas are often stripped so that police can investigate crimes in areas where the situation is hopeless) or on the basis of calls for service (despite the fact that most citizens do not call the police when they are merely frightened or annoyed). To allocate patrols wisely, the department must look at the neighborhoods and decide, from first-hand evidence, where an additional officer will make the greatest difference in promoting a sense of safety.

One way to stretch limited police resources is being tried in some public-housing projects. Tenant organizations hire off-duty police officers for patrol work in their buildings. The costs are not high (at least not per resident), the officer likes the additional income, and the residents feel safer. Such arrangements are probably more successful than hiring private watchmen, and the Newark experiment helps us understand why. A private security guard may deter crime or misconduct by his presence, and he may go to the aid of persons needing help, but he may well not intervene – that is, control or drive away someone challenging community standards. Being a sworn officer – a "real cop" – seems to give one the confidence, the sense of duty, and the aura of authority necessary to perform this difficult task.

Patrol officers might be encouraged to go to and from duty stations on public transportation and, while on the bus or subway car, enforce rules about smoking, drinking, disorderly conduct, and the like. The enforcement need involve nothing more than ejecting the offender (the offense, after all, is not one with which a booking officer or a judge wishes to be bothered). Perhaps the random but relentless maintenance of standards on buses would lead to conditions on buses that approximate the level of civility we now take for granted on airplanes.

But the most important requirement is to think that to maintain order in precarious situations is a vital job. The police know this is one of their functions, and they also believe, correctly, that it cannot be done to the exclusion of criminal investigation and responding to calls. We may have encouraged them to suppose, however, on the basis of our oft-repeated concerns about serious, violent crime, that they will be judged exclusively on their capacity as crime-fighters. To the extent that this is the case, police administrators will continue to concentrate police personnel in the highest-crime areas (though not necessarily in the areas most vulnerable to criminal invasion), emphasize their training in the law and criminal apprehension (and not their training in managing street life), and join too quickly in campaigns to decriminalize "harmless" behavior (though public drunkenness, street prostitution, and pornographic displays can destroy a community more quickly than any team of professional burglars).

Above all, we must return to our long-abandoned view that the police ought to protect communities as well as individuals. Our crime statistics and victimization surveys measure individual losses, but they do not measure communal losses. Just as physicians now recognize the importance of fostering health rather than simply treating illness, so the police – and the rest of us – ought to recognize the importance of maintaining, intact, communities without broken windows.

"The Competitive Advantage of the Inner City"

Harvard Business Review (1995)

Michael Porter

Editors' Introduction

Harvard Business School professor Michael Porter declares that the economic distress of inner-city neighborhoods may be the most pressing issue facing the United States. Like William Julius Wilson (p. 126), Porter sees lack of jobs as the root cause of drug abuse, crime, and other social problems. And like Wilson, Porter feels that government response to the problem of inner-city decline and lack of jobs has been ineffective.

Porter characterizes the current approach as based on a *social model* aimed at the individual. Governments have historically provided "relief" for the effects economic problems have had on individuals in the form of welfare payments, subsidized housing, food stamps, and free medical assistance. Corporate philanthropy has often done the same.

Porter feels that government programs to create jobs in inner-city neighborhoods and train local residents for them have been fragmented and inefficient. He argues that they have consisted of subsidies, preference programs, and expensive efforts to stimulate economic activity. Governments have dumped money on small marginal inner-city businesses, which could not turn a profit without government help. They have hired incompetent workers or required private sector contractors to hire them. Or they have spent large sums to get hopelessly blighted redevelopment areas and badly contaminated brownfields back into useable condition. Often the subsidized firms fail, the workers hired through preference programs are fired, and the brownfields and redevelopment project areas sit vacant or are "showcase" projects that burn up money better spent elsewhere.

Porter argues for a new *economic* model of inner-city revitalization. He favors private, for-profit initiatives based on economic self-interest, not artificial inducements, charity, and government mandates. Such an approach will only work, he says, if it takes advantage of the true competitive advantages of the inner city.

Porter punctures the myth that inner-city real estate or labor costs are sufficiently lower to make a compelling reason for firms to locate in inner-city neighborhoods rather than in suburban or exurban locations. Rather, he argues, there are four true advantages to inner-city locations: (1) their strategic location, (2) local market demand the areas themselves possess, (3) possibilities of integration with regional job clusters, and (4) an industrious labor force which is eager to work. Firms that can take advantage of these inner-city advantages may find inner-city locations the best place to do business. Unlike Wilson, who emphasizes the loss of jobs that people with limited education can perform, Porter argues that there are still firms that can use unskilled labor for warehouse and production-line workers, truck drivers, and other unskilled jobs.

Porter places a large part of the blame for high costs and the difficulties of doing business in inner cities on government regulation and anti-business attitudes. He argues that local governments can improve the economic climate and make inner-city neighborhoods more attractive for private investment by reducing regulation. This has been a theme in British Enterprise Zone and US Empowerment Zone/Enterprise Community Programs that have reduced environmental, occupational health, and other government regulations in distressed neighborhoods in order to lure in business.

What would his Harvard colleague William Julius Wilson (p. 126) think of Porter's characterization of the social model and preference programs? Given historic racism, bad schools, and the lack of skills many inner-city Black ghetto residents possess, aren't preference programs necesssary?

Michael Porter is the C. Roland Christensen Professor of Business Administration at the Harvard Business School. He is the author of *Competitive Advantage: Creating and Sustaining Superior Performance* (New York: Free Press, 1998), *Competitive Strategy: Techniques for Analyzing Industries and Competitors* (New York: Free Press, 1998), *On Competition* (Boston: Harvard Business School Press, 1998), *The Competitive Advantage of Nations* (New York: Free Press, 1990), an edited anthology, *Strategy: Seeking and Securing Competitive Advantage* (Boston: Harvard Business School Press, 1991), and many other books.

Books on community-based economic development – generally less oriented to private sector approaches than Porter – include Michael H. Shuman, *Going Local: Creating Self-reliant Communities in a Global Age* (New York: Free Press, 1998), Douglas Henton, John Melville, and Kimberly Walsh, *Grassroots Leaders for a New Economy* (San Francisco: Jossey-Bass, 1997), Richard P. Taub, *Community Capitalism* (Boston: Harvard Business School Press, 1994), Gregory D. Squires (ed.), *From Redlining to Reinvestment* (Philadelphia, PA: Temple University Press, 1992), Edward J. Blakeley, *Planning Local Economic Development* (Thousand Oaks, CA: Sage, 1994), and Julia Ann Parzen, Michael Hall Kieschnick (contributor), *Credit Where It's Due* (Philadelphia, PA: Temple University Press, 1993).

▪ ▪ ▪ ▪ ▪ ▪

The economic distress of America's inner cities may be the most pressing issue facing the nation. The lack of businesses and jobs in disadvantaged urban areas fuels not only a crushing cycle of poverty but also crippling social problems, such as drug abuse and crime. And, as the inner cities continue to deteriorate, the debate on how to aid them grows increasingly divisive.

The sad reality is that the efforts of the past few decades to revitalize the inner cities have failed. The establishment of a sustainable economic base – and with it employment opportunities, wealth creation, role models, and improved local infrastructure – still eludes us despite the investment of substantial resources.

Past efforts have been guided by a social model built around meeting the needs of individuals. Aid to inner cities, then, has largely taken the form of relief programs such as income assistance, housing subsidies, and food stamps, all of which address highly visible – and real – social needs.

Programs aimed more directly at economic development have been fragmented and ineffective. These piecemeal approaches have usually taken the form of subsidies, preference programs, or expensive efforts to stimulate economic activity in tangential fields such as housing, real estate, and neighborhood development. Lacking an overall strategy, such programs have treated the inner city as an island isolated from the surrounding economy and subject to its own unique laws of competition. They have encouraged and supported small, subscale businesses designed to serve the local community but ill equipped to attract the community's own spending power, much less export outside it. In short, the social model has inadvertently undermined the creation of economically viable companies. Without such companies and the jobs they create, the social problems will only worsen.

The time has come to recognize that revitalizing the inner city will require a radically different approach. While social programs will continue to play a critical role in meeting human needs and improving education, they must support – and not undermine – a coherent economic strategy. The question we should be asking is how inner-city-based businesses and nearby employment opportunities for inner-city residents can proliferate and grow. A sustainable economic base can be created in the inner city, but only as it has been created elsewhere: through private, for-profit initiatives and investment based on economic self-interest and genuine competitive advantage – not through artificial inducements, charity, or government mandates.

We must stop trying to cure the inner city's problems by perpetually increasing social investment and hoping for economic activity to follow. Instead, an economic model must begin with the premise that inner-city businesses should be profitable and should be positioned to compete on a regional, national, and even international scale. These businesses should be capable not only of serving the local community but also of exporting goods and services to the surrounding economy. The cornerstone of such a model is to identify and exploit the competitive advantages of inner cities that will translate into truly profitable businesses.

Our policies and programs have fallen into the trap of redistributing wealth. The real need – and the real opportunity – is to create wealth.

TOWARDS A NEW MODEL: LOCATION AND BUSINESS DEVELOPMENT

Economic activity in and around inner cities will take root if it enjoys a competitive advantage and occupies a niche that is hard to replicate elsewhere. If companies are to prosper, they must find a compelling competitive reason for locating in the inner city. A coherent strategy for development starts with that fundamental economic principle, as the contrasting experiences of the following companies illustrate.

Alpha Electronics (the company's name has been disguised), a 28-person company that designed and manufactured multimedia computer peripherals, was initially based in lower Manhattan. In 1987, the New York City Office of Economic Development set out to orchestrate an economic "renaissance" in the South Bronx by inducing companies to relocate there. Alpha, a small but growing company, was sincerely interested in contributing to the community and eager to take advantage of the city's willingness to subsidize its operations. The city, in turn, was happy that a high-tech company would begin to stabilize a distressed neighborhood and create jobs. In exchange for relocating, the city provided Alpha with numerous incentives that would lower costs and boost profits. It appeared to be an ideal strategy.

By 1994, however, the relocation effort had proved a failure for all concerned. Despite the rapid growth of its industry, Alpha was left with only 8 of its original 28 employees. Unable to attract high-quality employees to the South Bronx or to train local residents, the company was forced to outsource its manufacturing and some of its design work. Potential suppliers and customers refused to visit Alpha's offices. Without the city's attention to security, the company was plagued by theft.

What went wrong? Good intentions notwithstanding, the arrangement failed the test of business logic. Before undertaking the move, Alpha and the city would have been wise to ask themselves why none of the South Bronx's thriving businesses was in electronics. The South Bronx as a location offered no specific advantages to support Alpha's business, and it had several disadvantages that would prove fatal. Isolated from the lower Manhattan hub of computer-design and software companies, Alpha was cut off from vital connections with customers, suppliers, and electronic designers.

In contrast, Matrix Exhibits, a $2.2 million supplier of trade-show exhibits that has 30 employees, is thriving in Atlanta's inner city. When Tennessee-based Matrix decided to enter the Atlanta market in 1985, it could have chosen a variety of locations. All the other companies that create and rent trade show exhibits are based in Atlanta's suburbs. But the Atlanta World Congress Center, the city's major exhibition space, is just a six-minute drive from the inner city, and Matrix chose the location because it provided a real competitive advantage. Today Matrix offers customers superior response time, delivering trade-show exhibits faster than its suburban competitors. Matrix benefits from low rental rates for warehouse space – about half the rate its competitors pay for similar space in the suburbs – and draws half its employees from the local community. The commitment of local police has helped the company avoid any serious security problems. Today Matrix is one of the top five exhibition houses in Georgia.

Alpha and Matrix demonstrate how location can be critical to the success or failure of a business. Every location – whether it be a nation, a region, or a city – has a set of unique local conditions that underpin the ability of companies based there to compete in a particular field. The competitive advantage of a location does not usually arise in isolated companies but in clusters of companies – in other words, in companies that are in the same industry or otherwise linked together through customer, supplier, or similar relationships. Clusters represent critical masses of skill, information, relationships, and infrastructure in a given field. Unusual or sophisticated local demand gives companies insight into customers' needs. Take Massachusetts's highly competitive cluster of information-technology industries: it includes companies specializing in semiconductors, workstations, supercomputers, software, networking equipment, databases, market research, and computer magazines.

Clusters arise in a particular location for specific historical or geographic reasons – reasons that may cease to matter over time as the cluster itself becomes powerful and competitively self-sustaining. In successful clusters such as Hollywood, Silicon Valley, Wall Street, and Detroit, several competitors often push one another to improve products and processes. The presence of a group of competing companies contributes to the formation of new suppliers, the growth of companies in related fields, the formation of specialized training programs, and the emergence of technological centers of excellence in colleges and universities. The clusters also provide newcomers with access to expertise, connections, and infrastructure that they in turn can learn and exploit to their own economic advantage.

If locations (and the events of history) give rise to clusters, it is clusters that drive economic development. They create new capabilities, new companies, and new industries. I initially described this theory of location in *The Competitive Advantage of Nations* (Free Press, 1990), applying it to the relatively large geographic areas of nations and states. But it is just as relevant to smaller areas such as the inner city. To bring the theory to bear on the inner city, we must first identify the inner city's competitive advantages and the ways inner-city businesses can forge connections with the surrounding urban and regional economies.

THE TRUE ADVANTAGES OF THE INNER CITY

The first step toward developing an economic model is identifying the inner city's true competitive advantages. There is a common misperception that the inner

city enjoys two main advantages: low-cost real estate and labor. These so-called advantages are more illusory than real. Real estate and labor costs are often higher in the inner city than in suburban and rural areas. And even if inner cities were able to offer lower-cost labor and real estate compared with other locations in the United States, basic input costs can no longer give companies from relatively prosperous nations a competitive edge in the global economy. Inner cities would inevitably lose jobs to countries like Mexico or China, where labor and real estate are far cheaper.

Only attributes that are unique to inner cities will support viable businesses. My ongoing research of urban areas across the United States identifies four main advantages of the inner city: strategic location, local market demand, integration with regional clusters, and human resources. Various companies and programs have identified and exploited each of those advantages from time to time. To date, however, no systematic effort has been mounted to harness them.

Strategic location

Inner cities are located in what *should* be economically valuable areas. They sit near congested high-rent areas, major business centers, and transportation and communications nodes. As a result, inner cities can offer a competitive edge to companies that benefit from proximity to downtown business districts, logistical infrastructure, entertainment or tourist centers, and concentrations of companies.

Local market demand

The inner-city market itself represents the most immediate opportunity for inner-city-based entrepreneurs and businesses. At a time when most other markets are saturated, inner-city markets remain poorly served – especially in retailing, financial services, and personal services. In Los Angeles, for example, retail penetration per resident in the inner city compared with the rest of the city is 35% in supermarkets, 40% in department stores, and 50% in hobby, toy, and game stores.

The first notable quality of the inner-city market is its size. Even though average inner-city incomes are relatively low, high population density translates into an immense market with substantial purchasing power.

Boston's inner city, for example, has an estimated total family income of $3.4 billion.

Spending power per acre is comparable with the rest of the city despite a 21% lower average household income level than in the rest of Boston, and, more significantly, higher than in the surrounding suburbs. In addition, the market is young and growing rapidly, owing in part to immigration and relatively high birth rates.

Integration with regional clusters

The most exciting prospects for the future of inner-city economic development lie in capitalizing on nearby regional clusters: those unique-to-a-region collections of related companies that are competitive nationally and even globally. For example, Boston's inner city is next door to world-class financial-services and health-care clusters. South Central Los Angeles is close to an enormous entertainment cluster and a large logistical-services and wholesaling complex.

The ability to access competitive clusters is a very different attribute – and one much more far reaching in economic implication – than the more generic advantage of proximity to a large downtown area with concentrated activity. Competitive clusters create two types of potential advantages. The first is for business formation. Companies providing supplies, components, and support services could be created to take advantage of the inner city's proximity to multiple nearby customers in the cluster . . .

The second advantage of these clusters is the potential they offer inner-city companies to compete in downstream products and services. For example, an inner-city company could draw on Boston's strength in financial services to provide services tailored to inner-city needs – such as secured credit cards, factoring, and mutual funds – both within and outside the inner city in Boston and elsewhere in the country.

[. . .]

Human resources

The inner city's fourth advantage takes on a number of deeply entrenched myths about the nature of its residents. The first myth is that inner-city residents do not want to work and opt for welfare over gainful employment. Although there is a pressing need to deal with inner-city residents who are unprepared for work, most inner-city residents are industrious and eager to work. For moderate-wage jobs ($6 to $10 per hour) that require little formal education (for instance, warehouse workers, production-line workers, and truck drivers), employers report that they find hard-working, dedicated employees in the inner city. For example, a company in Boston's inner-city neighborhood of Dorchester bakes and decorates cakes sold to supermarkets throughout the region. It attracts and retains area residents at $7 to $8 per hour (plus contributions to pensions and health insurance) and has almost 100 local employees. The loyalty of its labor pool is one of the factors that has allowed the bakery to thrive.

Admittedly, many of the jobs currently available to inner-city residents provide limited opportunities for advancement. But the fact is that they are jobs; and the inner city and its residents need many more of them close to home. Proposals that workers commute to jobs in distant suburbs – or move to be near those jobs – underestimate the barriers that travel time and relative skill level represent for inner-city residents. Moreover, in deciding what types of businesses are appropriate to locate in the inner city, it is critical to be realistic about the pool of potential employees. Attracting high-tech companies might make for better press, but it is of little benefit to inner-city residents. Recall the contrasting experiences of Alpha Electronics and Matrix Exhibits. In the case of Alpha, there was a complete mismatch between the company's need for highly skilled professionals and the available labor pool in the local community. In contrast, Matrix carefully considered the available workforce when it established its Atlanta office. Unlike the Tennessee headquarters, which custom-designs and creates exhibits for each client, the Atlanta office specializes in rentals made from prefabricated components – work requiring less-skilled labor, which can be drawn from the inner-city. Given the workforce, low-skill jobs are realistic and economically viable: they represent the first rung on the economic ladder for many individuals who otherwise would be unemployed. Over time, successful job creation will trigger a self-reinforcing process that raises skill and wage levels.

The second myth is that the inner city's only entrepreneurs are drug dealers. In fact, there is a real capacity for legitimate entrepreneurship among inner-city residents, most of which has been channeled into the provision of social services. For instance, Boston's inner city has numerous social service providers as well as social, fraternal, and religious organizations. Behind the creation and building of those organizations is a whole cadre of local entrepreneurs who have responded to intense local demand for social services and to funding opportunities provided by government, foundations, and private sector sponsors. The challenge is to redirect some of that talent and energy toward building for-profit businesses and creating wealth.

The third myth is that skilled minorities, many of whom grew up in or near inner cities, have abandoned their roots. Today's large and growing pool of talented minority managers represents a new generation of potential inner-city entrepreneurs. Many have been trained at the nation's leading business schools and have gained experience in the nation's leading companies. Approximately 2,800 African Americans and 1,400 Hispanics graduate from M.B.A. programs every year compared with only a handful 20 years ago. Thousands of highly trained minorities are working at leading companies such as Morgan Stanley, Citibank, Ford, HewlettPackard, and McKinsey & Company. Many of these managers have developed

the skills, net-work, capital base, and confidence to begin thinking about joining or starting entrepreneurial companies in the inner city . . .

THE REAL DISADVANTAGES OF THE INNER CITY

The second step toward creating a coherent economic strategy is addressing the very real disadvantages of locating businesses in the inner city. The inescapable fact is that businesses operating in the inner city face greater obstacles than those based elsewhere. Many of those obstacles are needlessly inflicted by government. Unless the disadvantages are addressed directly, instead of indirectly through subsidies or mandates, the inner city's competitive advantages will continue to erode.

Land

Although vacant property is abundant in inner cities, much of it is not economically usable. Assembling

New Model	Old Model
Economic: create wealth	*Social: redistribute wealth*
Private sector	Government and social service organizations
Profitable businesses	Subsidized businesses
Integration with the regional economy	Isolation from the larger economy
Companies that are export oriented	Companies that serve the local community
Skilled and experienced minorities engaged in building businesses	Skilled and experienced minorities engaged in the social service sector
Mainstream, private sector institutions enlisted	Special institutions created
Inner-city disadvantages addressed directly	Inner-city disadvantages counterbalanced
Government focuses on improving the environment for business	Government directly involved with providing services or funding

Figure 1 Inner-city economic development

small parcels into meaningful sites can be prohibitively expensive and is further complicated by the fact that a number of city, state, and federal agencies each control land and fight over turf . . .

Building costs

The cost of building in the inner city is significantly higher than in the suburbs because of the costs and delays associated with logistics, negotiations with *community* groups, and strict urban regulations: restrictive zoning, architectural codes, permits, inspections, and government-required union contracts and minority setasides. Ironically, despite the desperate need for new projects, construction in inner cities is far more regulated than it is in the suburbs – a legacy of big city politics and entrenched bureaucracies.

[. . .]

Other costs

Compared with the suburbs, inner cities have high costs for water, other utilities, workers' compensation, health care, insurance, permitting and other fees, real estate and other taxes, OSHA compliance, and neighborhood hiring requirements. For example, Russer Foods, a manufacturing company located in Boston's inner city, operates a comparable plant in upstate New York. The Boston plant's expenses are 55% higher for workers' compensation, 50% higher for family medical insurance, 166% higher for unemployment insurance, 340% higher for water, and 67% higher for electricity. High costs like these drive away companies and hold down wages. Some costs, such as those for workers' compensation, apply to the state or region as a whole. Others, such as real estate taxes, apply citywide. Still others, such as property insurance, are specific to the inner city. All are devastating to maintaining fragile inner-city companies and to attracting new businesses.

It is an unfortunate reality that many cities – because they have a greater proportion of residents dependent on welfare, Medicaid, and other social programs – require higher government spending and, as a result, higher corporate taxes. The resulting tax burden feeds a vicious cycle – driving out more companies while requiring even higher taxes from those that remain. Cities have been reluctant to challenge entrenched bureaucracies and unions, as well as inefficient and outdated government departments, all of which unduly raise city costs.

Finally, excessive regulation not only drives up building and other costs but also hampers almost all facets of business life in the inner city, from putting up an awning over a shop window to operating a pushcart to making site improvements. Regulation also stunts inner-city entrepreneurship, serving as a formidable barrier to small and start-up companies. Restrictive licensing and permitting, high licensing fees, and archaic safety and health regulations create barriers to entry into the very types of businesses that are logical and appropriate for creating jobs and wealth in the inner city.

Security

Both the reality and the perception of crime represent profound impediments to urban economic development. First, crime against property raises costs. For example, the Shops at Church Square, an inner-city strip shopping center in Cleveland, Ohio, spends $2 per square foot more than a comparable suburban center for a full-time security guard, increased lighting, and continuous cleaning – raising overall costs by more than 20%. Second, crime against employees and customers creates an unwillingness to work in and patronize inner-city establishments and restricts companies' hours of operation. Fear of crime ranks among the most important reasons why companies opening new facilities failed to consider inner-city locations and why companies already located in the inner city left. Currently, police devote most of their resources to the security of residential areas, largely overlooking commercial and industrial sites.

Infrastructure

Transportation infrastructure planning, which today focuses primarily on the mobility of residents for shopping and commuting, should consider equally the mobility of goods and the ease of commercial transactions. The most critical aspects of the new economic model – the importance of the location of the inner city, the connections between inner-city businesses and regional clusters, and the development of export-oriented businesses – require the presence of strong logistical links between inner-city business sites and the surrounding economy. Unfortunately, the business infrastructure of the inner city has fallen into disrepair. The capacity of roads, the frequency and location of highway on-ramps and off-ramps, the links to downtown, and the access to railways, airports, and regional logistical networks are inadequate.

Employee skills

Because their average education levels are low, many inner-city residents lack the skills to work in any but the most unskilled occupations. To make matters worse, employment opportunities for less educated workers have fallen markedly. In Boston between 1970 and 1990, for example, the percentage of jobs held by people without high school diplomas dropped from 29% to 7%, while those held by college graduates climbed from 18% to 44%. And the unemployment rate for African-American men aged 16 to 64 with less than a high school education in major northeastern cities rose from 19% in 1970 to 57% in 1990.

Management skills

The managers of most inner-city companies lack formal business training. That problem, however, is not unique to the inner city; it is a characteristic of small businesses in general. Many individuals with extensive work histories but little or no formal managerial training start businesses. Inner-city companies without well trained managers experience a series of predictable problems that are similar to those that affect many small businesses: weaknesses in strategy development, market segmentation, customer-needs evaluation, introduction of information technology, process design, cost control, securing or restructuring financing, interaction with lenders and government regulatory agencies, crafting business plans, and employee training. Local community colleges often offer management courses, but their quality is uneven, and entrepreneurs are hard-pressed for time to attend them.

Capital

Access to debt and equity capital represents a formidable barrier to entrepreneurship and company growth in inner-city areas.

First, most inner-city businesses still suffer from poor access to debt funding because of the limited attention that mainstream banks paid them historically. Even in the best of circumstances, small business lending is only marginally profitable to banks because transaction costs are high relative to loan amounts. Many banks remain in small-business lending only to attract deposits and to help sell other more profitable products.

The federal government has made several efforts to address the inner city's problem of debt capital. As a result of legislation like the Community Reinvestment Act, passed in order to overcome bias in lending, banks have begun to pay much more attention to inner-city areas. In Boston, for example, leading banks are competing fiercely to lend in the inner city – and some claim to be doing so profitably. Direct financing efforts by government, however, have proved ineffective. The proliferation of government loan pools and quasi-public lending organizations has produced fragmentation, market confusion, and duplication of overhead. Business loans that would provide scale to private sector lenders are siphoned off by these organizations, many of which are high-cost, bureaucratic, and risk-averse. In the end, the development of high-quality private sector expertise in inner-city business financing has been undermined.

Second, equity capital has been all but absent. Inner-city entrepreneurs often lack personal or family savings and networks of individuals to draw on for capital. Institutional sources of equity capital are scarce for minority-owned companies and have virtually ignored inner-city business opportunities.

Attitudes

A final obstacle to companies in the inner city is anti-business attitudes. Some workers perceive businesses as exploitative, a view that guarantees poor relations between labor and management. Equally debilitating are the antibusiness attitudes held by community leaders and social activists. These attitudes are the legacy of a regrettable history of poor treatment of workers, departures of companies, and damage to the environment. But holding on to these views today is counterproductive. Too often, community leaders mistakenly view businesses as a means of directly meeting social needs; as a result, they have unrealistic expectations for corporate involvement in the community . . .

Demanding linkage payments and contributions and stirring up antibusiness sentiment are political tools that brought questionable results in the past when owners had less discretion about where they chose to locate their companies. In today's increasingly competitive business environment, such tactics will serve only to stunt economic growth.

Overcoming the business disadvantages of the inner city as well as building on its inherent advantages will require the commitment and involvement of business, government, and the nonprofit sector. Each will have to abandon deeply held beliefs and past approaches. Each must be willing to accept a new model for the inner city based on an economic rather than social perspective. The private sector, non-government or social service organizations, must be the focus of the new model.

The new role of the private sector

The economic model challenges the private sector to assume the leading role. First, however, it must adopt new attitudes toward the inner city. Most private sector initiatives today are driven by preference programs or charity. Such activities would never stand on their own merits in the marketplace. It is inevitable, then, that they contribute to growing cynicism. The private sector will be most effective if it focuses on what it does best: creating and supporting economically viable businesses built on true competitive advantage. It should pursue four immediate opportunities as it assumes its new role.

1. *Create and expand business activity in the inner city.* The most important contribution companies can make to inner cities is simply to do business there. Inner cities hold untapped potential for profitable businesses. Companies and entrepreneurs must seek out and seize those opportunities that build on the true advantages of the inner city. In particular, retailers, franchisers, and financial services companies have immediate opportunities. Franchises represent an especially attractive model for inner-city entrepreneurship because they provide not only a business concept but also training and support.

Businesses can learn from the mistakes that many outside companies have made in the inner city. One error is the failure of retail and service businesses to tailor their goods and services to the local market . . .

Another common mistake is the failure to build relationships within the community and to hire locally. Hiring local residents builds loyalty from neighborhood customers, and local employees of retail and service businesses can help stores customize their products. Evidence suggests that companies that were perceived to be in touch with the community had far fewer security problems, whether or not the owners lived in the community.

[. . .]

2. *Establish business relationships with inner city companies.* By entering into joint ventures or customer–supplier relationships, outside companies will help inner-city companies by encouraging them to export and by forcing them to be competitive. In the long run, both sides will benefit . . . Such relationships, based not on charity but on mutual self-interest, are sustainable ones; every major company should develop them.

3. *Redirect corporate philanthropy from social services to business-to-business efforts.* Countless companies give many millions of dollars each year to worthy inner-city social-service agencies. But philanthropic efforts will be more effective if they also focus on building business-to-business relationships that, in the long run, will reduce the need for social services.

First, corporations could have a tremendous impact on training. The existing system for job training in the United States is ineffective. Training programs are fragmented, overhead intensive, and disconnected from the needs of industry. Many programs train people for nonexistent jobs in industries with no projected growth. Although reforming training will require the help of government, the private sector must determine how and where resources should be allocated to ensure that the specific employment needs of local and regional businesses are met. Ultimately, employers, not government, should certify all training programs based on relevant criteria and likely job availability.

Training programs led by the private sector could be built around industry clusters located in both the inner city (for example, restaurants, food service, and food processing in Boston) and the nearby regional economy (for example, financial services and health care in Boston). Industry associations and trade groups, supported by government incentives, could sponsor their own training programs in collaboration with local training institutions.

[. . .]

Second, the private sector could make an equally substantial impact by providing management assistance to inner-city companies. As with training, current programs financed or operated by the government are inadequate. Outside companies have much to offer companies in the inner city: talent, know-how, and contacts. One approach to upgrading management skills is to emphasize networking with companies in the regional economy that either are part of the same cluster (customers, suppliers, and related businesses) or have expertise in needed areas. An inner-city company could team up with a partner in the region who provides management assistance; or a consortium of companies with a required expertise, such as information technology, could provide assistance to inner-city businesses in need of upgrading their systems.

[. . .]

4. *Adopt the right model for equity capital investments.* The investment community – especially venture capitalists – must be convinced of the viability of investing in the inner city. There is a small but growing number of minority-oriented equity providers (although none specifically focus on inner cities). A successful model for inner-city investing will probably not look like the familiar venture-capital model created primarily for technology companies. Instead, it may resemble the equity funds operating in the emerging economies of Russia or Hungary – investing in such mundane but potentially profitable projects as supermarkets and laundries. Ultimately, inner-city-based businesses that follow the principles of competitive advantage will generate appropriate returns to investors – particularly if aided by appropriate incentives, such as tax exclusions for capital gains and dividends for qualifying inner-city businesses.

The new role of government

To date, government has assumed primary responsibility for bringing about the economic revitalization of the inner city. Existing programs at the federal, state, and local levels designed to create jobs and attract businesses have been piecemeal and fragmented at best. Still worse, these programs have been based on subsidies and mandates rather than on marketplace realities. Unless we find new approaches, the inner city will continue to drain our rapidly shrinking public coffers.

Undeniably, inner cities suffer from a long history of discrimination. However, the way for government to move forward is not by looking behind. Government can assume a more effective role by supporting the private sector in new economic initiatives. It must shift its focus from direct involvement and intervention to creating a favorable environment for business. This is not to say that public funds will not be necessary. But subsidies must be spent in ways that do not distort business incentives, focusing instead on providing the infrastructure to support genuinely profitable businesses. Government at all levels should focus on four goals as it takes on its new role.

1. *Direct resources to the areas of greatest economic need.* The crisis in our inner cities demands that they be first in line for government assistance. This may seem an obvious assertion. But the fact is that many programs in areas such as infrastructure, crime prevention, environmental cleanup, land development, and purchasing preference spread funds across constituencies for political reasons. For example, most transportation infrastructure spending goes to creating still more attractive suburban areas. In addition, a majority of preference-program assistance does not go to companies located in low-income neighborhoods.

[. . .]

Unfortunately, the qualifying criteria for current government assistance programs are not properly designed to channel resources where they are most needed. Preference programs support business based on the race, ethnicity, or gender of their owners rather than on economic need. In addition to directing resources away from the inner city, such race-based or gender-based distinctions reinforce inappropriate stereotypes and attitudes, breed resentment, and increase the risk that programs will be manipulated to serve unintended populations. Location in an economically distressed area and employment of a significant percentage of its residents should be the qualification for government assistance and preference programs. Shifting the focus to economic distress in this way will help enlist all segments of the private sector in the solutions to the inner city's problems.

2. *Increase the economic value of the inner city as a business location.* In order to stimulate economic development, government must recognize that it is a part of the problem. Today its priorities often run counter to business needs. Artificial and outdated government-induced costs must be stripped away in the effort to make the inner city a profitable location for business.

Doing so will require rethinking policies and programs in a wide range of areas . . .

Indeed, there are numerous possibilities for reform. Imagine, for example, policy aimed at eliminating the substantial land and building cost penalties that businesses face in the inner city. Ongoing rent subsidies run the risk of attracting companies for which an inner-city location offers no other economic value. Instead, the goal should be to provide building-ready sites at market prices. A single government entity could be charged with assembling parcels of land and with subsidizing demolition, environmental cleanup, and other costs. The same entity could also streamline all aspects of building – including zoning, permitting, inspections, and other approvals.

That kind of policy would require further progress on the environmental front. A growing number of cities – including Detroit, Chicago, Indianapolis, Minneapolis, and Wichita, Kansas – have successfully developed so-called brownfield urban areas by making environmental cleanup standards more flexible depending on land use, indemnifying land owners against additional costs if contamination is found on a site after a cleanup, and using tax-increment financing to help fund cleanup and redevelopment costs.

Government entities could also develop a more strategic approach to developing transportation and communications infrastructures, which would facilitate the fluid movement of goods, employees, customers, and suppliers within and beyond the inner city. Two projects in Boston are prime examples: first, a new exit ramp connecting the inner city to the nearby Massachusetts Turnpike, which in turn connects to the surrounding region and beyond; and a direct access road to the harbor tunnel, which connects to Logan International Airport. Though inexpensive, both projects are stalled because the city does not have a clear vision of their economic importance.

3. *Deliver economic development programs and services through mainstream, private sector institutions.* There has been a tendency to rely on small community-based nonprofits, quasi-governmental organizations, and special-purpose entities, such as community development banks and specialized small-business investment corporations, to provide capital and business-related services. Social service institutions have a role, but it is not this. With few exceptions nonprofit and government organizations cannot provide the quality of training, advice, and support to substantial companies that mainstream, private sector

organizations can. Compared with private sector entities such as commercial banks and venture capital companies, special-purpose institutions and non-profits are plagued by high overhead costs; they have difficulty attracting and retaining high-quality personnel, providing competitive compensation, or offering a breadth of experience in dealing with companies of scale.

[. . .]

The most important way to bring debt and equity investment to the inner city is by engaging the private sector. Resources currently going to government or quasi-public financing would be better channeled through other private financial institutions or directed at recapitalizing minority-owned banks focusing on the inner city, provided that there were matching private sector investors. Minority-owned banks that have superior knowledge of the inner city market could gain a competitive advantage by developing business-lending expertise in inner-city areas.

As in lending, the best approach to increase the supply of equity capital to the inner cities is to provide private sector incentives consistent with building economically sustainable businesses. One approach would be for both federal and state governments to eliminate the tax on capital gains and dividends from long-term equity investments in inner-city-based businesses or subsidiaries that employ a minimum percentage of inner-city residents. Such tax incentives, which are based on the premise of profit, can play a vital role in speeding up private sector investment. Private sector sources of equity will be attracted to inner-city investment only when the creation of genuinely profitable businesses is encouraged.

4. *Align incentives built into government programs with true economic performance.* Aligning incentives with business principles should be the goal of every government program. Most programs today would fail such a test. For example, preference programs in effect guarantee companies a market. Like other forms of protectionism, they dull motivation and retard cost and quality improvement. A 1988 General Accounting Office report found that within six months of graduating from the Small Business Association's purchasing preference program, 30% of the companies had gone out of business. An additional 58% of the remaining companies claimed that the withdrawal of the SBA's support had had a devastating impact on business. To align incentives with economic performance, preference programs should be rewritten to require

an increasing amount of non-set-aside business over time.

Direct subsidies to businesses do not work. Instead, government funds should be used for site assembly, extra security, environmental clean-up, and other investments designed to improve the business environment. Companies then will be left to make decisions based on true profit.

The new role of community-based organizations

Recently, there has been renewed activity among community-based organizations (CBOs) to become directly involved in business development. CBOs can, and must, play an important supporting role in the process. But choosing the proper strategy is critical, and many CBOs will have to change fundamentally the way they operate. While it is difficult to make a general set of recommendations to such a diverse group of organizations, four principles should guide community-based organizations in developing their new role.

1. *Identify and build on strengths.* Like every other player, CBOs must identify their unique competitive advantages and participate in economic development based on a realistic assessment of their capabilities, resources, and limitations. Community-based organizations have played a much-needed role in developing low-income housing, social programs, and civic infrastructure. However, while there have been a few notable successes, the vast majority of businesses owned or managed by CBOs have been failures. Most CBOs lack the skills, attitudes, and incentives to advise, lend to, or operate substantial businesses. They were able to master low-income housing development, in which there were major public subsidies and a vacuum of institutional capabilities. But, when it comes to financing and assisting for-profit business development, CBOs simply can't compete with existing private sector institutions.

Moreover, CBOs naturally tend to focus on community entrepreneurship: small retail and service businesses that are often owned by neighborhood residents. The relatively limited resources of CBOs, as well as their focus on relatively small neighborhoods, is not well-suited to developing the more substantial companies that are necessary for economic vitality.

Finally, the competitive imperatives of for-profit business activity will raise inevitable conflicts for CBOs whose mission rests with the community. Turning down local residents in favor of better qualified outside entrepreneurs, supporting necessary layoffs or the dismissal of poorly performing workers, assigning prime sites for business instead of social uses, and approving large salaries to successful entrepreneurs and managers are only a handful of the necessary choices. Given these organizations' roots in meeting the social needs of neighborhoods, it will be difficult for them to put profit ahead of their traditional mission.

2. *Work to change workforce and community attitudes.* Community-based organizations have a unique advantage in their intimate knowledge of and influence within inner-city communities, and they can use that advantage to help promote business development. CBOs can help create a hospitable environment for business by working to change community and workforce attitudes and acting as a liaison with residents to quell unfounded opposition to new businesses . . .

3. *Create work-readiness and job-referral systems.* Community-based organizations can play an active role in preparing, screening, and referring employees to local businesses. A pressing need among many inner city residents is work-readiness training, which includes communication, self-development, and workplace practices. CBOs, with their intimate knowledge of the local community, are well equipped to provide this service in close collaboration with industry . . .

CBOs can also help inner-city residents by actively developing screening and referral systems. Admittedly, some inner-city-based businesses do not hire many local residents. The reasons are varied and complex but seem to revolve around a few bad experiences that owners have had with individual employees and their work attitudes, absenteeism, false injury claims, or drug use . . .

4. *Facilitate commercial site improvement and development.* Community-based organizations (especially community development corporations) can also leverage their expertise in real estate and act as a catalyst to facilitate environmental cleanup and the development of commercial and industrial property . . .

OVERCOMING IMPEDIMENTS TO PROGRESS

This economic model provides a new and comprehensive approach to reviving our nation's distressed urban communities. However, agreeing on and implementing it will not be without its challenges. The private sector, government, inner-city residents, and the public at large all hold entrenched attitudes and prejudices about the inner city and its problems. These will be slow to change. Rethinking the inner city in economic rather than social terms will be uncomfortable for many who have devoted years to social causes and who view profit and business in general with suspicion. Activists accustomed to lobbying for more government resources will find it difficult to embrace a strategy for fostering wealth creation. Elected officials used to framing urban problems in social terms will be resistant to changing legislation, redirecting resources, and taking on recalcitrant bureaucracies. Government entities may find it hard to cede power and control accumulated through past programs. Local leaders who have built social service organizations and merchants who have run mom-and-pop stores could feel threatened by the creation of new initiatives and centers of power. Local politicians schooled in old-style community organizing and confrontational politics will have to tread unfamiliar ground in facilitating co-operation between business and residents.

These changes will be difficult ones for both individuals and institutions. Nonetheless, they must be made. The private sector, government, and community-based organizations all have vital new parts to play in revitalizing the economy of the inner city. Businesspeople, entrepreneurs, and investors must assume a lead role; and community activists, social service providers, and government bureaucrats must support them. The time has come to embrace a rational economic strategy and to stem the intolerable costs of outdated approaches.

"The Center Cannot Hold"

from Michael Sorkin and Sharon Zukin (eds), *After the World Trade Center: Rethinking New York City* (2002)

Michael Sorkin

Editors' Introduction

No recent event has so shaken the world as the terrorist attack on the World Trade Center in New York City. In addition to killing thousands of people, the attack destroyed the area of the most intense economic activity in the pre-eminent modern city. Saskia Sassen notes that in global cities like New York new centers of intense economic activity are often located just outside the historic central business district because they require new, high-tech buildings. The twin towers of the World Trade Center were just outside New York City's Wall Street Area. They were the heart of the most advanced sector of the world economy.

Understandably this tragic event has led to an emotionally charged debate among both citizens and professionals about rebuilding at the site of the trade center. This selection by Michael Sorkin is a fitting final piece in this section on urban politics, governance, and economics because issues raised by the other authors in this section and elsewhere in *The City Reader* reappear in such a dramatic form.

John Mollenkopf (p. 235) described structualist theories of urban politics that argue that city governments are at the mercy of global economic forces beyond their control. Structuralists would argue that decision-makers in New York City would have to rebuild the World Trade Center site as a high-density center of intense economic activity because the economic logic of global capitalism makes this inevitable. On the other hand, Mollenkopf describes how pluralist political theorists believe that citizens and their elected representatives have more discretion to make political choices. In the case of the World Trade Center this would mean that New Yorkers could – despite the imperatives of the world economic order – dedicate much of the very expensive land where the twin towers once stood for a memorial to the attack victims, for a public park, or other uses. Sorkin begins his essay with a reference to a notable speech by then New York mayor Rudy Giuliani who, despite his pro-business beliefs, rises above the structural forces dictating rebuilding the area as a financial center and argues for special treatment of this symbolically important area. Mollenkopf concludes that urban political power is best analyzed by a synthesis of structuralist and pluralist approaches. Theory is important for practice and how the World Trade Center site is actually rebuilt – as a financial center, as a memorial, or as a blend of the two – is a good test of the competing political theories he discusses. If he is right, the World Trade Center site should blend both high-intensity economic functions related to New York's global city status, a memorial to the attack victims, and perhaps public space that transcends economic imperatives.

David Harvey's theoretical selection that began this section (p. 227) urges urbanists to focus on social *processes* as well as the *physical forms* and think of things as products of processes. Following Harvey's theoretical orientation, the social and political processes that will unfold as New York City recovers from the attack on the World Trade Center will be critical. The *physical forms* which will be built at the World Trade Center site and elsewhere in New York as a result of the attack will be the result of the processes Harvey urges us to consider. Sherry Arnstein's "ladder of citizen participation" (p. 244) provides insight on the processes that will unfold as the debate about rebuilding after the World Trade Center attack unfolds. The processes to involve New Yorkers in the decisions may be the largest and most complex citizen participation process of all time. The business-oriented

Lower Manhattan Development Corporation and the New York Port Authority have invited the local community and survivors' groups to participate in planning the rebuilding of the site. Given the emotional nature of the task the groups empowered to decide the future of the site cannot just ask the local community to rubber stamp their already-formulated plans for the area (manipulation) or talk with groups who lost family members to make them feel good (therapy). Participation at these lowest levels of Arnstein's ladder is unacceptable. But given the importance of this site to the city (and the world), decision-makers are not about to give citizens control of the ultimate decisions or to delegate power to them (the top rungs of Arnstein's ladder). The ultimate process for deciding the future of the World Trade Center site will probably fall somewhere in the middle of the ladder – information will certainly be shared and many groups consulted. Perhaps a true partnership between the multiple decision-makers will evolve. At this writing – fall of 2002 – the jury is still out.

Michael Sorkin raises fundamental questions about social equity and alternative visions for New York. Terrorism, he believes has opened new space for thinking about radical alternatives to urban planning questions. Sorkin's argument is related to Saskia Sassen's finding that globalization and information technology are creating both concentration and dispersal in global cities (p. 212). Sorkin notes that 15 million square feet of office space were lost in the World Trade Center attack and that New York City is projected to need four times that much new office space by 2020. Where should the investment to develop New York go? Rather than the all-eggs-in-one-basket approach of directing massive new investment to the World Trade Center site itself (*concentration*), Sorkin argues that the investment should be *dispersed* to multiple sites in the city. He would like to see the investment used to create a "village structure" with a "network of autonomous, comprehensible places" that make the city as a whole more like what people think of as New York rather than hyperconcentration at the World Trade Center site. Ethnic New York neighborhoods are in need of jobs, construction, and in some cases greater concentration. Sorkin sees the tragedy as a unique opportunity to join physical planning and investment to create a better city. Unlike Anthony Downs (p. 256), Sorkin's vision is based almost entirely within New York City. Downs is more of a regionalist and would like to see government investment break down the distinctions between poor ghetto areas of New York City and affluent suburbs.

There will be massive investment in the rebuilding of New York City by both the public and private sector. How might this funding benefit low-income and minority groups? How might community development corporations, nonprofits, and minority-owned private businesses relate to the rebuilding. Michael Porter (p. 277) would take a dim view of this reinvestment pursuing fragmented and inefficient government programs to create jobs in inner-city neighborhoods and train local residents. He would oppose subsidies, preference programs, and expensive efforts to stimulate small marginal inner-city businesses, which could not turn a profit without government help. He would see opportunities in the rebuilding process for New York City to pursue a new *economic* model of inner-city revitalization of private, for-profit initiatives based on economic self-interest, rather than artificial inducements, charity, and government mandates.

Michael Sorkin is an architect and principal of the Michael Sorkin Studio in New York City and a professor of urbanism and director of the Institute of Urbanism at the Academy of Fine Arts in Vienna. He has taught architecture at Harvard, Yale, Cornell, Columbia, University of North Carolina, Southern California Institute of Architecture, and other universities. He is a contributing editor at *Architectural Record*, *Metropolis*, and *I.D.* His recent architectural projects have included public housing in Vienna, a sustainable urban design scheme for San Francisco, and Shrooms, an experimental loft housing typology. For ten years he was the architectural critic for the *Village Voice*.

This selection is from Michael Sorkin and Sharon Zukin (eds), *After the World Trade Center: Rethinking New York City* (London and New York: Routledge, 2002). Other books by Michael Sorkin include *Wiggle* (New York: Monacelli Press, 1998), *Giving Ground: The Politics of Propinquity*, co-edited with Joan Copjec (London: Verso, 1998), *Local Code: The Constitution of a City at 42°N Latitude* (Princeton, NJ: Princeton Architectural Press, 1993), *Variations on a Theme Park* (New York: Hill and Wang, 1991), and *Exquisite Corpse: Writings on Buildings* (London: Verso, 1991).

For more about the attack on the World Trade Center see Bill Harris and Peter G. Goldmark, Jr., *The World Trade Center: A Tribute* (New York: Running Press, 2001).

In his farewell to office, [New York Mayor] Rudy Giuliani, standing in St Paul's Chapel, adjacent to the World Trade Center site, declared, "I really believe we shouldn't think about this site out there, right behind us, right here, as a site for economic development. We should think about a soaring, monumental, beautiful memorial that just draws millions of people here who just want to see it. We have to be able to create something here that enshrines this forever and that allows people to build on it and grow from it. And it's not going to happen if we just think about it in a very narrow way."

Giuliani's speech reminded me of Eisenhower's leave-taking from the presidency, in which he warned the nation against the growing antidemocratic power of the "military–industrial complex." In both cases, the cautionary appeals resonated because of their sources, one from a military man and architect of the Cold War, the other from a mayor whose leadership favored planning by the "market."

Giuliani's heartfelt call for restraint ran counter to the business-as-usual approach that has dominated official thinking since the tragedy. This has included obscene ambulance chasing on the part of the architectural community and robust talk about responding to the terror by rapid rebuilding, bigger than ever. The Lower Manhattan Development Corporation, empowered to decide the future of the site, is headed up by a patriarchal ex-director of Goldman Sachs whose credibility seems untainted by the spectacle of his own firm migrating over the river to Jersey. With the exception of a single community representative the board is comprised of the usual business crowd. Their initial consensus seems to favor the construction of a very large amount of office space on the cleared site of the fallen towers, with the memorial simply a modest component. And rumors grind on about the working drawings already in CAD at SOM.

Fortunately, the competition for authority over the site is both structural and complex. The Port Authority, Larry Silverstein (the ninety-nine-year leaseholder), the Development Corporation, the federal, state, and city governments, survivor groups, the local community, the business improvement district, the Battery Park City Authority, the Metropolitan Transit Authority, and myriad other civic and private interests are jostling to be heard and influential. If nothing else, this fog buys time for contention and for the serious consideration of alternatives.

What is clear is that despite the currently soft market, some of the 15 million square feet of lost space needs to be replaced sooner rather than later and downtown's dysfunctions repaired to allow the city's economy to reestablish jobs and networks lost in the attack. And the eventual need is to reestablish jobs and networks not simply for replacement space: the "Group of 35" – a business-heavy organization chaired by Charles Schumer and Robert Rubin – has predicted (in a report about the future of the city's commercial space) that an additional 60 million square feet of office space will be required by 2020.

The question is where to put it. Some will clearly go to Lower Manhattan. The relationships of propinquity that form the social and spatial substrate of commerce and culture downtown must be quickly reestablished by the restoration of infrastructure and the addition of new space. Railroading the restoration of the status quo by looking at the site as no more than its footprint, however, guarantees that we learn nothing from the tragedy and let the opportunity for better thinking slip away.

My studio is downtown, not far from the Western Union building at 60 Hudson Street, known to architects as the home of the New York City Building Department. Since September 11, this building has been the subject of unusual security, surrounded by concrete barriers and half a dozen police cars. It appears to be the only site outside the confines of Ground Zero to enjoy this level of fresh protection, and the reason seems to be the building's long-standing role as the nexus both for telecommunications cables coming into New York City, and for trunk lines to the nation and the world, a logical next target for terror, according to some scenarios.

Ironically, a system at the core of urban disaggregation depends on the joining of huge dispersed networks on a single site. The currently dominant pattern of our urbanism – enabled by the kinds of instantaneous artificial proximity that the electronic network of phones, faxes, e-mails, and other global systems allow – is the rapid growth of the "edge city," a sprawling realm that has become the antithesis of a traditional sense of place, the location to which security-conscious firms are now increasingly impelled to retreat.

Of course, this "nonplace" urban realm is the result of more than new communications technology. The suburbs were fertilized by massive government intervention in highway construction, by radical tax policy, by changes in the national culture of desire, by racism, by cheap, unencumbered land, and by an earlier fear of terror. The prospect of nuclear annihilation that made urban concentrations particularly vulnerable was on the minds of many planners during the Cold War, both in the United States and abroad. The massive deurbanization in Maoist China, for example, was the direct result of nuclear anxiety. The dispersal facilitated by the interstates – our erstwhile "National Defense Highway" – was likewise more than simply good for General Motors: both the company and the United States were playing at the same stratego-urban games.

Whatever the causes, however, the effects of this pattern of urbanization were in many ways antithetical to the presumptions behind the space of Lower Manhattan and its kith. Here, concentration has long been considered crucially advantageous. The possibility of conducting economic affairs face-to-face, the collective housing of related bureaucracies and businesses (the famous Finance Insurance Real Estate [FIRE] sectors that make up the majority of business downtown), the dense life of the streets, the convenience of having everything at hand, are the foundation for the viability of the main financial district for the planet. Its characteristic form – the superimposition of skyscrapers on the medieval street pattern left by the Dutch – has given Downtown its indelible shape.

Any changes reconstruction brings must deepen this formal singularity, expand the possibilities of exchange, and broaden the mix of uses supported. While it is now a bromide to apologize before suggesting that the tragedy can be turned to advantage, to have positive consequences, we must find a way to make things better. Life downtown is going to be refounded and the imperative is to do it well. Perhaps we can reply to the terror by showing that evil acts can have unintended positive consequences.

In fact, the enormous disruption in the life of the city has already had a number of constructive effects. Here in Tribeca, not far from Ground Zero, traffic is dramatically reduced on local streets, the polluted sewer of Canal Street is suddenly tractable, and deep civility abides many months later. The emergency carpooling and limited access instituted as the result of the disaster are equally positive and important contributions to a sustainable urban ecology.

The radical act of the terrorists opens a space for us to think radically as well, to examine alternatives for the future of all of New York City. It is no coincidence

that we have constructed a skyline in the image of a bar graph. This is not simply an abstraction but a multiplication, an utterly simple means of multiplying wealth: where land is scarce, make more. Lots more. There is a fantasy of Manhattan as driven simply by a pure and perpetual increase in density. But while our dynamism is surely a product of critical mass, not all arguments for concentration are the same. Viewed from the perspective of the city as a whole, the hyperconcentration of the World Trade Center was not necessarily optimal by any standard other than profit, and even that proved elusive.

Density has a downside in overcrowding and strained services, but this is not necessarily the result of the hyperscale of any particular building. More critical than specific effects on the ground are the consequences for densities elsewhere. While the anxiety over corporate and population flight to the suburbs comes from a general fear of both economic and social losses, the all-eggs-in-one-basket approach slights other areas of the city, themselves in need of jobs, construction, and greater concentration. Manhattan's gain has been the other boroughs' loss: the rise of the island's office towers historically makes the decline of industrial employment throughout the city and has obliged the respiratory pattern of one-directional commuting that marks its life. A new form of producing wealth with new spatial requirements has over the century completely supplanted its predecessor.

With thousands of jobs already relocated out of the city, a solution to the "practical" problems of reconstruction can and must engage possibilities well beyond the confines of the downtown site. While the billions that will be available for new building – from insurance, from federal aid, from city coffers, from developers – are certainly needed to restore health to the enterprises formerly in or servicing the World Trade Center, it seems reasonable to question – given the probable level of this investment – whether such massive expenditure should be focused exclusively here rather than throughout the city at additional sites of need and opportunity, places development could transform.

The majority of New York City's population and geography does not live in Manhattan: the island comprises only 8 percent of the city's land area and 19 percent of its inhabitants. Moreover, according to the 2000 census, the residential growth of the island since 1990 – slightly over 3 percent – lags far behind the explosive enlargement of Staten Island (17 percent) and Queens (nearly 15 percent), and the dramatic increases in the Bronx (10.7 percent) and Brooklyn (7.2 percent). Manhattan, however, remains the city's economic engine, producing up to 67 percent of its jobs and 46 percent of its retail sales.

The effect of these imbalances on the ground has fundamentally reshaped the city. The great infusions of capital and the artificial fortunes of the last decade have propelled the price of real estate in much of Manhattan to the stratosphere, accelerating the flight of the middle class and the poor and making Manhattan increasingly monochrome. We continue to revere our island as a place of thick, urbane interaction and cling to the fantasy of the great mixing engine of difference, of a place with many quarters housing many kinds of people. Increasingly, however, the differences in Manhattan's neighborhoods are merely physical. This uneven development and accelerated metamorphosis have had dramatic effects, distorting the character of our urbanity decisively.

Here in Tribeca, we are at the end of a familiar cycle in which a neighborhood moves from a mix of warehouses, manufacturing, offices, and housing, to an "artistic" neighborhood, and now to the climax form of gentrification, an extreme-high-end residential quarter. The corollary is that the jobs and people formerly employed here have either been eliminated or moved elsewhere: to the Hunts Point Market in the Bronx, to low-wage environments offshore, to the suburbs, or to the new bohemias of Williamsburg or Long Island City. We have scrupulously preserved the architectural character of Tribeca but at the expense of its human character.

With the exception of Chinatown, Manhattan south of 110th Street has become a faded mosaic of former ethnic enclaves and cultural variety. Increasingly, the city's ethnic and cultural quarters are being solidified outside the borough, in Flushing, Greenpoint, Dumbo (Down Underneath Manhattan Bridge), or Atlantic Avenue. Although the city remains a beacon for immigrants both from abroad and at home, the sites of intake and expression are not what they were, preserved to death. Manhattan is ceasing to be a place to get a start and becoming inhospitable to striving, less and less like New York.

But big changes can also suggest big opportunities for burgeoning neighborhoods struggling to find form or merely to keep up. Not all disaggregation leads to sprawl. Better, perhaps, to call it reaggregation, but it is also a notion that can be useful in cultivating

character and encouraging development within more traditional, compact cities like New York, itself the central place for an enormous region. The point is not to make New York more like Phoenix or Los Angeles, but to make the city as a whole more like New York.

Because of its dynamic population and superb movement infrastructure, New York City can become a model of a new kind of polycentric metropolis, with Manhattan remaining its centro di tutti centri, its concentrated vitality unsapped. In fact, Manhattan is itself already polycentric: the disaggregation represented by, for example, the easy movement of financial and legal services firms from Downtown to Midtown in recent years suggests that there is a certain fluidity to the idea of proximity within the city, that convenient movement and strong local character can substitute for immediate adjacency within an overall context of density.

Reinforcing New York's special polycentricity would return the city to something of its pre-twentieth-century character by restoring a network of autonomous, comprehensible places. Such a "village" structure – the origin of the great city of variegated neighborhoods – is again made possible by the technology behind the ephemeral and flexible nets and flows of the twenty-first century. Because it is aspatial, this malleability need not simply lead to generic sprawl but can fit within – and reinforce – preexisting infrastructures of neighborhood difference.

Cultivating this "natural" polycentricity would multiply opportunities for more self-sufficient neighborhoods where people walk to work, to school, to recreation, and to culture. Such places would also satisfy many of the needs that impel people to seek the densities and economies of the suburbs and edge cities. By regenerating local character, the energy of intracity reaggregation could reinforce the expressive singularity of each of these neighborhoods to which its energies were applied – the Asian flavor of Flushing, the Latin-American atmosphere of the Bronx Hub, the African cultures of Harlem.

This would be an advance on the wing-and-a-prayer style of current planning, in which good intentions are simultaneously frustrated by imprecise plans and the absence of economic drivers to set them in motion before changing times render them irrelevant. By joining physical planning to direct investment and to zoning and economic incentives, we can redistribute uses to a set of centers outside Manhattan where land and transit connections are available and economical, places like Flushing, Jamaica, Queens Plaza, Sunnyside, the Bronx Hub, St. George, and downtown Brooklyn, among others. These sites – also identified in the report of the Group of 35 – are not mysterious either in their needs or their suitability.

Planning comprehensively could help assure the mixed-use character of these places by including residential construction matched to the numbers of new workplaces, a pattern that has already begun downtown, where substantial office space has actually been eliminated by conversion to residential use. Indeed, in the last ten years, forty office buildings have been converted to residential use downtown, part of an 18 percent population growth in the area below Canal Street. The sense of locality that grows from a well-finessed mix would be further reinforced by the decentralization of cultural growth (the City Opera, the Guggenheim, the Whitney, the Met, and the Jets are all seeking space) and by encouraging the development of new cultural, health care, educational, and commercial institutions to enhance the variety and life of these neighborhood centers.

It is critical, however, that these centers be envisioned and planned as semiautonomous and not simply as ancillary. Downtown Brooklyn is already one of the largest "central places" in America but continues to be thought of as a back office for Manhattan. The key is zoning for sustainability and difference, not simply for a series of mini-Manhattans. Although the skyscraper is a preeminent symbol of twentieth-century technology, and of the culture of the corporation, other paradigms must now emerge as values change. The economic driver that has impelled these heights will be usefully moderated in smaller centers that foreground strong environmental values and in which land prices are restrainedly moderate.

Lower Manhattan is the commercial district with the highest public transportation usage in the country; 80 percent of those who come to work there – 350,000 people a day – arrive on mass transit. A comprehensive reexamination and reinforcement of this pattern are crucial to sustaining the city but must be approached noncentrifugally to facilitate movement, not simply in and out of Manhattan, but between the developing centers of lived life, reinforcing the repatterning. Our waterways, in particular, offer a tremendous opportunity for creating such links with great economy. In addition, the city's large areas of public greenspace and municipally owned property can be used to begin to create a third transport net – for pedestrians, bikers, and nonaggressive zero-emissions vehicles – to supplement the street grid and the subway.

Business as usual in New York City is more than the compulsion to repeat patterns of the past: our talent is creating the new. In the case of downtown Manhattan, however, it is also important to recognize that this is an area of the city that is near completion: its project of build-out and of formal invention is almost done. The construction of the World Trade Center, the isolation of Battery Park City by an overwide highway, the nasty scale of many newer high-rises, the abandonment of the piers, the elimination of manufacturing and small-scale commercial activity, and the elevation of the West Side Highway are all assaults on a satisfying paradigm of great scale contrasts, rich architectural textures, and pedestrian primacy that lies at the core of what's best about Downtown. Restoring this is the task at hand, and it cannot be accomplished in Lower Manhattan alone.

Plate 23 New Urbanist neo-traditionalism. In Seaside, Florida, designed by the firm of Duany–Plater-Zyberk, a postmodernist streetscape mimics the sense and scale of the traditional American small town.

Plate 24 The mall has it all. The suburban shopping center, sometimes called the "new downtown," is both an extravaganza of commercialism and a new kind of social and entertainment center. The Mall of America in Minneapolis, Minnesota, has shops, restaurants, and even a roller coaster!

Plate 25 **Architecture as symbolic power.** The Petronas Twin Towers in Kuala Lumpur, designed by Cesar Pelli and Associates, are the tallest buildings in the world and proclaim Malaysia to be one of the new power centers of the global economy. (Photograph by Roger Mellor.)

Plate 26 Luxury housing. This upscale housing development is for the managers of the new global enterprises that are transforming urban economies and social structures worldwide. The city is Shanghai, People's Republic of China. Note the private interior gardens. (Photograph by Lian Bloch.)

Plate 27 The persistence of poverty. While postmodernism and the global economy advance, streets in working-class neighborhoods continue to serve as market-places, playgrounds, and political venues. Here, a street in central Rome exhibits a lively blending of urban uses and purposes. Note the Communist Party wall poster. (Photograph by Patti Walters.)

Plate 28 The persistence of decay. Amid the splendor of new global power centers and the comfort of New Urbanist neighborhoods, inner-city decay persists, even dominates. Desperate graffiti, barred windows, and litter of all sorts mark this sidewalk in San Francisco, California. (Photograph by Derek Metzger.)

Plate 29 **The persistence of tradition.** Today, many Islamic cities are gleamingly modern, but others retain a traditional, almost medieval flavor. In Fez, Morocco, a crowd fills a walled plaza to watch a street entertainer. (Photograph by Paul V. Turner.)

Plate 30 The streets belong to the kids. Skateboarders may seem like a relatively recent phenomenon, but youth has always found a way to expropriate (adult) public space. (Photograph by Frederic Stout.)

Plate 31 **The streets belong to the people!** Street protests, like this demonstration in Seattle, Washington, have greeted the emergence of the new global economy. The political cause is new, but the use of public space for popular and insurgent politics is an ancient and honorable tradition. (Photograph by Jason Morrison.)

Plate 32 **Urban terror.** The World Trade Center, New York, September 11, 2001. A new kind of global politics renders the city suddenly vulnerable. (Photograph by Wanda McCormick, Readio.com.)

PART FIVE

Urban planning history and visions

THE CRISIS,

OR THE CHANGE FROM ERROR AND MISERY, TO TRUTH AND HAPPINESS

1832.

IF WE CANNOT YET   RECONCILE ALL OPINIONS,

LET US ENDEAVOUR   TO UNITE ALL HEARTS.

IT IS OF ALL TRUTHS THE MOST IMPORTANT, THAT THE CHARACTER OF MAN IS FORMED FOR — NOT BY HIMSELF.

Design of a Community of 2,000 Persons, founded upon a principle, recommended by Plato, Lord Bacon, Sir T. More, & R. Owen.

EDITED BY
ROBERT OWEN AND ROBERT DALE OWEN.

London:

PRINTED AND PUBLISHED BY J. EAMONSON, 15, CHICHESTER PLACE
GRAY'S INN ROAD.

STRANGE, PATERNOSTER ROW. PURKISS, OLD COMPTON STREET,
AND MAY BE HAD OF ALL BOOKSELLERS.

INTRODUCTION TO PART FIVE

The effects of urban planning are perhaps the greatest – and, at the same time, the most invisible – influences on human life and culture. In the words of Paul and Percival Goodman, the co-authors of *Communitas: Means of Livelihood and Ways of Life* (1947), we hardly realize as we go about the daily round of our lives "that somebody once drew some lines on a piece of paper who might have drawn otherwise" and that "now, as engineer and architect once drew, people have to walk and live."

When the Sumerian kings built the walls of Eridu and Uruk, they engaged in acts of urban planning and thus determined how their people would "walk and live." The walls provided safety and protection for the people of the city and also defined the new political unity of the city-state. The associated roads, bridges, irrigation systems, and centers for market and ceremonial functions – all served a dual function in that they met the practical social needs of the urban population in general and fulfilled the power aspirations of the urban elites in particular. The ancient citadels were centers of religious meaning, as well as economic and political power, and thus a third component of urban planning – an idealized, often spiritual vision of what constitutes the best possible state of human existence – was present at the very beginning of city building.

The origins of modern urban planning are complex. On one level, modern planning is a direct extension of the ancient and pre-modern models: imposing order on nature for the health, safety, and amenity of the urban masses, for the political benefit of the urban elites, and for a way of expressing each culture's highest spiritual ideals. On another level, however, modern planning is far more complex than anything that had ever gone on before. Modern planning operates, by and large, in a politically and economically pluralistic environment, making every alteration of the physical arrangements of the city a complex negotiation between competing interests. And the practice of modern urban planning also takes place at a stage of human development when the planner's defining goal is no longer merely to impose human order on nature, but to continuously impose order on the city itself.

All the goals and functions of planning – both the ancient hold-overs and the modern elaborations – are present in the first planning responses to the urban conditions associated with the Industrial Revolution. As Friedrich Engels (p. 58) and other contemporary observers described, the cities of the new industrialism were characterized by horrendous overcrowding, ubiquitous misery, and despair. There were daily threats to the public health and safety, not just for the impoverished working class but for the capitalist middle class as well. These conditions gave rise to movements for housing reform, to great advances in the technologies of water supply and sewage disposal, and to the emergence of middle-class suburbs. They also led to the construction of model "company towns" by various industrial firms in both Europe and America, and to the development of a modern urban planning profession.

Reviewing the history of urban planning in the nineteenth century, Richard LeGates and Frederic Stout (the co-editors of this volume) have written that "the classic texts of early urban planning history often seem surprisingly modern." An example of the surprising modernity of early urban planning is the nineteenth-century parks movement, especially the work of Frederick Law Olmsted, which gave rise to something very like comprehensive urban planning practice. Projects like Central Park in New York (Plate 33) represented a transplantation and democratization of European landscape gardening traditions, to be sure, but Olmsted's goal was not merely to bring nature into the city. Rather, Olmsted repeatedly appealed to the political and economic leadership of American cities to create parks that would achieve a whole range of public benefits: they would contribute to the public health by serving as the "lungs" of the city; they would be practical and necessary additions to the physical infrastructure of the metropolis, providing a general recreation ground; their ponds and reservoirs would serve as adjuncts to municipal water-supply systems; and they would soften and tame human nature, by providing wholesome alternatives to the vulgar street amusements that daily tempted poor and working-class youth.

Olmsted was a visionary and a reformer, but he was also a successful businessman and a canny political operative capable of offering his clients useful strategic advice on how to fund and build constituencies in favor of large municipal projects. Somewhat less practical, but even more visionary, were a group of architects, planners, and activists who may be termed, collectively, the utopian modernists. Three of these – Ebenezer Howard, Le Corbusier, and Frank Lloyd Wright – define the mainstream of that utopian tradition. Not one of them had his utopian vision realized in its entirety, but each had an enormous influence on the way contemporary cities, and city life, developed in the twentieth century. A fourth, the Spanish engineer/planner Arturo Soria y Mata, is influential for his vision of the relationship between transportation systems and land use. Plate 34 illustrates Soria y Mata's vision of a "linear city" developed along a central spine containing an electric streetcar line and other utilities.

Ebenezer Howard prided himself on being "the inventor of the Garden City idea," and his tireless devotion to the project of decongesting the modern metropolis by building small, self-contained, green-belted cities in the rural countryside is one of the marvels of modern urban planning history. Plate 35 from *Garden Cities of To-morrow* illustrates Howard's vision of "a group of slumless, smokeless cities." Howard originally wanted his Garden Cities to be cooperatively owned. He wanted the surrounding greenbelt to be much larger than the built-up part of the city itself. And he wanted his cities to be economically independent, not commuter suburbs. In the process of actually building Letchworth and Welwyn – the two Garden Cities constructed before his death in 1924 – Howard had to compromise many of his original goals. Building lots and businesses were privately owned; the greenbelt became more of a park than an extensive rural buffer zone; and neither of the original Garden Cities ever became a fully independent economic entity. Nonetheless as Plate 36, the plan for Welwyn illustrates, these were fully planned communities which embodied many of Howard's ideals. Still, the Garden City experiment gave rise to a larger movement of town planning, and disciples of Howard spread his ideas and his example worldwide.

Charles-Edouard Jeanneret, better known as Le Corbusier, was another utopian visionary who never saw his ideal plans fully developed but who was enormously influential nonetheless. Le Corbusier wanted his "Contemporary City of Three Million," illustrated in Plate 37, to be a series of exquisite towers, geometrically arranged in a surrounding park, and he spent years looking for governmental and industrial sponsors for his plan. Many "Corbusian" high-rise urban developments have been built throughout the world, of course. Indeed, the "International Style" of modern architecture and the principles of the International Congress of Modern Architecture (CIAM), which he pioneered, have become global standards of urban development. But in almost every case, the surrounding park has been compromised away in the process of realization. In case after case, the tower in the park has become the tower without the park or, even worse, the tower in the parking lot!

While Le Corbusier was issuing his manifestos and shocking the architectural and planning establishment with his modernist plans, American planners Clarence Stein and Henry Wright were also wrestling with the problem of how to adapt urban form to the automobile. In their influential plan for Radburn, New Jersey, illustrated in Plate 38, Stein and Wright invented and implemented a series of planning concepts including superblocks and the separation of pedestrian and vehicular traffic.

Frank Lloyd Wright, the originator of the visionary "Broadacre City" plan, was also responding to the automobile. Wright called for a city composed of family homesteads – one full acre per person – and the withering away of the dense and crowded traditional cities. Wright's 1935 plan for Broadacre City is illustrated in Plate 39. The private automobile, Wright thought, would virtually abolish distance and allow for a new kind of community based on individualism and self-reliance. What actually became of Wright's Broadacre was sprawl suburbia. One acre per person became one-eighth acre per family or less; the core cities refused to wither away; the transportation monoculture of the automobile became a new form of dependency, rather than a technology of liberation; and the family-oriented suburban community became problematic at best, an object of ridicule at worst.

The utopian visionaries were more that just planners, if they can be said to be planners at all. Even Ebenezer Howard, the most moderate of the group, was a dreamer. Together, the utopian modernists concerned themselves with great philosophical issues such as the connection between Mankind and Nature, the relationship of city plan to moral reform, and the role of urban design and technologies to the evolutionary transformation of society. It would be left to more practical men and women – the actual members of the urban planning profession as it developed in the twentieth century – to address the real-world problems of ever-changing cities and metropolitan regions. If the utopian modernists established the lofty goals, the professional planners attended to the details.

Still, the role of visionary projections of better lives through better urban planning persists as an important motivating force in contemporary urban planning. Establishing a good planning vision and sticking to it can have profound positive impacts. Plate 40 illustrates the famous "Paseo del Rio" of San Antonio, Texas. Like hundreds of other cities San Antonio had a blighted river-turned-drainage-ditch disfiguring the downtown. But unlike other cities San Antonio developed a vision of turning the problem area into a magnificent area of riverfront amenities and recreational activities. Today it provides a pleasant place to sit, stroll, paddle, and shop. Boston, Massachusetts worked the same kind of urban planning magic by collaborating with developer James Rouse to turn a seedy and obsolete market place around Quincy Market into a magnificent center for strolling, shopping, dining, and cultural events. Quincy Market today is illustrated in Plate 41.

The tradition of visionary urban planning is alive and well today. In addition to visions of sustainable urban development discussed throughout this anthology (p. 339 and p. 486), the contemporary planning movement called the "New Urbanism" is both a profit-making process of real-estate development that emphasizes small-town scale as the basis of new communities from Florida to California and a practical application of visionary ideas. Architect/planner Peter Calthorpe advocates a vision of a new "Regional City" in which land use and transportation systems are designed together in harmony with the natural environment to eliminate the blight of suburban sprawl and produce livable communities. One of Calthorpe's earliest visions was for "Pedestrian Pockets" such as those illustrated in Plate 42. Over time, these evolved into the kind of transit-oriented developments (TODs) that are currently being built not as a further extension of traditional suburbia but in and for what he has variously called "The Next American Metropolis" and "The Post-Suburban Metropolis." In one sense, Calthorpe's Pedestrian Pockets, transit-oriented developments, and Regional City conceptions look backward to the Garden Cities of Ebenezer Howard – especially in their use of greenbelting and light-rail mass-transit options – but, in another sense, they look forward to an entirely new relationship between city and region, between individual and community. In short, the New Urbanist vision is a response to a future already in the process of becoming, characterized by "a dramatic shift in the nature and location of our workplace and a fundamental change in the character of our increasingly diverse households."

F
I
V
E

"Public Parks and the Enlargement of Towns"

American Social Science Association (1870)

Frederick Law Olmsted

Editors' Introduction

Frederick Law Olmsted (1822–1903) has been called "America's great pioneer landscape architect," and, during his lifetime, he was widely recognized as one of the most influential public figures in the nation. Along with his business partner, the English-born architect Calvert Vaux, he originated and dominated the urban parks movement, pioneered the development of planned suburbs, and laid out scores of public and private institutions. If Central Park in New York, illustrated as it looked in 1863 in Plate 33, remains his best known masterpiece, the designs for Riverside, Illinois (outside Chicago), the Boston park system, the Capitol grounds in Washington, DC, the 1893 World's Fair, and the campus of Stanford University in California are equally impressive contributions to the built environment.

Olmsted began his career practicing and writing about farming, then turned his talents to journalism and, in the 1850s, published a series of books describing the society and economy of the slave states of the American South (collected into one volume as *The Cotton Kingdom* in 1861). With this background, it is hardly surprising that Olmsted thoroughly imbued his art of landscape architecture with a wide variety of social and political, as well as cultural, concerns.

"Public Parks and the Enlargement of Towns" was originally presented as an address to the American Social Science Association, meeting at the Lowell Institute, Boston, in 1870. In it, Olmsted provides a number of specific guidelines for parks and parkways and suggests ways to overcome political resistance to public funding for parks and planned urban growth. Most importantly, however, he lays out the political and philosophical case for public parks in terms of three great moral imperatives: first, the need to improve public health by sanitation measures and the use of trees to combat air and water pollution; second, the need to combat urban vice and social degeneration, particularly among the children of the urban poor; and third, the need to advance the cause of civilization by the provision of urban amenities that would be democratically available to all.

Both as a practitioner and as a theorist, Olmsted anticipated many of the principal concerns of urban planning, both infrastructural and social, down to the present day. Indeed, behind the somewhat convoluted Victorianisms of his prose lies a strikingly modern mind. In the design of the Garden City Ebenezer Howard (p. 309) borrowed directly from Olmsted, and even plans so fundamentally different as those of Frank Lloyd Wright (p. 325) and Le Corbusier (p. 317) owe a debt to Olmsted insofar as they recognize and address the central problem of the relationship between nature and the built urban environment. As the father of modern landscape architecture, Olmsted's work and thought invite comparison with all those, including J.B. Jackson (p. 164), who came after him in the profession, either as practitioners or critics.

A selection of Olmsted's most important writings may be found in S.B. Sutton (ed.), *Civilizing American Cities: Writings on City Landscapes by Frederick Law Olmsted* (Cambridge, MA: MIT Press, 1971). Johns Hopkins University has published most of Olmsted's work in the multi-volume *Collected Papers of Frederick Law Olmsted* (Baltimore, MD: Johns Hopkins University Press, 1977–1992).

Biographies of Olmsted and commentary on his work include Laura Wood Roper's *FLO: A Biography of Frederick Law Olmsted* (Baltimore, MD: Johns Hopkins University Press, 1973), Elizabeth Stevenson's *Park Maker: A Life of Frederick Law Olmsted* (New York: Macmillan, 1977), Charles E. Beveridge, Paul Rocheleau, and David Larkin's *Frederick Law Olmsted: Designing the American Landscape* (New York: Rizzoli, 1995), and Witold Rybczynski, *A Clearing in the Distance: Frederick Law Olmsted and America in the Nineteenth Century* (New York: Scribner, 1999).

Galen Cranz's *The Politics of Park Design: A History of Urban Parks in America* (Cambridge, MA: MIT Press, 1982) is a superb overview that places Olmsted's planning and landscape design achievements in the context of a larger movement for urban social reform. See also Cynthia Zaitzevsky, *Frederick Law Olmsted and the Boston Park System* (Cambridge: Belknap Press, 1992), Charles E. Beveridge, Paul Rocheleau, and David Larkin, *Frederick Law Olmsted: Designing the American Landscape* (New York: Rizzoli, 1995), and Susan L. Klaus, *Modern Arcadia: Frederick Law Olmsted, Jr. and the Plan for Forest Hill Gardens* (Cambridge, MA: MIT Press, 2002).

▓ ▓ ▓ ▓ ▓ ▓

We have reason to believe, then, that towns which of late have been increasing rapidly on account of their commercial advantages, are likely to be still more attractive to population in the future; that there will in consequence soon be larger towns than any the world has yet known, and that the further progress of civilization is to depend mainly upon the influences by which men's minds and characters will be affected while living in large towns.

Now, knowing that the average length of the life of mankind in towns has been much less than in the country, and that the average amount of disease and misery and of vice and crime has been much greater in towns, this would be a very dark prospect for civilization, if it were not that modern Science has beyond all question determined many of the causes of the special evils by which men are afflicted in towns, and placed means in our hands for guarding against them. It has shown, for example, that under ordinary circumstances, in the interior parts of large and closely built towns, a given quantity of air contains considerably less of the elements which we require to receive through the lungs than the air of the country or even of the outer and more open parts of a town, and that instead of them it carries into the lungs highly corrupt and irritating matters, the action of which tends strongly to vitiate all our sources of vigor – how strongly may perhaps be indicated in the shortest way by the statement that even metallic plates and statues corrode and wear away under the atmosphere influences which prevail in the midst of large towns, more rapidly than in the country.

The irritation and waste of the physical powers which result from the same cause, doubtless indirectly affect and very seriously affect the mind and the moral strength; but there is a general impression that a class of men are bred in towns whose peculiarities are not perhaps adequately accounted for in this way. We may understand these better if we consider that whenever we walk through the denser part of a town, to merely avoid collision with those we meet and pass upon the sidewalks, we have constantly to watch, to foresee, and to guard against their movements. This involves a consideration of their intentions, a calculation of their strength and weakness, which is not so much for their benefit as our own. Our minds are thus brought into close dealings with other minds without any friendly flowing toward them, but rather a drawing from them. Much of the intercourse between men when engaged in the pursuits of commerce has the same tendency – a tendency to regard others in a hard if not always hardening way. Each detail of observation and of the process of thought required in this kind of intercourse or contact of minds is so slight and so common in the experience of towns-people that they are seldom conscious of it. It certainly involves some expenditure nevertheless. People from the country are even conscious of the effect on their nerves and minds of the street contact – often complaining that they feel confused by it; and if we had no relief from it at all during our waking hours, we should all be conscious of suffering from it. It is upon our opportunities of relief from it, therefore, that not only our comfort in town life, but our ability to maintain a temperate, good-natured, and healthy state of mind, depends. This is one of many ways in which it happens that men who have been brought up, as the saying is, in the streets, who have been most directly and completely affected by town influences, so generally show, along with a remarkable quickness of apprehension, a peculiarly hard sort of selfishness. Every day of their lives they have seen thousands of their fellow-men, have met them face to face, have brushed against them, and yet have had no experience of anything in common with them.

[. . .]

It is practically certain that the Boston of today is the mere nucleus of the Boston that is to be. It is practically certain that it is to extend over many miles of country now thoroughly rural in character, in parts of which farmers are now laying out roads with a view to shortening the teaming distance between their wood-lots and a railway station, being governed in their courses by old property lines, which were first run simply with reference to the equitable division of heritages, and in other parts of which, perhaps, some wild speculators are having streets staked off from plans which they have formed with a rule and pencil in a broker's office, with a view, chiefly, to the impressions they would make when seen by other speculators on a lithographed map. And by this manner of planning, unless views of duty or of interest prevail that are not yet common, if Boston continues to grow at its present rate even for but a few generations longer, and then simply holds its own until it shall be as old as the Boston in Lincolnshire now is, more men, women, and children are to be seriously affected in health and morals than are now living on this Continent.

Is this a small matter – a mere matter of taste; a sentimental speculation?

It must be within the observation of most of us that where, in the city, wheel-ways originally twenty-feet wide were with great difficulty and cost enlarged to thirty, the present width is already less nearly adequate to the present business than the former was to the former business; obstructions are more frequent, movements are slower and oftener arrested, and the liability to collision is greater. The same is true of sidewalks. Trees thus have been cut down, porches, bow-windows, and other encroachments removed, but every year the walk is less sufficient for the comfortable passing of those who wish to use it.

It is certain that as the distance from the interior to the circumference of towns shall increase with the enlargement of their population, the less sufficient relatively to the service to be performed will be any given space between buildings.

In like manner every evil to which men are specially liable when living in towns, is likely to be aggravated in the future, unless means are devised and adapted in advance to prevent it.

Let us proceed, then, to the question of means, and with a seriousness in some degree befitting a question, upon our dealing with which we know the misery or happiness of many millions of our fellow-beings will depend.

We will for the present set before our minds the two sources of wear and corruption which we have seen to be remediable and therefore preventible. We may admit that commerce requires that in some parts of a town there shall be an arrangement of buildings, and a character of streets and of traffic in them which will establish conditions of corruption and of irritation, physical and mental. But commerce does not require the same conditions to be maintained in all parts of a town.

Air is disinfected by sunlight and foliage. Foliage also acts mechanically to purify the air by screening it. Opportunity and inducement to escape at frequent intervals from the confined and vitiated air of the commercial quarter, and to supply the lungs with air screened and purified by trees, and recently acted upon by sunlight, together with opportunity and inducement to escape from conditions requiring vigilance, wariness, and activity toward other men, – if these could be supplied economically, our problem would be solved.

In the old days of walled towns all tradesmen lived under the roof of their shops, and their children and apprentices and servants sat together with them in the evening about the kitchen fire. But now that the dwelling is built by itself and there is greater room, the inmates have a parlor to spend their evening in; they spread carpets on the floor to gain in quiet, and hang drapery in their windows and papers on their walls to gain in seclusion and beauty. Now that our towns are built without walls, and we can have all the room that we like, is there any good reason why we should not make some similar difference between parts which are likely to be dwelt in, and those which will be required exclusively for commerce?

Would trees, for seclusion and shade and beauty, be out of place, for instance, by the side of certain of our streets? It will, perhaps, appear to you that it is hardly necessary to ask such a question, as throughout the United States trees are commonly planted at the sides of streets. Unfortunately they are seldom so planted as to have fairly settled the question of the desirableness of systematically maintaining trees under these circumstances. In the first place, the streets are planned, wherever they are, essentially alike. Trees are planted in the space assigned for sidewalks, where at first, while they are saplings and the vicinity is rural or suburban, they are not much in the way, but where, as they grow larger, and the vicinity becomes urban, they take up more and more space, while space is more and more required for passage. That is not all. Thousands and tens of thousands are planted every year in a manner and under conditions as nearly certain as possible either to kill them outright, or to so lessen their vitality as to prevent their natural and beautiful development, and to cause premature decrepitude. Often, too, as their lower limbs are found inconvenient, no space having been provided for trees in laying out the street, they are deformed by butcherly amputations. If by rare good fortune they are suffered to become beautiful, they still stand subject to be condemned to death at any time, as obstructions in the highway.

What I would ask is, whether we might not with economy make special provision in some of our streets – in a twentieth or a fiftieth part, if you please, of all – for trees to remain as a permanent furniture of the city? I mean, to make a place for them in which they would have room to grow naturally and gracefully. Even if the distance between the houses should have to be made half as much again as it is required to be in our commercial streets, could not the space be afforded? Out of town space is not costly when measures to secure it are taken early. The assessments for benefit where such streets were provided for, would, in nearly all cases, defray the cost of the land required. The strips of ground required for the trees, six, twelve, twenty feet wide, would cost nothing for paving or flagging.

The change both of scene and of air which would be obtained by people engaged for the most part in the necessarily confined interior commercial parts of the town, on passing into a street of this character after the trees have become stately and graceful, would be worth a good deal. If such streets were made still broader in some parts, with spacious malls, the advantage would be increased. If each of them were given the proper capacity, and laid out with laterals and connections in suitable directions to serve as a convenient trunk line of communication between two large districts of the town or the business centre and the suburbs, a very great number of people might thus be placed every day under influences counteracting those with which we desire to contend.

These, however, would be merely very simple improvements upon arrangements which are in common use in every considerable town. Their advantages would be incidental to the general uses of streets as they are. But people are willing very often to seek recreations as well as receive it by the way. Provisions may indeed be made expressly for public recreations, with certainty that if convenient they will be resorted to.

We come then to the question: what accommodations for recreation can we provide which shall be so agreeable and so accessible as to be efficiently attractive to the great body of citizens, and which, while giving decided gratification, shall also cause those who resort to them for pleasure to subject themselves, for the time being, to conditions strongly counteractive to the special, enervating conditions of the town?

In the study of this question all forms of recreation may, in the first place, be conveniently arranged under two general heads. One will include all of which the predominating influence is to stimulate exertion of any part or parts needing it; the other, all which cause us to receive pleasure without conscious exertion. Games chiefly of mental skill, as chess, or athletic sports, as baseball, are examples of means of recreation of the first class, which may be termed that of *exertive* recreation; music and the fine arts generally of the second or *receptive* division.

Considering the first by itself, much consideration will be needed in determining what classes of exercises may be advantageously provided for. In the Bois de Boulogne there is a race-course; in the Bois de Vincennes a ground for artillery target-practice. Military parades are held in Hyde Park. A few cricket clubs are accommodated in most of the London parks, and swimming is permitted in the lakes at certain hours. In the New York Park, on the other hand, none of these exercises are provided for or permitted, except that the boys of the public schools are given the use on holidays of certain large spaces for ball playing. It is considered that the advantage to individuals which would be gained in providing for them would not compensate for the general inconvenience and expense they would cause.

I do not propose to discuss this part of the subject at present, as it is only necessary to my immediate purpose to point out that if recreations requiring large spaces to be given up to the use of a comparatively small number, are not considered essential, numerous small grounds so distributed through a large town that some one of them could be easily reached by a short walk from every house, would be more desirable than a single area of great extent, however rich in landscape attractions it might be. Especially would this be the case if the numerous local grounds were connected and supplemented by a series of trunk-roads or boulevards such as has already been suggested.

Proceeding to the consideration of receptive recreations, it is necessary to ask you to adopt and bear in mind a further subdivision, under two heads, according to the degree in which the average enjoyment is greater when a large congregation assembles for a purpose of receptive recreation, or when the number coming together is small and the circumstances are favorable to the exercise of personal friendliness.

The first I shall term *gregarious*; the second, *neighborly*. Remembering that the immediate matter in hand is a study of fitting accommodations, you will, I trust, see the practical necessity of this classification.

Purely gregarious recreation seems to be generally looked upon in New England society as childish and savage, because, I suppose, there is so little of what we call intellectual gratification in it. We are inclined to engage in it indirectly, furtively, and with complication. Yet there are certain forms of recreation, a large share of the attraction of which must, I think, lie in the gratification of the gregarious inclination, and which, with those who can afford to indulge in them, are so popular as to establish the importance of the requirement.

If I ask myself where I have experienced the most complete gratification of this instinct in public and out of doors, among trees, I find that it has been in the promenade of the Champs-Élysées. As closely following it I should name other promenades of Europe, and our own upon the New York parks. I have studiously watched the latter for several years. I have several times seen fifty thousand people participating in them; and the more I have seen of them, the more highly have I been led to estimate their value as means of counteracting the evils of town life.

Consider that the New York Park and the Brooklyn Park are the only places in those associated cities where, in this eighteen hundred and seventieth year after Christ, you will find a body of Christians coming together, and with an evident glee in the prospect of coming together, all classes largely represented, with a common purpose, not at all intellectual, competitive with none, disposing to jealousy and spiritual or intellectual pride toward none, each individual adding by his mere presence to the pleasure of all others, all

helping to the greater happiness of each. You may thus often see vast numbers of persons brought closely together, poor and rich, young and old, Jew and Gentile. I have seen a hundred thousand thus congregated, and I assure you that though there have been not a few that seemed a little dazed, as if they did not quite understand it, and were, perhaps, a little ashamed of it, I have looked studiously but vainly among them for a single face completely unsympathetic with the prevailing expression of good nature and light-heartedness.

Is it doubtful that it does men good to come together in this way in pure air and under the light of heaven, or that it must have an influence directly counteractive to that of the ordinary hard, hustling working hours of town life?

You will agree with me, I am sure, that it is not, and that opportunity, convenient, attractive opportunity, for such congregation, is a very good thing to provide for, in planning the extension of a town.

[. . .]

Think that the ordinary state of things to many is at this beginning of the town. The public is reading just now a little book in which some of your streets of which you are not proud are described. Go into one of those red cross streets any fine evening next summer, and ask how it is with their residents. Oftentimes you will see half a dozen sitting together on the door-steps or, all in a row, on the curb-stones, with their feet in the gutter; driven out of doors by the closeness within; mothers among them anxiously regarding their children who are dodging about at their play, among the noisy wheels on the pavement.

Again, consider how often you see young men in knots of perhaps half a dozen in lounging attitudes rudely obstructing the sidewalks, chiefly led in their little conversation by the suggestions given to their minds by what or whom they may see passing in the street, men, women, or children, whom they do not know and for whom they have no respect or sympathy. There is nothing among them or about them which is adapted to bring into play a spark of admiration, of delicacy, manliness, or tenderness. You see them presently descend in search of physical comfort to a brilliantly lighted basement, where they find others of their sort, see, hear, smell, drink, and eat all manner of vile things.

Whether on the curb-stones or in the dram-shops, these young men are all under the influence of the same impulse which some satisfy about the tea-table with neighbors and wives and mothers and children, and all things clean and wholesome, softening, and refining.

If the great city to arise here is to be laid out little by little, and chiefly to suit the views of land-owners, acting only individually, and thinking only of how what they do is to affect the value in the next week or the next year of the few lots that each may hold at the time, the opportunities of so obeying this inclination as at the same time to give the lungs a bath of pure sunny air, to give the mind a suggestion of rest from the devouring eagerness and intellectual strife of town life, will always be few to any, to many will amount to nothing.

But is it possible to make public provision for recreation of this class, essentially domestic and secluded as it is?

It is a question which can, of course, be conclusively answered only from experience. And from experience in some slight degree I shall answer it. There is one large American town, in which it may happen that a man of any class shall say to his wife, when he is going out in the morning: "My dear, when the children come home from school, put some bread and butter and salad in a basket, and go to the spring under the chestnut-tree where we found the Johnsons last week. I will join you there as soon as I can get away from the office. We will walk to the dairy-man's cottage and get some tea, and some fresh milk for the children, and take our supper by the brook-side"; and this shall be no joke, but the most refreshing earnest.

There will be room enough in the Brooklyn Park, when it is finished, for several thousand little family and neighborly parties to bivouac at frequent intervals through the summer, without discommoding one another, or interfering with any other purpose, to say nothing of those who can be drawn out to make a day of it, as many thousand were last year. And although the arrangements for the purpose were yet very incomplete, and but little ground was at all prepared for such use, besides these small parties, consisting of one or two families, there came also, in companies of from thirty to a hundred and fifty, somewhere near twenty thousand children with their parents, Sunday-school teachers, or other guides and friends, who spent the best part of a day under the trees and on the turf, in recreations of which the predominating element was of this neighborly receptive class. Often they would bring a fiddle, flute, and harp, or other music. Tables, seats, shade, turf, swings, cool spring-water, and a pleasing rural prospect, stretching off half a mile or more each way, unbroken by a carriage road or the slightest

evidence of the vicinity of the town, were supplied them without charge and bread and milk and ice-cream at moderate fixed charges. In all my life I have never seen such joyous collections of people. I have, in fact, more than once observed tears of gratitude in the eyes of poor women, as they watched their children thus enjoying themselves.

The whole cost of such neighborly festivals, even when they include excursions by rail from the distant parts of the town, does not exceed for each person, on an average, a quarter of a dollar; and when the arrangements are complete, I see no reason why thousands should not come every day where hundreds come now to use them; and if so, who can measure the value, generation after generation, of such provisions for recreation to the over-wrought, much-confined people of the great town that is to be?

For this purpose neither of the forms of ground we have heretofore considered are at all suitable. We want a ground to which people may easily go after their day's work is done, and where they may stroll for an hour, seeing, hearing, and feeling nothing of the bustle and jar of the streets, where they shall, in effect, find the city put far away from them. We want the greatest possible contrast with the streets and the shops and the rooms of the town which will be consistent with convenience and the preservation of good order and neatness. We want, especially, the greatest possible contrast with the restraining and confining conditions of the town, those conditions which compel us to walk circumspectly, watchfully, jealously, which compel us to look closely upon others without sympathy. Practically, what we most want is a simple, broad, open space of clean greensward, with sufficient play of surface and a sufficient number of trees about it to supply a variety of light and shade. This we want as a central feature. We want depth of wood enough about it not only for comfort in hot weather, but to completely shut out the city from our landscapes.

The word *park*, in town nomenclature, should, I think, be reserved for grounds of the character and purpose thus described.

[...]

A park fairly well managed near a large town, will surely become a new center of that town. With the determination of location, size, and boundaries should therefore be associated the duty of arranging new trunk routes of communication between it and the distant parts of the town existing and forecasted.

These may be either narrow informal elongations of

the park, varying say from two to five hundred feet in width, and radiating irregularly from it, or if, unfortunately, the town is already laid out in the unhappy way that New York and Brooklyn, San Francisco and Chicago, are, and, I am glad to say, Boston is not, on a plan made long years ago by a man who never saw a spring-carriage, and who had a conscientious dread of the Graces, then we must probably adopt formal Park-ways. They should be so planned and constructed as never to be noisy and seldom crowded, and so also that the straightforward movement of pleasure-car carriages need never be obstructed, unless at absolutely necessary crossings, by slow-going heavy vehicles used for commercial purposes. If possible, also, they should be branched or reticulated with other ways of a similar class, so that no part of the town should finally be many minutes' walk from some one of them; and they should be made interesting by a process of planting and decoration, so that in necessarily passing through them, whether in going to or from the park, or to and from business, some substantial recreative advantage may be incidentally gained. It is a common error to regard a park as something to be produced complete in itself, as a picture to be painted on canvas. It should rather be planned as one to be done in fresco, with constant consideration of exterior objects, some of them quite at a distance and even existing as yet only in the imagination of the painter.

I have thus barely indicated a few of the points from which we may perceive our duty to apply the means in our hands to ends far distant, with reference to this problem of public recreations. Large operations of construction may not soon be desirable, but I hope you will agree with me that there is little room for question, that reserves of ground for the purposes I have referred to should be fixed upon as soon as possible, before the difficulty of arranging them, which arises from private building, shall be greatly more formidable than now.

"Author's Introduction" and "The Town–Country Magnet"

from *Garden Cities of To-morrow* (1898)

Ebenezer Howard

Editors' Introduction

A stenographer by trade, Ebenezer Howard (1850–1928) was a quiet, modest, self-effacing man – "a man without credentials or connections," as one biographer put it – who nevertheless managed to change the world.

Born in London, Howard early experienced the pollution, congestion, and social dislocations of the modern industrial metropolis. After a year in America (as a homesteader in Nebraska!), he returned to England in 1876 and became involved in political movements and discussion groups addressing what was then termed "the Social Question." Howard was influenced by a number of radical theorists and visionaries including the social reformer Robert Owen, the utopian novelist Edward Bellamy, and the "single tax" advocate Henry George. He published *To-morrow: a Peaceful Path to Real Reform* in 1898 (now better known under its 1902 title, *Garden Cities of To-morrow*) and methodically set about convincing people of the beauty and utility of "the Garden City idea."

Although Howard's plan may seem quaintly Victorian to the modern reader, the ideas he put forth were nothing short of revolutionary at the time. Indeed, Howard's ideas of urban decentralization, zoning for different uses, the integration of nature into cities, greenbelting, and the development of self-contained "New Town" communities outside crowded central cities illustrated in Plate 35 laid the groundwork for the entire tradition of modern city planning. Unlike many other utopian dreamers, Howard lived to see his plans (if in a somewhat compromised form) actually put into action. In his own lifetime, the Garden Cities of Letchworth and Welwyn were built in England. Later, the Garden City idea spread to continental Europe, to America by way of the New Deal, and to much of the rest of the world.

Howard's argument begins with a protest against urban over-crowding; the one issue upon which, he writes, "men of all parties" are "well-nigh universally agreed." He then explains why "the people continue to stream into the already over-crowded cities" by reference to "the town magnet," that combination of jobs and amenities that characterizes the modern metropolis. Arrayed against this urban magnetic force is "the country magnet," the appealing features of the more natural, but increasingly desolate, rural districts. Finally, Howard describes his own plan, a new kind of human community based on "the town–country magnet," which is the best of both worlds. As detailed in his famous concentric-ring diagram (which, he is careful to warn, is "a diagram only," not an actual site plan), the center of Garden City is to be a central park containing important public buildings and surrounded by a "Crystal Palace" ring of retail stores. The entire city of approximately 1,000 acres is to be encircled by a permanent agricultural greenbelt of some 5,000 acres, and the new cities are to be connected with central "Social Cities" and each other by a system of railroad lines.

Howard's ideas about the evils of over-crowding are similar to those of Friedrich Engels (p. 58), and his solution to the problem invites comparison with the very different solutions proposed by Le Corbusier (p. 317) and Frank Lloyd Wright (p. 325). Direct followers of Howard include Patrick Geddes and Lewis Mumford (p. 92), who helped to spread the Garden City idea throughout Europe and America. More recently, Peter Calthorpe

(p. 331) has effectively re-invented the Garden City idea in California as the Regional City in the form of greenbelted, suburban "Pedestrian Pockets" linked to central cities (and each other) by a network of light-rail transportation systems.

Garden Cities of To-morrow remains a readable and relevant book. It is available as the second volume of Richard T. LeGates and Frederic Stout (eds), *Early Urban Planning* (nine volumes, London: Routledge/Thoemmes, 1998) and earlier editions by Attic Books (1985), Eastbourne (1985), MIT Press (1965), Faber and Faber (1960, 1951, and 1946), Sonnenschein (1902), and the original edition which appeared under the title *To-morrow: A Peaceful Path to Real Reform* (Sonnenschein, 1898).

Biographies of Ebenezer Howard include Robert Beevers, *The Garden City Utopia: A Critical Biography of Ebenezer Howard* (New York: St. Martin's, 1988), and Dugald Macfadyen, *Sir Ebenezer Howard and the Town Planning Movement* (Manchester: Manchester University Press, 1933; reprinted Cambridge, MA: MIT Press, 1970).

Excellent accounts of Howard and the Garden City movement may be found in Robert Fishman's *Urban Utopias in the Twentieth Century* (New York: Basic Books, 1977) and Peter Hall's *Cities of Tomorrow* (Oxford: Basil Blackwell, 1988).

Additional books about Ebenezer Howard and the Garden City movement include Standish Meacham, *Regaining Paradise: Englishness and the Early Garden City Movement* (New Haven, CT: Yale, 1999), Peter Geoffrey Hall and Colin Ward, *Sociable Cities: The Legacy of Ebenezer Howard* (New York: John Wiley and Sons, 1998), Stephen V. Ward (ed.), *The Garden City: Past, Present and Future* (London and New York: E & FN Spon, 1992), Stanley Buder, *Visionaries and Planners: The Garden City Movement and the Modern Community* (Oxford: Oxford University Press, 1990), and Kermit Parsons and David Schuyler (eds), *From the Garden City to Green Cities: The Legacy of Ebenezer Howard* (Baltimore, MD: Johns Hopkins University Press, 2002).

■ ■ ■ ■ ■ ■

AUTHOR'S INTRODUCTION

In these days of strong party feeling and of keenly contested social and religious issues, it might perhaps be thought difficult to find a single question having a vital bearing upon national life and well-being on which all persons, no matter of what political party, or of what shade of sociological opinion, would be found to be fully and entirely agreed . . .

[. . .]

There is, however, a question in regard to which one can scarcely find any difference of opinion . . . It is wellnigh universally agreed by men of all parties, not only in England, but all over Europe and America and our colonies, that it is deeply to be deplored that the people should continue to stream into the already over-crowded cities, and should thus further deplete the country districts.

All . . . are agreed on the pressing nature of this problem, all are bent on its solution, and though it would doubtless be quite Utopian to expect a similar agreement as to the value of any remedy that may be proposed, it is at least of immense importance that, on a subject thus universally regarded as of supreme importance, we have such a consensus of opinion at the outset. This will be the more remarkable and the more hopeful sign when it is shown, as I believe will be conclusively shown in this work, that the answer to this, one of the most pressing questions of the day, makes of comparatively easy solution many other problems which have hitherto taxed the ingenuity of the greatest thinkers and reformers of our time. Yes, the key to the problem how to restore the people to the land – that beautiful land of ours, with its canopy of sky, the air that blows upon it, the sun that warms it, the rain and dew that moisten it – the very embodiment of Divine love for man – is indeed a *Master Key*, for it is the key to a portal through which, even when scarce ajar, will be seen to pour a flood of light on the problems of intemperance, of excessive toil, of restless anxiety, of grinding poverty – the true limits of Governmental interference, ay, and even the relations of man to the Supreme Power.

It may perhaps be thought that the first step to be taken towards the solution of this question – how to restore the people to the land – would involve a careful consideration of the very numerous causes which have hitherto led to their aggregation in large cities. Were this the case, a very prolonged enquiry would be necessary at the outset. Fortunately, alike for writer and for reader, such an analysis is not, however, here requisite, and for a very simple reason, which may be stated thus: Whatever may have been the causes which have operated in the past, and are operating now, to draw the people into the cities, those causes may all be summed up as "attractions"; and it is obvious, therefore, that no remedy can possibly be effective which will not present to the people, or at least to considerable portions of them, greater "attractions" than our cities now possess, so that the force of the old "attractions" shall be overcome by the force of new "attractions" which are to be created. Each city may be regarded as a magnet, each person as a needle; and, so viewed, it is at once seen that nothing short of the discovery of a method for constructing magnets of yet greater power than our cities possess can be effective for redistributing the population in a spontaneous and healthy manner.

So presented, the problem may appear at first sight to be difficult, if not impossible, of solution. "What", some may be disposed to ask, "can possibly be done to make the country more attractive to a workaday people than the town – to make wages, or at least the standard of physical comfort, higher in the country than in the town; to secure in the country equal possibilities of social intercourse, and to make the prospects of advancement for the average man or woman equal, not to say superior, to those enjoyed in our large cities?" The issue one constantly finds presented in a form very similar to that. The subject is treated continually in the public press, and in all forms of discussion, as though men, or at least working men, had not now, and never could have, any choice or alternative, but either, on the one hand, to stifle their love for human society – at least in wider relations than can be found in a straggling village – or, on the other hand, to forgo almost entirely all the keen and pure delights of the country. The question is universally considered as though it were now, and for ever must remain, quite impossible for working people to live in the country and yet be engaged in pursuits other than agricultural; as though crowded, unhealthy cities were the last word of economic science; and as if our present form of industry, in which sharp lines divide agricultural from industrial pursuits, were necessarily an enduring one. This fallacy is the very common one of ignoring altogether the possibility of alternatives other than those presented to the mind. There are in reality not only, as is so constantly assumed, two alternatives – town life and country life – but a third alternative, in which all the advantages of the most energetic and active town life, with all the beauty and delight of the country, may be secured in perfect combination; and the certainty of being able to live this life will be the magnet which will produce the effect for which we are all striving – the spontaneous movement of the people from our crowded cities to the bosom of our kindly mother earth, at once the source of life, of happiness, of wealth, and of power. The town and the country may, therefore, be regarded as two magnets, each striving to draw the people to itself – a rivalry which a new form of life, partaking of the nature of both, comes to take part in. This may be illustrated by a diagram (Figure 1) of "The Three Magnets", in which the chief advantages of the Town and of the Country are set forth with their corresponding drawbacks, while the advantages of the Town–Country are seen to be free from the disadvantages of either.

The Town magnet, it will be seen, offers, as compared with the Country magnet, the advantages of high wages, opportunities for employment, tempting prospects of advancement, but these are largely counterbalanced by high rents and prices. Its social opportunities and its places of amusement are very alluring, but excessive hours of toil, distance from work, and the "isolation of crowds" tend greatly to reduce the value of these good things. The well lit streets are a great attraction, especially in winter, but the sunlight is being more and more shut out, while the air is so vitiated that the fine public buildings, like the sparrows, rapidly become covered with soot, and the very statues are in despair. Palatial edifices and fearful slums are the strange, complementary features of modern cities.

The Country magnet declares herself to be the source of all beauty and wealth; but the Town magnet mockingly reminds her that she is very dull for lack of society, and very sparing of her gifts for lack of capital. There are in the country beautiful vistas, lordly parks, violet-scented woods, fresh air, sounds of rippling water; but too often one sees those threatening words, "Trespassers will be prosecuted". Rents, if estimated by the acre, are certainly low, but such low rents are the natural fruit of low wages rather than a cause of substantial comfort; while long hours and lack of amusements forbid the bright sunshine and the pure air to gladden the hearts of the people. The one industry, agriculture, suffers frequently from excessive rainfalls; but this wondrous harvest of the clouds is seldom properly in-gathered, so that, in times of drought,

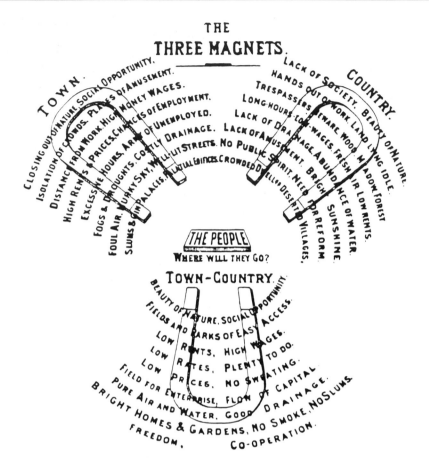

Figure 1

there is frequently, even for drinking purposes, a most insufficient supply. Even the natural healthfulness of the country is largely lost for lack of proper drainage and other sanitary conditions, while, in parts almost deserted by the people, the few who remain are yet frequently huddled together as if in rivalry with the slums of our cities.

But neither the Town magnet nor the Country magnet represents the full plan and purpose of nature. Human society and the beauty of nature are meant to be enjoyed together. The two magnets must be made one. As man and woman by their varied gifts and faculties supplement each other, so should town and country. The town is the symbol of society – of mutual help and friendly co-operation, of fatherhood, motherhood, brotherhood, sisterhood, of wide relations between man and man – of broad, expanding sympathies – of science, art, culture, religion. And the country! The country is the symbol of God's love and care for man. All that we are and all that we have comes from

it. Our bodies are formed of it; to it they return. We are fed by it, clothed by it, and by it are we warmed and sheltered. On its bosom we rest. Its beauty is the inspiration of art, of music, of poetry. Its forces propel all the wheels of industry. It is the source of all health, all wealth, all knowledge. But its fullness of joy and wisdom has not revealed itself to man. Nor can it ever, so long as this unholy, unnatural separation of society and nature endures. Town and country *must be married*, and out of this joyous union will spring a new hope, a new life, a new civilization. It is the purpose of this work to show how a first step can be taken in this direction by the construction of a Town–Country magnet; and I hope to convince the reader that this is practicable, here and now, and that on principles which are the very soundest, whether viewed from the ethical or the economic standpoint.

I will undertake, then, to show how in "Town–Country" equal, nay better, opportunities of social intercourse may be enjoyed than are enjoyed in any crowded city, while yet the beauties of nature may encompass and enfold each dweller therein; how higher wages are compatible with reduced rents and rates; how abundant opportunities for employment and bright prospects of advancement may be secured for all; how capital may be attracted and wealth created; how the most admirable sanitary conditions may be ensured; how beautiful homes and gardens may be seen on every hand; how the bounds of freedom may be widened, and yet all the best results of concert and co-operation gathered in by a happy people.

The construction of such a magnet, could it be effected, followed, as it would be, by the construction of many more, would certainly afford a solution of the burning question set before us by Sir John Gorst, "how to back the tide of migration of the people into the towns, and to get them back upon the land".

[. . .]

THE TOWN–COUNTRY MAGNET

The reader is asked to imagine an estate embracing an area of 6,000 acres, which is at present purely agricultural, and has been obtained by purchase in the open market at a cost of £40 an acre, or £240,000. The purchase money is supposed to have been raised on mortgage debentures, bearing interest at an average rate not exceeding 4 per cent. The estate is legally vested in the names of four gentlemen of responsible position and of undoubted probity and honour, who hold it in trust, first, as a security for the debenture-holders, and, secondly, in trust for the people of Garden City, the Town–Country magnet, which it is intended to build thereon. One essential feature of the plan is that all ground rents, which are to be based upon the annual value of the land, shall be paid to the trustees, who, after providing for interest and sinking fund, will hand the balance to the Central Council of the new municipality, to be employed by such Council in the creation and maintenance of all necessary public works – roads, schools, parks, etc. The objects of this land purchase may be stated in various ways, but it is sufficient here to say that some of the chief objects are these: To find for our industrial population work at wages of *higher purchasing power*, and to secure healthier surroundings and more regular employment. To enterprising manufacturers, co-operative societies, architects, engineers, builders, and mechanicians

of all kinds, as well as to many engaged in various professions, it is intended to offer a means of securing new and better employment for their capital and talents, while to the agriculturists at present on the estate as well as to those who may migrate thither, it is designed to open a new market for their produce close to their doors. Its object is, in short, to raise the standard of health and comfort of all true workers of whatever grade – the means by which these objects are to be achieved being a healthy, natural, and economic combination of town and country life, and this on land owned by the municipality.

Garden City, which is to be built near the centre of the 6,000 acres, covers an area of 1,000 acres, or a sixth part of the 6,000 acres, and might be of circular form, 1,240 yards (or nearly three-quarters of a mile) from centre to circumference. (Figure 2 is a ground plan of the whole municipal area, showing the town in the centre; and Figure 3, which represents one section or ward of the town, will be useful in following the description of the town itself – *a description which is, however, merely suggestive, and will probably be much departed from . . .*)

Six magnificent boulevards – each 120 feet wide – traverse the city from centre to circumference, dividing it into six equal parts or wards. In the centre is a circular space containing about five and a half acres, laid out as a beautiful and well-watered garden; and, surrounding this garden, each standing in its own ample grounds, are the larger public buildings – town hall, principal concert and lecture hall, theatre, library, museum, picture-gallery, and hospital.

The rest of the large space encircled by the "Crystal Palace" is a public park, containing 145 acres, which includes ample recreation grounds within very easy access of all the people.

Running all round the Central Park (except where it is intersected by the boulevards) is a wide glass arcade called the "Crystal Palace", opening on to the park. This building is in wet weather one of the favourite resorts of the people, whilst the knowledge that its bright shelter is ever close at hand tempts people into Central Park, even in the most doubtful of weathers. Here manufactured goods are exposed for sale, and here most of that class of shopping which requires the joy of deliberation and selection is done. The space enclosed by the Crystal Palace is, however, a good deal larger than is required for these purposes, and a considerable part of it is used as a Winter Garden – the whole forming a permanent exhibition of a most

Figure 2

attractive character, whilst its circular form brings it near to every dweller in the town – the furthest removed inhabitant being within 600 yards.

Passing out of the Crystal Palace on our way to the outer ring of the town, we cross Fifth Avenue – lined, as are all the roads of the town, with trees – fronting which, and looking on to the Crystal Palace, we find a ring of very excellently built houses, each standing in its own ample grounds; and, as we continue our walk, we observe that the houses are for the most part built either in concentric rings, facing the various avenues (as the circular roads are termed), or fronting the boulevards and roads which all converge to the centre of the town. Asking the friend who accompanies us on our journey what the population of this little city may be, we are told about 30,000 in the city itself, and about 2,000 in the agricultural estate, and that there are in the town 5,500 building lots of an *average* size of 20 feet × 130 feet – the minimum space allotted for the purpose being 20 × 100. Noticing the very varied architecture and design which the houses and groups of houses display – some having common gardens

and co-operative kitchens – we learn that general observance of street line or harmonious departure from it are the chief points as to house building, over which the municipal authorities exercise control, for, though proper sanitary arrangements are strictly enforced, the fullest measure of individual taste and preference is encouraged.

Walking still toward the outskirts of the town, we come upon "Grand Avenue". This avenue is fully entitled to the name it bears, for it is 420 feet wide, and, forming a belt of green upwards of three miles long, divides that part of the town which lies outside Central Park into two belts. It really constitutes an additional park of 115 acres – a park which is within 240 yards of the furthest removed inhabitant. In this splendid avenue six sites, each of four acres, are occupied by public schools and their surrounding playgrounds and gardens, while other sites are reserved for churches, of such denominations as the religious beliefs of the people may determine, to be erected and maintained out of the funds of the worshippers and their friends. We observe that the houses fronting on Grand

Figure 3

Avenue have departed (at least in one of the wards – that of which Figure 3 is a representation) – from the general plan of concentric rings, and, in order to ensure a longer line of frontage on Grand Avenue, are arranged in crescents – thus also to the eye yet further enlarging the already splendid width of Grand Avenue.

On the outer ring of the town are factories, warehouses, dairies, markets, coal yards, timber yards, etc., all fronting on the circle railway, which encompasses the whole town, and which has sidings connecting it with a main line of railway which passes through the estate. This arrangement enables goods to be loaded direct into trucks from the warehouses and workshops, and so sent by railway to distant markets, or to be taken direct from the trucks into the warehouses or factories; thus not only effecting a very great saving in regard to packing and cartage, and reducing to a minimum loss from breakage, but also, by reducing the traffic on the roads of the town, lessening to a very marked extent the cost of their maintenance. The smoke fiend is kept well within bounds in Garden City; for all machinery is driven by electric energy, with the result that the cost of electricity for lighting and other purposes is greatly reduced.

The refuse of the town is utilized on the agricultural portions of the estate, which are held by various individuals in large farms, small holdings, allotments, cow pastures, etc.; the natural competition of these various methods of agriculture, tested by the willingness of occupiers to offer the highest rent to the municipality, tending to bring about the best system of husbandry, or, what is more probable, the best systems adapted for various purposes. Thus it is easily conceivable that it may prove advantageous to grow wheat in very large fields, involving united action under a capitalist farmer, or by a body of co-operators; while the cultivation of vegetables, fruits, and flowers, which requires closer and more personal care, and more of the artistic and inventive faculty, may possibly be best dealt with by individuals, or by small groups of individuals having a common belief in the efficacy and value of certain dressings, methods of culture, or artificial and natural surroundings.

This plan, or, if the reader be pleased to so term it, this absence of plan, avoids the dangers of stagnation or dead level, and, though encouraging individual initiative, permits of the fullest co-operation, while the increased rents which follow from this form of competition are common or municipal property, and by far the larger part of them are expended in permanent improvements.

While the town proper, with its population engaged in various trades, callings, and professions, and with a store or depot in each ward, offers the most natural market to the people engaged on the agricultural estate, inasmuch as to the extent to which the towns-people demand their produce they escape altogether any railway rates and charges; yet the farmers and others are not by any means limited to the town as their only market, but have the fullest right to dispose of their produce to whomsoever they please. Here, as in every feature of the experiment, it will be seen that it is not the area of rights which is contracted, but the area of choice which is enlarged.

This principle of freedom holds good with regard to manufacturers and others who have established themselves in the town. These manage their affairs in their own way, subject, of course, to the general law of the land, and subject to the provision of sufficient space for workmen and reasonable sanitary conditions. Even in regard to such matters as water, lighting, and telephonic communication – which a municipality, if efficient and honest, is certainly the best and most natural body to supply – no rigid or absolute monopoly is sought; and if any private corporation or any body of individuals proved itself capable of supplying on more advantageous terms, either the whole town or a section of it, with these or any commodities the supply of which was taken up by the corporation, this would be allowed. No really sound system of *action* is in more need of artificial support than is any sound system of *thought*. The area of municipal and corporate action is probably destined to become greatly enlarged; but, if it is to be so, it will be because the people possess faith in such action, and that faith can be best shown by a wide extension of the area of freedom.

Dotted about the estate are seen various charitable and philanthropic institutions. These are not under the control of the municipality, but are supported and managed by various public-spirited people who have been invited by the municipality to establish these institutions in an open healthy district, and on land let to them at a pepper-corn rent, it occurring to the authorities that they can the better afford to be thus generous, as the spending power of these institutions greatly benefits the whole community. Besides, as those persons who migrate to the town are among its most energetic and resourceful members, it is but just and right that their more helpless brethren should be able to enjoy the benefits of an experiment which is designed for humanity at large.

"A Contemporary City"

from *The City of Tomorrow and its Planning* (1929)

Le Corbusier

Editors' Introduction

Le Corbusier (1887–1965) was one of the founding fathers of the Modernist movement and of what has come to be known as the International Style in architecture. Painter, architect, city planner, philosopher, author of revolutionary cultural manifestos – Le Corbusier exemplified the energy and efficiency of the Machine Age. His was the bold, nearly mystical rationality of a generation that was eager to accept the scientific spirit of the twentieth century on its own terms and to throw off all pre-existing ties – political, cultural, conceptual – with what it considered an exhausted, out-moded past.

Born Charles-Edouard Jeanneret, Le Corbusier grew up in the Swiss town of La Chaux-de-Fonds, noted for its watch-making industry. He took his famous pseudonym after he moved to Paris to pursue a career in art and architecture. From the first, his designs for modern houses – he called them "machines for living" – were strikingly original, and many people were shocked by the spare cubist minimalism of his designs. The real shock, however, came in 1922 when Le Corbusier presented the public with his plan for "A Contemporary City of Three Million People." Laid out in a rigidly symmetrical grid pattern, the city consisted of neatly spaced rows of identical, strictly geometrical skyscrapers as illustrated in Plate 37. This was not the city of the future, Le Corbusier insisted, but the city of today. It was to be built on the Right Bank, after demolishing several hundred acres of the existing urban fabric of Paris!

The "Contemporary City" proposal certainly caught the attention of the public, but it did not win Le Corbusier many actual urban planning commissions. Throughout the 1920s, 1930s, and 1940s, he sought out potential patrons wherever he could find them – the industrial capitalists of the Voisin automobile company, the communist rulers of the Soviet Union, and the fascist Vichy government of occupied France – mostly without success. Le Corbusier's real impact came not from cities he designed and built himself but from cities that were built by others incorporating the planning principles that he pioneered. Most notable among these was the notion of "the skyscraper in the park," an idea that is today ubiquitous. Whether in relatively complete examples like Brasilia and Chandigar, India (where new cities were built from scratch), or in partial examples such as the skyscraper parks and the high-rise housing blocks that have been built in cities worldwide, the Le Corbusier vision has truly transformed the global urban environment.

Le Corbusier's "Contemporary City" plan has often been contrasted to Frank Lloyd Wright's "Broadacre" (p. 325), and the comparison of a thoroughly centralized versus a thoroughly decentralized plan is indeed striking. Le Corbusier's boldness invites comparison with the original optimism of the post-World War II reconstruction and redevelopment efforts and even with the work of such visionary megastructuralists as Paolo Soleri. Jane Jacobs (p. 114) may be counted as one of the severest critics and grassroots opponents of Corbusian city planning principles, and Allan Jacobs and Donald Appleyard's "Urban Design Manifesto" (p. 437) deliberately takes the form of a Le Corbusier pronouncement but rejects his program, opting instead for lively streets, participatory planning, and the integration of old buildings into the new urban fabric. Beneath all the sparkling clarity of Le

Corbusier's urban designs are questions, which must forever remain conjectural: How would democratic politics be practiced in a Corbusian city? What would social relationships be like amid the gleaming towers?

Le Corbusier's writings include *The City of Tomorrow and its Planning* (New York: Dover, 1987) (translated by Frederich Etchells from *Urbanisme* [1929]), *Concerning Town Planning* (translated by Clive Entwistle from *Propos d'Urbanisme* [1946], New Haven, CT: Yale University Press, 1948), and *L'Urbanisme des Trois Etablissements Humaines* (Paris: Editions de Minuit, 1959).

Excellent accounts of Le Corbusier's ideas may be found in Robert Fishman's *Urban Utopias in the Twentieth Century* (New York: Basic Books, 1977), Peter Hall's *Cities of Tomorrow* (Oxford: Blackwell, 1988), and Kenneth Frampton's *Le Corbusier: Architect of the Twentieth Century* (New York: Abrams, 2002, with photographs by Roberto Schezen). For a closer look at some of Le Corbusier's most important urban planning projects, consult Vikramaditya Prakash's *Chandigarh's Le Corbusier: The Struggle for Modernity in Postcolonial India* (Seattle: University of Washington Press, 2002) and Klaus-Peter Gast and Arthur Ruegg, *Le Corbusier: Paris-Chandigarh* (Birkhauser, 2000).

■ ■ ■ ■ ■ ■

The existing congestion in the centre must be eliminated.

The use of technical analysis and architectural synthesis enabled me to draw up my scheme for a contemporary city of three million inhabitants. The result of my work was shown in November 1922 at the Salon d'Automne in Paris. It was greeted with a sort of stupor; the shock of surprise caused rage in some quarters and enthusiasm in others. The solution I put forward was a rough one and completely uncompromising. There were no notes to accompany the plans, and, alas! not everybody can read a plan. I should have had to be constantly on the spot in order to reply to the fundamental questions which spring from the very depths of human feelings. Such questions are of profound interest and cannot remain unanswered. When at a later date it became necessary that this book should be written, a book in which I could formulate the new principles of Town Planning, I resolutely decided *first of all* to find answers to these fundamental questions. I have used two kinds of argument: first, those essentially human ones which start from the mind or the heart or the physiology of our sensations as a basis; secondly, historical and statistical arguments. Thus I could keep in touch with what is fundamental and at the same time be master of the environment in which all this takes place.

In this way I hope I shall have been able to help my reader to take a number of steps by means of which he can reach a sure and certain position. So that when I unroll my plans I can have the happy assurance that his astonishment will no longer be stupefaction nor his fears mere panic.

[. . .]

A CONTEMPORARY CITY OF THREE MILLION INHABITANTS

Proceeding in the manner of the investigator in his laboratory, I have avoided all special cases, and all that may be accidental, and I have assumed an ideal site to begin with. My object was not to overcome the existing state of things, but *by constructing a theoretically watertight formula to arrive at the fundamental principles of modern town planning*. Such fundamental principles, if they are genuine, can serve as the skeleton of any system of modern town planning; being as it were the rules according to which development will take place. We shall then be in a position to take a special case, no matter what: whether it be Paris, London, Berlin, New York or some small town. Then, as a result of what we have learnt, we can take control and decide in what direction the forthcoming battle is to be waged. For the desire to rebuild any great city in a modern way is to engage in a formidable battle. Can you imagine people engaging in a battle without knowing their objectives? Yet that is exactly what is happening. The authorities are compelled to do something, so they give the police white sleeves or set them on horseback, they invent sound signals and light signals, they propose to put bridges over streets or moving pavements under the streets; more garden cities are suggested, or it is decided to suppress the tramways, and so on. And these decisions are reached in a sort of frantic haste in order, as it were, to hold a wild beast at bay. That beast is the great city. It is infinitely more powerful than all these devices. And it is just beginning to wake. What will to-morrow bring forth to cope with it?

We must have some rule of conduct.

We must have fundamental principles for modern town planning.

Site

A level site is the ideal site [for the contemporary city (Figure 1)]. In all those places where traffic becomes over-intensified the level site gives a chance of a normal solution to the problem. Where there is less traffic, differences in level matter less.

The river flows far away from the city. The river is a kind of liquid railway, a goods station and a sorting house. In a decent house the servants' stairs do not go through the drawing room – even if the maid is charming (or if the little boats delight the loiterer leaning on a bridge).

Population

This consists of the citizens proper; of suburban dwellers; and of those of a mixed kind.

(a) Citizens are of the city: those who work and live in it.
(b) Suburban dwellers are those who work in the outer industrial zone and who do not come into the city: they live in garden cities.
(c) The mixed sort are those who work in the business parts of the city but bring up their families in garden cities.

To classify these divisions (and so make possible the transmutation of these recognized types) is to attack the most important problem in town planning, for such a classification would define the areas to be allotted to these three sections and the delimitation of their boundaries. This would enable us to formulate and resolve the following problems:

1 The *City*, as a business and residential centre.
2 The *Industrial City* in relation to the *Garden Cities* (i.e. the question of transport).
3 The *Garden Cities* and the *daily transport* of the workers.

Figure 1

Our first requirement will be an organ that is compact, rapid, lively and concentrated: this is the City with its well organized centre. Our second requirement will be another organ, supple, extensive and elastic; this is the *Garden City* on the periphery. Lying between these two organs, we must *require the legal establishment* of that absolute necessity, a protective zone which allows of extension, *a reserved zone* of woods and fields, a fresh-air reserve.

Density of population

The more dense the population of a city is the less are the distances that have to be covered. The moral, therefore, is that we must *increase the density of the centres of our cities, where business affairs are carried on.*

Lungs

Work in our modern world becomes more intensified day by day, and its demands affect our nervous system in a way that grows more and more dangerous. Modern toil demands quiet and fresh air, not stale air.

The towns of to-day can only increase in density at the expense of the open spaces which are the lungs of a city.

We must *increase the open spaces and diminish the distances to be covered.* Therefore the centre of the city must be constructed *vertically.*

The city's residential quarters must no longer be built along "corridor-streets", full of noise and dust and deprived of light.

It is a simple matter to build urban dwellings away from the streets, without small internal courtyards and with the windows looking on to large parks; and this whether our housing schemes are of the type with "set-backs" or built on the "cellular" principle.

The street

The street of to-day is still the old bare ground which has been paved over, and under which a few tube railways have been run.

The modern street in the true sense of the word is a new type of organism, a sort of stretched-out workshop, a home for many complicated and delicate organs, such as gas, water and electric mains. It is contrary to all economy, to all security, and to all sense to bury these important service mains. They ought to be accessible throughout their length. The various storeys of this stretched-out workshop will each have their own particular functions. If this type of street, which I have called a "workshop", is to be realized, it becomes as much a matter of construction as are the houses with which it is customary to flank it, and the bridges which carry it over valleys and across rivers.

The modern street should be a masterpiece of civil engineering and no longer a job for navvies.

The "corridor-street" should be tolerated no longer, for it poisons the houses that border it and leads to the construction of small internal courts or "wells".

Traffic

Traffic can be classified more easily than other things.

To-day traffic is not classified – it is like dynamite flung at hazard into the street, killing pedestrians. Even so, *traffic does not fulfil its function.* This sacrifice of the pedestrian leads nowhere.

If we classify traffic we get:

(a) Heavy goods traffic.
(b) Lighter goods traffic, i.e. vans, etc., which make short journeys in all directions.
(c) Fast traffic, which covers a large section of the town.

Three kinds of roads are needed, and in superimposed storeys:

(a) Below-ground there would be the street for heavy traffic. This storey of the houses would consist merely of concrete piles, and between them large open spaces which would form a sort of clearing-house where heavy goods traffic could load and unload.
(b) At the ground floor level of the buildings there would be the complicated and delicate network of the ordinary streets taking traffic in every desired direction.
(c) Running north and south, and east and west, and forming the two great axes of the city, there would be great *arterial roads for fast one-way traffic* built on immense reinforced concrete bridges 120 to 180 yards in width and approached every half-mile or so by subsidiary roads from ground level. These

arterial roads could therefore be joined at any given point, so that even at the highest speeds the town can be traversed and the suburbs reached without having to negotiate any cross-roads.

The number of existing streets should be diminished by two-thirds. The number of crossings depends directly on the number of streets; and *cross-roads are an enemy to traffic*. The number of existing streets was fixed at a remote epoch in history. The perpetuation of the boundaries of properties has, almost without exception, preserved even the faintest tracks and footpaths of the old village and made streets of them, and sometimes even an avenue . . . The result is that we have cross-roads every fifty yards, even every twenty yards or ten yards. And this leads to the ridiculous traffic congestion we all know so well.

The distance between two bus stops or two tube stations gives us the necessary unit for the distance between streets, though this unit is conditional on the speed of vehicles and the walking capacity of pedestrians. So an average measure of about 400 yards would give the normal separation between streets, and make a standard for urban distances. My city is conceived on the gridiron system with streets every 400 yards, though occasionally these distances are subdivided to give streets every 200 yards.

This triple system of superimposed levels answers every need of motor traffic (lorries, private cars, taxis, buses) because it provides for rapid and *mobile* transit.

Traffic running on fixed rails is only justified if it is in the form of a convoy carrying an immense load; it then becomes a sort of extension of the underground system or of trains dealing with suburban traffic. *The tramway has no right to exist in the heart of the modern city.*

If the city thus consists of plots about 400 yards square, this will give us sections of about 40 acres in area, and the density of population will vary from 50,000 down to 6,000, according as the "lots" are developed for business or for residential purposes. The natural thing, therefore, would be to continue to apply our unit of distance as it exists in the Paris tubes to-day (namely, 400 yards) and to put a station in the middle of each plot.

Following the two great axes of the city, two "storeys" below the arterial roads for fast traffic, would run the tubes leading to the four furthest points of the garden city suburbs, and linking up with the metropolitan network . . . At a still lower level, and again following these two main axes, would run the one-way loop systems for suburban traffic, and below these again the four great main lines serving the provinces and running north, south, east and west. These main lines would end at the Central Station, or better still might be connected up by a loop system.

The station

There is only one station. The only place for the station is in the centre of the city. It is the natural place for it, and there is no reason for putting it anywhere else. The railway station is the hub of the wheel.

The station would be an essentially subterranean building. Its roof, which would be two storeys above the natural ground level of the city, would form the aerodrome for aero-taxis. This aerodrome (linked up with the main aerodrome in the protected zone) must be in close contact with the tubes, the suburban lines, the main lines, the main arteries and the administrative services connected with all these . . .

The plan of the city

The basic principles we must follow are these:

1 We must de-congest the centres of our cities.
2 We must augment their density.
3 We must increase the means for getting about.
4 We must increase parks and open spaces.

At the very centre we have the *station* with its landing stage for aero-taxis.

Running north and south, and east and west, we have the *main arteries* for fast traffic, forming elevated roadways 120 feet wide.

At the base of the sky-scrapers and all round them we have a great open space 2,400 yards by 1,500 yards, giving an area of 3,600,000 square yards, and occupied by gardens, parks and avenues. In these parks, at the foot of and round the sky-scrapers, would be the restaurants and cafes, the luxury shops, housed in buildings with receding terraces: here too would be the theatres, halls and so on; and here the parking places or garage shelters.

The sky-scrapers are designed purely for business purposes.

On the left we have the great public buildings, the museums, the municipal and administrative offices. Still further on the left we have the "Park" (which is available for further logical development of the heart of the city).

On the right, and traversed by one of the arms of the main arterial roads, we have the warehouses, and the industrial quarters with their goods stations.

All around the city is the *protected zone* of woods and green fields.

Further beyond are the *garden cities*, forming a wide encircling band.

Then, right in the midst of all these, we have the *Central Station*, made up of the following elements:

(a) The landing-platform; forming an aerodrome of 200,000 square yards in area.
(b) The entresol or mezzanine; at this level are the raised tracks for fast motor traffic: the only crossing being gyratory.
(c) The ground floor where are the entrance halls and booking offices for the tubes, suburban lines, main line and air traffic.
(d) The "basement": here are the tubes which serve the city and the main arteries.
(e) The "sub-basement": here are the suburban lines running on a one-way loop.
(f) The "sub-sub-basement": here are the main lines (going north, south, east and west).

The city

Here we have twenty-four sky-scrapers capable each of housing 10,000 to 50,000 employees; this is the business and hotel section, etc., and accounts for 400,000 to 600,000 inhabitants.

The residential blocks, of the two main types already mentioned, account for a further 600,000 inhabitants.

The garden cities give us a further 2,000,000 inhabitants, or more.

In the great central open space are the cafes, restaurants, luxury shops, halls of various kinds, a magnificent forum descending by stages down to the immense parks surrounding it, the whole arrangement providing a spectacle of order and vitality.

Density of population

(a) The sky-scraper: 1,200 inhabitants to the acre.
(b) The residential blocks with set-backs: 120 inhabitants to the acre. These are the luxury dwellings.

(c) The residential blocks on the "cellular" system, with a similar number of inhabitants.

This great density gives us our necessary shortening of distances and ensures rapid inter-communication.

Note. The average density to the acre of Paris in the heart of the town is 146, and of London 63; and of the over-crowded quarters of Paris 213, and of London 169.

Open spaces

Of the area (a), 95 per cent of the ground is open (squares, restaurants, theatres).

Of the area (b), 85 per cent of the ground is open (gardens, sports grounds).

Of the area (c), 48 per cent of the ground is open (gardens, sports grounds).

Educational and civic centres, universities, museums of art and industry, public services, county hall

The "Jardin anglais". (The city can extend here, if necessary.)

Sports grounds: Motor racing track, Racecourse, Stadium, Swimming baths, etc.

The protected zone (which will be the property of the city), with its aerodrome

A zone in which all building would be prohibited; reserved for the growth of the city as laid down by the municipality: it would consist of woods, fields, and sports grounds. The forming of a "protected zone" by continual purchase of small properties in the immediate vicinity of the city is one of the most essential and urgent tasks which a municipality can pursue. It would eventually represent a tenfold return on the capital invested.

Industrial quarters: types of buildings employed

For business: sky-scrapers sixty storeys high with no internal wells or courtyards . . .

Residential buildings with "set-backs", of six double storeys; again with no internal wells: the flats looking on either side on to immense parks.

Residential buildings on the "cellular" principle, with "hanging gardens", looking on to immense parks; again no internal wells. These are "service-flats" of the most modern kind.

Garden cities: their aesthetic, economy, perfection and modern outlook

A simple phrase suffices to express the necessities of tomorrow: WE MUST BUILD IN THE OPEN.

The lay-out must be of a purely geometrical kind, with all its many and delicate implications.

[. . .]

The city of to-day is a dying thing because it is not geometrical. To build in the open would be to replace our present haphazard arrangements, *which are all we have to-day*, by a *uniform* lay-out. Unless we do this *there is no salvation*.

The result of a true geometrical lay-out is *repetition*. The result of repetition is a *standard*, the perfect form (i.e. the creation of standard types). A geometrical lay-out means that mathematics play their part.

There is no first-rate human production but has geometry at its base. It is of the very essence of Architecture. To introduce uniformity into the building of the city we must *industrialize building*. Building is the one economic activity which has so far resisted industrialization. It has thus escaped the march of progress, with the result that the cost of building is still abnormally high.

The architect, from a professional point of view, has become a twisted sort of creature. He has grown to love irregular sites, claiming that they inspire him with original ideas for getting round them. Of course he is wrong. For nowadays the only building that can be undertaken must be either for the rich or built at a loss (as, for instance, in the case of municipal housing schemes), or else by jerry-building and so robbing the inhabitant of all amenities. A motor-car which is achieved by mass production is a masterpiece of comfort, precision, balance and good taste. A house built to order (on an "interesting" site) is a masterpiece of incongruity – a monstrous thing.

If the builder's yard were reorganized on the lines of standardization and mass production we might have gangs of workmen as keen and intelligent as mechanics.

The mechanic dates back only twenty years, yet already he forms the highest caste of the working world.

The mason dates . . . from time immemorial! He bangs away with feet and hammer. He smashes up everything round him, and the plant entrusted to him falls to pieces in a few months. The spirit of the mason must be disciplined by making him part of the severe and exact machinery of the industrialized builder's yard.

The cost of building would fall in the proportion of 10 to 2.

The wages of the labourers would fall into definite categories; to each according to his merits and service rendered.

The "interesting" or erratic site absorbs every creative faculty of the architect and wears him out. What results is equally erratic: lopsided abortions; a specialist's solution which can only please other specialists.

We must build *in the open*: both within the city and around it.

Then having worked through every necessary technical stage and using absolute ECONOMY, we shall be in a position to experience the intense joys of a creative art which is based on geometry.

THE CITY AND ITS AESTHETIC

(The plan of a city which is here presented is a direct consequence of purely geometric considerations.)

A new unit *on a large scale* (400 yards) inspires everything. Though the gridiron arrangement of the streets every 400 yards (sometimes only 200) is uniform (with a consequent ease in finding one's way about), no two streets are in any way alike. This is where, in a magnificent contrapuntal symphony, the forces of geometry come into play.

Suppose we are entering the city by way of the Great Park. Our fast car takes the special elevated motor track between the majestic sky-scrapers: as we approach nearer there is seen the repetition against the sky of the twenty-four sky-scrapers; to our left and right on the outskirts of each particular area are the municipal and administrative buildings; and enclosing the space are the museums and university buildings.

Then suddenly we find ourselves at the feet of the first sky-scrapers. But here we have, not the meagre shaft of sunlight which so faintly illumines the dismal streets of New York, but an immensity of space. The whole city is a Park. The terraces stretch out over lawns and into groves. Low buildings of a horizontal kind lead the eye on to the foliage of the trees. Where are now the trivial *Procuracies*? Here is the *city* with its crowds living in peace and pure air, where noise is smothered under the foliage of green trees. The chaos of New York is overcome. Here, bathed in light, stands the modern city [Figure 2].

Our car has left the elevated track and has dropped its speed of sixty miles an hour to run gently through the residential quarters. The "set-backs" permit of vast architectural perspectives. There are gardens, games and sports grounds. And sky everywhere, as far as the eye can see. The square silhouettes of the terraced roofs stand clear against the sky, bordered with the verdure of the hanging gardens. The uniformity of the units that compose the picture throw into relief the firm lines on which the far-flung masses are constructed. Their outlines softened by distance, the sky-scrapers

raise immense geometrical facades all of glass, and in them is reflected the blue glory of the sky. An overwhelming sensation. Immense but radiant prisms.

And in every direction we have a varying spectacle: our "gridiron" is based on a unit of 400 yards, but it is strangely modified by architectural devices! (The "set-backs" are in counterpoint, on a unit of 600 × 400.)

The traveller in his airplane, arriving from Constantinople or Pekin it may be, suddenly sees appearing through the wavering lines of rivers and patches of forests that clear imprint which marks a city which has grown in accordance with the spirit of man: the mark of the human brain at work.

As twilight falls the glass sky-scrapers seem to flame.

This is no dangerous futurism, a sort of literary dynamite flung violently at the spectator. It is a spectacle organized by an Architecture which uses plastic resources for the modulation of forms seen in light.

A city made for speed is made for success.

Figure 2 A Contemporary city

"Broadacre City: A New Community Plan"

Architectural Record (1935)

Frank Lloyd Wright

Editors' Introduction

For more than half a century, the question "Who is the greatest American architect?" could have only one answer: Frank Lloyd Wright (1867–1959). First with his revolutionary "prairie houses" that seemed to grow directly out of the Midwest landscape with their long, low cantilevered rooflines, and later with such masterpieces as the Imperial Hotel in Tokyo, the Guggenheim Museum in New York, and the breathtaking "Fallingwater" in Western Pennsylvania, Wright became the spokesman for "organic architecture" and a style of building that expressed "the nature of the materials."

To many, Wright's architecture and "the architecture of American democracy" were synonymous. As an unabashed egotist and a pioneer in the field of media celebrity, Wright encouraged the popular identification of himself with the American spirit. He cultivated an Imperious image of plain speaking, anti-collectivist democracy and sought personally to embody the notion of radical individualism. As an artistic genius, Wright despised the popular philistinism of his day and attributed the observable decline of American popular culture to "the mobocracy" and to the unprincipled bankers and politicians who served its interests. By the 1920s and 1930s, Wright had become a social revolutionary but not, characteristically, of the socialist Left. Rather, Wright called for a radical transformation of American society to restore earlier Emersonian and Jeffersonian virtues. The physical embodiment of that utopian vision was Broadacre City.

Wright unveiled his model of Broadacre City, illustrated in Plate 39, at Rockefeller Center, New York, in 1935. The article reprinted here represents his first and clearest statement of the revolutionary proposal whereby every citizen of the United States would be given a minimum of one acre of land per person, with the family homestead being the basis of civilization, and with government reduced to nothing more than a county architect who would be in charge of directing land allotments and the construction of basic community facilities. Many at the time thought the idea was totally outlandish, but Broadacre (and the small, efficient "Usonian" house) proved to be prophetic as sprawling suburban regions transformed the American landscape during the last half of the twentieth century.

Wright believed that two inventions – the telephone and the automobile – made the old cities "no longer modern," and he fervently looked forward to the day when dense, crowded conglomerations like New York and Chicago would wither and decay. In their place, Americans would reinhabit the rural landscape (and re-acquire the rural virtues of individual freedom and self-reliance) with a "city" of independent homesteads in which people would be isolated enough from one another to insure family stability but connected enough, through modern telecommunications and transportation, to achieve a real sense of community. Borrowing an idea from the anarchist philosopher Kropotkin, Wright believed that the citizens of Broadacre would pursue a combination of manual and intellectual work every day, thus achieving a human wholeness that modern society and the modern city had destroyed. He also believed that a system of personal freedom and dignity through land ownership was the way to guarantee social harmony and avoid class struggle.

Broadacre City invites immediate comparison with the very different models of Ebenezer Howard's Garden City (p. 309) and with Le Corbusier's cities based on towers in a park (p. 317). Intriguingly, the overall population density of Broadacre, on the one hand, and the Garden City and Corbusian visions, on the other, were not all that different, depending on the actual acreage of the surrounding parkland or greenbelt. And both Wright's and Le Corbusier's plans are wedded to the automobile, one vision seeing a centralizing, the other a decentralizing, effect. But the most revealing comparisons are with Robert Fishman's description of the now-emerging "technoburbs" (p. 77) and Melvin Webber's prediction of a "post-urban age" (p. 470). One cannot help but wonder whether what seemed impossible in 1935 may actually be realized, with the help of computer-based telecommunications and the possibility of "telecommuting" to work over the Internet in the twenty-first century.

This selection is from *Architectural Record*, vol. 77 (April 1935). For more on Broadacre City see Robert Fishman's *Urban Utopias of the Twentieth Century* (New York: Basic Books, 1977). John Sergeant's *Frank Lloyd Wright's Usonian Houses: The Case for Organic Architecture* (New York: Whitney Library of Design, 1984) is also useful, and William Allin Storer's *A Frank Lloyd Wright Companion* (Chicago: University of Chicago Press, 1994) is an impressive and definitive reference book.

For a biography of Wright see Meryle Secrest, *Frank Lloyd Wright: A Biography* (Chicago: University of Chicago Press, 1998). For good overviews of Wright's work see David Larkin and Bruce Brooks Pfeiffer (eds), *Frank Lloyd Wright: The Masterworks* (New York: Rizzoli, 1993) and Neil Levine, *The Architecture of Frank Lloyd Wright* (Princeton, NJ: Princeton University Press, 1996). But the very best sources on Wright are Wright himself, although his writing style is often quirky and hyperbolic. Of particular interest are *When Democracy Builds* (Chicago: University of Chicago Press, 1945), *Genius and the Mobocracy* (New York: Duell, Sloan & Pearce, 1949), and *The Living City* (New York: Horizon, 1958).

■ ■ ■ ■ ■ ■

Given the simple exercise of several inherently just rights of man, the freedom to decentralize, to redistribute and to correlate the properties of the life of man on earth to his birthright – the ground itself – and Broadacre City becomes reality.

As I see Architecture, the best architect is he who will devise forms nearest organic as features of human growth by way of changes natural to that growth. Civilization is itself inevitably a form but not, if democracy is sanity, is it necessarily the fixation called "academic." All regimentation is a form of death which may sometimes serve life but more often imposes upon it. In Broadacres all is symmetrical but it is seldom obviously and never academically so.

Whatever forms issue are capable of normal growth without destruction of such pattern as they may have. Nor is there much obvious repetition in the new city. Where regiment and row serve the general harmony of arrangement both are present, but generally, both are absent except where planting and cultivation are naturally a process or walls afford a desired seclusion.

Rhythm is the substitute for such repetitions everywhere. Wherever repetition (standardization) enters, it has been modified by inner rhythms either by art or by nature as it must, to be of any lasting human value.

The three major inventions already at work building Broadacres, whether the powers that over-built the old cities otherwise like it or not, are:

1 The motor car: general mobilization of the human being.
2 Radio, telephone and telegraph: electrical inter-communication becoming complete.
3 Standardized machine-shop production: machine invention plus scientific discovery.

The price of the major three to America has been the exploitation we see everywhere around us in waste and in ugly scaffolding that may now be thrown away. The price has not been so great if by way of popular government we are able to exercise the use of three inherent rights of any man:

1 His social right to a direct medium of exchange in place of gold as a commodity: some form of social credit.
2 His social right to his place on the ground as he has had it in the sun and air: land to be held only by use and improvements.
3 His social right to the ideas by which and for which he lives: public ownership of invention and scientific discoveries that concern the life of the people.

The only assumption made by Broadacres as ideal is that these three rights will be the citizen's so soon as the folly of endeavoring to cheat him of their democratic values becomes apparent to those who hold (feudal survivors or survivals), as it is becoming apparent to the thinking people who are held blindly abject or subject against their will.

The landlord is no happier than the tenant. The speculator can no longer win much at a game about played out. The present success-ideal, placing, as it does, premiums upon the wolf, the fox and the rat in human affairs and above all, upon the parasite, is growing more evident every day as a falsity just as injurious to the "successful" as to the victims of such success. Well – sociologically, Broadacres is release from all that fatal "success" which is, after all, only excess. So I have called it a new freedom for living in America. It has thrown the scaffolding aside. It sets up a new ideal of success.

In Broadacres, by elimination of cities and towns the present curse of petty and minor officialdom, government, has been reduced to one minor government for each county. The waste motion, the back and forth haul, that today makes so much idle business is gone. Distribution becomes automatic and direct, taking place mostly in the region of origin. Methods of distribution of everything are simple and direct. From the maker to the consumer by the most direct route.

Coal (one-third the tonnage of the haul of our railways) is eliminated by burning it at the mines and transferring that power, making it easier to take over the great railroad rights of way; to take off the cumbersome rolling stock and put the right of way into general service as the great arterial on which truck traffic is concentrated on lower side lanes, many lanes of speed traffic above and monorail speed trains at the center, continuously running. Because traffic may take off or take on at any given point, these arterials are traffic not dated but fluescent. And the great arterial as well as all the highways become great architecture, auto-matically affording within their structure all necessary storage facilities of raw materials, the elimination of all unsightly piles of raw material.

In the hands of the state, but by way of the county, is all redistribution of land – a minimum of one acre going to the childless family and more to the larger family as effected by the state. The agent of the state in all matters of land allotment or improvement, or in matters affecting the harmony of the whole, is the architect. All building is subject to his sense of the whole as organic architecture. Here architecture is landscape and landscape takes on the character of architecture by way of the simple process of cultivation.

All public utilities are concentrated in the hands of the state and county government as are matters of administration, patrol, fire, post, banking, license and record, making politics a vital matter to everyone in the new city instead of the old case where hopeless indifference makes "politics" a grafter's profession.

In the buildings for Broadacres no distinction exists between much and little, more and less. Quality is in all, for all, alike. The thought entering into the first or last estate is of the best. What differs is only individuality and extent. There is nothing poor or mean in Broadacres.

Nor does Broadacres issue any dictum or see any finality in the matter either of pattern or style.

Organic character is style. Such style has myriad forms inherently good. Growth is possible to Broad-acres as a fundamental form, not as mere accident of change but as integral pattern unfolding from within.

Here now may be seen the elemental units of our social structure [Figure 1]: the correlated farm, the factory – its smoke and gases eliminated by burning coal at places of origin, the decentralized school, the various conditions of residence, the home offices, safe traffic, simplified government. All common interests take place in a simple coordination wherein all are employed: *little* farms, *little* homes for industry, *little* factories, *little* schools, a *little* university going to the people mostly by way of their interest in the ground, *little* laboratories on their own ground for professional men. And the farm itself, notwithstanding its animals, becomes the most attractive unit of the city. The husbandry of animals at last is in decent association with them and with all else as well. True farm relief.

To build Broadacres as conceived would auto-matically end unemployment and all its evils forever. There would never be labor enough nor could under-consumption ever ensue. Whatever a man did would

Figure 1

be done – obviously and directly – mostly by himself in his own interest under the most valuable inspiration and direction: under training, certainly, if necessary. Economic independence would be near, a subsistence certain; life varied and interesting.

Every kind of builder would be likely to have a jealous eye to the harmony of the whole within broad limits fixed by the county architect, an architect chosen by the county itself. Each county would thus naturally develop an individuality of its own. Architecture – in the broad sense – would thrive.

In an organic architecture the ground itself predetermines all features; the climate modifies them; available means limit them; function shapes them.

Form and function are one in Broadacres. But Broadacres is no finality! The model shows four square miles of a typical countryside developed on the acre as unit according to conditions in the temperate zone and accommodating some 1,400 families. It would swing north or swing south in type as conditions, climate and topography of the region changed.

In the model the emphasis has been placed upon diversity in unity, recognizing the necessity of cultivation as a need for formality in most of the planting. By a simple government subsidy certain specific acres or groups of acre units are, in every generation, planted to useful trees, meantime beautiful, giving privacy and various rural divisions. There are no rows of trees alongside the roads to shut out the view. Rows where they occur are perpendicular to the road or the trees are planted in groups. Useful trees like white pine, walnut, birch, beech, fir, would come to maturity as well as fruit and nut trees and they would come as a profitable crop meantime giving character, privacy and comfort to the whole city. The general park is a flowered meadow beside the stream and is bordered with ranks of trees, tiers gradually rising in height above the flowers at the ground level. A music-garden is sequestered from noise at one end. Much is made of general sports and festivals by way of the stadium, zoo, aquarium, arboretum and the arts.

The traffic problem has been given special attention, as the more mobilization is made a comfort and a facility the sooner will Broadacres arrive. Every Broadacre citizen has his own car. Multiple-lane highways make travel safe and enjoyable. There are no grade crossings nor left turns on grade. The road system and construction is such that no signals nor any lamp-posts need be seen. No ditches are alongside the roads. No curbs either. An inlaid purfling over which the car cannot come without damage to itself takes its place to protect the pedestrian.

In the affair of air transport Broadacres rejects the present airplane and substitutes the self-contained mechanical unit that is sure to come: an aerator capable of rising straight up and by reversible rotors able to travel in any given direction under radio control at a maximum speed of, say, 200 miles an hour, and able to descend safely into the hexacomb from which it arose or anywhere else. By a doorstep if desired.

The only fixed transport trains kept on the arterial are the long-distance monorail cars traveling at a speed (already established in Germany) of 220 miles per hour. All other traffic is by motor car on the twelve lane levels or the triple truck lanes on the lower levels which have on both sides the advantage of delivery direct to warehousing or from warehouses to consumer. Local trucks may get to warehouse-storage on lower levels under the main arterial itself. A local truck road parallels the swifter lanes.

Houses in the new city are varied: make much of fireproof synthetic materials, factory-fabricated units adapted to free assembly and varied arrangement, but do not neglect the older nature-materials wherever they are desired and available. House-holders' utilities are nearly all planned in prefabricated utility stacks or units, simplifying construction and reducing building costs to a certainty. There is the professional's house with its laboratory, the minimum house with its workshop, the medium house ditto, the larger house and the house of machine-age luxury. We might speak of them

as a one-car house, a two-car house, a three-car house and a five-car house. Glass is extensively used as are roofless rooms. The roof is used often as a trellis or a garden. But where glass is extensively used it is usually for domestic purposes in the shadow of protecting overhangs.

Copper for roofs is indicated generally on the model as a permanent cover capable of being worked in many appropriate ways and giving a general harmonious color effect to the whole.

Electricity, oil and gas are the only popular fuels. Each land allotment has a pit near the public lighting fixture where access to the three and to water and sewer may be had without tearing up the pavements.

The school problem is solved by segregating a group of low buildings in the interior spaces of the city where the children can go without crossing traffic. The school building group includes galleries for loan collections from the museum, a concert and lecture hall, small gardens for the children in small groups and well-lighted cubicles for individual outdoor study: there is a small zoo, large pools and green playgrounds.

This group is at the very center of the model and contains at its center the higher school adapted to the segregation of the students into small groups.

This tract of four miles square, by way of such liberal general allotment determined by acreage and type of ground, including apartment buildings and hotel facilities, provides for about 1,400 families at, say, an average of five or more persons to the family.

To reiterate: the basis of the whole is general decentralization as an applied principle and architectural reintegration of all units into one fabric; free use of the ground held only by use and improvements; public utilities and government itself owned by the people of Broadacre City; privacy on one's own ground for all and a fair means of subsistence for all by way of their own work on their own ground or in their own laboratory or in common offices serving the life of the whole.

There are too many details involved in the model of Broadacres to permit complete explanation. Study of the model itself is necessary study. Most details are explained by way of collateral models of the various types of construction shown: highway construction, left turns, crossovers, underpasses and various houses and public buildings.

Anyone studying the model should bear in mind the thesis upon which the design has been built by the Taliesin Fellowship, built carefully not as a finality in any sense but as an interpretation of the changes inevitable to our growth as a people and a nation.

Individuality established on such terms must thrive. Unwholesome life would get no encouragement and the ghastly heritage left by over-crowding in overdone ultra-capitalistic centers would be likely to disappear in three or four generations. The old success ideals having no chance at all, new ones more natural to the best in man would be given a fresh opportunity to develop naturally.

"Designing the Region" and "Designing the Region Is Designing the Neighborhood"

from *The Regional City: Planning for the End of Sprawl* (2001)

Peter Calthorpe and William Fulton

Editors' Introduction

What is the most appropriate scale for urban planning? The home? The street? The neighborhood? The city? Or the entire metropolitan region? In *The Regional City: Planning for the End of Sprawl*, Peter Calthorpe and William Fulton argue that every element of the planned environment interacts with every other and that planners embracing the New Urbanist conception should follow a holistic, ecological approach that harmonizes the intimate scale of the neighborhood with the metropolitan scale of the region.

Peter Calthorpe – principal of the architectural firm of Calthorpe Associates based in Berkeley, California – is an urban futurist very much rooted in both the realities of the present and the traditions of the past. An author, a practicing architect, and a former academic who taught at the University of California, Berkeley, Calthorpe is one of the founders of the Congress for the New Urbanism. William Fulton is president of the Solimar Research Group of Ventura, California, a firm that provides governments, foundations, and planning agencies with empirical research on land use, metropolitan growth, and related policy concerns. Founded in 1999, Solimar specializes in urban growth boundaries and what it calls "urban containment policy."

In *The Regional City*, Calthorpe and Fulton present a subtly nuanced exploration of the New Urbanism at its best, a unified vision that combines earlier ideas about neighborhoods as "Pedestrian Pockets" and "transit-oriented developments" in a sweeping regional conception of contemporary urban form.

As a leading proponent of ecology and environmentalism as applied to urban design, Calthorpe has long been a prophet of a new kind of twenty-first-century community: the intimate, human-scale model that is descended directly from Ebenezer Howard's Garden Cities (p. 309) and that quickly became an ubiquitous element of New Urbanist planning and design, first in America and later in Europe and elsewhere. Calthorpe began his career by writing and lecturing extensively on the necessity of ecologically sensitive design and energy-efficient building based on the application of passive solar techniques. For a time, he was a partner of Sim Van der Ryn, the California state architect in the administration of progressive governor Jerry Brown. Together, they published *Sustainable Communities*, an influential volume that helped spread the ideas of solar energy, recycling, appropriate technology, and environmentalist approaches to urban planning and design.

In Doug Kelbaugh's *The Pedestrian Pocket Book* (1989), Calthorpe examined the "profound mismatch between the old suburban patterns . . . and the post-industrial culture in which we now find ourselves." The similarities between his solution and the one proposed by Howard in 1898 were striking. Both the Garden Cities and the Pedestrian Pockets were to be surrounded by greenbelts of permanent agricultural land. Both were relatively dense developments, allowing residents to walk to the urban center in a short period of time. Both combined residential, commercial, and workplace elements; and where the Garden City was served by a railroad connection, the Pedestrian Pocket avoided the typical suburban monoculture of the automobile by a system of light-rail transit connectors.

But although an important source of the New Urbanism style of community development is the Garden City, the real application of the Calthorpe plan is in the blossoming new ring of suburban development currently springing up around the old metropolitan cores, the area that journalist Joel Garreau has dubbed "Edge City" and which Robert Fishman calls the "technoburbs" (p. 77). In *The Next American Metropolis*, Calthorpe refined and matured the Pedestrian Pocket idea to fit the emerging realities of Edge City technoburbia with what he calls TODs or "transit-oriented developments." And in *The Regional City*, he and Fulton present nothing less than "a new paradigm of community . . . that leads from the Edge City to the Regional City." This, they argue, offers a conception that presents a fuller, richer social, economic, and civic life than is possible in the cultural sterilities of Edge City as it has developed in the final decades of the twentieth century. Signaling a clear break with traditional modernism as exemplified by the theories of Le Corbusier (p. 317), Calthorpe and Fulton argue that the Regional City expresses a "set of principles rooted more in ecology than in mechanics," focusing on issues of human scale, diversity, and conservation rather than rationality and mechanical efficiency.

Calthorpe is the most practical of urban visionaries because his visions represent "a response to a transformation that has already expressed itself: the transformation from the industrial forms of segregation and centralization to the decentralized and integrated forms of the post industrial era." This, he writes, is the result of "a culture adjusting itself" to new realities. Moreover, Calthorpe and Fulton are comfortable recognizing that the creation of a new urban paradigm will require the clear and careful application of state power operating through local and regional planning agencies. They write that "the idea that a region can be 'designed' is central to the Regional City" and call for an active role on the part of central governments, especially in the provision of transit systems, open-space, and development financing.

Andres Duany and Elizabeth Plater-Zyberk (p. 207), Jaime Correa, Steven Peterson and Barbara Littenberg, Mark Schimmenti, Daniel Solomon, and a number of other architects and planners are designing human-scale communities with design aspects similar to the ones Calthorpe and Fulton espouse. Because their architecture draws on traditional small town elements, these New Urbanist architects are sometimes referred to as neotraditionalists, and some of their most famous designs – notably the town of Seaside and the Disney-sponsored community of Celebration in Florida – have been roundly criticized as artificial enclaves for the well-to-do that care more for nostalgia and middle-class amenity than the needs of working people and depressed inner-city neighborhoods.

This selection is from *The Regional City: Planning for the End of Sprawl* (Washington, DC: Island Press, 2001). Other books by Peter Calthorpe include *Sustainable Communities* (San Francisco: Sierra Club Books, 1986) with Sym Van der Ryn, and *The Next American Metropolis* (New York: Princeton Architectural Press, 1993). Other books by William Fulton include *The Reluctant Metropolis* (Baltimore, MD: Johns Hopkins University Press, 2001) and Fulton's *Guide to California Planning* (Pt. Arena, CA: Solano Press, 1999).

Peter Katz, *The New Urbanism: Towards an Architecture of Community* (New York: McGraw-Hill, 1994) and Doug Kelbaugh, *Common Place: Toward Neighborhood and Regional Design* (Seattle: University of Washington Press, 1997) provide overviews of designs by Peter Calthorpe and other New Urbanists. See also James Howard Kunstler's *The Geography of Nowhere* (New York: Simon & Schuster, 1993) and *Home from Nowhere: Remaking Our Everyday World for the Twenty-first Century* (New York: Simon & Schuster, 1998) for lively analyses of Calthorpe and other New Urbanists' work. Also of interest are Michael Leccese (ed.), *Charter of the New Urbanism* (New York: McGraw-Hill, 1999) and Todd W. Bressi (ed.), *The Seaside Debates: A Critique of the New Urbanism* (New York: Rizzoli, 2002).

DESIGNING THE REGION

At the heart of creating concrete visions for the Regional City is the notion that they can be "designed." We use the term "design" not in the typical sense of artistically configuring a physical form but to imply a process that synthesizes many disciplines. Regional design is an act that integrates multiple facets at once: the demands of the region's ecology, its economy, its history, its politics, its regulations, its culture, and its social structure. And its results are specific physical forms as well as abstract goals and policies – regional maps and neighborhood urban design standards as well as implementation strategies, governmental policies, and financing mechanisms.

Too often we plan and engineer rather than design. Engineering tends to optimize isolated elements without regard for the larger system, whereas planning tends to be ambiguous, leaving the critical details of place making to chance. If we merely plan and engineer, we forfeit the possibility of developing a "whole systems" approach or a "design" that recognizes the trade-offs between isolated efficiencies and integrated parts.

The engineering mentality often reduces complex, multifaceted problems to one measurable dimension. For example, traffic engineers optimize road size for auto capacity without considering the trade-off of neighborhood scale, walkability, or beauty. Civil engineers efficiently channelize our streams without considering recreational, ecological, or esthetic values. Commercial developers optimize the delivery of goods without balancing the social need of neighborhoods for local identity and meeting places. Again and again we sacrifice the synergy of the whole for the efficiency of the parts.

The idea that a region or even a neighborhood could or should be "designed" is central to the Regional City. We need to acknowledge that we can direct our growth and that such action can include complex trade-offs as well as unexpected synergies. The common impression is that our neighborhoods, towns, or regions evolve organically (and somewhat mysteriously). They are the product of invisible market forces or the summation of technical imperatives. There also is the illusion that these forces cannot and should not be tampered with. Planning failed in the past; therefore it will fail in the future.

The real illusion, of course, is that we cannot control the form of our communities. Historically, design played a large role in shaping our forms of settlement. The template that underlies much of our suburban growth was designed in the thirties by Frank Lloyd Wright with his Broadacre Cities plans and Clarence Stein's Greenbelt towns. These were then bastardized and codified by the HUD minimum property standards of the 1950s. The template for urban redevelopment was developed about the same time by Le Corbusier and a European group of architects called CIAM (Congrès Internationaux d'Architecture Moderne). Their vision of superblocks and high-rise development became the basis of our urban renewal programs of the 1960s.

The problem is not that our suburbs and cities are lacking design but that they are designed according to failed principles with flawed implementation. They are designed in accord with modernist principles and implemented by specialists. The modernist principles of specialization, standardization, and mass production in emulating our industrial economy had a severe effect on the character of our neighborhoods and regions.

At the neighborhood scale, specialization meant that each land use – residential, retail, commercial, or civic – was isolated and developed by "experts" who optimized their particular zones without any responsibility for the whole. Regional specialization meant that each area within the region could play an independent role: suburbs for the middle class and new businesses, cities for the poor and declining industries, and countryside for nature and agriculture.

As a complement to specialization, standardization led to the homogenization of our communities, a blindness to history and the demise of unique ecological systems. A "one size fits all" mentality of efficiency overrode the special qualities of place and community.

Mass production (in housing, transportation, offices, and so forth) upends the delicate balance between local enterprise, regional systems, and global networks. The logic of mass production moves relentlessly toward ever-increasing scales, which in turn reinforces the specialization and standardization of everyday life.

Against this modern alliance of specialization, standardization, and mass production stands a set of principles rooted more in ecology than in mechanics. They are the principles of diversity, conservation, and human scale. Diversity at each scale calls for more complex, differentiated communities shaped from the unique qualities of place and history. Conservation implies care for existing resources whether natural, social, or institutional. And the principle of human scale brings the individual back into a picture increasingly fashioned around remote and mechanistic concerns.

These alternative principles apply equally to the social, economic, and physical dimensions of communities. For example, the social implications of human scale may mean more police officers walking a beat rather than hovering overhead in a helicopter; the economic implications of human scale may mean economic policies that support small local business rather than major industries and corporations; and the physical implications of human scale may be realized in the form and detail of buildings as they relate to the street. Unlike the standard governmental categories of economic development, housing, education, and health services, each of these design principles incorporates physical design, social programs, and economic strategies. These principles, then, are the ones that we believe should form the foundation of a new regional and neighborhood design ethic.

Human scale

For several generations, the design of buildings, the planning of communities, and the growth of our institutions have exemplified the view that "bigger is better." Efficiency was correlated with large, centralized organizations and processes. Now the idea of decentralized networks of smaller working groups and more personalized institutions is gaining currency in both government and business. Efficiency is correlated with nimble, small working groups, not large hierarchical institutions.

Certainly, the reality of our time is a complex mix of both of these trends. For example, we have ever-larger retail outlets at the same time that Main Streets are making a comeback. Some businesses are growing larger and more centralized while the "new economy" is bursting with small-scale start-ups and intimate working groups. Housing production is diversifying home types at the same time that it consolidates into larger financing packages. Both directions are evolving at the same time, and the shape of our communities will have to accommodate this complex reality.

Yet people are reacting to an imbalance between these two forces. The building blocks of our communities – schools, local shopping areas, housing

subdivisions, apartment complexes, and office parks – have all grown into forms that defy human scale. And we are witnessing a reaction to this lack of scale in many ways. People uniformly long for an architecture that puts detail and identity back into what have too often become generic, if functional, buildings. They desire the character and scale of a walkable street, complete with shade trees and buildings that orient windows and entries their way. They idealize Main Street shopping areas and historic urban districts.

Human scale is a design principle that responds simultaneously to simple human desires and the emerging ethos of the new decentralized economics. The focus on human scale represents a shift away from top-down social programs, from characterless housing projects, and from more and more remote institutions. In its most concrete expression, human scale is the stoop of a townhouse or the front porch of a home rather than the stairwell of a high-rise or the garage door of a tract home. Human scale in economics means supporting individual entrepreneurs and local businesses. Human scale in community means a strong neighborhood focus and an environment that encourages everyday interaction.

Diversity

Diversity has multiple meanings and profound implications. It has the most challenging implications for the social, environmental, and economic dimensions of community planning. Perhaps its most obvious outcome is the creation of communities that are diverse in use and in population. As a planning axiom, it calls for a return to mixed-use neighborhoods that contain a broad range of uses as well as a broad range of housing types and people.

The four fundamental elements of community – civic places, commercial uses, housing opportunities, and natural systems – define the physical elements of diversity at any scale. As a physical principle, diversity in neighborhoods ensures that destinations are close at hand and that the shared institutions of community are integrated. It also implies an architecture rich in character and streetscapes that vary with place and use.

As a social principle, diversity is controversial and perhaps the most challenging of all. It implies creating neighborhoods that provide for a large range in age group, household type, income, and race. As already stated, neighborhoods have always (to a greater or lesser degree) been defined by commonalties even if energized by differences. But today we have reached an extreme: age, income, family size, and race are all divided into discrete market segments and locations that are built independently. Complete housing integration may be a distant goal, but inclusive neighborhoods that broaden the economic range, expand the mix of age and household types, and open the door to racial integration are feasible and desirable.

Diversity is a principle with significant economic implications. Gone are the days when economic-revitalization efforts focused on a single industry or a major governmental program. A more ecological understanding of industry clusters has emerged. This sensibility validates the notion that a range of complementary but differing enterprises (large and small, local, regional, and global) are important to maintaining a robust economy, and that now more than ever, the quality of life and the urbanism of a place, as well as the more traditional economic factors, play a significant role in the emerging economy.

Finally, diversity is a fundamental principle that can help to guide the preservation of local and regional ecologies. Clearly, understanding the complex nature of the existing or stressed habitats and watershed systems mandates a different approach to open-space planning. Active recreation, agriculture, and habitat preservation are often at odds. A broad range of open-space types, from the most active to the most passive, must be integrated in neighborhood and regional designs. Diversity in use, diversity in population, diversity in enterprise, and diversity in natural systems are fundamental to the Regional City.

Conservation

Conservation implies many things in community design beyond husbanding resources and protecting natural systems; it implies preserving and restoring the cultural, historic, and architectural assets of a place as well. Conservation calls for designing communities and buildings that require fewer resources – less energy, less land, less waste, and fewer materials – but it also implies caring for what we have and developing an ethic of reuse and repair – in both our physical and our social realms.

The principle of conservation and its complements, restoration and preservation, should be applied to the built environment as well as to the natural environment – not only to our historic building stock and neighborhood institutions, but also to human resources and human history. Communities should strive to conserve their cultural identity, physical history, and unique natural systems. Restoration and conservation are more than environmental themes; they are an approach to the way that we think about community at the regional and local levels.

Conserving resources has many obvious implications in community planning. Foremost are the quantities of farmlands and natural systems displaced by sprawling development and the quantity of auto travel required to support it. Even within more compact, walkable communities, conservation of resources can lead to new design strategies. The preservation of waterways and on-site water-treatment systems can add identity and natural amenities at the same time that they conserve water quality. Energy-conservation strategies in buildings often lead to environments that are climate responsive and unique to place.

Conserving the historic buildings and institutions of a neighborhood can preserve the icons of community identity. Restoring and enhancing the vernacular architecture of a place can simultaneously reduce energy costs, reestablish local history, and create jobs. Although the preservation movement has made great strides with landmark buildings, it is correct now in extending its agenda beyond building facades to the social fabric of neighborhoods and to the ecology of the communities that are the lifeblood of historic districts.

Conserving human resources is another implication of this fundamental principle. In too many of our communities, poverty, lack of education, and declining job opportunities lead to a tragic waste of human potential. As we have begun to see, communities are not viable when concentrations of poverty turn them into a wasteland of despair and crime. In this context, "conservation" takes on a larger meaning – the restoration and rehabilitation of human potential wherever it is being squandered and overlooked. There should be no natural or cultural environments that are disposable or marginalized. Conservation and restoration are practical undertakings that can be economically strengthening and socially enriching.

FIVE

DESIGNING THE REGION IS DESIGNING THE NEIGHBORHOOD

What happens to regions or neighborhoods if they are "designed" according to these principles? An interesting set of parallel design strategies emerges at both the regional and the neighborhood levels. First and foremost, the region and its elements – the city, suburbs, and their natural environment – should be conceived as a unit, just as the neighborhood and its elements – housing, shops, open space, civic institutions, and businesses – should be designed as a unit. Treating each element separately is endemic to many of the problems that we now face. Just as a neighborhood needs to be developed as a whole system, the region must be treated as a human ecosystem, not a mechanical assembly.

Seen as this integrated whole, the region can be designed in much the same way as we would design a neighborhood. That the whole, the region, would be similar to its most basic pieces, its neighborhoods, is an important analogy. Both need protected natural systems, vibrant centers, human-scale circulation systems, a common civic realm, and integrated diversity. Developing such an architecture for the region creates the context for healthy neighborhoods, districts, and city centers. Developing such an architecture for the neighborhood creates the context for regions that are sustainable, integrated, and coherent. The two scales have parallel features that reinforce one another.

Major open-space corridors within the region, such as rivers, ridge lands, wetlands, or forests, can be seen as a "village green" at a megascale – the commons of the region. These natural commons establish an ecological identity as the basis of a region's character. Similarly, the natural systems and shared open spaces at the neighborhood scale are fundamental to its identity and character. A neighborhood's natural systems, like the region's, are as much a part of its commons as its civic institutions or commercial center.

Just as a neighborhood needs a vital center to serve as the crossroads of a local community, the region needs a vital central city to serve as its cultural heart and as a link to the global economy. In the Edge City metropolis, both types of centers are failing. In the suburbs, what were village centers of human proportions are overcome by remote discount centers and relentless commercial strips. In the central cities, poverty and disinvestment erode historic neighborhood communities. Both fall prey to specialized enterprises oriented to mass distribution rather than the local community. Like the commons, healthy centers, both urban and suburban, are fundamental to local and regional coherence.

Regional and neighborhood design has other parallels. Pedestrian scale within the neighborhood – walkable streets and nearby destinations – has a partner in transit systems at the regional scale. Transit can organize the region in much the same way as a street network orders a neighborhood. Transit lines focus growth and redevelopment in the region just as main streets can focus a neighborhood. Crossing local and metropolitan scales, transit supports the life of the pedestrian within each neighborhood and district by providing access to regional destinations. In a complementary fashion, pedestrian-friendly neighborhoods support transit by providing easy access for riders, not cars. The two scales, if designed as parallel strategies, reinforce each other.

As we have pointed out, diversity is a fundamental design principle for both the neighborhood and the region. A diverse population and job base within a region supports a resilient economy and a rich culture in much the same way that diverse uses and housing in a neighborhood support a complex and active community. The suburban trend to segregate development by age and income translates at the regional level into an increasing spatial and economic polarization . . . Both trends can be countered by policies that support inclusionary housing and mixed-use environments.

These parallels across scales are not merely coincidence. The fundamental nature of a culture and economy expresses itself at many scales simultaneously. Sprawl and our lack of regional structure is a manifestation of an older and quite different paradigm. Since World War II, our economy and culture have accelerated their movement toward the industrial qualities of mass production, standardization, and specialization. The massive suburbanization that marks this period is the direct expression of these qualities. As a counterpoint, the principles and concurrences just outlined define a new paradigm of community and growth, one that leads from the Edge City to the Regional City.

PART SIX

Urban planning theory and practice

INTRODUCTION TO PART SIX

Contemporary urban planning has come a long way from its origins in the Olmsted-inspired parks movement, the visionary plans of Ebenezer Howard and the Garden Cities that more or less followed his ideas, Daniel Burnham's monumental City Beautiful projects, and the plans of eccentric Scottish biologist Patrick Geddes, Le Corbusier and his modernist followers, the brilliant Broadacre City vision of Frank Lloyd Wright, and a host of other imagined and implemented plans discussed in Part Five, above, on "Urban Planning History and Visions." Today city and regional planning (or town and country planning as it is called in the UK) has matured into an important profession with its own body of theory and set of professional practices. This section focuses on the theory and practice of urban planning today.

Large local governments may employ dozens or even hundreds of professionally trained planners. Even most small and medium-sized cities and towns employ city planners and have explicit plans for their future development. In addition to professionals whose formal education is in city and regional (or town and country) planning, planning staffs are likely to include architects and urban designers, geographers, economists, civil engineers, transportation experts, environmental professionals, computer experts, and staff trained in negotiation and other communicative planning skills.

City plans are grounded in *analysis* of local conditions and contain a *vision* of an urban future the citizens, local elected officials, planning staff, and consultants consider desirable. Urban plans vary greatly in approach, content, sophistication, comprehensiveness, time frame, and format. Plans developed for the Los Angeles City Planning Commission or Curitiva, Brazil, run to many volumes built on mountains of data and sophisticated analysis. The town plan for a small town like J.B. Jackson's hypothetical Optimo City (p. 164) may contain a common-sense description of the town's situation and some practical suggestions worked out by the residents under the direction of a part-time planning consultant. The substance of city plans varies as widely as the planning culture of the cities that produce them. The plan for Optimo City might advance a narrow, business-as-usual vision for its future that envisions the city tearing down the courthouse and building a parking garage to attract more off-highway business. By contrast, green city plans in some of the most environmentally sensitive European cities that Timothy Beatley describes (p. 399) are filled with imaginative ideas for use of public bicycle depots, wind and solar power generation, community gardens, and recycling, and the city and regional plans Peter Calthorpe and William Fulton (p. 331) and Andres Duany and Elizabeth Plater-Zyberk (p. 207) have developed reflect their "New Urbanist" and "sustainable urban development" values.

The first two selections in this section pick up where Part Five leaves off. Sir Peter Hall's "The City of Theory" (p. 342) discusses the evolution and current status of twentieth-century urban planning theory, and Edward J. Kaiser and David R. Godschalk's "Twentieth Century Land Use Planning" (p. 354) describes current American land use planning practice. Both selections touch on planning in the first half of the twentieth century, but focus on how planning has evolved during the last half century and what it is like today.

In the first half of the twentieth century, urban planning was elitist and paid little attention to implementation. According to Hall, planning theory in the first half of the twentieth century was preoccupied with how to create stable cities "geared to a static world." During this "golden age," the planner was "free from political interference" and "serenely sure of his technical capacities." He (male) was also generally ineffective. Heavily influenced by the Garden Cities and City Beautiful movements, early planners worked out theoretically perfect physical plans in excruciating detail. Slowly a more conceptual, visionary, and flexible approach to creating city "general plans" emerged. By the 1950s, according to Kaiser and Godschalk (p. 354), there was a general consensus that urban general plans should be long-term, visionary documents, charting the desired physical form of the city. While an advance over earlier static architectural city planning, his ivory tower general planning approach was never very practical and was no longer defensible by the 1950s, as the pace of urban development and urban change accelerated.

In the last half century, urban planning theory has been buffeted by a whole series of conflicting approaches proposed by Marxists, advocacy planners, equity planners, systems analysts, pluralists, disjointed incrementalists, probabilistic planners, green urbanists, ecological designers, feminist planners, and communicative action theorists. In practice the general plan "trunk" of the urban land use planning "tree" has branched into management and policy as well as physical design. Planning no longer takes place in ivory towers. As planning has become more relevant it has also become more conflictual as Paul Davidoff (p. 388) and John Forester (p. 375) describe. Citizens and decision-makers did not care much about unrealistic static physical designs or utopian general plans developed in the first half of the twentieth century. But they care a great deal about plans that propose using public regulatory power and spending money to create or destroy neighborhoods, retail areas, open space, and office parks.

Cornell planning professor John Forester describes how planners actually work with neighborhood residents, local elected officials, interest groups, and private developers. Forester provides perhaps the best view we have of what current urban planning is really like as perceived by planners themselves. Based on interviews with dozens of practicing planners Forester describes planners' day-to-day activities, the nature of and limitations on their power, and the strategies they actually employ to get things done. Planners negotiate, mediate, resolve conflict, and serve as diplomats. Forester is interested in equity planning and applauds the efforts of many planners to redirect market forces and to empower people and communities poorly served by the private market. Planning practitioners who study Forester's theoretical writings on planning in the face of conflict have much to learn that can help them be more effective.

Planner/lawyer Paul Davidoff proposes an approach to urban planning that recognizes that different groups compete in the planning process. Davidoff envisioned a kind of planning practice in which city planners would act as *advocates* for poor people and disenfranchised groups. Davidoff and many planners whom he inspired saw "advocacy planning" as one way to bring about non-violent social change. This concern with social justice is an enduring one. From the early efforts of nineteenth-century reformers advocating on behalf of slum residents in the new industrial slums, through the New Deal, New Frontier, and Great Society programs of the last century, some progressives have always made a connection between urban planning and social justice. The most recent incarnation of this tradition is "equity planning," a subfield developed by Cleveland state professor and former Cleveland city planning director Norman Krumholz.

Another enduring value in urban planning has been a concern to harmonize the built environment with the natural environment. By the middle of the nineteenth century, park planners like Joseph Paxton in England and Frederick Law Olmsted in the United States were working hard to bring nature into crowded cities. In the early twentieth century, Scottish biologist/planner Patrick Geddes had worked out an elaborate scheme for regional planning that reflected different ecosystems. Similar thinking informed Ian McHarg's theories in his classic *Design with Nature* (1969). Today "sustainable" urban planning as described by Stephen Wheeler (p. 486), ecological design, and green urbanism have moved to the fore. The final selection in Part Six describes green urbanism in Europe and what planners worldwide can learn from it. Planning professor Timothy Beatley describes what environmentally conscious cities in Europe have actually done to substitute renewable energy sources for non-renewable energy sources, reduce auto dependency, make themselves more compact, produce more of their own food, and weave green space into the urban fabric. Beatley identifies core aspects of the European green urbanism agenda, provides some specific examples of exemplary green practices that some European cities have implemented, and eloquently argues that green urbanism is possible and can produce livable cities that respect the natural environment of the Earth. Beatley's selection combines theory and practice. It is both realistic about urban environmental challenges and optimistic about what can be done. It is a fitting conclusion to this section on "Urban Planning Theory and Practice."

What makes urban planning so fascinating is the tremendous variety of issues planning theory and practice must confront. Issues of aesthetics, design, economic feasibility, decision-making theory, conflict resolution, advocacy, race, class and gender equity, and sustainability are just some of the core concerns which planning must confront in the twenty-first century.

S
I
X

"The City of Theory"

from *Cities of Tomorrow: An Intellectual History of Urban Planning and Design in the Twentieth Century*, 3rd edn (2001)

Peter Hall

Editors' Introduction

In this selection British geographer/planner Sir Peter Hall reviews the evolution of city planning theory in the United States and Europe during the last half century.

Hall points out that before World War II, city/town planning was defined as the craft of physical planning. Professors who had been educated as architects, landscape architects, and other design professionals dominated teaching and writing. They taught students to prepare self-contained, end-state physical plans like Raymond Unwin's plan for the garden city of Letchworth, England: architectural drawings extended to city scale. Once a well-crafted, aesthetically pleasing, functional plan was complete on paper the early planning theorists believed that a city could be built just as a house is built from architectural drawings.

But, as Hall notes, few planners today (or indeed in the past) actually get to plan whole new towns from scratch. Rather they have to decide how to tie new housing, street, retail districts, industrial areas, parks, and infrastructure into existing cities. And city planning does not end the way house construction does with a finished product.

During the 1960s, systems analysts who had been educated as economists, computer scientists, mathematical geographers, and engineers developed a competing paradigm of what city planning should be. They argued that city planning should be a science, not a craft. They felt planners could and should use quantitative and exact methodologies to plan urban transportation, utility, or other "systems" on an ongoing basis. They were pioneers in using computers at a time when computers were new, costly, and extraordinarily hard to use. The systems theorists thought plans should be mathematical models rather than end-state architectural drawings. It is helpful to imagine a planner from the old school saying, "Here is my plan," and unrolling a series of large, carefully drafted blueprints and design sketches confronting a systems planner saying, "Here's my plan," as she types an equation into a huge mainframe computer!

But planning for what kind of city and for whom? Normative theorists, shaken by 1960s racial and class conflict in cities, felt city planning was too important to leave to either elitist designers or technocratic systems analysts. They were more concerned with who the city was being built for and how planning might benefit poor and powerless people than with either design excellence or elegant mathematical modeling.

During the 1970s, Marxist theorists developed a coherent body of urban theory based on class conflict. Twentieth-century Marxists could not apply the categories Friedrich Engels applied to mid-nineteenth-century Manchester, England – the stark class conflict between an oppressed industrial proletariat and wealthy owners of the means of production had become far more complicated. But Marxist theorists felt updated class analysis was highly appropriate. Marxist urban theory dominated much academic discourse throughout the 1970s.

Hall ends his tour of planning theory with some critical comments on the current divorce between planning theory and practice. He argues that as graduate programs in city and regional planning (called town and country planning in the UK) have grown in number and size and as a formal body of planning theory has developed, too often academics today merely debate each other's academic theories with little attention to actual practice or the needs of planning practitioners on the front lines. Finding academic planning theory irrelevant, city planning practitioners concern themselves only with the nuts-and-bolts of planning practice without the deeper understanding relevant theory could offer. Hall calls for an improved, reciprocal relationship between the two: theory that is informed by and relevant to planning practice and planning practice informed and improved by theory. John Forester's empirical research (p. 375) is a good example of what Peter Hall feels academic planning theorists should do. Forester interviewed dozens of practicing planners so that he was very knowledgeable about actual planning practice and then developed theory based on what they told him.

Until assuming the Bartlett Professorship of City Planning at University College, London, in 1994, Peter Hall shuttled between the University of California Berkeley's Department of City and Regional Planning and the University of Reading's Geography Department. He has traveled widely and written prolifically on urban geography and city planning. In addition to his academic accomplishments, Peter Hall has worked with actual planning projects as varied as long-term plans for the London region and bringing high-speed bullet trains to California. His theory is informed by the experience of actual planning. Practicing planners as well as academics use his writings.

This selection is from *Cities of Tomorrow: An Intellectual History of Urban Planning and Design in the Twentieth Century*, 3rd edn (Oxford: Blackwell, 2001). Other books by Peter Hall include *Urban and Regional Planning*, 4th edn (London and New York: Routledge, 2002), *Urban Future 21: A Global Agenda for 21st Century Cities* (London: E & F Spon, 2001) co-authored with Ulrich Pfeiffer, *Cities in Civilization* (New York: Pantheon, 1998), *Sociable Cities: The Legacy of Ebenezer Howard* co-authored with Colin Ward (London: John Wiley & Sons, 1998), *The World Cities*, 3rd edn (London: Weidenfeld and Nicolson; New York: St. Martin's, 1984), and *Great Planning Disasters* (Berkeley: University of California Press, 1982).

Other overviews of planning theory include Susan Fainstein and Scott Campbell, *Readings in Urban Theory* (London: Blackwell, 2001), Seymour Mandelbaum (ed.), *Explorations in Planning Theory* (Rutgers, NJ: CUPR Press, 1996), and John Friedmann, *Planning in the Public Domain: From Knowledge to Action* (Princeton, NJ: Princeton University Press, 1987).

PLANNING AND THE ACADEMY: PHILADELPHIA, MANCHESTER, CALIFORNIA, PARIS, 1955–1987

... about 1955 ... city planning at last became legitimate; but in doing so, it began to sow the seeds of its own destruction. All too quickly, it split into two separate camps: the one, in the schools of planning, increasingly and exclusively obsessed with the theory of the subject; the other, in the offices of local authorities and consultants, concerned only with the everyday business of planning in the real world. That division was not at first evident; indeed, during the late 1950s and most of the 1960s, it seemed that at last a complete and satisfactory link had been forged between the world of theory and the world of practice. But all too soon, illusion was stripped aside: honeymoon was followed in quick succession during the 1970s by tiffs and temporary reconciliations, in the 1980s by divorce. And, in the process, planning lost much of its new-found legitimacy.

The prehistory of academic city planning 1930–1955

It was not that planning was innocent of academic influence before the 1950s. On the contrary: in virtually every urbanized nation, universities and polytechnics had created courses for the professional education of planners; professional bodies had come into existence to define and protect standards, and had forged links with the academic departments. Britain took an early lead when in 1909 ... the soap magnate William Hesketh Lever, founder of Port Sunlight, won a libel action against a newspaper and used the proceeds to endow his local University of Liverpool with a Department of Civic Design. Stanley Adshead, the first professor, almost immediately created a new journal, the *Town Planning Review*, in which theory and good practice were to be firmly joined; its first editor was a young faculty recruit, Patrick Abercrombie, who was later to succeed Adshead in the chair first at Liverpool, then at Britain's second school of planning: University

College London, founded in 1914. The Town Planning Institute – the Royal accolade was conferred only in 1959 – was founded in 1914 on the joint initiative of the Royal Institute of British Architects, the Institution of Civil Engineers and the Royal Institution of Chartered Surveyors; by the end of the 1930s, it had recognized seven schools whose examinations provided an entry to membership.

The United States was slower: though Harvard had established a planning course in 1909, neck and neck with Liverpool, it had no separate department until 1929. Nevertheless, by the 1930s America had schools also at MIT, Cornell, Columbia and Illinois, as well as courses taught in other departments at a great many universities across the country. And the American City Planning Institute, founded in 1917 as a breakaway from the National Conference on City Planning, ten years later became – mainly through the insistence of Thomas Adams – a full-fledged professional body on TPI lines, a status it retained when in 1938 it broadened to include regional planning and renamed itself the American Institute of Planners.

The important point about these, and other, initiatives was this: stemming as they did from professional needs, often through spin-offs from related professions like architecture and engineering, they were from the start heavily suffused with the professional styles of these design-based professions.

The job of the planners was to make plans, to develop codes to enforce these plans, and then to enforce those codes; relevant planning knowledge was what was needed for that job; planning education existed to convey that knowledge together with the necessary design skills. So, by 1950, the utopian age of planning . . . was over; planning was now institutionalized into comprehensive land-use planning. All this was strongly reflected in the curricula of the planning schools down to the mid-1950s, and often for years after that; and these in turn were reflected in the books and articles that academic planners wrote. Land-use planning, Keeble told his British audience in 1959 and Kent reminded the American counterpart in 1964, was a distinct and tightly bounded subject, quite different from social or economic planning. And these texts reflected the fact that "city planners early adopted the thoughtways and the analytical methods that engineers developed for the design of public works, and they then applied them to the design of cities."

The result, as Michael Batty has put it, was a subject that for the ordinary citizen was "somewhat mystical"

or arcane, as law or medicine were, but that was – in sharp contrast to education for these older professions – not based on any consistent body of theory; rather, in it, "scatterings of social science bolstered the traditional architectural determinism." Planners acquired a synthetic ability not through abstract thinking, but by doing real jobs; in them, they used first creative intuition, then reflection. Though they might draw on bits and pieces of theory about the city – the Chicago school's sociological differentiation of the city, the land economists' theory of urban land rent differentials, the geographers' concepts of the natural region – these were employed simply as snippets of useful knowledge. In the important distinction later made by a number of writers, there was some theory in planning but there was no theory of planning. The whole process was very direct, based on a single-shot approach: survey (the Geddesian approach) was followed by analysis (an implicit learning approach), followed immediately by design.

True, as Abercrombie's classic text of 1933 argued, the making of the plan was only half the planner's job; the other half consisted of planning, that is implementation, but it was nowhere assumed that some kind of continuous learning process was needed. True, too, the 1947 Act provided for plans – and the surveys on which they were based – to be quinquennially updated; the assumption was still that the result would be a fixed land-use plan. And, a decade after that, though Keeble's equally classic text referred to the planning process, by this he simply meant the need for a spatial hierarchy of related plans from the regional to the local, and the need at each scale for survey before plan. Nowhere is found a discussion of implementation or updating. Thus – apart from extremely generalized statements like Abercrombie's famous triad of 'beauty, health and convenience' – the goals were left implicit; the planner would develop them intuitively from his own values, which by definition were "expert" and apolitical.

So, in the classic British land-use planning system created by the 1947 Town and Country Planning Act, no repeated learning process was involved, since the planner would get it right first time:

The process was therefore not characterized by explicit feedback as the search "homed in" on the best plan, for the notion that the planner had to learn about the nature of the problem was in direct conflict with his assumed infallibility as an expert, a professional . . . The assumed certainty of the process was such that possible links back to the reality in the form of new surveys were rarely if ever considered . . . This certainty, based on the infallibility of the expert, reinforced the apolitical, technical nature of the process. The political environment was regarded as totally passive, indeed subservient to the "advice" of the planners and in practice, this was largely the case.

(Batty, 1979)

It was, as Batty calls it, the golden age of planning: the planner, free from political interference, serenely sure of his technical capacities, was left to get on with the job. And this was appropriate to the world outside, with which planning had to deal: a world of glacially slow change – stagnant population, depressed economy – in which major planning interventions would come only seldom and for a short time, as after a major war. Abercrombie, in the plan for the West Midlands he produced with Herbert Jackson in 1948, actually wrote that a major objective of the plan should be to slow down the rate of urban change, thus reducing the rate at which built structures became obsolescent: the ideal city would be a static, stable city:

Let us assume . . . that a maximum population has been decided for a town, arrived at after consideration of all the factors appearing to be relevant . . . Allowance has been made for proper space for all conceivable purposes in the light of present facts and the town planner's experience and imagination. Accordingly, an envelope or green belt has been prescribed, outside which the land uses will be those involving little in the way of resident population. The town planner is now in the happy position for the first time of knowing the limits of his problem. He is able to address himself to the design of the whole and the parts in the light of a basic overall figure for population. The process will be difficult enough in itself, but at least he starts with one figure to reassure him.

(Abercrombie and Jackson, 1948)

American planning was never quite like that. Kent's text of 1964 on the urban general plan, though it deals with the same kind of land-use planning, reminds its students of end-directions which are continually adjusted as time passes. And, because the planner's basic understanding of the interrelationship between socio-economic forces and the physical environment was largely intuitive and speculative, Kent warned his student readers,

In most cases it is not possible to know with any certainty what physical design measures should be taken to bring about a given social or economic objective, or what social and economic consequences will result from a given physical-design proposal. Therefore, the city council and the city-planning commission, rather than professional city planners, should make the final value judgements upon which the plan is based.

(Kent, 1964)

But even Kent was certain that, despite all this, it was still possible for the planner to produce some kind of optimal land use plan; the problem of objectives was just shunted off.

The systems revolution

It was a happy, almost dream-like, world. But increasingly, during the 1950s, it did not correspond to reality. Everything began to get out of hand. In every industrial country, there was an unexpected baby boom, to which the demographers reacted with surprise, the planners with alarm; only its timing varied from one country to another, and everywhere it created instant demands for maternity wards and child-care clinics, only slightly delayed needs for schools and playgrounds. In every one, almost simultaneously, the great postwar economic boom got under way, bringing pressures for new investment in factories and offices. And, as boom generated affluence, these countries soon passed into the realms of high mass-consumption societies, with unprecedented demands for durable consumer goods: most notable among these, land-hungry homes and cars. The result everywhere – in America, in Britain, in the whole of western Europe – was that the pace of urban development and urban change began to accelerate to an almost superheated level. The old planning system, geared to a static world, was overwhelmed.

These demands in themselves would force the system to change; but, almost coincidentally, there were changes on the supply side too. In the mid-1950s there occurred an intellectual revolution in the whole cluster of urban and regional social studies, which provided planners with much of their borrowed intellectual baggage. A few geographers and industrial economists discovered the works of German theorists of location, such as Johann Heinrich von Thünen (1826) on agriculture, Alfred Weber (1909) on industry, Walter Christaller (1933) on central places, and August Lösch (1940) on the general theory of location; they began to summarize and analyse these works, and where necessary to translate them. In the United States, academics coming from a variety of disciplines began to find regularities in many distributions, including spatial ones. Geographers, beginning to espouse the tenets of logical positivism, suggested that their subject should cease to be concerned with descriptions of the detailed differentiation of the earth's surface, and should instead begin to develop general hypotheses about spatial distributions, which could then be rigorously tested against reality: the very approach which these German pioneers of location theory had adopted. These ideas, together with the relevant literature, were brilliantly synthesized by an American economist, Walter Isard, in a text that became immediately influential. Between 1953 and 1957, there occurred an almost instant revolution in human geography and the creation, by Isard, of a new academic discipline uniting the new geography with the German tradition of locational economics. And, with official blessing – as in the important report of Britain's Schuster Committee of 1950, which recommended a greater social science content in planning education – the new locational analysis began to enter the curricula of the planning schools.

The consequences for planning were momentous: with only a short timelag, "the discipline of physical planning changed more in the 10 years from 1960 to 1970, than in the previous 100, possibly even 1000 years" (Batty, 1979).

The subject changed from a kind of craft, based on personal knowledge of a rudimentary collection of concepts about the city, into an apparently scientific activity in which vast amounts of precise information were garnered and processed in such a way that the planner could devise very sensitive systems of guidance and control, the effects of which could be monitored and if necessary modified. More precisely, cities and regions were viewed as complex systems – they were, indeed, only a particular spatially based subset of a whole general class of systems – while planning was seen as a continuous process of control and monitoring of these systems, derived from the then new science of cybernetics developed by Norbert Wiener.

There was thus, in the language later used in the celebrated work of Thomas Kuhn, a "paradigm shift." It affected city planning as it affected many other related areas of planning and design. Particularly, its main early applications – already in the mid-1950s – concerned defence and aerospace; for these were the Cold War years, when the United States was engaging in a crash programme to build new and complex electronically controlled missile systems. Soon, from that field, spun off another application. Already in 1954, Robert Mitchell and Chester Rapkin – colleagues of Isard at the University of Pennsylvania – had published a book suggesting that urban traffic patterns were a direct and measurable function of the pattern of activities – and thus land uses – that generated them. Coupled with earlier work on spatial interaction patterns, and using for the first time the data-processing powers of the computer, this produced a new science of urban transportation planning, which for the first time claimed to be able scientifically to predict future urban-traffic patterns. First applied in the historic Detroit Metropolitan Area transportation study of 1955, further developed in the Chicago study of 1956, it soon became a standardized methodology employed in literally hundreds of such studies, first across the United States, then across the world.

Heavily engineering-based in its approach, it adopted a fairly standardized sequence. First, explicit goals and objectives were set for the performance of the system. Then, inventories were taken of the existing state of the system: both the traffic flows, and the activities that gave rise to them. From this, models were derived which sought to establish these relationships in precise mathematical form. Then, forecasts were made of the future state of the system, based on the relationships obtained from the models. From this, alternative solutions could be designed and evaluated in order to choose a preferred option. Finally, once implemented the network would be continually monitored and the system modified as necessary.

At first, these relationships were seen as operating in one direction: activities and land uses were given; from these, the traffic patterns were derived. So the resulting methodology and techniques were part of a new field, transportation planning, which came to exist on one side of traditional city planning. Soon, however, American regional scientists suggested a crucial modification: the locational patterns of activities – commercial, industrial, residential – were in turn influenced by the available transportation opportunities; these relationships, too, could be precisely modelled and used for prediction; therefore the relationship was two-way, and there was a need to develop an interactive system of land-use–transportation planning for entire metropolitan or subregional areas. Now, for the first time, the engineering-based approach invaded the professional territory of the traditional land-use planner. Spatial interaction models, especially the Garin–Lowry model – which, given basic data about employment and transportation links, could generate a resulting pattern of activities and land uses – became part of the planner's stock in trade. As put in one of the classic systems texts:

In this general process of planning we particularise in order to deal with more specific issues: that is, a specific real world system or subsystem must be represented by a specific conceptual system or subsystem within the general conceptual system. Such a particular representation of a system is called a *model* . . . the use of models is a means whereby the high variety of the real world is reduced to a level of variety appropriate to the channel capacities of the human being.

(Chadwick, 1971)

This involved more than a knowledge of computer applications – novel as that seemed to the average planner of the 1960s. It meant also a fundamentally different concept of planning. Instead of the old master-plan or blueprint approach, which assumed that the objectives were fixed from the start, the new concept was of planning as a process, "whereby programmes are adapted during their implementation as and when incoming information requires such changes". And this planning process was independent of the thing that was planned; as Melvin Webber put it, it was "a special way of deciding and acting" which involved a constantly recycled series of logical steps: goal-setting, forecasting of change in the outside world, assessment of chains of consequences of alternative courses of action, appraisal of costs and benefits as a basis for action strategies, and continuous monitoring. This was the approach of the new British textbooks of systems planning, which started to emerge at the end of the 1960s, and which were particularly associated with a group of younger British graduates, many teaching or studying at the University of Manchester. It was also the approach of a whole generation of subregional studies, made for fast-growing metropolitan areas in

Britain during that heroic period of growth and change, 1965–75: Leicester–Leicestershire, Nottinghamshire–Derbyshire, Coventry–Warwickshire–Solihull, South Hampshire. All were heavily suffused with the new approach and the new techniques; in several, the same key individuals – McLoughlin in Leicester, Batty in Notts–Derby – played a directing or a crucial consulting role.

But the revolution was less complete – at least, in its early stages – than its supporters liked to argue: many of these "systems" plans had a distinctly blueprint tint, in that they soon resulted in all-too-concrete proposals for fixed investments like freeway systems. Underlying this, furthermore, were some curious metaphysical assumptions, which the new systems planners shared with their blueprint elders: the planning system was seen as active, the city system as purely passive; the political system was regarded as benign and receptive to the planner's expert advice. In practice, the systems planner was involved in two very different kinds of activity: as a social scientist, he or she was passively observing and analysing reality; as a designer, the same planner was acting on reality to change it – an activity inherently less certain, and also inherently subject to objectives that could only be set through a complex, often messy, set of dealings between professionals, politicians and public.

The core of this problem was a logical paradox: despite the claims of the systems planners, the urban planning system was different from (say) a weapons system. In this latter kind of system, to which the "systems approach" had originally and successfully been applied, the controls were inside the system; but here, the urban-regional system was inside its own system of control. Related to this were other crucial differences: in urban planning, there was not just one problem and one overriding objective, but many, perhaps contradictory; it was difficult to move from general goals to specific operational ones; not all were fully perceived; the systems to be analysed did not self-evidently exist, but had to be synthesized; most aspects were not deterministic, but probabilistic; costs and benefits were difficult to quantify. So the claims of the systems school to scientific objectivity could not readily be fulfilled. Increasingly, members of the school came to admit that in such "open" systems, systematic analysis would need to play a subsidiary role to intuition and judgement; in other words, the traditional approach. By 1975 Britton Harris, perhaps the most celebrated of all the systems planners, could write that he no longer believed that the more difficult problems of planning could be solved by optimizing methods.

The search for a new paradigm

All this, in the late 1960s, came to focus in an attack from two very different directions, which together blew the ship of systems planning at least half out of the water. From the philosophical right came a series of theoretical and empirical studies from American political scientists, arguing that – at least in the United States – crucial urban decisions were made within a pluralist political structure in which no one individual or group had total knowledge or power, and in which, consequently, the decision-making process could best be described as "disjointed incrementalism" or "muddling through". Meyerson and Banfield's classic analysis of the Chicago Housing Authority concluded that it engaged in little real planning, and failed because it did not correctly identify the real power structure in the city; its elitist view of the public interest was totally opposed to the populist view of the ward politicians, which finally prevailed. Downs theorized about such a structure, suggesting that politicians buy votes by offering bundles of policies, rather as in a market. Lindblom contrasted the whole rational-comprehensive model of planning with what he found to be the actual process of policy development, which was characterized by a mixture of values and analysis, a confusion of ends and means, a failure to analyse alternatives, and an avoidance of theory. Altshuler's analysis of Minneapolis–St Paul suggested that the professional planner carried no clout against the political machine, which backed the highway-building engineers against him; they won by stressing expertise and concentrating on narrow goals, but theirs was a political game; the conclusion was that planners should recognize their own weakness, and devise strategies appropriate to that fact.

All these analyses arose from study of American urban politics, which is traditionally more populist, more pluralist, than most. Even there, Rabinowitz's study of New Jersey cities suggested that they varied greatly in style, from the highly fragmented to the very cohesive; while Etzioni, criticizing Lindblom, suggested that recent United States history showed several important examples of non-incremental decision-making, especially in defence. But, these reservations taken, the studies did at least suggest that planning in

actuality was a very long way indeed from the cool, rational, Olympian style envisaged in the systems texts. Perhaps it might have been better if it had been closer; perhaps not. The worrisome point was that in practice, local democracy proved to be an infinitely messier business than the theory would have liked. Some theorists accordingly concluded that if this was the way planning was, this was the way it should be encouraged to be: partial, experimental, incremental, working on problems as they arose.

That emerged even more clearly, because – as so often seems to happen – in America the left-wing criticism was reaching closely similar conclusions. By the late 1960s, fuelled by the civil-rights movement and war on poverty, the protests against the Vietnam war and the campus free-speech movement, it was this wing that was making all the running. Underlying the general current of protest were three key themes, which proved fatal to the legitimacy of the systems planners. One was a widespread distrust of expert, top-down planning generally – whether for problems of peace and war, or for problems of the cities. Another, much more specific, was an increasing paranoia about the systems approach, which in its military applications was seen as employing pseudo-science and incomprehensible jargon to create a smokescreen, behind which ethically reprehensible policies could be pursued. And a third was triggered by the riots that tore through American cities starting with Birmingham, Alabama, in 1963 and ending with Detroit, in 1967. They seemed to prove the point: systems planning had done nothing to ameliorate the condition of the cities; rather, by assisting or at least conniving in the dismemberment of inner-city communities, it might actually have contributed to it. By 1967 one critic, Richard Bolan, could argue that systems planning was old-fashioned comprehensive planning, dressed up in fancier garb; both, alike, ignored political reality.

The immediate left-wing reaction was to call on the planners themselves to turn the tables, and to practise bottom-up planning by becoming advocate-planners. Particularly, in this way they would make explicit the debate about the setting of goals and objectives, which both the blueprint and systems approaches had bypassed by means of their comfortable shared assumption that this was the professional planner's job. Advocacy planners would intervene in a variety of ways, in a variety of groups; diversity should be their keynote. They would help to inform the public of alternatives; force public planning agencies to compete for support; help critics to generate plans that were superior to official ones; compel consideration of underlying values. The resulting structure was highly American: democratic, locally grounded, pluralistic, but also legalistic in being based on institutionalized conflict. But, interestingly, while demoting the planner in one respect, it enormously advanced his or her power in another: the planner was to take many of the functions that the locally elected official had previously exercised. And, in practice, it was not entirely clear how it would all work; particularly, how the process would resolve the very real conflicts of interest that could arise within communities, or how it could avoid the risk that the planners, once again, would become manipulators.

At any rate, there is more than a passing resemblance between the planner as a disjointed incrementalist, and the planner-advocate; and, indeed, between either of these and a third model set out in Bolan's paper of 1967, the planner as informal co-ordinator and catalyst, which in turn shades into a fourth: Melvin Webber's probabilistic planner, who uses new information systems to facilitate debate and improve decision-making. All are assumed to work within a pluralist world, with very many different competing groups and interests, where the planner has at most (and, further, should have) only limited power or influence; all are based, at least implicitly, on continued acceptance of logical positivism. As Webber put it, at the conclusion of his long two-part paper of 1968–9:

> The burden of my argument is that city planning failed to adopt the planning method, choosing instead to impose input bundles, including regulatory constraints, on the basis of ideologically defined images of goodness. I am urging, as an alternative, that planning tries out the planning idea and the planning method.

In turn, Webber's view of planning – which flatly denied the possibility of a stable predictable future or agreed goals – provided some of the philosophical underpinnings of the Social Learning or New Humanist approach of the 1970s, which stressed the importance of learning systems in helping cope with a turbulent environment. But finally, this approach divorced itself from logical positivism, returning to a reliance on personal knowledge which was strangely akin to old-style blueprint planning; and, as developed by John Friedmann of the University of California at Los Angeles, it finally resulted in a demand for all political activity to be decomposed into decision by minute political groups: a return to the anarchist roots of planning, with a vengeance.

So these different approaches diverged, sometimes in detailed emphasis, sometimes more fundamentally. What they shared was the belief that – at any rate in the American political system – the planner did not have much power and did not deserve to have much either; within a decade, from 1965 to 1975, these approaches together neatly stripped the planner of whatever priestly clothing, and consequent mystique, s/he may have possessed. Needless to say, this view powerfully communicated itself to the professionals themselves. Even in countries with more centralized, top-down political systems, such as Great Britain, young graduating planners increasingly saw their roles as rather like barefoot doctors, helping the poor down on the streets of the inner city, working either for a politically acceptable local authority, or, failing that, for community organizations battling against a politically objectionable one.

Several historical factors, in addition to the demolition job on planning by the American theorists, contributed to this change: planners and politicians belatedly discovered the continued deprivation of the inner-city poor; then, it was seen that the areas where these people lived were suffering depopulation and deindustrialization; in consequence, planners progressively moved away from the merely physical, and into the social and the economic. The change can be caricatured thus: in 1955, the typical newly graduated planner was at the drawing board, producing a diagram of desired land uses; in 1965, s/he was analysing computer output of traffic patterns; in 1975, the same person was talking late into the night with community groups, in the attempt to organize against hostile forces in the world outside.

It was a remarkable inversion of roles. For what was wholly or partly lost, in that decade, was the claim to any unique and useful expertise, such as was possessed by the doctor or the lawyer. True, the planner could still offer specialized knowledge on planning laws and procedures, or on how to achieve a particular design solution; though often, given the nature of the context and the changed character of planning education, s/he might not have enough of either of these skills to be particularly useful. And, some critics were beginning to argue, this was because planning had extended so thinly over so wide an area that it became almost meaningless; in the title of Aaron Wildavsky's celebrated paper, "If Planning is Everything, Maybe it's Nothing".

The fact was that planning, as an academic discipline, had theorized about its own role to such an extent that it was denying its own claim to legitimacy. Planning, Faludi pointed out in his text of 1973, could be merely *functional*, in that the goals and objectives are taken as given; or *normative*, in that they are themselves the object of rational choice. The problem was whether planning was really capable of doing that latter job. As a result, by the mid-1970s planning had reached the stage of a "paradigm crisis"; it had been theoretically useful to distinguish the planning process as something separate from what is planned, yet this had meant a neglect of substantive theory, pushing it to the periphery of the whole subject. Consequently, new theory is needed which attempts to bridge current planning strategies and the urban physical and social systems to which strategies are applied.

The Marxist ascendancy

That became ever clearer in the following decade, when the logical positivists retreated from the intellectual field of battle and the Marxists took possession. As the whole world knows, the 1970s saw a remarkable resurgence – indeed a veritable explosion – of Marxist studies. This could not fail to affect the closely related worlds of urban geography, sociology, economics and planning. True, like the early neo-classical economists, Marx had been remarkably uninterested in questions of spatial location – even though Engels had made illuminating comments on the spatial distribution of classes in mid-Victorian Manchester. The disciples now reverently sought to extract from the holy texts, drop by drop, a distillation that could be used to brew the missing theoretical potion. At last, by the mid-1970s, it was ready; then came a flood of new work. It originated in various places and in various disciplines: in England and the United States the geographers David Harvey and Doreen Massey helped to explain urban growth and change in terms of the circulation of capital; in Paris, Manuel Castells and Henri Lefebvre developed sociologically based theories. In the endless debates that followed among the Marxists themselves, a critical question concerned the role of the state. In France, Lokjine and others argued that it was mainly concerned, through such devices as macroeconomic planning and related infrastructure investment, directly to underpin and aid the direct productive investments of private capital. Castells, in contrast, argued that its main function had been to provide collective consumption – as in public housing, or schools, or transportation – to help guarantee the reproduction of the labour force and to dampen class conflict, essential for the maintenance of the system. Clearly, planning might play a very large role in both these state functions; hence, by the mid-1970s French Marxist urbanists were engaging in major studies of this role in the industrialization of such major industrial areas as Dieppe.

At the same time, a specifically Marxian view of planning emerged in the English-speaking world. To describe it adequately would require a course in Marxist theory. But, in inadequate summary, it states that the structure of the capitalist city itself, including its land-use and activity patterns, is the result of capital in pursuit of profit. Because capitalism is doomed to recurrent crises, which deepen in the current stage of late capitalism, capital calls upon the state, as its agent, to assist it by remedying disorganization in commodity production, and by aiding the reproduction of the labour force. It thus tries to achieve certain necessary objectives: to facilitate continued capital accumulation, by ensuring rational allocation of resources; by assisting the reproduction of the labour force through provision of social services, thus maintaining a delicate balance between labour and capital and preventing social disintegration; and by guaranteeing and legitimating capitalist social and property relations. As Dear and Scott put it: "In summary, planning is an historically-specific and socially-necessary response to the self-disorganizing tendencies of *privatized capitalist* social and property relations as these appear in urban space." In particular, it seeks to guarantee collective provision of necessary infrastructure and certain basic urban services, and to reduce negative externalities whereby certain activities of capital cause losses to other parts of the system.

But, since capitalism also wishes to circumscribe state planning as far as possible, there is an inbuilt contradiction: planning, because of this inherent inadequacy, always solves one problem only by creating another. Thus, say the Marxists, nineteenth-century clearances in Paris created a working-class housing problem; American zoning limited the powers of industrialists to locate at the most profitable locations.

And planning can never do more than modify some parameters of the land development process; it cannot change its intrinsic logic, and so cannot remove the contradiction between private accumulation and collective action. Further, the capitalist class is by no means homogenous; different fractions of capital may have divergent, even contradictory interests, and complex alliances may be formed in consequence; thus, latter-day Marxist explanations come close to being pluralist, albeit with a strong structural element. But in the process, "the more that the State intervenes in the urban system, the greater is the likelihood that different social groups and fractions will contest the legitimacy of its decisions. *Urban life as a whole becomes progressively invaded by political controversies and dilemmas.*"

Because traditional non-Marxian planning theory has ignored this essential basis of planning, so Marxian commentators argue, it is by definition vacuous: it seeks to define what planning ideally ought to be, devoid of all context; its function has been to depoliticize planning as an activity, and thus to legitimate it. It seeks to achieve this by representing itself as the force which produces the various facets of real-world planning. But in fact, its various claims – to develop abstract concepts that rationally represent real-world processes, to legitimate its own activity, to explain material processes as the outcome of ideas, to present planning goals as derived from generally shared values, and to abstract planning activity in terms of metaphors drawn from other fields like engineering – all these are both very large and quite unjustified. The reality, Marxists argue, is precisely the opposite: viewed objectively, planning theory is nothing other than a creation of the social forces that bring planning into existence.

It makes up a disturbing body of coherent criticism: yes, of course, planning cannot simply be an independent self-legitimating activity, as scientific inquiry may claim to be; yes, of course, it is a phenomenon that – like all phenomena – represents the circumstances of its time. As Scott and Roweis put it:

> . . . there is a definite mismatch between the world of current planning theory, on the one hand, and the real world of practical planning intervention on the other hand. The one is the quintessence of order and reason in relation to the other which is full of disorder and unreason. Conventional theorists then set about resolving this mismatch between theory

and reality by introducing the notion that planning theory is in any case not so much an attempt to explain the world as it is but as it ought to be. Planning theory then sets itself the task of rationalizing irrationalities, and seeks to materialize itself in social and historical reality (like Hegel's World Spirit) by bringing to bear upon the world a set of abstract, independent, and transcendent norms.

(Scott and Roweis, 1977)

It was powerful criticism. But it left in turn a glaringly open question, both for the unfortunate planner – whose legitimacy is now totally torn from him, like the epaulette from the shoulder of a disgraced officer – and, equally, for the Marxist critic: what, then, is planning theory about? Has it any normative or prescriptive content whatsoever? The answer, logically, would appear to be no. One of the critics, Philip Cooke, is uncompromising:

> The main criticism that tends to have been made, justifiably, of planning is that it has remained stubbornly normative . . . in this book it will be argued that [planning theorists] should identify mechanisms which cause changes in the nature of planning to be brought about, rather than assuming such changes to be either the creative idealizations of individual minds, or mere regularities in observable events.

(Cooke, 1983)

This is at least consistent: planning theory should avoid all prescription; it should stand right outside the planning process, and seek to analyse the subject – including traditional theory – for what it is, the reflection of historical forces. Scott and Roweis, a decade earlier, seem to be saying exactly the same thing: planning theory cannot be normative, it cannot assume "transcendent operational norms". But then, they stand their logic on its own head, saying that "a viable theory of urban planning should not only tell us what planning is, but also what we can, and must, do as progressive planners".

This, of course, is sheer rhetoric. But it nicely displays the agony of the dilemma. Either theory is about unravelling the historical logic of capitalism, or it is about prescription for action. Since the planner-theorist – however sophisticated – could never hope to divert the course of capitalist evolution by more than a millimetre or a millisecond, the logic would seem to demand that s/he sticks firmly to the first and abjures the second. In other words, the Marxian logic is strangely quietist; it suggests that the planner retreats from planning altogether into the academic ivory tower.

Some were acutely conscious of the dilemma. John Forester tried to resolve it by basing a whole theory of planning action on the work of Jürgen Habermas. Habermas, perhaps the leading German social theorist of the post-World War Two era, had argued that latter-day capitalism justified its own legitimacy by spinning around itself a complex set of distortions in communication, designed to obscure and prevent any rational understanding of its own workings. Thus, he argued, individuals became powerless to understand how and why they act, and so were excluded from all power to influence their own lives,

> as they are harangued, pacified, mislead [*sic*], and ultimately persuaded that inequality, poverty, and ill-health are either problems for which the victim is responsible or problems so "political" and "complex" that they can have nothing to say about them. Habermas argues that democratic politics or planning requires the consent that grows from processes of collective criticism, not from silence or a party line.
>
> (Forester, 1980)

But, Forester argues, Habermas's own proposals for communicative action provide a way for planners to improve their own practice:

> By recognizing planning practice as normatively role-structured communication action which distorts, covers up, or reveals to the public the prospects and possibilities they face, a critical theory of planning aids us practically and ethically as well. This is the contribution of critical theory to planning: pragmatics with vision – to reveal true alternatives, to correct false expectations, to counter cynicism, to foster inquiry, to spread political responsibility, engagement, and action. Critical planning practice, technically skilled and politically sensitive, is an organizing and democratizing practice.
>
> (Forester, 1980)

Fine. The problem is that – stripped of its Germanic philosophical basis, which is necessarily a huge over-simplification of a very dense analysis – the practical prescription all comes out as good old-fashioned democratic common sense, no more and no less than Davidoff's advocacy planning of fifteen years before: cultivate community networks, listen carefully to the people, involve the less-organized groups, educate the citizens in how to join in, supply information and make sure people know how to get it, develop skills in working with groups in conflict situations, emphasize the need to participate, compensate for external pressures. True, if in all this planners can sense that they have penetrated the mask of capitalism, that may help them to help others to act to change their environment and their lives; and, given the clear philosophical impasse of the late 1970s, such a massive metaphysical underpinning may be necessary.

The world outside the tower: practice retreats from theory

Meanwhile, if the theorists were retreating in one direction, the practitioners were certainly reciprocating. Whether baffled or bored by the increasingly scholastic character of the academic debate, they lapsed into an increasingly untheoretical, unreflective, pragmatic, even visceral style of planning. That was not entirely new: planning had come under a cloud before, as during the 1950s, and had soon reappeared in a clear blue sky. What was new, strange, and seemingly unique about the 1980s was the divorce between the Marxist theoreticians of academe – essentially academic spectators, taking grandstand seats at what they saw as one of capitalism's last games – and the anti-theoretical, anti-strategic, anti-intellectual style of the players on the field down below. The 1950s were never like that; then, the academics were the coaches, down there with the team.

The picture is of course exaggerated. Many academics did still try to teach real-life planning through simulation of real-world problems. The Royal Town Planning Institute enjoined them to become ever more practice-minded. The practitioners had not all shut their eyes and ears to what comes out of the academy; some even returned there for refresher courses. And if all this was true in Britain, it was even more so of America, where the divorce had never been so evident. Yet the picture does describe a clear and unmistakable trend; and it was likely to be more than a cyclical one.

The reason is simple: as professional education of any kind becomes more fully absorbed by the academy, as its teachers become more thoroughly socialized within it, as careers are seen to depend on academic peer judgements, then its norms and values – theoretical, intellectual, detached – will become ever more pervasive; and the gap between teaching and practice will progressively widen. One key illustration: of the huge output of books and papers from the planning schools in the 1980s, there were many – often, those most highly regarded within the academic community – that were simply irrelevant, even completely incomprehensible, to the average practitioner.

Perhaps, it might be argued, that was the practitioner's fault; perhaps too we need fundamental science, with no apparent payoff, if we are later to enjoy its technological applications. The difficulty with that argument was to find convincing evidence that – not merely here, but in the social sciences generally – such payoff eventually comes. Hence the low esteem into which the social sciences had everywhere fallen, not least in Britain and the United States: hence too the diminished level of support for them, which – at any rate in Britain – had directly redounded on the planning schools. The relationship between planning and the academy had gone sour, and that is the major unresolved question that must now be addressed.

REFERENCES

Abercrombie, P. and Jackson, H. (1948) *West Midlands Plan*. Interim Confidential Edition. 5 vols. London: Ministry of Town and Country Planning.

Batty, M. (1979) "On Planning Processes", in: B. Goodall and A. Kirby (eds) *Resources and Planning*. Oxford: Pergamon.

Chadwick, G. (1971) *A Systems View of Planning: Towards a Theory of the Urban and Regional Planning Process*. Oxford: Pergamon.

Cooke, P.N. (1983) *Theories of Planning and Spatial Development*. London: Hutchinson.

Forester, J. (1980) "Critical Theory and Planning Practice", *Journal of the American Planning Association*, 46, 275–86.

Kent, T.J. (1964) *The Urban General Plan*. San Francisco: Chandler.

Scott, A.J. and Roweis, S.T. (1977) "Urban Planning in Theory and Practice: An Appraisal", *Environment and Planning*, 9, 1097–1119.

Webber, M.M. (1968–9) "Planning in an Environment of Change", *Town Planning Review*, 39, 179–95, 277–95.

S I X

"Twentieth Century Land Use Planning: A Stalwart Family Tree"

Journal of the American Planning Association (1995)

Edward J. Kaiser and David R. Godschalk

Editors' Introduction

Much of what city planning departments do is *physical planning* related to land use, transportation, capital improvements, and infrastructure. In the following selection, Edward J. Kaiser and David R. Godschalk describe how concepts of what land use plans should be like have evolved over time and what land use planning practice in the United States is like today.

Kaiser and Godschalk liken the development of land use planning to a tree. From disparate roots in planning theory and practice, a sturdy trunk based on the 1950s vision of a general plan for cities' long-term physical development has grown over time, with periodic branches, to a rich foliage of "hybrid" contemporary plans which typically blend aspects of design, policy, and management.

Kaiser and Godschalk begin their account of twentieth-century land use planning by describing the elitist, architecturally based, and often unrealistic plans that Peter Hall noted were common prior to World War II (p. 342). These have evolved into contemporary hybrid urban land use plans developed through participatory processes and which incorporate elements of policy and management that make them far more realistic than plans developed during planning's comfortable, but ineffective "golden age."

Kaiser and Godschalk trace the roots of their tree to mid-century. By that time, they say, there was a consensus that city plans should be focused on long-term physical development – what is called a "master plan" or "general plan."

At mid-century, city general plans tended to be elitist, inspirational, long-range visions developed with little attention to implementation. In contrast, Kaiser and Godschalk argue that today, urban land use plans have become frameworks for community consensus on future growth supported by fiscally grounded actions to manage change. Plans are becoming more sensitive to the green planning issues raised by Timothy Beatley (p. 399) and sustainable urban development as discussed by Stephen Wheeler (p. 486).

Most modern city plans still contain maps showing projected long-range urban form for the city's land uses, transportation, community facilities, and other infrastructure. But today's plans do not consist of architecturally complete renderings of an entire static new town or section of a city like many pre-war plans. In addition to or in place of maps, the city plan's vision may be expressed in words and visual images.

At the beginning of the twenty-first century it appears that the "sturdy tree" of urban land use planning will continue to grow. Kaiser and Godschalk anticipate that the next generation of development plans will mature and adapt without abandoning the urban physical plan heritage. Edward J. Kaiser and David R. Godschalk are professors of City Planning at the University of North Carolina. They are co-authors (with Stuart Chapin) of *Urban Land Use Planning*, 4th edn (Chicago: University of Illinois Press, 1995), a leading American text on land use planning.

Other books on city planning in the United States include Eugenie Birch (ed.), *The Urban and Regional Planning Reader* (London and New York: Routledge, 2003), J. Barry Cullingworth and Roger Caves, *Planning in the USA: Policies, Issues, and Process*, 2nd edn (London: Routledge, 2003), John M. Levy, *Contemporary Urban Planning*, 6th edn (New York: Prentice-Hall, 2001), Barbara Becker, Eric Damian Kelley, and Frank So, *Community Planning: An Introduction to the Comprehensive Plan* (Washington, DC: Milldale Press, 2000), International City Management Association (ICMA), *The Practice of Local Government Planning* (Washington, DC: ICMA, 1999), Alexander Garvin, *The American City: What Works, What Doesn't* (New York: McGraw-Hill, 1995), and Jay Stein (ed.), *Classic Readings in Urban Planning* (New York: McGraw-Hill, 1995).

Leading books on European urban planning include Peter Hall, *Urban and Regional Planning*, 4th edn (London and New York: Routledge, 2002), Barry Cullingworth, J. Barry Cullingworth, and Vincent Nadin, *Town and Country Planning in the UK*, 13th edn (London and New York: Routledge, 2002) and Patsy Healey and Stephen Graham (eds), *Managing Cities* (London: Wiley, 1995). J. Barry Cullingworth has written additional books which provide an overview of city planning in Canada and Britain.

■ ■ ■ ■ ■ ■

How a city's land is used defines its character, its potential for development, the role it can play within a regional economy and how it impacts the natural environment.
(Seattle Planning Commission, 1993)

During the twentieth century, community physical development plans have evolved from elite, City Beautiful designs to participatory, broad-based strategies for managing urban change. A review of land use planning's intellectual and practice history shows the continuous incorporation of new ideas and techniques. The traditional mapped land use design has been enriched with innovations from policy plans, land classification plans, and development management plans. Thanks to this flexible adaptation, local governments can use contemporary land use planning to build consensus and support decisions on controversial issues about space, development, and infrastructure. If this evolution persists, local plans should continue to be mainstays of community development policy into the twenty-first century.

Unlike the more rigid, rule-oriented modern architecture, contemporary local planning does not appear destined for deconstruction by a postmodern revolution. Though critics of comprehensive physical planning have regularly predicted its demise, the evidence demonstrates that spatial planning is alive and well in hundreds of United States communities. A 1994 tabulation found 2,742 local comprehensive plans prepared under state growth management regulations in twelve states. (See table 1.) This figure of course significantly understates the overall nationwide total, which would include all those plans prepared in the other thirty-eight states and in the noncoastal areas of California and North Carolina. It is safe to assume that most, if not all, of these plans contain a mapped land use element. Not only do such plans help decision-makers to manage urban growth and change, they also provide a platform for the formation of community consensus about land use issues, now among the most controversial items on local government agendas.

This article looks back at the history of land use planning and forward to its future. It shows how planning ideas, growing from turn-of-the-century roots, culminated in a midcentury consensus on a general concept – the traditional land use design plan. That consensus was stretched as planning branched out to deal with public participation, environmental protection, growth management, fiscal responsibility, and effective implementation under turbulent conditions. To meet these new challenges, new types of plans arose: verbal policy plans, land classification plans, and growth management plans. These in turn became integrated into today's hybrid comprehensive plans, broadening and strengthening the traditional approach.

Future land use planning will continue to evolve in certain foreseeable directions, as well as in ways unforeseen. Among the foreseeable developments are even more active participation by interest groups, calling for planners' skills at consensus building and managing conflict; increased use of computers and electronic media, calling for planners' skills in information management and communication; and continuing concerns over issues of diversity, sustainability, and quality of life, calling for planners' ability to analyze and seek creative solutions to complex and interdependent problems.

State	Number of comprehensive plans				Source
	Cities/towns	Counties	Regions	Total	
California (coastal)	97	7	0	104	Coastal Commission
Florida	377	49	0	426	Department of Community Affairs
Georgia	298	94	0	392	Department of Community Affairs
Maine	270	0	0	270	Dept. of Economic and Community Development
Maryland	1	1	0	2	Planning Office
New Jersey	567	0	0	567	Community Affairs Department
North Carolina (coastal)	70	20	0	90	Division of Coastal Management
Oregon	241	36	1	278	Department of Local Community Development
Rhode Island	39	0	1	40	Department of Planning and Development
Vermont	235	0	10	245	Department of Housing and Community Affairs
Virginia	211	94	0	305	Department of Housing and Community Affairs
Washington	23	9	9	23	Office of Growth Management
Total	**2429**	**301**	**12**	**2742**	

Table 1 Local comprehensive plans in growth-managing states and coastal areas as of 1994
Source: Compiled from telephone survey of state sources

THE LAND USE PLANNING FAMILY TREE

We liken the evolution of the physical development plan to a family tree. The early genealogy is represented as the roots of the tree (figure 1). The general plan, constituting consensus practice at midcentury, is represented by the main trunk. Since the 1970s this traditional "land use design plan" has been joined by several branches – the verbal policy plan, the land classification plan, and the development management plan. These branches connect to the trunk although springing from different planning disciplines, in a way reminiscent of the complex structure of a Ficus tree. The branches combine into the contemporary, hybrid comprehensive plan integrating design, policy, classification, and management, represented by the foliage at the top of the tree.

As we discuss each of these parts of the family tree, we show how plans respond both to social climate changes and to "idea genes" from the literature. We also draw conclusions about the survival of the tree and the prospects for new branches in the future.

The focus of the article is the plan prepared by a local government – a county, municipality, or urban region – for the long-term development and use of the land.

ROOTS OF THE FAMILY TREE: THE FIRST 50 YEARS

New World city plans certainly existed before this century. They included L'Enfant's plan for Washington, William Penn's plan for Philadelphia, and General Oglethorpe's plan for Savannah. These plans, however, were blueprints for undeveloped sites, commissioned by unitary authorities with power to implement them unilaterally.

In this century, perhaps the most influential early city plan was Daniel Burnham's plan for Chicago, published by the Commercial Club of Chicago (a civic, not a government entity) in 1909. The archetypical plan-as-inspirational-vision, it focuses only on design of public spaces as a City Beautiful effort.

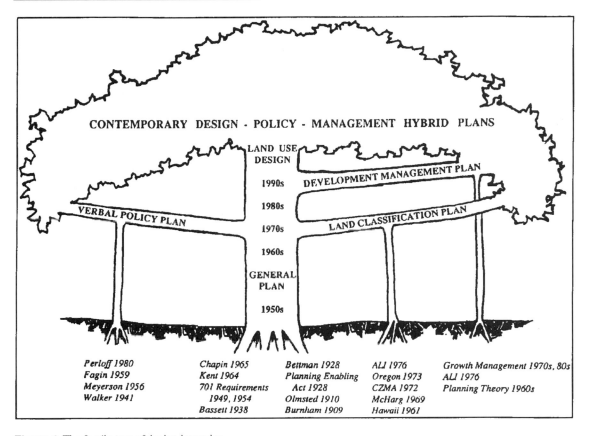

CONTEMPORARY DESIGN · POLICY · MANAGEMENT HYBRID PLANS

LAND USE DESIGN

DEVELOPMENT MANAGEMENT PLAN

VERBAL POLICY PLAN

LAND CLASSIFICATION PLAN

GENERAL PLAN

1990s
1980s
1970s
1960s
1950s

Perloff 1980	Chapin 1965	Bettman 1928	ALI 1976	Growth Management 1970s, 80s
Fagin 1959	Kent 1964	Planning Enabling	Oregon 1973	ALI 1976
Meyerson 1956	701 Requirements	Act 1928	CZMA 1972	Planning Theory 1960s
Walker 1941	1949, 1954	Olmsted 1910	McHarg 1969	
	Bassett 1938	Burnham 1909	Hawaii 1961	

Figure 1 The family tree of the land use plan

The City Beautiful approach was soon broadened to a more comprehensive view. At the 1911 National Conference on City Planning, Frederick Law Olmsted, Jr., son of the famous landscape architect and in his own right one of the fathers of planning, defined a city plan as encompassing all uses of land, private property, public sites, and transportation. Alfred Bettman, speaking at the 1928 National Conference of City Planning, envisioned the plan as a master design for the physical development of the city's territory, including "the general location and extent of new public improvements . . . and in the case of private developments, the general distribution amongst various classes of land uses, such as residential, business, and industrial uses . . . designed for . . . the future, twenty-five to fifty years" (Black 1968, 352–3). Together, Olmsted and Bettman anticipated the development of the midcentury land use plan.

Another early influence, the federal Standard City Planning Enabling Act of 1928, shaped enabling acts passed by many states. However, the Act left many planners and public officials confused about the difference between a master plan and a zoning ordinance, so that hundreds of communities adopted "zoning plans" without having created comprehensive plans as the basis for zoning (Black 1968, 353). Because the Act also did not make clear the importance of comprehensiveness or define the essential elements of physical development, no consensus about the essential content of the plan existed.

Ten years later, Edward Bassett's book, *The Master Plan* (1938), spelled out the plan's subject matter and format – supplementing the 1928 Act, and consistent with it. He argued that the plan should have seven elements, all relating to land areas (not buildings) and capable of being shown on a map: streets, parks, sites for public buildings, public reservations, routes for public utilities, pier-head and bulkhead lines (all public facilities), and zoning districts for private lands. Bassett's views were incorporated in many state enabling laws.

The physical plans of the first half of the century were drawn by and for independent commissions, reflecting the profession's roots in the Progressive

Reform movement, with its distrust of politics. The 1928 Act reinforced that perspective by making the planning commission, not the legislative body, the principal client of the plan, and purposely isolating the commission from politics. Bassett's book reinforced the reliance on an independent commission. He conceived of the plan as a "plastic" map, kept within the purview of the planning commission, capable of quick and easy change. The commission, not the plan, was intended to be the adviser to the local legislative body and to city departments.

By the 1940s, both the separation of the planning function from city government and the plan's focus on physical development were being challenged. Robert Walker, in *The Planning Function in Local Government*, argued that the "scope of city planning is properly as broad as the scope of city government." The central planning agency might not necessarily do all the planning, but it would coordinate departmental planning in the light of general policy considerations – creating a comprehensive plan but one without a physical focus. That idea was not widely accepted. Walker also argued that the independent planning commission should be replaced by a department or bureau attached to the office of mayor or city manager. That argument did take hold, and by the 1960s planning in most communities was the responsibility of an agency within local government, though planning boards still advised elected officials on planning matters.

This evolution of ideas over 50 years resulted at midcentury in a consensus concept of a plan as focused on long-term physical development; this focus was a legacy of the physical design professions. Planning staff worked both for the local government executive officer and with an appointed citizen planning board, an arrangement that was a legacy of the Progressive insistence on the public interest as an antidote to governmental corruption. The plan addressed both public and private uses of the land, but did not deal in detail with implementation.

THE PLAN AFTER MIDCENTURY: NEW GROWTH INFLUENCES

Local development planning grew rapidly in the 1950s, for several reasons. First, governments had to contend with the postwar surge of population and urban growth, as well as a need for the capital investment in infrastructure and community facilities that had been postponed during the depression and war years. Second, municipal legislators and managers became more interested in planning as it shifted from being the responsibility of an independent commission to being a function within local government. Third, and very important, Section 701 of the Housing Act of 1954 required local governments to adopt a long-range general plan in order to qualify for federal grants for urban renewal, housing, and other programs, and it also made money available for such comprehensive planning. The 701 program's double-barreled combination of requirements and financial support led to more urban planning in the United States in the latter half of the 1950s than at any previous time in history.

At the same time, the plan concept was pruned and shaped by two planning educators. T. J. Kent, Jr. was a professor at the University of California at Berkeley, a planning commissioner, and a city councilman in the 1950s. His book, *The Urban General Plan* (1964), clarified the policy role of the plan. F. Stuart Chapin, Jr. was a TVA planner and planning director in Greensboro, NC in the 1940s, before joining the planning faculty at the University of North Carolina at Chapel Hill in 1949. His contribution was to codify the methodology of land use planning in the various editions of his book, *Urban Land Use Planning* (1957, 1965).

What should the plan look like? What should it be about? What is its purpose (besides the cynical purpose of qualifying for federal grants)? The 701 program, Kent, and Chapin all offered answers.

The "701" program comprehensive plan guidelines

In order to qualify for federal urban renewal aid and, later, for other grants – a local government had to prepare a general plan that consisted of plans for physical development, programs for redevelopment, and administrative and regulatory measures for controlling and guiding development. The 701 program specified what the content of a comprehensive development plan should include:

- A land use plan, indicating the locations and amounts of land to be used for residential, commercial, industrial, transportation, and public purposes
- A plan for circulation facilities

- A plan for public utilities
- A plan for community facilities

T.J. Kent's urban general plan

Kent's view of the plan's focus was similar to that of the 701 guidelines: long-range physical development in terms of land use, circulation, and community facilities. In addition, the plan might include sections on civic design and utilities, and special areas, such as historic preservation or redevelopment areas. It covered the entire geographical jurisdiction of the community, and was in that sense comprehensive. The plan was a vision of the future, but not a blueprint; a policy statement, but not a program of action; a formulation of goals, but not schedules, priorities, or cost estimates. It was to be inspirational, uninhibited by short-term practical considerations.

Kent (1964, 65–89) believed the plan should emphasize policy, serving the following functions:

- Policy determination – to provide a process by which a community would debate and decide on its policy
- Policy communication – to inform those concerned with development (officials, developers, citizens, the courts, and others) and educate them about future possibilities
- Policy effectuation – to serve as a general reference for officials deciding on specific projects
- Conveyance of advice – to furnish legislators with the counsel of their advisors in a coherent, unified form

The format of Kent's proposed plan included a unified, comprehensive, but general physical design for the future, covering the whole community and represented by maps. (See figure 2.) It also contained goals and policies (generalized guides to conduct, and the most important ingredients of the plan), as well as summaries of background conditions, trends, issues, problems, and assumptions. (See figure 3.) So that the plan would be suitable for public debate, it was to be a complete, comprehensible document, containing factual data, assumptions, statements of issues, and goals, rather than merely conclusions and recommendations. The plan belonged to the legislative body and was intended to be consulted in decision-making during council meetings.

Kent recommended overall goals for the plan:

- Improve the physical environment of the community to make it more functional, beautiful, decent, healthful, interesting, and efficient
- Promote the overall public interest, rather than the interests of individuals or special groups within the community
- Effect political and technical coordination in community development
- Inject long-range considerations into the determination of short-range actions
- Bring professional and technical knowledge to bear on the making of political decisions about the physical development of the community

F. Stuart Chapin, Jr.'s urban land use plan

Chapin's ideas, though focusing more narrowly on the land use plan, were consistent with Kent's in both the 1957 and 1965 editions of *Urban Land Use Planning*, a widely used text and reference work for planners. Chapin's concept of the plan was of a generalized, but scaled, design for the future use of land, covering private land uses and public facilities, including the thoroughfare network.

Chapin conceived of the land use plan as the first step in preparing a general or comprehensive plan. Upon its completion, the land use plan served as a temporary general guide for decisions, until the comprehensive plan was developed. Later, the land use plan would become a cornerstone in the comprehensive plan, which also included plans for transportation, utilities, community facilities, and renewal, only the general rudiments of which are suggested in the land use plan. Purposes of the plan were to guide government decisions on public facilities, zoning, subdivision control, and urban renewal, and to inform private developers about the proposed future pattern of urban development.

The format of Chapin's land use plan included a statement of objectives, a description of existing conditions and future needs for space and services, and finally the mapped proposal for the future development of the community, together with a program for implementing the plan (customarily including zoning, subdivision control, a housing code, a public works expenditure program, an urban renewal program, and other regulations and development measures).

Figure 2　Example of the land use design map featured in the 1950s General Plan
Source: Kent 1991, 111

THE URBAN GENERAL PLAN

INTRODUCTION: Reasons for G.P.; roles of council, CPC, citizens; historical background and context of G.P.

SUMMARY OF G.P.: Unified statement including (a) basic policies, (b) major proposals, and (c) one schematic drawing of the physical design.

BASIC POLICIES

1. CONTEXT OF THE G.P.:
Historical background; geographical and physical factors; social and economic factors; major issues, problems, and opportunities. } facts trends assumptions forecasts

2. SOCIAL OBJECTIVES AND URBAN PHYSICAL-STRUCTURE CONCEPTS: Value judgments concerning social objectives; professional judgments concerning major physical-structure concepts adopted as basis for G.P.

3. BASIC POLICIES OF THE G.P.: Discussion of the basic policies that the general physical design is intended to implement.

GENERAL PHYSICAL DESIGN

Description of plan proposals in relation to large-scale G.P. drawing and citywide drawings of:

1. Working-and-living-areas section.
2. Community-facilities section.
3. Civic-design section.
4. Circulation section.
5. Utilities section.

} These drawings must remain general. They are needed because single G.P. drawing is too complex to enable each element to be clearly seen.

(Plus regional, functional, and district drawings that are needed to explain G.P.)

This diagram also suggests the contents of the official G.P. and publication as a single document.

Figure 3 Components of the 1950s–1960s General Plan
Source: Kent 1964, 93

The typical general plan of the 1950s and 1960s

Influenced by the 701 program, Kent's policy vision, and Chapin's methods, the plans of the 1950s and 1960s were based on a clear and straightforward concept: The plans' purposes were to determine, communicate, and effectuate comprehensive policy for the private and public physical development and redevelopment of the city. The subject matter was long-range physical development, including private uses of the land, circulation, and community facilities. The standard format included a summary of existing and emerging conditions and needs; general goals; and a long-range urban form in map format, accompanied by consistent development policies. The coverage was comprehensive, in the sense of addressing both public and private development and covering the entire planning jurisdiction, but quite general. The tone was

typically neither as "inspirational" as the Burnham plan for Chicago, nor as action-oriented as today's plans. Such was the well-defined trunk of the family tree in the 1950s and 1960s, in which today's contemporary plans have much of their origin.

CONTEMPORARY PLANS: INCORPORATING NEW BRANCHES

Planning concepts and practice have continued to evolve since midcentury, maturing in the process. By the 1970s, a number of new ideas had taken root. Referring back to the family tree in figure 1, we can see a trunk and several distinct branches:

■ *The land use design*, a detailed mapping of future land use arrangements, is the most direct descendant of the 1950s plan. It still constitutes the trunk of the tree. However, today's version is more likely to be accompanied by action strategies, also mapped, and to include extensive policies.

- *The land classification plan*, a more general map of growth policy areas rather than a detailed land use pattern, is now also common, particularly for counties, metropolitan areas, and regions that want to encourage urban growth in designated development areas and to discourage it in conservation or rural areas. The roots of the land classification plan include McHarg's *Design With Nature* (1969), the 1976 American Law Institute (ALI) Model Land Development Code, the 1972 Coastal Zone Management Act, and the 1973 Oregon Land Use Law.
- *The verbal policy plan* de-emphasizes mapped policy or end-state visions and focuses on verbal action policy statements, usually quite detailed; sometimes called a strategic plan, it is rooted in Meyerson's middle-range bridge to comprehensive planning, Fagin's policies plan, and Perloff's strategies and policies general plan.
- *The development management plan* lays out a specific program of actions to guide development, such as a public investment program, a development code, and a program to extend infrastructure and services; and it assumes public sector initiative for influencing the location, type, and pace of growth. The roots of the development management plan are in the environmental movement, and the movements for state growth management and community growth control, as well as in ideas from Fagin (1959) and the ALI Code.

We looked for, but could not find, examples of land use plans that could be termed purely prototypical "strategic plans," in the sense of Bryson and Einsweller. Hence, rather than identifying strategic planning as a separate branch on the family tree of the land use plan, we see the influence of strategic planning showing up across a range of contemporary plans. We tend to agree with the planners surveyed by Kaufman and Jacobs that strategic planning differs from good comprehensive planning more in emphasis (shorter range, more realistically targeted, more market-oriented) than in kind.

The land use design plan

The land use design plan is the most traditional of the four prototypes of contemporary plans and is the most direct descendent of the Kent–Chapin–701 plans of the 1950s and 1960s. It proposes a long-range future urban form as a pattern of retail, office, industrial, residential, and open spaces, and public land uses and a circulation system. Today's version, however, incorporates environmental processes, and sometimes agriculture and forestry, under the "open space" category of land use. Its land uses often include a "mixed use" category, honoring the neotraditional principle of closer mingling of residential, employment, and shopping areas. In addition, it may include a development strategy map, which is designed to bring about the future urban form and to link strategy to the community's financial capacity to provide infrastructure and services. The plans and strategies are often organized around strategic themes or around issues about growth, environment, economic development, transportation, or neighborhood/community scale change.

Like the other types of plans in vogue today, the land use design plan reflects recent societal issues, particularly the environmental crisis, the infrastructure crisis, and stresses on local government finance. Contemporary planners no longer view environmental factors as development constraints, but as valuable resources and processes to be conserved. They also may question assumptions about the desirability and inevitability of urban population and economic growth, particularly as such assumptions stimulate demand for expensive new roads, sewers, and schools. While at midcentury plans unquestioningly accommodated growth, today's plans cast the amount, pace, location, and costs of growth as policy choices to be determined in the planning process.

The 1990 Howard County (Maryland) General Plan, winner of an American Planning Association (APA) award in 1991 for outstanding comprehensive planning, exemplifies contemporary land use design. (See figure 4.) While clearly a direct descendent of the traditional general plan, the Howard County plan adds new types of goals, policies, and planning techniques. To enhance communication and public understanding, it is organized strategically around six themes/chapters (responsible regionalism, preservation of the rural area, balanced growth, working with nature, community enhancement, and phased growth), instead of the

Figure 4 Howard County, Maryland, General Plan, Land Use 2010
Source: Adapted from Howard County 1990

customary plan elements. Along with the traditional land use design, the plan includes a "policy map" (strategy map) for each theme and an overall policies map for the years 2000 and 2010. A planned service area boundary is used to contain urban growth within the eastern urbanized part of the county, home to the well-known Columbia New Town. The plan lays out specific next steps to be implemented over the next two years, and defines yardsticks for measuring success. An extensive public participation process for formulating the plan involved a 32-member General Plan Task Force, public opinion polling to discover citizen concerns, circulation of preplan issue papers on development impacts, and consideration of six alternative development scenarios.

The land classification plan

Land classification, or development priorities mapping, is a proactive effort by government to specify where and under what conditions growth will occur. Often, it also regulates the pace or timing of growth. Land classification addresses environmental protection by designating "nondevelopment" areas in especially vulnerable locations. Like the land use design, the land classification plan is spatially specific and map-oriented. However, it is less specific about the pattern of land uses within areas specified for development, which results in a kind of silhouette of urban form. On the other hand, land classification is more specific about development strategy, including timing. Counties, metropolitan areas, and regional planning agencies are more likely than cities to use a land classification plan.

The land classification plan identifies areas where development will be encouraged (called urban, transition, or development areas) and areas where development will be discouraged (open space, rural, conservation, or critical environmental areas). For each designated area, policies about the type, timing, and density of allowable development, extension of infrastructure, and development incentives or constraints apply. The planning principle is to concentrate financial resources, utilities, and services within a limited, prespecified area suitable for development, and to relieve pressure on nondevelopment areas by withholding facilities that accommodate growth.

Ian McHarg's (1969) approach to land planning is an early example of the land classification concept. He divides planning regions into three categories: natural use, production, and urban. Natural use areas, those with valuable ecological functions, have the highest priority. Production areas, which include agriculture, forestry, and fishing uses, are next in priority. Urban areas have the lowest priority and are designated after allocating the land suitable to the two higher-priority uses. McHarg's approach in particular, and land classification generally, also reflect the emerging environmental consciousness of the 1960s and 1970s.

As early as 1961, Hawaii had incorporated the land classification approach into its state growth management system. The development framework plan of the Metropolitan Council of the Twin Cities Area defined "planning tiers," each intended for a different type and intensity of development. The concepts of the "urban service area," first used in 1958 in Lexington, Kentucky, and the "urban growth boundary," used throughout Oregon under its 1973 statewide planning act, classify land according to growth management policy. Typically, the size of an urban growth area is based on the amount of land necessary to accommodate development over a period of ten or twenty years.

Vision 2005: A Comprehensive Plan for Forsyth County, North Carolina exemplifies the contemporary approach to land classification plans. The plan, which won honorable mention from APA in 1989, employs a six category system of districts, plus a category for activity centers. It identifies both short- and long-range growth areas (4A and 4B in figure 5). Policies applicable to each district are detailed in the plan.

The verbal policy plan: shedding the maps

The verbal policy plan focuses on written statements of goals and policy, without mapping specific land use patterns or implementation strategy. Sometimes called a policy framework plan, a verbal policy plan is more easily prepared and flexible than other types of plans, particularly for incorporating nonphysical development policy. Some claim that such a plan helps the planner to avoid relying too heavily on maps, which

are difficult to keep up to date with the community's changes in policy. The verbal policy plan also avoids falsely representing general policy as applying to specific parcels of property. The skeptics, however, claim that verbal statements in the absence of maps provide too little spatial specificity to guide implementation decisions.

The verbal policy plan may be used at any level of government, but is especially common at the state level, whose scale is unsuited to land use maps. The plan usually contains goals, facts and projections, and general policies corresponding to its purposes – to understand current and emerging conditions and issues, to identify goals to be pursued and issues to be addressed, and to formulate general principles of action. Sometimes communities do a verbal policy plan as an interim plan or a first step in the planning process. Thus, verbal policies are included in most land use design plans, land classification plans, and development management plans.

The Calvert County, MD, Comprehensive Plan (Calvert County 1983), winner of a 1985 APA award, exemplifies the verbal policy plan. Its policies are concise, easy to grasp, and grouped in sections corresponding to the six divisions of county government responsible for implementation. It remains a policy plan, however, because it does not specify a program of specific actions for development management. Though the plan clearly addresses physical development and discusses specific spatial areas, it contains no land use map. (See figure 6 for an illustrative page from the Calvert County plan.)

The development management plan

The development management plan features a coordinated program of actions, supported by analyses and goals, for specific agencies of local government to undertake over a three-to-ten-year period. The program of actions usually specifies the content, geographic coverage, timing, assignment of responsibility, and coordination among the parts. Ideally, the plan includes most or all of the following components:

- Description of existing and emerging development conditions, with particular attention to development processes, the political-institutional context, and a critical review of the existing systems of development management

Figure 5 Example of a land classification plan
Source: Adapted from Forsyth County City–County Planning Board 1988

- Statement of goals and/or legislative intent, including management-oriented goals
- Program of actions – the heart of the plan – including:
  1 Outline of a proposed development code, with: (a) procedures for reviewing development permits; (b) standards for the type of development, density, allowable impacts and/or performance standards; (c) site plan, site engineering, and construction practice requirements; (d) exactions and impact fee provisions and other incentives/disincentives; and (e) delineation of districts where various development standards, procedures, exactions, fees, and incentives apply

  2 Program for the expansion of urban infrastructure and community facilities and their service areas
  3 Capital improvement program
  4 Property acquisition program
  5 Other components, depending on the community situation, for example, a preferential taxation program, an urban revitalization program for specific built-up neighborhoods, or a historic preservation program
- Official maps, indicating legislative intent, which may be incorporated into ordinances, with force of law – among them, goal-form maps (e.g., land classification plan or land use design); maps

---

INDUSTRIAL DISTRICTS

Industrial Districts are intended to provide areas in the county which are suitable for the needs of industry. They should be located and designed to be compatible with the surrounding land uses, either due to existing natural features or through the application of standards.

> **RECOMMENDATIONS:**
>
> 1. Identify general locations for potential industrial uses.
> 2. Permit retail sales as an accessory use in the Industrial District.

SINGLE-FAMILY RESIDENTIAL DISTRICTS

Single-Family Residential Districts are to be developed and promoted as neighborhoods free from any land usage which might adversely affect them.

> **RECOMMENDATIONS:**
>
> 1. For new development, require buffering for controlling visual, noise, and activity impacts between residential and commercial uses.
> 2. Encourage single-family residential development to locate in the designated towns.
> 3. Allow duplexes, triplexes, and fourplexes as a conditional use in the "R-1" Residential Zone so long as the design is compatible with the single-family residential development.
> 4. Allow home occupations (professions and services, but not retail sales) by permitting the employment of one full-time equivalent individual not residing on the premises.

MULTIFAMILY RESIDENTIAL DISTRICTS

Multifamily Residential Districts provide for townhouses and multifamily apartment units. Areas designated in this category are those which are currently served or scheduled to be served by community or multi-use sewerage and water supply systems.

> **RECOMMENDATIONS:**
>
> 1. Permit multifamily development in the Solomons, Prince Frederick, and Twin Beach Towns.
> 2. Require multifamily projects to provide adequate recreational facilities— equipment, structures, and play surfaces.
> 3. Evaluate the feasibility of increasing the dwelling unit density permitted in the multifamily Residential Zone (R-2).

Figure 6 An excerpt from a verbal policy plan
Source: Calvert County, Maryland 1983

---

of zoning districts, overlay districts, and other special areas for which development types, densities, and other requirements vary; maps of urban services areas; maps showing scheduled capital improvements; or other maps related to development management standards and procedures

The development management plan is a distinct type, emphasizing a specific course of action, nor general policy. At its extreme the management plan actually incorporates implementation measures, so that the plan becomes part of a regulative ordinance.

Although the spatial specifications for regulations and other implementation measures are included, a land use map may not be.

One point of origin for development management plans is Henry Fagin's concept of the "policies plan," whose purpose was to coordinate the actions of line departments and provide a basis for evaluating their results, as well as to formulate, communicate, and implement policy (the traditional purpose). Such a plan's subject matter was as broad as the responsibilities of the local government, including but not limited to physical development. The format included a "state of the community" message, a physical plan, a financial plan, implementation measures, and detailed sections for each department of the government.

A more recent point of origin is *A Model Land Development Code* (American Law Institute 1976), intended to replace the 1928 Model Planning Enabling Act as a model for local planning and development management. The model plan consciously retains an emphasis on physical development (unlike Fagin's broader concept), but stresses a short-term program of action, rather than a long-term, mapped goal form. The ALI model plan contains a statement of conditions and problems; objectives, policies, and standards; and a short-term (from one to five years) program of specific public actions. It may also include land acquisition requirements, displacement impacts, development regulations, program costs and fund sources, and environmental, social, and economic consequences. More than other plan types, the development management plan is a "course of action" initiated by government to control the location and timing of development.

The Sanibel, Florida, Comprehensive Land Use Plan (1981) exemplifies the development management plan. The plan outlines the standards and procedures of regulations (i.e., the means of implementation), as well as the analyses, goals, and statements of intent normally presented in a plan. Thus, when the local legislature adopts the plan, it also adopts an ordinance for its implementation. Plan and implementation are merged into one instrument, as can be seen in the content of its articles:

Article 1: Preamble: including purposes and objectives, assumptions, coordination with surrounding areas, and implementation

Article 2: Elements of the Plan: Safety, Human Support Systems, Protection of Natural Environmental, Economic and Scenic Resources, Intergovernmental Coordination, and Land Use

Article 3: Development Regulations: Definitions, Maps, Requirements, Permitted Uses, Subdivisions, Mobile Home and Recreation Vehicles, Flood and Storm Proofing, Site Preparation, and Environmental Performance Standards

Article 4: Administrative Regulations (i.e., procedures): Standards, Short Form Permits, Development Permits, Completion Permits, Amendments to the Plan, and Notice, Hearing and Decision Procedures on Amendments

Figure 7 shows the Sanibel plan's map of permitted uses, which is more like a zoning plan than a land use design plan, because it shows where regulations apply, and boundaries are exact.

THE CONTEMPORARY HYBRID PLAN: INTEGRATING DESIGN, POLICY, AND MANAGEMENT

The rationality of practice has integrated the useful parts of each of the separate prototypes reviewed here into contemporary hybrid plans that not only map and classify land use in both specific and general ways, but also propose policies and management measures. For example, Gresham, Oregon, combined land use design (specifying residential, commercial, and industrial areas, and community facilities and public lands) with an overlay of land classification districts (developed, developing, rural, and conservation), and also included standards and procedures for issuing development permits (i.e., a development code). Prepared with considerable participation by citizens and interest groups, such plans usually reflect animated political debates about the costs and benefits of land use alternatives.

The states that manage growth have created new land use governance systems whose influence has broadened the conceptual arsenals of local planners. DeGrove identifies the common elements of these systems:

■ Consistency – intergovernmentally and internally (i.e., between plan and regulations)
■ Concurrency – between infrastructure and new development
■ Compactness – of new growth, to limit urban sprawl affordability – of new housing
■ Economic development, or "managing to grow"
■ Sustainability – of natural systems

DeGrove attributes the changes in planning under growth management systems to new hard-nosed concerns for measurable implementation and realistic funding mechanisms. For example, Florida local governments must adopt detailed capital improvement programs as part of their comprehensive plans, and substantial state grants may be withheld if their plans do not meet consistency and concurrency requirements.

Figure 7 Map of permitted uses, Sanibel

Source: City of Sanibel 1981

Another important influence on contemporary plans is the renewed attention to community design. The neotraditional and transit-oriented design movements have inspired a number of proposals for mixed-use villages in land use plans. *Toward a Sustainable Seattle: A Plan for Managing Growth* (1994) exemplifies a city approach to the contemporary hybrid plan. Submitted as the Mayor's recommended comprehensive plan, it attempted to muster political support for its proposals. Three core values – social equity, environmental stewardship, and economic security and opportunity – underlie the plan's overall goal of sustainability. This goal is to be achieved by integrating plans for land use and transportation, healthy and affordable housing, and careful capital investment in a civic compact based on a shared vision. Citywide population and job growth targets, midway between growth completely by regional sprawl and growth completely by infill, are set forth within a 20-year time frame. The plan is designed to meet the requirements of the Washington State Growth Management Act.

The land use element designates urban center villages, hub urban villages, residential urban villages, neighborhood villages, and manufacturing/industrial centers, each with specific design guidelines (figure 8). The city's capacity for growth is identified, and then allocated according to the urban village strategy. Future development is directed to mixed-use neighborhoods, some of which are already established; existing single-family areas are protected. Growth is shaped to build community, promote pedestrian and transit use, protect natural amenities and existing residential and employment areas, and ensure diversity of people and activities. Detailed land use policies carry out the plan.

Loudoun County Choices and Changes: General Plan (1991), which won APA's 1994 award for comprehensive planning in small jurisdictions, exemplifies a county approach to the contemporary hybrid plan. Its goals are grouped into three categories:

1 Natural and cultural resources goals seek to protect fragile resources by limiting development or mitigating disturbances, while at the same time not unduly diminishing land values.
2 Growth management: goals seek to accommodate and manage the county's fair share of regional growth, guiding development into the urbanized eastern part of the county or existing western towns and their urban growth areas, and conserving

agriculture and open space areas in the west. (See figure 9.)
3 Community design goals seek to concentrate growth in compact, urban nodes to create mixed-use communities with strong visual identities, human-scale street networks, and a range of housing and employment opportunities utilizing neo-traditional design concepts (illustrated in figure 9).

Three time horizons are addressed: the "ultimate" vision through 2040, the 20-year, long-range development pattern, and the five-year, short-range development pattern. The plan uses the concept of community character areas as an organizing framework for land use management. Policies are proposed for the overall county, as well as for the eastern urban growth areas, town urban growth areas, rural areas, and existing rural village areas. Implementation tools include capital facility and transportation proffers by developers, density transfers, community design guidelines, annexation guidelines, and an action schedule of next steps.

SUMMARY OF THE CONTEMPORARY SITUATION

Since midcentury, the nature of the plan has shifted from an elitist, inspirational, long-range vision that was based on fiscally innocent implementation advice, to a framework for community consensus on future growth that is supported by fiscally grounded actions to manage change. Subject matter has expanded to include the natural as well as the built environment. Format has shifted from simple policy statements and a single large-scale map of future land use, circulation, and community facilities, to a more complex combination of text, data, maps, and timetables. In a number of states plans are required by state law, and their content is specified by state agencies. Table 2 compares the general plan of the 1950s–1960s with the four contemporary prototype plans and the new 1990s hybrid design-policy management plan, which combines aspects of the prototype plans.

Today's prototype land use design continues to emphasize long-range urban form for land uses, community facilities, and transportation systems as shown by a map; but the design is also expressed in general policies. Land use design is still a common form of development plan, especially in municipalities.

S I X

Figure 8 Seattle urban villages strategy
Source: Seattle Planning Department 1993

Figure 9 Neotraditional community schematic and generalized policy planning areas, Loudoun County, Virginia General Plan

Source: *Planning* 60, 3:10 (1994)

		Contemporary prototype plans				
Features of plans	1950s general plan	Land use design	Land classification plan	Verbal policy plan	Development management plan	1990s Hybrid design policy management plan
Land use maps	Detailed	Detailed	General	No	By growth areas	General *and* area specific
Nature of recommendations	General community goals	Land use policies & objectives	Growth locations & incentives	Variety of community policies	Specific management actions	Policy *and* actions
Time horizon	Long range	Long range	Long range	Intermediate range	Short range	Short *and* long range
Link to implementation	Very weak	Weak to moderate	Moderate	Moderate	Strong	Moderate to strong
Public participation	Pro-forma	Active	Moderate	Moderate	Active	Active
Capital improvements	Advisory	Recommended	Recommended	Recommended	Required	Recommended to required
Land use/transportation linkage	Moderate	Strong	Weak	Varies	Strong	Strong
Environmental protection	Weak	Moderate	Strong	Varies	Varies	Strong
Social policy linkage	Weak	Weak	Weak	Moderate to strong	Weak	Moderate

Table 2 Comparison of plan types

The land classification plan also still emphasizes mapping, but of development policy rather than policy about a pattern of urban land uses. Land classification is more specific about development management and environmental protection, but less specific about transportation, community facilities, and the internal arrangement of the future urban form. County and regional governments are more likely than are municipalities to use land classification plans.

The verbal policies plan eschews the spatial specificity of land use design and land classification plans and focuses less on physical development issues. It is more suited to regions and states, or may serve as an interim plan for a city or county while another type of plan is being prepared.

The development management plan represents the greatest shift from the traditional land use plan. It embodies a short-to-intermediate-range program of governmental actions for ongoing growth management rather than for long-range comprehensive planning.

In practice, these four types of plans are not mutually exclusive. Communities often combine aspects of them into a hybrid general plan that has policy sections covering environmental/social/economic/housing/infrastructure concerns, land classification maps defining spatial growth policy, land use design maps specifying locations of particular land uses, and development management programs laying out standards and procedures for guiding and paying for growth. Regardless of the type of plan used, the most progressive planning programs today regard the plan as but one part of a coordinated growth management program, rather than, as in the 1950s, the main planning product. Such a program incorporates a capital improvement program, land use controls, small area plans, functional plans, and other devices, as well as a general plan.

THE ENDURING LAND USE FAMILY TREE AND ITS FUTURE BRANCHES

For the first 50 years of this century, planning responded to concerns about progressive governmental reform, the City Beautiful, and the "City Efficient." Plans were advisory, specifying a future urban form, and were developed by and for an independent commission. By midcentury this type of plan, growing out of the design tradition, had become widespread in local practice. During the 1950s and 1960s the 701 program, T. J. Kent, and F. Stuart Chapin, Jr. further articulated the plan's content and methodology. Over the last 30 years, environmental and infrastructure issues have pushed planning toward growth management. As citizen activists and interest groups have taken more of a role, land use politics have become more heated. Planning theorists, too, have questioned the midcentury approach to planning, and have proposed changes in focus, process, subject matter, and format, sometimes challenging even the core idea of rational planning. As a result, practice has changed, though not to a monolithic extent and without entirely abandoning the traditional concept of a plan. Instead, at least four distinct types of plans have evolved, all descending more or less from the midcentury model, but advocating very different concepts of what a plan should be. With a kind of self-correcting common sense, the plans of the 1990s have subsequently incorporated the useful parts of each of these prototypes to create today's hybrid design/policy/management plans.

To return to our analogy of the plan's family tree: Roots for the physical development plan became well established during the first half of this century. By 1950, a sturdy trunk concept had developed. Since then, new roots and branches have appeared – land classification plans, verbal policy plans, and development management plans. Meanwhile development of the main trunk of the tree – the land use design – has continued. Fortunately, the basic gene pool has been able to combine with new genes in order to survive as a more complex organism – the 1990s design-policy management hybrid plan. The present family tree of planning reflects both its heredity and its environment.

The next generation of physical development plans also should mature and adapt without abandoning their heritage. We expect that by the year 2000, plans will be more participatory, more electronically based, and concerned with increasingly complex issues. An increase in participation seems certain, bolstered by interest groups' as well as governments' use of expert systems and computer databases. A much broader consideration of alternative plans and scenarios, as well as a more flexible and responsive process of plan amendment, will become possible. These changes will call upon planners to use new skills of consensus building and conflict management, as more groups articulate their positions on planning matters, and government plans and interest group plans compete, each backed by experts.

With the advent of the "information highway," plans are more likely to be drafted, communicated, and debated through electronic networks and virtual reality images. The appearance of plans on CD ROM and cable networks will allow more popular access and input, and better understanding of plans' three-dimensional consequences. It will be more important than ever for planners to compile information accurately and ensure it is fairly communicated. They will need to compile, analyze, and manage complex databases, as well as to translate abstract data into understandable impacts and images.

Plans will continue to be affected by dominant issues of the times: aging infrastructure and limited public capital, central city decline and suburban growth, ethnic and racial diversity, economic and environmental sustainability, global competition and interdependence, and land use/transportation/air quality spillovers. Many of these are unresolved issues from the last 30 years, now grown more complex and interrelated. Some are addressed by new programs like the Intermodal Surface Transportation Efficiency Act (ISTEA) and HUD's Empowerment Zones and Enterprise Communities. To cope with others,

planners must develop new concepts and create new techniques.

One of the most troubling new issues is an attempt by conservative politicians (see the Private Property Protection Act of 1995 passed by the US House of Representatives) and "wise use" groups to reverse the precedence of the public interest over individual private property rights. These groups challenge the use of federal, state, and local regulations to implement land use plans and protect environmental resources when the result is any reduction in the economic value of affected private property. Should their challenge succeed and become widely adopted in federal and state law, growth management plans based on regulations could become toothless. Serious thinking by land use lawyers and planners would be urgently needed to create workable new implementation techniques, setting in motion yet another planning evolution.

We are optimistic, however, about the future of land use planning. Like democracy, it is not a perfect institution but works better than its alternatives. Because land use planning has adapted effectively to this century's turbulence and become stronger in the process, we believe that the twenty-first century will see it continuing as a mainstay of strategies to manage community change.

"Planning in the Face of Conflict"

Journal of the American Planning Association (1987)

John Forester

Editors' Introduction

While good city planning needs to be inspired by a vision of the end results, and should be informed by theory, in democracies planning is never achieved without conflict. Planners, citizens, local elected officials, developers, and others invariably have different views on what a city *should* be like and *how* to build. Passions run high at important city planning commission meetings.

John Forester, the chair of the Department of City and Regional Planning at Cornell University, got down in the trenches with practicing city planners and others involved in city development to study what the practice of city planning is really like in the face of conflict. The following selection summarizes what he learned about the process and his ideas on how planners can be effective in the face of conflict. It is valuable for the actual lessons Forester learned. But equally important Forester helps point the way out of the impasse Peter Hall describes in much academic planning theory today (p. 342). Unlike some ivory tower theorists, Forester listened carefully to practicing city planners and learned from them. His work synthesizes what he learned and develops theoretical concepts that are highly relevant to planning practice.

Forester found that city planners guide both developers and neighborhood residents through the complexities of the planning process using a variety of strategies. Planners have to be attentive to timing. Successful planners handle conflicts both through both formal and informal channels. They must respond to complex and contradictory duties – tugged this way by local politicians, that way by legal mandates, and yet another way by citizen demands. Through all of this, successful city planners must be true to professional norms and hold fast to their own visions of high-quality city development. City planners who retreat to their planning offices to draw beautiful plans or create elegant computerized models of how cities will develop are doomed to fail. They need to wed these professional skills to practical skills for managing conflict among competing groups.

There are many lessons in Forester's work. People who want to be effective translating city plans into action need to expect opposition and should not be surprised or worn down by what often seems an endless and frustrating process. They need to be aware of their own power and also its limitations. They have to be sensitive to and understand the interests of other actors in the city development process and form alliances.

Forester argues that city planners can self-consciously follow any of a number of strategies to keep projects on track and achieve success – as rule-enforcers, negotiators and mediators, resource people, or shuttle diplomats.

Compare Forester's insights with David Harvey's description of social conflict around urban spatial issues (p. 227) and John Mollenkopf's observations on how to study urban power (p. 235). Consider the kind of conflicts you would expect if you were trying to implement the different types of plans that Edward J. Kaiser and David R. Godschalk describe (p. 354).

A related article by John Forester is "Planning in the Face of Power," *Journal of the American Planning Association* 48, 1 (1982). Forester teamed up with former Cleveland city planning director Norman Krumholz to write an account of Krumholz's experience implementing socially responsible equity planning in *Making Equity*

Planning Work (Philadelphia, PA: Temple University Press, 1992). He is the author of *The Deliberative Practitioner: Encouraging Participatory Planning Processes* (Cambridge, MA: MIT Press, 1999) and *Critical Theory, Public Choice, and Planning Practice: Towards a Critical Pragmatism* (Buffalo: State University of New York Press, 1993). He edited *Critical Theory and Public Life* (Cambridge, MA: MIT Press, 1987), a book of essays about Jürgen Habermas's critical communications theory of society.

Other books on what city planning is actually like include Allan Jacobs, *Making City Planning Work* (Chicago: American Society of Planning Officials, 1976) and Bruce W. McClendon (ed.), *Planners on Planning* (San Francisco: Jossey-Bass, 1996).

In the face of local land-use conflicts, how can planners mediate between conflicting parties and at the same time negotiate as interested parties themselves? To address that question, this article explores planners' strategies to deal with conflicts that arise in local processes of zoning appeals, subdivision approvals, special permit applications, and design reviews.

Local planners often have complex and contradictory duties. They may seek to serve political officials, legal mandates, professional visions, and the specific requests of citizens' groups, all at the same time. They typically work in situations of uncertainty, of great imbalances of power, and of multiple, ambiguous, and conflicting political goals. Many local planners, therefore, may seek ways both to negotiate effectively, as they try to satisfy particular interests, and to mediate practically, as they try to resolve conflicts through a semblance of a participatory planning process.

But these tasks – negotiating and mediating – appear to conflict in two fundamental ways. First, the negotiator's interest in the subject threatens the independence and the presumed neutrality of a mediating role. Second, although a negotiating role may allow planners to protect less powerful interests, a mediating role threatens to undercut this possibility and thus to leave existing inequalities of power intact. How can local planners deal with these problems? I discuss their strategies in detail below.

This article first presents local planners' own accounts of the challenges they face as simultaneous negotiators and mediators in local land-use permitting processes. Planning directors and staff in New England cities and towns, urban and suburban, shared their viewpoints with me during extensive open-ended interviews. The evidence reported here, therefore, is qualitative, and the argument that follows seeks not generalizability but strong plausibility across a range of planning settings.

The article next explores a repertoire of mediated negotiation strategies that planners use as they deal with local land-use permitting conflicts. It assesses the emotional complexity of mediating roles and asks: What skills are called for? Why do planners often seem reticent to adopt face-to-face mediating roles?

Finally, the article turns to the implications of these discussions. How might local planning organizations encourage both effective negotiation and equitable, efficient mediation? How might mediated-negotiation strategies empower the relatively powerless instead of simply perpetuating existing inequalities of power?

ELEMENTS OF LOCAL LAND-USE CONFLICTS

Consider first the settings in which planners face local permitting conflicts. Private developers typically propose projects. Formal municipal boards – typically planning boards and boards of zoning appeals – have decision-making authority to grant variances, special permits, or design approvals. Affected residents often have a say – but sometimes little influence – in formal public hearings before these boards. Planning staff report to these boards with analyses of specific proposals. When the reports are positive, they often recommend conditions to attach to a permit or suggest design changes to improve the final project. When the reports are negative, there are arguments to be made, reasons to be given.

Some municipalities have elected permit-granting boards; some have appointed boards. Some municipal ordinances mandate design review; others do not. Some local by-laws call for more than one planning board hearing on "substantial" projects, but others do not. Nevertheless, for several reasons, planners' roles in these different settings may be more similar than dissimilar.

Common planning responsibilities

First, planners must help both developers and neighborhood residents to navigate a potentially complex review process; clarity and predictability are valued goods. Second, the planners need to be concerned with timing. When a developer or neighborhood resident is told about an issue may be even more important than the issue itself. Third, planners typically need to deal with conflicts between project developers and affected neighborhood residents that usually concern several issues at once: scale, the income of tenants, new traffic, existing congestion, the character of a street, and so on. Such conflicts simultaneously involve questions of design, social policy, safety, transportation, and neighborhood character as well. Fourth, how much planners can do in the face of such conflicts depends not only upon their formal responsibilities, but also upon their informal initiatives. A zoning by law, for example, can specify a time by which a planning board is to hold a public hearing, but it usually will not tell a planner how much information to give a developer or a neighbor, when to hold informal meetings with either or both, how to do it, just whom to invite, or how to negotiate with either party. So within the formal guidelines of zoning appeals, special permit applications, site plan and design reviews, planning staff can exercise substantial discretion and exert important influence as a result.

Planners' influence

The complexity of permitting processes is a source of influence for planning staff. Complexity creates uncertainties for everyone involved. Some planners eagerly use the resulting leverage, as an associate planning director explains, beginning with a truism but then elaborating:

> Time is money for developers. Once the money is in, the clock is ticking. Here we have some influence. We may not be able to stop a project that we have problems with, but we can look at things in more or less detail, and slow them down. Getting back to [the developers] can take two days or two

months, but we try to make it clear, "We're people you can get along with." So many developers will say, "Let's get along with these people and listen to their concerns . . ."

He continues,

> But we have influence in other ways too. There are various ways to interpret the ordinance, for example. Or I can influence the building commissioner. He used to work in this office and we have a good relationship . . . his staff may call us about a project they're looking over and ask, "Hey, do you want this project or not?"

Planners think strategically about timing not only to discourage certain projects but to encourage or capture others. The associate director explains,

> On another project, we waited before pushing for changes. We wanted to let the developer get fully committed to it; then we'd push. If we'd pushed earlier, he might have walked away . . .

A director in another municipality echoes the point:

> Take an initial meeting with the developer, the mayor, and me. Depending on the benefits involved – fiscal or physical – the mayor might kick me under the table; "Not now," he's telling me. He doesn't want to discourage the project . . . and so I'll be able to work on the problems later . . .

For the astute, it seems, the complexity of the planning process creates more opportunities than headaches. For the novice, no doubt, the balance shifts the other way.

But isn't everything, in the last analysis, all written down in publicly available documents for everyone to see? Hardly. Could all the procedures ever be made entirely clear? Consider the experience of an architect planner who grappled with these problems in several planning positions. The following conversation took place toward the end of my interview with this planner. The planner pulled a diagram from a folder and said, "Here's the new flow-chart I just drew up that shows how our design review process works. If you have any questions, let's talk. I think it's still pretty cryptic."

"If you think it's cryptic," interjected the zoning appeals planner, who was standing nearby and had overheard this, "just think what developers and neighborhood people will think!"

Both planners shook their heads and laughed, since the problem was all too plain: the arrows on the design review flow chart seemed to run everywhere. The chart was no doubt correct, but it did look complicated.

I recalled my first interview with the zoning appeals planner. Probing with a deliberately leading question, I had asked, "But what influence can you have in the process if everything's written down as public information, if it's all clear there on the page?"

The zoning appeals planner had grinned: "But that's just it! The process is not clear! And that's where I come in . . ." The architect-planner developed the point further:

> Where I worked before, the planning director wanted to adopt a new "policy and procedures" document that would have every last item defined. We were going to get it all clear. The whole staff spent a lot of time writing that, trying to get all the elements and subsections and so on clearly defined . . . But it was chaos. Once we had the document, everyone fought about what each item meant . . .

So clarity, apparently, has its limits!

Different actors, different strategies

Planning staff point almost poignantly to the different issues that arise as they work with developers and neighborhood residents. The candor of one planning director is worth quoting at length:

> It's easy to sit down with developers or their lawyers. They're a known quantity. They want to meet. There's a common language – say, of zoning – and they know it, along with the technical issues. And they speak with one voice (although that's not to say that we don't play off the architect and the developer at times – we'll push the developer, for example, and the architect is happy because he agrees with us) . . .
>
> But then there's the community. With the neighbors, there's no consistency. One week one group comes in, and the next week it's another. It's hard if there's no consistent view. One group's worried about traffic; the other group's not worried

about traffic but about shadows. There isn't one point of view there. They also don't know the process (though there are cases where there are too many experts).
>
> So at the staff level (as opposed to planning board meetings) we usually don't deal with both developers and neighbors simultaneously.

Although these comments may distress advocates of neighborhood power, they say much about the practical situation in which the director finds himself.

All people may be created equal, but when they walk into the planning department, they are simply not all the same. This director suggests that getting all the involved parties together around the table in the planners' conference room is not an obviously good idea, for several reasons. (It is, however, an idea we shall consider more closely below.)

First, the director suggests, planners generally know what to expect from developers; the developers' interests are often clearer than the neighbors', and project proponents may actually want to meet with the staff. Neighborhood residents may be less likely to treat planners as potential allies; after all, the planners are not the decision makers, and the decision makers can often easily ignore the planners' recommendations. Because developers may cultivate good relations with planning staff (this is in part their business, after all), while neighborhood groups do not, local planning staff may find meetings with developers relatively cordial and familiar, but meetings with neighborhood activists more guarded and uncertain.

Second, the planning director suggests that planners and developers often share a common professional language. They can pinpoint technical and regulatory issues and know that both sides understand what is being said. But on any given project, he implies, he may need to teach the special terms of the local zoning code to affected neighbors before they can really get to the issues at hand.

The planning director makes a third point. Developers speak with one voice; neighbors do not. When planners listen to developers talk, they know whom they're listening to, and they know what they're likely to hear repeated, elaborated, defended, or qualified next week. When planners listen to neighborhood residents, though, this director suggests, they can't be so sure how strongly to trust what they hear. "Who really speaks for the neighborhood?" the director wonders.

Planners must make practical judgments about who represents affected residents and about how to interpret their concerns. This director implies, therefore, that until planners find a way to identify "the neighborhood's voice," the problems of conducting joint mediated negotiations between developers and neighbors are likely to seem insurmountable. We return to this issue of representation below.

Inequalities of information, expertise, and financing

What about imbalances of power? Developers, typically, initiate site developments. Planners respond. Neighbors, if they are involved at all, then try to respond to both. Developers have financing and capital to invest; neighbors have voluntary associations and not capital, but lungs. Developers hire expertise; neighborhood groups borrow it. Developers typically have economic resources; neighbors often have time, but not always the staying power to turn that time into real negotiating power.

Where power relations are unbalanced, must mediated negotiation simply lead to coopting the weaker party? No, because, as we shall see below, mediated negotiation is not a gimmick or a recipe; it is a practical and political strategy to be applied in ways that address the specific relations of power at hand.

When either developers or neighborhood groups are so strong that they need not negotiate, mediated negotiation is irrelevant, and other political strategies are more appropriate. But when both developers and neighbors want to negotiate, planners can act both as mediators, assisting the negotiations, and as interested negotiators themselves. But how is this possible? What strategies can planners use?

PLANNERS' STRATEGIES: SIX WAYS TO MEDIATE LOCAL LAND-USE CONFLICTS

Consider the following six mediated-negotiation strategies that planning staff can utilize in the face of local land-use conflicts. They are *mediated* strategies because planners employ them to assure that the interests of the major parties legitimately come into play. They are *negotiation* strategies because (except for the first) they focus attention on the informal negotiations that may produce viable agreements even before formal decision-making boards meet.

Strategy one: The facts! The rules! (The planner as regulator)

The first strategy is a traditional response, pristine in its simplicity, but obviously more complex in practice. A young planner who handles zoning appeals and design review says:

> I see my role often as a fact finder so that the planning board can evaluate this project and form a recommendation; whether it's design review, special permits, or variances, you still need lots of facts . . .

Here of course is the clearest echo of the planner as technician and bureaucrat; the planner processes information and someone else takes responsibility for making decisions. But the echo quickly fades. A moment later, this planner continues,

> Our role is to listen to the neighbors, to be able to say to the board, "Okay, this project meets the technical requirements but there will be impacts . . ." The relief will usually then be granted, but with conditions . . . We'll ask for as much in the way of conditions as we think necessary for the legitimate protection of the neighborhood. The question is, is there a legitimate basis for complaint? And it's not just a matter of complaint, but of the merits.

This planner's role is much more complex than that of fact finder; it is virtually judicial in character. He implies, essentially, "I'm not just a bureaucrat, I'm a professional. I need to think not only about the technical requirements, but about what's legitimate protection for the neighbors. Now I have to think about the merits!" Thinking about the merits, though, does not yet mean thinking about politics, the feelings of other agencies, the chaos at community meetings – it means making professional judgments and then recommending to the planning board the conditions that should be attached to the permits.

Consider now a slightly more complex strategy.

Strategy two: Pre-mediate and negotiate – representing concerns

When developers meet with planners to discuss project proposals, neighborhood representatives rarely join them. Yet planners might nevertheless speak *for* neighborhood concerns as well as *about* them. A planning director in a municipality where neighborhood groups are well-organized, vocal, and influential notes,

> We temper our recommendations to developers. While we might accept A, the neighbors want D, and so we'll tell the developers to think about something in the middle – if they can make it work.

Here, the planner anticipates the concerns of affected residents and changes the informal staff recommendation accordingly to search for an acceptable compromise with the developers. He explains,

> What we do is premediate rather than mediate after the fact. We project people's concerns and then raise them; so we do more before [explicit conflict arises] . . . The only other way we step in and mediate, later, is when we support changes to be made in a project, changes that consider the neighbors' views; but that's later, after the public hearing . . .

Unlike the planner-regulator quoted above, this planning director relies on far more than his professional judgment when he meets with a developer. He will negotiate to reach project outcomes that satisfy local statutes, professional standards, and the interests of affected residents as well. His calculation is not only judicial, but explicitly political. He anticipates the concerns of interested community members. So he seeks to represent neighborhood interests – without neighborhood representatives.

Such premediation – articulating others' concerns well before they can erupt into overt conflict – involves a host of political, strategic, and ethical issues. What relationships does the planner have with neighborhood groups? In what senses can the planner "know what the community wants"? To which "key actors" might the planner "steer" the developer? How much information and how much advice should the planner give, or withhold?

Such questions arise whether or not project developers ever meet with neighbors. In many cases, where "neighborhoods" are sprawling residential areas, and where "the interests of the neighbors" seem most difficult to represent through actual neighborhood representatives, the planners' premediation may be the only mediation that takes place.

Strategy three: Let them meet – the planner as a resource

The planner's influence might be used in still other ways. The director continues:

> Regardless of how our first meeting with a developer goes, we recommend to them that they meet with neighbors and the neighbors' representatives [on the permit-granting board]. We usually can give the developer a good inkling about what to expect both professionally and politically. The same elected representative might say that a project is "okay" professionally, but not "okay" for them in their elected capacity. We try to encourage back and forth meetings . . .

The director, then, regularly takes the pulse of neighborhood groups and elected representatives. Working in city hall has its advantages: "We'll discuss a project with the representatives; we see them so much here, just in the halls, and they ask us to let them know what's happening in their parts of the city." So the director listens to the developers, listens to the neighbors, and "encourages back and forth meetings."

A planning director who seldom met jointly with neighbors and developers had an acute sense of other strategies he used:

> We . . . urge the applicants, the developers, to deal directly with the neighborhood for several reasons: First, if the neighbors are confronted at a hearing with glossy plans, they'll think it's all a *fait accompli*; so they'll just adopt the "guns blazing, full charge ahead" strategy, since they think it'll just be a "yea" or "nay" decision. Second, we tell them to talk to the neighbors since if they can come up with something that the neighbors will "okay," it'll be easier at the board of appeals. Third, we try to get them to meet one on one, or maybe as a group, but in as deinstitutionalized a way as possible, informally. We try to get the developers to sell their case that way; it'll get a much better hearing than at the big formal public hearings.

But why should planners be reluctant to convene joint negotiating sessions between developers and neighbors, yet still be willing to encourage both parties to meet on their own? Why don't these planners embrace opportunities to mediate local land-use conflicts face to face? One planner could hardly imagine such a mediating role:

Work as a neutral between developers and neighbors? I don't know how I'd approach it. I'd just answer questions, suggest what could be done, and so on. That's what our role should be – although we should reach compromises between developers and neighbors. But we have to work within the rules – that's my reference point – to say what the rules of the game are; that's the job.

This planner's image of a "neutral" between disputing parties is less that of a mediator facilitating agreement than it is of a referee in a boxing match. The referee assures that the rules are followed, but the antagonists might still kill each other. No wonder planners might find this image of mediation unattractive!

A senior planner envisions further complications:

If I could be assured I could be wholly independent, then I could mediate – but I still have to pay my bills . . . The planning department always has some vested interests, as much as we try to stay objective, independent . . . I work for a mayor, for the elected representatives, for 14 committees . . . So there's always the question of compromise on my part: if the mayor says, "Tell me how to make this project work," for example. It took me a long time before I was able to say, "I'm going to have to say no." We have a very strong mayor . . .

Strategy four: Perform shuttle diplomacy – probe and advise both sides

A planning director proposes another way to facilitate developer–neighbor negotiations:

I feel more comfortable in shuttle diplomacy, if you will; trying to get the neighbors' concerns on the table, to get the developers to deal with them . . . I'd rather bounce ideas off each side individually than be caught in the middle if they're both there. If both sides are there, I'm less likely to give my own ideas than if I'm alone with each of them.

Shuttle diplomacy, this director suggests, allows planners to address the concerns of each party in a professionally effective way. He explains:

If I'm with the developer, I feel I can make a much more extreme proposal – "knock off three stories" – but I wouldn't dare say that if neighbors were there. The neighbors would be likely to pick up and run with it, and it could damage the negotiations rather than help them . . . I'm willing to back off on an issue if the developer has a good argument, but the neighborhood might not, and then they might use my point as a club to hit the developer with: "Well, the planning director suggested that; it must be a good idea" . . . and then I can't unsay it . . .

This planning director is as concerned about how his suggestions, proposals, queries, and arguments will be understood and used as he is about what ought to be altered in the project at hand. He recognizes clearly that when he talks he acts politically and inevitably fuels one argument or another. He not only conveys information in talking, but he acts practically, influentially. He focuses attention on specific problems, shapes future agendas, legitimates a point of view, and suggests lines of further argument.

The director continued,

I might not want to concede to a developer that there won't be a traffic problem, because I want to push him to relieve a problem or a perceived problem . . . but I could say to the neighbors aside, "Look, this will be no big deal; it'll be five trips, not fifty." I can say that in a private meeting, but in a public meeting if I say it to a neighborhood representative I'm insulting him, even if the developer snickers silently . . . So I lose my ability to be frank with both sides if we're all together. Not that this should be completely shuttle diplomacy, but it has its place.

These comments suggest that planning staff can certainly mediate conflicts in local permitting processes, if not in ways that mediators are thought typically to act. The planners may not be independent third parties who assist developers and neighbors in face-to-face meetings to reach development agreements – but they might still mediate such conflicts as "shuttle diplomats."

Strategy five: Active and interested mediation – thriving as a non-neutral

We can consider a case that involves not a zoning appeal but a rezoning proposal. One planner, who had earlier worked as a community organizer, had convened a working group of five community representatives and five local business representatives to draft a rezoning proposal for a large stretch of the major arterial street in their municipality. She considers her work on that project a kind of mediation and reflects about how she as a planner acts as a mediator, dealing with substantive and affective issues alike:

> Am I in a position of having to think about everyone's interest and yet being trusted by no one? Sure, all the time. But I've been in this job for seven years, and I have a reputation that's good, fortunately . . . Trust is an issue of your integrity and planning process. I talk to people a lot; communication is a big part of it . . . My approach is to let people let off steam – let them say negative things about other people to me, and then in a different conversation at another time, I'll be sure to say something positive about that person – to try to let them feel that they can say whatever they want to me, and to try to confront them with the fact that the other person isn't just out to ruin the process. But I'd do that in another conversation; I let them let off steam if they're angry.

This planner is well aware that distrust on all sides is an abiding issue, so she tries to build trust as she works. She works to assure others that she will listen to them and more; that she will acknowledge and respect their thoughts and feelings, whatever they have to say. She pays attention first to the person, then to the words. Then, as she establishes trust with her committee members and with others, she can also make sure, carefully, that real evidence is not ignored.

She realizes that anger makes its own demands, so she responds with an interested patience. She seeks throughout to mediate the conflicting interests of the groups with whom she's working:

> I also make a point to tell each side the other's concerns – categorically, not with names, but all the other sides' concerns . . . Why's that important? I like to let people anticipate the arguments and prepare a defense, either to stand or fall on its own

merits. For people to be surprised is unfortunate. It's better to let people know what's coming so they can build a case. They can hear an objection, if you can retain credibility, and absorb it; but in another setting they might not be able to hear it . . . If they hear an objection first as a surprise, you're likely to get blamed for it. If concerns are raised in an emotional setting, people concentrate more on the emotion than on the substance. This is a concern of mine. In emotional settings, lots gets thrown out, and lots is peripheral, but possibly also central later . . .

This planner is keenly aware that emotion and substance are interwoven, and that planners who focus only upon substance and try to ignore or wish away emotion do so at their own practical peril. Yet she is saying even more.

She knows that in some settings disputing groups can hear objections, understand the points at stake, and address them, while in other settings those points may be lost. She tries to present each side's concerns to the other so that they can be understood and addressed. Anticipating issues is central; learning of important objections late in the process will be mostly emotionally and financially, and planning staff are likely to share the blame. "Why didn't you tell us sooner . . .?" the refrain is likely to sound.

Consider next, then, this planner-mediator's thoughts about the sort of mediation role she is performing. She continues,

> But what I do is different from the independent mediator model. In a job like mine, you have an on-going relationship with parties in the city. You have more information than a mediator does about the history of various individuals, about participating organizations, about the political history of city agencies, and so on. You also have a vested interest in what happens. You want the process to be credible. You want the product to be successful; in my case I want the city council to adopt the committee's proposal. And you're invested . . . both professionally and emotionally. And then you have an opinion about particular proposals; you're a professional, you should have one, you should be able to look at a proposal and have an opinion.

Thus, she suggests, mediation has its place in local land-use conflicts, but the "rules of the game" will not be those that labor mediators follow. Indeed, planners who now mediate local land-use conflicts are not waiting for someone else to write the rules of the game, they are writing them themselves.

Strategy six: Split the job – you mediate, I'll negotiate

Consider finally a planning strategy that promotes face-to-face mediation with planning staff at the table – but as negotiators or advisors, not as mediators. A planning director explains:

> There's another way we deal with these conflicts; we might involve a local planning board member. For example, if there's a sophisticated neighborhood group that's well organized, we've brought in an architect from the board who's as good with words as he is with his pencil . . . The chair of the board might ask the board member to be a liaison to the neighborhood, say, and sometimes he'll talk just to the neighbors, sometimes with both . . .

Here the "process manager" comes from the planning board with highly developed "communications skills." How does the planner feel in these situations?

> It's more comfortable from my point of view, and the citizens', to have a board member in the convening role. I'm still a hired hand. It seems more appropriate in a negotiating situation to have a citizen in that role and not an employee . . . Since they've come from the neighborhoods, a board member is in a better position to bring neighbors and developers together – if they behave properly. Some board members are good communicators; some are more dynamic than others in pressing for specific solutions.

This planner identifies so strongly with the professional and political mandate of his position that he cannot imagine a role as neutral convener or mediator of neighborhood–developer negotiations. But that does not prevent mediation; it means rather that the planner retains a substantively interested posture while another party, here a planning board member, convenes informal, but organized, project negotiations between developers and neighbors. This planner's example makes the point:

Take the example of the Mayfair Hospital site. The hospital was going to close, and the neighbors and the planning board were concerned about what might happen with the site. So Jan from the planning board got involved with the hospital and the neighborhood to look at the possibilities. Both the neighbors and the hospital set up re-use committees, and Jan and I went to the meetings. There was widespread agreement that the best use of the site would be residential – the neighbors definitely preferred that to an institutional use – but then there was a lot of haggling over scale, density, and so on. Ultimately, a special zoning district was proposed that included the site; the neighbors supported it, and it went to [the elected representatives] where they voted to rezone the several acres involved . . .

When local planners feel they cannot mediate disputes themselves, then one strategy may be to search for informal, most likely volunteer, mediators. These ad hoc mediators might be "borrowed" from respected local institutions, and their facilitation of meetings between disputing parties might allow planning staff to participate as professionally interested parties concerned with the site in question.

Table 1 summarizes the six approaches presented. Together, these approaches form a repertoire of strategies that land-use planners can use to encourage mediated negotiations in the face of conflicts in local zoning, special permit, and design review processes. To refine these strategies, local planning staff can build upon several basic theories and techniques of conflict resolution. Consider now the distinctive competences and sensitivities required by these strategies.

1 The Facts! The Rules! (The Planner as Regulator)
2 Pre-Mediate and Negotiate: Representing Concerns
3 Let Them Meet: The Planner as a Resource . . .
4 Perform Shuttle Diplomacy: Probe and Advise Both Sides
5 Active and Interested Mediation: Thriving as a Non-Neutral
6 Split the Job: You Mediate, I'll Negotiate

Table 1 A repertoire of mediated-negotiation strategies used by local land-use planners in permitting processes

THE EMOTIONAL COMPLEXITY OF MEDIATED-NEGOTIATION STRATEGIES

More than a lack of independence keeps planners from easily adopting roles as mediators. The emotional complexity of the mediating role makes quite different demands upon planners than those that they have traditionally been prepared to meet. The community-organizer-turned-planner makes the point brilliantly:

> In the middle, you get all the flak. You're the release valve. You're seen as having some power, and you do have some . . .Look, if you have a financial interest in a project, or an emotional one, you want the person in the middle to care about your point of view, and if you don't think they do, you'll be angry!
>
> ["So when planners try to be professional by appearing detached, objective, does it get people angry at them?" I asked.]
>
> Sure!

This comment cuts to the heart of planners' professional identities. Must "professional," "objective," and "detached" be synonymous? If so, this planner suggests, then planners' own striving for an independent professionalism will fuel the anger, resentment, and suspicion of the same people those planners presume to serve!

Thus we can understand the caution with which a planner speaks of his way of handling emotional participants in public hearings:

> How do I deal with people's anger? I try to keep cool, but occasionally I get irritated. But that's how we're expected to behave, to be rational. It's all right for citizens to be irrational, but not the staff!

How does one keep cool, be rational, and still respond to the claims of an emotional public at formal hearings? This planner elaborates:

> It's one thing to begin the discussion of a project [to present our analysis] and anticipate problems. But it's another thing to *rebut* a neighborhood resident in public in a gentle way . . . Part of the problem is that if you antagonize people it'll haunt you in the future . . . We're here for the long haul, and we have to try to maintain our credibility . . .

The planner's problem here is precisely *not* the facts of the case: the facts themselves may be clear enough.

But how should the planner present the analysis that he feels must be made and how should he decide which arguments to make and which to hold back at a given time?

> The biggest problem I have in the board meetings is when to respond and when to keep quiet. In a hearing, for example, I can't possibly respond to all the accusations and issues that come up. So I have to pick a direction, to deal with a generally felt concern. It's just not effective to enter into a debate on each point in turn; it's better to clarify things, to explain what's misunderstood . . .

This planner does much more than simply recapitulate facts. He tries to avoid an adversarial posture, even when he feels the situation is quite conflictual. He listens as much to the individuals and their concern as he does to each point. He knows that points and demands and positions may change as issues are clarified, but that if he cannot respond to people's concerns, he's in some trouble. Because he and his staff are there "for the long haul," he wants to be able to work with neighbors, community leaders, and elected representatives alike not just now but in the future as well. How he relates to the parties involved in local disputes, he suggests, is as important as what he has to say.

Another planner points to the skills involved:

> Whom would I try to hire to deal with such conflicts? I'd look for someone who's a careful listener, someone who's good at explaining a position coherently, succinctly, quietly, in a calm tone . . . someone who could hear a point, understand it whether he or she agreed with it or not, and then verbalize a clear, concise response. Most people though – myself included – try to jump the gun and answer before it's appropriate. So I want someone who's able to stay cool and stay on the issues . . .

A community development director first mentions "a good listener" and then elaborates:

> [To deal with these conflicting situations I'd want to hire staff] who won't say, "I know best," who won't get people's backs up just by their style. I'd want someone with some openness, with a sense of how things work who won't accept everything, but who won't offend people. They have to have

critical judgment – to leave doors open, to give people a sense of involvement and a sense of the feasible – [someone who] can't be convinced of something that's not likely to work, just for the sake of getting agreement . . .

This planning director also points to the balance necessary between what planners say and how they say it. The "how" counts; he doesn't want staff who will "get people's backs up," "offend people," and not communicate an openness to others' concerns. Nor does he want someone who will sacrifice project viability for the temporary comfort of agreement. He asks for substantive judgment and the skills to manage a process.

Referring to the demands of working and negotiating with developers as they navigate the approval process, the director stresses the role of diplomacy:

We [planners] have access to information, to resources, to skills . . . so developers usually want to work with us. They have certain problems getting through the process . . . so we'll go to them and ask, "What do you want?" and we'll start a process of meetings . . . It's diplomacy; that's the real work. You have to have the technical skill . . . but that's the first 25 percent. The next 75 percent is diplomacy, working through the process.

Percentages aside, the point remains. To the extent that planning practitioners and educators focus predominantly upon facts, rules, likely consequences, and mitigation measures, they may fail to attend to the pressing emotional and communicative dimensions of local land-use conflicts. Because the planning profession has not traditionally embraced the diplomat's skills, it should surprise no one that practicing planners envision mediating roles with more reticence than relish.

In the next section, we turn to administrative and political questions. What, initially, can be done in planning organizations to improve planners' abilities to mediate local land-use negotiations successfully? What about imbalances of power?

ADMINISTRATIVE IMPLICATIONS FOR PLANNING ORGANIZATIONS

What does this analysis imply for policymakers and planners who wish to build options for mediation into local review processes? Mediation may offer several opportunities, under conditions of interdependent power: a shift from adversarial to collaborative problem-solving; voluntary development controls and agreements; improved city–developer–neighborhood relationships enabling early and effective reviews of future projects; more effective neighborhood voice; and joint gains ("both gain" outcomes) for the municipality, neighbors, and developers alike. Such opportunities present themselves *only* when no single party is so dominant that it need not negotiate at all, that it is likely simply to get what it wants in any case.

Planners already use the strategies reviewed in diverse settings. Which strategy a planner uses, and at which times, depends largely on practical judgment: What skills does the planner have? How willing are developers or neighbors, or other agency staff, to meet jointly? Does enough time exist to allow early, joint meetings? Are the practical and political alternatives of any one party so attractive that they see no point in mediated negotiations?

No strategy is likely to be desirable in all circumstances, so no one approach will provide the model to formalize into new zoning or permitting procedures. But to say that we should not formalize these strategies does not mean that we cannot regularly use them. How, then, can planners apply the mediated-negotiation strategies in local zoning, permitting, and design review processes?

First, planning staff must distinguish clearly the two complementary but distinct mandates they typically must serve: to press professionally, and thus to negotiate, for particular substantive goals (design quality or affordable housing, for example), and to enable a participatory process that gives voice to affected parties; thus, like mediators, to facilitate negotiations between disputants.

Second, planning staff need to adopt, administratively if not formally, a goal of supplementing (not substituting for) formal permitting processes with mediated negotiations: attempting to craft workable and voluntary tentative agreements before formal hearing dates.

Third, planning staff should examine each of the strategies reviewed here. They need to determine how each could work, given the size of their agency, their zoning and related by-laws, the political and institutional history of elected officials, neighborhood groups, and other agencies. Planning staff must ask which skills and competencies they need to develop to employ each of these strategies appropriately.

Fourth, planning staff must be able to show others – developers, neighborhood groups, public works department staff, elected and appointed officials – how and when mediated negotiations can lead to "both gain" outcomes and so improve the local land-use planning and development process. Planners also have to be clear about what mediated negotiation will not do: it will not solve problems of radically unbalanced power, for example. It can, however, refine an adversarial process into a partially collaborative one. It will not solve problems of basic rights, but it can often expand the range of affected parties' interests that developers will take into account. Mediated negotiations will neither necessarily co-opt project opponents (as skeptical neighborhood residents might suspect) nor stall proposals and projects (as skeptical developers and builders might suspect). Yet when each side can effectively threaten the other, when each side's interests depend upon the other's actions, then mediated negotiations may enable voluntary agreements, incorporate measures of control on both sides, allow "both gain" trades to be achieved, and do so more efficiently for all sides than pursuing alternative strategies (e.g. going to court or, sometimes, community organizing).

Fifth, planners need administratively to create an organized process to match incoming projects with one or more of the mediated-negotiation strategies and to review their progress as they go along. With staff training in negotiation and mediation principles and techniques, planning departments would be better able to carry out these strategies effectively once they have organized administratively to promote them.

DEALING WITH POWER IMBALANCES: CAN THE SIX STRATEGIES MAKE A DIFFERENCE?

The six strategies we have considered are hardly "neutral." Planners who adopt them inevitably either perpetuate or challenge existing inequalities of information, expertise, political access, and opportunity. Consider each approach, briefly, in turn.

To provide only the facts, or information about procedures, to whomever asks for them seems to treat everyone equally. Yet where severe inequalities exist, to treat the strong and the weak alike only ensures that the strong remain strong, the weak remain weak. The planner who pretends to act as a neutral regulator may sound egalitarian but nevertheless act, ironically, to perpetuate and ignore existing inequalities.

The premediation strategy can involve substantial discretion on the part of the planning staff. If the staff fail to put the interests of weaker parties "on the negotiating table," then here, too, inequalities will be perpetuated, not mitigated. If the staff do defend neighborhood interests in the development negotiations, they may challenge existing inequalities. But which "neighborhood interests" should the planning staff identify? How should neighborhoods – especially weakly organized ones – be represented? These questions are both practical and theoretical and they have no purely technical, "recipe"-like answers.

At first glance, the strategy of letting developers and neighbors meet without an active staff presence seems only to reproduce the initial strengths of the parties. Yet depending on how the planning staff intervene, one party or another may be strengthened or weakened. At times planners have helped developers anticipate and ultimately evade the concerns of citizens who opposed projects. Yet planners may also provide expertise, access, information, and so on to strengthen weaker citizens' positions.

The same discretion exists for planning staff who act as shuttle diplomats. Here a planner may counsel weaker parties to help them both before and during actual negotiations by identifying concerns that might effectively be raised, experts or other influentials who might be called upon, prenegotiation strategies and tactics to be employed, and so on. The shuttle diplomat need not appear neutral to all parties but he or she does need to appear useful to, or needed by, those parties. Planners who act as "interested mediators" face many of the same problems and opportunities that shuttle diplomats confront. In addition, though, the activist mediator may risk being perceived by planning board members, officials, or elected representatives as making deals that preempt their own formal authority. Thus the invisibility of the shuttle diplomat has its advantages; the planners can give counsel discreetly, suggesting packages and "deals" but avoiding the glare – and the heat – of the limelight.

Finally, the strategy of separating mediation and negotiation functions also involves substantial staff discretion. Here, too, the ways that mediators and negotiators consider the interests and enable the voice of weaker parties will affect existing power imbalances.

Because negotiations always involve questions of relative power, they depend heavily upon the parties' *prenegotiation* work of marshalling resources, developing options, and organizing support. Thus politically astute planners need both organizing and mediated-negotiation skills if conflicts are to be addressed without pretending that structural power imbalances just do not exist. Finally, note that a planner who explicitly calls everyone's attention to class-based power imbalances, for example, may not obviously do better in any practical sense of the word than an activist mediator who knows the same thing and acts on it in just the same ways without explicitly framing the planning negotiations in those terms.

CONCLUSION

The repertoire of mediated-negotiation strategies inevitably requires that planners exercise practical judgment, both politically and ethically. These judgments involve who is and who is not invited to meetings; where, when, and which meetings are held;

what issues should and should not appear on agendas; whose concerns are and are not acknowledged; how interventionist the planner's role is; and so on.

In local planning processes, then, planners often have the administrative discretion not only to mediate among conflicting parties, but to negotiate as interested parties themselves. Planning staff can routinely engage in the complementary tasks of supporting organizing efforts, negotiating, and mediating. In these ways, local planners can use a range of mediated-negotiation strategies to address practically existing power imbalances of access, information, class, and expertise that perpetually threaten the quality of local planning outcomes.

Mediated negotiations in local permitting processes will, of course, not resolve the structural problems of our society. Yet when local conflicts involve multiple issues, when differences in interests can be exploited by trading to achieve joint gains, and when diverse interests rather than fundamental rights are at stake, mediated-negotiation strategies for planners make good sense, politically, ethically, and practically.

SIX

"Advocacy and Pluralism in Planning"

Journal of the American Institute of Planners (1965)

Paul Davidoff

Editors' Introduction

Each year at the annual meeting of the Association of Collegiate Schools of Planning (ACSP), professors of urban planning present the Paul Davidoff Award to a city planning scholar whose work exemplifies the practice and ideals of the professor/lawyer/activist who is the author of this selection. It is an honor to receive the Davidoff award, because Davidoff exemplified professional commitment to vigorous advocacy on behalf of the less fortunate members of society.

During the 1960s, Davidoff, a lawyer and city planner, taught city planning students at Hunter College and simultaneously fought successfully to get racially integrated low-income housing built in exclusive white suburbs. This experience as an advocate for low-income minority residents shaped his view of what city planning should be like.

Unlike the "advocacy planning" Davidoff proposes, most city or town planning is performed by a single local government agency which develops plans which it feels will best serve the welfare of the whole community, not of individual interest groups such as organizations of homeless people, merchants, environmentalists, or bicycle enthusiasts. While city planning commissions may explore many alternatives and consider conflicting interest group demands before finalizing plans, generally they end up with a single unitary plan.

Davidoff's vision for how planning might be structured is quite different. He argues that different groups in society have different interests, which would result in fundamentally different plans if they were recognized. Business elites and other articulate, wealthy, and powerful groups have the skill and resources to shape city plans to serve their interests. But what about the poor and powerless? Davidoff argued that there should be planners acting as *advocates* articulating the interests of these and other groups much as a lawyer represents a client. For example, a planner might develop and advocate for a plan, which would meet the needs of poor West Indian residents of London's Brixton neighborhood. Another planner might develop a different plan representing the point of view of shopkeepers in the same area. And yet another might work with Brixton environmentalists to develop and advocate for a plan for the Brixton area incorporating the kind of sustainable urban development urged by Stephen Wheeler (p. 486). Confronted with these different "visions" (and the empirical data which makes the most compelling case for them), the local planning commission could weigh the merits of the competing plans much as a court weighs evidence and conflicting characterizations of a legal case by competing lawyers. Davidoff believed that the plan which would emerge from such a process would be better than a plan prepared by planning department staff without the interplay of competing advocate planners. And, Davidoff reasoned, the needs of the poor and powerless would be better met in city plans if – a big if – they were adequately represented by advocacy planners speaking on their behalf.

Davidoff's view of planning profoundly influenced activist planners of the 1960s and 1970s, many of whom defined themselves as advocacy planners and developed plans to meet the needs and interests of under-represented groups, with some notable successes. "Equity planners" today continue this tradition.

Compare Davidoff's humanistic, grassroots, pluralistic approach to city planning with Le Corbusier's brilliant, but elitist vision of an elite cadre of CIAM architects to impose the forms they felt modern machine culture demanded on the fabric of cities (p. 317). Compare Davidoff's views with John Forester's comments on how planners working within the system can use their influence to empower stakeholders in the planning process (p. 375). Compare the advocacy planning approach to strategies to empower communities to reach the highest possible level on the "ladder" of citizen participation that Sherry Arnstein has developed (p. 244). Review John Mollenkopf's description of how urban political power is actually exercised to understand how stakeholders represented by advocacy planners might better fit into pluralist local decision-making structures and urban regimes (p. 235).

While he was a planning director Norman Krumholz worked hard to make city planning responsive to the needs of the poor and powerless. As a planning professor at Cleveland State University he continues to develop the theory and practice of equity planning. Krumholtz describes his experience practicing equity planning as the planning director of Cleveland, Ohio, in *Making Equity Planning Work* (Philadelphia, PA: Temple University Press, 1990), co-authored with John Forester and the experience of other equity planners in *Reinventing Cities: Equity Planners Tell Their Stories* co-authored with Pierre Clavel (Philadelphia, PA: Temple University Press, 1994). Chester Hartman's *Cities for Sale: The Transformation of San Francisco* (Berkeley: University of California Press, 2002) describes how advocacy planners and lawyers fought to make urban renewal more responsive to very low-income residents of the area. For an application of advocacy planning to women, see Jacqueline Levitt, "Feminist Advocacy Planning in the 1980s" in Barry Checkoway (ed.), *Strategic Perspectives in Planning Practice* (Lexington, MA: Lexington Books, 1986). For a radical critique of advocacy planning, see Frances Fox Piven, "Whom Does the Advocate Planner Serve?" in Richard A. Cloward and Frances Fox Piven, *The Politics of Turmoil* (New York: Vintage, 1965). Piven sees advocacy planners as unwitting dupes of the system. She argues that angry and potentially violent groups will obtain more political leverage bargaining directly for themselves without professional intermediaries. Piven feels the poor and powerless need power, not plans.

The present can become an epoch in which the dreams of the past for an enlightened and just democracy are turned into a reality. The massing of voices protesting racial discrimination have roused this nation to the need to rectify racial and other social injustices. The adoption by Congress of a host of welfare measures and the Supreme Court's specification of the meaning of equal protection by law both reveal the response to protest and open the way for the vast changes still required.

The just demand for political and social equality on the part of the Negro and the impoverished requires the public to establish the bases for a society affording equal opportunity to all citizens. The compelling need for intelligent planning, for specification of new social goals and the means for achieving them, is manifest. The society of the future will be an urban one, and city planners will help to give it shape and content.

The prospect for future planning is that of a practice which openly invites political and social values to be examined and debated. Acceptance of this position means rejection of prescriptions for planning which would have the planner act solely as a technician. It has been argued that technical studies to enlarge the information available to decision makers must take precedence over statements of goals and ideals:

> We have suggested that, at least in part, the city planner is better advised to start from research into the functional aspects of cities than from his own estimation of the values which he is attempting to maximize. This suggestion springs from a conviction that at this juncture the implications of many planning decisions are poorly understood, and that no certain means are at hand by which values can be measured, ranked, and translated into the design of a metropolitan system.

While acknowledging the need for humility and openness in the adoption of social goals, this statement amounts to an attempt to eliminate, or sharply reduce, the unique contribution planning can make: understanding the functional aspects of the city and recommending appropriate future action to improve the urban condition.

Another argument that attempts to reduce the importance of attitudes and values in planning and other policy sciences is that the major public questions are themselves matters of choice between technical methods of solution. Dahl and Lindblom put forth this position at the beginning of their important text-book *Politics, Economics, and Welfare*:

In economic organization and reform, the "great issues" are no longer the great issues, if they ever were. It has become increasingly difficult for thoughtful men to find meaningful alternatives posed in the traditional choices between socialism and capitalism, planning and the free market, regulation and laissez faire, for they find their actual choices neither so simple nor so grand. Not so simple, because economic organization poses knotty problems that can only be solved by pain-staking attention to technical details – how else, for example, can inflation be controlled? Nor so grand, because, at least in the Western world, most people neither can nor wish to experiment with the whole pattern of socio-economic organization to attain goals more easily won. If, for example, taxation will serve the purpose, why "abolish the wages system" to ameliorate income inequality?

These words were written in the early 1950s and express the spirit of that decade more than that of the 1960s. They suggest that the major battles have been fought. But the "great issues" in economic organization, those revolving around the central issue of the nature of distributive justice, have yet to be settled. The world is still in turmoil over the way in which the resources of nations are to be distributed. The justice of the present social allocation of wealth, knowledge, skill, and other social goods is clearly in debate. Solutions to questions about the share of wealth and other social commodities that should go to different classes cannot be technically derived; they must arise from social attitudes.

Appropriate planning action cannot be prescribed from a position of value neutrality, for prescriptions are based on desired objectives. One conclusion drawn from this assertion is that "values are inescapable elements of any rational decision-making process" and that values held by the planner should be made clear. The implications of that conclusion for planning have been described elsewhere and will not be considered in this article. Here I will say that the planner should do more than explicate the values underlying his prescriptions

for courses of action; he should affirm them; he should be an advocate for what he deems proper.

Determinations of what serves the public interest, in a society containing many diverse interest groups, are almost always of a highly contentious nature. In performing its role of prescribing courses of action leading to future desired states, the planning profession must engage itself thoroughly and openly in the contention surrounding political determination. Moreover, planners should be able to engage in the political process as advocates of the interests both of government and of such other groups, organizations, or individuals who are concerned with proposing policies for the future development of the community.

The recommendation that city planners represent and plead the plans of many interest groups is founded upon the need to establish an effective urban democracy, one in which citizens may be able to play an active role in the process of deciding public policy. Appropriate policy in democracy is determined through a process of political debate. The right course of action is always a matter of choice, never of fact. In a bureaucratic age great care must be taken that choices remain in the area of public view and participation.

Urban politics, in an era of increasing government activity in planning and welfare, must balance the demands for ever-increasing central bureaucratic control against the demands for increased concern for the unique requirements of local, specialized interests. The welfare of all and the welfare of minorities are both deserving of support; planning must be so structured and so practiced as to account for this unavoidable bifurcation of the public interest.

The idealized political process in a democracy serves the search for truth in much the same manner as due process in law. Fair notice and hearings, production of supporting evidence, cross-examination, reasoned decision are all means employed to arrive at relative truth: a just decision. Due process and two-(or more) party political contention both rely heavily upon strong advocacy by a professional. The advocate represents an individual, group, or organization. He affirms their position in language understandable to his client and to the decision makers he seeks to convince.

If the planning process is to encourage democratic urban government then it must operate so as to include rather than exclude citizens from participating in the process. "Inclusion" means not only permitting the citizen to be heard. It also means that he be able to become well informed about the underlying reasons for planning proposals, and be able to respond to them in the technical language of professional planners.

A practice that has discouraged full participation by citizens in plan making in the past has been based on what might be called the "*unitary plan*." This is the idea that only one agency in a community should prepare a comprehensive plan; that agency is the city planning commission or department. Why is it that no other organization within a community prepares a plan? Why is only one agency concerned with establishing both general and specific goals for community development, and with proposing the strategies and costs required to effect the goals? Why are there not plural plans?

If the social, economic, and political ramifications of a plan are politically contentious, then why is it that those in opposition to the agency plan do not prepare one of their own? It is interesting to observe that "rational" theories of planning have called for consideration of alternative courses of action by planning agencies. As a matter of rationality it has been argued that all of the alternative choices open as means to the ends ought be examined. But those, including myself, who have recommended agency consideration of alternatives have placed upon the agency planner the burden of inventing "a few representative alternatives." The agency planner has been given the duty of constructing a model of the political spectrum, and charged with sorting out what he conceives to be worthy alternatives. This duty has placed too great a burden on the agency planner, and has failed to provide for the formulation of alternatives by the interest groups who will eventually be affected by the completed plans.

Whereas in a large part of our national and local political practice contention is viewed as healthy, in city planning where a large proportion of the professionals are public employees, contentious criticism has not always been viewed as legitimate. Further, where only government prepares plans, and no minority plans are developed, pressure is often applied to bring all professionals to work for the ends espoused by a public agency. For example, last year a Federal official complained to a meeting of planning professors that the academic planners were not giving enough support to Federal programs. He assumed that every planner should be on the side of the Federal renewal program. Of course government administrators will seek to gain the support of professionals outside of government, but such support should not be expected as a matter of loyalty. In a democratic system opposition to a public agency should be just as normal and appropriate as support. The agency, despite the fact that it is concerned with planning, may be serving undesired ends.

In presenting a plea for plural planning I do not mean to minimize the importance of the obligation of the public planning agency. It must decide upon appropriate future courses of action for the community. But being isolated as the only plan maker in the community, public agencies as well as the public itself may have suffered from incomplete and shallow analysis of potential directions. Lively political dispute aided by plural plans could do much to improve the level of rationality in the process of preparing the public plan.

The advocacy of alternative plans by interest groups outside of government would stimulate city planning in a number of ways. First, it would serve as a means of better informing the public of the alternative choices open, *alternatives strongly supported by their proponents*. In current practice those few agencies which have portrayed alternatives have not been equally enthusiastic about each. A standard reaction to rationalists' prescription for consideration of alternative courses of action has been "it can't be done; how can you expect planners to present alternatives which they don't approve?" The appropriate answer to that question has been that planners, like lawyers, may have a professional obligation to defend positions they oppose. However, in a system of plural planning, the public agency would be relieved of at least some of the burden of presenting alternatives. In plural planning the alternatives would be presented by interest groups differing with the public agency's plan. Such alternatives would represent the deep-seated convictions of their proponents and not just the mental exercises of rational planners seeking to portray the range of choice.

A second way in which advocacy and plural planning would improve planning practice would be in forcing the public agency to compete with other planning groups to win political support. In the absence of opposition or alternative plans presented by interest groups the public agencies have had little incentive to improve the quality of their work or the rate of production of plans. The political consumer has been offered a yes–no ballot in regard to the comprehensive plan; either the public agency's plan was to be adopted or no plan would be adopted.

A third improvement in planning practice which might follow from plural planning would be to force those who have been critical of "establishment" plans to produce superior plans, rather than only to carry out the very essential obligation of criticizing plans deemed improper.

THE PLANNER AS ADVOCATE

Where plural planning is practiced, advocacy becomes the means of professional support for competing claims about how the community should develop. Pluralism in support of political contention describes the process; advocacy describes the role performed by the professional in the process. Where unitary planning prevails, advocacy is not of paramount importance, for there is little or no competition for the plan prepared by the public agency. The concept of advocacy as taken from legal practice implies the opposition of at least two contending viewpoints in an adversary proceeding.

The legal advocate must plead for his own and his client's sense of legal propriety or justice. The planner as advocate would plead for his own and his client's view of the good society. The advocate planner would be more than a provider of information, an analyst of current trends, a simulator of future conditions, and a detailer of means. In addition to carrying out these necessary parts of planning, he would be a *proponent* of specific substantive solutions.

The advocate planner would be responsible to his client and would seek to express his client's views. This does not mean that the planner could not seek to persuade his client. In some situations persuasion might not be necessary, for the planner would have sought out an employer with whom he shared common views about desired social conditions and the means toward them. In fact one of the benefits of advocate planning is the possibility it creates for a planner to find employment with agencies holding values close to his own. Today the agency planner may be dismayed by the positions affirmed by his agency, but there may be no alternative employer.

The advocate planner would be above all a planner. He would be responsible to his client for preparing plans and for all of the other elements comprising the planning process. Whether working for the public agency or for some private organization, the planner would have to prepare plans that take account of the

arguments made in other plans. Thus the advocate's plan might have some of the characteristics of a legal brief. It would be a document presenting the facts and reasons for supporting one set of proposals, and facts and reasons indicating the inferiority of counter-proposals. The adversary nature of plural planning might, then, have the beneficial effect of upsetting the tradition of writing plan proposals in terminology which makes them appear self-evident.

A troublesome issue in contemporary planning is that of finding techniques for evaluating alternative plans. Technical devices such as cost–benefit analysis by themselves are of little assistance without the use of means for appraising the values underlying plans. Advocate planning, by making more apparent the values underlying plans, and by making definitions of social costs and benefits more explicit, should greatly assist the process of plan evaluation. Further, it would become clear (as it is not at present) that there are no neutral grounds for evaluating a plan; there are as many evaluative systems as there are value systems.

The adversary nature of plural planning might also have a good effect on the uses of information and research in planning. One of the tasks of the advocate planner in discussing the plans prepared in opposition to his would be to point out the nature of the bias underlying information presented in other plans. In this way, as critic of opposition plans, he would be performing a task similar to the legal technique of cross-examination. While painful to the planner whose bias is exposed (and no planner can be entirely free of bias) the net effect of confrontation between advocates of alternative plans would be more careful and precise research.

Not all the work of an advocate planner would be of an adversary nature. Much of it would be educational. The advocate would have the job of informing other groups, including public agencies, of the conditions, problems, and outlook of the group he represented. Another major educational job would be that of informing his clients of their rights under planning and renewal laws, about the general operations of city government, and of particular programs likely to affect them.

The advocate planner would devote much attention to assisting the client organization to clarify its ideas and to give expression to them. In order to make his client more powerful politically the advocate might also become engaged in expanding the size and scope of his client organization. But the advocate's most important function would be to carry out the planning process for the organization and to argue persuasively in favor of its planning proposals.

Advocacy in planning has already begun to emerge as planning and renewal affect the lives of more and more people. The critics of urban renewal have forced response from the renewal agencies, and the ongoing debate has stimulated needed self-evaluation by public agencies. Much work along the lines of advocate planning has already taken place, but little of it by professional planners. More often the work has been conducted by trained community organizers or by student groups. In at least one instance, however, a planner's professional aid led to the development of an alternative renewal approach, one which will result in the dislocation of far fewer families than originally contemplated.

Pluralism and advocacy are means for stimulating consideration of future conditions by all groups in society. But there is one social group which at present is particularly in need of the assistance of planners. This group includes organizations representing low-income families. At a time when concern for the condition of the poor finds institutionalization in community action programs, it would be appropriate for planners concerned with such groups to find means to plan with them. The plans prepared for these groups would seek to combat poverty and would propose programs affording new and better opportunities to the members of the organization and to families similarly situated. The difficulty in providing adequate planning assistance to organizations representing low-income families may in part be overcome by funds allocated to local antipoverty councils. But these councils are not the only representatives of the poor; other organizations exist and seek help. How can this type of assistance be financed? This question will be examined below, when attention is turned to the means for institutionalizing plural planning.

THE STRUCTURE OF PLANNING

Planning by special interest groups

The local planning process typically includes one or more "citizens'" organizations concerned with the nature of planning in the community. The Workable Program requirement for "citizen participation" has enforced this tradition and brought it to most large communities. The difficulty with current citizen participation programs is that citizens are more often *reacting* to agency programs than proposing their concepts of appropriate goals and future action.

The fact that citizens' organizations have not played a positive role in formulating plans is to some extent a result of both the enlarged role in society played by government bureaucracies and the historic weakness of municipal party politics. There is something very shameful to our society in the necessity to have organized "citizen participation." Such participation should be the norm in an enlightened democracy. The formalization of citizen participation as a required practice in localities is similar in many respects to totalitarian shows of loyalty to the state by citizen parades.

Will a private group interested in preparing a recommendation for community development be required to carry out its own survey and analysis of the community? The answer would depend upon the quality of the work prepared by the public agency, work which should be public information. In some instances the public agency may not have surveyed or analyzed aspects the private group thinks important; or the public agency's work may reveal strong biases unacceptable to the private group. In any event, the production of a useful plan proposal will require much information concerning the present and predicted conditions in the community. There will be some costs associated with gathering that information, even if it is taken from the public agency. The major cost involved in the preparation of a plan by a private agency would probably be the employment of one or more professional planners.

What organizations might be expected to engage in the plural planning process? The first type that comes to mind are the political parties; but this is clearly an aspirational thought. There is very little evidence that local political organizations have the interest, ability, or concern to establish well-developed programs for their communities. Not all the fault, though, should be placed upon the professional politicians, for the registered members of political parties have not demanded very much, if anything, from them as agents.

Despite the unreality of the wish, the desirability for active participation in the process of planning by the political parties is strong. In an ideal situation local parties would establish political platforms which would contain master plans for community growth and both the majority and minority parties in the legislative branch of government would use such plans as one basis for appraising individual legislative proposals. Further, the local administration would use its planning agency to carry out the plans it proposed to the electorate. This dream will not turn to reality for a long time. In the interim other interest groups must be sought to fill the gap caused by the present inability of political organizations.

The second set of organizations which might be interested in preparing plans for community development are those that represent special interest groups having established views in regard to proper public policy. Such organizations as chambers of commerce, real estate boards, labor organizations, pro- and anti-civil rights groups, and anti-poverty councils come to mind. Groups of this nature have often played parts in the development of community plans, but only in a very few instances have they proposed their own plans.

It must be recognized that there is strong reason operating against commitment to a plan by these organizations. In fact it is the same reason that in part limits the interests of politicians and which limits the potential for planning in our society. The expressed commitment to a particular plan may make it difficult for groups to find means for accommodating their various interests. In other terms, it may be simpler for professionals, politicians, or lobbyists to make deals if they have not laid their cards on the table.

There is a third set of organizations that might be looked to as proponents of plans and to whom the foregoing comments might not apply. These are the ad hoc protest associations which may form in opposition to some proposed policy. An example of such a group is a neighborhood association formed to combat a renewal plan, a zoning change, or the proposed location of a public facility. Such organizations may seek to develop alternative plans, plans which would, if effected, better serve their interests.

From the point of view of effective and rational planning it might be desirable to commence plural planning at the level of city-wide organizations, but a more realistic view is that it will start at the neighborhood level. Certain advantages of this outcome should be noted. Mention was made earlier of tension in government between centralizing and decentralizing forces. The contention aroused by conflict between the central planning agency and the neighborhood organization may indeed be healthy, leading to clearer

definition of welfare policies and their relation to the rights of individuals or minority groups.

Who will pay for plural planning? Some organizations have the resources to sponsor the development of a plan. Many groups lack the means. The plight of the relatively indigent association seeking to propose a plan might be analogous to that of the indigent client in search of legal aid. If the idea of plural planning makes sense, then support may be found from foundations or from government. In the beginning it is more likely that some foundation might be willing to experiment with plural planning as a means of making city planning more effective and more democratic. Or the Federal Government might see plural planning, if carried out by local anti-poverty councils, as a strong means of generating local interest in community affairs.

Federal sponsorship of plural planning might be seen as a more effective tool for stimulating involvement of the citizen in the future of his community than are the present types of citizen participation programs. Federal support could only be expected if plural planning were seen, not as a means of combating renewal plans, but as an incentive to local renewal agencies to prepare better plans.

The public planning agency

A major drawback to effective democratic planning practice is the continuation of that non-responsible vestigial institution, the planning commission. If it is agreed that the establishment of both general policies and implementation policies are questions affecting the public interest and that public interest questions should be decided in accord with established democratic practices for decision making, then it is indeed difficult to find convincing reasons for continuing to permit independent commissions to make planning decisions. At an earlier stage in planning the strong arguments of John T. Howard and others in support of commissions may have been persuasive. But it is now more than a decade since Howard made his defense against Robert Walker's position favoring planning as a staff function under the mayor. With the increasing effect planning decisions have upon the lives of citizens the Walker proposal assumes great urgency.

Aside from important questions regarding the propriety of independent agencies which are far removed from public control determining public policy, the failure to place planning decision choices in the hands of elected officials has weakened the ability of professional planners to have their proposals effected. Separating planning from local politics has made it difficult for independent commissions to garner influential political support. The commissions are not responsible directly to the electorate and in turn the electorate is, at best, often indifferent to the planning commission.

During the last decade, in many cities power to alter community development has slipped out of the hands of city planning commissions, assuming they ever held it, and has been transferred to development co-ordinators. This has weakened the professional planner. Perhaps planners unknowingly contributed to this by their refusal to take concerted action in opposition to the perpetuation of commissions.

Planning commissions are products of the conservative reform movement of the early part of this century. The movement was essentially anti-populist and pro-aristocracy. Politics was viewed as dirty business. The commissions are relics of a not-too-distant past when it was believed that if men of good will discussed a problem thoroughly, certainly the right solution would be forthcoming. We know today, and perhaps it was always known, that there are no right solutions. Proper policy is that which the decision-making unit declares to be proper.

Planning commissions are responsible to no constituency. The members of the commissions, except for their chairman, are seldom known to the public. In general the individual members fail to expose their personal views about policy and prefer to immerse them in group decision. If the members wrote concurring and dissenting opinions, then at least the commissions might stimulate thought about planning issues. It is difficult to comprehend why this aristocratic and undemocratic form of decision making should be continued. The public planning function should be carried out in the executive or legislative office and perhaps in both. There has been some question about which of these branches of government would provide the best home, but there is much reason to believe that both branches would be made more cognizant of planning issues if they were each informed by their own planning staffs. To carry this division further, it would probably be advisable to establish minority and majority planning staffs in the legislative branch.

At the root of my last suggestion is the belief that there is or should be a Republican and Democratic way of viewing city development; that there should be

conservative and liberal plans, plans to support the private market, and plans to support greater government control. There are many possible roads for a community to travel and many plans should show them. Explication is required of many alternative futures presented by those sympathetic to the construction of each such future. As indicated earlier, such alternatives are not presented to the public now. Those few reports which do include alternative futures do not speak in terms of interest to the average citizen. They are filled with professional jargon and present sham alternatives. These plans have expressed technical land use alternatives rather than social, economic, or political value alternatives. Both the traditional unitary plans and the new ones that present technical alternatives have limited the public's exposure to the future states that might be achieved. Instead of arousing healthy political contention as diverse comprehensive plans might, these plans have deflated interest.

The independent planning commission and unitary plan practice certainly should not co-exist. Separately they dull the possibility for enlightened political debate; in combination they have made it yet more difficult. But when still another hoary concept of city planning is added to them, such debate becomes practically impossible. This third of a trinity of worn-out notions is that city planning should focus only upon the physical aspects of city development.

AN INCLUSIVE DEFINITION OF THE SCOPE OF PLANNING

The view that equates physical planning with city planning is myopic. It may have had some historic justification, but it is clearly out of place at a time when it is necessary to integrate knowledge and techniques in order to wrestle effectively with the myriad of problems afflicting urban populations.

The city planning profession's historic concern with the physical environment has warped its ability to see physical structures and land as servants to those who use them. Physical relations and conditions have no meaning or quality apart from the way they serve their users. But this is forgotten every time a physical condition is described as good or bad without relation to a specified group of users. High density, low density, green belts, mixed uses, cluster developments, centralized or decentralized business centers are per se neither good nor bad. They describe physical relations or conditions, but take on value only when seen in terms of their social, economic, psychological, physiological, or aesthetic effects upon different users.

The profession's experience with renewal over the past decade has shown the high costs of exclusive concern with physical conditions. It has been found that the allocation of funds for removal of physical blight may not necessarily improve the overall physical condition of a community and may engender such harsh social repercussions as to severely damage both social and economic institutions. Another example of the deficiencies of the physical bias is the assumption of city planners that they could deal with the capital budget as if the physical attributes of a facility could be understood apart from the philosophy and practice of the service conducted within the physical structure. This assumption is open to question. The size, shape, and location of a facility greatly interact with the purpose of the activity the facility houses. Clear examples of this can be seen in public education and in the provision of low cost housing. The racial and other socioeconomic consequences of "physical decisions" such as location of schools and housing projects have been immense, but city planners, while acknowledging the existence of such consequences, have not sought or trained themselves to understand socioeconomic problems, their causes or solutions.

The city planning profession's limited scope has tended to bias strongly many of its recommendations toward perpetuation of existing social and economic practices. Here I am not opposing the outcomes, but the way in which they are developed. Relative ignorance of social and economic methods of analysis has caused planners to propose solutions in the absence of sufficient knowledge of the costs and benefits of proposals upon different sections of the population.

Large expenditures have been made on planning studies of regional transportation needs, for example, but these studies have been conducted in a manner suggesting that different social and economic classes of the population did not have different needs and different abilities to meet them. In the field of housing, to take another example, planners have been hesitant to question the consequences of locating public housing in slum areas. In the field of industrial development, planners have seldom examined the types of jobs the community needs; it has been assumed that one job was about as useful as another. But this may not be the case where a significant sector of the population finds it difficult to get employment.

"Who gets what, when, where, why, and how" are the basic political questions which need to be raised about every allocation of public resources. The questions cannot be answered adequately if land use criteria are the sole or major standards for judgment.

The need to see an element of city development, land use, in broad perspective applies equally well to every other element, such as health, welfare, and recreation. The governing of a city requires an adequate plan for its future. Such a plan loses guiding force and rational basis to the degree that it deals with less than the whole that is of concern to the public.

The implications of the foregoing comments for the practice of city planning are these. First, state planning enabling legislation should be amended to permit planning departments to study and to prepare plans related to any area of public concern. Second, planning education must be redirected so as to provide channels of specialization in different parts of public planning and a core focused upon the planning process. Third, the professional planning association should enlarge its scope so as to not exclude city planners not specializing in physical planning.

A year ago at the AIP convention it was suggested that the AIP Constitution be amended to permit city planning to enlarge its scope to all matters of public concern. Members of the Institute in agreement with this proposal should seek to develop support for it at both the chapter and national level. The Constitution at present states that the Institute's "particular sphere of activity shall be the planning of the unified development of urban communities and their environs and of states, regions and the nation *as expressed through determination of the comprehensive arrangement of land and land occupancy and regulation thereof.*" It is time that the AIP delete the words in my italics from its Constitution. The planner limited to such concerns is not a city planner, he is a land planner or a physical planner. A city is its people, their practices, and their political, social, cultural and economic institutions as well as other things. The city planner must comprehend and deal with all these factors.

The new city planner will be concerned with physical planning, economic planning, and social planning. The scope of his work will be no wider than that presently demanded of a mayor or a city councilman. Thus, we cannot argue against an enlarged planning function on grounds that it is too large to handle. The mayor needs assistance; in particular he needs the assistance of a planner, one trained to examine needs and aspirations in terms of both short- and long-term perspectives. In observing the early stages of development of Community Action Programs, it is apparent that our cities are in desperate need of the type of assistance trained planners could offer. Our cities require for their social and economic programs the type of long-range thought and information that have been brought forward in the realm of physical planning. Potential resources must be examined and priorities set.

What I have just proposed does not imply the termination of physical planning, but it does mean that physical planning be seen as part of city planning. Uninhibited by limitations on his work, the city planner will be able to add his expertise to the task of coordinating the operating and capital budgets and to the job of relating effects of each city program upon the others and upon the social, political, and economic resources of the community.

An expanded scope reaching all matters of public concern will make planning not only a more effective administrative tool of local government but it will also bring planning practice closer to the issues of real concern to the citizens. A system of plural city planning probably has a much greater chance for operational success where the focus is on live social and economic questions instead of rather esoteric issues relating to physical norms.

THE EDUCATION OF PLANNERS

Widening the scope of planning to include all areas of concern to government would suggest that city planners must possess a broader knowledge of the structure and forces affecting urban development. In general this would be true. But at present many city planners are specialists in only one or more of the functions of city government. Broadening the scope of planning would require some additional planners who specialize in one or more of the services entailed by the new focus.

A prime purpose of city planning is the coordination of many separate functions. This coordination calls for men holding general knowledge of the many elements comprising the urban community. Educating a man for performing the coordinative role is a difficult job, one not well satisfied by the present tradition of two years

of graduate study. Training of urban planners with the skills called for in this article may require both longer graduate study and development of a liberal arts undergraduate program affording an opportunity for holistic understanding of both urban conditions and techniques for analyzing and solving urban problems.

The practice of plural planning requires educating planners who would be able to engage as professional advocates in the contentious work of forming social policy. The person able to do this would be one deeply committed to both the process of planning and to particular substantive ideas. Recognizing that ideological commitments will separate planners, there is tremendous need to train professionals who are competent to express their social objectives.

The great advances in analytic skills, demonstrated in the recent May issue of this journal [*Journal of the American Institute of Planners*] dedicated to techniques of simulating urban growth processes, portend a time when planners and the public will be better able to predict the consequences of proposed courses of action. But these advances will be of little social advantage if the proposals themselves do not have substance. The contemporary thoughts of planners about the nature of man in society are often mundane, unexciting or gimmicky. When asked to point out to students the planners who have a developed sense of history and philosophy concerning man's situation in the urban world one is hard put to come up with a name. Sometimes Goodman or Mumford might be mentioned. But planners seldom go deeper than acknowledging the goodness of green space and the soundness of proximity of linked activities. We cope with the problems of the alienated man with a recommendation for reducing the time of the journey to work.

CONCLUSION

The urban community is a system comprised of interrelated elements, but little is known about how the elements do, will, or should interrelate. The type of knowledge required by the new comprehensive city planner demands that the planning profession be comprised of groups of men well versed in contemporary philosophy, social work, law, the social sciences, and civic design. Not every planner must be knowledgeable in all these areas, but each planner must have a deep understanding of one or more of these areas and he must be able to give persuasive expression to his understanding. As a profession charged with making urban life more beautiful, exciting, and creative, and more just, we have had little to say.

Our task is to train a future generation of planners to go well beyond us in its ability to prescribe the future urban life.

"Green Urbanism and the Lessons of European Cities: 'The Vision of Green Urbanism' and 'Creating Green and Sustainable Cities in the United States: Lessons from Europe'"

from *Green Urbanism: Learning from European Cities* (2000)

Timothy Beatley

Editors' Introduction

The urbanization of the human population described by Kingsley Davis (p. 25), particularly in the last hundred years, and industrialization without regard for the environmental consequences in both capitalist and communist countries, have not only produced serious problems of pollution and congestion in cities throughout the world, but also raise serious concerns about planet Earth's capacity to sustain urban life.

Today "green politics" has established itself as a serious political movement worldwide. "Green parties" have a political agenda to protect the environment and manage Earth's resources more responsibly. In Europe green parties have been successful in electing enough representatives to be serious political forces in the governing coalitions of a number of countries. In the United States some local green party candidates have been elected and green politics is stimulating new thinking about the nature of urban development.

Some European cities where the green movement is strongest work hard to live within their ecological limits, reduce their ecological footprints, and implement policies that are sensitive to the needs of other cities in their regions, countries, and the world. In this selection American urban planning professor Timothy Beatley describes lessons from the experience of green urbanism in European cities. The examples he gives and the generalizations he reaches are highly relevant for readers everywhere in the world.

Green urbanism is not a new idea. The idea of building cities in harmony with the natural environment is expressed in other current intellectual movements in architecture and planning today. Before World War I eccentric Scottish biologist Patrick Geddes had classified the environmental needs of different ecological systems and developed a systematic approach to building cities that respected natural systems. Ian McHarg, another Scot, wrote a seminal book titled *Design with Nature* that inspired the environmentally conscious generation of the 1960s. Today there is a movement in ecological design among architects, many urban planners pursue sustainable urban development strategies as described by Stephen Wheeler (p. 486), and respect for the natural environment is a cornerstone of the New Urbanism movement described by Andres Duany and Elizabeth Plater-Zyberk (p. 207) and Peter Calthorpe and William Fulton (p. 331). Green urbanism is well established in urban *theory*, but the practice of green urbanism is most advanced in Europe. Beatley argues that while no European city has achieved all of the goals of green urbanism many European cities have successfully implemented green urbanism practices. He believes that the recent experience in progressive European cities demonstrates that alternative patterns of urban growth and development are possible. Compact, walkable, green communities can be created; cities that are sustainable, economically viable, and livable.

Hard-headed (hearted?) decision-makers often dismiss alternative green visions as hopelessly unrealistic. In their view it is easy to dream of cities where renewable solar and wind power rather than non-renewable resources like coal, oil, and natural gas produce much of the communities' energy needs. It is pleasant to fantasize about communities where people walk and bicycle to work rather than drive cars. And growing enough vegetables near or in cities sounds appealing enough. But, these critics of green urbanism say that alternative energy sources can never really produce nearly enough energy to run cities. Bicycle and pedestrian paths are nice, but people need cars and super highways to get around. And agribusiness, not urban farms, is the only way to grow enough food cheaply enough to feed Earth's six plus billion inhabitants. Beatley disagrees. He produces hard evidence from Europe that green communities are possible, economically feasible, and can be very livable.

Many European cities are more compact than their American counterparts because citizens accept much higher density development than their suburb-loving, auto-dependent American counterparts. Tough European land use planning and a culture that values urban community reduce sprawl and keep cities compact. Americans who have chosen to live in low-density suburbs travel to Paris or Rome to enjoy the energy and street life!

Many European cities invest much more per capita in public transportation systems such as buses, trains, and subways than cities in the United States. This makes it possible for households to get around with one car or no car at all. If there is a really good bus system serving a compact city, as is the case in many European cities, people will use buses rather than cars for many trips. This in turn generates revenue to keep the system viable.

In contrast, if there is a terrible bus system that functions (poorly) as the transit system of last resort it will not attract riders who have other alternatives, and without enough revenue will deteriorate further. In many European countries gasoline taxes are two or three times as high as in the United States. This discourages people from driving and further reinforces public transit. Amsterdam and other European cities work hard to make bicycling possible. Some even provide free public bicycle depots. This bundle of carrots and sticks makes for much more environmentally friendly transit in Europe than in North America.

Timothy Beatley is a professor of Urban and Environmental Planning at the University of Virginia, Charlottesville. This selection is from *Green Urbanism: Learning from European Cities* (Washington, DC: Island Press, 2000). Professor Beatley's other books include *The Sustainable Urban Development Reader* (London and New York: Routledge, 2003) co-edited with Stephen Wheeler, *The Ecology of Place* (Washington, DC: Island Press, 1997), and *An Introduction to Coastal Zone Management* (Washington, DC: Island Press, 1993).

Other readings on green urbanism include Judy Corbett, Michael Corbett, and Robert Thayer, *Designing Sustainable Communities: Learning from Village Homes* (Washington, DC: Island Press, 2000), Andrew Dobson (ed.), *The Green Reader: Essays Toward a Sustainable Society* (San Francisco: Mercury House, 1991), and Alex Wilson, Jen Uncapher, Lisa McManigal, and L. Hunter Lovins, *Green Development: Integrating Ecology and Real Estate* (New York: John Wiley, 1998).

For more on green politics in Europe see John S. Dryzek (ed.), *Green States and Social Movements: Environmentalism in the United States, United Kingdom, Germany and Norway* (London and New York: Oxford University Press, 2002) and Michael O'Neill, *Green Parties and Political Change in Contemporary Europe: New Politics, Old Predicaments* (Aldershot, UK and Brookfield, VT: Ashgate, 1997). For one view on green politics in America see John Rensenbrink, *Against All Odds: The Green Transformation of American Politics* (Raymond, ME: Leopold Press, 1999).

Classic statements that anticipate the green urbanism movement are Patrick Geddes, *Cities in Evolution* (London: Williams and Norgate, 1915, reprinted in Richard LeGates and Frederic Stout, *Early Urban Planning 1870–1940* (London: Routledge/Thoemmes, 1998), and Ian McHarg, *Design with Nature* (Garden City, NY: Doubleday & Company, 1969).

THE VISION OF GREEN URBANISM

There are . . . many different terms used today in discussing efforts to reduce environmental impacts and to live more lightly on the land. *Sustainable development*, *sustainable communities*, and *sustainable cities* are a few of these terms, and each captures much of the agenda of this book. *Green urbanism* effectively captures both the central urban and environmental dimensions of the agenda I will be discussing. It emphasizes the important role of cities and positive urbanism in shaping more sustainable places, communities, and lifestyles. And, it implicitly emphasizes that our old approaches to urbanism – our old views of cities, towns, and communities – are incomplete and must be substantially expanded to incorporate ecology and more ecologically responsible forms of living and settlement. This need for a revised approach has been an ongoing concern with the so-called *new urbanism*, so enthusiastically endorsed by many American architects and planners . . . What we need today are cities that reflect a *different* new urbanism, a new urbanism that is dramatically more ecological in design and functioning and that has ecological limits at its core.

Precisely what green urbanism implies is evolving and unclear, but in the programs, policies, and creative design ideas found in many European cities, we begin to sharpen our sense of what might be possible. To elaborate on what the vision of green urbanism includes (and incorporating and extending the thinking of others), there are several important design qualities or characteristics. Cities that exemplify green urbanism are:

■ *Cities that strive to live within their ecological limits, fundamentally reduce their ecological footprints, and acknowledge their connections with and impacts on other cities and communities and the larger planet.*

Green urbanism accepts that public (and private) decisions about how cities grow, the kinds of transportation systems they employ, and the ways they generate and supply energy and food for their populations have tremendous environmental impacts. Green urbanism takes as a primary goal the need to greatly reduce the ecological footprints of cities, to live within the limit of local and regional ecosystems, and to acknowledge that in a host of ways the decisions in one city affect the quality of environment and life in other places, as well as the overall health of the planet.

Efforts by cities to reduce carbon dioxide emissions and to reduce the impact of urban consumption patterns, for example, reflect all of these goals.

[. . .]

■ *Cities that are green and that are designed for and function in ways analogous to nature.*

Green urbanism requires us to overcome our traditional view of the polarity of cities and nature. Cities, to many, are indeed the very antithesis of nature – places of gray, where one finds concrete and asphalt, buildings and cars, things that could not be natural. Yet, nature does exist in cities, and cities are fundamentally embedded within a larger natural setting.

Ecological architect William McDonough frequently ponders whether towns and cities might function like forests. Indeed, nature is a profoundly helpful paradigm for cities. Cities must strive to be places of nature; they should be sheltering; cleansing of air, water, and spirit; and restorative and replenishing of the planet, rather than fundamentally extractive and damaging.

Moreover, innumerable ways of restoring, replenishing, and nurturing urban ecology exist, such as daylighting streams, planting green rooftops, bringing forests and greenspaces into the very heart of cities, as well as many other creative planning approaches.

■ *Cities that strive to achieve a circular rather than a linear metabolism, which nurtures and develops positive symbiotic relationships with and between its hinterland (whether that be regional, national, or international).*

Nothing in nature is wasted. Wastes become productive inputs for other natural processes. In hundreds of ways, the same principle could apply to the functioning of cities. Wastewater treatment systems can extract biogas to fuel community heating systems; organic household waste can become fertilizer returned to urban populations in the form of food. Industries can feed off each other, with each company's wastes becoming the productive inputs to production processes of the others.

Green urbanism calls for a circular metabolism for cities, rather than the prevailing linear approach. Green urbanism calls for the balancing of the ecocycles of a city, so that its inputs and outputs are harmonized, complementary, and fundamentally in balance.

■ *Cities that strive toward local and regional self-sufficiency and take full advantage of and nurture local/regional food production, economy, power production, and many other activities that sustain and support their populations.*

Green urbanism demands cities that assume responsibility for the environmental and other impacts of lifestyle and consumption decisions. Historically, it has been easier to ignore the devastating impacts of conventional energy production and production of food, for example, because these impacts have typically been externalized. They happen far away, in distant places, out of sight and mind. Bringing these many functions closer means greater scrutiny and presents more likelihood that responsible actions and choices will result.

What will also result are more healthful lives, for example by avoiding the consumption of chemicals and level of processing needed to transport food hundreds of miles.

■ *Cities that facilitate (and encourage) more sustainable, healthful lifestyles.*

An important measure of a sustainable city, and an important goal of green urbanism, is that such places should make it easier for people to live richer, fuller lives. Most Americans, for instance, have few options besides the automobile. Green urbanism emphasizes giving individuals the ability to walk or ride bicycles if they choose. It gives them the option and the ability to grow food, to live with fewer consumer goods, to live without a car if they choose to, and so on. The benefits again extend beyond the environmental. Such conditions empower individuals and families to change in meaningful ways (if they so desire) the directions of their lives and to emphasize the quality of their relationships, rather than the size of their home or their possessions.

■ *Cities that emphasize a high quality of life and the creation of highly livable neighborhoods and communities.*

Green urbanism gives centrality to creating (and strengthening) neighborhoods and places where people enjoy being, places that are emotionally uplifting and aesthetically inspirational. It is an agenda that emphasizes the provision of adequate housing and services for all members of society and that seeks to be socially and economically inclusive. It is as much about

creating highly livable cities as it is about creating ecological cities (indeed, the goals are mutually reinforcing). Nature in cities is important to livability. Green urbanism presumes that connections to nature are important to personal health and well-being.

Are there actually cities in the world that satisfy these ambitious criteria? The insertion of the words strive to is important to note. The many examples of European cities described in the pages that follow, I will argue, often come the closest to meeting the vision. But, in many ways, even the most exemplary individual cities fall short. Nevertheless, the following text contains a menu of creative tools, design concepts, and tangible examples that illustrate how cities – particularly American cities – might move closer to these ideals.

CREATING GREEN AND SUSTAINABLE CITIES IN THE UNITED STATES: LESSONS FROM EUROPE

It might be wondered whether what has been accomplished in these most exemplary European cities can also occur in their American counterparts. There is no question that economic, political, geographic, and other circumstances make many of the programs and initiatives described here more feasible in European countries than in the United States. The background context is clearly different in many significant ways. For example, the density of population and the scarcity of land in countries such as the Netherlands make compact city policies much more of a necessity. The much higher price of gasoline, mostly a result of historically higher government taxes, helps to make automobile use more expensive and public transit more attractive. And, the structure and more extensive role of government, and its stronger land use planning function, is also helpful in achieving many of the outcomes described here. Nevertheless, there are many lessons and broad themes that emerge from these European cities, which American planners and public officials should consider. These lessons include the following:

■ *A more sustainable urban development model is possible: the promise of ecological living environments.*

Perhaps the clearest lesson from European urban practice is that alternative patterns of growth and development do exist, and it is possible to organize space and public investments in ways that create compact, walkable, green communities – places that exhibit many highly attractive qualities. Public transit, walking, and bicycle use, under the right circumstances and with the appropriate long-term public policies, can serve as viable alternatives to reliance on the automobile. Substantially less energy and fewer resources can be consumed by citizens, and at the same time personal choice and mobility can be expanded. In short, these European cities, while clearly not perfect and clearly confronting a host of their own problems, show that different, more sustainable future paths do exist. Moreover, collectively the cases have provided substantial guidance about the tools, techniques, methods, and strategies that can be employed in the pursuit of more sustainable paths

These cities also strongly demonstrate that compact urban form and other sustainable design measures need not conflict with a high quality of life. Indeed, just the opposite is the case. Compact, denser, walkable communities offer tremendous amenities, social and environmental, and are life-enhancing as well as ecologically restorative. A major lesson can be seen in the multiple objectives achieved and in reinforcing agendas possible in these green-urban cities.

[. . .]

Many of the cities studied here have taken amazing steps to experiment with and promote ecological housing and development. From the scale of a single home or building to entire new neighborhoods and urban districts, these experiences demonstrate that new urban development can fundamentally reduce its ecological footprint. Moreover, and very importantly, these new ecological neighborhoods and developments can occur in urban settings and can indeed contribute to a broader, more sustainable urban form . . .

■ *Sustainable mobility.*

Many of the European cities are grappling with American trends of decentralization and increasing auto use and thus provide special insights and examples. In the face of these trends, cities such as Freiburg, Zürich, and Amsterdam have made major commitments and investments in other, more sustainable modes of mobility and transportation. These European cities provide important models for how American cities can address their transportation

problems. The solution is not a simple one, not a silver bullet, but a package of initiatives, which includes commitment to and investment in comprehensive, integrated public transit systems, coordination of land use and transport investment decisions, increasing opportunities for bicyclists and pedestrians, and looking for creative ways to minimize or reduce automobility and reliance. Although some of the models are still experimental, these European cities offer some important ideas about handling mobility in the future, including the use of car-sharing, comprehensive mobility packages, car-free development and housing areas, and financial incentives to reduce car use (e.g., road pricing), among others.

■ *Integrated strategies and solutions.*

The hallmark of many of these successful city efforts is the attempt to take an integrated, holistic approach. This is especially clear in the transportation area, where the most impressive cities have adopted and implemented a variety of reinforcing policies and measures. These measures include simultaneously investing in and expanding public transit, expanding bicycle networks and facilities, pedestrianization and traffic calming, aggressive parking restrictions, and locating major new developments near public transit and where walking and bicycling can take place.

[. . .]

■ *Urban metabolism: the city as an organic whole.*

These European cases illustrate the importance of looking at cities as organic wholes – whether through the lens of urban metabolism, ecological footprints, or ecocycle balancing. Cities require substantial environmental inputs (e.g., water, energy, and food) and generate substantial outputs (e.g., waste). Conventional thinking sees this process as essentially linear – inputs are drawn in, outputs flow out. Many of the cities profiled, however, are attempting (many admittedly in the early stages) to connect these inputs and outputs with the promise of reducing long-term ecological impacts, using resources more efficiently, and reducing economic costs. Stockholm's efforts to promote a balanced ecocycles approach is one such promising example.

An equally powerful metaphor is to see the city as a living environment – fundamentally like a forest, or a wetland, or a prairie. These themes have been increasingly important in Europe. The EU's *Green Paper on the Urban Environment* and later the important report of European Sustainable Cities both strongly emphasize an ecosystem's view of cities. The cities profiled here provide a variety of powerful ideas and tangible examples of how to acknowledge and put into practice this view of cities. They range from the impressive urban-ecological structures of Scandinavian cities (such as Helsinki), which bring wild nature almost to the very city-center, to the very important use of greenroofs (eco-roofs) in Germany and the Netherlands, which create new habitats and nature in the most urban and developed of city spaces. American cities, perhaps because the option of horizontal expansion (ever in pursuit of the elusive nature beyond the boundaries of cities) has always existed, have yet to face squarely the challenge of naturalizing or eco-logizing urbanity. The Europeans have not always done it completely or perfectly but they are, nevertheless, on the better course.

■ *The shift toward renewable energy cities.*

Although there is certainly much still to be done, there are now many positive examples of cities (and national governments) actively promoting a shift toward renewable forms of energy. The support for solar energy is particularly instructive, with governments such as Germany's and the Netherlands' providing significant financial support (as is the European Commission). Development projects such as Nieuwland in Amersfoort, in which a very high percentage of the homes will incorporate some combination of active and passive solar, are impressive and should be emulated here in the United States. Through a mixture of energy conservation and support of renewables, cities and settlements can significantly reduce their overall ecological footprint and make a significant contribution to addressing global climate change.

The integration of and support for solar energy, and a sense of the need to fundamentally move away from dependence on local fuels, are especially impressive in many of these European cities . . . There is a great missed opportunity in American cities to seriously embrace the renewables agenda. Cities in the American Southwest, for example, whose economies and infrastructure could be essentially driven by solar power, seem to have little interest in this (ubiquitous) renewable energy source.

■ *The importance of reinforcing actions at each scale; the importance of connections and linkages.*

The cases illustrate the importance of reinforcing actions and policies at different jurisdictional levels and geographical scales. Ecological design features at the building or project scale can significantly reduce overall energy or resource demands (e.g., through insulation and passive solar). At a regional or extra-regional level, however, efforts are needed to develop renewable energy resources (e.g., wind energy in Bologna). More sustainable forms of transportation and mobility depend on actions at both a project or neighborhood level (e.g., designing for pedestrians and bicycles and de-emphasizing automobiles) and metropolitan and regional levels (e.g., public transit investments and improvements). Creative attention at each scale is critical to the outcome of creating sustainable places.

Almost all of the cities are working to enhance mobility through a strategy of linkages and connections, which take several forms. One form is the spatial connections between different neighborhoods and districts in the city. In most of the bicycle-friendly cities studied, the urban bike network links up with recreational routes, making it relatively easy to get to more open expanses of land and nature. An emphasis placed on connectedness pays tremendous dividends in these cities.

Attention to detail is another part of the story. For example, while there are many larger locational decisions and land use and infrastructural policies that explain more sustainable transportation systems, one is struck by the attention to detail at every planning level and the important cumulative effect this has. As the tram systems in cities such as Freiburg effectively show, enticing people to use public transit is often a result of the cumulative effect of lots of actions that make it easy and enjoyable to use this mode. Positive design details include comfortable seat configurations, easy ticket purchase and use, real-time reporting of when the next tram will arrive, low-carriage cars for easy loading and unloading, and so forth.

Facilitating the use of bicycles is similarly a matter of detailing, including the many individual design considerations that make bicyclists feel safe, such as the physical separation of cars and bikes, separate traffic lighting for bicycles, special painted surfaces (e.g., the red bike lanes in Heidelberg and elsewhere), and actions to slow and calm auto traffic . . .

■ *Advances in sustainability through partnerships.*

Many of the city case studies, and perhaps the European approach more generally, reflect a spirit of collaboration between the industrial and business communities, and government. There appears to be less antagonism and divisiveness between these groups and a sense of the mutuality of their objectives and agendas. This certainly is partly cultural (consider the experiences of the Dutch and the consensus-based nature of their society), but local governments can clearly make the choice to include and collaborate with business in pursuit of sustainability.

Cities such as Leicester have successfully built partnerships among the city, community groups, and the media. Leicester's Turning the Tide Campaign is actually guided by a so-called "Partnership Board." The close involvement of the media has been especially important in helping to get the word out about more sustainable lifestyles and practices. Leicester's Quality Bus Partnership, a collaborative effort with bus operators to improve service, is yet another example. Universities can also be important partners (e.g., consider the positive role played by De Montfort University's Institute for Energy and Sustainable Development, in Leicester). Many of the case cities provide similar positive examples. Successful partnering is clearly a touchstone in these sustainable cities.

[. . .]

■ *Engaging target groups; collaborative, interdisciplinary approaches.*

The case studies here illustrate that movement toward a more sustainable society requires efforts and actions by a variety of groups in society, including industry, government, and consumers, among others. Different initiatives and creative approaches are needed to involve and engage these different target groups. Programs such as Graz's ECOPROFIT, for instance, recognize the importance of engaging the industrial and business sectors. The use of eco-teams, a concept pioneered and promoted in a number of cities by Global Action Plan, addresses the changes needed in personal and household consumption. Each of these different groups in society will need to be involved in crafting a more sustainable future.

Many of the most successful urban sustainability initiatives documented in this study, moreover, are the direct result of collaboration among and between different actors and groups in the community, and of the collaborative involvement of different professions and disciplines. The sustainable housing projects being built in the Netherlands, Denmark, and other countries are possible in large part because they explicitly involve environmental planners and local environmental departments early in their design, as well as engineers, developers, and housing associations. There is a creativity and attention to multiple factors that results from these interdisciplinary processes.

[. . .]

■ *A shift from products to services in the urban environment.*

A significant lesson illustrated by many of these innovative city programs is the importance of a shift in thinking from the provision of *products* or material goods to the provision of the actual *services* or benefits resulting from them. This is a commonly heralded shift among environmentalists in the United States, but rarely do we see evidence of this at an urban level as we do in Europe. The result can clearly be less environmental damage and greater cost-efficiency when focusing on the desired service or benefit, rather than on a specific product. Examples include the promising approach of focusing on the provision of "mobility" rather than on accommodating auto traffic, as well as any specific form of transportation service. Projects such as Amsterdam's Ijburg development have the potential of satisfying through mobility packages the movement needs of residents, while minimizing the use of automobiles. The car-sharing services that have developed in many cities in Europe are a similar example. It can be argued that it is not specifically an automobile that people desire or need, but the mobility services provided by them. Car-sharing companies provide the service without product ownership per se. Yet another example can be seen in the innovative public bike programs in Copenhagen, Amsterdam, and other cities.

[. . .]

■ *Making sustainability profitable.*

Related to the importance of emphasizing the multiple benefits of sustainable city initiatives is the idea that cities can structure these initiatives in ways that are perhaps not always economically profitable but that at least pay for themselves in the long or even the short term. Examples of such programs include Heidelberg's creative approach to contracting for energy reduction in its city buildings. The idea of hiring a mobility manager (as in several London boroughs) whose salary is paid for out of savings in transportation costs is another example.

Another method is to create positive incentives for users to reduce resource consumption. Several cities, including Heidelberg and Münster, have school energy conservation programs that create incentives for school administrators to find ways to reduce consumption and waste by allowing them to keep a portion of the savings. In Münster, for example, schools can keep half of the energy savings that result, using these funds for books, computers, or other school-related needs. These experiences provide important insights about how to structure similar green-urban initiatives in the United States.

[. . .]

■ *Ecological governance.*

The cities described in this book effectively illustrate that municipal governments can reform *themselves* – their purchasing, building, employee management, and other policies – to significantly reduce use of resources and promote sustainability . . . A common first step is to conduct an extensive audit of the environmental effects of local government policies and decisions and to prepare a local state-of-the-environment study.

[. . .]

A lesson repeatedly stated by many interviewees was the importance of local governments setting positive examples at the local level, along with the need for local authorities to take their own serious actions to promote sustainability before proselytizing about what its citizens, businesses, and private organizations should or must do. Such actions are seen as an essential step in ensuring the credibility of the municipality, as well as a matter of fairness.

What is also especially impressive is the more proactive stance assumed by the best of these three green-urban cities. Saarbrücken has taken major proactive initiatives to grow and support a solar energy industry, for example. Stockholm's environmental vehicle initiatives is in large part about stimulating and strengthening a local market for such vehicles. Graz has made exemplary strides in proactively working with local industries to help them become more energy

efficient and less polluting. In a host of ways, these cities show the virtues of creative, proactive policies. Government in these cities is not seen as laissez-faire or caretaking in nature, but as an entity exercising important proactive leadership; it is a pacesetter, not a follower or a spectator.

■ *The importance of economic signals and incentives for green-urban living.*

The green-urban cities profiled here also highlight the important role played by economic signals and incentives, and the direction and type of public investments made. Higher gasoline prices, higher costs associated with auto ownership and use, and the increasing use of green taxes in Europe all help to reinforce more sustainable patterns of land and resource use in these cities. The willingness by these cities (and countries) to invest substantially in public transit is a further example. Such investment also fundamentally shapes individual, family, and corporate decisions.

The lessons for the United States are perhaps difficult to embrace politically, but they do suggest that, to a large degree, what is necessary is the shifting of economic incentives and signals (e.g., at least a modest increase in gas taxes, perhaps to more fully fund transit and high-speed rail). These actions represent the conscious decisions of a society to ensure that prices more accurately reflect the true costs of personal choices and behavior and create positive inducements for more sustainable lifestyles.

■ *The role of demonstration and experimentation.*

The cities examined here have in common a history of creative experimentation with new ideas, new tools, and new strategies. Together, these green/urban cities show how powerful a spirit of experimentation can be, as well as how important this experimentation can be in setting new directions. This study has uncovered numerous demonstration projects sponsored by national and local governments. These demonstration projects, taken as a whole, appear to have played an extremely important role in shaping perception about what actions and policies are possible and feasible. Examples of valuable demonstration projects include Ecolonia (an innovative residential housing project in the Netherlands, sponsored in part by a national group committed to promoting housing innovation),

Stockholm's environmental vehicles program, and Heidelberg's minimum energy social housing project, among many others. Not only do such efforts effectively illustrate the technical feasibility of particular sustainability ideas, but they also help to constructively push private companies and markets along in the direction of such new ideas. They can serve as both examples and market catalysts.

[. . .]

■ *Technology and people.*

The experiences of European cities in grappling with sustainability and the green-urban agenda suggest that the challenge is both about technology and people. Attention to both will be needed and, indeed, as these many European initiatives show, one form of solution will not be successful without the other. European cities, as we have seen, are investing in, utilizing, and experimenting with a host of sustainability technologies, including photovoltaics, wind energy, sustainable building techniques, and public bicycle-depot systems, among many others. These innovative techniques, products, and technologies are potentially powerful in their ability to fundamentally reduce environmental impacts while at the same time making life easier and more enjoyable. The Europeans demonstrate convincingly the importance of underwriting and supporting the development and application of these technological innovations.

[. . .]

Many of the European cities' experiences further demonstrate that technology requires attention to the human users for it to be effective. Numerous examples have emerged, such as in the area of innovative sustainable building technology. Creating special areas for car-washing in Ecolonia (thus separating out this harmful waste stream from water that would normally drain into the community lake) is an idea that sounds good on paper, but if residents do not understand the purpose or need for this device, they are not likely to use it (which was indeed the case). As a further example, research suggests that residents of the Dutch ecological housing project Morra Park did not understand the purpose of the serres (sunrooms) attached to their homes. As a consequence, many residents attempted to heat these spaces and to treat them as year-round living areas. Perhaps partly faulty design, this passive design strategy clearly failed at the level of human use. Indeed, the emphasis on educating resi-

dents in new ecological housing projects reflects this understanding that human use and behavior are essential elements if the environmental goals behind technology and design innovation are to be realized.

■ *City networks and peer recognition.*

European cities benefit greatly from the existence of a number of supportive networks or associations of communities and cities that deal with local sustainability issues. Among these are the International Council for Local Environmental Initiatives (ICLEI, with its European Secretariat in Freiburg), Eurocities, and Car Free Cities, among others. These organizations serve a number of useful functions, including lobbying, technical assistance, and publication and dissemination of information about what different communities are doing. And there are also a number of EU-funded programs that further support information dissemination and collaboration between cities and the development of demonstration programs.

[. . .]

■ *The role of national programs, leadership, and coordination.*

A major factor helping to explain the creative and bold initiatives in the cities examined has been the pull and the push of governmental units at higher levels, and especially the existence of strong national initiatives. The examples are numerous and can be seen in virtually every sector (such as transportation, energy, and waste management). The financial and other forms of support and encouragement are substantial and should not be underestimated. European national governments, especially those of northern European nations, are not timid in taking the initiative in solving problems of unsustainable development, and they frequently have developed ambitious national goals, plans, and programs, from which this assistance flows.

Examples of national sustainability initiatives that have stimulated or facilitated local programs are numerous. Most countries have prepared national sustainable development plans under Agenda 21, laying a foundation for local actions. The Netherlands government, for example, has national-level strategies and action plans that are backed by substantial funding and that promote a host of sustainable practices and technologies. As mentioned, there is an ambitious strategy for promoting sustainable building nationwide.

Similarly, programs exist for promoting bicycle use (see Welleman, undated). In Denmark, tremendous national leadership can be seen in promoting and investing in wind power and other renewable energy technologies of the future. Numerous examples of critical national leadership can be cited.

Important financial incentives are often provided at the national level. The Netherlands has creatively harnessed the power of the tax code by, for example, providing investors in green projects with tax-free status, as well as generating substantial new capital to support sustainable building and other practices. European national governments have adopted carbon and other eco-taxes that encourage more sustainable use of land and water and eco-cycle balancing in cities and households.

The adoption of strong national codes and standards also explains some of the good progress that has been made. Countries such as Denmark, Germany, and the Netherlands have adopted strong energy codes for new housing, for example.

In the area of spatial planning, local initiatives are helped greatly by the existence in many countries of a national development strategy or plan. In the Netherlands, a national compact cities strategy has been effectively advanced, providing a broad structure in which provincial and local planning occurs . . .

In turn, national spatial plans serve as the basis for more logical, coordinated investment decisions at the national level, as in the form of new, high-speed rail lines or a new justice complex. In numerous ways, the examples in this book show the power and impact of strong national leadership in setting out basic conditions and a supportive climate in which many green-urban cities and green-urban initiatives can occur.

CENTRAL PARK.

Published by JOHN BACHMANN, 76 Nassau St. New York.

Plate 33 Central Park, New York, 1863. Frederick Law Olmsted and his partner Calvert Vaux conceived and executed a park, which was not only a masterpiece of design excellence, but also articulated a philosophy about what urban parks were for. Central Park provided the illusion of nature in the city. It was an oasis of calm and an intended meeting place for different classes. Central Park provided areas for quiet contemplation, boating, strolling, riding, baseball, Sunday school picnics, and countless other activities. (Courtesy of the Museum of the City of New York.)

Plate 34 Arturo Soria y Mata's plan for a linear city around Madrid, 1894. Spanish engineer/planner Arturo Soria y Mata envisaged the liberating force of new transportation technology. He conceived of "linear cities" built along electric streetcar lines which would provide for access to nature, large and relatively inexpensive lots, quick transportation, and efficient provision of infrastructure. Portions of the linear city around Madrid pictured here were built. (From Arturo Soria y Mata, *The Linear City*.)

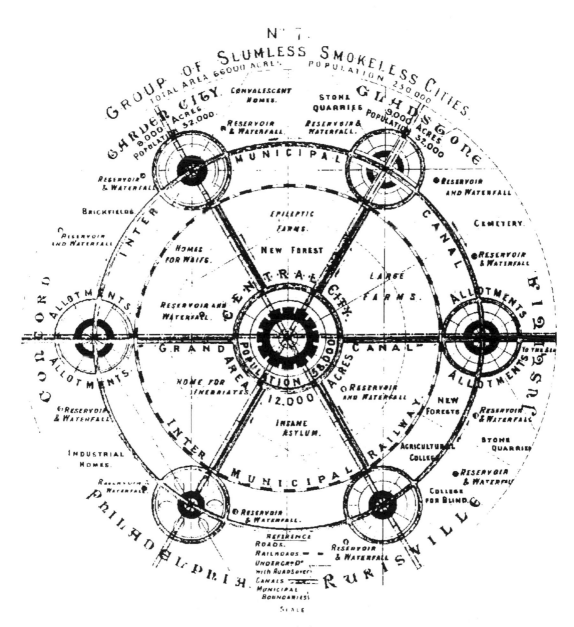

Plate 35 Ebenezer Howard's plan for a Garden City, 1898. From Ebenezer Howard, *To-morrow: A Peaceful Path for Real Reform* (London: Swan Sonnenschein, 1898). Reacting to the squalor of the nineteenth-century city, Ebenezer Howard proposed self-sufficient "Garden Cities" of about 32,000 people, carefully planned and surrounded by a permanent green belt. The Garden City movement spread worldwide and continues to inspire city planners.

Plate 36 Plan for Welwyn Garden City, 1909. Welwyn, the second of Britain's Garden Cities, closely followed Howard's vision. Howard lived to see Welwyn built and spent the last years of his life there.

Plate 37 Le Corbusier's "Plan Voisin" for a hypothetical city of three million people, 1925. Visionary modernist Le Corbusier planned huge new cities of steel and concrete dominated by highways and large modern buildings in park-like settings. This 1925 plan contemplates an entire new city of three million people built on these principles.

Plate 38 Plan for Radburn, New Jersey, 1929. Architects Clarence Stein and Henry Wright responded to the automobile by designing cities with a separation between pedestrian and traffic arteries, superblocks, and separate city areas with strong neighborhood identity. Their plan for Radburn, New Jersey, has inspired generations of planners.

Plate 39 Frank Lloyd Wright's plan for Broadacre City, 1935. Frank Lloyd Wright's extreme decentralist vision of the modern city contemplated a new form of Jeffersonian democracy with decentralized households living with at least one acre of land for each person.

Plate 40 **Paseo del Rio, San Antonio, Texas.** Good city planning can turn the most mundane landscape into a good urban space. Through adopting a bold vision and seeing it through to completion, San Antonio, Texas, turned a blighted riverfront into a magnificent area of parkland, water-oriented activities, and retail shopping. (Photograph by Alexander Garvin.)

Plate 41 **Quincy Market, Boston, Massachusetts.** Suburban malls and shopping centers may be "the new downtown" in many metropolitan areas, but some cities like Boston, Massachusetts are rebuilding their urban core into attractive shopping and entertainment areas. Boston's Quincy Market has been transformed from a seedy and economically marginal market area into a bustling and successful commercial area. (Photograph by Alexander Garvin.)

Plate 42 Peter Calthorpe's plan for a Pedestrian Pocket. Some modern architects and planners are incorporating elements from earlier small towns into their plans. California architect Peter Calthorpe advocates and has designed a series of "Pedestrian Pockets" which link land use and transportation together. Calthorpe's Pedestrian Pockets are dense enough to support light rail systems, environmentally sensitive, and pedestrian friendly. Calthorpe's designs emphasize front porches and de-emphasize garages. (Photograph courtesy of Peter Calthorpe Associates.)

PART SEVEN

Perspectives on urban design

INTRODUCTION TO PART SEVEN

This section focuses on urban *design* – the way in which humans actually shape the built environment. Urban designers are usually trained as architects with further training in urban design, city and regional planning, or both. They focus on the design of sites larger than individual buildings – neighborhoods, park systems, highway corridors, or even entire new towns. Professionals from the related field of landscape architecture are trained to provide advice on the relationship of the natural environment to the built environment. As in other areas where academics study cities or professionals work to build cities, material from many disciplines and professional fields is relevant to urban design. The best designers are also social scientists who study how people use the environments they are designing.

The selections in this section by architect Camillo Sitte, urban designer Kevin Lynch, sociologist William Whyte, city planners Allan Jacobs and Donald Appleyard, and historian/architect/planner Dolores Hayden are examples of the way different approaches and perspectives can enrich urban design.

Urban designers may disagree on what makes for a good design, but they share a belief in the value of design itself. They believe professionals should consciously think about physical relationships in the creation of urban space. Design is often expressed through drawings – generated by hand or computer – but design ideas may be expressed in many other ways.

Good urban design usually begins with intensive observation. Camillo Sitte (p. 413) observed squares and plazas in cities all over Europe in order to develop his principles for designing cities. William Whyte (p. 429) and his students spent hundreds of hours watching people use New York City parks and plazas – filming them, analyzing the films, and quantifying their behavior in order to reach principles for good park design. Kevin Lynch and his students (p. 424) surveyed people to see how they perceived the cities where they lived and asked them to draw maps to help understand what parts were and were not clear to them. Allan Jacobs and Donald Appleyard's "urban design manifesto" grew out of many hours surveying, sketching, counting, mapping, photographing, measuring, and simply walking and looking at San Francisco.

Urban designers often criticize how badly the built environment fits human needs. Ugly, impersonal, dirty, dangerous, dysfunctional, race- and gender-segregated areas dominate many large cities today. Designers may have different priorities with respect to the value of improving traffic flow versus making pedestrian-friendly streets, economically revitalizing an area versus retaining historic buildings, or protecting the natural environment versus keeping the city competitive in the global economy, but urban design is never value-free. The design of the built environment may not *determine* human behavior, but implicit in the following selections is the notion that bad design can numb the human spirit and good design can have powerful, positive influences on human beings.

The section begins with Austrian architect Camillo Sitte's theories on "the art of building cities," written at the end of the nineteenth century. Sitte is often viewed as the father of modern urban design. Dissatisfied with the impersonality of building projects in his own city of Vienna, Sitte set out to rediscover the artistic principles that guided classical Greek and Roman city builders and their medieval and renaissance successors. Sitte emphasizes the aesthetic, artistic character of city design. Underlying Sitte's theories is a profound belief in the importance of public places as venues to celebrate civic life. Like Lewis Mumford (p. 92) Sitte values human-scale development, the retention of historical elements, the joys of irregularity, and cities as theaters for the display of culture and civilization. Sitte's writings ran counter to the grand building schemes the new industrial wealth in Europe was creating during his own day. But he touched a responsive chord, and he quickly developed followers throughout the world during his lifetime. In the 1930s modernists, enamored of industrial technology and efficiency, dismissed Sitte. But in today's postmodern era his ideas are enjoying a resurgence of interest. Sitte's perspective – the aesthetic character of cities – did not deal with many of the difficult and important issues raised in prior sections of this Reader. Issues of class and race, economics and governance, and how to implement city plans are nowhere discussed in his work. But within the sphere of aesthetics and design he made an important contribution.

Seventy years after *The Art of Building Cities* was published another slim volume on urban design appeared. Massachusetts Institute of Technology planning professor Kevin Lynch's *The Image of the City* quickly established itself as the foundation of contemporary urban design. Lynch asked basic questions: How do people perceive the built environment? What are underlying elements common to human perception of the city? Armed with a better theoretical understanding of how people perceive the city image, what can urban designers actually do to design cities better? The selection by Lynch reprinted here describes the city image and the characteristics of five elements of urban form that Lynch considers to be fundamental. It is rich in suggestions about how these findings can shape better urban design. Urban planners like Allan Jacobs and Donald Appleyard (p. 437) have put Lynch's ideas into practice all over the world.

Sociologist William Whyte's writing on the design of spaces summarizes ideas he developed studying the way in which New Yorkers use urban parks and plazas. Whyte was a sharp observer and a fine writer. His writings are exemplary of the way in which social scientists can link understanding of human behavior to good urban design. Whyte's work also illustrates how urban research can lead directly to changes in city policy. The New York City Planning Department and organizations in New York involved in planning and managing parks have incorporated Whyte's ideas directly into policy. Other cities have followed his lead.

Urban planners Allan Jacobs and Donald Appleyard also illustrate the connection between theory and practice. Jacobs is an emeritus professor of city planning at the University of California, Berkeley, where Appleyard also taught until his death. Both were students of Kevin Lynch. Jacobs served as the director of the San Francisco City Planning Department, and Appleyard worked with him on notable studies of street livability and urban design. Under Jacobs's leadership, the San Francisco City Planning Department produced an award-winning urban design plan, which draws heavily on Lynch's ideas and the insights of Appleyard's studies. That work has profoundly shaped the development of San Francisco over the last quarter century.

The final selection in this section raises important questions about the relationship between design and gender and work roles and the relationship between design of the natural and built environments. Dolores Hayden (p. 448) condemns the isolated, wasteful, sexist world that male architects and planners have built reflecting outdated gender and work roles. She asks what a non-sexist city would be like and imagines groups reflecting the gender, age, and occupational structure of America sharing physical space, appliances, child rearing, and wage earning.

Designing the urban environment requires sensitivity to human needs and the natural environment. Urban planners, architects, landscape architects, and other design professions must be sensitive to biological and other natural systems, physical form and function, aesthetics, economics, transportation systems, and all the myriad way in which people use urban space. Each of the major subdivisions of this anthology can contribute to sensitive urban design.

"Author's Introduction," "The Relationship Between Buildings, Monuments, and Public Squares," and "The Enclosed Character of the Public Square"

from *The Art of Building Cities* (1889)

Camillo Sitte

Editors' Introduction

Camillo Sitte (1843–1903), an Austrian architect, felt that much was lost in the transformation of his native city of Vienna in the latter part of the nineteenth century. Sitte witnessed the old city walls – no longer useful against modern artillery – torn down, a "Ringstrasse" with new electric streetcars built to encircle the city, and old areas in the city leveled for monumental boulevards and impressive new buildings. Sitte felt nostalgia for the oddly shaped cathedral squares and narrow streets of Vienna, Salzburg, and other European cities that had evolved over time. He mourned the loss of structures built to human scale and public spaces embellished with statues, fountains, and other "municipal art" that adorned cites in classical Greece and Rome, the Middle Ages, and the Renaissance.

Sitte knew instinctively that he enjoyed qualities in cities he had visited that had been built before the modern era. He also admired the urban form of Greek and Roman cities of classical antiquity discernible in ruins and fragments of existing cities. But what exactly were these qualities? And how might modern-day designers, architects, and city planners incorporate the principles that Sitte and so many others enjoyed into new building projects? To answer these questions, Sitte embarked on a careful study of the built environment of notable European cities. Armed with a sketchbook he visited Athens, Rome, Florence, Venice, Paris, Pisa, Salzburg, Rothenburg on the Tauber, Dresden, and dozens of other European cities. Everywhere he went, Sitte carefully sketched the physical form of squares and plazas, the outline of cathedrals and public buildings, the location of statues and fountains. He thought about scale and building materials, views and elevations, the integration of ornamental features with functional buildings. He imagined what civic life in these urban spaces must have been like at the time of Pericles and Julius Caesar, of Lorenzo the Magnficent, and Louis XI of France. Sitte reflected on how the architects and city planners designed aesthetically pleasing spaces that reinforced civic culture. The result was a masterful little book that set in motion the modern study of urban design. Sitte produced a volume that was passionate in its advocacy of human-scale building and consideration of "artistic principles" in city building.

Sitte celebrated public space and was particularly enamoured of public squares and plazas. He applauded the practice in ancient Greece and Rome and during the Italian Renaissance of concentrating outstanding buildings around a single public square or plaza and ornamenting this center of community life with fountains, monuments, and statues.

Sitte had limited influence on the rebuilding of his native Vienna, but enormous and continuing impact elsewhere. "Sitte Schülen" (Sitte schools) sprang up all over Europe as young architects and planners read his book and discussed how to implement his ideas. *The Art of Building Cities* was translated into other languages. In the United States it was the Bible of the turn-of-the-century municipal arts movement. Dozens of local committees inspired by Sitte and the American writer Charles Mulford Robinson formed to "embellish and adorn" American cities.

Sitte fell out of favor in the interwar period when Le Corbusier and the insurgent young architects of the Congrès International de Architecture Moderne (CIAM) developed plans to raze and rebuild what they saw as the obsolete cities using modern materials, monumental scale, and designs inspired by industrial society (p. 317). Today there is a renewed interest in human-scale post-modernist designs, and "New Urbanist" architects and planners like Andres Duany and Elizabeth Plater-Zyberk (p. 207) and Peter Calthorpe and William Fulton (p. 331) are rediscovering Sitte's writings and find much of value in the principles he developed more than a century ago.

Note the connections between Sitte's celebration of plazas and public squares in classical Greece and Rome and in medieval and renaissance European cities with Lewis Mumford's notion that cities above all should be theaters in which humans could display their culture (p. 92) and William Whyte's views on how urban parks and plazas contribute to urban life (p. 429).

Camillo Sitte's treatise is available in many languages and editions. The most recent English language edition is Camillo Sitte, *The Art of Building Cities: City Building According to its Artistic Fundamentals* (Westport, CT: Hyperion Press, 1979).

An exhaustive study of the form of European (and non-European) cities before the Industrial Revolution is A.E.J. Morris, *The History of Urban Form Before the Industrial Revolutions* (New York: John Wiley, 1994). Another study of public spaces in European cities is Paul Zucker's *Town and Square from the Agora to the Village Green*, 2nd edn (Cambridge, MA: MIT Press, 1959). Erwin Anton Gutkind's eight-volume *International History of City Development* (New York: Free Press, 1964) has hundreds of illustrations of public spaces in European cities.

Spiro Kostoff's monumental studies of urban form, *The City Shaped* (New York: Little Brown, 1991), and *The City Assembled* (New York: Little Brown, 1992) contain illustrations of many of the features Sitte discusses. An account of the transformation of Vienna at the time that Sitte lived and wrote is contained in Carl Shorske, *Fin-De-Siècle Vienna: Politics and Culture* (New York: Vintage, 1981).

A biography and study of Sitte's work is George and Christiane Collins, *Camillo Sitte: The Birth of Modern City Planning* (New York: Rizzoli, 1981).

AUTHOR'S INTRODUCTION

Memory of travel is the stuff of our fairest dreams. Splendid cities, plazas, monuments, and landscapes thus pass before our eyes, and we enjoy again the charming and impressive spectacles that we have formerly experienced. If we could but stop again at those places where beauty never satiates, we could bear many dreary hours with a light heart and pursue life's long struggle with new energies. Assuredly the imperturbable lightheartedness of the South, on the Hellenic coast, in lower Italy and other favored climes, is above all a gift of nature. And the old cities of these countries, built after the beauty of nature itself, continue to augment nature's gentle and irresistible influence upon the soul of man. Only the person who has never understood the beauty of an ancient city could contradict this assertion.

Let him go ramble on the ruins of Pompeii to convince himself of it. If, after a day of patient investigation there, he walks across the bare Forum, he will be drawn, in spite of himself, to the summit of the monumental staircase toward the terrace of Jupiter's temple. On this platform, which dominates the entire place, he will sense, rising within him, waves of harmony like the pure, full tones of sublime music. Under this influence he will truly understand the words of Aristotle, who thus summarized all principles of city building: "A city should be built to give its inhabitants security and happiness."

The science of the technician will not suffice to accomplish this. We need, in addition, the talent of the artist. Thus it was in ancient times, in the Middle Ages, and in the Renaissance, wherever fine arts were held in esteem. It is only in our mathematical century that the construction and extension of cities has become a purely technical matter. Perhaps, then, it is not beside the point to recall that these problems have diverse aspects, and that he who has been given the least attention in our time is perhaps not the least important.

The object of this study, then, is clear. It is not our purpose to republish ancient and trite ideas, nor to reopen sterile complaints against the already proverbial banality of modern streets. It is useless to hurl general condemnations and to put everything that has been done in our time and place once more to the pillory. That kind of purely negative effort should be left to the critic who is never satisfied and who can only contradict. Those who have enough enthusiasm and faith in good causes should be convinced that our own era can create works of beauty and worth. We shall examine the plan of a number of cities, but neither as historian nor as critic. We wish to seek out, as technician and artist, the elements of composition which formerly produced such harmonious effects, and those which today produce only loose and dull results. Perhaps this study will permit us to find the means of satisfying the three principal requirements of practical city building: to rid the modern system of blocks and regularly aligned houses; to save as much as possible of that which remains from ancient cities; and in our creation to approach more closely the ideal of the ancient models.

This standpoint of practical art will lead us to consider especially the cities of the Middle Ages and the Renaissance. We shall be content, on recalling examples from Greek and Roman conceptions, either to explain the creations of following epochs, or to support the ideas that we propose to develop. For the principal architectural elements of cities have greatly changed since antiquity. Public squares (Forum, market, etc.) are used in our times not so much for great popular festivals or for the daily needs of our life. The sole reason for their existence is to provide more air and light, and to break the monotony of oceans of houses. At times they also enhance a monumental edifice by freeing its walls. It was quite different in ancient times. Public squares, or plazas, were then of prime necessity, for they were theaters for the principal scenes of public life, which today take place in enclosed halls. Under the open sky, on the agora, the council of the ancient Greeks gathered.

The market place, a further center of activity for our ancestors, has persisted, it is true, to the present time, but more and more it is being replaced by vast enclosed halls. And how many other scenes of public life have totally disappeared? Sacrifices before the temples, games, and theatrical presentations of all kinds. The temples themselves were scarcely covered, and the principal part of dwellings, around which were grouped large and small rooms, consisted of an open court. In a word, the distinction between the public square and other structures was so slight that it is amazing to our modern minds, accustomed to a very different state of things.

A review of the writings of the period proves to us that the ancients themselves sensed this similarity. Thus Vitruvius does not discuss the Forum in connection with the placement of public buildings or the arrangement of streets in his account of Dinocrates and his plan of Alexandria. But he does mention it in the same chapter which discusses the Basilica, and in the same book (1, 5) he deals with the theaters, palaces, the circus, and the baths. That is to say, all gathering places under the open sky constituted architectural works. The ancient Forum corresponds exactly to this definition, and Vitruvius logically places it in this group. This close relationship between the Forum and a public hall enhanced architecturally by statues and paintings is brought out clearly by the Latin writer's description, and more clearly still by an examination of the Forum of Pompeii. Vitruvius writes again on this subject:

The Greeks arrange their market places in the form of a square and surround them by vast double column supporting stone or marble architraves above which run the promenades. In Italian cities the Forum takes another aspect, for from time immemorial it has been the theater of gladiatorial combats. The columns, therefore, must be less densely grouped. They shelter the stalls of the silversmiths, and their upper floors have projections in the form of balconies which are advantageously placed for frequent use and for public revenue.

This description illustrates well the correspondence between theater and Forum. This relationship appears still more striking when we examine the plan of the Forum of Pompeii (Figure 1). The square is surrounded on all sides by public buildings. The temple of Jupiter alone rises in isolation. And the two-story colonnade

Figure 1 Forum of Pompeii
I Temple of Jupiter, II Enclosed Market, III Temple of Household Gods, IV Temple of Vespasianus, V Eumachia, VI Comitium, VII–IX Public Buildings, X Basilica, XI Temple of Apollo, XII Market Hall

which surrounds the entire space is interrupted only by the peristyle of the temple of the household gods, which makes a greater projection than the other buildings. The center of the Forum remains free, but its periphery is occupied by numerous monuments, the pedestals of which, covered with inscriptions, are still visible.

What a grandiose impression this place must have made! To our modern point of view its effect is like that of a great concert hall without ceiling. In every direction the eye fell upon edifices which in no respect resembled our files of modern houses, and there were

far fewer streets opening directly on the plaza. Streets ran behind buildings III, IV, and V, but they did not extend as far as the Forum. Streets C, D, E, and F were closed by grilles, and even those on the north side passed under the monumental portals, A and B.

Forum Romanum (Figure 2) was conceived according to the same principles. It is surrounded, of course, by buildings more varied in type but all monumental. The streets which open onto it were arranged to avoid too frequent openings in the frame of the plaza. Monuments are located around its sides rather than in

Figure 2 Forum Romanum

its center. In brief, the place of the forum in cities corresponds to that of the principal room of a house. It is to the city, so to speak, the principal hall, as well arranged as it is richly furnished. There stand assembled in immense bulk the columns, the statues, the monuments, and everything that can contribute to the splendor of the place. The art treasures of some of them were said to be numbered in hundreds and thousands. As they did not encumber the midst of the plaza, but were always located at the periphery, it was possible to encompass them all with a single glance; and the spectacle must have been imposing. This concentration of plastic and architectural masterpieces at a single point was a stroke of genius. Aristotle had taught it. He advocated grouping the temples of the gods with public buildings. Pausanias wrote similarly, "A city without public edifices and squares is not worthy of its name."

The market place of Athens is arranged in its principal features according to the same rules, as well as may be judged from the restoration projects. They are applied on a still grander scale in the consecrated cities of Hellenic antiquity (Olympia, Delphi, Eleusis) (Figure 3). Masterpieces of architecture, painting, and sculpture are found there in a superb and imposing union capable of rivaling the most powerful tragedies and the most majestic symphonies. The Acropolis of Athens (Figure 4) is the most finished creation of this character. A high plateau surrounded by high walls is the base of it. The lower entrance portal, the enormous flight of steps, and monumental vestibules constitute the first phrase of this symphony in marble, gold, ivory, bronze, and color. The interior temples and monuments are the stone myths of the Greek people. The highest poetry and thought are embodied in them. It is truly the center of a considerable city, an expression of the feelings of a great people. It is no longer a simple square in the ordinary sense of the term, but the work of several centuries grown to the maturity of pure art.

It is impossible to establish a higher aim in this style, and it is difficult to imitate successfully this splendid model, but it should always remain before our eyes in all our works as the most sublime ideal to attain. In the progress of our study we shall see that the principles which have inspired such building are not entirely lost, but that they remain to us.

Figure 3 The Temple of Zeus and Plaza of Olympia

Figure 4 The Acropolis of Athens in the age of Pericles

THE RELATIONSHIP BETWEEN BUILDINGS, MONUMENTS, AND PUBLIC SQUARES

In the South of Europe, and especially in Italy, where ancient cities and ancient public customs have remained alive for ages, even to the present in some places, public squares still follow the type of the ancient forum. They have preserved their role in public life. Their natural relationships with the buildings which enclose them may still be readily discerned. The distinction between the forum, or agora, and the market place also remains. As before, we find the tendency to concentrate outstanding buildings at a single place, and to ornament this center of community life with fountains, monuments, and statues which can bring back historical memories and which, during the Middle Ages and the Renaissance, constituted the glory and pride of each city.

It was there that traffic was most intense. That is where public festivals and theatrical presentations were held. There it was that official ceremonies were conducted and laws promulgated. In Italy, according to varying circumstances, two or three public places, rarely a single one, served these practical purposes.

The existence of two powers, temporal and spiritual, required two distinct centers: one, the cathedral square (Figure 5) dominated by the campanile, the baptistry, and the palace of the bishop; the other, the Signoria, or manor place, which is a kind of vestibule to a royal residence. It is enclosed by houses of the country's great and adorned with monuments. Sometimes we see there a loggia, or open gallery, used by a military guard, or a high terrace from which laws and public statements were promulgated. The Signoria of Florence (Figure 6) is the finest example of this. The market square, rarely lacking even in cities of northern Europe, is the meeting place of the citizens. There stand the City Hall and the more or less richly decorated traditional fountain, the sole vestige of the past that has been conserved since the lively activity of merchants and traders has been moved within to iron cages and glass market places.

The important function of the public square in the community life of past ages is evident. The period of the Renaissance saw the birth of masterpieces in the manner of the Acropolis of Athens, where everything concurred to produce a finished artistic effect. The cathedral place at Pisa, an Acropolis of Pisa (Figure 5), is the proof of this. It includes everything that the people of the City have been able to create in building religious edifices of unparalleled richness and grandeur. The splendid cathedral, the campanile, the baptistry, the incomparable Campo Santo are not depreciated by profane or banal surroundings of any kind. The effect produced by such a place, removed from the world of baseness while rich in the noblest works of the human spirit, is overpowering. Even those with a poorly developed sensitiveness to art are unable to escape the power of this impression. There is nothing there to distract our thoughts or to intrude our daily affairs. The esthetic enjoyment of those who look upon

Figure 5 Pisa: Cathedral Square
Key
a. Saint-Jean b. Cathedral c. Campanile d. Campo Santo (Cemetery)

Figure 6 Florence: Piazza of the Signoria

the noble facade of the Cathedral is not spoiled by the sight of a modern haberdashery, by the cries of drivers and porters, or by the tumult of a cafe. Peace reigns over the place. It is thus possible to give full attention to the artwork assembled there.

This situation is almost unique, although that of Saint Francis of Assisi and the arrangement of the Certosa de Pavia closely approach it. In general, the modern period does not encourage the formation of such perfect groupings. Cities, even in the fatherland of art, undergo the fate of palaces and dwellings. They no longer have distinct character. They present a mixture of motifs borrowed as much from the architecture of the north as from that of the southern countries. Ideas and tastes have been mingled as the people themselves have been interchanged. Local characteristics are gradually disappearing. The market place alone, with its City Hall and fountain, has here and there remained intact.

In passing we should like to remark that our intention is not to suggest a sterile imitation of the beauties spoken of as "picturesque" in the ancient cities for our present needs. The proverb, "Necessity breaks even iron," is fully applicable here. Changes made necessary by hygiene or other requirements must be carried out, even if the picturesque suffers from it. But that does not prevent us from examining the work of our forebears at close range to determine how much of it may be adapted to modern conditions. In this way alone can we resolve the esthetic part of the practical problem of city building, and determine what can be saved from the heritage of our ancestors.

Before determining the question in a positive manner, we state the principle that during the Middle Ages and Renaissance public squares were often used for practical purposes, and that they formed an entirety with the buildings which enclosed them. Today they serve at best as places for stationing vehicles, and they have no relation to the buildings which dominate them. Our parliament buildings have no agora enclosed by columns. Our universities and cathedrals have lost their atmosphere of peace. Surging throngs no longer circulate on market days before our City Halls. In brief, activity is lacking precisely in those places where, in ancient times, it was most intense – near public structures. Thus, to a great extent, we have lost that which contributed to the splendor of public squares.

And the fabric of their very splendor, the numerous statues, is almost entirely lacking today. What have we to compare to the richness of ancient forums and to works of majestic style like the Signoria of Florence and its Loggia dei Lanzi?

A few years ago there flourished at Vienna a remarkable school of sculpture whose works of merit cannot be scorned. They were generally used to adorn buildings. In only a few exceptional cases were their works used in public squares. Statues adorn the two museums, the palace of Parliament, the two Court theaters, the City Hall, the new university, the Votive Church. But there is no interest in adorning public open spaces. And that is true not only in Vienna, but nearly everywhere.

Buildings lay claim to so many statues that commissions are needed to find new subjects to be represented. It is often necessary to wait for years to find a suitable place for a statue although many appropriate places remain empty in the meantime. After long efforts we have reconciled ourselves to modern public squares as vast as they are deserted, and the monument, without a place of refuge, becomes stranded on some small and ancient space. That is even more strange, yet true. After much groping about, this fortunate result occurs, for it is thus that a work of art derives its value and produces a more powerful impression. Indifferent artists who neglect to provide for such effects must bear the entire responsibility of it.

The story of Michelangelo's *David* at Florence shows how mistakes of this kind are perpetrated in modern times. This gigantic marble statue stands close to the walls of the Palazzo Vecchio, to the left of its principal entrance, in the exact place chosen by Michelangelo. The idea of erecting a statue on this place of ordinary appearance would have appeared to moderns as absurd if not insane. Michelangelo chose it, however, and without doubt deliberately; for all those who have seen the masterpiece in this place testify to the extraordinary impression that it makes. In contrast to the relative scantiness of the place, affording an easy comparison with human stature, the enormous statue seems to swell even beyond its actual dimensions. The sombre and uniform, but powerful, walls of the palace provide a background on which we could not wish to improve to make all the lines of the figure stand out.

Today the David is moved into one of the academy's halls under a glass cupola in the midst of plaster reproductions, photographs, and engravings. It serves as a model for study and an object of research for historians and critics. A special mental preparation is needed now to resist the morbid influences of an art prison that we call a museum, and to have the ability to enjoy the imposing work. Moreover, the spirit of the times, which believed that it was perfecting art, and which was still not satisfied with this innovation, had a bronze cast made of the David in its original grandeur and put it up on a vast plaza (naturally in its mathematical center) far from Florence at the Via dei Colli. It has a superb horizon before it; behind it, cafes; on one side, a carriage station, a corso; and from all sides the murmurs of Baedeker readers ascend to it. In this setting the statue produces no effect at all. The opinion that its dimensions do not exceed human stature is often heard. Michelangelo thus understood best the

kind of placement that would be suitable for his work, and, in general, the ancients were abler than we are in these matters.

The fundamental difference between the procedures of former times and those of today rests in the fact that we constantly seek the largest possible space for each little statue. Thus we diminish the effect that it could produce, instead of augmenting it with the assistance of a neutral background such as painters have used in their portraits.

This explains why the ancients erected their monuments by the sides of public places, as is shown in the view of the Signoria of Florence. In this way, the number of statues could increase indefinitely without obstructing the circulation of traffic, and each of them had a fortunate background. Contrary to this, we hold the middle of a public place as the sole spot worthy to receive a monument. Thus no esplanade, however magnificent, can have more than one. If by misfortune it is irregular and if its center cannot be located geometrically we become confused and allow the space to remain empty for eternity.

THE ENCLOSED CHARACTER OF THE PUBLIC SQUARE

The old practice of setting churches and palaces back against other buildings brings to mind the ancient forum and its unbroken frame of public buildings. In examining the public squares that came into being during the Middle Ages and the Renaissance, especially in Italy, it is seen that this pattern has been retained for ages by tradition. The old plazas produce a collective harmonious effect because they are uniformly enclosed. In fact, the public square owes its name to this characteristic in an expanse at the center of a city. It is true that we now use the term to indicate any parcel of land bounded by four streets on which all construction has been renounced.

That can satisfy the public health officer and the technician, but for the artist these few acres of ground are not yet a public square. Many things must be done to embellish the area to give it character and importance. For just as there are furnished and unfurnished rooms, we could speak of complete and incomplete squares. The essential thing of both room and square is the quality of enclosed space. It is the most essential condition of any artistic effect, although it is ignored by those who are now elaborating on city plans.

SEVEN

The ancients, on the contrary, employed the most diverse methods of fulfilling this condition under the most diverse circumstances. They were, it is true, supported by tradition and favored by the usual narrowness of streets and less active traffic movement. But it is precisely in cases where these aids were lacking that their talent and artistic feeling is displayed most conspicuously.

A few examples will assist in accounting for this. The following is the simplest. Directly facing a monumental building a large gap was made in a mass of masonry, and the square thus created, completely surrounded by buildings, produced a happy effect. Such is the Piazza S. Giovanni at Brescia. Often a second street opens on to a small square, in which case care is taken to avoid an excessive breach in the border, so that the principal building will remain well enclosed. The methods used by the ancients to accomplish this were so greatly varied that chance alone could not have guided them. Undoubtedly they were often assisted by circumstance, but they also knew how to use circumstances admirably.

Today in such cases all obstructions would be taken down and large breaches in the border of the public place would be opened, as is done when we decide to "modernize" a city. Ancient streets would be found to open on the square in a manner precisely contrary to the methods of modern city builders, and mere chance would not account for this. Today the practice is to join two streets that intersect at right angles at each corner of the square, probably to enlarge as much as possible the opening made in the enclosure and to destroy every impression of cohesion. Formerly the procedure was entirely different. There was an effort to have only one street at each angle of the square. If a second artery was needed in a direction at right angles to the first it was designed to terminate at a sufficient distance from the square to remain out of view from the square. And better still, the three or four streets which came in at the corners each ran in a different direction. This interesting arrangement was reproduced so frequently, and more or less completely, that it can be considered as one of the conscious or subconscious principles of ancient city building.

Careful study shows that there are many advantages to an arrangement of street openings in the form of turbine arms. From any part of the square there is but one exit on the streets opening into it, and the enclosure of buildings is not broken. It even seems to enclose the square completely, for the buildings set at an angle conceal each other, thanks to perspective, and unsightly impressions which might be made by openings are avoided. The secret of this is in having streets enter the square at right angles to the visual lines instead of parallel to them. Joiners and carpenters have followed this principle since the Middle Ages when, with subtle art, they sought to make joints of wood and stone inconspicuous if not invisible.

The Cathedral Square at Ravenna shows the purest type of the arrangement just described. The square of

Figure 7 Pistoia: Cathedral Square
Key
a. Cathedral b. Baptistery
c. Residence d. Palais de la Commune
e. Palais du Podestat

Figure 8 Parma
Key
a. Pal. del Commune b. Madonna della Steccata
c. Pal. della Podesteria
I. Piazza d. Steccata II. Piazza Grande

Pistoia (Figure 7) is in the same manner; as is . . . the Piazza Grande at Parma (Figure 8).

The ancients had recourse to still other means of closing in their squares. Often they broke the infinite perspective of a street by a monumental portal or by several arcades of which the size and number were determined by the intensity of traffic circulation. This splendid architectural pattern has almost entirely disappeared, or, more accurately, it has been suppressed. Again Florence gives us one of the best examples in the portico of the Uffizi with its view of the Arno in the distance. Every Italian city of average importance has its portico, and this is also true north of the Alps. We mention only the Langasser Thor at Danzig, the entrance portal of the City Hall and Chancellery at Bruges, the Kerkboog at Nimeguen, the great Bell Tower at Rouen, the monumental Portals of Nancy, and the windows of the Louvre.

More or less ornate portals like those that simply but effectively frame the Piazza dei Signori at Verona (Figure 9) are to be found in all the royal residences, in the chateaux and city halls, and they are used as much for vehicular traffic as by pedestrians. While ancient architects used this pattern wherever possible with infinite variations, our modern builders seem to ignore its existence. Let us recall, to demonstrate again the persistence of ancient traditions, that at Pompeii, too, there is an Arc de Triomphe at the entrance to the Forum.

Columns were used with porticos to form enclosures for public squares. Saint Peter's in Rome is the best example of this . . .

Arcades were used to embellish monumental buildings more frequently in former times than at present, either on the higher stories, as in the City Halls of Halle (1548) and Cologne (1568), or on the ground level . . .

All of these above-mentioned architectural forms in former times made up a complete system of enclosing public squares. Today there is a contrary tendency to open them on all sides. It is easy to describe the results that have come about. It has tended to destroy completely the old public squares. Wherever these openings have been made the cohesive effect of the square has been completely nullified.

Figure 9 Verona: Piazza dei Signori

"The City Image and its Elements"

from *The Image of the City* (1960)

Kevin Lynch

Editors' Introduction

Kevin Lynch (1918–1989) is the towering figure of twentieth-century urban design. *The Image of the City*, from which this selection is taken, is the most widely read urban design book of all time. Lynch was a professor of urban studies and planning at the Massachusetts Institute of Technology where he taught courses in urban design.

As a young student Kevin Lynch apprenticed himself to Frank Lloyd Wright. Drawing widely on material from psychology and the humanities, Lynch sought to understand how people perceive their environments and how design professionals can respond to the deepest human needs. Lynch's rambling, profoundly humane writings weave together a unique blend of theory and practical design suggestions drawn from his voluminous reading in history, anthropology, architecture, art, literature, and a host of other areas.

This influential chapter on "The City Image and its Elements" presents Lynch's best known concepts on how people perceive cities. Lynch argues that people perceive cities as consisting of underlying city form "elements" such as *paths* (along which movement flows) and *edges* (which differentiate one part of the urban fabric from another). If they understand how people perceive these elements and design to make cities more imageable, Lynch argues, urban designers can create more psychologically satisfying urban environments.

Urban designers throughout the world today sketch out the elements of cities or parts of cities they are designing as paths, edges, nodes, landmarks, and districts – the underlying elements of city form that Lynch identified – and draw on his theories and practical suggestions to strengthen the city image. Planners in cities as diverse as San Francisco, Cairo, Havana, and Ciudad Guyana in Venezuela, have used Lynch's concepts to inform their urban planning and design strategies.

Compare Lynch's ideas about what people find psychologically satisfying and aesthetically appealing about cities with Camillo Sitte's ideas (p. 413). Contrast his practical suggestions with William Whyte's applied principles and standards for park and plaza design (p. 429), and Frederick Law Olmsted's vision of urban parks (p. 362).

In addition to *The Image of the City* (Cambridge, MA: MIT Press, 1960), Lynch's many books include a textbook on site design co-authored with Gary Hack, *Site Planning*, 3rd edn (Cambridge, MA: MIT Press, 1984), *What Time Is This Place* (Cambridge, MA: MIT Press, 1979), a book on historic preservation, *Managing the Sense of a Region* (Cambridge, MA: MIT Press, 1976), and his magnum opus, *Good City Form* (Cambridge, MA: MIT Press, 1991). Other of Lynch's writings are contained in Kevin Lynch, Tridib Banerjee and Michael Southworth (eds), *City Sense and City Design: Writings and Projects of Kevin Lynch* (Cambridge, MA: MIT Press, 1995).

Other books on the way in which people perceive urban space include Anthony Hiss, *The Experience of Place* (New York: Knopf, 1990) and Robert Sommer, *Personal Space* (Englewood Cliffs, NJ: Prentice-Hall, 1969). For more on urban design see Edmund Bacon, *The Design of Cities* (New York: Viking Press, 1976), Spiro Kostoff's *The City Shaped* (New York: Little Brown, 1991) and *The City Assembled* (New York: Little Brown, 1992), Mike Greenberg, *The Poetics of Cities: Designing Neighborhoods that Work* (Columbus: Ohio State University Press,

1995), and Doug Kelbaugh, *Common Place: Toward Neighborhood and Regional Design* (Seattle: University of Washington Press, 1997).

■ ■ ■ ■ ■ ■

There seems to be a public image of any given city which is the overlap of many individual images. Or perhaps there is a series of public images each held by some significant number of citizens. Such group images are necessary if an individual is to operate successfully within his environment and to cooperate with his fellows. Each individual picture is unique, with some content that is rarely or never communicated, yet it approximates the public image, which in different environments is more or less compelling, more or less embracing.

[. . .]

The contents of the city images so far studied, which are referable to physical forms, can conveniently be classified into five types of elements: paths, edges, districts, nodes, and landmarks . . . These elements may be defined as follows:

1 *Paths*. Paths are the channels along which the observer customarily, occasionally, or potentially moves. They may be streets, walkways, transit lines, canals, railroads. For many people, these are the predominant elements in their image. People observe the city while moving through it, and along these paths the other environmental elements are arranged and related.

2 *Edges*. Edges are the linear elements not used or considered as paths by the observer. They are the boundaries between two phases, linear breaks in continuity: shores, railroad cuts, edges of development, walls. They are lateral references rather than coordinate axes. Such edges may be barriers, more or less penetrable, which close one region off from another; or they may be seams, lines along which two regions are related and joined together. These edge elements, although probably not as dominant as paths, are for many people important organizing features, particularly in the role of holding together generalized areas, as in the outline of a city by water or wall.

3 *Districts*. Districts are the medium-to-large sections of the city, conceived of as having two-dimensional extent, which the observer mentally enters "inside of,"

and which are recognizable as having some common, identifying character. Always identifiable from the inside, they are also used for exterior reference if visible from the outside. Most people structure their city to some extent in this way, with individual differences as to whether paths or districts are the dominant elements. It seems to depend not only upon the individual but also upon the given city.

4 *Nodes*. Nodes are points, the strategic spots in a city into which an observer can enter, and which are the intensive foci to and from which he is traveling. They may be primarily junctions, places of a break in transportation, a crossing or convergence of paths, moments of shift from one structure to another. Or the nodes may be simply concentrations, which gain their importance from being the condensation of some use or physical character, as a street-corner hangout or an enclosed square. Some of these concentration nodes are the focus and epitome of a district, over which their influence radiates and of which they stand as a symbol. They may be called cores. Many nodes, of course, partake of the nature of both junctions and concentrations. The concept of node is related to the concept of path, since junctions are typically the convergence of paths, events on the journey. It is similarly related to the concept of district, since cores are typically the intensive foci of districts, their polarizing center. In any event, some nodal points are to be found in almost every image, and in certain cases they may be the dominant feature.

5 *Landmarks*. Landmarks are another type of point-reference, but in this case the observer does not enter within them, they are external. They are usually a rather simply defined physical object: building, sign, store, or mountain. Their use involves the singling out of one element from a host of possibilities. Some landmarks are distant ones, typically seen from many angles and distances, over the tops of smaller elements, and used as radial references. They may be within the city or at such a distance that for all practical purposes they symbolize a constant direction. Such are isolated towers, golden domes, great hills. Even a mobile point,

like the sun, whose motion is sufficiently slow and regular, may be employed. Other landmarks are primarily local, being visible only in restricted localities and from certain approaches. These are the innumerable signs, store fronts, trees, doorknobs, and other urban detail, which fill in the image of most observers. They are frequently used clues of identity and even of structure, and seem to be increasingly relied upon as a journey becomes more and more familiar.

[. . .]

PATHS

For most people interviewed, paths were the predominant city elements, although their importance varied according to the degree of familiarity with the city. People with the least knowledge of Boston tended to think of the city in terms of topography, large regions, generalized characteristics, and broad directional relationships. Subjects who knew the city better had usually mastered parts of the path structure; these people thought more in terms of specific paths and their interrelationships. A tendency also appeared for the people who knew the city best of all to rely more upon small landmarks and less upon either regions or paths.

The potential drama and identification in the highway system should not be underestimated. One Jersey City subject, who can find little worth describing in her surroundings, suddenly lit up when she described the Holland Tunnel. Another recounted her pleasure:

> You cross Baldwin Avenue, you see all of New York in front of you, you see the terrific drop of land [the Palisades] . . . and here's this open panorama of Lower Jersey City in front of you and you're going downhill, and there you know: there's the tunnel, there's the Hudson River and everything . . . I always look to the right to see if I can see the . . . Statue of Liberty . . . Then I always look up to see the Empire State Building, see how the weather is . . . I have a real feeling of happiness because I'm going someplace, and I love to go places.

[. . .]

Concentration of special use or activity along a street may give it prominence in the minds of observers. Washington Street in Boston is the outstanding Boston example: subjects consistently associated it with shopping and theaters . . . People seemed to be sensitive to variations in the amount of activity they encountered and sometimes guided themselves largely by following the main stream of traffic. Los Angeles' Broadway was recognized by its crowds and its street cars; Washington Street in Boston was marked by a torrent of pedestrians. Other kinds of activity at ground level also seemed to make places memorable, such as construction work near South Station, or the bustle of the food markets.

Characteristic spatial qualities were able to strengthen the image of particular paths. In the simplest sense, streets that suggest extremes of either width or narrowness attracted attention . . .

[. . .]

Where major paths lacked identity, or were easily confused one for the other, the entire city image was in difficulty . . . Boston's Longfellow Bridge was not infrequently confused with the Charles River Dam, probably since both carry transit lines and terminate in traffic nodes . . .

[. . .]

People tended to think of path destinations and origin points: they liked to know where paths came from and where they led. Paths with clear and well-known origins and destinations had stronger identities, helped tie the city together, and gave the observer a sense of his bearings whenever he crossed them. Some subjects thought of general destinations for paths, to a section of the city, for example, while others thought of specific places. One person, who made rather high demands for intelligibility upon the city environment, was troubled because he saw a set of railroad tracks, and did not know the destination of trains using them.

[. . .]

EDGES

Edges are the linear elements not considered as paths: they are usually, but not quite always, the boundaries between two kinds of areas. They act as lateral references. They are strong in Boston and Jersey City but weaker in Los Angeles. Those edges seem strongest which are not only visually prominent, but also continuous in form and impenetrable to cross movement. The Charles River in Boston is the best example and has all of these qualities . . .

[. . .]

It is difficult to think of Chicago without picturing Lake Michigan. It would be interesting to see how many Chicagoans would begin to draw a map of their city by putting down something other than the line of the lake shore. Here is a magnificent example of a visible edge, gigantic in scale, that exposes an entire metropolis to view. Great buildings, parks, and tiny private beaches all come down to the water's edge, which throughout most of its length is accessible and visible to all. The contrast, the differentiation of events along the line, and the lateral breadth are all very strong. The effect is reinforced by the concentration of paths and activities along its extent. The scale is perhaps unrelievedly large and coarse, and too much open space is at times interposed between city and water, as at the Loop. Yet the facade of Chicago on the Lake is an unforgettable sight.

DISTRICTS

Districts are the relatively large city areas which the observer can mentally go inside of, and which have some common character. They can be recognized internally, and occasionally can be used as external reference as a person goes by or toward them. Many persons interviewed took care to point out that Boston, while confusing in its path pattern even to the experienced inhabitant, has, in the number and vividness of its differentiated districts, a quality that quite makes up for it. As one person put it: "Each part of Boston is different from the other. You can tell pretty much what area you're in."

[. . .]

Subjects, when asked which city they felt to be a well-oriented one, mentioned several, but New York (meaning Manhattan) was unanimously cited. And this city was cited not so much for its grid, which Los Angeles has as well, but because it has a number of well-defined characteristic districts, set in an ordered frame of rivers and streets. Two Los Angeles subjects even referred to Manhattan as being "small" in comparison to their central area! Concepts of size may depend in part on how well a structure can be grasped.

[. . .]

Usually the typical features were imaged and recognized in a characteristic cluster, the thematic unit. The Beacon Hill image, for example, included steep narrow streets; old brick row houses of intimate scale; inset, highly maintained, white doorways; black trim;

cobblestones and brick walks; quiet; and upper-class pedestrians. The resulting thematic unit was distinctive by contrast to the rest of the city and could be recognized immediately . . .

NODES

Nodes are the strategic foci into which the observer can enter, typically either junctions of paths, or concentrations of some characteristic. But although conceptually they are small points in the city image, they may in reality be large squares, or somewhat extended linear shapes, or even entire central districts when the city is being considered at a large enough level. Indeed, when conceiving the environment at a national or international level, then the whole city itself may become a node.

The junction, or place of a break in transportation, has compelling importance for the city observer. Because decisions must be made at junctions, people heighten their attention at such places and perceive nearby elements with more than normal clarity. This tendency was confirmed so repeatedly that elements located at junctions may automatically be assumed to derive special prominence from their location. The perceptual importance of such locations shows in another way as well. When subjects were asked where on a habitual trip they first felt a sense of arrival in downtown Boston, a large number of people singled out break-points of transportation as the key places . . .

LANDMARKS

Landmarks, the point reference considered to be external to the observer, are simple physical elements which may vary widely in scale. There seemed to be a tendency for those more familiar with a city to rely increasingly on systems of landmarks for their guides – to enjoy uniqueness and specialization, in place of the continuities used earlier.

Since the use of landmarks involves the singling out of one element from a host of possibilities, the key physical characteristic of this class is singularity, some aspect that is unique or memorable in the context. Landmarks become more easily identifiable, more likely to be chosen as significant, if they have a clear form; if they contrast with the background; and if there is some prominence of spatial location.

SEVEN

Figure–background contrast seems to be the principal factor. The background against which an element stands out need not be limited to immediate surroundings: the grasshopper weathervane of Faneuil Hall, the gold dome of the State House, or the peak of the Los Angeles City Hall are landmarks that are unique against the background of the entire city.

[. . .]

"The Design of Spaces"

from *City: Rediscovering the Center* (1988)

William H. Whyte

Editors' Introduction

Puzzled by why some of New York's parks and plazas were well used while others were almost empty, the New York City Planning Commission asked sociologist William Whyte (1918–1999) to study park and plaza use and help draft a comprehensive design plan for the city.

Whyte's lucid writing on planning and design gave him great credibility. Hunter College appointed him a Distinguished Professor, and the National Geographic Society gave him the first domestic "expedition grant" they had ever made.

Whyte worked with bright young designers and planners at the New York City Planning Department, Hunter College sociology students, and other talented people he drew to "The Street Life Project." This team produced an exceptional study of how people use urban space and a set of urban design guidelines for New York that have been widely praised and used in New York and many other cities.

The Street Life Project is an excellent example of how to do urban research. Whyte formed hypotheses about how people use urban space. Then he tested each hypothesis by filming people using different plazas and parks in New York City and carefully analyzing the films. His results were often startling. Some initial hypotheses were validated, but Whyte found that he had to reject or modify many others that had seemed intuitively obvious. For example, Whyte hypothesized that the number of people using a plaza would be related to the *amount* of space or its *shape*. Big parks should have more people than little ones. A long skinny strip park should have fewer people than a rectangular one. But Chart 1 (Figure 1) shows that is not the case. New Yorkers use tiny Greenacre Park much more than the much larger J.C. Penny Park. One of New York's most popular parks is just a long, narrow indentation in a building. Whyte eventually concluded that the amount of *sittable* space in a park or plaza was much more important than either the total space or its shape.

Most writing on urban design ignores gender differences or is written from a male perspective with a separate section on design implications for women. Whyte is one of the few authors to notice gender and to weave its significance into the fabric of his study. He noticed that women are more discriminating than men as to where they will sit and are more sensitive to annoyances. He concluded that if a plaza has a high proportion of women, it is probably a good and well-managed one.

William Whyte's ideas have had a wide impact. The New York City Planning Commission held hearings on his recommendations and, after much debate, adopted many of his suggestions as requirements or guidelines for new development. Sharon Zukin (p. 136) describes how a public–private partnership inspired by Whyte's ideas tore down walls isolating Bryant Park from the street, put in sittable space and food, and transformed a dangerous, little-used park into one bustling with life, while at the same time evicting many homeless park "residents" and raising troubling questions about who should control public space and prescribe how it is to be used.

Note the similarity between Whyte's description of Seagram's Plaza as "the best of stages" and Lewis Mumford's emphasis on the city as theater (p. 92). Mike Davis (p. 201) found designers in Los Angeles consciously design public spaces to keep *people away*. As Whyte comments, "it takes real work to create a lousy place." Note the importance of good public spaces in Camillo Sitte's work on *The Art of Building Cities* (p. 413) and in H.D.F. Kitto's description of the Greek polis (p. 43).

This selection is from *City: Rediscovering the Center* (New York: Anchor Books, 1988). In other chapters of *City*, Whyte explores water, wind, trees, light, steps and entrances, undesirables, walls, sun and shadows, and many other factors. Whyte produced a delightful film titled *Public Spaces/Human Places* based on his research (available from Direct Cinema Limited in Los Angeles). The original report of Whyte's classic street life project has recently been reprinted as *The Social Life of Small Urban Spaces* (New York: Project for Public Spaces Inc., 2001). An anthology of William Whyte's most important writings is Albert LaFarge (ed.), *The Essential William Whyte* (New York: Fordham University Press, 2000). Whyte's other main book on planning is *The Last Landscape* (Garden City, NY: Doubleday, 1968). His best known book is *The Organization Man* (New York: Simon and Schuster, 1956) – a study of corporate culture that sold more than two million copies.

There are many books concerning human aspects of design. Spiro Kostoff's monumental *The City Shaped: Urban Patterns and Meanings Through History* (London: Thames and Hudson, 1991) and *The City Assembled* (London: Thames and Hudson, 1992) synthesize a vast amount of material from around the world and contain excellent illustrations. Oscar Newman's influential *Defensible Space* (New York: Macmillan, 1972) is a study of the way in which low-rent public housing project residents use space with suggestions to architects and planners on how to meet their security concerns. Clare Cooper and Wendy Sarkissien's *Housing as if People Mattered* (Berkeley: University of California Press, 1986) provides practical suggestions for designing housing responsive to the needs of all its residents, particularly working women and children. Allan Jacobs's *Looking at Cities* (Cambridge, MA: MIT Press, 1985) provides a stimulating discussion of how close observation like Whyte undertook can inform city planning and how to do it.

▪ ▪ ▪ ▪ ▪ ▪ ▪

. . . Since 1961 New York City had been giving incentive bonuses to developers who would provide plazas . . . Every new office building qualified for the bonus by providing a plaza or comparable space; in total, by 1972 some twenty acres of the world's most expensive open space.

Some plazas attracted lots of people . . .

But on most plazas there were few people. In the middle of the lunch hour on a beautiful day the number of people sitting on plazas averaged four per thousand square feet of space – an extraordinarily low figure for so dense a center . . .

. . . The city was being had. For the millions of dollars of extra floor space it was handing out to developers, it had every right to demand much better spaces in return.

I put the question to the chairman of the city planning commission, Donald Elliott . . . He felt tougher zoning was in order. If we could find out why the good places worked and the bad ones didn't and come up with tight guidelines, there could be a new code . . .

We set to work. We began studying a cross section of spaces – in all, sixteen plazas, three small parks, and a number of odds and ends of space . . .

[. . .]

We started by charting how people used plazas. We mounted time-lapse cameras at spots overlooking the plaza . . . and recorded the dawn-to-dusk patterns. We made periodic circuits of the plazas and noted on sighting maps where people were sitting, their gender, and whether they were alone or with others . . . We also interviewed people and found where they worked, how frequently they used the plaza, and what they thought of it. But mostly we watched what they did.

Most of them were young office workers from nearby buildings. Often there would be relatively few from the plaza's own building. As some secretaries confided, they would just as soon put a little distance between themselves and the boss come lunchtime. In most cases the plaza users came from a building within a three-block radius. Small parks, such as Paley and Greenacre, had a somewhat more varied mix of people – with more upper-income older people – but even here office workers predominated.

This uncomplicated demography underscores an elemental point about good spaces: supply creates demand. A good new space builds a new constituency.

It gets people into new habits – such as alfresco lunches – and induces them to use new paths . . .

The best-used plazas are sociable places, with a higher proportion of couples and groups than you will find in less-used places. At the plazas in New York, the proportion of people in twos or more runs about 50–62 percent; in the least-used, 25–30 percent. A high proportion is an index of selectivity. If people go to a place in a group or rendezvous there, it is most often because they decided to beforehand. Nor are these places less congenial to the individual. In absolute numbers, they attract more individuals than do the less-used spaces. If you are alone, a lively place can be the best place to be.

The best-used places also tend to have a higher than average proportion of women. The male–female ratio of a plaza reflects the composition of the work force and this varies from area to area. In midtown New York it runs about 60 percent male, 40 percent female. Women are more discriminating than men as to where they will sit, they are more sensitive to annoyances, and they spend more time casing a place. They are also more likely to dust off a ledge with their handkerchief.

The male–female ratio is one to watch. If a plaza has a markedly low proportion of women, something is wrong. Conversely, if it has a high proportion, the plaza is probably a good and well-managed one and has been chosen as such.

The rhythms of plaza life are much alike from place to place. In the morning hours, patronage will be sporadic . . .

Around noon the main clientele begins to arrive. Soon activity will be near peak and will stay there until a little before two . . .

Some 80 percent of the people activity on plazas comes during the lunchtime, and there is very little of any kind after five-thirty . . .

During the lunch period, people will distribute themselves over space with considerable consistency, with some sectors getting heavy use day in and day out, others much less so. We also found that off-peak use often gives the best clues to people's preferences. When a place is jammed, people sit where they can; this may or may not be where they most want to. After the main crowd has left, however, the choices can be significant. Some parts of the plaza become empty; others continue to be use . . .

Men show a tendency to take the front row seats and if there is a kind of gate they will be the guardians of it. Women tend to favor places slightly secluded. If

there are double-sided ledges parallel to the street, the inner side will usually have a higher proportion of women; the outer, of men.

Of the men up front the most conspicuous are the girl watchers. As I have noted, they put on such a show of girl watching as to indicate that their real interest is not so much the girls as the show. It is all machismo. Even in the Wall Street area, where girl watchers are especially demonstrative you will hardly ever see one attempt to pick up a girl.

Plazas are not ideal places for striking up acquaintances. Much better is a very crowded street with lots of eating and quaffing going on. An outstanding example is the central runway of the South Street Seaport. At lunch sometimes, one can hardly move for the crush. As in musical chairs, this can lead to interesting combinations. On most plazas, however, there isn't much mixing. If there are, say, two smashing blondes on a ledge, the men nearby will usually put on an elaborate show of disregard. Look closely, however, and you will see them giving away the pose with covert glances.

Lovers are to be found on plazas, but not where you would expect them. When we first started interviewing, people would tell us to be sure to see the lovers in the rear places. But they weren't usually there. They would be out front. The most fervent embracing we've recorded on film has taken place in the most visible of locations, with the couple oblivious of the crowd. (In a long clutch, however, I have noted that one of the lovers may sneak a glance at a wristwatch.)

Certain locations become rendezvous points for groups of various kinds. The south wall of the Chase Manhattan Plaza was, for a while, a gathering point for camera bugs, the kind who are always buying new lenses and talking about them. Patterns of this sort may last no more than a season – or persist for years. A black civic leader in Cincinnati told me that when he wants to make contact, casually, with someone, he usually knows just where to look at Fountain Square . . .

Standing patterns on the plazas are fairly regular. When people stop to talk they will generally do so athwart one of the main traffic flows, as they do on streets. They also show an inclination to station themselves near objects, such as a flagpole or a piece of sculpture. They like well-defined places, such as steps or the border of a pool. What they rarely choose is the middle of a large space.

There are a number of explanations. The preference might be ascribed to some primeval instinct: you have a full view of all comers but your rear is covered. But this doesn't explain the inclination men have for lining up at the curb. Typically, they face inward, with their backs exposed to the vehicle traffic of the street.

Whatever their cause, people's movements are one of the great spectacles of a plaza. You do not see this in architectural photographs, which are usually devoid of human beings and are taken from a perspective that few people share. It is a misleading one. Looking down on a bare plaza, one sees a display of geometry, done almost in monochrome. Down at eye level the scene comes alive with movement and color – people walking quickly, walking slowly, skipping up steps, weaving in and out on crossing patterns, accelerating and retarding to match the moves of others. Even if the paving and the walls are gray, there will be vivid splashes of color – in winter especially, thanks to women's fondness for red coats and colored umbrellas.

There is a beauty that is beguiling to watch, and one senses that the players are quite aware of this themselves. You can see this in the way they arrange themselves on ledges and steps. They often do so with a grace that they must appreciate themselves. With its brown-gray setting, Seagram is the best of stages – in the rain, too, when an umbrella or two puts spots of color in the right places, like Corot's red dots.

Let us turn to the factors that make for such places. The most basic one is so obvious it is often overlooked: people. To draw them, a space should tap a strong flow of them. This means location, and, as the old adage has it, location and location. The space should be in the heart of downtown, close to the 100 percent corner – preferably right on top of it.

Because land is cheaper further out, there is a temptation to pick sites away from the center. There may also be some land for the asking – some underused spaces, for example, left over from an ill-advised civic center campus of urban renewal days. They will be poor bargains. A space that is only a few blocks too far might as well be ten blocks for all the people who will venture to walk to it.

People *ought* to walk to it, perhaps; the exercise would do them good. But they don't. Even within the core of downtown the effective radius of a good place is about three blocks. About 80 percent of the users will have come from a place within that area. This does indicate a laziness on the part of pedestrians and this may change a bit, just as the insistence on close-in parking may. But there is a good side to the constrained radius. Since usage is so highly localized, the addition of other good open spaces will not saturate demand. They will increase it.

Given a fine location, it is difficult to design a space that will not attract people. What is remarkable is how often this has been accomplished. Our initial study made it clear that while location is a prerequisite for success, it in no way assures it. Some of the worst plazas are in the best spots . . .

All of the plazas and small parks that we studied had good locations; most were on the major avenues, some on attractive side-streets. All were close to bus-stops or subway stations and had strong pedestrian flows on the sidewalks beside them. Yet when we rated them according to the number of people sitting at peak time, there was a wide range: from 160 people at 77 Water Street to 17 at 280 Park Avenue (Figure 1).

How come? The first factor we studied was the sun. We thought it might well be the critical one, and our first time-lapse studies seemed to bear this out. Subsequent studies did not. As I will note later they show that sun was important but did not explain the differences in popularity of plazas.

Nor did aesthetics . . . The elegance and purity of a complex's design, we had to conclude, had little relationship to the usage of the spaces around it.

[. . .]

Another factor we considered was the shape of spaces. Members of the commission's urban design group believed this was very important and hoped our findings would support tight criteria for proportions and placement. They were particularly anxious to rule out strip plazas: long, narrow spaces that were little more than enlarged sidewalks, and empty of people more times than not . . .

Our data did not support such criteria. While it was true that most strip plazas were little used, it did not follow that their shape was the reason. Some squarish plazas were little used too, and, conversely, several of the most heavily used spaces were in fact long, narrow strips. One of the five most popular sitting places in New York is essentially an indentation in a building, long and narrow. Our research did not prove shape unimportant or designers' instincts misguided. As with the sun, however, it proved that other factors were more critical.

If not the shape of the space, what about the *amount* of it? Some conservationists believed this would be the key factor. In their view, people seek open space as a relief from overcrowding and it would follow that places with the greatest sense of space and light and air

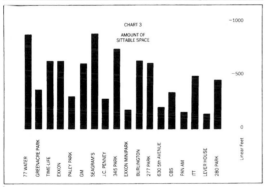

Figure 1

. . . No matter how many other variables we checked, one basic point kept coming through. We at last recognized that it was the major one.

People tend to sit most where there are places to sit.

This may not strike the reader as an intellectual bombshell, and now that I look back on our study I wonder why it was not more apparent to us from the beginning . . . Whatever the attractions of a space, it cannot induce people to come and sit if there is no place to sit.

INTEGRAL SEATING

The most basic kind of seating is the kind that is built into a place, such as steps and ledges. Half the battle is seeing to it that these features are usable by people. And there is a battle. Another force has been diligently at work finding ways to deny these spaces. Here are some of the ways:

Horizontal metal strip with sawtooth points.
Jagged rocks set in concrete (Southbridge House, New York City).
Spikes imbedded in ledges (Peachtree Plaza Hotel).
Railing placed to hit you in small of back (GM Plaza, New York City).
Canted ledges of slippery marble (Celanese Building, New York City).

It takes real work to create a lousy place. In addition to spikes and metal objects, there are steps to be made steep, additional surveillance cameras to be mounted, walls to be raised high. Just not doing such things can produce a lot of sitting space.

It won't be the most comfortable kind but it will have the great advantage of enlarging choice. The more sittable the inherent features are made, the more freedom people have to sit up front, in the back, to the side, in the sun, or out of it. This means designing ledges and parapets and other flat surfaces so they can do double duty as seating, tables, and shelves. Since most building sites have some slope in them, there are bound to be opportunities for such features, and it is no more trouble to leave them sittable than not.

[. . .]

would draw the best. If we ranked plazas by the amount of space they provided, there surely would be a positive correlation between space and people.

Once again we found no clear relationship. Several of the smallest spaces had the largest number of people, and several of the largest spaces had the least number of people . . .

What about the amount of *sittable* space? Here we began to get close. As we tallied the number of linear feet of sitting space, we could see that the plazas with the most tended to be among the most popular . . .

SITTING HEIGHT

One guideline we thought would be easy to establish was for sitting heights. It seemed obvious enough that somewhere around sixteen to seventeen inches would probably be the optimum. But how much higher or lower could a surface be and still be sittable? Thanks to slopes, several of the most popular ledges provided a range of continuously variable heights. The front ledge at Seagram, for example started at seven inches at one corner and rose to forty-four inches at the other. Here was an opportunity for a definitive study, we thought; by recording over time how many people sat at what heights, we would get a statistical measure of preferences.

We didn't . . . We had to conclude that people will sit almost anywhere between a height of one foot and three, and this was the range that was to be specified in the zoning. People will sit lower or higher, of course, but there are apt to be special conditions – a wall too high for most adults to mount but just right for teenagers.

A dimension that is truly important is the human backside. It is a dimension many architects ignore. Rarely will you find a ledge or bench that is deep enough to be sittable on both sides . . . Most frustrating are the ledges just deep enough to tempt people to sit on both sides, but too shallow to let them do so comfortably. At peak times people may sit on both sides but they won't be comfortable doing it. They will be sitting on the forward edge, awkwardly.

Thus to another of our startling findings: ledges and spaces two backsides deep seat more people than those that are not as deep . . .

[. . .]

Steps work for the same reason. They afford an infinity of possible groupings, and the excellent sight lines make all the seating great for watching the theatre of the street . . .

[. . .]

Circulation and sitting, in sum, are not antithetical but complementary. I stress this because a good many planners think that the two should be kept separate. More to the point, so do some zoning codes. New York's called for "pedestrian circulation areas" separate from "activity areas" for sitting. People ignore such boundaries.

We felt that pedestrian circulation through and within plazas should be encouraged. Plazas that are sunken or elevated tend to attract low flows, and for that reason the zoning specifies that plazas be not more than three feet below street level or above it. The easier the flow between street and plaza the more likely they are to come in and tarry and sit.

This is true of the handicapped also. If a place is planned with their needs in mind, the place is apt to function more easily for everyone. Drinking fountains that are low enough for wheelchair users are low enough for children. Walkways that are made easier for the handicapped by ramps, handrails, and steps of gentle pitch are easier for all. The guidelines make such amenities mandatory . . . For the benefit of the handicapped, it is required that at least 5 percent of the seating spaces have backrests. These are not segregated for the handicapped. No facilities are segregated. The idea is to make all of the place useful for everyone.

BENCHES

Benches are design artifacts the purpose of which is to punctuate architectural photographs. They are most often sited in modular form, spaced equidistant from one another in a symmetry that is pleasing in plan view. They are not very good, however, for sitting. There are usually too few of them; they are too short and too narrow; they are isolated from other benches and from what action there is to look at.

[. . .]

Watch how benches fill up. The first arrival will usually take the end of a bench, not the middle. The next arrival will take the end of another bench. Subsequent arrivals head for whatever end spots are not taken. Only when there are few other places left will people sit in the middle of the bench, and some will elect to stand.

Since it's the ends of the benches that do most of the work, it could be argued that benches ought to be shortened so they're all end and no middle. But the unused middles are functional for not being used. They provide buffer space. They also provide choice, and if it is the least popular choice, that does not negate its utility.

[. . .]

CHAIRS

We come now to a wonderful invention: the movable chair. Having a back, it is comfortable, and even more so if it has armrests as well. But the big asset is movability. Chairs enlarge choice: to move into the sun, out of it; to move closer to someone, further away from another.

The possibility of choice is as important as the exercise of it. If you know you can move if you want to, you can feel all the more comfortable staying put. This is why, perhaps, people so often approach a chair and then, before sitting on it, move the chair a few inches this way or that, finally ending up with the chair just about where it was in the first place. These moves are functional. They are a declaration of one's free will to oneself, and rather satisfying. In this one small matter you are the master of your fate.

Small moves can say things to other people. If a newcomer chooses a chair next to a couple in a crowded situation, he may make several moves with the chair. He is conveying a message: Sorry about the closeness, but it can't be helped and I am going to respect your privacy as you will mine. A reciprocal shift of a chair may signal acknowledgment.

Chair arranging by groups is a ritual worth watching. In a group of three or four women, one may be dominant and direct the sitting, including the fetching of an extra chair. More times, the members of the group work it out themselves, often with false starts and second choices. The chair arranging can take quite a bit of time on occasion – it is itself a form of recreation – but people enjoy it. Watching these exercises in civility is one of the pleasures of a good place.

Fixed individual seats deny choice. They may be good to look at, and in the form of stools, metal love seats, granite cubes, and the like, they make interesting decorative elements. That is their primary function. For sitting, however, they are inflexible and socially uncomfortable.

[. . .]

Where space is at a premium – in theatres, stadia – fixed seats are a necessity. In open spaces, however, they are uncalled for; there is so much space around them that the compression makes for awkard sitting . . . On one campus a group of metal love seats was cemented to the paving with epoxy glue; in short order they were wrenched out of position by students. The designer is unrepentant. His love seats have won several design awards.

[. . .]

A salute to grass is in order. It is a wonderfully adaptable substance, and while it is not the most comfortable seating, it is fine for napping, sunbathing, picnicking, and Frisbee throwing. Like movable chairs, it also has the great advantage of offering people the widest possible choice of sitting arrangements. There are an infinity of possible groupings, but you will note that the most frequent has people self-positioned at oblique angles from each other.

Grass offers a psychological benefit as well. A patch of green is a refreshing counter to granite and concrete, and when people are asked what they would like to see in a park, trees and grass usually are at the top of the list . . .

RELATIONSHIP TO THE STREET

Let us turn to a more difficult consideration. With the kind of amenities we have been discussing, there are second chances. If the designers have goofed on seating, more and better seating can be provided. If they have been too stingy with trees, more trees can be planted. If there is no food, a food cart can be put in – possibly a small pavilion or gazebo. If there is no water feature, a benefactor might be persuaded to donate a small pool or fountain. Thanks to such retrofitting, spaces regarded as hopeless dogs have been given new life.

What is most difficult to change, however, is what is most important: the location of the space and its relationship to the street. The real estate people are right about location, location, location. For a space to function truly well it must be central to the constituency it is to serve – and if not in physical distance, in visual accessibility . . .

The street functions as part of the plaza or square; indeed, it is often hard to tell where the street leaves off and the plaza begins. The social life of the spaces flows back and forth between them. And the most vital space of all is often the street corner. Watch one long enough and you will see how important it is to the life of the large spaces. There will be people in 100 percent conversations or prolonged goodbyes. If there is a food vendor at the corner, like Gus at Seagram, people will be clustered around him, and there will be a brisk traffic between corner and plaza.

It is a great show, and one of the best ways to make the most of it is, simply, not to wall off the plaza from it. Frederick Law Olmsted spoke of an "interior park" and an "outer park," and he argued that the latter – the surrounding streets – was vital to the enjoyment of the former. He thought it an abomination to separate the two with walls or, worse yet, with a spiked iron fence. "In expression and association," he said, "it is in the most distinct contradiction and discord with all the sentiment of a park. It belongs to a jail or to the residence of a despot who dreads assassination."

But walls are still being put up, usually in the mistaken notion that they will make the space feel safer. They do not . . . they make a space feel isolated and gloomy. Lesser defensive measures can work almost as much damage. The front rows of a space – whether ledges or steps or benches – are the best of sitting places, yet they are often modified against human use. At the General Motors Building on Fifth Avenue, the front ledges face out on one of the greatest of promenades. But you cannot sit on the ledges for more than a minute or so. There is a fussy little railing that catches you right in the small of your back. I do not think it was deliberately planned to do so. But it does and you cannot sit for more than a few moments before your back hurts. Another two inches of clearance for the railing and you would be comfortable. But day after day, year after year, one of the great front rows goes scarcely used, for want of two inches. Canted ledges, especially ones of polished marble, are another nullifying feature. You can almost sit on them if you keep pressing down on your heel hard enough.

[. . .]

A good space beckons people in, and the progression from street to interior is critical in this respect. Ideally, the transition should be such that it's hard to tell where one ends and the other begins. You shouldn't have to make a considered decision to enter; it should be almost instinctive . . .

[. . .]

"Toward an Urban Design Manifesto"

Journal of the American Institute of Planners (1987)

Allan Jacobs and Donald Appleyard

Editors' Introduction

Allan Jacobs and Donald Appleyard deplore many of the same aspects of Los Angeles, London, New York, and other large postmodern cities that Mike Davis (p. 201), Ali Madanipour (p. 181), Edward Soja (p. 189), and David Harvey (p. 227) criticize: vast anonymous areas developed by giant public and private developers; dangerous, polluted, noisy, anonymous living environments; fortress-like buildings which present windowless façades to the street; and pervasive semiotics that tell "outsiders" they are not welcome in subtle and not-so-subtle ways. But "Toward an Urban Design Manifesto" moves beyond observation and critique to set out goals for urban life and advance ideas for how the urban fabric of cities might be designed to encourage livable urban environments.

Allan Jacobs is an emeritus professor of city and regional planning at the University of California, Berkeley. Donald Appleyard also taught at Berkeley until his death shortly before this selection was published. Both Jacobs and Appleyard worked closely with Kevin Lynch, whose *Image of the City* (p. 424) and other writings strongly influenced their work.

Jacobs alternated between careers as a practicing city planner (in Pittsburgh, Philadelphia, New Delhi, and San Francisco) and teaching at the University of Pennsylvania and the University of California, Berkeley's Department of City and Regional Planning. While he was San Francisco's City Planning Director, Jacobs enlisted Appleyard to work on studies of street livability in San Francisco and to help develop an award-winning citywide urban design plan reflecting Lynch's ideas.

Jacobs and Appleyard title their piece a "manifesto" and model it on the celebrated Charter of Athens adopted by the International Congress of Modern Architecture (CIAM) – the organization which advanced ideas for building contemporary cities based on Le Corbusier's principles (p. 317).

Jacobs and Appleyard do not like the vast clearance projects, highways, and high-rise buildings surrounded by enormous open space that have resulted from the CIAM's design ideology. They acknowledge that the Garden City ideas of Ebenezer Howard (p. 309) have produced some pleasant communities, but dismiss them as more like suburbs than true *cities*. Their manifesto suggests an approach that is more subtle and humane than the CIAM's and more truly urban than Howard's approach.

Jacobs and Appleyard's manifesto is grounded in both a command of academic theory and their own practical experience in city design. In this manifesto they propose urban development at densities higher than Howard proposed for Garden City designs – high enough to qualify as truly urban. But they do not endorse urban densities nearly as great as the CIAM theorists do, particularly in megastructures surrounded by open space. (Le Corbusier deliberately shocked the architectural establishment by producing a plan to tear down all of central Paris and replace it with modern concrete and steel high-rise buildings!)

While Jacobs and Appleyard favor *reasonable* standards for decibel levels and street widths, they oppose excessive engineering standards that destroy the texture of urban life. Like Jane Jacobs (p. 114), William Whyte

(p. 429), and Lewis Mumford (p. 92) they relish some of the disorder that makes urban life enjoyable including noise, smell, and jumbled uses that some engineers and many modernist architects wanted to separate into orderly districts. Jacobs and Appleyard value pedestrians and public space. Unlike the elitist CIAM theorists, they argue that participatory planning of the kind Sherry Arnstein (p. 244) and John Forester (p. 375) describe is essential.

In *Making City Planning Work* (Chicago: Planner's Press, 1976), Jacobs alternates chapters describing the practical aspects of a city planning director's job with case studies on successful and not so successful projects he undertook during his tenure as city planning director of San Francisco. Jacobs's *Looking at Cities* (Cambridge, MA: MIT Press, 1985) grew out of a class he taught at Berkeley in which students took him to an unfamiliar neighborhood, left him to carefully observe it, and then compared what he found out from observation with what they learnt by examining data and city planning reports on the same neighborhood. *Looking at Cities* reminds professionals to open their eyes and experience the areas they are planning. It outlines a methodology for reading clues in the built environment that can improve urban planning practice. Jacobs's most recent books are *Great Streets* (Cambridge, MA: MIT Press, 1995) and *The Boulevard Book: History, Evolution, Design of Multiway Boulevards* (Cambridge, MA: MIT Press, 2001) co-authored with Elizabeth MacDonald and Yodan Rofe. Both of these recent books reflect many of the values expressed in this selection.

Donald Appleyard's *The View from the Road* co-authored with Kevin Lynch and John Myer (Cambridge, MA: MIT Press, 1963) and *Livable Streets* (Berkeley: University of California Press, 1981) show how ideas expressed here can be translated into action in street design.

For an overview of urban design see Jonathon Barnett, *An Introduction to Urban Design* (New York: Harper & Row, 1982). Other works on urban design include Edmund N. Bacon, *Design of Cities* (New York: Penguin, 1976), Vincent Scully, *American Architecture and Urbanism* (New York: Praeger, 1969), and Gordon Cullen, *Townscape* (New York: Penguin, 1976).

▫ ▫ ▫ ■ ▫ ▫

We think it's time for a new urban design manifesto. Almost 50 years have passed since Le Corbusier and the International Congress of Modern Architecture (CIAM) produced the Charter of Athens, and it is more than 20 years since the first Urban Design Conference, still in the CIAM tradition, was held (at Harvard in 1957). Since then the precepts of CIAM have been attacked by sociologists, pecently by architects themselves. But it is still a strong influence, and we will take it as our starting point. Make no mistake: the charter was, simply, a manifesto – a public declaration that spelled out the ills of industrial cities as they existed in the 1930s and laid down physical requirements necessary to establish healthy, humane, and beautiful urban environments for people. It could not help but deal with social, economic, and political phenomena, but its basic subject matter was the physical design of cities. Its authors were (mostly) socially concerned architects, determined that their art and craft be responsive to social realities as well as to improving the lot of man. It would be a mistake to write them off as simply elitist designers and physical determinists.

So the charter decried the medium-size (up to six storys) high-density buildings with high land coverage that were associated so closely with slums. Similarly, buildings that faced streets were found to be detrimental to healthy living. The seemingly limitless horizontal expansion of urban areas devoured the countryside, and suburbs were viewed as symbols of terrible waste. Solutions could be found in the demolition of unsanitary housing, the provision of green areas in every residential district, and new high-rise, high-density buildings set in open space. Housing was to be removed from its traditional relationship facing streets, and the whole circulation system was to be revised to meet the needs of emerging mechanization (the automobile). Work areas should be close to but separate from residential areas. To achieve the new city, large land holdings, preferably owned by the public, should replace multiple small parcels (so that projects could be properly designed and developed).

Now thousands of housing estates and redevelopment projects in socialist and capitalist countries the world over, whether built on previously undeveloped land or developed as replacements for old urban areas, attest to the acceptance of the charter's dictums. The design notions it embraced have become part of a world design language, not just the intellectual property of an enlightened few, even though the principles have been devalued in many developments.

Of course, the Charter of Athens has not been the only major urban philosophy of this century to influence the development of urban areas. Ebenezer Howard, too, was responding to the ills of the nineteenth-century industrial city, and the Garden City movement has been at least as powerful as the Charter of Athens. New towns policies, where they exist, are rooted in Howard's thought. But you don't have to look to new towns to see the influence of Howard, Olmsted, Wright, and Stein. The superblock notion, if nothing else, pervades large housing projects around the world, in central cities as well as suburbs. The notion of buildings in a park is as common to garden city designs as it is to charter-inspired development. Indeed, the two movements have a great deal in common: superblocks, separate paths for people and cars, interior common spaces, housing divorced from streets, and central ownership of land. The garden city-inspired communities place greater emphasis on private outdoor space. The most significant difference, at least as they have evolved, is in density and building type: the garden city people preferred to accommodate people in row houses, garden apartments, and maisonettes, while Corbusier and the CIAM designers went for high-rise buildings and, inevitably, people living in flats and at significantly higher densities.

We are less than enthralled with what either the Charter of Athens or the Garden City movement has produced in the way of urban environments. The emphasis of CIAM was on buildings and what goes on within buildings that happen to sit in space, not on the public life that takes place constantly in public spaces. The orientation is often inward. Buildings tend to be islands, big or small. They could be placed anywhere. From the outside perspective, the building, like the work of art it was intended to be, sits where it can be seen and admired in full. And because it is large it is best seen from a distance (at a scale consistent with a moving auto). Diversity, spontaneity, and surprise are absent, at least for the person on foot. On the other hand, we find little joy or magic or spirit in the charter cities. They are not urban, to us, except according to some definition one might find in a census. Most garden cities, safe and healthy and even gracious as they may be, remind us more of suburbs than of cities. But they weren't trying to be cities. The emphasis has always been on "garden" as much as or more than on "city."

Both movements represent overly strong design reactions to the physical decay and social inequities of industrial cities. In responding so strongly, albeit understandably, to crowded, lightless, airless, "utilitiless," congested buildings and cities that housed so many people, the utopians did not inquire what was good about those places, either socially or physically. Did not those physical environments reflect (and maybe even foster) values that were likely to be meaningful to people individually and collectively, such as publicness and community? Without knowing it, maybe these strong reactions to urban ills ended up by throwing the baby out with the bathwater.

In the meantime we have had a lot of experience with city building and rebuilding. New spokespeople with new urban visions have emerged. As more CIAM-style buildings were built people became more disenchanted. Many began to look through picturesque lenses back to the old preindustrial cities. From a concentration on the city as a kind of sculpture garden, the townscape movement, led by the *Architectural Review*, emphasized "urban experience." This phenomenological view of the city was espoused by Rasmussen, Kepes, and ultimately Kevin Lynch and Jane Jacobs. It identified a whole new vocabulary of urban form – one that depended on the sights, sounds, feels, and smells of the city, its materials and textures, floor surfaces, facades, style, signs, lights, seating, trees, sun, and shade all potential amenities for the attentive observer and user. This has permanently humanized the vocabulary of urban design, and we enthusiastically subscribe to most of its tenets, though some in the townscape movement ignored the social meanings and implications of what they were doing.

The 1960s saw the birth of community design and an active concern for the social groups affected, usually negatively, by urban design. Designers were the "soft cops," and many professionals left the design field for social or planning vocations, finding the physical environment to have no redeeming social value. But at the beginning of the 1980s the mood in the design professions is conservative. There is a withdrawal from social engagement back to formalism. Supported by semiology and other abstract themes, much of architecture has become a dilettantish and narcissistic pursuit, a chic component of the high art consumer culture, increasingly remote from most people's everyday lives, finding its ultimate manifestation in the art gallery and the art book. City planning is too immersed in the administration and survival of housing,

environmental, and energy programs and in responding to budget cuts and community demands to have
any clear sense of direction with regard to city form.

While all these professional ideologies have
been working themselves out, massive economic,
technological, and social changes have taken place
in our cities. The scale of capitalism has continued
to increase, as has the scale of bureaucracy, and the
automobile has virtually destroyed cities as they once
were.

In formulating a new manifesto, we react against
other phenomena than did the leaders of CIAM 50
years ago. The automobile cities of California and the
Southwest present utterly different problems from
those of nineteenth-century European cities, as do the
CIAM-influenced housing developments around
European, Latin American, and Russian cities and the
rash of squatter settlements around the fast-growing
cities of the Third World. What are these problems?

PROBLEMS FOR MODERN URBAN DESIGN

Poor living environments

While housing conditions in most advanced countries
have improved in terms of such fundamentals as
light, air, and space, the surroundings of homes are
still frequently dangerous, polluted, noisy, anonymous
wastelands. Travel around such cities has become
more and more fatiguing and stressful.

Giantism and loss of control

The urban environment is increasingly in the hands of
the large-scale developers and public agencies. The
elements of the city grow inexorably in size, massive
transportation systems are segregated for single travel
modes, and vast districts and complexes are created
that make people feel irrelevant.

People, therefore, have less sense of control over
their homes, neighborhoods, and cities than when they
lived in slower-growing locally based communities.
Such giantism can be found as readily in the housing
projects of socialist cities as in the office buildings and
commercial developments of capitalist cities.

Large-scale privatization and the loss of public life

Cities, especially American cities, have become
privatized, partly because of the consumer society's
emphasis on the individual and the private sector,
creating Galbraith's "private affluence and public
squalor," but escalated greatly by the spread of the
automobile. Crime in the streets is both a cause and a
consequence of this trend, which has resulted in a new
form of city: one of closed, defended islands with blank
and windowless facades surrounded by wastelands of
parking lots and fast-moving traffic. As public transit
systems have declined, the number of places in
American cities where people of different social groups
actually meet each other has dwindled. The public
environment of many American cities has become
an empty desert, leaving public life dependent for its
survival solely on planned formal occasions, mostly in
protected internal locations.

Centrifugal fragmentation

Advanced industrial societies took work out of the
home, and then out of the neighborhood, while
the automobile and the growing scale of commerce
have taken shopping out of the local community. Fear
has led social groups to flee from each other into
homogeneous social enclaves. Communities themselves have become lower in density and increasingly
homogeneous. Thus the city has spread out and separated to form extensive monocultures and specialized
destinations reachable often only by long journeys – a
fragile and extravagant urban system dependent on
cheap, available gasoline, and an effective contributor
to the isolation of social groups from each other.

Destruction of valued places

The quest for profit and prestige and the relentless
exploitation of places that attract the public have led
to the destruction of much of our heritage, of historic
places that no longer turn a profit, of natural amenities
that become overused. In many cases, as in San
Francisco, the very value of the place threatens its
destruction as hungry tourists and entrepreneurs flock
to see and profit from it.

Placelessness

Cities are becoming meaningless places beyond their citizens' grasp. We no longer know the origins of the world around us. We rarely know where the materials and products come from, who owns what, who is behind what, what was intended. We live in cities where things happen without warning and without our participation. It is an alien world for most people. It is little surprise that most withdraw from community involvement to enjoy their own private and limited worlds.

Injustice

Cities are symbols of inequality. In most cities the discrepancy between the environments of the rich and the environments of the poor is striking. In many instances the environments of the rich, by occupying and dominating the prevailing patterns of transportation and access, make the environments of the poor relatively worse. This discrepancy may be less visible in the low density modern city, where the display of affluence is more hidden than in the old city; but the discrepancy remains.

Rootless professionalism

Finally, design professionals today are often part of the problem. In too many cases, we design for places and people we do not know and grant them very little power or acknowledgment. Too many professionals are more part of a universal professional culture than part of the local cultures for whom we produce our plans and products. We carry our "bag of tricks" around the world and bring them out wherever we land. This floating professional culture has only the most superficial conception of particular place. Rootless, it is more susceptible to changes in professional fashion and theory than to local events. There is too little inquiry, too much proposing. Quick surveys are made, instant solutions devised, and the rest of the time is spent persuading the clients. Limits on time and budgets drive us on, but so do lack of understanding and the placeless culture. Moreover, we designers are often unconscious of our own roots, which influence our preferences in hidden ways.

At the same time, the planning profession's retreat into trendism, under the positivist influence of social science, has left it virtually unable to resist the social pressures of capitalist economy and consumer sovereignty. Planners have lost their beliefs. Although we believe citizen participation is essential to urban planning, professionals also must have a sense of what we believe is right, even though we may be vetoed.

GOALS FOR URBAN LIFE

We propose, therefore, a number of goals that we deem essential for the future of a good urban environment: livability; identity and control; access to opportunity, imagination, and joy; authenticity and meaning; open communities and public life; self-reliance; and justice.

Livability

A city should be a place where everyone can live in relative comfort. Most people want a kind of sanctuary for their living environment, a place where they can bring up children, have privacy, sleep, eat, relax, and restore themselves. This means a well managed environment relatively devoid of nuisance, overcrowding, noise, danger, air pollution, dirt, trash, and other unwelcome intrusions.

Identity and control

People should feel that some part of the environment belongs to them, individually and collectively, some part for which they care and are responsible, whether they own it or not. The urban environment should be an environment that encourages people to express themselves, to become involved, to decide what they want and act on it. Like a seminar where everybody has something to contribute to communal discussion, the urban environment should encourage participation. Urbanites may not always want this. Many like the anonymity of the city, but we are not convinced that the freedom of anonymity is a desirable freedom. It would be much better if people were sure enough of themselves to stand up and be counted. Environments should therefore be designed for those who use them or are affected by them, rather than for those who own them. This should reduce alienation and anonymity (even if people want them); it should increase people's sense of identity and rootedness and encourage more care and responsibility for the physical environment of cities.

Respect for the existing environment, both nature and city, is one fundamental difference we have with the CIAM movement. Urban design has too often assumed that new is better than old. But the new is justified only if it is better than what exists. Conservation encourages identity and control and, usually, a better sense of community, since old environments are more usually part of a common heritage.

Access to opportunity, imagination, and joy

People should find the city a place where they can break from traditional molds, extend their experience, meet new people, learn other viewpoints, have fun. At a functional level, people should have access to alternative housing and job choices; at another level, they should find the city an enlightening cultural experience. A city should have magical places where fantasy is possible, a counter to and an escape from the mundaneness of everyday work and living. Architects and planners take cities and themselves too seriously; the result too often is deadliness and boredom, no imagination, no humor, alienating places. But people need an escape from the seriousness and meaning of the everyday. The city has always been a place of excitement; it is theater, a stage upon which citizens can display themselves and see others. It has magic, or should have, and that depends on a certain sensuous, hedonistic mood, on signs, on night lights, on fantasy, color, and other imagery. There can be parts of the city where belief can be suspended, just as in the experience of fiction. It may be that such places have to be framed so that people know how to act. Until now such fantasy and experiment have been attempted mostly by commercial facilities, at rather low levels of quality and aspiration, seldom deeply experimental. One should not have to travel as far as the Himalayas or the South Sea Islands to stretch one's experience. Such challenges could be nearer home. There should be a place for community utopias; for historic, natural, and anthropological evocations of the modern city, for encounters with the truly exotic.

Authenticity and meaning

People should be able to understand their city (or other people's cities), its basic layout, public functions, and institutions; they should be aware of its opportunities. An authentic city is one where the origins of things and places are clear. All this means an urban environment should reveal its significant meanings; it should not be dominated only by one type of group, the powerful; neither should publicly important places be hidden. The city should symbolize the moral issues of society and educate its citizens to an awareness of them.

That does not mean everything has to be laid out as on a supermarket shelf. A city should present itself as a readable story, in an engaging and, if necessary, provocative way, for people are indifferent to the obvious, overwhelmed by complexity. A city's offerings should be revealed or they will be missed. This can affect the forms of the city, its signage, and other public information and education programs.

Livability, identity, authenticity, and opportunity are characteristics of the urban environment that should serve the individual and small social unit, but the city has to serve some higher social goals as well. It is these we especially wish to emphasize here.

Community and public life

Cities should encourage participation of their citizens in community and public life. In the face of giantism and fragmentation, public life, especially life in public places, has been seriously eroded. The neighborhood movement, by bringing thousands, probably millions of people out of their closed private lives into active participation in their local communities, has begun to counter that trend, but this movement has had its limitations. It can be purely defensive, parochial, and self-serving. A city should be more than a warring collection of interest groups, classes, and neighborhoods; it should breed a commitment to a larger whole, to tolerance, justice, law, and democracy. The structure of the city should invite and encourage public life, not only through its institutions, but directly and symbolically through its public spaces. The public environment, unlike the neighborhood, by definition should be open to all members of the community. It is where people of different kinds meet. No one should be excluded unless they threaten the balance of that life.

Urban self-reliance

Increasingly cities will have to become more self-sustaining in their uses of energy and other scarce resources. "Soft energy paths" in particular not only will reduce dependence and exploitation across regions and countries but also will help reestablish a stronger sense of local and regional identity, authenticity, and meaning.

An environment for all

Good environments should be accessible to all. Every citizen is entitled to some minimal level of environmental livability and minimal levels of identity, control, and opportunity. Good urban design must be for the poor as well as the rich. Indeed, it is more needed by the poor.

We look toward a society that is truly pluralistic, one where power is more evenly distributed among social groups than it is today in virtually any country, but where the different values and cultures of interest- and place-based groups are acknowledged and negotiated in a just public arena.

These goals for the urban environment are both individual and collective, and as such they are frequently in conflict. The more a city promises for the individual, the less it seems to have a public life; the more the city is built for public entities, the less the individual seems to count. The good urban environment is one that somehow balances these goals, allowing individual and group identity while maintaining a public concern, encouraging pleasure while maintaining responsibility, remaining open to outsiders while sustaining a strong sense of localism.

AN URBAN FABRIC FOR AN URBAN LIFE

We have some ideas, at least, for how the fabric or texture of cities might be conserved or created to encourage a livable urban environment. We emphasize the structural qualities of the good urban environment – qualities we hope will be successful in creating urban experiences that are consonant with our goals.

Do not misread this. We are not describing all the qualities of a city. We are not dealing with major transportation systems, open space, the natural environment, the structure of the large-scale city, or even the structure of neighborhoods, but only the grain of the good city.

There are five physical characteristics that must be present if there is to be a positive response to the goals and values we believe are central to urban life. They must be designed, they must exist, as prerequisites of a sound urban environment. All five must be present, not just one or two. There are other physical characteristics that are important, but these five are essential: livable streets and neighborhoods; some minimum density of residential development as well as intensity of land use; an integration of activities – living, working, shopping – in some reasonable proximity to each other; a manmade environment, particularly buildings, that defines public space (as opposed to buildings that, for the most part, sit in space); and many, many separate, distinct buildings with complex arrangements and relationships (as opposed to few, large buildings).

Let us explain, keeping in mind that all five of the characteristics must be present. People, we have said, should be able to live in reasonable (though not excessive) safety, cleanliness, and security. That means livable streets and neighborhoods: with adequate sunlight, clean air, trees, vegetation, gardens, open space, pleasantly scaled and designed buildings; without offensive noise; with cleanliness and physical safety. Many of these characteristics can be designed into the physical fabric of the city.

The reader will say, "Well of course, but what does that mean?" Usually it has meant specific standards and requirements, such as sun angles, decibel levels, lane widths, and distances between buildings. Many researchers have been trying to define the qualities of a livable environment. It depends on a wide array of attributes, some structural, some quite small details. There is no single right answer. We applaud these efforts and have participated in them ourselves. Nevertheless, desires for livability and individual comfort by themselves have led to fragmentation of the city. Livability standards, whether for urban or for suburban developments, have often been excessive.

Our approach to the details of this inclusive physical characteristic would center on the words "reasonable, though not excessive . . ." Too often, for example, the requirement of adequate sunlight has resulted in buildings and people inordinately far from each other, beyond what demonstrable need for light would dictate. Safety concerns have been the justifications for ever wider streets and wide, sweeping curves rather than narrow ways and sharp corners. Buildings are removed from streets because of noise considerations when there might be other ways to deal with this concern. So although livable streets and neighborhoods are a primary requirement for any good urban fabric – whether for existing, denser cities or for new development – the quest for livable neighborhoods, if pursued obsessively, can destroy the urban qualities we seek to achieve.

A *minimum density* is needed. By density we mean the number of people (sometimes expressed in terms of housing units) living on an area of land, or the number of people using an area of land.

Cities are not farms. A city is people living and working and doing the things they do in relatively close proximity to each other.

We are impressed with the importance of density as a perceived phenomenon and therefore relative to the beholder and agree that, for many purposes, perceived density is more important than an "objective" measurement of people per unit of land. We agree, too, that physical phenomena can be manipulated so as to render perceptions of greater or lesser density. Nevertheless, a narrow, winding street, with a lot of signs and a small enclosed open space at the end, with no people, does not make a city. Cities are more than stage sets. Some minimum number of people living and using a given area of land is required if there is to be human exchange, public life and action, diversity and community.

Density of people alone will account for the presence or absence of certain uses and services we find important to urban life. We suspect, for example, that the number and diversity of small stores and services – for instance, groceries, bars, bakeries, laundries and cleaners, coffee shops, secondhand stores, and the like – to be found in a city or area is in part a function of density. That is, that such businesses are more likely to exist, and in greater variety, in an area where people live in greater proximity to each other ("higher" density). The viability of mass transit, we know, depends partly on the density of residential areas and partly on the size and intensity of activity at commercial and service destinations. And more use of transit, in turn, reduces parking demands and permits increases in density. There must be a critical mass of people, and they must spend a lot of their time in reasonably close proximity to each other, including when they are at home, if there is to be an urban life. The goal of local control and community identity is associated with density as well. The notion of an optimum density is elusive and is easily confused with the health and livability of urban areas, with lifestyles, with housing types, with the size of area being considered (the building site or the neighborhood or the city), and with the economics of development. A density that might be best for child rearing might be less than adequate to support public transit. Most recently, energy efficiency has emerged as a concern associated with density, the notion being that conservation will demand more compact living arrangements.

Our conclusion, based largely on our experience and on the literature, is that a minimum net density (people or living units divided by the size of the building site, excluding public streets) of about 15 dwelling units (30–60 people) per acre of land is necessary to support city life. By way of illustration, that is the density produced with generous town houses (or row houses). It would permit parcel sizes up to 25 feet wide by about 115 feet deep. But other building types and lot sizes also would produce that density. Some areas could be developed with lower densities, but not very many. We don't think you get cities at 6 dwellings to the acre, let alone on half-acre lots. On the other hand, it is possible to go as high as 48 dwelling units per acre (96 to 192 people) for a very large part of the city and still provide for a spacious and gracious urban life. Much of San Francisco, for example, is developed with three-story buildings (one unit per floor) above a parking story, on parcels that measure 25 feet by 100 or 125 feet. At those densities, with that kind of housing, there can be private or shared gardens for most people, no common hallways are required, and people can have direct access to the ground. Public streets and walks adequate to handle pedestrian and vehicular traffic generated by these densities can be accommodated in rights-of-way that are 50 feet wide or less. Higher densities, for parts of the city, to suit particular needs and lifestyles, would be both possible and desirable. We are not sure what the upper limits would be but suspect that as the numbers get much higher than 200 people per net residential acre, for larger parts of the city, the concessions to less desirable living environments mount rapidly.

Beyond residential density, there must be a minimum intensity of people using an area for it to be urban, as we are defining that word. We aren't sure what the numbers are or even how best to measure this kind of intensity. We are speaking here, particularly, of the public or "meeting" areas of our city. We are confident that our lowest residential densities will provide most meeting areas with life and human exchange, but are not sure if they will generate enough activity for the most intense central districts.

There must be an *integration of activities* – living, working, and shopping as well as public, spiritual, and recreational activities – reasonably near each other.

The best urban places have some mixtures of uses. The mixture responds to the values of publicness and diversity that encourage local community identity. Excitement, spirit, sense, stimulation, and exchange are more likely when there is a mixture of activities than when there is not. There are many examples that we all know. It is the mix, not just the density of people and uses, that brings life to an area, the life of people going about a full range of normal activities without having to get into an automobile.

We are not saying that every area of the city should have a full mix of all uses. That would be impossible. The ultimate in mixture would be for each building to have a range of uses from living, to working, to shopping, to recreation. We are not calling for a return to the medieval city. There is a lot to be said for the notion of "living sanctuaries," which consist almost wholly of housing. But we think these should be relatively small, of a few blocks, and they should be close and easily accessible (by foot) to areas where people meet to shop or work or recreate or do public business. And except for a few of the most intensely developed office blocks of a central business district or a heavy industrial area, the meeting areas should have housing within them. Stores should be mixed with offices. If we envision the urban landscape as a fabric, then it would be a salt-and-pepper fabric of many colors, each color for a separate use or a combination. Of course, some areas would be much more heavily one color than another, and some would be an even mix of colors. Some areas, if you squinted your eyes, or if you got so close as to see only a small part of the fabric, would read as one color, a red or a brown or a green. But by and large there would be few if any distinct patterns, where one color stopped and another started. It would not be patchwork quilt, or an even-colored fabric. The fabric would be mixed.

In an urban environment, *buildings* (and other objects that people place in the environment) *should be arranged in such a way as to define and even enclose public space, rather than sit in space.* It is not enough to have high densities and an integration of activities to have cities. A tall enough building with enough people living (or even working) in it, sited on a large parcel, can easily produce the densities we have talked about and can have internally mixed uses, like most "mixed use" projects. But that building and its neighbors will be unrelated objects sitting in space if they are far enough apart, and the mixed uses might be only privately available. In large measure that is what the Charter of Athens, the garden cities, and standard suburban development produce.

Buildings close to each other along a street, regardless of whether the street is straight, or curved, or angled, tend to define space if the street is not too wide in relation to the buildings. The same is true of a plaza or a square. As the spaces between buildings become larger (in relation to the size of the buildings, up to a point), the buildings tend more and more to sit in space. They become focal points for few or many people, depending on their size and activity. Except where they are monuments or centers for public activities (a stadium or meeting hall), where they represent public gathering spots, buildings in space tend to be private and inwardly oriented. People come to them and go from them in any direction. That is not so for the defined outdoor environment. Avoiding the temptation to ascribe all kinds of psychological values to defined spaces (such as intimacy, belonging, protection – values that are difficult to prove and that may differ for different people), it is enough to observe that spaces surrounded by buildings are more likely to bring people together and thereby promote public interaction. The space can be linear (like streets) or in the form of plazas of myriad shapes. Moreover, interest and interplay among uses is enhanced. To be sure, such arrangements direct people and limit their freedom – they cannot move in just any direction from any point – but presumably there are enough choices (even avenues of escape) left open, and the gain is in greater potential for sense stimulation, excitement, surprise, and focus. Over and over again we seek out and return to defined ways and spaces as symbolic of urban life emphasizing the public space more than the private building.

It is important for us to emphasize *public places* and a *public* way system. We have observed that the central value of urban life is that of publicness, of people from different groups meeting each other and of people acting in concert, albeit with debate. The most important public places must be for *pedestrians*, for no public life can take place between people in automobiles. Most public space has been taken over by the automobile, for travel or parking. We must fight to restore more for the pedestrian. Pedestrian malls are not simply to benefit the local merchants. They have an essential public value. People of different kinds meet each other directly. The level of communication may be only visual, but that itself is educational and can encourage tolerance. The revival of street activities, street vending, and street theater in American cities may be the precursor of a more flourishing public environment, if the automobile can be held back.

There also must be symbolic, public meeting places, accessible to all and publicly controlled. Further, in order to communicate, to get from place to place, to interact, to exchange ideas and goods, there must be a healthy public circulation system. It cannot be privately controlled. Public circulation systems should be seen as significant cultural settings where the city's finest products and artifacts can be displayed, as in the piazzas of medieval and renaissance cities.

Finally, *many different buildings and spaces with complex arrangements and relationships* are required. The often elusive notion of human scale is associated with this requirement – a notion that is not just an architect's concept but one that other people understand as well.

Diversity, the possibility of intimacy and confrontation with the unexpected, stimulation, are all more likely with many buildings than with few taking up the same ground areas.

For a long time we have been led to believe that large land holdings were necessary to design healthy, efficient, aesthetically pleasing urban environments. The slums of the industrial city were associated, at least in part, with all those small, overbuilt parcels. Socialist and capitalist ideologies alike called for land assembly to permit integrated, socially and economically useful developments. What the socialist countries would do via public ownership the capitalists would achieve through redevelopment and new fiscal mechanisms that rewarded large holdings. Architects of both ideological persuasions promulgated or were easily convinced of the wisdom of land assembly. It's not hard to figure out why. The results, whether by big business or big government, are more often than not inward-oriented, easily controlled or controllable, sterile, large-building projects, with fewer entrances, fewer windows, less diversity, less innovation, and less individual expression than the urban fabric that existed previously or that can be achieved with many actors and many buildings. Attempts to break up facades or otherwise to articulate separate activities in large

buildings are seldom as successful as when smaller properties are developed singly.

Health, safety, and efficiency can be achieved with many smaller buildings, individually designed and developed. Reasonable public controls can see to that. And, of course, smaller buildings are a lot more likely if parcel sizes are small than if they are large. With smaller buildings and parcels, more entrances must be located on the public spaces, more windows and a finer scale of design diversity emerge. A more public, lively city is produced. It implies more, smaller groups getting pieces of the public action, taking part, having a stake. Other stipulations may be necessary to keep public frontages alive, free from the deadening effects of offices and banks, but small buildings will help this more than large ones. There need to be large buildings, too, covering large areas of land, but they will be the exception, not the rule, and should not be in the centers of public activity.

ALL THESE QUALITIES . . . AND OTHERS

A good city must have all those qualities. Density without livability could return us to the slums of the nineteenth century. Public places without small-scale, fine-grain development would give us vast, overscale cities. As an urban fabric, however, those qualities stand a good chance of meeting many of the goals we outlined. They directly attend to the issue of livability though they are aimed especially at encouraging public places and a public life. Their effects on personal and group identity are less clear, though the small-scale city is more likely to support identity than the large-scale city. Opportunity and imagination should be encouraged by a diverse and densely settled urban structure. This structure also should create a setting that is more meaningful to the individual inhabitant and small group than the giant environments now being produced. There is no guarantee that this urban structure will be a more just one than those presently existing. In supporting the small against the large, however, more justice for the powerless may be encouraged.

Still, an urban fabric of this kind cannot by itself meet all these goals. Other physical characteristics are important to the design of urban environments. Open space, to provide access to nature as well as relief from the built environment, is one. So are definitions, boundaries if you will, that give location and identity to neighborhoods (or districts) and to the city itself.

There are other characteristics as well: public buildings, educational environments, places set aside for nurturing the spirit, and more. We still have work to do.

MANY PARTICIPANTS

While we have concentrated on defining physical characteristics of a good city fabric, the process of creating it is crucial. As important as many buildings and spaces are many participants in the building process. It is through this involvement in the creation and management of their city that citizens are most likely to identify with it and, conversely, to enhance their own sense of identity and control.

AN ESSENTIAL BEGINNING

The five characteristics we have noted are essential to achieving the values central to urban life. They need much further definition and testing. We have to know more about what configurations create public space: about maximum densities, about how small a

community can be and still be urban (some very small Swiss villages fit the bill, and everyone knows some favorite examples), about what is perceived as big and what small under different circumstances, about landscape material as a space definer, and a lot more. When we know more we will be still further along toward a new urban design manifesto.

We know that any ideal community, including the kind that can come from this manifesto, will not always be comfortable for every person. Some people don't like cities and aren't about to. Those who do will not be enthralled with all of what we propose.

Our urban vision is rooted partly in the realities of earlier, older urban places that many people, including many utopian designers, have rejected, often for good reasons. So our utopia will not satisfy all people. That's all right. We like cities. Given a choice of the kind of community we would *like* to live in – the sort of choice earlier city dwellers seldom had – we would choose to live in an urban, public community that embraces the goals and displays the physical characteristics we have outlined. Moreover, we think it responds to what people want and that it will promote the good urban life.

"What Would a Non-sexist City Be Like? Speculations on Housing, Urban Design, and Human Work"

from Catharine Stimpson *et al.* (eds), *Women and the American City* (1981)

Dolores Hayden

Editors' Introduction

Dolores Hayden is a professor of architecture, urbanism, and American studies at Yale University where she teaches courses on American urban history and urban design. A social historian and architect, Hayden has proposed ways to make the built environment and society more responsive to women's needs, particularly the needs of working women with children. She is committed to new ways of building cities for a more egalitarian society, with more community interaction, which are more responsive to both social and environmental concerns, and reflect a concern for public history, cultural diversity, and urban preservation.

This and other of Hayden's writings have struck a deeply responsive cord with architects and planners who want to design new forms of housing to better fit contemporary family structures and with feminists favoring changes in gender and work roles. Like other of Hayden's writings this selection is grounded in the *ideas* of earlier feminist writers, provides evidence that changes can work based on *actual projects* that have been successfully tried, and forcefully states Hayden's own feminist and egalitarian *values*.

Today only a small percentage of people live in nuclear households consisting of a married husband and wife with one or more children where only the husband is gainfully employed outside of the household. Single individuals without children, unmarried couples, households consisting of unrelated individuals, gay couples, and multi-generational households together account for a much larger proportion of the total population. Yet arguably most existing housing and the entire organization of the rest of the built environment was built to meet the needs of nuclear families with a sole (male) breadwinner. Hayden argues that the existing built environment does not work well for many households – particularly for working women with children. Accordingly she urges architects and planners to design new urban developments and retrofit existing ones to meet the needs of the varied household types that exist today and the reality that two or more people in many households work outside of the home.

In this visionary selection Hayden imagines what a non-sexist city would be like and proposes the creation of HOMES groups (Homemakers Organization for a More Egalitarian Society). A HOMES group would be a forty-household hologram of existing society. It would have men and women, young and old people, professionals and non-professionals in proportion to their presence in the general population. Members of the HOMES group would live in newly built or retrofitted building complexes with common space and shared appliances. A HOMES group would involve both men and women equally in unpaid housekeeping and child care labor and in the paid labor force; eliminate residential segregation by class, race, and age; eliminate government programs and laws that reinforce the unpaid role of female homemakers and minimize unpaid domestic labor; minimize wasteful energy consumption by sharing domestic electrical appliances; and maximize real choices for households for recreation and sociability. This is a radical socialist/feminist vision for an alternative society.

Life in a HOMES group would be somewhat like life in a particularly egalitarian Israeli kibbutz, a Chinese commune during the great proletarian cultural revolution, or an egalitarian religious society like the Shakers. While the idea might appeal to some people, it is questionable whether most people anywhere in the world today would personally want to live in a HOMES group. But by advancing such a thorough critique of existing social organization and the built environment and proposing such a radically different vision Hayden has deliberately stimulated a very productive debate. Many architects and planners agree with much of her critique and are incorporating new design features into their plans in response to them. Generally their proposals are less extreme than what would be required in a pure HOMES group. Dolores Hayden's work has inspired better design of buildings, neighborhoods, and cities adapted to families that do not consist of an employed husband, non-working wife, and children; where men and women can more easily share both child-rearing and work outside the home; and where diverse households can interact in mutually helpful and personally satisfying ways.

"Co-housing" is a current movement to build housing with more shared space than is common in subdivisions of single-family detached homes. Co-housing developments typically consist of both privately owned units and commonly owned spaces such as a common kitchen, daycare center, and perhaps a bicycle repair shop. Members of a co-housing project might pool use of appliances just as Hayden suggests. The idea of mutual support in cooking, child-rearing, gardening, and social activities is fundamental to the co-housing philosophy. The movement proposes much more interaction among members living in a co-housing project than is common in typical suburban subdivisions of single-family detached homes, but a less visionary restructuring of gender and work roles than Hayden proposes. A number of co-housing projects have been built and more are under way both in Europe and the United States. The best source of information on the co-housing movement is Kathryn McCamant and Charles Durrett, *Cohousing*, 2nd edn (Berkeley, CA: Ten Speed Press, 1992).

Dolores Hayden's books include *The Power of Place: Urban Landscapes as Public History* (Cambridge, MA: MIT Press, 1995), *Redesigning the American Dream: The Future of Housing, Work, and Family Life* (New York: W.W. Norton, 1984), *The Grand Domestic Revolution: A History of Feminist Designs for American Houses, Neighborhoods and Cities* (Cambridge, MA: MIT Press, 1981), *Seven American Utopias: The Architecture of Communitarian Socialism: 1780–1975* (Cambridge, MA: MIT Press, 1976), and *Preserving Cultural Landscapes in America* (Baltimore, MD: Johns Hopkins University Press, 2000) co-edited with Arnold R. Alanen and Robert Melrick.

Writings about women and cities include Catharine Stimpson (ed.), *Women and the American City* (Chicago: University of Chicago Press, 1980) from which this selection in reprinted, Caroline Andrew and Beth Moore Milroy (eds), *Life Spaces: Gender, Household, Employment* (Vancouver: University of British Columbia Press, 1988), Leslie Kanes Weisman, *Discrimination by Design* (Urbana and Chicago: University of Illinois Press, 1992), Daphne Spain's *Gendered Space* (Chapel Hill: University of North Carolina Press, 1992) and *How Women Saved the City* (Minneapolis: University of Minnesota Press, 2002), and Clara Greed's *Women and Planning: Creating Gendered Realities* (London: Routledge, 1993). Leonie Sandercock and Ann Forsythe's bibliography on gender issues in planning theory in "A Gender Agenda: New Directions for Planning Theory," *Journal of the American Planning Association*, Vol. 58, No. 1 (Winter 1992) contains further bibliography of related writings.

■ ■ ■ ■ ■ ■

"A woman's place is in the home" has been one of the most important principles of architectural design and urban planning in the United States for the last century. An implicit rather than explicit principle for the conservative and male-dominated design professions, it will not be found stated in large type in textbooks on land use. It has generated much less debate than the other organizing principles of the contemporary American city in an era of monopoly capitalism, which include the ravaging pressure of private land development, the fetishistic dependence on millions of private automobiles, and the wasteful use of energy.[1] However, women have rejected this dogma and entered the paid labor force in larger and larger numbers. Dwellings, neighborhoods, and cities designed for homebound women constrain women physically, socially, and economically. Acute frustration occurs when women defy these constraints to spend all or part of the work day in the paid labor force. I contend that the only remedy for this situation is to develop a new paradigm of the home, the neighborhood, and the city; to begin to describe the physical, social, and economic design of a human settlement that would support, rather than restrict, the activities of employed women and their families. It is essential to recognize such needs in order to begin both the rehabilitation of the existing housing stock and the construction of new housing to meet the needs of a new and growing majority of Americans – working women and their families.

When speaking of the American city in the last quarter of the twentieth century, a false distinction between "city" and "suburb" must be avoided. The urban region, organized to separate homes and workplaces, must be seen as a whole. In such urban regions, more than half of the population resides in the sprawling suburban areas, or "bedroom communities." The greatest part of the built environment in the United States consists of "suburban sprawl": single-family homes grouped in class-segregated areas, crisscrossed by freeways and served by shopping malls and commercial strip developments. Over 50 million small homes are on the ground. About two-thirds of American families "own" their homes on long mortgages; this includes over 77 percent of all AFL-CIO members.[2] White, male skilled workers are far more likely to be homeowners than members of minority groups and women, long denied equal credit or equal access to housing. Workers commute to jobs either in the center or elsewhere in the suburban ring. In metropolitan areas studied in 1975 and 1976, the journey to work, by public transit or private car, averaged about nine miles each way. Over 100 million privately owned cars filled two- and three-car garages (which would be considered magnificent housing by themselves in many developing countries). The United States, with 13 percent of the world's population, uses 41 percent of the world's passenger cars in support of the housing and transportation patterns described.[3]

The roots of this American settlement form lie in the environmental and economic policies of the past. In the late nineteenth century, millions of immigrant families lived in the crowded, filthy slums of American industrial cities and despaired of achieving reasonable living conditions. However, many militant strikes and demonstrations between the 1890s and 1920s made some employers reconsider plant locations and housing issues in their search for industrial order.[4] "Good homes make contented workers" was the slogan of the Industrial Housing Associates in 1919. These consultants and many others helped major corporations plan better housing for white, male skilled workers and their families in order to eliminate industrial conflict. "Happy workers invariably mean bigger profits, while unhappy workers are never a good investment," they chirruped.[5] Men were to receive "family wages," and become home "owners" responsible for regular mortgage payments, while their wives became home "managers" taking care of spouse and children. The male worker would return from his day in the factory or office to a private domestic environment, secluded from the tense world of work in an industrial city characterized by environmental pollution, social degradation, and personal alienation. He would enter a serene dwelling whose physical and emotional maintenance would be the duty of his wife. Thus the private suburban house was the stage set for the effective sexual division of labor. It was the commodity par excellence, a spur for male paid labor and a container for female unpaid labor. It made gender appear a more important self-definition than class, and consumption more involving than production. In a brilliant discussion of the "patriarch as wage slave," Stuart Ewen has shown how capitalism and anti-feminism fused in campaigns for homeownership and mass consumption: the patriarch whose home was his "castle" was to work year in and year out to provide the wages to support this private environment.[6]

Although this strategy was first boosted by corporations interested in a docile labor force, it soon appealed to corporations that wished to move from World War I defense industries into peacetime production of domestic appliances for millions of families. The development of the advertising industry, documented by Ewen, supported this ideal of mass consumption and promoted the private suburban dwelling, which maximized appliance purchases.[7] The occupants of the isolated household were suggestible. They bought the house itself, a car, stove, refrigerator, vacuum cleaner, washer, carpets. Christine Frederick, explaining it in 1929 as Selling Mrs. Consumer, promoted homeownership and easier consumer credit and advised marketing managers on how to manipulate American women.[8] By 1931 the Hoover Commission on Home

Ownership and Home Building established the private, single-family home as a national goal, but a decade and a half of depression and war postponed its achievement. Architects designed houses for Mr. and Mrs. Bliss in a competition sponsored by General Electric in 1935; winners accommodated dozens of electrical appliances in their designs with no critique of the energy costs involved.[9] In the late 1940s the single-family home was boosted by FHA and VA mortgages, and the construction of isolated, overprivatized, energy-consuming dwellings became commonplace. "I'll Buy That Dream" made the postwar hit parade.[10]

Mrs. Consumer moved the economy to new heights in the 1950s. Women who stayed at home experienced what Betty Friedan called the "feminine mystique" and Peter Filene renamed the "domestic mystique."[11] While the family occupied its private physical space, the mass media and social science experts invaded its psychological space more effectively than ever before.[12] With the increase in spatial privacy came pressure for conformity in consumption. Consumption was expensive. More and more married women joined the paid labor force, as the suggestible housewife needed to be both a frantic consumer and a paid worker to keep up with the family's bills. Just as the mass of white male workers had achieved the "dream houses" in suburbia where fantasies of patriarchal authority and consumption could be acted out, their spouses entered the world of paid employment. By 1975, the two-worker family accounted for 39 percent of American households. Another 13 percent were single-parent families, usually headed by women. Seven out of ten employed women were in the work force because of financial need. Over 50 percent of all children between the ages of one and seventeen had employed mothers.[13]

How does a conventional home serve the employed woman and her family? Badly. Whether it is in a suburban, exurban, or inner-city neighborhood, whether it is a split-level ranch house, a modern masterpiece of concrete and glass, or an old brick tenement, the house or apartment is almost invariably organized around the same set of spaces: kitchen, dining room, living room, bedrooms, garage or parking area. These spaces require someone to undertake private cooking, cleaning, child care, and usually private transportation if adults and children are to exist within it. Because of residential zoning practices, the typical dwelling will usually be physically removed from any shared community space – no commercial or communal day-care facilities, or laundry facilities, for example, are likely to be part of the dwelling's spatial domain. In many cases these facilities would be illegal if placed across property lines. They could also be illegal if located on residentially zoned sites. In some cases sharing such a private dwelling with other individuals (either relatives or those unrelated by blood) is also against the law.[14]

Within the private spaces of the dwelling, material culture works against the needs of the employed woman as much as zoning does, because the home is a box to be filled with commodities. Appliances are usually single-purpose, and often inefficient, energy-consuming machines, lined up in a room where the domestic work is done in isolation from the rest of the family. Rugs and carpets that need vacuuming, curtains that need laundering, and miscellaneous goods that need maintenance fill up the domestic spaces, often decorated in "colonial," "Mediterranean," "French Provincial," or other eclectic styles purveyed by discount and department stores to cheer up that bare box of an isolated house. Employed mothers usually are expected to, and almost invariably do, spend more time in private housework and child care than employed men; often they are expected to, and usually do, spend more time on commuting per mile traveled than men, because of their reliance on public transportation. One study found that 70 percent of adults without access to cars are female.[15] Their residential neighborhoods are not likely to provide much support for their work activities. A "good" neighborhood is usually defined in terms of conventional shopping, schools, and perhaps public transit, rather than additional social services for the working parent, such as day care or evening clinics.

While two-worker families with both parents energetically cooperating can overcome some of the problems of existing housing patterns, households in crisis, such as subjects of wife and child battering, for example, are particularly vulnerable to its inadequacies. According to Colleen McGrath, every thirty seconds a woman is being battered somewhere in the United States. Most of these batterings occur in kitchens and bedrooms. The relationship between household isolation and battering, or between unpaid domestic labor and battering, can only be guessed, at this time, but there is no doubt that America's houses and households are literally shaking with domestic violence.[16] In addition, millions of angry and upset women are treated with tranquilizers in the private home – one drug company advertises to doctors: "You can't change her environment but you can change her mood."[17]

The woman who does leave the isolated, single-family house or apartment finds very few real housing alternatives available to her.[18] The typical divorced or battered woman currently seeks housing, employment, and child care simultaneously. She finds that matching her complex family requirements with the various available offerings by landlords, employers, and social services is impossible. One environment that unites housing, services, and jobs could resolve many difficulties, but the existing system of government services, intended to stabilize households and neighborhoods by ensuring the minimum conditions for a decent home life to all Americans, almost always assumes that the traditional household with a male worker and an unpaid homemaker is the goal to be achieved or simulated. In the face of massive demographic changes, programs such as public housing, AFDC, and food stamps still attempt to support an ideal family living in an isolated house or apartment, with a full-time homemaker cooking meals and minding children many hours of the day.

By recognizing the need for a different kind of environment, far more efficient use can be made of funds now used for subsidies to individual households. Even for women with greater financial resources, the need for better housing and services is obvious. Currently, more affluent women's problems as workers have been considered "private" problems – the lack of good day care, their lack of time. The aids to overcome an environment without child care, public transportation, or food service have been "private," commercially profitable solutions: maids and baby-sitters by the hour; franchise day care or extended television viewing; fast food service; easier credit for purchasing an automobile, a washer, or a microwave oven. Not only do these commercial solutions obscure the failure of American housing policies; they also generate bad conditions for other working women. Commercial day-care and fast-food franchises are the source of low-paying nonunion jobs without security. In this respect they resemble the use of private household workers by bourgeois women, who may never ask how their private maid or child-care worker arranges care for her own children. They also resemble the insidious effects of the use of television in the home as a substitute for developmental child care in the neighborhood. The logistical problems which all employed women

face are not private problems, and they do not succumb to market solutions.

The problem is paradoxical: women cannot improve their status in the home unless their overall economic position in society is altered; women cannot improve their status in the paid labor force unless their domestic responsibilities are altered. Therefore, a program to achieve economic and environmental justice for women requires, by definition, a solution that overcomes the traditional divisions between the household and the market economy, the private dwelling and the workplace. One must transform the economic situation of the traditional homemaker, whose skilled labor has been unpaid but economically and socially necessary to society; one must also transform the domestic situation of the employed woman. If architects and urban designers were to recognize all employed women and their families as a constituency for new approaches to planning and design and were to reject all previous assumptions about "woman's place" in the home, what could we do? Is it possible to build non-sexist neighborhoods and design non-sexist cities? What would they be like?

Some countries have begun to develop new approaches to the needs of employed women. The Cuban Family Code of 1974 requires men to share housework and child care within the private home. The degree of its enforcement is uncertain, but in principle it aims at men's sharing what was formerly "women's work," which is essential to equality. The Family Code, however, does not remove work from the house, and relies upon private negotiation between husband and wife for its day-to-day enforcement. Men feign incompetence, especially in the area of cooking, with tactics familiar to any reader of Patricia Mainardi's essay, "The Politics of Housework," and the sexual stereotyping of paid jobs for women outside the home, in day-care centers for example, has not been successfully challenged.[19]

Another experimental approach involves the development of special housing facilities for employed women and their families. The builder Otto Fick first introduced such a program in Copenhagen in 1903. In later years it was encouraged in Sweden by Alva Myrdal and by the architects Sven Ivar Lind and Sven Markelius. Called "service houses" or "collective houses," such projects (Figures 1 and 2) provide child care and cooked food along with housing for employed women and their families.[20] Like a few similar projects in the USSR in the 1920s, they aim at offering services,

TRADGÅRDSTERASS   DAGHEMMETS LEKPLATS

RESTAURANG   ENTRÉ   DAGHEM

Figure 1 Sven Ivar Lind, Marieberg collective house, Stockholm, 1944, plan of entrance (entré), restaurant (restaurang), and day nursery (deghem): (1) entrance hall, (2) doorman's office, (3) restaurant delivery room, (4) real estate office, (5) connecting walkway to Swedberg House, (6) restaurant anteroom, (7) main dining room, (8) small dining room, (9) restaurant kitchen, (10) to day nursery's baby carriage room, (11) day nursery's baby changing room, (12) office for day nursery's directress, (13) to Wennerberg House's cycle garage

Figure 2 Plan of residential floors. Type 2A contains two rooms, bath, and kitchenette. Types 1C and 4D are efficiency units with bath and kitchenette. Type 4S includes four rooms with bath and full kitchen

2 BR ///.	1 BR ≡
1 BR S |||||	2 BR \\\
1 BR S ≡	2 BR ¦¦¦¦

Figure 3 "Urbanes Wohnen" (urban living) Steilshoop, north of Hamburg, public housing for 206 tenants, designed by the tenant association in collaboration with Rolf Spille, 1970–3. Instead of 72 conventional units, they built 20 multifamily units and 2 studios. Twenty-six mental patients were included in the project, of whom 24 recovered. Partial floor plan. Units include private bedrooms (br), living rooms (lr), and some studios (s). They share a collective living room, kitchen, dining room, and playroom. Each private apartment can be closed off from the collective space and each is different. Key: (1) storage room, (2) closets, (3) wine cellar, (4) buanderie, (5) fire stairs

either on a commercial basis or subsidized by the state, to replace formerly private "women's work" performed in the household. The Scandinavian solution does not sufficiently challenge male exclusion from domestic work, nor does it deal with households' changing needs over the life cycle, but it recognizes that it is important for environmental design to change.

Some additional projects in Europe extend the scope of the service house to include the provision of services for the larger community or society. In the Steilshoop Project, in Hamburg, Germany, in the early 1970s, a group of parents and single people designed public housing with supporting services (Figure 3).[21] The project included a number of former mental patients as residents and therefore served as a halfway house for them, in addition to providing support services for the public-housing tenants who organized it. It suggests the extent to which current American residential stereotypes can be broken down – the sick, the aged, the unmarried can be integrated into new types of households and housing complexes, rather than segregated in separate projects.

Another recent project was created in London by Nina West Homes, a development group established in 1972, which has built or renovated over sixty-three units of housing on six sites for single parents. Children's play areas or day-care centers are integrated with the dwellings; in their Fiona House project the housing is designed to facilitate shared baby-sitting, and the day-care center is open to the neighborhood residents for a fee (Figure 4). Thus the single parents can find jobs as day-care workers and help the neighborhood's working parents as well.[22] What is most exciting here is the hint that home and work can be reunited on one site for some of the residents, and home and child-care services are reunited on one site for all of them.

In the United States, we have an even longer history of agitation for housing to reflect women's needs. In the late nineteenth century and early twentieth century there were dozens of projects by feminists, domestic scientists, and architects attempting to develop community services for private homes. By the late 1920s, few such experiments were still functioning.[23] In general, feminists of that era failed to recognize the problem of exploiting other women workers when providing services for those who could afford them. They also often failed to see men as responsible parents and workers in their attempts to socialize "women's" work. But feminist leaders had a very strong

Figure 4 A, Fiona House, second-floor plan, main building, showing corridor used as a playroom, with kitchen windows opening into it; first-floor plan, rear building, showing nursery school. B, Axonometric drawing, Fiona House, Nina West Homes, London, 1972, designed by Sylvester Bone. Twelve two-bedroom units for divorced or separated mothers with additional outdoor play space and neighborhood nursery school facility. Flats can be linked by intercom system to provide an audio substitute for babysitting

sense of the possibilities of neighborly cooperation among families and of the economic importance of "women's" work.

In addition, the United States has a long tradition of experimental utopian socialist communities building model towns, as well as the example of many communes and collectives established in the 1960s and 1970s which attempted to broaden conventional definitions of household and family.[24] While some communal groups, especially religious ones, have often demanded acceptance of a traditional sexual division of labor, others have attempted to make nurturing activities a responsibility of both women and men. It is important to draw on the examples of successful projects of all kinds, in seeking an image of a non-sexist settlement. Most employed women are not interested in taking themselves and their families to live in communal families, nor are they interested in having state bureaucracies run family life. They desire, not an end to private life altogether, but community services to support the private household. They also desire solutions that reinforce their economic independence and maximize their personal choices about child rearing and sociability.

What, then, would be the outline of a program for change in the United States? The task of reorganizing both home and work can only be accomplished by organizations of homemakers, women and men dedicated to making changes in the ways that Americans deal with private life and public responsibilities. They must be small, participatory organizations with members who can work together effectively. I propose calling such groups HOMES (Homemakers Organization for a More Egalitarian Society). Existing feminist groups, especially those providing shelters for battered wives and children, may wish to form HOMES to take over existing housing projects and develop services for residents as an extension of those offered by feminist counselors in the shelter. Existing organizations supporting cooperative ownership of housing may wish to form HOMES to extend their housing efforts in a feminist direction. A program broad enough to transform housework, housing, and residential neighborhoods must: (1) involve both men and women in the unpaid labor associated with housekeeping and child care on an equal basis; (2) involve both men and women in the paid labor force on an equal basis; (3) eliminate residential segregation by class, race, and age; (4) eliminate all federal, state, and local programs and laws that offer implicit or explicit reinforcement of the unpaid role of the female homemaker; (5) minimize unpaid domestic labor and wasteful energy consumption; (6) maximize real choices for households concerning recreation and sociability. While many partial reforms can support these goals, an incremental strategy cannot achieve them. I believe that the establishment of experimental residential centers, which in their architectural design and economic organization transcend traditional definitions of home, neighborhood, city, and workplace, will be necessary to make changes on this scale. These centers could be created through renovation of existing neighborhoods or through new construction.

Suppose forty households in a US metropolitan area formed a HOMES group and that those households, in their composition, represented the social structure of the American population as a whole. Those forty households would include: seven single parents and their fourteen children (15 percent); sixteen two-worker couples and their twenty-four children (40 percent); thirteen one-worker couples and their twenty-six children (35 percent); and four single residents, some of them "displaced homemakers" (10 percent). The residents would include sixty-nine adults and sixty-four children. There would need to be forty private dwelling units, ranging in size from efficiency to three bedrooms, all with private, fenced outdoor space. In addition to the private housing, the group would provide the following collective spaces and activities: (1) a day-care center with landscaped outdoor space, providing day care for forty children and after-school activities for sixty-four children; (2) a laundromat providing laundry service; (3) a kitchen providing lunches for the day-care center, take-out evening meals, and "meals-on-wheels" for elderly people in the neighborhood; (4) a grocery depot, connected to a local food cooperative; (5) a garage with two vans providing dial-a-ride service and meals-on-wheels; (6) a garden (or allotments) where some food can be grown; (7) a home help office providing helpers for the elderly, the sick, and employed parents whose children are sick. The use of all these collective services should be voluntary; they would exist in addition to private dwelling units and private gardens.

To provide all of the above services, thirty-seven workers would be necessary: twenty day-care workers; three food-service workers; one grocery-depot worker; five home helpers; two drivers of service vehicles; two laundry workers; one maintenance worker; one gardener; two administrative staff. Some of these may be part-time workers, some full-time. Day care, food services, and elderly services could be organized as producers' cooperatives, and other workers could be employed by the housing cooperative as discussed below.

Because HOMES is not intended as an experiment in isolated community buildings but as an experiment in meeting employed women's needs in an urban area, its services should be available to the neighborhood in which the experiment is located. This will increase demand for the services and insure that the jobs are real ones. In addition, although residents of HOMES should have priority for the jobs, there will be many who choose outside work. So some local residents may take jobs within the experiment.

In creating and filling these jobs it will be important to avoid traditional sex stereotyping that would result from hiring only men as drivers, for example, or only women as food-service workers. Every effort should be made to break down separate categories of paid work for women and men, just as efforts should be made to recruit men who accept an equal share of domestic responsibilities as residents. A version of the Cuban Family Code should become part of the organization's platform.

Similarly, HOMES must not create a two-class society with residents outside the project making more money than residents in HOMES jobs that utilize some of their existing domestic skills. The HOMES jobs should be paid according to egalitarian rather than sex-stereotyped attitudes about skills and hours. These jobs must be all classified as skilled work rather than as unskilled or semiskilled at present, and offer full social security and health benefits, including adequate maternity leave, whether workers are part-time or full-time.

Many federal Housing and Urban Development programs support the construction of nonprofit, low- and moderate-cost housing, including section 106b, section 202, and section 8. In addition, HUD section 213 funds are available to provide mortgage insurance for the conversion of existing housing of five or more units to housing cooperatives. HEW programs also fund special facilities such as day-care centers or meals-on-wheels for the elderly. In addition, HUD and HEW offer funds for demonstration projects which meet community needs in new ways.[25] Many trade unions, churches, and tenant cooperative organizations are active as nonprofit housing developers. A limited-equity housing cooperative offers the best basis for economic organization and control of both physical design and social policy by the residents.

Many knowledgeable nonprofit developers could aid community groups wishing to organize such projects, as could architects experienced in the design of housing cooperatives. What has not been attempted is the reintegration of work activities and collective services into housing cooperatives on a large enough scale to make a real difference to employed women. Feminists in trade unions where a majority of members are women may wish to consider building cooperative housing with services for their members. Other trade unions may wish to consider investing in such projects. Feminists in the co-op movement must make strong, clear demands to get such services from existing housing cooperatives, rather than simply go along with plans for conventional housing organized

on a cooperative economic basis. Feminists outside the cooperative movement will find that cooperative organizational forms offer many possibilities for supporting their housing activities and other services to women. In addition, the recently established national Consumer Cooperative Bank has funds to support projects of all kinds that can be tied to cooperative housing.

In many areas, the rehabilitation of existing housing may be more desirable than new construction. The suburban housing stock in the United States must be dealt with effectively. A little bit of it is of architectural quality sufficient to deserve preservation; most of it can be aesthetically improved by the physical evidence of more intense social activity. To replace empty front lawns without sidewalks, neighbors can create blocks where single units are converted to multiple units; interior land is pooled to create a parklike setting at the center of the block; front and side lawns are fenced to make private outdoor spaces; pedestrian paths and sidewalks are created to link all units with the central open space; and some private porches, garages, tool sheds, utility rooms, and family rooms are converted to community facilities such as children's play areas, dial-a-ride garages, and laundries.

Figure 5A shows a typical bleak suburban block of thirteen houses, constructed by speculators at different times, where about four acres are divided into plots of one-fourth to one-half acre each. Thirteen driveways are used by twenty-six cars; ten garden sheds, ten swings, thirteen lawn mowers, thirteen outdoor dining tables, begin to suggest the wasteful duplication of existing amenities. Yet despite the available land there are no transitions between public streets and these private homes. Space is either strictly private or strictly public. Figure 6A shows a typical one-family house of 1,400 square feet on this block. With three bedrooms and den, two-and-a-half baths, laundry room, two porches, and a two-car garage, it was constructed in the 1950s at the height of the "feminine mystique."

To convert this whole block and the housing on it to more efficient and sociable uses, one has to define a zone of greater activity at the heart of the block, taking a total of one and one half to two acres for collective use (Figure 5B). Essentially, this means turning the block inside out. The Radburn plan, developed by Henry Wright and Clarence Stein in the late 1920s, delineated this principle very clearly as correct land use in "the motor age," with cars segregated from residents' green spaces, especially spaces for children.

Figure 5 Diagram showing some of the possibilities of reorganizing a typical suburban block through rezoning, rebuilding, and landscaping. A, Ten single-family houses (1) on ten private lots (2); B, the same houses (1) with smaller private lots (2) after a backyard rehabilitation progam has created a new village green (3) at the heart of the block; C, the same houses (1) and many small private gardens (2) with a new village green (3) surrounded by a zone for new services and accessory apartments (4) connected by a new sidewalk or arcade (5) and surrounded by a new border of street trees (6). In C, (4) can include space for such activities as day care, elderly care, laundry, and food service as well as housing, while (3) can accommodate a children's play area, vegetable or flower gardens, and outdoor seating. (5) may be a sidewalk, a vine-covered trellis, or a formal arcade. The narrow ends of the block can be emphasized as collective entrances with gates (to which residents have keys), leading to new accessory apartments entered from the arcade or sidewalk. In the densest possible situations, (3) may be alley and parking lot, if existing street parking and public transit are not adequate

Figure 6A Suburban single-family house, plan, three bedrooms plus den

Figure 6B Proposed HOMES revitalization, same house converted to three units (two bedroom, one bedroom, and efficiency), plus dial-a-ride garage and collective outdoor space

In Radburn, New Jersey, and in the Baldwin Hills district of Los Angeles, California, Wright and Stein achieved remarkably luxurious results (at a density of about seven units to the acre) by this method, since their multiple-unit housing always bordered a lush parkland without any automobile traffic. The Baldwin Hills project demonstrates this success most dramat-ically, but a revitalized suburban block with lots as small as one-fourth acre can be reorganized to yield something of this same effect.[26] In this case, social amenities are added to aesthetic ones as the interior park is designed to accommodate community day care, a garden for growing vegetables, some picnic tables, a playground where swings and slides are grouped, a grocery depot connected to a larger neighborhood food cooperative, and a dial-a-ride garage.

Large single-family houses can be remodeled quite easily to become duplexes and triplexes, despite the "open plans" of the 1950s and 1960s popularized by many developers. The house in Figure 6A becomes, in Figure 6B, a triplex, with a two-bedroom unit (linked to a community garage); a one-bedroom unit; and an efficiency unit (for a single person or elderly person). All three units are shown with private enclosed gardens. The three units share a front porch and entry hall. There is still enough land to give about two-fifths of the original lot to the community. Particularly striking is the way in which existing spaces such as back porches or garages lend themselves to conversion to social areas or community services. Three former private garages out of thirteen might be given over to collective uses – one as a central office for the whole block, one as a grocery depot, and one as a dial-a-ride garage. Is it possible to have only twenty cars (in ten garages) and two vans for twenty-six units in a rehabilitated block? Assuming that some residents switch from outside employment to working within the block, and that for all residents, neighborhood shopping trips are cut in half by the presence of day care, groceries, laundry, and cooked food on the block, as well as aided by the presence of some new collective transportation, this might be done.

What about neighbors who are not interested in such a scheme? Depending on the configuration of lots, it is possible to begin such a plan with as few as three or four houses. In Berkeley, California, where neighbors on Derby Street joined their backyards and created a cooperative day-care center, one absentee landlord refused to join – his entire property is fenced in and the community space flows around it without difficulty. Of course, present zoning laws must be changed, or variances obtained, for the conversion of single-family houses into duplexes and triplexes and the introduction of any sort of commercial activities into a residential block. However, a community group that is able to organize or acquire at least five units could become a HUD housing cooperative, with a nonprofit corporation owning all land and with producers' cooperatives running the small community services. With a coherent plan for an entire block, variances could be obtained much more easily than on a lot-by-lot basis. One can also imagine organizations that run halfway houses – for ex-mental patients, or runaway teenagers, or battered women – integrating their activities into

such a block plan, with an entire building for their activities. Such groups often find it difficult to achieve the supportive neighborhood context such a block organization would offer.

I believe that attacking the conventional division between public and private space should become a socialist and feminist priority in the 1980s. Women must transform the sexual division of domestic labor, the privatized economic basis of domestic work, and the spatial separation of homes and workplaces in the built environment if they are to be equal members of society. The experiments I propose are an attempt to unite the best features of past and present reforms in our own society and others, with some of the existing social services available in the United States today. I would like to see several demonstration HOMES begun, some involving new construction following the program I have laid out, others involving the rehabilitation of suburban blocks. If the first few experimental projects are successful, homemakers across the United States will want to obtain day-care, food, and laundry services at a reasonable price, as well as better wages, more flexible working conditions, and more suitable housing. When all homemakers recognize that they are struggling against both gender stereotypes and wage discrimination, when they see that social, economic, and environmental changes are necessary to overcome these conditions, they will no longer tolerate housing and cities, designed around the principles of another era, that proclaim that "a woman's place is in the home."

NOTES

This paper comprised part of the text of a talk for the conference "Planning and Designing a Non-Sexist Society," University of California, Los Angeles, April 21, 1979. I would like to thank Catharine Stimpson, Peter Marris, S.M. Miller, Kevin Lynch, Jeremy Brecher, and David Thompson for extensive written comments on drafts of this paper.

1. There is an extensive Marxist literature on the importance of spatial design to the economic development of the capitalist city, including Henri Lefebre, *La Production de l'espace* (Paris: Editions Anthropos, 1974); Manuel Castells, *The Urban Question* (Cambridge, Mass.: MIT Press, 1977); David Harvey, *Social Justice and the City* (London: Edward Arnold, 1974); and David Gordon, "Capitalist Development and the History of American

Cities," in *Marxism and the Metropolis*, ed. William K. Tabb and Larry Sawyers (New York: Oxford University Press, 1978). None of this work deals adequately with the situation of women as workers and homemakers, nor with the unique spatial inequalities they experience. Nevertheless, it is important to combine the economic and historical analysis of these scholars with the empirical research of non-Marxist feminist urban critics and sociologists who have examined women's experience of conventional housing, such as Gerda Wekerle, "A Woman's Place Is in the City" (paper for the Lincoln Institute of Land Policy, Cambridge, Mass., 1978); and Suzanne Keller, "Women in a Planned Community" (paper for the Lincoln Institute of Land Policy, Cambridge, Mass., 1978). Only then can one begin to provide a socialist-feminist critique of the spatial design of the American city. It is also essential to develop research on housing similar to Sheila B. Kamerman, "Work and Family in Industrialized Societies," *Signs: Journal of Women in Culture and Society* 4, no. 4 (Summer 1979): 632–50, which reviews patterns of women's employment, maternity provisions, and child-care policies in Hungary, East Germany, West Germany, France, Sweden, and the United States. A comparable study of housing and related services for employed women could be the basis for more elaborate proposals for change.

Many attempts to refine socialist and feminist economic theory concerning housework are discussed in an excellent article by Ellen Malos, "Housework and the Politics of Women's Liberation," *Socialist Review* 37 (January–February 1978): 41–7. A most significant theoretical piece is Movimento di Lotta Femminile, "Programmatic Manifesto for the Struggle of Housewives in the Neighborhood," *Socialist Revolution* 9 (May–June 1972): 85–90.

2. *Survey of AFL-CIO Members Housing 1975* (Washington, DC: AFL-CIO, 1975), p. 16. I am indebted to Allan Heskin for this reference.

3. *Transit Fact Book*, 1977–78 ed. (Washington, DC: American Public Transit Association, 1978), p. 29); *Motor Vehicle Facts and Figures* (Detroit, Mich.: Motor Vehicle Manufacturers Association, 1977), pp. 29, 31, 53.

4. Gordon, pp. 48–50, discusses suburban relocation of plants and housing.

5. Industrial Housing Associates, "Good Homes Make Contented Workers," 1919, Edith Elmer Wood Papers, Avery Library, Columbia University. Also see Barbara Ehrenreich and Deirdre English, "The Manufacture of Housework," *Socialist Revolution* 5 (1975): 16. They quote an unidentified corporate official (ca. 1920): "Get them to invest their savings in homes and own them. Then they won't leave and they won't strike. It ties them down so they have a stake in our prosperity."

6. Stuart Ewen, *Captains of Consciousness: Advertising and the Social Roots of the Consumer Culture* (New York: McGraw-Hill Book Co., 1976).

7. Richard Walker, "Suburbanization in Passage," unpublished draft paper (Berkeley: University of California, Berkeley, Department of Geography, 1977).

8. Christine Frederick, *Selling Mrs. Consumer* (New York: Business Bourse, 1929).

9. Carol Barkin, "Home, Mom, and Pie-in-the-Sky" (M. Arch. thesis, University of California, Los Angeles, 1979), pp. 120–24, gives the details of this competition; Ruth Schwartz Cowan, in an unpublished lecture at MIT in 1977, explained GE's choice of an energy-consuming design for its refrigerator in the 1920s, because this would increase demand for its generating equipment by municipalities.

10. Peter Filene, *Him/Her/Self: Sex Roles in Modern America* (New York: Harcourt Brace Jovanovich, 1974), p. 189.

11. Betty Friedan, *The Feminine Mystique* (1963; New York: W.W. Norton & Co., 1974), p. 307, somewhat hysterically calls the home a "comfortable concentration camp"; Filene, p. 194, suggests that men are victimized by ideal homes too, thus "domestic" mystique.

12. Eli Zaretsky, *Capitalism, the Family, and Personal Life* (New York: Harper & Row, 1976), develops Friedman's earlier argument in a more systematic way. This phenomenon is misunderstood by Christopher Lasch, *Haven in a Heartless World* (New York: Alfred A. Knopf, 1977), who seems to favor a return to the sanctity of the patriarchal home.

13. Rosalyn Baxandall, Linda Gordon, and Susan Reverby, eds., *America's Working Women: A Documentary History, 1600 to the Present* (New York: Vintage Books, 1976). For more detail, see Louise Kapp Howe, *Pink Collar Workers: Inside the World of Woman's Work* (New York: Avon Books, 1977).

14. Recent zoning fights on the commune issue have occurred in Santa Monica, Calif.; Wendy Schuman, "The Return of Togetherness," *New York Times* (March 20, 1977), reports frequent illegal down zoning by two-family groups in one-family residences in the New York area.

15. Study by D. Foley, cited in Wekerle (see n. 1 above).

16. Colleen McGrath, "The Crisis of Domestic Order," *Socialist Review* 9 (January–February 1979): 12, 23.

17. Research by Malcolm MacEwen, cited in *Associate Collegiate Schools of Architecture Newsletter* (March 1973), p. 6.

18. See, for example, Carol A. Brown, "Spatial Inequalities and Divorced Mothers" (paper delivered at the annual meeting of the American Sociological Association, San Francisco, 1978); Susan Anderson-Khleif, research report for HUD on single-parent families and their housing, summarized in "Housing for Single Parents," *Research Report, MIT-Harvard Joint Center for Urban Studies* (April 1979), pp. 3–4.

19. Patricia Mainardi, "The Politics of Housework," in *Sisterhood Is Powerful*, ed. Robin Morgan (New York: Vintage Books, 1970). My discussion of the Cuban Family Code is based on a visit to Cuba in 1978; a general review is Carollee Bengelsdorf and Alice Hageman, "Emerging from Underdevelopment: Women and Work in Cuba," in *Capitalist Patriarchy and the Case for Socialist Feminism*, ed. Z. Eisenstein (New York: Monthly Review Press, 1979). Also see Geoffrey E. Fox, "Honor, Shame and Women's Liberation in Cuba: Views of Working-Class Emigré Men," in *Female and Male in Latin America*, ed. A. Pescatello (Pittsburgh: University of Pittsburgh Press, 1973).

20. Erwin Muhlestein, "Kollektives Wohnen gestern und heute," *Archithese* 14 (1975): 3–23.

21. This project relies on the "support structures" concept of John Habraken to provide flexible interior partitions and fixed mechanical core and structure.

22. "Bridge over Troubled Water," *Architects' Journal* (September 27, 1972), pp. 680–84; personal interview with Nina West, 1978.

23. Dolores Hayden, *A "Grand Domestic Revolution": Feminism, Socialism and the American Home, 1870–1930* (Cambridge, Mass.: MIT Press, 1980); "Two Utopian Feminists and Their Compaigns for Kitchenless Houses," *Signs: Journal of Women in Culture and Society* 4, no. 2 (Winter 1979): 274–90; "Melusina Fay Peirce and Cooperative Housekeeping," *International Journal of Urban and Regional Research* 2 (1978): 401–20; "Challenging the American Domestic Ideal," and "Catharine Beecher and the Politics of Housework," in *Women in American Architecture*, ed. S. Torre (New York: Whitney Library of Design, 1977), pp. 22–39, 40–49; "Charlotte Perkins Gilman: Domestic Evolution or Domestic Revolution," *Radical History Review*, vol. 21 (Winter 1979–80), in press.

24. Dolores Hayden, *Seven American Utopias: The Architecture of Communitarian Socialism, 1790–1975* (Cambridge, Mass.: MIT Press, 1976), discusses historical examples and includes a discussion of communes of the 1960s and 1970s, "Edge City, Heart City, Drop City: Communal Building Today," pp. 320–47.

25. I am indebted to Geraldine Kennedy and Sally Kratz, whose unpublished papers, "Toward Financing Cooperative Housing," and "Social Assistance Programs Whose Funds Could Be Redirected to Collective Services," were prepared for my UCLA graduate seminar in spring 1979.

26. See also the successful experience of Zurich, described in Hans Wirz, "Back Yard Rehab: Urban Microcosm Rediscovered," *Urban Innovation Abroad* 3 (July 1979): 2–3.

SEVEN

PART EIGHT

The future of the city

PARK

PROMENADE

COMMERCIAL

CITY CENTER

DWELLINGS

PUBLIC

INTRODUCTION TO PART EIGHT

The desire to peer into the future is a human trait as old as the Biblical prophets and the oracle at Delphi. And the desire to project *urban* futures is at least as old as Plato's description of the ideal city-state in *The Republic*. But the pace of futurist predictions seems to quicken at times when great cultural and historic shifts are taking place. Such was the case during the Industrial Revolution of the nineteenth century when fantasists like Jules Verne and political idealists like Edward Bellamy (author of *Looking Backward, 2000–1887*) captured the popular imagination and when the utopian visionaries discussed in Part Five of this volume helped to establish the theoretical basis of urban planning practice. And such is the case today as the realization becomes every day more clear that the advanced economies of the world are entering a new global, information-based, post-industrial stage of development that promises to reveal new forms of urban civilization and human community.

Clearly, a new urban paradigm is now emerging. But in order to predict the shape of the emerging postmodern city, one must have a clear sense of the probable direction of the world urbanization process described by Kingsley Davis in "The Urbanization of the Human Population," the essay with which the "Evolution of Cities" section of this volume begins. Will the percentage of the world's total population living in cities, now at about fifty percent, continue to increase in the decades and centuries to come? Will the "townward drift," as Frederick Law Olmsted called it, continue? To what eventual level? Eighty percent? Ninety? One hundred percent? Or has urbanization reached its peak, ready to stabilize at more or less the present level? Perhaps the S-curve on the chart representing world urbanization will prove to be a bell curve, with the percentage of the human population living in cities going into a long, gradual decline until only a small number of the total population remains urbanized. This last possibility has spawned an intriguing, if highly conjectural body of literature. Some, pondering the possible effects of modern transportation and telecommunications technologies, see a gradual withering away of cities. Others, particularly environmentalists, talk about urbanization reaching its natural limits and beginning to reverse in an age of ecological and economic constraints, or suggest that unrestrained growth is sure to result in widespread economic and environmental collapse.

Today, the variety of possible futures from which to choose is extraordinarily diverse. Each option mirrors some of our deepest hopes and fears. Marshall McLuhan, the 1960s guru of communications theory, suggested that the whole world would one day become a "global village," with every member of humanity interacting with every other in a real-time simulacrum of the neolithic community. Some of the more radical members of the environmentalist movement have gone "back to nature" by establishing rural communes along the fringes of urbanized civilization, while other equally radical social activists have established "urban kibbutzim" in the very hearts of the inner cities. Techno-optimists see Earth's future in space colonization projects while techno-pessimists envision post-apocalypse cities like the ones depicted in *Terminator*, *Bladerunner*, *The Matrix*, and other popular science fiction films.

One possible urban future is post-urbanism. Melvin Webber, author of "The Post-City Age" (1968), argues that certain technological developments will result in an end to traditional cities and the emergence of a post-urban period of human development. Since he wrote his prophetic essay in the 1960s, the technologies he referred to were air travel and telephones. Today, the advent of computers and telecommunications make his predictions even more plausible. Even if we reject the basic idea of a reversal of the course of urbanization, Webber's insight into the way the new technological world will leave some people informationally rich and others informationally poor is a troubling reality that cannot be ignored.

Webber is something of a utopian visionary engaged in brilliant, if somewhat mystical, speculation about humankind's future. It is perhaps more useful, when considering the future of the city, to come down to earth, to shorten the futurist perspective to the near term, and to project immediate futures based on present and observable trends. Among the most important such trends today are the parallel emergence of (1) a global, post-industrial, telecommunications-based economy, (2) a new, far-flung suburban ring of development that journalist Joel Garreau has dubbed "Edge City" and that Robert Fishman has analyzed as "technoburbia" (p. 77), (3) transnational environmental concerns such as global warming and atmospheric pollution, and (4) the persistence, even intensification, of the classic social conflicts along racial and class lines that have been a feature of urban life for hundreds of years. These influences are likely to determine the course of early twenty-first-century urban development.

Manuel Castells is a critical analyst of urban life who has carefully analyzed the near-term futures of urban society – both for the United States and Western Europe – and suggested possible directions for urban development strategies in ways that combine the influences of new technologies and old social inequalities based on uneven access to economic empowerment. No one thinks that solving the problem of inner-city poverty and race-and-class-based segregation will be an easy task, nor that one solution is the only one worth considering. Clearly, there are many contentious issues of politics and urban governance, and even deeper concerns at the level of social structure and cultural development that will affect the future of cities. But as the neo-suburban technoburbs and the Pedestrian Pockets beckon, and as local economies and national sovereignties give way to global interdependence, the ongoing and sometimes explosive violence of racial and class discrimination is still there at the core of many cities, threatening to destroy urban civilization and to undermine the human community. Defusing those possible sources of conflict remains part of the unfinished business of the city of the future.

In "European Cities, the Informational Society, and the Global Economy" (1993), Manuel Castells surveys the urban future of Western Europe. Recognizing that racial and class conflicts – in this case exacerbated by immigration from the underdeveloped world – will be a troubling feature of future urban development, Castells worries about the effects of economic globalism and telecommunications technologies on the democratic values of equality and social justice. He argues that technology-based cities represent "a new industrial space," located in centers throughout Europe, Asia, and America and tied to a global "informational" economy, that are fundamentally different from anything that has ever come before. A two-tier economy and a widening gulf between the educated elites and the ghettoized, marginalized urban populations is intensifying rather than diminishing in the cities of the information economy and the global market-place. The new urban order, abetted by new economic realities, seems cut off from its past, from history and cultural tradition as sources of communal meaning and individual identity.

In *The Rise of the Network Society* (1998), Castells argued that the implicit tendency of the work-styles of the post-industrial economy is to detach themselves from traditional cultures, values, and communities. Information flows through networks and across vast distances, and "the historical emergence of the space of flows," he writes, supersedes "the meaning of the space of places." For Castells, then, the real challenge of the new informational cities of Western Europe is to reconcile the "new techno-economic paradigm" and "place-based social meaning" in a way that will avoid what he calls "urban schizophrenia." "At the cultural level," he writes, "local societies, territorially defined, must preserve their identities, and build upon their historical roots, regardless of their economic and functional dependence upon the space of flows."

If Castells is one of the great theoreticians of the techno-urbanism of the future, Stephen Wheeler recognizes that the new telecommunications technologies are just another aspect of the man-made infrastructures of production, management, and control that have always been a part of urban life. Increasingly, however, another factor will greatly influence the future of cities: the natural environment, now seen not just as a regional rural hinterland but as a global context in which cities can either flourish or die. To use the popular sports metaphor: Nature bats last!

By the middle of the nineteenth century, park planners like Joseph Paxton in England and Frederick Law Olmsted in the United States were working hard to bring nature into crowded cities. In the early twentieth century, Patrick Geddes had worked out an elaborate scheme for regional planning that reflected different ecosystems. Similar thinking informed Ian McHarg's theories in his classic *Design with Nature* (1965). Today "sustainable" urban planning has moved to the fore, and issues of urban sustainability – the ability of cities to grow and develop without overwhelming the environmental carrying capacity of the earth – have become an important feature of contemporary thinking about the future of cities. In "Planning Sustainable and Livable Cities," Wheeler summarizes the current sustainable urban planning literature and provides his own definition of sustainable urban development and the operating principles needed to achieve it. Wheeler illustrates his ideas with actual examples of good sustainable planning practice and thereby points the way toward desirable urban futures that can actually be realized. Wheeler's vision of a sustainable urban future is not narrowly ecological, in the sense of placing natural above human values. "In the long run," he writes, "sustainable development will require systemic cultural change that builds democracy and social capital."

But can social capital, or even the accepted norms of urban social life, withstand the underlying implications of the telecommunications revolution? William Mitchell thinks not and, in "The Teleserviced City" from his imaginative book *E-topia*, suggests that the future of urban social life will be totally and fundamentally transformed by the very nature of digital electronics. Not only will urban space and urban economic relationships be transformed by computers and the daily online transactions that already characterize our lives, but urban communities and even individual urban personalities, human identity itself, are on the verge of a major new historic and cultural revolution, equal in effect to the Neolithic, Urban, and Industrial Revolutions that came before. Mitchell's vision of the inevitable future is not without promise, but many will find it deeply challenging. And it is how we will adapt to the challenges of the future – population growth, the new globalism, the emerging technologies, environmental constraints, and the ongoing struggles for social justice – that will determine how cities and citizens in the years to come will achieve, or fail to achieve, a meaningful sense of personal identity and social community, and thereby give ongoing justification or final refutation to the validity of the urban project.

"The Post-City Age"

Daedalus (1968)

Melvin M. Webber

Editors' Introduction

In the introduction to Kingsley Davis's "The Urbanization of the Human Population" (p. 25), we asked what the future of urbanization might be. Would the urban population of the earth stabilize at fifty percent or seventy percent? Might it increase to one hundred percent? Or might the S-curve of historic urbanization become a standard bell curve in the future, charting a steady decline in the proportion of the total human population that lives in cities?

As long ago as 1968, Melvin Webber thought through the possibility of a gradual decline in urbanization in his brilliant and prophetic essay "The Post-City Age." Prophetic? some will ask. Major urban centers have not decreased in density and population during the last thirty years. To the contrary, major global cities – as well as newly emerging mega-cities in Asia, Africa, and Latin America – have grown tremendously in the decades since Webber made his bold and counter-intuitive prediction. Still, Webber was at least prescient in pointing to two major technological developments – air transportation and the telephone – that even in 1968 were eliminating the traditional space–time constraints on human interaction and bringing about stunning cultural changes in the nature of an increasingly global human civilization.

Webber argues that the widespread availability of commercial air transportation and global telephonic communication permits an educated and affluent class of people to live anywhere – in suburbs, in rural districts, on mountain-tops, for that matter – and still be thoroughly "urban," participating fully in intellectual, professional, and economic life. At the same time, it is precisely the poor populations that are trapped in inner cities and deprived of access to technology that are becoming increasingly "rural" in the sense that they are non-participants in global community affairs.

Are these insights still applicable, perhaps even more applicable, now that networks of instantaneous global telecommunications exist to serve the needs of technologically advanced individuals and populations? Is the society that Webber foresaw in 1968 something like the contemporary "technoburbs" (p. 77) that Robert Fishman describes? Might not computers and modern telecommunications make the Broadacre City that Frank Lloyd Wright prophesied (p. 325) even more possible today and in the future? And might Webber's vision of a "post-city age" prefigure the kind of "teleserviced city" that William Mitchell describes in *E-topia* (p. 497)?

Melvin Webber is professor emeritus of planning at the University of California, Berkeley. He is the author of *Explorations into Urban Structure* (Philadelphia: University of Pennsylvania Press, 1964) and has written extensively on issues of urban transit and cyberspace citizenship.

We are passing through a revolution that is unhitching the social processes of urbanization from the locationally fixed city and region. Reflecting the current explosion in science and technology, employment is shifting from the production of goods to services; increasing ease of transportation and communication is dissolving the spatial barriers to social intercourse; and Americans are forming social communities comprised of spatially dispersed members. A new kind of large scale urban society is emerging that is increasingly independent of the city. In turn, the problems of the city place generated by early industrialization are being supplanted by a new array different in kind. With but a few remaining exceptions (the new air pollution is a notable one), the recent difficulties are not place type problems at all. Rather, they are the transitional problems of a rapidly developing society-economy-and-polity whose turf is the nation. Paradoxically, just at the time in history when policy-makers and the world press are discovering the city, "the age of the city seems to be at an end."

Our failure to draw the rather simple conceptual distinction between the spatially defined city or metropolitan area and the social systems that are localized there clouds current discussions about the "crisis of our cities." The confusion stems largely from the deficiencies of our language and from the anachronistic thoughtways we have carried over from the passing era. We still have no adequate descriptive terms for the emerging social order, and so we use, perforce, old labels that are no longer fitting. Because we have named them so, we suppose that the problems manifested inside cities are, therefore and somehow, "city problems." Because societies in the past had been spatially and locally structured, and because urban societies used to be exclusively city-based, we seem still to assume that territorial is a necessary attribute of social systems.

The error has been a serious one, leading us to seek local solutions to problems whose causes are not of local origin and hence are not susceptible to municipal treatment. We have been tempted to apply city-building instruments to correct social disorders, and we have then been surprised to find that they do not work. (Our experience with therapeutic public housing, which was supposed to cure "social pathologies," and urban renewal, which was supposed to improve the lives of the poor, may be our most spectacular failures.) We have lavished large investments on public facilities, but neglected the quality and the distribution of the social services. And we have defended and reinforced home-rule prerogatives of local and state governments with elaborate rhetoric and protective legislation.

Neither crime-in-the-streets, poverty, unemployment, broken families, race riots, drug addiction, mental illness, juvenile delinquency, nor any of the commonly noted "social pathologies" marking the contemporary city can find its causes or its cure there. We cannot hope to invent local treatments for conditions whose origins are not local in character, nor can we expect territorially defined governments to deal effectively with problems whose causes are unrelated to territory or geography. The concepts and methods of civil engineering and city planning suited to the design of unitary physical facilities cannot be used to serve the design of social change in a pluralistic and mobile society. In the novel society now emerging – with its sophisticated and rapidly advancing science and technology, its complex social organization, and its internally integrated societal processes – the influence and significance of geographic distance and geographic place are declining rapidly.

This is, of course, a most remarkable change. Throughout virtually all of human history, social organization coincided with spatial organization. In preindustrial society, men interacted almost exclusively with geographic neighbors. Social communities, economies, and polities were structured about the place in which interaction was least constrained by the frictions of space. With the coming of large-scale industrialization during the latter half of the nineteenth century, the strictures of space were rapidly eroded, abetted by the new ease of travel and communication that the industrialization itself brought.

The initial counterparts of industrialization in the United States were, first, the concentration of the nation's population into large settlements and, then, the cultural urbanization of the population. Although these changes were causally linked, they had opposite spatial effects. After coming together at a common place, people entered larger societies tied to no specific place. Farming and village peoples from throughout the continent and the world migrated to the expanding cities, where they learned urban ways, acquired the occupational skills that industrialization demanded, and became integrated into the contemporary society.

In recent years, rising societal scale and improvements in transportation and communications systems have loosed a chain of effects robbing the city of its once unique function as an urbanizing instrument of society. Farmers and small-town residents, scattered throughout the continent, were once effectively removed from the cultural life of the nation. City folks visiting the rural areas used to be treated as strangers, whose styles of living and thinking were unfamiliar. News of the rest of the world was hard to get and then had little meaning for those who lived the local life. Country folk surely knew there was another world out there somewhere, but little understood it and were affected by it only indirectly. The powerful anti-urban traditions in early American thought and politics made the immigrant city dweller a suspicious character whose crude ways marked him as un-Christian (which he sometimes was) and certainly un-American. The more sophisticated urban upper classes – merchants, landowners, and professional men – were similarly suspect and hence rejected. In contrast, the small-town merchant and the farmer who lived closer to nature were the genuine Americans of pure heart who lived the simple, natural life. Because the contrasts between the rural and the urban ways-of-life were indeed sharp, antagonisms were real, and the differences became institutionalized in the conduct of politics. America was marked by a diversity of regional and class cultures whose followers interacted infrequently, if ever.

By now this is nearly gone. The vaudeville hick-town and hayseed characters have left the scene with the vaudeville act. Today's urbane farmer watches television documentaries, reads the national news magazines, and manages his acres from an office (maybe located in a downtown office building), as his hired hands ride their tractors while listening to the current world news broadcast from a transistor. Farming has long since ceased to be a handicraft art; it is among the most highly technologized industries and is tightly integrated into the international industrial complex.

During the latter half of the nineteenth century and the first third of the twentieth, the traditional territorial conception that distinguished urbanites and ruralites was probably valid: The typical rural folk lived outside the cities, and the typical urbanites lived inside. By now this pattern is nearly *reversed*. Urbanites no longer reside exclusively in metropolitan settlements, nor do ruralites live exclusively in the hinterlands. Increasingly, those who are least integrated into modern society – those who exhibit most of the attributes of rural folk – are concentrating within the highest-density portions of the large metropolitan centers. This profoundly important development is only now coming to our consciousness, yet it points up one of the major policy issues of the next decades.

Cultural diffusion is integrating immigrants, city residents, and hinterland peoples into a national urban society, but it has not touched all Americans evenly. At one extreme are the intellectual and business elites, whose habitat is the planet; at the other are the lower-class residents of city and farm who live in spatially and cognitively constrained worlds. Most of the rest of us, who comprise the large middle class, lie somewhere in-between, but in some facets of our lives we all seem to be moving from our ancestral localism toward the unbounded realms of the cosmopolites.

High educational attainments and highly specialized occupations mark the new cosmopolites. As frequent patrons of the airlines and the long-distance telephone lines, they are intimately involved in the communications networks that tie them to their spatially dispersed associates. They contribute to and consume the specialized journals of science, government, and industry, thus maintaining contact with information resources of relevance to their activities, whatever the geographic sources or their own locations. Even though some may be employed by corporations primarily engaged in manufacturing physical products, these men trade in information and ideas. They are the producers of the information and ideas that fuel the engines of societal development. For those who are tuned into the international communications circuits, cities have utility precisely because they are rich in information. The way such men use the city reveals its essential character most clearly, for to them the city is essentially a massive communications switchboard through which human interaction takes place.

Indeed, cities exist *only* because spatial agglomeration permits reduced costs of interaction. Men originally elected to locate in high-density settlements precisely because space was so costly to overcome. It is still cheaper to interact with persons who are nearby, and so men continue to locate in such settlements. Because there *are* concentrations of associates in city places, the new cosmopolites establish their offices there and then move about from city to city conducting their affairs. The biggest settlements attract the most long-distance telephone and airline traffic and have undergone the most dramatic growth during the era of city-building.

The recent expansion of Washington, DC is the most spectacular evidence of the changing character of metropolitan development. Unlike the older settlements whose growth was generated by expanding manufacturing activities during the nineteenth and early-twentieth centuries, Washington produces almost no goods whatsoever. Its primary products are information and intelligence, and its fantastic growth is a direct measure of the predominant roles that information and the national government have come to play in contemporary society.

This terribly important change has been subtly evolving for a long time, so gradually that it seems to have gone unnoticed. The preindustrial towns that served their adjacent farming hinterlands were essentially alike. Each supplied a standardized array of goods and services to its neighboring market area. The industrial cities that grew after the Civil War and during the early decades of this century were oriented to serving larger markets with the manufacturing products they were created to produce. As their market areas widened, as product specialization increased, and as the information content of goods expanded, establishments located in individual cities became integrated into the spatially extensive economies. By now, the large metropolitan centers that used to be primarily goods-producing loci have become interchange junctions within the international communications networks. Only in the limited geographical, physical sense is any modern metropolis a discrete, unitary, identifiable phenomenon. At most, it is a localized node within the integrating international networks, finding its significant identity as contributor to the workings of that larger system. As a result, the new cosmopolites belong to none of the world's metropolitan areas, although they use them. They belong, rather, to the national and international communities that merely maintain information exchanges at these metropolitan junctions.

Their capacity to interact intimately with others who are spatially removed depends, of course, upon a level of wealth adequate to cover the dollar costs of long-distance intercourse, as well as upon the cognitive capacities associated with highly skilled professional occupations. The intellectual and business elites are able to maintain continuing and close contact with their associates throughout the world because they are rich not only in information, but also in dollar income.

As the costs of long-distance interaction fall in proportion to the rise in incomes, more and more people are able and willing to pay the transportation and communication bills. As expense-account privileges are expanded, those costs are being reduced to

zero for ever larger numbers of people. As levels of education and skill rise, more and more people are being tied into the spatially extensive communities that used to engage only a few.

Thus, the glue that once held the spatial settlement together is now dissolving, and the settlement is dispersing over ever widening terrains. At the same time, the pattern of settlement upon the continent is also shifting (moving toward long strips along the coasts, the Gulf, and the Great Lakes). These trends are likely to be accelerated dramatically by cost-reducing improvements in transportation and communications technologies now in the research-and-development stages. (The SST, COMSAT communications, high-speed ground transportation with speeds up to 500 mph, TV and computer-aided educational systems, no-toll long-distance telephone service, and real-time access to national computer-based information systems are likely to be powerful ones.) Technological improvements in transport and communications reduce the frictions of space and thereby ease long-distance intercourse. Our compact, physical city layouts directly mirror the more primitive technologies in use at the time these were built. In a similar way, the locational pattern of cities upon the continent reflects the technologies available at the time the settlements grew. If currently anticipated technological improvements prove workable, each of the metropolitan settlements will spread out in low-density patterns over far more extensive areas than even the most frightened future-mongers have yet predicted. The new settlement-form will little resemble the nineteenth-century city so firmly fixed in our images and ideologies. We can also expect the large junction points will no longer have the communications advantage they now enjoy, and smaller settlements will undergo a major spurt of growth in all sorts of now isolated places where the natural amenities are attractive.

Moreover, as ever larger percentages of the nation's youth go to college and thus enter the national and international cultures, attachments to places of residence will decline dramatically. This prospect, rather than the spatial dispersion of metropolitan areas, portends the functional demise of the city. The signs are already patently clear among those groups whose worlds are widest and least bounded by parochial constraints.

Consider the extreme cosmopolite, if only for purposes of illustrative cartooning. He might be engaged in scientific research, news reporting, or international

EIGHT

business, professions exhibiting critical common traits. The astronomer, for example, maintains instantaneous contact with his colleagues around the world; indeed, he is a day-to-day collaborator with astronomers in all countries. His work demands that he share information and that he and his colleagues monitor stellar events jointly, as the earth's rotation brings men at different locales into prime viewing position. Because he is personally committed to their common enterprise, his social reference group is the society of astronomers. He assigns his loyalties to the community of astronomers, since their work and welfare matter most to him.

To be sure, as he plays out other roles – say, as citizen, parent, laboratory director, or grocery shopper – he is a member of many other communities, both interest-based and place-defined ones. But the striking thing about our astronomer, and the millions of people like him engaged in other professions, is how little of his attention and energy he devotes to the concerns of place-defined communities. Surely, as compared to his grandfather, whose life was largely bound up in the affairs of his locality, the astronomer, playwright, newsman, steel broker, or wheat dealer lives in a life-space that is not defined by territory and deals with problems that are not local in nature. For him, the city is but a convenient setting for the conduct of his professional work; it is not the basis for the social communities that he cares most about.

"European Cities, the Informational Society, and the Global Economy"

Journal of Economic and Social Geography (1993)

Manuel Castells

Editors' Introduction

The information revolution sweeping the world today has profound implications for the future of cities. The power of computers is increasing and the cost of computing is dropping at astonishing rates. Fax machines, modems, fiber-optic cable, communication satellites, a global electronic (e-mail) system, videoconferencing, the Internet, information highways, virtual reality, and multi-media – all are transforming traditional urban space and communities in ways that are still only partially understood.

Manuel Castells was born in Spain and educated as a sociologist in France. He has taught at more than fifteen universities in Europe, North America, Latin America, and Asia, including the University of Paris and the Universidad Autonoma de Madrid, and is currently Professor of Sociology and Professor of City and Regional Planning at the University of California, Berkeley. As a young man, Castells fled Franco's authoritarian and intellectually stifling Spain for the freedom and intellectual excitement of Paris. He became a neo-Marxian and crafted sophisticated theories on the role of the capitalist state and grassroots urban protest movements. More recently, Castells has turned his attention to the implications of high technology and the information revolution for community life and urban development.

Castells sees information technologies as the fundamental instrument of the new organizational logic transforming the world today. Accordingly he uses the adjective "informational" as a type of city just as "industrial" or "colonial" city might have been used in the nineteenth century. In his magisterial three-volume *The Rise of the Network Society* (1998) Castells describes how information technology will restructure relationships between rich and poor regions, labor and capital, centralization and decentralization of services, governments and non-governmental entities, the individual and society.

Castells argues that what he calls the "space of flows" will increasingly govern the actions of power-holding organizations rather than territorially based institutions operating in the "space of places." Industry and services will be organized worldwide around the operation of their information-generating units. He envisions power-ful, secretive, multi-national institutions not tied to any particular place as the dominant institutions of the future.

In "European Cities, the Informational Society, and the Global Economy," Castells predicts how the basic elements of the new informational society will affect the cities of Western Europe. Among the main influences, he lists the inevitability of European Union political and economic integration, the cultural pressures of Third World immigration, social conflicts arising out of gentrification and the "polarized occupational structure," and the increasing importance of environmentalism and gender-based social restructuring. Thus, he argues that the "fundamental urban dualism of our time . . . opposed the cosmopolitanism of the elite, living on a daily connection to the whole world to the tribalism of local communities, retrenched in their spaces that they try to control as their last stand against the macro forces that shape their lives out of their reach."

Castells's analysis raises many questions. Will local and even national governments wither in importance with the information dominance of powerful multinational institutions? Will the information revolution lead to greater social inequality between rich and poor nations? Are "Dual Cities" inevitable where rich and powerful professional managers manipulate impoverished masses, both blocks and continents away? May his vision of cosmopolitan elites versus retrenched local communities be too dualistic, too either–or, when the broad mass of middle-class urban residents may feel torn between both lifestyles? Will mass urban social movements emerge to reassert popular power against such a techno-Orwellian future? And can they succeed against so powerful and subtle a global system? Thirty years ago, radicals could identify and work to overthrow a territorially based dictator like Francisco Franco. But who and where is the enemy in the emerging space of flows?

Many of the visionary writers in this anthology predicted that the information (and transportation) technology of their times would undermine old, and make possible new, city patterns. Ebenezer Howard argued that Garden Cities were possible because they could communicate by telegraph and send goods by railroad (p. 309). Frank Lloyd Wright foresaw instant communication and rapid transportation in Broadacre City by pneumatic tube and helicopter (p. 325). And Melvin Webber predicted a "post-city age" as a result of telephones and air transportation (p. 470).

Other books on information technology and cities by Castells are *The Rise of the Network Society*, three volumes (Oxford: Blackwell, 1998; 2nd edn, 2000), *The Informational City: Information Technology, Economic Restructuring, and the Urban-Regional Process* (Oxford and Cambridge, MA: Blackwell, 1991), an edited anthology *High Tech, Space, and Society* (Beverly Hills, CA: Sage, 1985), and *Technopoles of the World* (London: Routledge, 1994) jointly authored with Sir Peter Hall.

Other books by Castells include *The Power of Identity: The Information Age – Economy, Society, and Culture* (Oxford: Blackwell, 1997), *The Rise of the Network Society* (Oxford: Blackwell, 1996), *Dual City: Restructuring New York* (New York: Russell Sage Foundation, 1991) edited with John Mollenkopf, *The City and The Grassroots* (Berkeley: University of California Press, 1983) and *The Urban Question* (London: Edward Arnold, 1977). An excellent anthology of works from throughout his career is Ida Susser (ed.), *The Castells Reader on Cities and Social Theory* (Oxford: Blackwell, 2002).

▨ ▨ ▨ ■ ▨ ▨

An old axiom in urban sociology considers space as a reflection of society. Yet life, and cities, are always too complex to be captured in axioms. Thus the close relationship between space and society, between cities and history, is more a matter of expression than of reflection. The social matrix expresses itself into the spatial pattern through a dialectical interaction that opposes social contradictions and conflicts as trends fighting each other in an endless supersession. The result is not the coherent spatial form of an overwhelming social logic be it the capitalist city, the preindustrial city or the ahistorical utopia but the tortured and disorderly, yet beautiful patchwork of human creation and suffering.

Cities are socially determined in their forms and in their processes. Some of their determinants are structural, linked to deep trends of social evolution that transcend geographic or social singularity. Others are historically and culturally specific. And all are played out, and twisted, by social actors that impose their interests and their values, to project the city of their dreams and to fight the space of their nightmares.

Sociological analysis of urban evolution must start from the theoretical standpoint of considering the complexity of these interacting trends in a given time–space context. The last twenty years of urban sociology have witnessed an evolution of thinking (including my own) from structuralism to subjectivism, then to an attempt, however imperfect, at integrating both perspectives into a structural theory of urban change that, if a label rooted in an intellectual tradition is necessary, I would call Marxian, once history has freed the Marxian theoretical tradition from the terrorist tyranny of Marxism-Leninism.

I intend to apply this theoretical perspective to the understanding of the fundamental transformations that are taking place in West European cities at the end of the second millennium. In order to understand such transformations we have to refer to major social trends that are shaking up the foundations of our existence: the coming of a technological revolution centered on information technologies, the formation of a global economy, and transition to a new society, the informational society, that without ceasing to be capitalist or statist replaces the industrial society as the framework of social institutions.

But this analysis has to be at the same time general and structural (if we accept that a historical transformation is under way) and specific to a given social and cultural context, such as Western Europe (with all due acknowledgement to its internal differentiation).

In recent years, a new trademark has become popular in urban theory: capitalist restructuring. Indeed it is most relevant to pinpoint the fundamental shift in policies that both governments and corporations have introduced in the 1980s to steer capitalist economies out of their 1970s crises. Yet more often than not, the theory of capitalist restructuring has missed the specificity of the process of transformation in each area of the world, as well as the variation of the cultural and political factors that shape the process of economic restructuring, and ultimately determine its outcome.

Thus the deindustrialization processes of New York and London take place at the same time that a wave of industrialization of historic proportions occurs in China and in the Asian Pacific. The rise of the informal economy and of urban dualism takes place in Los Angeles, as well as in Madrid, Miami, Moscow, Bogota and Kuala Lumpur, but the social paths and social consequences of such similarly structural processes are so different as to induce a fundamental variegation of each resulting urban structure.

I will try therefore to analyze some structural trends underlying the current transformation of European cities, while accounting for the historical and social specificity of the processes emerging from such structural transformation.

THE THREAD OF THE NEW HISTORY

Urban life muddles through the pace of history. When this pace accelerates, cities and their people become confused, spaces turn threatening, and meaning escapes from experience. In such disconcerting yet magnificent times, knowledge becomes the only source to restore meaning, and thus meaningful action.

At the risk of schematism, and for the sake of clarity, I will summarize what seem to be the main trends that, together and in their interaction, provide the framework of social, economic and political life for European cities in this particular historical period.

First of all, we live in the midst of a fundamental technological revolution that is characterized by two features:

As all major technological revolutions in history, the effects are pervasive. They are not limited to industry, or to the media, or to telecommunications or transportation. New technologies, that have emerged in their applications in full strength since the mid 1970s, are transforming production and consumption, management and work, life and death, culture and warfare, communication and education, space and time. We have entered a new technological paradigm.

As the industrial revolution was based on energy (although it embraced many other technological fields) the current revolution is based upon information technologies, in the broadest sense of the concept, which includes genetic engineering (the decoding and reprogramming of the codes of living matter).

This technological informational revolution is the backbone (although not the determinant) of all other major structural transformations:

It provides the basic infrastructure for the formation of a functionally interrelated world economic system.

It becomes a crucial factor in competitiveness and productivity for countries throughout the world, ushering in a new international division of labour.

It allows for the simultaneous process of centralization of messages and decentralization of their reception, creating a new communications world made up at the same time of the global village and of the incommunicability of those communities that are switched off from the global network. Thus an asymmetrical space of communication flows emerges from the uneven appropriation of a global communication system.

It creates a new, intimate linkage between the productive forces of the economy and the cultural capacity of society. Because knowledge generation and information processing are at the roots of the new productivity, the ability of a society to accumulate knowledge and manipulate symbols translates into economic productivity and political military might, anchoring the sources of wealth and power in the informational capacity of each society.

While this technological revolution does not determine per se the emergence of a social system, it is an essential component of the new social structure that characterizes our world: the informational society. By this concept, I understand a social structure where the sources of economic productivity, cultural hegemony and political military power depend, fundamentally, on the capacity to retrieve, store, process and generate information and knowledge. Although information and knowledge have been critical for economic accumulation and political power throughout history, it is only under the current technological, social, and cultural parameters that they become directly productive forces. In other words, because of the interconnection of the whole world and the potential automation of most standard production and management functions, the generation and control of knowledge, information and technology is a necessary and sufficient condition to organize the overall social structure around the interests of the information holders. Information becomes the critical raw material of which all social processes and social organizations are made. Material production, as well as services, become subordinate to the handling of information in the system of production and in the organization of society. Empirically speaking, an ever growing majority of employment in Western European cities relates to information processing jobs. The growing proportion of employment in service activities is not the truly distinctive feature, because of the ambiguity of the notion of 'services' (e.g. in Third World cities a majority of the population also works in 'services', although these are indeed very different kinds of activities). What is truly fundamental is the growing quantitative size and qualitative importance of information processing activities in both goods production and services delivery. The contradictory but ineluctable emergence of the informational society shapes European cities as the onset of the industrial era marked forever the urban and rural spaces of the nineteenth century.

A third major structural trend of our epoch is the formation of a global economy. The global economy concept must be distinguished from the notion of a world economy, which reflects a very old historical reality for most European nations, and particularly for the Netherlands, which emerged as a nation through its role as one of the nodal centres of the sixteenth century's world economy. Capitalism has accumulated, since its beginnings, on a worldwide scale. This is not to say that the capitalist economy was a global economy. It is only now becoming such.

By global economy we mean an economy that works as a unit in real time on a planetary scale. It is an economy where capital flows, labour markets, commodity markets, information, raw materials, management, and organization are internationalized and fully interdependent throughout the planet, although in an asymmetrical form, characterized by the uneven integration into the global system of different areas of the planet. Major functions of the economic system are fully internationalized and interdependent on a daily basis. But many others are segmented and unevenly structured depending upon functions, countries, and regions. Thus the global economy embraces the whole planet, but not all regions and all people on the planet. In fact, only a minority of people are truly integrated into the global economy, although all the dominant economic and political centres on which people depend are indeed integrated into the global economic networks (with the possible exception of Bhutan . . .). With the disintegration of the Soviet empire, the last area of the planet that was not truly integrated into the global economy, it is restructuring itself in the most dramatic conditions to be able to reach out to the perceived avenues of prosperity of our economic model. (China already started its integration in the global capitalist economy in December 1979, while trying to preserve its statist political regime.)

This global economy increasingly concentrates wealth, technology, and power in 'the North', a vague geopolitical notion that replaces the obsolete West–East differentiation, and that roughly corresponds to the OECD countries. The East has disintegrated and is quickly becoming an economic appendage of the North. Or at least such is the avowed project of its new leaders. The 'South' is increasingly differentiated. East Asia is quickly escaping from the lands of poverty and underdevelopment to link up, in fact, with the rising sun of Japan, in a model of development that Japanese writers love to describe as 'the flying geese pattern', with Japan of course leading the way, and the other Asian nations taking off harmoniously under the technological guidance and economic support of Japan. China is at the crossroads of a potential process of substantial economic growth at a terrible human cost as hundreds of millions of peasants are uprooted without structures able to integrate them into the new urban industrial world. South and Southeast Asia struggle to survive the process of change, looking for a subordinate yet livable position in the new world order. Most of Africa, on the contrary, finds itself increasingly disconnected from the new global economy, reduced to piecemeal, secondary functions that see the continent deteriorate, with the world only waking up from time to time to the structural genocide taking place there when television images strike the moral consciousness of public opinion and affect the political interests of otherwise indifferent policy makers. Latin America, and many regions and cities around the world, struggle in the in-between land of being only partially integrated into the global economy, and then submitted to the tensions between the promise of full integration and the daily reality of a marginal existence.

In this troubled world, Western Europe has, in fact, become a fragile island of prosperity, peace, democracy, culture, science, welfare and civil rights. However, the selfish reflex of trying to preserve this heaven by erecting walls against the rest of the world may undermine the very fundamentals of European culture and democratic civilization, since the exclusion of the other is not separable from the suppression of civil liberties and a mobilization against alien cultures. Major European cities have become nodal centres of the new global economy, but they have also seen themselves transformed into magnets of attraction for millions of human beings from all around the world who want to share the peace, democracy, and prosperity of Europe in exchange for their hard labour and their commitment to a promised land. But the overcrowded and aged Western Europe of the late twentieth century does not seem as open to the world as was the young, mostly empty America of the beginning of the century. Immigrants are not welcome as Europe tries to embark on a new stage of its common history, building the supranational Europe without renouncing its national identities. Yet the cultural isolation of the pan-European construction is inseparable from the affirmation of ethnic nationalism that will eventually turn not only against the 'alien immigrants' but against European foreigners as well. European cities will have to cope with this new global economic role while accommodating to a multi-ethnic society that emerges from the same roots that sustain the global economy.

The fourth fundamental process under way in European cities is the process of European integration, into what will amount in the twenty-first century to some form of confederation of the present nation states. This is an ineluctable process for at least fifteen countries (the current twelve EC countries plus Sweden, Austria and Switzerland) regardless of the fate of the symbolic Maastricht Treaty. If, as is generally accepted, the European Community is heading toward a common market, a common resident status for all its citizens, a common technology policy, a common currency, a common defense and a common foreign policy, all the basic prerogatives of the national state will be shifted to the European institutions by the end of the century. This will certainly be a tortuous path, with the nostalgics of the past, neofascists, neo-communists and fundamentalists of all kinds fighting the tide of European solidarity, fueling the fears of ignorance among people, building upon demagogy and opportunism. Yet however difficult the process, and with whatever substantial modifications to the current technocratic blueprints, Europe will come into existence: there are too many interests and too much political will at stake to see the project destroyed after having come this far.

The process of European integration will cause the internationalization of major political decision-making processes, and thus will trigger the fear of subordination of specific social interests to supranational institutions. But most of these specific interests express themselves on a regional or local basis rather than at the national level. Thus we are witnessing the renewal of the role of regions and cities as locuses of autonomy and political decision. In particular, major cities throughout Europe constitute the nervous system of both the economy and political system of the continent. The more national states fade in their role, the more cities emerge as a driving force in the making of the new European society.

IDENTITY AND SOCIAL MOVEMENTS

The process of historical transition experienced by Europe's cities leads to an identity crisis in its cultures and among its people, that becomes another major element of the new urban experience. This identity crisis is the result of two above mentioned processes that, however contradictory among themselves, jointly contribute to shake up the foundations of European national and local cultures. On the one hand, the march to supranationality blurs national identities and makes people uncertain about the power holders of their destiny, thus pushing them into withdrawal, either individualistic (neolibertarianism) or collective (neonationalism). On the other hand, the arrival of millions of immigrants and the consolidation of multi-ethnic, multicultural societies in most West European countries, confronts Europe head on with the reality of a nonhomogeneous culture, precisely at the moment when national identity is most threatened. There follows a crisis of cultural identity (with the corollary of collective alienation) that will mark urban processes in Europe for years to come.

More to the point: major cities will concentrate the overwhelming proportion of immigrants and ethnic minority citizens (the immigrants' sons and daughters). Thus they will also be at the forefront of the waves of racism and xenophobia that will shake up the institutions of the new Europe even before they come into existence. As a reaction to the national identity crisis we observe the emergence of territorially defined identities at the level of the region, of the city, of the neighbourhood. European cities will be increasingly oriented toward their local culture, while increasingly distrustful of higher order cultural identities. The issue then is to know if cities can reach out to the whole world without surrendering to a localistic, quasi-tribal reaction that will create a fundamental divide between local cultures, European institutions, and the global economy.

European cities are also affected by the rise of the social movements of the informational society, and in particular by the two central movements of the informational society: the environmental movement, and the women's movement.

The environmental movement is at the origin of the rise of the ecological consciousness that has substantially affected urban policies and politics. The issue of sustainable development is indeed a fundamental theme of our civilization and a dominant topic on today's political agendas. Because major cities in Europe are at the same time nodal centres of economic growth and the living places for the most environmentally conscious segment of the population, the battles for integration will be fought in the streets and institutions of these major cities.

The structural process of transformation of women's condition, in dialectical interaction with the rise of the feminist movement, has completely changed the social fabric of cities. Labour markets have been massively feminized, resulting in a change in the conditions of work and management, of struggle and negotiation, and ultimately in the weakening of a labour movement that could not overcome its sexist tradition. This also points to the possibility of a new informational labour movement that because it will have to be based on women's rights and concerns, as well as on those of men, will be historically different from its predecessor.

At the same time, the transformation of households and the domestic division of labour is fundamentally changing the demands on collective consumption, and thus urban policy. For instance, childcare is becoming as important an issue as housing in today's cities. Transportation networks have to accommodate the demands of two workers in the family, instead of relying on the free driving service provided by the suburban housewife in the not so distant past.

Some of the new social movements, the most defensive, the most reactive, have taken and will be taking the form of territorially based countercultures, occupying a given space to cut themselves off from the outside world, hopeless of being able to transform the society they refuse. Because such movements are likely to occur in major cities that concentrate a young, educated population, as well as marginal cultures that accommodate themselves in the cracks of the institutions, we will be witnessing a constant struggle over the occupation of meaningful space in the main European cities, with business corporations trying to appropriate the beauty and tradition for their noble quarters, and urban countercultures making a stand on the use value of the city, while local residents try to get on with their living, refusing to be bent by the alien wind of the new history.

Beyond the territorial battles between social movements and elite interests, the new marginality, unrelated to such social movements, is spreading over the urban space. Drug addicts, drug dealers and drug victims populate the back alleys of European cities, creating the unpredictable, awakening our own psychic terrors, and tarnishing the shine of civilized prosperity at the daily coming of darkness. The 'black holes' of our society, those social conditions from where there is no return, take their territory too, making cities tremble at the fear of their unavowed misery.

The occupation of urban space by the new poverty and the new marginality takes two forms: the tolerated ghettoes where marginalized people are permitted to stay, out of sight of the mainstream society; the open presence in the core area of cities of 'street people', a risky strategy, but at the same time a survival technique since only there do they exist, and thus only there can they relate to society, either looking for a chance or provoking a final blow.

Because the informational society concentrates wealth and power, while polarizing social groups according to their skills, unless deliberate policies correct the structural tendencies we are also witnessing the emergence of a social dualism that could ultimately lead to the formation of a dual city, a fundamental concept that I will characterize below, when considering the spatial consequences of the structural trends and social processes that I have proposed as constituting the framework that underlies the new historical dynamics of European cities.

THE SPATIAL TRANSFORMATION OF MAJOR CITIES

From the trends we have described stem a number of spatial phenomena that characterize the current structure of major metropolitan centres in Western Europe. These centres are formed by the uneasy articulation of various sociospatial forms and processes that I find useful to specify in their singularity, although it is obvious that they cannot be understood without reference to each other.

First of all, the national-international business centre is the economic engine of the city in the informational global economy. Without it, there is no wealth to be appropriated in a given urban space, and the crisis overwhelms any other project in the city, as survival becomes the obvious priority. The business centre is made up of an infrastructure of telecommunications, communications, urban services and office space, based upon technology and educational institutions. It thrives through information processing and control functions. It is sometimes complemented by tourism and travel facilities. It is the node of the space of flows that characterizes the dominant space of informational societies. That is, the abstract space constituted in the networks of exchange of capital flows, information flows, and decisions that link directional centres among themselves throughout the planet. Because the space

of flows needs nodal points to organize its exchange, business centres and their ancillary functions constitute the localities of the space of flows. Such localities do not exist by themselves but by their connection to other similar localities organized in a network that forms the actual unit of management, innovation, and power.

Secondly, the informational society is not disincarnated. New elites make it work, although they do not necessarily base their power and wealth on majority ownership of the corporations. The new managerial technocratic political elite does however create exclusive spaces, as segregated and removed from the city at large as the bourgeois quarters of the industrial society. In European cities, unlike in America, the truly exclusive residential areas tend to appropriate urban culture and history, by locating in rehabilitated areas of the central city, emphasizing the basic fact that when domination is clearly established and enforced, the elite does not need to go into a suburban exile, as the weak and fearful American elite did to escape from the control of the urban population (with the significant exceptions of New York, San Francisco, and Boston).

Indeed, the suburban world of European cities is a socially diversified space, which is segmented in different peripheries around the central city. There are the traditional working class suburbs (either blue collar or white collar) of the well kept subsidized housing estates in home ownership. There are the new towns, inhabited by a young cohort of lower middle class, whose age made it difficult for them to penetrate the expensive housing market of the central city. And there are also the peripheral ghettoes of the older public housing estates where new immigrant populations and poor working families experience their exclusion from the city.

Suburbs are also the locus of industrial production in European cities, both for traditional manufacturing and for the new high technology industries that locate in new peripheries of the major metropolitan areas, close enough to the communication centres but removed from older industrial districts.

Central cities are still shaped by their history. Thus traditional working class neighbourhoods, increasingly populated by service workers rather than by industrial workers, constitute a distinctive space, a space that, because it is the most vulnerable, becomes the battleground between the redevelopment efforts of business and the upper middle class, and the invasion attempts of the countercultures trying to reappropriate the use value of the city. They often become defensive spaces for workers who have only their home to fight for, while at the same time meaningful popular neighbourhoods and likely bastions of xenophobia and localism.

The new professional middle class is torn between attraction to the peaceful comfort of the boring suburbs and the excitement of a hectic, and often too expensive, urban life. The structure of the household generally determines the spatial choice. The larger the role women play in the household, the more the proximity to jobs and urban services in the city makes central urban space attractive to the new middle class, triggering the process of gentrification of the central city. On the contrary, the more patriarchal is the middle class family, the more we are likely to observe the withdrawal to the suburb to raise children, all economic conditions being equal.

The central city is also the locus for the ghettoes of the new immigrants, linked to the underground economy, and to the networks of support and help needed to survive in a hostile society. Concentration of immigrants in some dilapidated urban areas in European cities is not the equivalent however to the experience of the American ghettoes, because the overwhelming majority of European ethnic minorities are workers, earning their living and raising their families, thus counting on a support structure that makes their ghettoes strong, family oriented communities, unlikely to be taken over by street crime.

It is in the core administrative and entertainment districts of European cities that urban marginality makes itself present. Its pervasive occupation of the busiest streets and public transport nodal points is a survival strategy deliberately designed so they can receive public attention or private business, be it welfare assistance, a drug transaction, a prostitution deal, or the customary police care.

Major European metropolitan centres present some variation around the structure of urban space we have outlined, depending upon their differential role in the European economy. The lower their position in the new informational network, the greater the difficulty of their transition from the industrial stage, and the more traditional will be their urban structure, with old established neighbourhoods and commercial quarters playing the determinant role in the dynamics of the city. On the other hand, the higher their position in the competitive structure of the new European economy, the greater the role of their advanced services in the business district, and the more intense will be the restructuring of the urban space. At the same time, in those cities where the new European society reallocates functions and people throughout the space, immigration, marginality and countercultures will be the most present, fighting over control of the territory, as identities become increasingly defined by the appropriation of space.

The critical factor in the new urban processes is, however, the fact that urban space is increasingly differentiated in social terms, while being functionally interrelated beyond physical contiguity. There follows a separation between symbolic meaning, location of functions, and the social appropriation of space in the metropolitan area. The transformation of European cities is inseparable from a deeper, structural transformation that affects urban forms and processes in advanced societies: the coming of the Informational City.

THE INFORMATIONAL CITY

The spatial evolution of European cities is a historically specific expression of a broader structural transformation of urban forms and processes that expresses the major social trends that I have presented as characterizing our historical epoch: the rise of the Informational City. By this concept I do not refer to the urban form resulting from the direct impact of information technologies on space. The Informational City is the urban expression of the whole matrix of determinations of the Informational Society, as the Industrial City was the spatial expression of the Industrial Society. The processes constituting the form and dynamics of this new urban structure, the Informational City, will be better understood by referring to the actual social and economic trends that are restructuring the territory. Thus the new international and interregional division of labour ushered in by the informational society leads, at the world level, to three simultaneous processes:

> The reinforcement of the metropolitan hierarchy exercised throughout the world by the main existing nodal centres, which use their informational potential and the new communication technologies to extend and deepen their global reach.
>
> The decline of the old dominant industrial regions that were not able to make a successful transition to the informational economy. This does not imply however that all traditional manufacturing cities are forced to decline: the examples

of Dortmund or Barcelona show the possibility to rebound from the industrial past into an advanced producer services economy and high technology manufacturing.

The emergence of new regions (such as the French Midi or Andalusia) or of new countries (e.g. the Asian Pacific) as dynamic economic centres, attracting capital, people, and commodities, thus recreating a new economic geography.

In the new economy, the productivity and competitiveness of regions and cities is determined by their ability to combine informational capacity, quality of life, and connectivity to the network of major metropolitan centres at the national and international levels.

The new spatial logic, characteristic of the Informational City, is determined by the preeminence of the space of flows over the space of places. By space of flows I refer to the system of exchanges of information, capital, and power that structures the basic processes of societies, economies and states between different localities, regardless of localization. I call it 'space' because it does have a spatial materiality: the directional centres located in a few selected areas of a few selected localities; the telecommunication system, dependent upon telecommunication facilities and services that are unevenly distributed in the space, thus marking a telecommunicated space; the advanced transportation system, that makes such nodal points dependent on major airports and airlines services, on freeway systems, on high speed trains; the security systems necessary to the protection of such directional spaces, surrounded by a potentially hostile world; and the symbolic marking of such spaces by the new monumentality of abstraction, making the locales of the space of flows meaningfully meaningless, both in their internal arrangement and in their architectural form. The space of flows, superseding the space of places, epitomizes the increasing differentiation between power and experience, the separation between meaning and function.

The Informational City is at the same time the Global City, as it articulates the directional functions of the global economy in a network of decision-making and information processing centres. Such globalization of urban forms and processes goes beyond the functional and the political, to influence consumption patterns, lifestyles, and formal symbolism.

Finally, the Informational City is also the Dual City. This is because the informational economy has a structural tendency to generate a polarized occupational structure, according to the informational capabilities of different social groups. Informational productivity at the top may incite structural unemployment at the bottom or downgrading of the social conditions of manual labour, particularly if the control of labour unions is weakened in the process and if the institutions of the welfare state are undermined by the concerted assault of conservative politics and libertarian ideology.

The filling of downgraded jobs by immigrant workers tends to reinforce the dualization of the urban social structure. In a parallel movement the age differential between an increasingly older native population in European cities and a younger population of newcomers and immigrants forms two extreme segments of citizens polarized simultaneously along lines of education, ethnicity, and age. There follows a potential surge of social tensions.

The necessary mixing of functions in the same metropolitan area leads to the attempt to preserve social segregation and functional differentiation through planning of the spatial layout of activities and residence, sometimes by public agencies, sometimes by the influence of real estate prices. There follows a formation of cities made up of spatially coexisting, socially exclusive groups and functions, that live in an increasingly uneasy tension vis-à-vis each other. Defensive spaces emerge as a result of the tension.

This leads to the fundamental urban dualism of our time. It opposes the cosmopolitanism of the elite, living on a daily connection to the whole world (functionally, socially, culturally), to the tribalism of local communities, retrenched in their spaces that they try to control as their last stand against the macro forces that shape their lives out of their reach. The fundamental dividing line in our cities is the inclusion of the cosmopolitans in the making of the new history while excluding the locals from the control of the global city to which ultimately their neighbourhoods belong.

Thus the Informational City, the Global City, and the Dual City are closely interrelated, forming the background of urban processes in Europe's major metropolitan centres. The fundamental issue at stake is the increasing lack of communication between the directional functions of the economy and the informational elite that performs such functions, on the one hand, and the locally oriented population that experiences an ever deeper identity crisis, on the other. The separation between function and meaning, translated into the tension between the space of flows and the space of places, could become a major destabilizing force in European cities, potentially ushering in a new type of urban crisis.

BACK TO THE FUTURE?

The most important challenge to be met in European cities, as well as in major cities throughout the world, is the articulation of the globally oriented economic functions of the city with the locally rooted society and culture. The separation between these two levels of our new reality leads to a structural urban schizophrenia that threatens our social equilibrium and our quality of life. Furthermore, the process of European integration forces a dramatic restructuring of political institutions, as national states see their functions gradually voided of relevance, pulled from the top toward supranational institutions and from the bottom toward increasing regional and local autonomy. Paradoxically, in an increasingly global economy and with the rise of the supranational state, local governments appear to be at the forefront of the process of management of the new urban contradictions and conflicts. National states are increasingly powerless to control the global economy, and at the same time they are not flexible enough to deal specifically with the problems generated in a given local society. Local governments seem to be equally powerless vis-à-vis the global trends but much more adaptable to the changing social, economic, and functional environment of cities.

The effectiveness of the political institutions of the new Europe will depend more on their capacity for negotiation and adaptation than on the amount of power that they command, since such power will be fragmented and shared across a variety of decision-making processes and organizations. Thus instead of trying to master the whole complexity of the new European society, governments will have to deal with specific sets of problems and goals in specific local circumstances. This is why local governments, in spite of their limited power, could be in fact the most adequate instances of management of these cities, working in the world economy and living in the local cultures. The strengthening of local governments is thus a precondition for the management of European cities. But local governments can only exercise such management potential if they engage in at least three fundamental policies:

The fostering of citizen participation, on the basis of strong local communities that feed local government with information, present their demands, and lay the ground for the legitimacy of local government so that they can become respected partners of the global forces operating in their territory.

The interconnection and cooperation between local governments throughout Europe, making it difficult for the global economic forces to play one government against the other, thus forcing the cooperation of global economy and local societies in a fruitful new social contract.

New information technologies should make possible a qualitative upgrading of the cooperation between local governments. A European Municipal Data Bank, and a network of instant communication between local leaders, could allow the formation of a true association of interests of the democratic representatives of the local populations. An electronically connected federation of quasi-free communes could pave the way for restoring social and political control over global powers in the informational age.

Managing the new urban contradictions at the local level by acting on the social trends that underlie such contradictions requires a vision of the new city and new society we have entered into, including the establishment of cooperative mechanisms with national governments and European institutions, beyond natural and healthy competition of parties. The local governments of the new Europe will have to do their homework in understanding their cities, if they are to assume the historical role that the surprising evolution of society could offer them.

Thus the historical specificity of European cities may be a fundamental asset in creating the conditions for managing the contradictions between the global and the local in the new context of the informational society. Because European cities have strong civil societies, rooted in an old history and a rich, diversified culture, they could stimulate citizen participation as a fundamental antidote against tribalism and alienation. And because the tradition of European cities as city states leading the pace to the modern age in much of Europe is engraved in the collective memory of their people, the revival of the city state could be the necessary complement to the expansion of a global economy and the creation of a European state. The old urban tradition of Amsterdam as a centre of politics, trade, culture and innovation, suddenly becomes more strategically important for the next stage of urban civilization than the meaningless suburban sprawl of high technology complexes that characterize the informational space in other areas of the world.

European cities, because they are cities and not just locales, could manage the articulation between the space of flows and the space of places, between function and experience, between power and culture, thus recreating the city of the future by building on the foundations of their past.

"Planning Sustainable and Livable Cities"
(1998)

Stephen Wheeler

Editors' Introduction

The enormous and exponential growth of the world's population described by Kingsley Davis in the "Evolution of Cities" section of this anthology (p. 25) has had profound and often catastrophic effects on the natural environment of planet Earth. Non-renewable energy sources have been consumed, forests cleared, buffalo slaughtered, species extinguished. Rainforests are disappearing. Today, many argue that global warming caused by population growth, urbanization, and irresponsible fossil fuels consumption threatens to inundate low-lying cities and irreparably damage the climate of the entire world.

In the following selection on "Planning Sustainable and Livable Cities," Stephen Wheeler reviews the evolution of sustainable urban development thinking and ties it to the new concern with urban livability. He defines sustainable development as "development that improves long-term health of human and ecological systems." He dismisses as inadequate recent debates about "needs" (which are hard to distinguish from wants), the "carrying capacity" of areas (which are tough to pin down, particularly for people), and "sustainable end states" (since it is virtually impossible to decide on end states).

Wheeler lays out a helpful compendium of core themes in the sustainable urban development literature, a list of specific approaches that can guide planning practice, and numerous examples to learn from. Like John Forester's theoretical description of how city planners actually make decisions (p. 375), Wheeler's theoretical framework provides real guidance to planners about how to carry out plans to foster sustainable development.

Whatever their specifics, Wheeler argues, sustainable urban development strategies need to be *long term* – plans for twenty, fifty, one hundred years or longer rather than year-by-year plans that optimize short-term present enjoyment at the expense of future welfare. Wheeler emphasizes that a core theme in urban sustainability planning must be attention to the natural environment. And finally urban sustainability planning requires holistic and interdisciplinary approaches connecting the insights of biologists and transit planners, agronomists, economists, and many other disciplines. Sustainable urban development planning requires that land use, transportation, housing, community development, economic development, and environmental planning all be woven together.

Wheeler outlines eight directions sustainable urban development should take to move towards his definition of sustainable urban development. He illustrates each suggested direction with specific, concrete examples of actual good things some cities have done. In contrast to unplanned sprawl that consumes land and leads to costly and inefficient infrastructure, for example, Wheeler urges planners to move in the direction of compact, efficient land use. Wheeler illustrates each of his other suggested directions – fewer automobiles, more efficient resource use, restoring natural systems, good housing, healthy social ecology, and sustainable economics – with specific suggestions and concrete examples of where sustainability has been successfully implemented.

Compare Wheeler's ideas on planning sustainable and livable cities to Timothy Beatley's ideas about green urbanism (p. 399). For ingenious and influential ideas on how to combine land use, housing, and transit design in environmentally sensitive ways, see Peter Calthorpe and William Fulton's conception of the Regional City.

Stephen Wheeler is an assistant professor of city and regional planning at the University of New Mexico. He was a board member of Urban Ecology and for eight years edited their publication *The Urban Ecologist*. Mr. Wheeler is active in Bay Area transportation issues, serves the city of Berkeley as a Transportation Commissioner, and was the co-founder of the Bay Area Transportation Choices Forum.

Among the literature on the destructive effects of urbanization on the natural environment see particularly William Cronon, *Nature's Metropolis* (New York: W.W. Norton, 1992) and Mark Reisner, *Cadillac Desert* (New York: Penguin, 1993). An early and influential report on sustainability is the World Commission on Environment and Development, *Our Common Future* (New York: Oxford University Press, 1987) commonly referred to as the Brundtland Report. Other writings on sustainable urban development include Graham Haughton and Colin Hunter, *Sustainable Cities* (London: Regional Studies Association, 1994), the President's Council on Sustainable Development, *Sustainable America* (Washington, DC: Government Printing Office, 1996), and William Rees, *Our Ecological Footprint: Reducing Human Impact on Earth* (Philadelphia, PA: New Society Publishers, 1996). Ian McHarg's classic book, *Design with Nature* (Garden City, NY: Doubleday & Company, 1969) and Patrick Geddes, *Cities in Evolution* (London: Williams and Norgate, 1915, reprinted in Richard LeGates and Frederic Stout, *Early Urban Planning 1870–1940*, London: Routledge/Thoemmes, 1998) are classic works that anticipate today's sustainable urban development, regional planning, and ecological design debates.

INTRODUCTION

The term "sustainable" is now widely used to describe a world in which both human and natural systems can continue to exist long into the future. The concept of "sustainable development" is used to refer to alternatives to traditional patterns of physical, social and economic development that can avoid problems such as exhaustion of natural resources, ecosystem destruction, pollution, overpopulation, growing inequality, and the degradation of human living conditions.

However, the notion of sustainability is still very recent, and understandings of what it means and how to apply it are still evolving. Many questions surround this concept. Does it in fact express a coherent and meaningful philosophy? Can it be defined in ways that the general public can relate to? Can it be used to generate consensus around specific urban planning directions? Can it avoid cooptation by existing political forces?

In the following pages I will propose a framework for thinking about sustainable development in the metropolitan context. After examining the origins of the sustainability concept, different definitions, and key themes, I will outline implications for urban and regional planning, and suggest processes through which sustainable development planning might be achieved. I will also consider the related concept of "livability," another broad notion that is often used in the same breath as "sustainability." These terms are closely related, in that they both promote urban planning that enhances long-term community well-being. It is no accident that these concepts have come to the fore in urban planning discussions recently, because they address important unmet needs arising from the nature of twentieth-century urban development.

ORIGINS OF THE "SUSTAINABILITY" CONCEPT

The verb "sustain" has been used in English since the year 1290 or before, and comes from the Latin roots "sub" + "tenere," meaning "to uphold" or "to keep". The *Oxford English Dictionary* traces the adjective "sustenable" to around 1400 and the modern form "sustainable" to 1611. However, this term appears to have been used mainly in legal contexts until recently, as in "The Defendant has taken several technical objections to the order, none of which . . . are sustainable" (1884). Many other variants of "sustain" have existed for centuries, but only in the past several decades has the word "sustainable" emerged with its current meaning, perhaps most simply defined as "that which can be maintained into the future."

It is far from clear who was the first to use the term "sustainable development" in its current sense. Rather, it seems one of those inevitable expressions that so neatly encapsulates what many people are thinking that once the words are mentioned they quickly become ubiquitous. The birth of the sustainability concept in the 1970s can be seen as the logical outgrowth of a new consciousness about global problems related to environment and development, fueled in part by 1960s environmentalism, publications such as *The Limits to Growth*, and the first United Nations Conference on Environment and Development held in Stockholm in 1972. However, the notion of sustainability is also rooted in older environmental traditions, particularly "sustained yield" techniques of forest management developed by nineteenth-century German foresters. These concepts influenced American policy makers such as Gifford Pinchot, Theodore Roosevelt's chief forester, and natural resource scientists such as Aldo Leopold. Leopold's notion of a "land ethic" – a human responsibility to care for particular lands and ecosystems, discussed most fully in *A Sand County Almanac* (1948) – represents a fundamental shift from the view that natural resources should be seen in terms of their utility for human beings, toward the perspective that species and ecosystems have intrinsic value in their own right and should be stewarded and sustained indefinitely into the future.

In the post-World War II period a long line of environmentalist works such as William Vogt's best-selling *Road to Survival* (1948), Fairfield Osborn's *Our Plundered Planet* (1948), Rachel Carson's *Silent Spring* (1962), and Barry Commoner's *The Closing Circle* (1971) sounded a note of alarm about the global ecological situation, and helped tie the rise of ecological problems to ongoing patterns of industrial development. Particular events also helped change consciousness around development issues, such as the 1973 oil embargo during which millions of people suddenly realized that their fossil fuel use could not continue to expand forever. Social critics, futurists, feminists, and new age writers further prepared the way for discussions of sustainability by critiquing existing notions of development and proposing alternative paradigms that would emphasize the spiritual, the natural, and the human over values of profit and economic progress as

traditionally conceived. At the same time humanistic and transpersonal psychologists pointed out ways in which human potential is shaped by the surrounding environment, and ways in which it can perhaps be shaped in healthier directions in the future. The implication of such work is that people and perhaps entire societies can evolve towards more conscious, compassionate, and sustainable modes of existence, given the right conditions.

The earliest specific reference to sustainability that I have been able to document occurs in the 1972 book *Limits to Growth*, in which Donella Meadows and other MIT researchers describe computer models showing a collapse of global systems in the mid-twenty-first century, but state optimistically that "It is possible to alter these growth trends and to establish a condition of ecological and economic stability that is sustainable far into the future." A 1974 conference of the World Council of Churches then issued a call for a "sustainable society," and the earliest book specifically discussing sustainability appeared in 1976, a volume entitled *The Sustainable Society: Ethics and Economic Growth* by Lutheran theologian Robert L. Stivers.

In the late 1970s the number of writings on sustainability grew rapidly. The sustainability literature got one of its strongest pushes from Lester Brown and others at the Worldwatch Institute, a group which began publishing an extensive series of papers and books related to global sustainability, including annual State of the World reports. The tide of literature expanded in the 1980s with the International Union for the Conservation of Nature's influential World Conservation Strategy (1980), the President's Council on Environmental Quality's Global 2000 Report (1981), and above all the 1987 report of the World Commission on Environment and Development, chaired by Norwegian Prime Minister Gro Harlem Brundtland. These documents warned about global environmental problems and critiqued notions of "development," although generally accepting the desirability of continued economic growth. The influence of the IUCN and Brundtland reports in particular stemmed from the broad participation of mainstream governmental officials and academics within these bodies, which gave their findings an air of authority going beyond the "alarmist" reports of the *Limits to Growth* researchers, Global 2000, or the Worldwatch Institute. The Brundtland Commission in particular received input from literally thousands of individuals and organizations from around the world. Initiated at the request of the United Nations Secretary-General, it followed in the footsteps of two other highly respected UN-sponsored commissions, the Brandt Commission on North–South Issues and the Palme Commission on Security and Disarmament Issues. A more authoritative body to explore the topic would have been hard to find.

With the release of the Brundtland Commission report *Our Common Future* in 1987 and the United Nations "Earth Summit" conference in 1991, calls for sustainable development entered the mainstream internationally. "Sustainable city" programs emerged in many parts of the world, some resulting from grass-roots activism, some based on municipal initiative, some benefiting from the support of national governments, and some facilitated by multilateral entities such as the European Community, the World Bank, and the UN. The 1996 UN Habitat II "City Summit" in Istanbul took slow but significant steps toward establishing global consensus on how the sustainability agenda can be applied to urban planning. National reports such as that of the President's Commission on Sustainable Development (PCSD) in 1996 attempted to establish sustainable development directions for particular countries. As the 1990s went on academics in fields such as urban planning began to delve into the subject. Although actual implementation of sustainability programs remains difficult, the persistence and spread of the concept over three decades indicates that sustainability is a notion of lasting importance.

DEFINITIONS

Unfortunately, no perfect definition of sustainable development has emerged. The most widely used is that of the Brundtland Commission: "development that meets the needs of the present without compromising the ability of future generations to meet their own needs." However, this formulation is open to criticism for being anthropocentric and for raising the difficult-to-define concept of needs. (Does every household really need two cars? A VCR? A 2,000-square-foot house on a 5,000-square-foot lot? What happens if every household worldwide has these things?)

Other definitions include that given by the World Conservation Union in 1991: "improving the quality of human life while living within the carrying capacity of supporting ecosystems." This version raises the problematic notion of "carrying capacity," which is useful to think about for educational purposes but extremely hard to pin down in practice. It is one thing to say that the carrying capacity of a given watershed is a certain number of white-tailed deer; it is far more difficult to say that it is a certain number of human beings, when humans readily transport themselves and the resources they use over vast distances, and can substitute some resources for others if necessary. Although William Rees at the University of British Columbia argues that it is useful to try to calculate the "ecological footprint" of cities in terms of their resource use, my own belief is that attempts to actually define "carrying capacity" are best avoided.

Still other writers prefer to define sustainability in terms of preserving existing stocks of "ecological capital" and "social capital." This approach builds on the economic wisdom of living on the interest of an investment – in this case the earth's stock of natural resources – rather than the principal. For example, British economist David Pearce argues that "Sustainability requires at least a constant stock of natural capital, construed as the set of all environmental assets."

Most sustainability advocates throw up their hands when faced with the definition question, and fall back on Brundtland. However, my own preference is to move instead towards a relatively simple, process-oriented definition emphasizing long-term systemic welfare: "Sustainable development is development that improves the long-term health of human and ecological systems." This strategy avoids fruitless debates over "carrying capacity," "needs," or sustainable end states, while emphasizing the process of continually moving towards healthier human and natural communities. In theory the directions of this process can be agreed upon through participatory processes in which all relevant stakeholders are represented, and progress can be measured by means of various performance indicators.

CORE THEMES

The widespread use of the sustainability concept testifies to the strength and relevance of its underlying themes, both for urban planning and other fields. Foremost among these is a concern for the long-term perspective. Though it seems only common sense that planning and building should be for the long-term, this is manifestly often not the case in practice. In particular, long-term patterns of metropolitan growth, land use,

EIGHT

resource use, and infrastructure development demand attention, giving new impetus to old quests such as halting suburban sprawl. From this perspective it is very important to think about expanding planning horizons from a year-by-year approach or even a 20-year horizon, to think instead about the effects of urban development over 50 years, 100 years, or longer.

A second main theme is concern about the earth's natural environment. As widespread as this sentiment is these days, it is remarkably recent within industrial society and should not be taken for granted. Also new is a recognition that current development patterns are leading to ecological and social problems on a global scale. Problems such as the greenhouse effect and damage to the earth's ozone layer were only taken seriously starting in the late 1980s. In its simplest formulation, concern about environmental problems often focuses on worries about global collapse or large-scale disaster. In more complex, systems-oriented formulations, environmental and social problems can be seen not so much as leading to specific disasters in specific time frames, but as contributing to an increasingly unstable and unhealthy global system, which could be plagued by any number of unpredictable catastrophes as well as by a generally increased level of suffering on the part of human beings and natural ecosystems. Either way, the environmental costs and risks of current development patterns are viewed as unacceptable by increasing numbers of observers. Hence the search for "sustainable" alternatives.

Lastly, the sustainable development discourse can be seen as based on a new recognition of the complex web of interconnections between different issues, fields, disciplines, and actors. This holistic and inter-disciplinary perspective, based on the ecological metaphor of the world as an organic system, has huge implications for planning. Among other things, it means that different specialties having to do with transportation, land use, housing, community development, economic development, and environmental protection should not be handled in isolation from one another, but should be integrated to the extent possible even while specific tasks are carried out. Coordination of economic, environmental, and social goals within planning is also necessary. Indeed, it is widely believed that social dimensions of sustainable development should be given equal weight to environmental goals.

In some ways the related concept of "livability" is a bit simpler, since it focuses less on abstract themes and more on specific human needs and people's subjective reactions to places. Dictionary definitions such as "fit to live in" or "conducive to comfortable living" actually work fairly well. Of course human needs are to a large extent culturally determined and are open to extensive debate. However, there is widespread agreement on basic elements that make cities and towns livable – a healthy environment, decent housing, safe public places, uncongested roads, parks and recreational opportunities, vibrant social interaction, and so on. Such elements obviously contribute to sustainability as well.

Livability themes are becoming more and more important to modern societies in which the basic problems of food, shelter, public health and sanitation that plagued nineteenth century cities have long since been solved, at least for most people. Instead, in the postindustrial world the emphasis is increasingly on "quality of life." It is no longer enough just to throw up cities and suburbs that are ugly, uncoordinated, automobile-dominated, and lacking in parks, sidewalks, local shops, community vitality, and sense of place. The question becomes How do we make these places green, safe, convenient, and human-oriented? How do we turn mass-produced urban landscapes into places that have character and nurture community? How do we make cities attractive and comfortable to all groups within society, including women, children, the elderly, and minorities? All of these concerns touch upon the task of making urban places "livable" in the long run.

IMPLICATIONS FOR URBAN DEVELOPMENT

Exactly what constitutes a "sustainable city" is impossible to determine, given the extent to which cities are embedded in the global context. To be absolutely self-sustaining, an urban region would need to wall itself off from the rest of the world and produce all food, energy, and materials locally. Such an autarkic model is generally infeasible and would be seen as undesirable by most residents.

Rather, it is more useful to speak of cities as moving *toward* sustainability. A metropolitan area might seek to move toward greater resource efficiency, environmental quality, social equity, and community vitality, while moving away from automobile dependency, non-renewable resource consumption, hazardous waste generation, and inequity. While local self-sufficiency may indeed offer many benefits for sustainability, this can be set forth as a value to be enhanced rather than as an absolute goal.

Until the early 1990s very little of the sustainable development literature focused on cities or patterns of urban development. Instead, writers addressed topics such as the global environmental crisis, ecological economics, critiques of prevailing models of international development, and the need for a transformation of values and mindsets. However, in recent years architects and planners have begun looking more specifically at what sustainability means for patterns of metropolitan development. Some authors have emphasized urban design and physical planning. Others have focused on environmental planning concerns having to do with the quality of air, water, and natural ecosystems. A number have stressed the need to address social problems and inequities within the urban community, and emphasize that environmental and social issues are inextricably linked. In all of these categories, urban sustainability advocates can be seen as building on the work of past planning visionaries such as Patrick Geddes, Ebenezer Howard, Lewis Mumford, Jane Jacobs, and Ian McHarg.

Although authors may have different emphases, there is substantial agreement on many dimensions of sustainable urban development. Such directions have been endorsed by documents such as UN's Agenda 21 and Habitat Agenda, professional manifestos such as the Charter of the Congress for the New Urbanism and the Local Government Commission's Ahwahnee Principles, and publications of the European Community and the President's Council on Sustainable Development.

"Sustainable urban development" might be defined as "development that improves the long-term social and ecological health of cities and towns". Based on this definition and sources such as those mentioned above, main directions for urban sustainability can be seen to include the following:

1. Compact, efficient land use. Land is perhaps our most important limited resource, and current urban development patterns are clearly consuming the landscape in unsustainable ways. Land is also often divided very inequitably, and in many parts of the world those inequities are increasing. A wide range of devices can help lead to more sustainable land use. For example, urban growth boundaries (UGBs) have been adopted by Portland, Oregon, and eleven San Francisco Bay Area cities to restrain sprawl. To be effective in the long run, UGBs need to be coupled with policies to increase the efficiency of land use within already built-up areas, and to make these places more green, safe, attractive, and livable. Other types of land use controls can help preserve farm land, ecological habitat, and open space near cities. Meanwhile, urban park systems can be expanded and property tax systems changed to promote equity. Beyond specific land use changes, sustainable patterns of development are likely to involve alterations to the relationship between people and the land. In particular, a new balance between private property rights and human responsibilities toward the land is needed, as Leopold urged. The view of land as a commodity for human use and profit needs to shift towards a respect for the landscape as a thing of value in its own right, and a renewed sense of connection between human beings and the land they live on.

2. Less automobile use, better access. Current transportation systems contribute to a complex web of urban problems such as air pollution, congestion, blight, suburban sprawl, ecosystem destruction, and social fragmentation. Transport in more sustainable cities will most likely be based on several key principles: access by proximity, an inversion of the current transportation hierarchy, and demand reduction. Together these are likely to reduce the total amount that people need to travel, while allowing them to travel by far cleaner, more resource efficient, and more community-enhancing modes.

"Access by proximity" means solving transportation problems by bringing people closer to the places they need to go everyday. This is done primarily through land use changes, for example by promoting mixed-use development and the creation of neighborhood centers and "urban villages" which contain homes, workplaces, shops, and recreational facilities in close proximity to one another. Not coincidentally, fine-grained, mixed-use development also tends to create more interesting and livable places. "Inverting the transportation hierarchy" means placing the heaviest emphasis on the pedestrian, who represents the most energy efficient form of transportation and adds a much-needed human presence to the city. Bicycle planning should also be near the top of the priority list, followed by public transit. The automobile should be given lowest priority in the new hierarchy, and existing automobile subsidies reduced, although it is enormously difficult to do this politically. Finally, efforts to develop more sustainable transportation systems should also include looking at the "demand side" of the equation. By providing incentives to reduce the amount that people travel (the "demand side"), congestion problems can be solved and quality of life improved without building new roads or other infrastructure (the "supply side"). Pricing mechanisms are particularly useful to do this, such as higher parking charges, gas taxes, tolls, and vehicle registration fees.

3. Efficient resource use, less pollution and waste. Moving toward sustainability means paying greater attention to flows of energy and materials through human society, and planning for wiser use of resources. The overall challenge can be seen as one of moving from open-ended resource flows, in which nonrenewable resources are harvested, used once by human systems, and then discarded (often creating pollution and toxic waste problems in the process), toward closed-loop flows in which resources are reused and recycled.

A great number of mechanisms are available to urban and regional authorities to encourage this transition toward more sustainable resource flows. Energy conservation and materials recycling are two areas in which ordinary citizens can most directly take action through small daily initiatives, and so are good subjects for public involvement efforts. Municipal recycling programs are one of the most obvious areas in which cities can demonstrate their commitment to sustainable resource use. As in the transportation field, demand-side management programs offer great potential to reduce resource consumption. Since the late 1970s, for example, many utility companies around the US have offered consumers free or reduced-price compact fluorescent light bulbs, which typically use about one-fifth the electricity of incandescent models of similar wattage, as well as rebates on energy-efficient refrigerators, air conditioners, and water heaters. This saves both the consumer and the utility money, while reducing energy consumption. Unfortunately such programs were de-emphasized in the 1990s as many states shifted their attention to deregulating the utility industry.

Tougher energy conservation codes in building construction have also produced large energy savings in many cities and states. Pollution prevention programs are being developed to try to eliminate pollution before it is created rather than cleaning it up afterwards. "Industrial ecosystem" projects try to take a systematic look at manufacturing processes to see if the wastes from one industry can be used as inputs to another. And in the economic realm, attempts are being made to shift

the costs of pollution from society as a whole onto the individual or group who produces it – what has become known as the "polluter pays principle."

4. Restoration of natural systems. Even though many urban areas may seem entirely artificial – full of pavement and buildings, and often landscaped with non-native plants – still in almost every location there are many elements of the original ecosystem that can be reclaimed. Such restoration efforts add to the livability, ecological health, and overall sustainability of the urban region. Creek restoration, for example, is an idea that is catching on rapidly in many parts of the U.S. as well as overseas. Restoring a natural water-course provides corridors and habitat for wildlife as well as walkways and open space for people. It also helps reconnect urban dwellers to the bioregion, reminding them that they live in a natural world with cycles of rainfall and waterflow. Existing urban parks and areas of open space can benefit from restoration activities as well. Volunteer site restoration programs and stewardship approaches to watershed manage-ment can help this happen. Urban agriculture is another area in which nature is being brought back into the city and urban sustainability enhanced. Biointensive methods make it possible for urbanites to grow sub-stantial amounts of food on very small areas of land. In both real and symbolic ways, urban gardening helps reconnect city dwellers with the earth. Finally, ecological restoration is urgently needed in many inner city areas which are often home to lower-income neighborhoods and communities of color. Abandoned or contaminated industrial land can be reclaimed and restored, while vacant lots can be recycled into parks, housing, and community gardens.

Cities are often seen as unlivable because they have lost any sense of connection between people and the natural world. People move to suburbs in search of trees and nature, and there is a widespread belief that densely settled areas cannot also be green. Restoring urban ecosystems can lead to healthier and more livable cities, while providing important amenities that can help entice residents back from suburbia.

5. Good housing and living environments. One of the main purposes of cities and towns is to create decent places for people to live, and if these do not exist or are not affordable, the urban system is bound to suffer. Housing affordability is a recurrent crisis in many North American cities and suburbs. Steps to address the affordability problem include active government construction of housing – which has a checkered history in the US, although a somewhat better record in European countries – support of nonprofit housing developers, tenant subsidies, and requirements that developers include a certain number of affordable units in any market-rate project. Also, many urban areas are characterized by ugly, homogenous, pedestrian-unfriendly development which writer James Howard Kunstler has called a "geography of nowhere." The design of housing and neighborhoods needs to be rethought in many cases to ensure that people have access to open space, meeting areas, shared facilities, shops, offices, public transportation, child care facilities, and other essentials which can make urban communities more livable.

6. A healthy social ecology. The health of human communities within an urban region is more difficult to grasp than the natural ecology. Certain social problems such as homelessness are quite obvious to anyone who walks down an urban street. Other deeply entrenched social problems, which help to decrease overall sustainability and livability in the long run, are often more hidden. Racism, for example, has been an enormous factor shaping American cities for many decades. Expressed in particular through the denial of housing, financing, insurance, or other necessities to persons of color, this factor has done as much as any other to contribute to the decline of many central city areas.

Promoting a healthy and sustainable social ecology means looking for every opportunity to enhance human community, opportunity and empowerment. It requires planners in particular to advocate on behalf of those groups who do not have access to power or expertise, and to fight for equity and justice. On a personal level, it requires an ability to put oneself in the shoes of any resident of an urban region and ask, What are the opportunities available to him or her? What is the environment like in which he or she lives? What public policies, design improvements, and social programs could help improve this environment?

7. A sustainable economics. Developing an economy that values the long-term health of human and natural systems is one of the biggest challenges related to sustainability. It would be a mistake to say that any one economic model holds all the answers, but in general a sustainable regional economy is likely be oriented around three principles. First, it is likely to be what Paul Hawken terms a "restoration economy" – one which helps restore the environmental and social damage done in the past, and that prevents new

problems from occurring. Second, it is likely to be a "human-centered economy," one which meets real human needs and provides meaningful work to people at decent pay. Third, it is likely to be a locally-oriented economy, one which emphasizes local ownership, local control, local investment, use of local resources, production for local markets. This does not mean that economic development policies should totally downplay export-oriented industries, but that they should encourage as much economic activity as possible to be rooted in particular communities and regions.

The sustainable economy is likely to meet these goals through a mixture of market mechanisms, government action, and incentives for social and environmental responsibility in economic decision-making. One important step toward a more sustainable economy will be to phase out industries or processes that consume large amounts of nonrenewable resources and produce large quantities of pollutants and toxics. In addition, economic sectors based on large-scale extraction of natural resources such as minerals and oil are unlikely to be sustainable in the long run, since these resources will eventually run out after enduring various price and supply shocks along the way. Likewise, sections of the regional economy based on government subsidy – such as military industries – are not particularly sustainable either, since these handouts may cease. Somewhat more arguably, sectors of the economy which support the automobile should not be considered sustainable, since cars and car-dependent patterns of suburban sprawl cause a great deal of ecological and social damage. In contrast businesses engaged in environmental cleanup, recycling, public transportation, affordable housing, organic food production and the like contribute to sustainability, in that these improve the social and environmental health of the region. But while some industries may dwindle in a sustainable economy and others grow, many will simply find ways of doing the same things better. This can have advantages for businesses as well as the environment.

Many writers have argued that a cooperative, locally-oriented economics, emphasizing worker, producer and consumer co-ops and small, locally-owned businesses, is healthiest for local communities. Such a system promotes economic democracy, local control, diversity of ownership, and social responsibility, and offers an alternative model to the global market economy envisioned under free trade agreements such as GATT. Such a free-market global economy led by large corporations has many negative effects on cities and towns, in that it tends to undermine local ownership and control, replace a diversity of small retailers with a few standardized chains, and export capital from local communities to financial centers in other parts of the world.

8. Community participation and involvement. One of the most important components of urban sustainability will be creation of more functional local and regional democracy, which in turn can bring about other positive changes. There is no single best way to promote this. But a package of policies aimed at opening up local political processes, insulating them from money and special interests, producing an educated and informed electorate, and promoting responsible local decision-making can help. Community participation in local planning and design is important, as is the broad-minded leadership of officials at local, state and federal levels of government, who must demonstrate that it is possible at every level to make decisions with global, regional, and local sustainability in mind.

9. Preservation of local culture and wisdom. Much of the strength of any particular urban region lies in its cultural traditions and the unique relationships that its residents develop with each other and with the land. This uniqueness gives a region vitality, helps it take advantage of particular local contexts, and makes it an interesting place to live. Local culture, history and wisdom can add to sustainability, and their best aspects should be preserved. Such preservation will often take conscious action by governments to encourage traditional crafts, languages, rituals, cultural practices, and building techniques; to protect important local products from mass-produced imports; to protect local farmland and resource stocks; and to integrate vernacular architecture and materials into local development.

HOW CAN SUSTAINABLE AND LIVABLE CITIES BE CREATED?

How can sustainability goals be put into practice for cities and towns? On the one hand, progress will depend on sustainability themes diffusing into all existing planning and development processes. On the other hand, specific planning efforts and changes to planning procedures may be necessary as well. Sustainability-oriented planning processes could focus on particular urban issues or problems, such as air quality or watershed management (as long as these are undertaken with an awareness of other related issues). Or city leaders and public participants could take a more comprehensive look at the sustainability of a city or region. Cities such as San Francisco, Seattle, Santa Monica, and Leicester, England, have taken this latter approach.

In this age of entrenched economic and political forces opposing sustainability, no single planning effort is going to set cities on a path towards a healthy long-term future. Rather, the need is for a long-term strategy emphasizing consensus processes, public education, political organizing, policy tools such as indicators and performance standards, development of vision documents and "best practices" examples, and the creation of institutions that can more effectively address physical planning and equity issues. Together, such efforts can develop the knowledge, political will, and institutional capacity to bring about change.

A systematic planning approach to promote urban sustainability might first seek to get a wide range of parties involved in efforts to improve the long-term health of a particular city or region. Participants would then attempt to reach agreement on particular values and goals that might move the community in more sustainable directions. Vision statements and review of "best practices" examples worldwide are often helpful in giving stakeholders ideas about the range of possible approaches. Performance standards and sustainability indicators could be developed to help measure whether or not the community is making progress toward long-term goals, and to allow policies and programs to be revised to better achieve their objectives.

The most difficult challenge comes in implementing sustainability visions, policies, and programs, and in modifying institutions so as to be able to do this. This process depends on effective political organizing, and on the development of a coalition of interests supporting common objectives around sustainability and quality of life. Participants should expect the process to be a long one. Most cities and towns contain entrenched political and economic forces with an interest in continuing unsustainable patterns of development, and so progress will be slow.

In the long run, sustainable development will require systemic cultural change that builds democracy and social capital (accumulations of trust and cooperation between people). The problems created by concentrations of economic power must be addressed, as well as the tendency of capitalist systems to reinforce values oriented toward short-term private profit rather than long-term social or ecological well-being. Ultimately, moving towards sustainable cities will require a different mix of values than dominates urban development at present. There will still be opportunities for individuals to make a profit. However, the emphasis must be instead on caring deeply about the community, the region, and each other. Each development or planning decision must be evaluated in terms of its effects on the health of human and ecological communities. In *A New Theory of Urban Design* (1987), architect Christopher Alexander sets forth a single overriding rule of city development: "Every increment of construction must be made in such a way as to heal the city." This is not a bad guideline for sustainable development planning.

CONCLUSION

Planning for urban sustainability is still in the very early stages. As of now little progress has been made on turning today's huge, resource-consumptive metropolises and sprawling, even more resource-intensive suburbs into sustainable communities. But the seeds are being laid for future change, in terms of emerging consensus on what more sustainable and livable cities would be like, and on some processes and institutions that can help implement these planning directions.

Different approaches will have to be found for different cities. The Third World's rapidly growing megacities face the challenge of providing basic water, sewer, utility, and transportation services for their residents in ways that will be sustainable in the long run and that will avoid some of the problems that industrialized cities in the "developed" world have gotten into. First World cities face a different task: redeveloping urban areas that have plenty of infrastructure but fail to provide an ecologically or socially healthy urban environment. Sometimes the needs of cities in different parts of the world will be exactly opposite. For example, planners in the U.S. often seek higher residential densities to support transit, conserve open land, and promote community interaction. Yet in high-rise Hong Kong – the world's densest city – efficient land use is not an issue, and densities may fall without affecting sustainability.

It is important to remember that the current city is very recent. Its form and environment are heavily determined by technological innovations such as the automobile and the elevator which have only existed since the late nineteenth century. The creation of megacities of more than ten million people is an even more recent phenomenon, occurring only in the latter part of the twentieth century. Just as recent patterns of suburban development are now layered on top of nineteenth-century streetcar grids and eighteenth-century walking cities, so new and more ecological patterns of development may someday be layered on top of these, gradually bringing cities back into a better balance with the ecological limits of regions and the planet as a whole.

Twentieth-century suburbanization was in large part a reaction against the dirty, crowded, unhealthy cities of the industrial revolution, in which people were crammed into terrible housing virtually without amenities, to serve the needs of ruthless early forms of industrial capitalism. In a similar manner sustainable city initiatives of the next century may form a reaction against the excesses of twentieth-century culture, which is dominated by economic rather than environmental or social values. The transition toward more sustainable cities will not happen overnight. But through a growing ecological and social consciousness, the development of innovative models and examples, and better understandings of the policies, programs and designs appropriate to urban sustainability, new, more sustainable forms of urban development can come about.

RECOMMENDED READINGS

Beatley, Timothy and Kristy Manning, *The Ecology of Place: Planning for Environment, Economy, and Community*, Island Press, Washington, D.C., 1997.

Braidotti, Rosi *et al.*, *Women, the Environment and Sustainable Development: Towards a Theoretical Synthesis*, Zed Books, London, 1994.

Brown, Lester R., *Building a Sustainable Society*, Norton, New York, 1981.

Calthorpe, Peter, *The Next American Metropolis*, Princeton Architectural Press, Princeton, 1993.

Elkin, Tim and Duncan McLaren, with Mayer Hillman, *Reviving the City: Toward Sustainable Urban Development*, Friends of the Earth, London, 1990.

Girardet, Herbert, *The GAIA Atlas of Cities: New Directions in Sustainable Urban Living*, Anchor Books/Doubleday, New York, 1992.

Haughton, Graham and Colin Hunter, *Sustainable Cities*, Regional Studies Association, London, 1994.

Hawken, Paul, *The Ecology of Commerce*, HarperCollins, New York, 1993.

Holmberg, Johan, ed., *Making Development Sustainable: Redefining Institutions, Policy, and Economics*, Island Press, Washington, D.C., 1992.

Kelbaugh, Douglas, *Common Place: Toward Neighborhood and Regional Design*, University of Washington Press, Seattle, 1997.

Local Government Commission, *Land Use Strategies for More Livable Places*, Sacramento, 1992.

Lowe, Marcia, *Shaping Cities: The Environmental and Human Dimensions*, Worldwatch Paper 105, The Worldwatch Institute, Washington, DC, 1991.

Lyle, John Tillman, *Regenerative Design for Sustainable Development*, John Wiley & Sons, New York, 1994.

Maclaren, Virginia W., "Urban Sustainability Reporting," *Journal of the American Planning Association*, Vol. 62, No. 2, Spring 1996, 184–202.

Mitlin, Diana, "Sustainable Development: A Guide to the Literature," *Environment and Urbanization*, Vol. 4, No. 1, April 1992.

Norgaard, Richard B., *Development Betrayed: The End of Progress and a Coevolutionary Revisioning of the Future*, Routledge, New York, 1994.

Pearce, David, Edward Barbier, and Anil Markandya, *Blueprint for a Green Economy*, Earthscan Publications, London, 1989.

President's Council on Sustainable Development, *Sustainable America*, Government Printing Office, Washington, D.C. 1996.

Rees, William, *Our Ecological Footprint: Reducing Human Impact on Earth*, New Society Publishers, Philadelphia, 1996

Stren, Richard, Rodney White, and Joseph Whitney, eds., *Sustainable Cities: Urbanization and the Environment in International Perspective*, Westview Press, Boulder, CO, 1992.

United Nations, *Agenda 21: Program of Action for Sustainable Development*, New York, 1992.

Van der Ryn, Sim and Stuart Cowan, *Ecological Design*, Island Press, Washington, D.C., 1995.

World Commission on Environment and Development, *Our Common Future*, Oxford University Press, New York, 1987.

Zuckerman, Wolfgang, *End of the Road: The World Car Crisis and How We Can Solve It*, Chelsea Green Publishing, Post Mills, VT, 1991.

"The Teleserviced City"

from *E-topia: Urban Life, Jim –
But Not as we Know it* (2001)

William J. Mitchell

Editors' Introduction

Population growth, environmental constraints, and new communications technologies are widely seen as the key determinants of urban futures. But according to William J. Mitchell, the city of the future has already arrived. Many elements of the "e-topia" that he predicts will become the new urban paradigm based on new ways of organizing urban space, urban communities, and urban personalities that flow from the demands of digitalized electronic communications technologies and have already become accepted parts of everyday life. ATMs, Mitchell points out, have largely replaced banks with human tellers, we order everything from books to clothing from online companies like Amazon.com and LandsEnd, and even going to the doctor has become intermixed with the delivery of digitalized information through electronic channels. Indeed, the worldwide digital network has become a "new form of urban infrastructure – one that will change the forms of our cities as dramatically as railroads, highways, electric power supply, and telephone networks did in the past."

William Mitchell studied at the University of Melbourne, Yale, and Cambridge and has taught at Yale, Cambridge, Carnegie-Mellon, the University of Virginia, the UCLA Graduate School of Architecture and Urban Planning, and the Harvard Graduate School of Design. He is currently Professor of Architecture and Media Arts and Dean of the School of Architecture and Planning at the Massachusetts Institute of Technology. For years he has studied the relationship between architecture and the built environment and the new technologies of mass communications, including photography, cinema, and the new telecommunications technologies.

Mitchell argues that virtual urban places have become as important as physical ones. Cities of the future, he writes, will be "characterized by live/work dwellings, twenty-four-hour pedestrian-scale neighborhoods rich in social relationships, and vigorous community life, complemented by far-flung configurations of electronic meeting places and decentralized production, marketing, and distribution systems."

In *E-topia*, Mitchell explores all aspects of communal virtuality in ways that illuminate both the present and the future. Dispassionately, without assuming an advocacy position either for or against the new technological realities, he details how computers have already revolutionized – and will continue to revolutionize – the way we live in our homes, entertain ourselves, work, consume, define ourselves, and interact with both neighbors and the natural environment. Personal identity, he writes, is destined to become "an electronically metered commodity." And although many may want to opt out of "the vast electronic Panoptikon" that the world of online space has become, the changes already made are changes that are here to stay. As to the "scoffers" who question whether the new technologies are really so influential, they, according to Mitchell, "are just not getting it."

Mitchell's e-topia invites immediate comparison with the other writings in this section, especially Manuel Castells's "informational cities" (p. 475) and the 1968 predictions of technology-driven "post-urbanism" of Melvin Webber (p. 470).

Mitchell's earlier works include *City of Bits: Space, Place and the Infobahn* (Cambridge, MA: MIT Press, 1996), *The Reconfigured Eye: Visual Truth in the Post-photographic Era* (Cambridge, MA: MIT Press, 1994), *The Logic of Architecture: Design, Computation, and Cognition* (Cambridge, MA: MIT Press, 1990), and with Donald Schön and Bish Sayal, *High Technology and Low-income Communities* (Cambridge, MA: MIT Press, 1999).

■ ■ ■ ■ ■ ■

In ancient Rome you got better military protection and better spectacles than in the provinces. In Manhattan you get better medical care, restaurants, and haircuts than in a cow town. As everyone knows, availability of high-quality services is a major attraction of urban areas.

In the emergent computer-networked world, though, this reduces to a half-truism. Some services still depend on the local presence of the providers, but others can effectively be summoned and delivered from a distance. As a result, new patterns of service distribution are superimposing themselves on cities and rapidly displacing some older ones.

A TYPOLOGY OF SERVICE SYSTEMS

Service systems consist, in their barest and most obvious essentials, of service providers, service consumers, and effective means to connect the two. The various possible patterns of connection define an elementary typology of service systems. And the effects of digital telecommunications vary by type.

Before telecommunications, the privileged surrounded themselves by servants or slaves and summoned them verbally as required. Members of large service staffs were classified and named according to their roles – butlers, chambermaids, cooks, chauffeurs, gamekeepers, viziers, broom-wallahs, personal trainers, scribes, amahs, corporate lawyers, whatever. It mostly depended on maintaining close physical proximity, even when primitive systems of bells and buzzers began to augment direct verbal communication. And it reflected itself architecturally in provision of servants' quarters, janitors' cubbyholes, service stairs, outer offices, and so on.

As large modern cities grew, so did an alternative system of centralized service points – particularly for sophisticated, specialized services. These allowed economies of scale and could serve large populations at relatively low cost, but service consumers had to travel to them. Medical care, education, and many

commercial services followed this pattern, and associated building types such as modern hospitals and schools emerged as a result.

One way to resolve the inconvenient contradiction between gaining economies of scale through centralization and remaining close to consumers through decentralization was to develop distributed systems of branches. Thus, for example, in the nineteenth and earlier twentieth centuries, large banking organizations had head offices in prominent downtown locations, back offices for centralized processing activities in low-rent suburban space, and huge numbers of branches to provide customer service in local communities. Retailing followed a similar pattern. The general result was that Main Streets, commercial strips, and shopping malls became collections of branches and franchises. And the downtown office towers in all but the largest global cities mostly contained branches of national and international organizations.

Yet another strategy was to service distributed populations by means of mobile providers. This has its roots in the ancient tradition of itinerant healers, teachers, peddlers, and cops on the beat. The disadvantages are that the mobile providers have to carry their tools of trade with them, and that it is hard to achieve economies of scale in this way.

Finally, in efforts to maximize the advantages and minimize the disadvantages of these basic patterns, all manner of hybrids developed. A large, central medical facility might be combined with a system of local clinics, mobile care units, and home care providers. A retailer might make use both of downtown showrooms and traveling salesmen.

SUMMONING ASSISTANCE

In the nineteenth century, early telecommunications technology was quickly adapted to the task of summoning mobile service providers from central locations as required. This speeded response times, and so made centralized services far more effective.

In 1852, for example, Boston began constructing a system of telegraphic call boxes connected to fire stations, and other cities soon followed. Together with the replacement of hand-drawn firefighting apparatus by horse-drawn fire engines, then by motorized fire trucks, this allowed stations to serve wider areas and larger populations.

Successive waves of telecommunication and transportation technology extended and elaborated the idea. By the 1880s, telephones were being installed in police stations, and police forces eventually began to make combined use of the telephone, the two-way radio, and patrol cars to service extensive areas. From 1928, the Royal Flying Doctor Service began providing medical care to the vast, sparsely populated Australian outback through use of light aircraft summoned by means of pedal-powered Morse code radio transmitter-receivers. Now, in the era of cellular telephones and pagers, service providers of every kind – from drain cleaners to brain surgeons – can continually be on call.

KEEPING TABS

In all of these systems, somebody still has to take the action of calling the cops, the medics, the firefighters, the plumbers, or the caterers. But by adding sensors to remote summoning capabilities, the tasks of monitoring need and summoning services as required can be automated.

Thus it is now routine to install fire and smoke detectors in buildings; these not only sound a local alarm, but in many cases they automatically call firefighters to the scene. Burglar alarms that sense opening of doors, breakage of windows, or motion within interior spaces function in much the same way. Continuous electronic monitoring, based on embedded sensors, is beginning to revolutionize the maintenance of structures such as bridges and dams. In industry, it has long been commonplace to embed sensors in plant and machinery to signal malfunctions, and in a ubiquitously networked world this idea will increasingly be extended to automobiles and to domestic appliances of all kinds.

Consider automobile and truck tires, for example. Traditionally, it has been up to drivers and service mechanics to check the pressure manually, and to adjust it as required; neglect results in poor performance and excessive wear. A smart vehicle might perform these routine service tasks itself, making use of onboard pressure monitors, computers, and controllable pumps and valves to maintain constant pressure. But even smarter logging trucks in Alaska and British Columbia now link their onboard computers, via satellite, to geographic information systems and weather information systems and dynamically adjust tire pressures to current conditions. Overkill? Not if the performance payoff is there.

What works for structures and machines can also work for our own bodies. We are also likely to see proliferation of sophisticated medical monitoring devices hooked up to health care providers; once available only in hospital beds, they will increasingly take the form of unobtrusive wearable devices and of systems that provide continuous monitoring in the homes of those who require it.

[. . .]

SURVEILLANCE AND SECLUSION

All this, of course, superimposes yet another layer of electronically mediated social relationships on daily life. Wherever such electronic monitoring is carried out, it adds what are sometimes called quaternary social relationships – those that exist between observed and anonymous observer – to our primary, secondary, and tertiary relationships. And as vigilant civil libertarians have been quick to point out, we could well end up imprisoning ourselves in a vast electronic Panopticon.

More subtly, we will increasingly face tradeoffs between maintaining our privacy and getting better service by giving some of it up. If an online bookstore or CD store keeps track of your purchases, for example, it can automatically compare these with the purchases of other customers and use these comparisons to tell you what customers with similar interests to yours have been buying. It is a very effective collaborative filtering and recommendation mechanism, and it adds considerable value to the bookseller's service – but you might want to opt out of it if you discovered that the purchase profiles were also being sold to direct-mail marketeers. And you might be very upset if you discovered that nosy journalists were poking around in them.

What if you were checking into a fancy hotel? If the hotel could electronically access a detailed profile of your needs and preferences, it could organize the space and the menu to your liking. But is it worth it? Would you want to reveal so much in order to gain this benefit? Then, what about checking into a hospital? Does the greater criticality of the situation change things? Would you be prepared to reveal much more about yourself if it made a significant difference in the quality of your health care?

[. . .]

Under the most pessimistic scenarios, the mechanisms will inevitably fail, the powerful will always get whatever information they want, and we will end up with no privacy left. Under more optimistic ones, we will find effective ways to treat identity as an electronically metered commodity. We will turn it up or down, depending on the context.

DELIVERY AT A DISTANCE

Remote monitoring and summoning significantly change service systems – particularly medical care and emergency services – but it is remote delivery that really makes a difference. If you can pump a service out through a network, then you can extend the service area to wherever that network reaches – potentially globally. This creates large service markets, promises greater distributional equity, and is particularly good news for inhabitants of remote and underdeveloped areas, and for those who are immobilized by age or infirmity. Furthermore, the service agent on the end of the line can often become an unsleeping piece of software rather than a human operator.

In the most elementary case, as with audio and video entertainment, news, and some educational services, delivery reduces to transmission and display of an information stream. It may be synchronous, as with radio and television broadcasts, or it may be asynchronous, as with Web news servers. Either way, the network simply provides one-way pipes; the logic is little different from that of water supply systems.

With two-way telecommunication, remote delivery becomes an attractive option for service businesses that pursue the strategy of informing clients about their options, advising them on their choices, then eventually executing a purchase transaction of some kind. It works even in cases where the product or service purchased is eventually delivered in an utterly conventional way.

Travel is typical. Once, you had to go to a railway station, steamship office, or local travel agency to get information and advice and to purchase your tickets. Then, when the telephone came along, you could get

much the same service by dialing in; airlines and other transportation companies began to rely on their call center operations, and travel agents began to spend most of their time on the phone. More recently, interactive Web sites have offered a third alternative; you can use them to consult comprehensive online databases of travel information, to conduct sophisticated searches to find flights and rates that meet your requirements, then immediately to make reservations and ticket purchases by executing online transactions. This has made it tougher for travel agents to survive (as they traditionally have) on commissions from ticket sales, and has forced them to compete on the quality of the information and advice that they can offer.

Some areas of retailing are going the same way. Online book and CD stores such as Amazon.com do not just offer convenience and twenty-four/seven service; they also compete with traditional bookstores by offering increasingly sophisticated information and advice. Their catalogs are extensive and detailed, contain summaries, reviews, and cross references, and can be searched in a variety of ways . . .

Banks and financial services vendors have also been affected dramatically, Deposits, withdrawals, and balance queries have become high-volume, low-price commodity transactions; they are increasingly handled remotely and automatically by ATM machines and electronic home banking systems, rather than by tellers behind counters as in the past. Bills get paid online rather than through the mail. And an increasing number of investors use inexpensive online trading sites instead of calling live brokers.

[. . .]

The scoffers who think these online services are little different from mail-supported and phone-supported services of the past, and that nothing can replace face-to-face interaction with a knowledgeable human specialist, are just not getting it. It is the ubiquity and speed of electronic delivery capabilities, combined with the possibility of effectively integrating electronic intelligence, that makes the crucial difference . . .

EXPANDING THE WEB OF INDIRECT RELATIONSHIPS

The general social effect of these new sorts of tele-service systems is to eliminate familiar middlemen. They are replaced by electronic systems and software.

Instead of going to a branch bank and meeting with a teller – perhaps someone you have come to know through regular contact – you interact with a faceless ATM machine or an electronic home banking system. Instead of buying a theater ticket at the box office (or from a scalper), you surf into a Web site, select a seat from an onscreen plan, and prepay by credit card. Rather than visit your friendly local retailer, you run searches on catalogs and click "order" buttons. Where you once had to line up at the Registry of Motor Vehicles to renew your driver's license, you now perform that task online.

So indirect, anonymous, electronically enabled relationships are proliferating in our daily lives, while certain kinds of face-to-face transactions (and the secondary social relationships with familiar intermediaries that these have fostered) are correspondingly being reduced. Society as a whole is becoming more and more dependent on a vast, complex web of automated, electronic intermediation – our new, all-purpose go-between. The reductions in transaction costs, and the gains in market efficiency, are potentially enormous; not surprisingly, Bill Gates has written lip-smackingly of a coming era of "friction-free capitalism."

Many people, understandably enough, fear reduction to a bitsucking subspecies of homo economicus, and the attendant loss of human contacts and relationships. But let's ask the hard questions. What are these particular social relationships really worth? And what will replace them? For my part, I can well live without the human contact that I used to get from the bored and overworked clerks at the RMV, and I can put the time that I spent in line to much better use. And I doubt that I am alone in this.

[. . .]

TELEROBOTICS

All this applies to services that can successfully be disembodied. But what about those that traditionally have required not just exchange of information but also human hands, right there, on the spot? Can you get your car fixed remotely?

Well, you can certainly get your computer fixed – under some circumstances, at least. If you let a skilled technician log in remotely, you may be able to get software problems resolved without a site visit or taking the machine to a service center; large-scale networks would, in fact, be very difficult to maintain

without this sort of remote servicing. As more and more machines get embedded software and network connections, they will also be serviced in this fashion. If they cannot be fixed directly, their problems will be diagnosed remotely, and technicians dispatched with appropriate parts and tools.

Where this strategy does not suffice, telerobots may, at least in principle, be brought in to do the job. Telerobots are remotely controlled machines that are capable of performing various physical tasks. They may be fixed in place, like industrial robots, or they may be mobile, like delivery vehicles. They may be wired up to telecommunications networks, or they may be linked wirelessly. Their every move may explicitly be controlled, or they may be equipped with some autonomous decision-making capability.

[. . .]

In general, telerobotics sounds complicated, expensive, and a bit perverse. Indeed it often is, but it can make practical sense under conditions where distances and travel costs are very high, where services must be delivered to dangerous localities, or where demand is widespread but skilled providers are confined to a few locations. Consider specialized surgery, for example. No doubt it normally makes eminent sense to have the surgeon in the room, with the patient. But is it necessarily better to transport a sick patient or busy surgeon over very long distances if a combination of telerobotics and digital imaging, delivered via a smart surgical suite, might substitute? What about battlefield or disaster situations, where surgeons may be too precious to put at risk on the front lines? And what if the demand for some specialized procedure is widely scattered around the world, but the necessary skills are only available in a couple of major centers? The need to provide service under these sorts of conditions has motivated intensive research into the possibility of telesurgery, and the development of some impressive prototype systems.

So, yes, you can sometimes use telerobotics to be in touch – literally in touch – with distant service providers. But don't get too excited about robotic limbs, tactile feedback devices, and shaking the booty electric. Not yet, anyway.

THE TELESERVICE PARADOX

The limitations of telerobotics are instructive. Whatever the successes of new teleservice systems, it remains true that some services – including many of the most humble – still depend on local presence of the providers. Teleworkers need to get their dry cleaning done, and don't want to go too far for it. Electronic traders may operate globally, but the janitors who empty their wastebaskets and vacuum their floors have to be right there on the spot . . . Chefs have to get the food to the table while it's hot. Telehairdressing and teledentistry seem very far off in the future. Add it all up and you can quickly see that the goods and services produced within a city for local consumption – as distinct from what regional economists term the "export base" – are likely to remain a very significant percentage of the total.

Thus concentrations of population and economic activity, once established, still have some powerful glue holding them together. Digital delivery of educational, entertainment, medical, retail, financial, and other services can be expected to produce new patterns of service availability within and among these concentrations, but it will certainly not dissolve them. Indeed, something of a paradox emerges; hot spots of electronically mediated activity – such as Manhattan's Financial District, the City of London, or the teleworking-millionaire enclave of Aspen – become magnets for the low-wage service workers who do the sorts of things that computers and electronically controlled machinery cannot. And, of course, these concentrations of service workers become part of the attraction of such localities for the more privileged. It is a dirty little semisecret, then, that all of these high-flying places have large, far less interesting and attractive, low-rent counterparts somewhere nearby.

On this one, though, the fat lady is still clearing her vocal chords. Paul Krugman has suggested – and he is probably right – that the wrong side of the tracks will eventually have its revenge. As networks spread, smart places proliferate, and software becomes more and more capable, the prices of information-related services will be driven down. At the same time, the value of manually performed services that cannot readily be automated or remotely delivered will correspondingly rise. Cooks, gardeners, nannies, and plumbers will do increasingly well.

[. . .]

ELECTRONIC FRONTS, ARCHITECTURAL BACKS

Architecturally, the most striking consequence of tele-service is transformation of the traditional relationship between facade and back room. Many organizations are beginning to acquire electronic fronts and architectural backs.

Think, for example, of a retail store on an old-fashioned shopping street. The storefront presents the enterprise to the public, and the floor space immediately behind is where customers browse the stock, interact with sales staff, and make purchases. Behind that again is backroom stock and administration space which is not open to the public. Still further in the background, perhaps, is remote warehouse and head office space. Overall, there is a very clear hierarchy of visibility and public presence.

In the electronic equivalent of this store, though, the online interface takes over the functions of the street facade, the signage and display windows, and the retail floor; the software carries the full burden of mediating the organization's interactions with its customers. The backroom space remains; the need to store stock and accommodate administrative staff does not disappear. Locational constraints are loosened, however, and this backroom space can freely be distributed in whatever new patterns make practical sense. Furthermore, the buildings that provide this backroom space do not necessarily need to be in prominent, high-rent urban locations, nor do they need to perform representational roles. They can be remote and anonymous.

[. . .]

Not surprisingly, the character and distribution of electronically administered backroom space varies according to the natures of the products and services that organizations provide. Highly perishable items that require very quick delivery, like hot food, need backroom space distributed throughout service areas; you cannot have a national pizza delivery center! Online supermarkets need warehouses located to support same-day delivery within metropolitan areas. Online bookstores, which rely on air and truck delivery, need large concentrations of backroom space at national and international transportation hubs. And financial services organizations, which do not deliver anything physically, can locate wherever rents and available labor forces make it attractive. Where backroom workers do not handle physical items, they can even be located in highly distributed telecommuter workspaces that have no spatial connection to customers whatsoever.

At the same time, an organization's electronic front changes the style and granularity of its public representation. Once, for example, banks were represented by branches located on Main Streets. Now, they are represented by much larger numbers of ATM machines, minibranches, and electronic home banking screens scattered in a very different, much more diffuse pattern.

SERVED AND SERVING SPACES REVISITED

Most importantly, though, teleservice demands a new way of thinking about the organization of architectural space at both building and urban scales.

Back in the 1960s, Louis Kahn drew an influential distinction between the served and serving spaces of a building. Served spaces were the sites of important human activities, while serving spaces accommodated the support activities and equipment that the served spaces required. Thus a laboratory floor might be a served space, with adjacent plant rooms and exhaust shafts as serving spaces.

Since then, network technologists have learned to think in similar ways – and even to reinvent similar terminology. The Web, and comparable network structures, consist of client sites and server sites. Your home office might be a client site, for example, with your employer's intranet server supporting it.

Today, in the digital network era, the two traditions are beginning to converge. You might still relate smart served and serving spaces in the traditional way, by making them adjacent, but you might also establish their functional linkage through distant electronic connection. You can still partially read the functional organization of architectural space from floor plans and land use maps, but you must now look, as well, at the networking and the software.

ILLUSTRATION CREDITS

Every effort has been made to contact copyright holders for their permission to reprint plates in this book. The publishers would be grateful to hear from any copyright holder who is not here acknowledged and will undertake to rectify any errors or omissions in future editions of this book. Following is copyright information for the plates that appear in this book.

1 Palace of Sargon II of Khorsabad. Jim Harter (ed.), *Images of World Architecture* (New York: Bonanza Books, 1997). Public domain.

2 Theater of Dionysus, Athens, Greece, fifth century BCE. Jim Harter (ed.), *Images of World Architecture* (New York: Bonanza Books, 1997). Public domain.

3 Medieval market-place, Montpazier, France. © Jay Vance. Used by permission of Tiffany Vance.

4 The nineteenth-century industrial city. Augustus Welby Pugin, *Contrasts: Or A Parallel Between the Noble Edifices of the Middle Ages and the Corresponding Buildings of the Present Day Shewing the Present Decay of Taste* (London: Charles Dolman, 1841). Public domain.

5 Bird's-eye view of La Crosse, Wisconsin, showing the principal business street. *Frank Leslie's Illustrated Newspaper* (April 30, 1887). Public domain.

6 Levittown, New York, 1947. © Levittown, New York Public Library. Used by permission of the New York Public Library.

7 The auto-centered metropolis. Security Pacific Collection. Los Angeles Public Library (© Los Angeles Public Library). Used by permission of Los Angeles Public Library.

8 Technoburbia. Microsoft Corporate Headquarters, Redmond, Washington. Used by permission of Microsoft.

9 Unknown artist, "Music in the Street/Music in the Parlor" (1868, *The Illustrated News*). Public domain.

10 Sol Eytinge, Jr., "The Hearth-stone of the Poor" (1876, *Harper's Weekly*). Public domain.

11 C.A. Barry, "City Sketches" (1855, *Ballou's Pictorial Drawing-Room Companion*). Public domain.

12 Unknown artist, "Bicycles and Tricycles – How They Are Made" (1887, *Frank Leslie's Illustrated Newspaper*). Public domain.

13 George N. Barnard, "Burning Mills, Oswego, New York" (1853, *Oswego Daily Times*). Used by permission of George Eastman House.

14 Edward Anthony, "A Rainy Day on Broadway" (1859). Used by permission of George Eastman House.

15 Jacob Riis, "Bandits' Roost, 39½ Mulberry Street," New York (1889). The Jacob A. Riis collection # 101. © Museum of the City of New York. Used by permission of the Museum of the City of New York.

16 Lewis Hine, Ellis Island immigrant portraits (ca. 1910). In Grove S. Dow, *Social Problems of Today* (New York: Thomas Y. Crowell, 1925). Public domain.

17 Alfred Stieglitz, "The Steerage" (1907). © Museum of Modern Art, New York. Used by permission of the Museum of Modern Art, New York.

18 Dorothea Lange, untitled (roadside sign) (1939). Farm Security Administration, US Department of Agriculture. Printed in C.E. Lively and Conrad Tauber, *Rural Migration in the United States* (Washington, DC: Works Progress Administration, 1939). Public domain.

19 Charles Sheeler, "Ford Plant, Detroit" (1927). © Museum of Modern Art, New York. Used by permission of the Museum of Modern Art, New York.

20 Weegee (Arthur Fellig), "The Critic" (1943). © 1994, International Center of Photography, New York. Bequest of Wilma Wilcox. Used by permission of the International Center of Photography.

21 King Vidor, still from *The Crowd* (1926). Museum of Modern Art, New York, Film Stills Archive. © Museum of Modern Art, New York. Used by permission of the Museum of Modern Art, New York.

22 Fritz Lang, still from *Metropolis* (1929). Museum of Modern Art, New York, Film Stills Archive. © Museum of Modern Art, New York. Used by permission of the Museum of Modern Art, New York.

23 Street in Seaside, Florida. © Duany–Plater-Zyberk. Used by permission of Duany–Plater-Zyberk.

24 North Garden, Mall of America, Minneapolis. © Mall of America/Simon MOA Management Co, Inc. Used by permission of Mall of America.

25 The Petronas Twin Towers, Kuala Lumpur. © Roger Mellor. Used by permission of the photographer.

26 Luxury housing development, Shanghai. © Lian Bloch. Used by permission of the photographer.

27 Neighborhood market in Rome. © Patti Walters. Used by permission of the photographer.

28 Sumner Street, San Francisco. © Derek Metzger. Used by permission of the photographer.

29 Walled plaza in Fez. © Paul V. Turner. Used by permission of the photographer.

30 Skateboarder along the San Francisco waterfront. © StoutFoto, San Francisco. Used by permission of the photographer.

31 Anti-globalization demonstration, Seattle, Washington. © Jason Morrison, Pacific Institute. Used by permission of the photographer.

32 World Trade Center under attack. © Wanda McCormick, photographer, Readio.com. Used by permission of the photographer.

33 Central Park, New York, 1863. © Museum of the City of New York. Used by permission of the Museum of the City of New York.

34 Arturo Soria y Mata's plan for a linear city around Madrid, 1894. From Arturo Soria y Mata, *The Linear City* in Richard LeGates and Frederic Stout (eds), *Early Urban Planning 1870–1940* (London: Routledge/Thoemmes, 1998).

35 Ebenezer Howard's plan for a Garden City, 1898. Ebenezer Howard, *Garden Cities of To-morrow* in Richard LeGates and Frederic Stout (eds), *Early Urban Planning 1870–1940* (London: Routledge/Thoemmes, 1998).

36 Plan for Welwyn Garden City, 1909. Royal Town Planning Institute. Public domain.

37 Le Corbusier's "Plan Voisin" for a hypothetical city of three million people, 1925. Image Le Corbusier, *Urbanisme* (London: John Rodher, 1929). Public domain.

38 Plan for Radburn, New Jersey, 1929. Clarence Stein and Henry Wright architects. Public domain.

39 Frank Lloyd Wright's plan for Broadacre City, 1935. Frank Lloyd Wright, *When Democracy Builds* (Chicago: University of Chicago Press, 1945).

40 Paseo del Rio, San Antonio, Texas. © Alexander Garvin. Used by permission of Alexander Garvin.

41 Quincy Market, Boston, Massachusetts. © Alexander Garvin. Used by permission of Alexander Garvin.

42 Peter Calthorpe's plan for a Pedestrian Pocket. © Peter Calthorpe Associates. Used by permission of Peter Calthorpe.

PART PAGE ILLUSTRATIONS

PROLOGUE Lisa Ryan, 1998. © Lisa Ryan. Used by permission of Lisa Ryan.

PART 1 Hartman *Schedel*, WeltChronik (Nuremberg, 1493). Public domain.

PART 2 Gustave Doré, "London Traffic," 1872. Public domain.

COPYRIGHT INFORMATION

PROLOGUE

1 THE EVOLUTION OF CITIES

Index